ORGANOBISMUTH CHEMISTRY

Photograph on cover: Crystal of bismuth that the Great Architect makes.

Organobismuth Chemistry

Edited by

Hitomi Suzuki

Department of Chemistry
School of Science
Kwansei Gakuin University

Yoshihiro Matano

Department of Chemistry
Graduate School of Science
Kyoto University

2001

ELSEVIER

AMSTERDAM – LONDON – NEW YORK – OXFORD – PARIS – SHANNON – TOKYO

ELSEVIER SCIENCE B.V.
P.O. Box 211, 1000 AE Amsterdam
The Netherlands

First edition 2001

Library of Congress Cataloging in Publication Data

Organobismuth chemistry / edited by Hitomi Suzuki.
p. cm.
Includes bibliographical references and index.
ISBN 0-444-20528-4
1. Organobismuth compounds. I. Suzuki, Hitomi.

QD412.B5 .O74 1999
547'.05718--dc21

99-046573

ISBN 0-444-20528-4

Transferred to digital printing 2005

Printed and bound by Antony Rowe Ltd, Eastbourne

Foreword

Some aspects of bismuth chemistry have been known and studied for many decades but in recent years there has been a resurgence of interest in the properties of this element not least because of its application in a number of important yet often disparate fields. These fields include bismuth containing pharmaceuticals, superconductors, ferroelectrics, heterogeneous catalysts and, of particular relevance to this book, reagents for organic synthesis.

Bismuth is present in its compounds almost exclusively in two oxidation states namely (III) and (V), but as a result of the inert pair effect and relativistic effects, bismuth(V) is strongly oxidising in contrast to the lighter congeneric antimony(V). This feature is reflected in the organisation of the book such that, after a brief introduction, Chapter 2 deals with organobismuth(III) chemistry whilst Chapter 3 focuses on compounds containing bismuth(V) many of which are useful oxidising agents. After a look at bismuth heterocycles in Chapter 4, a detailed summary of the use of bismuth compounds in organic transformations is given in Chapter 5 and Chapter 6 deals with structural aspects.

Hitomi Suzuki and his coauthors have done an excellent job in preparing this work which besides being very comprehensive and well organised includes a number of novel and useful features. Thus within the text of Chapters 2–5 a large number of representative experimental preparations are given. Not only do these serve to illustrate the practical side of bismuth chemistry, they also provide a convenient alternative to looking up the original literature which can save a lot of time. Likewise, the structural data presented in chapter six offers the reader a convenient access to many key structures with both visual and numerical data clearly presented. The references are also given in a handy and easy to use format.

In summary, I have no doubt that this work will be a valuable resource for all groups working the field of organobismuth chemistry; I will certainly be happy to have a copy on my bookshelf.

N.C. Norman
1999

Preface

Bismuth in human history dates back about two millennia when it was used as an additive to ancient metal products. However, its identity as a distinct metal element was only established by the middle of the eighteenth century. In marked contrast to the history of inorganic bismuth chemistry which started more than two centuries ago, the chemistry of organic bismuth compounds was born only two decades ago, when Professor D.H.R. Barton and coworkers began a systematic study of organobismuth compounds and laid the foundation for further development.

In the last two decades, a large number of organic bismuth compounds have been synthesized and their properties obtained by laboratory measurements. At the same time, many new methods have been developed to synthesize a variety of organic bismuth compounds. In view of the rapidly expanding field of organobismuth chemistry as well as increasing attention in this area from both academia and industry, we are of the opinion that there is a need for a handy monograph on this topic. The main objective of this book is to provide an information resource to researchers who need to become better acquainted with a specific topic of organobismuth chemistry or to obtain information related to a particular organobismuth compound. The material includes only stoichiometric bismuth compounds containing Bi–C bond(s), which are covered from early 1999. With some exceptions, inorganic compounds, minerals, metal alloys, and non-stoichiometric materials are not included.

This book covers comprehensively all aspects of the subject related to organobismuth chemistry in a self-contained manner. However, it is focused on academic aspects and consequently the references include a limited numbers of patents that are mentioned only where pertinent to the description. Chapter 1 serves as an introduction to bismuth as the element. In Chapters 2–4, organobismuth compounds are classified according to the types of compounds and dealt with in detail. Chapter 5 is devoted to the use of bismuth and derivatives in organic transformations. In these four chapters, brief to moderate descriptions for selected experimental procedures are included; they are intended to inform the readers of relevant protocols and should serve in preparative studies which are based on analogies. In the final chapter the X-ray data of fundamental and/or structurally interesting organobismuth(III) and (V) compounds are given. The table of contents of the book is detailed and from there it should be possible to quickly locate information on a specific organobismuth compound.

Much effort has been made to include all factual data pertinent to an understanding of each class of organobismuth compounds. The main attention is paid to methods of synthesis, molecular structure, and chemical behavior of organobismuth compounds, although some knowledge of spectroscopy and other physical properties are also included. The format for presenting the information is both descriptive and numerical. Numerical data are mostly presented in tabular form. Tables of known compounds in each chapter are organized to give the readers easy access to the most relevant data source of a compound. The nomenclature does not follow strictly the recommendations of IUPAC, but usage is mostly consistent with common practice in the current literature. In order to help the readers to save time in looking for appropriate spectral data, an effort has also been made to provide the IR, MS, NMR and UV spectral data sources in tabular form. All references for chapters are collected together in a list at the end of this book. In the list, references are given chronologically both in code and in full form, with authors' names.

This book has been written with the close cooperation of six chemists from our bismuth research group. Collection of the literature and selection of the material took much time and energy. The book will appeal to academic and industrial researchers alike; it is useful especially to chemists engaged in bench work. It is hoped that this book will provide a stimulus as well as the basis for further development of organobismuth chemistry.

It is a privilege and pleasure for us to dedicate this book to the memory of late Professor Derek H.R. Barton, who was the pioneer in the area of organobismuth chemistry. H.S. is indebted to him for encouragement in producing this volume. He also thanks Professor Noboru Sonoda of Kansai University for a gift of a beautiful specimen of bismuth crystal. Most of the works from our group cited in this book, both published and unpublished, were supported by Grants-in-Aid from the Ministry of Education, Science, Culture and Sports, Japan. Finally, we express our sincere gratitude to Dr. Guido Zosimo-Landolfo, Elsevier Science, for his kind help in bringing out the first monograph on organobismuth chemistry that is both comprehensive and up-to-date.

Hitomi Suzuki
Kyoto, May 1999

About the Authors

Hitomi Suzuki

Department of Chemistry, School of Science, Kwansei Gakuin University, Nishinomiya 662-8501, Japan

Born in 1935 in Kyoto. Graduated from Kyoto University in 1958 and in 1963 obtained his first Ph.D. from the same university. Associated with Professor P.B.D. de la Mare at the University of London as a Ramsay Memorial Fellow from 1940 to 1942 and obtained his second Ph.D. there. Appointed Associate Professor, Hiroshima University, in 1971. Moved to Ehime University as Full Professor in 1978 and then to Kyoto in 1990. Became Emeritus Professor of Kyoto University in 1999 and moved to Kwansei Gakuin University as Full Professor. His research interests mainly involve the development of new synthetic methodologies and related mechanistic studies.

Takuji Ogawa

Department of Chemistry, Faculty of Science, Ehime University, Matsuyama 790-8577, Japan

Born in 1955 in Osaka. Graduated from Kyoto University in 1979 and received his Ph.D. in 1984 from the same university. Appointed Assistant Professor, Ehime University, in 1984 and promoted to Associate Professor in 1989. From 1995 to 1996, he joined the group of Professor M.S. Wrighton at Massachusetts Institute of Technology for postdoctoral work. His interests are focused on material science.

Naoki Komatsu

Department of Chemistry, Graduate School of Science, Kyoto University, Kyoto 606-8502, Japan

Born in 1962 in Nagoya. Graduated from Kyoto University in 1986 and obtained his Ph.D. in 1993 from the same university. Following postdoctoral work at Okayama University, he was appointed Assistant Professor, Kyoto University, in 1995. From 1996 to 1997, he joined the group of Professor R.A. Holton at Florida State University. His research is mainly concerned with the development of new synthetic methodologies.

Yoshihiro Matano

Department of Chemistry, Graduate School of Science, Kyoto University, Kyoto 606-8502, Japan

Born in 1963 in Tokyo. Graduated from Kyoto University in 1987 and obtained his Ph.D. from the same university in 1994. Appointed Assistant Professor, Kyoto University, in 1990. From 1996 to 1997, he joined the group of Professor J.M. Mayer at University of Washington, Seattle, for postdoctoral work. His research interests are focused on organometallic and heavy heteroatom chemistry.

Toshihiro Murafuji

Department of Chemistry, Faculty of Science, Yamaguchi University, Yamaguchi 753-8512, Japan

Born in 1966 at Mikame, a fishing town in western Shikoku Island. Graduated in chemistry at Ehime University in 1988 and obtained his Ph.D. in 1993 from Kyoto University. Appointed Associate Professor, Yamaguchi University, in 2000. His research interests are mainly concerned with organometallic and heteroatom chemistry.

Tohru Ikegami

Department of Polymer Science and Engineering, Faculty of Textile Science, Kyoto Institute of Technology, Matsugasaki, Kyoto 606-8585, Japan

Born in 1969 at Tsuna in Awaji Island. Graduated in chemistry at Ehime University in 1992 and obtained his Ph.D. in 1997 from Kyoto University. Appointed Assistant Professor, Kyoto Institute of Technology, in 1997. His research interests are mainly concerned with molecular recognition and development of bismuth-based fungicides.

Contents

Chapter 1

Introduction

1.1. ELEMENTAL BISMUTH

Bismuth was first mentioned in the 1450s as *Wismutton* or *Bisemutum*, probably derived from the old German word *Weissmuth* or *Wismut*, meaning *white substance*. Although not yet recognized as a distinct metallic element, in the past it had been included as a minor component, in a variety of metal products, for specific metallurgical purposes. However, its identity as a specific metallic element was only established by the middle of the eighteenth century.

Bismuth, Bi, the 83rd element of the periodic table is the most metallic and the least abundant of the elements in the nitrogen family (group 15). It has an atomic mass of 208.980 and a ground state electronic configuration of [Xe] $4f^{14}5d^{10}6s^{2}6p^{3}$. The bismuth atom usually utilizes the three $6p$ electrons in bond formation and retains the two $6s$ electrons as an inert pair, hence the oxidation state +3 exhibited by bismuth in the vast majority of its compounds. However, a variety of organobismuth compounds can contain the element in the +5 oxidation state. Coordination numbers are 2, 3, 4, 5 and 6. Bismuth not only has metallic characteristics but also exhibits many properties similar to those of semiconductors and insulators. Consequently, it is often classified as a *semi-metal* or *metalloid*. Bismuth compounds are usually colorless unless the metal is bound to a chromophore.

Natural bismuth consists of only one stable isotope, ^{209}Bi, with a nuclear spin of $-9/2$, but there are many isotopes which exhibit a wide range of half-lives, from ^{189}Bi of $t_{1/2} < 1.5$ s up to ^{215}Bi of $t_{1/2}$ 7 min. Natural isotope ^{209}Bi decays by α-emission with a half-life of more than 10^{18} years to form ^{207}Tl. The second and third stable isotopes, ^{208}Bi and ^{207}Bi, decay by electron capture with a half-life of 3.86×10^{5} and 32.2 years, respectively. Isotopes with a mass number of more than 210 are found in natural decay chains of some radioactive elements such as ^{232}Th, ^{235}U, ^{238}U and ^{237}Np [78B-LS]. The atmospheric concentration of ^{214}Bi ($t_{1/2}$, 19.27 min; α-emitter), resulting from the radiation decay of ^{222}Rn, is used to assess the exposure of miners and households to airborne radioactivity.[1]

1.2. OCCURRENCE AND PRODUCTION

Bismuth is a relatively rare element and ranks 64th in abundance in the earth's crust, which is estimated to contain 0.17–0.2 ppm of bismuth. This is comparable in abundance to silver and cadmium. Bismuth content increases from ultrabasic (1×10^{-7}%) to acid magmatic rock (1×10^{-6}%) and is not concentrated in the magmatic phase. Native bismuth is rarely found in nature in cobalt and silver veins, but usually occurs as the compounds, mostly sulfide (bismuthinite or bismuth glance, Bi_2S_3) together with complex minerals asso-

---

[1] For general references of bismuth and its chemistry, see: [58B-P(11), 64GH(19), 70OCR(A)283, 73B-BENT547, 73S328, 78KOE(3)912, 79B-WSA1111, 82COMC(2)681, 84B-GE, 85UE(A4)171, 87CCC(3)237, 88B-CW, 94EIC(1)280, 95EAS(1)365, 98B-CN]. For organic aspects of bismuth chemistry, see: [42CRV(30)281, 70B-D, 70B-DF, 77GH(47), 78MOC(13-8), 82CRV15, 94B-AY, 94EIC(1)292, 95COMC(2)321, 96F1501]. For structural chemistry of bismuth compounds, refer to [99CRV3277]. For medicinal aspects of bismuth compounds, refer to [97CCR(163)345, 97CB669, 99CRV2601]. Review articles devoted to specific topics in organobismuth chemistry are: [83TCC1, 90B-S, 92AOC(34)207] for pentaarylbismuth and related compounds, [90AOC(30)77, 92AOC(34)207] for dibismuthines, [72OMR73, 96BCJ2673] for chemistry of bismuthonium compounds, [86ACR406, 88CSR269, 90AOC(31)91, 94CCR(133)115] for transition metal complexes, and [95CT167, 98AOC(42)1] for bismuth alkoxides. The use of bismuth and its derivatives as reagents for organic transformations was reviewed in [89B-FD(5)397, 89YGK425, 95COMC(11)502, 96CCR(155)127, 97S249, 98B-SM, 99YGK689], particularly for arylation in [88T3039, 89CRV1487, 90B-B], and for oxidation in [86B-K, 87PAC937]. Progress of organobismuth chemistry is reviewed on an annual basis in the 1980s and irregularly in the 1970s and 1990s in *Journal of Organometallic Chemistry* [71JOM(48)292, 74JOM(68)411, 75JOM(89)183, 76JOM(106)255, 78JOM(163)141, 79JOM(180)143], 80JOM(203)359, 82JOM(237)51, 84JOM(261)59, 85JOML(17)353, 86JOM(298)67, 87JOM(324)39, 88JOM(351)63, 89JOM(360)297, 90JOM(380)35, 91JOM(404)87, 92JOM(442)61, 94JOM(477)31, 95JOM(485)1, 95JOM(496)1], in *Organometallic Chemistry* [82OMC129, 83OMC157, 84OMC127, 85OMC133, 86OMC141, 87OMC138, 88OMC142, 89OMC130, 90OMC124], and in *Organometallic Chemistry Reviews* [68OCR(B)426, 69OCR(B)186, 70OCR(A)283, 70OCR(B)656, 71OCR(B)247]. Information on all aspects of the bismuth chemistry is available as *CA SELECTS, Bismuth Chemistry* from Chemical Abstracts Service (P.O. Box 3012, 2450 Olentangy River Road, Columbus, OH 43210-0012, USA). *The Bulletin of the Bismuth Institute*, a publication from a non-profit organization, Bismuth Institute (301 Borgtstraat, B. 1850, Grimbergen, Belgium), contains wide aspects of up-to-date information on bismuth. The first issue appeared in 1973. Back issues and the cumulative table of contents of the bulletin will be supplied on request.

ciated with lead, copper, tin, antimony, molybdenum, tungsten, silver and gold [95EAS(1)365]. Other bismuth minerals include bismite (α-Bi_2O_3), bismutite ($Bi_2(O_2/CO_3)\cdot H_2O$), bismuth ochre ($Bi_2O_3\cdot 3H_2O$), guanajuatite (Bi_2Se_3), tetradymite (Bi_2Te_2S), matildite ($AgBiS_2$), eulytite ($Bi_4Si_3O_{12}$), galenobismutite ($PbBi_2S_4$), bismutotantolite ($BiTaO_4$), and pucherite ($Bi_2V_2O_8$). The former three are the weathering products of bismuthinite. These natural bismuth compounds are found mostly as minerals in Australia, Bolivia, Canada, Peru and China.

Bismuth is produced, in small amounts, from ores; large amounts are recovered as a byproduct from lead, copper, tin and zinc smelting in Western countries, and from tungsten and molybdenum production in China and Bolivia. The matte, slag, flue dust and fume discharged from the smelting process of base-metal ores contain appreciable amounts of bismuth compounds extracted or evaporated. The slime formed during the electrorefining of copper and lead also contains considerable amounts of bismuth. The slag, dust, and slime are treated by lead-smelting in order to collect bismuth into a lead bullion. The resulting bullions are then subjected to two important methods for recovering bismuth: the Betterton–Kroll and the Betts process. The former process involves the reduction of bismuth oxide with metallic magnesium or calcium in a reverberatory furnace, while the latter involves the electrolysis of the bismuth-containing lead bullion in a solution of $PbSiF_6/H_2SiF_6$. The crude

TABLE 1.1
Estimated world production of bismuth

Country	1980 (ton)[a]	1996 (ton)[b,c]
Australia	907	–
Belgium	–	1000
Mexico	748	1071
Peru	522	939
USA	400	200
Republic of China	259	1375
Canada	171	200
South Korea	90	75
Japan	80	560
Romania	80	80
Others	465	–
Total	3722	5500

[a] Source: [85UE(A4)171].
[b] Source: [96MJR].
[c] Fickling, M., personal communication.

TABLE 1.2
Bismuth consumption in 1996 in USA and Japan[a]

Use	USA (ton)	Japan (ton)
Fusible alloys	400	32
Metallurgical additives	230	431[b]
Chemicals[c]	855	77
Others	35	189
Total	1520	729

[a] Source: [96MJR, 99BBI(74)].
[b] Includes 273 tons for ferrite.
[c] Includes pharmaceuticals, cosmetics, catalysts and materials for electronics.

TABLE 1.3
US bismuth statistics for 1997–1998[a]

	1997 (ton)	1998 (ton)
Domestic consumption	1531	2025
Chemicals[b]	655	908
Alloys	593	745
Metallurgical additives	252	340
Other	31	32
Exports[c]	206	245
Imports, total[d]	2164	2712
From Belgium	691	739
Canada	99	155
China	161	385
Germany	–	45
Japan	–	1
Macao	20	–
Malaysia	4	–
Mexico	601	807
Peru	163	69
Russia	–	19
UK	425	492
Price/lb. (NY dealer) (US$)	3.33	3.15[e]
Stocks, consumer	213	173

[a] [99BBI(73)].
[b] Includes industrial and laboratory chemicals, cosmetics, and pharmaceuticals.
[c] Comprises Bi metal and the Bi content of alloys and waste and scrap.
[d] Source: Bureau of the Census.
[e] US$3.53 (April 1999).

bismuth obtained is further refined by electrolysis, but more modern techniques such as zone-melting and volatilization are also used [85UE(A4)171]. Bismuth is produced and sold at a purity of 99.99–99.999%.

Current world production of bismuth is estimated at about 5500 tons per year. The estimated world productions in 1980 and 1996 are shown in Table 1.1. The joint US Bureau of Mines/Geological Survey study in 1992 estimated the economic reserves of bismuth at 107 000 tons (48% in Asia/Australia, 37% in USA, and 15% in Europe).

The USA and Japan together represent almost half of the world consumer demand. The details of the US and Japanese consumption of bismuth in 1996 are shown in Table 1.2. The US domestic consumption, exports and imports statistics in 1997–1998 is shown in Table 1.3.

1.3. PHYSICAL PROPERTIES OF BISMUTH

Bismuth is a soft, heavy, lustrous, silver-white metal with a slight pink tinge. It forms large brittle crystals with a highly intricate feature of step dislocation when the melt is left to cool slowly (see the color photograph on the front cover). Large pieces of bismuth crystals with a multi-colored sheen are sometimes sold at gem or souvenir shops. Of several known allotropic forms of bismuth, the α-form is the most stable. Under high pressures, bismuth undergoes several allotropic transformations involving changes in volume; pronounced diminution in volume (3–5%) is observed at 2.56 and 2.71×10^8 kg/m^2 and further smaller changes at 4.8, 6.5 and 9.0×10^8 kg/m^2 at room temperature. Bismuth and gallium are the only metals that expand slightly on solidification, the expansion of the former element being 3.32%. Alloys with more than 55% bismuth content expand, those with less than 48% contract, and mixtures in the remaining gap are dimensionally stable. This property renders bismuth alloys particularly suitable to the manufacture of sharply detailed metal castings. Such bismuth-enhanced castability and wear resistance of print types brought a great improvement to old press works.

The electrical resistance of solid bismuth is greater than that of liquid bismuth, the ratio of liquid to solid resistivity being approximately 0.5–0.8; for most metals this ratio is 2.0. The linear dependence of resistivity on temperature does not hold for temperatures below 10 K. High thermoelectric effects are produced when bismuth is coupled with certain other metals. Of all metals, it is the most diamagnetic and the increase in resistivity in a magnetic field (the Hall effect) is the greatest. Its thermal conductivity is the lowest among all metals, with the exception of mercury. Both the negative Hall effect and the negative thermoelectric power of bismuth indicate that electrons, and not holes, are mainly responsible for electrical conductivity. The effective number of free electrons in bismuth is approximately 3×10^{17} cm^{-3}, the mean free path being 2×10^{-4}–4×10^{-4} cm. Superconductivity has not been detected. The major physical and mechanical properties of elemental bismuth are summarized in Table 1.4.

TABLE 1.4
Some physical and mechanical properties of bismuth

Property	Value
Atomic weight (amu)[a]	208.980374(5)
Radius (Å)[b]	
Atomic	1.55
Covalent	1.52
van der Waals	2.40
Ionic	
Bi^{3+}	0.96
Bi^{5+}	0.74
Ionization potentials (kJ/mol)[b]	
1st	703.2
2nd	1609
3rd	2466
4th	4370
5th	5403
Electron affinity (eV)[a]	0.946
Electronegativity[b]	
Pauling	2.02
Allred-Rochow	1.67
Electrode reduction potential Bi^{3+} (V vs NHE (25°C))[a]	0.308
Electrochemical equivalent Bi^{5+} (mg/coulomb)[a]	0.43316
Atomic volume (cm^3/g-atom)[a]	21.3
Atomic heat (cal/°C)[c]	6.17
Crystal system[a]	Rhombohedral
Space group[a]	$R\bar{3}m$ (hexagonal setting)
Lattice constant[a]	$a = 4.5459$ Å
	$c = 11.86225$
	$\alpha = 57.24°$
	$u = 0.237$
	$Z = 6$
Melting point (°C)[c]	271.40
Boiling point (°C)[c]	1564
Latent heat of fusion (kJ/mol)[b]	10.48
Latent heat of vaporization (kJ/mol)[b]	179.1

TABLE 1.4 (*continued*)

Property	Value
Vapor pressure (mmHg)[d]	
917°C	1
1257°C	100
Density (g/cm^3)[e]	
20°C	9.808
300°C	10.03
600°C	9.66
Mean specific heat (cal/g)[d]	
0–270°C	0.0294
300–1000°C	0.0373
Coefficient of linear expansion (K^{-1})[c]	13.45×10^{-6}
Thermal conductivity (cal/s per cm^2)[d]	
100°C	0.018
300°C	0.041
400°C	0.037
Electrical resistivity (nΩ·cm)[e]	
0°C	106.5
300°C	128.9
Surface tension (dyne/cm)[c]	
300°C	376
500°C	363
Viscosity (centipoise)[d]	
300°C	1.662
600°C	0.996
Modulus of elasticity (GPa)[c]	338
Shear modulus (Pa)[f]	12.4106×10^9
Poisson's ratio[c]	0.33
Hardness[c]	
Brinell (N/mm^2)	184
Mohs	2.5
Magnetic moment (μ)[a]	+4.1106
Quadruple moment (m^2)[a]	-0.4×10^{-28}
Magnetic induction (T)[c]	−0.393
Magnetic susceptibility (cgs unit)[a]	
Solid	-1.35×10^{-6}
Liquid	-0.08×10^{-6}
Hall coefficient (m^3/A.s (18°C))[c]	-6.33×10^{-7}
Thermal neutron absorption cross section (barn/atom)[c]	0.034
Thermal neutron activation cross section[c]	0.019

[a] Source: [92B-L].
[b] Source: [94EIC(1)280].
[c] Source: [85UE(A4)171].
[d] Source: [64GH(19)].
[e] Source: [78KOE(3)912].
[f] Source: [95EAS(1)365].

1.4. CHEMICAL PROPERTIES OF BISMUTH

Bismuth is the heaviest member of group 15 and the heaviest stable element on the periodic table. Although it belongs to the group 15 family, the chemistry of bismuth differs considerably from that of other lighter members, such as phosphorus, arsenic and antimony. Decreasing availability and increasing diffuseness of the *s* electrons makes the +5 oxidation state less stable when compared with phosphorus, arsenic and antimony. Due to its more pronounced metallic character than antimony and arsenic, bismuth forms formal stoichiometric compounds with other metals. Typical examples include M_3Bi and MBi (M = Li, Na, K), M_3Bi_2 (M = Mg, Ca), and MBi (M = La, Ce, Nd, Gd, Sm, Y, etc.).

Elemental bismuth is inert in dry air at room temperature, but oxidizes slowly to become covered with a thin film of the oxide, which gives it a beautiful multi-colored luster. Above its melting point, it oxidizes rapidly to form an oxide film. At red heat in air, it burns with a bluish flame to give a yellow fume of bismuth oxide (Bi_2O_3). It can also be attacked by super-heated water vapor to form the yellow oxide. Heating with sulfur produces bismuth sulfide Bi_2S_3 as a dark brown to grayish black solid.

Bismuth does not dissolve in non-oxidizing acids, but does dissolve in nitric acid to give bismuth trinitrate $Bi(NO_3)_3{\cdot}5H_2O$ with the evolution of nitrogen oxides, and in hot sulfuric acid to evolve sulfur dioxide, forming bismuth sulfate $Bi_2(SO_4)_3$. Prolonged heating of bismuth with concentrated hydrochloric acid in the presence of air gives bismuth chloride $BiCl_3$ which, on sublimation in air, is in part converted to bismuth oxychloride (BiOCl). Hydrogen chloride reacts with bismuth chloride in dry ether to form an acid $H^+BiCl_4^-{\cdot}(Et_2O)_n$.

The reaction of elemental bismuth with molecular halogens at elevated temperatures leads to the production of the corresponding trivalent bismuth halides BiX_3. Above 100°C, bismuth fluoride (BiF_3) can add an additional molecule of fluorine to form bismuth fluoride (BiF_5) as moisture sensitive crystals subliming at 120°C, which is highly toxic and the only known homoleptic inorganic bismuth(V) compound. Pentavalent bismuth fluoride is a powerful oxidizing and fluorinating agent. It reacts with xenon tetrafluoride (XeF_4) in liquid hydrogen fluoride to give a fluorobismuthate salt $[XeF_3^+][BiF_6^-]$.

Treatment of a solution of bismuth chloride in hydrochloric acid with hypophosphorous acid gives free bismuth as a dull gray powder with a diameter not greater than 15 μm. Its purity is more than 98.5% and it is readily dispersed in water.

The most sensitive method for the determination of bismuth is atomic absorption spectrometry in conjunction with graphite tube atomization. Bismuth has at least five atom lines suitable for determination: strong lines at 278, 289.8, 293.8 and 306.9 nm and a weak line at 267.9 nm. A detection limit of 0.05 ppm has been reported for this procedure.

1.5. BISMUTH COMPOUNDS COMMERCIALLY AVAILABLE

Bismuth metal and several inorganic bismuth compounds are important as starting materials for industrial products and they are commercially available at a cheap price. A few inorganic bismuth compounds are also employed as reagents in organic synthesis. More than ten inorganic bismuth compounds of reagent grade are available commercially at a moderate price. However, commercially available organic bismuth compounds are quite limited in number and generally expensive. For the convenience of intending researchers, some information on the properties of commercial bismuth compounds, as well as their current prices, taken from *Aldrich Catalogue/Handbook of Fine Chemicals 1998–1999* (Aldrich Chemical Co., Inc., 1001, West Saint Paul Avenue, Milwaukee, WI 53233, USA; dollar-based for US customers), are recorded in this section. The yen-based prices are shown in parentheses.

Metalic bismuth of 99.9–99.999% purity is available commercially in the form of ingot, elongated pieces, shot (1–2 mm), and powder (~100 mesh). Prices of powdered bismuth per 100 g are US$114.40 (¥28 300) for 4 nine bismuth and US$193.70 (¥47 900) for 5 nine bismuth.

Bismuth oxide (Bi_2O_3) is available as a pale yellow powder at a purity ranging of 98–99.99%. Highly pure material is prepared best by either igniting pure bismuth nitrate or basic carbonate, or dehydrating bismuth hydroxide (hydrated bismuth oxide). This compound exists in four modifications, α–δ; the α form is monoclinic, stable below 710°C, the β form is tetragonal, stable above 710°C, the γ form is body-centered cubic, metastable between room temperature and melting point, and the δ form is highly disordered and has about 75% of the disorder of the molten compound. Bismuth oxide is practically insoluble both in water and in organic solvents, but it dissolves readily in acid solution to give the corresponding salts or complex salts. The oxide is slightly soluble in concentrated alkali solution to form the bismuthate ion,

$Bi(OH)_6^-$ or $[Bi_2O_3{\cdot}3H_2O]^-$ and related species. Fusion of this oxide with potassium chlorate is reported to form a very unstable pentavalent oxide, Bi_2O_5, however, questions arise as to the exact nature of this compound. Other oxides of bismuth such as BiO, BiO_2, Bi_2O_4, Bi_3O_5 and Bi_4O_9 have been reported, but evidence for their existence as distinct chemical entities is meager. Bismuth oxide of a 99.999% purity costs US$184.25 (¥45 600) per 250 g. Mixed metal oxides of bismuth and other metals such as *titanate* ($Bi_2O_3{\cdot}2TiO_2$) and *zirconate* ($2Bi_2O_3{\cdot}3ZrO_2$) are available at prices of US$10.60 (¥2700) and US$18.05 (¥4500) per 100 g. A lead-doped mixture of Bi–Sr–Ca–Cu oxides of an approximate composition $Bi_{1.6}Pb_{0.4}Sr_{1.6}Ca_{2.0}Cu_{2.8}O_{9.2+x}$ ($x = 0.45$) is also commercially available in the form of fine powder (below 5 μm) at a price of US$345.60 (¥85 400) per 50 g. The undoped sample of a composition $Bi_2Sr_2CaCu_2O_{8+x}$ ($x = 0.15$–0.20) costs US$293.55 (¥72 600) per 50 g.

Sodium bismuthate ($NaBiO_3$) is a moisture-sensitive yellow to brown powder of ill-defined composition, which has long been known as an analytical reagent for oxidizing Mn^{2+} ion to Mn^{7+} ion, used for the qualitative test of manganese. The commercial product is thought to be a mixture of bismuth pentaoxide, sodium carbonate and sodium peroxide with about two molecules of water per bismuth atom. It is insoluble in water and organic solvents, but dissolves in strong acids to give an as yet uncharacterized Bi(V) species which is stable for a few days in solution. Sodium bismuthate of a 85% purity costs US$91.60 (¥21 500) per 100 g.

All four trivalent bismuth halides (BiX_3) are commercially available. The chloride and bromide are especially moisture-sensitive and corrosive, and are readily converted into oxyhalides, especially (BiOX). Some physical properties are shown in Table 1.5.

TABLE 1.5
Some physical properties of bismuth halides[a]

Halide	Appearance	m.p. (°C)	b.p. (°C)	d (g/cm^3)
BiF_3	White powder	649[b], 725[c]	900[d]	7.90
BiF_5	White needles	151.4	230	5.40
$BiCl_3$	White crystals, deliquescent	233.5	440.1	4.76
$BiBr_3$	Golden yellow crystals, deliquescent	218.5[d]	460.9[d]	5.72
BiI_3	Greenish black crystals	408.5	542	5.80

[a] Sources: [85UE(A4)171]. [b][78KOE(3)912]. [c][94EIC(1)280]. [d][64GH(19)].

Among these halides, the most common is bismuth chloride ($BiCl_3$), which is used as the starting material for a variety of organobismuth compounds. The commercial product is a white powder, more or less wet with atmospheric moisture, and therefore, refluxing with thionyl chloride ($SOCl_2$) is recommended for dehydrating the salt before use, especially for the Grignard reaction. Anhydrous bismuth chloride is also obtained by distilling commercial dihydrate $BiCl_3{\cdot}2H_2O$, which is a white to pale yellow, deliquescent crystalline solid, soluble in aromatic hydrocarbons, and donating organic solvents such as diethyl ether and tetrahydrofuran (THF). In an ethereal solution, it is monomeric, presumably in the form of $BiCl_3{\cdot}(OEt_2)_n$, but association occurs at a concentration greater than 0.1 N. It shows a considerable tendency to form addition products with organic molecules, and a wide variety of organic $BiCl_3$ adducts have been reported (Table 2.29). Bismuth chloride of 98 and 99.999% purity costs US$28.35 (¥7100) and US$75.05 (¥18 600) per 25 g, respectively.

Bismuth fluoride (BiF_3) is found in the form of white or gray dimorphic crystals, practically insoluble in water, but soluble in concentrated hydrofluoric acid with the formation of complexes. It volatilizes slowly with partial decomposition at high temperatures. Bismuth bromide ($BiBr_3$) is found as yellowish crystals, soluble in aqueous alkali halides and dilute hydrochloric acid, but practically insoluble in alcohols. It is readily decomposed by water to give bismuth oxybromide BiOBr. Bismuth iodide (BiI_3) is a black fine crystalline solid with a metallic sheen, practically insoluble in water, but slowly decomposes in hot water. It dissolves in liquid ammonia, aqueous potassium iodide, hydriodic acid and hydrochloric acid, but not so much (ca. 3.5%) in absolute ethanol. When exposed to air for a prolonged time, it is slowly converted to bismuth iodate ($Bi(IO_3)_3$). Bismuth fluoride of 98% purity costs US$24.30 (¥6200) per 25 g, bromide of 98% purity US$24.90 (¥6200), and iodide of 99% US$34.50 (¥8600). These halides are all corrosive and moisture-sensitive.

Bismuth oxychloride (BiOCl) is a colorless fine powder, which melts at 232°C and boils at 447°C. It is practically insoluble in water and alcohols, but readily soluble in hydrochloric acid and nitric acid. Above 570°C, it liberates $BiCl_3$ to be transformed into a $Bi_4O_5Cl_3$ compound. Among bismuth oxyhalides, only the oxychloride is available commercially at a cost of US$30.00 (¥7500) for a sample of 99% purity and US$98.60 (¥24 400) for a sample of 99.99% purity per 50 g. However, other oxyhalides can readily be prepared by partial hydrolysis of trivalent bismuth halides or by the interaction of basic bismuth nitrate ($BiO(NO_3)$) with the corresponding sodium halide in aqueous solution. These oxyhalides are crystalline solids or amorphous powders, practically insoluble in water and common organic solvents, but soluble in strong inorganic acids. They are destroyed by aqueous alkali. Bismuth oxyiodide BiOI is brick-red in color. Bismuth oxyperchlorate hydrate ($BiOClO_4 \cdot xH_2O$) of 99.99% purity, m.p. 39°C, is also available at a cost of US$60.50 (¥5400) per 100 g.

Bismuth sulfide (Bi_2S_3) is a dark brown to grayish black solid, readily obtained by heating bismuth with sulfur or by adding an alkali sulfide to the aqueous solution of bismuth(III) salts. It is used as a high temperature lubricant. The sulfide is almost insoluble in water, but dissolves in concentrated nitric acid and hot hydrochloric acid. Concentrated aqueous potassium sulfide dissolves bismuth sulfide to form bismuth thiobismuthate $KBiS_2$, which is readily oxidized in air. Fusion of the sulfide with bismuth halide (BiX_3) gives the corresponding halosulfide (BiXS) as an air-stable solid. Bismuth sulfide of 99% purity costs US$153.30 (¥37 900) per 250 g.

Commercial bismuth selenide (Bi_2Se_3) (m.p. 710°C) and telluride Bi_2Te_3 (m.p. 585°C) are both a gray to black crystalline powder. The powder is insoluble in water and decomposed by concentrated nitric acid and aqua regia. A variety of non-stoichiometric compounds can be derived from this powder by replacing part of a selenium or tellurium atom by another chalcogen element and are used as semi-conducting materials. Bismuth selenide and telluride cost US$55.40 (¥13 700) and US$75.30 (¥18 600) per 25 g, respectively. Both bismuth chalcogenides are moisture-sensitive and the former, especially, is highly toxic.

Bismuth sulfate ($Bi_2(SO_4)_3$) is found in the form of white hygroscopic crystals, which readily decompose in water to give basic subsulfate and dissolve without decomposition in aqueous sulfuric acid. Above 465°C, it is converted to the oxide under the liberation of sulfur trioxide. This salt costs ¥1900 per 25 g; taken from *Wako Catalogue 1998* (Wako Pure Chemicals, Ltd.; 3-1-2, Dosho-machi, Chuo-ku, Osaka 540-8605, Japan).

Bismuth nitrate pentahydrate ($Bi(NO_3)_3 \cdot 5H_2O$) is found as lustrous hygroscopic crystals and smells like nitric acid. It is readily decomposed by water to subnitrate $BiO(NO_3)$, but dissolves without decomposition in water containing nitric acid. It is soluble in organic solvents such as acetone, acetic acid and glycerol, but practically insoluble in ethanol and ethyl acetate. By heating above 590°C, it is converted to bismuth oxide. Bismuth subnitrate is a white, odorless and tasteless powder of non-stoichiometric composition; its properties and composition vary considerably depending on the conditions of preparation. The subnitrate is practically insoluble in water and organic solvents, but soluble in dilute inorganic acids. Bismuth nitrate pentahydrate and subnitrate monohydrate of both 98% purity are available commercially priced at US$15.20 (¥4000) and US$21.20 (¥5300) per 100 g, respectively.

Bismuth acetate ($Bi(OAc)_3$) and citrate ($Bi[O_2CCH_2CHOH(CO_2)CH_2CO_2]$) both 99.99% purity cost US$64.90 (¥16 100) per 50 g for the former and US$12.00 (¥3000) per 100 g for the latter. A variety of bismuth carboxylates such as subcitrate, subgallate, subsalicylate, etc. are also available commercially priced at around US$16–17 (¥3000–4000) per 100 g, but these compounds mostly lack chemical integrity and are not suitable for use as synthetic reagents. Bismuth subgallate and subsalicylate of 99.9% purity cost US$16.00 (¥4000) and US$12.00 (¥3000) per 100 g, respectively. Bismuth oxalate of 99.9% purity is available commercially at ¥3670 per 25 g [98B-W].

From 1999, the Aldrich catalogue menu became quite rich with organobismuth compounds [98B-A]. Triphenylbismuthine (Ph_3Bi) is a white, easy-to-handle, stable, crystalline solid melting at 78–80°C, readily soluble in benzene, dichloromethane, ether, THF and ethyl acetate, but poorly soluble in hexane and insoluble in water. It is not toxic (LD_{50} 180g/kg for dogs (oral)). The commercial product of 99% purity costs US$55.00 (¥13 600) per 25 g. Other trivalent organobismuth compounds include tributylbismuthine (US$59.00 per 5 g), methyldiphenylbismuthine (US$89.00 per 10 g), and tris(2-methoxyphenyl)bismuthine (US$120.00 per 10 g). The purity of these products ranges from 95 to 97%. Triphenylbismuth dichloride (Ph_3BiCl_2), diacetate ($Ph_3Bi(OAc)_2$), and carbonate (Ph_3BiCO_3) are probably the only pentavalent organobismuth compounds that we can find in common laboratory reagent catalogues. They cost US$55.60 (¥13 800), US$98.00, and US$42.20 (¥10 500) per 10 g, respectively.

1.6. PRESENT-DAY USE OF BISMUTH

Currently, bismuth is used mainly in the manufacture of medicines, cosmetics, low-melting alloys and metallurgical additives [78KOE(3)912, 85UE(A4)171, 93BBI(suppl.)].

1.6.1. Metallurgical use

Bismuth combines with one to several elements such as antimony, cadmium, indium, gallium, lead and tin to form a number of fusible alloys of various eutectic composition; the melting point ranges from 47 to 262°C. In manipulating the composition to meet a requirement, a wide variety of fusible or low-melting alloys are produced and used in special solders as cylinders, heat transfer medium, and many other safety devices such as electric fuses, water sprinkler plugs, and safety valves. Due to its low melting point, high boiling point and low neutron absorption cross section, bismuth is also used as a carrier for the fuels ^{235}U and ^{233}U, as well as a coolant in nuclear reactors. The nuclear fuel is dissolved to 0.15% in liquid bismuth and pumped through graphite modulators. Typical examples of bismuth-based alloys are given in Table 1.6.

The addition of minute amounts of bismuth improves the characteristics of steels and iron castings, by increasing the wear resistance and life of parts used for machine tools and automobiles. Bismuth also improves the machinability of free-cutting aluminum, increasing the corrosion and stress resistance. In recent years, bismuth has been replacing lead more and more, for example in shots and cartridges for waterfowl shooting, plumbing fittings for drinking water systems, and fire assay for gold mining and refining.

TABLE 1.6
Eutectic and non-eutectic fusible alloys[a]

Eutectic alloys	Eutectic temperature (°C)	Composition (%)					
		Bi	Pb	Sn	Cd	In	Tl
Bi–Pb–Sn–Cd–In	47	44.7	22.6	8.3	5.3	19.1	–
Bi–In	72	33.7	–	–	–	66.3	–
Bi–Sn–Cd–Tl	94.6	49.1	–	23.5	18.2	–	9.2
Bi–Pb	125	56.5	43.5	–	–	–	–
Bi–Cd	144	60	–	–	40	–	–
Bi–Tl	188	47.5	–	–	–	–	52.5
Non-eutectic alloys	Melting range (°C)	Composition (%)					
		Bi	Pb	Sn	Cd	In	Hg
Bi–Pb–Sn–Cd–In	61–65	48.0	25.63	12.77	9.6	4.0	–
Rose metal	94–110	50	28	22	–	–	–
Newton metal	96–97	50	31.25	18.75	–	–	–
Wood metal	70–72	50	24	14	12	–	–
Lipowitz metal	70–72	50	27	13	10	–	–
Cerro alloy 105	38–43	42.91	21.70	7.97	5.09	18.33	4.0

[a] Source: [74BBI(suppl.), 85UE(A4)171, 95EAS(1)365].

1.6.2. Use in electronics, ceramics and plastics

Thermoelectric devices for refrigeration based on the Peltier cooling effect uses the intermetallic compounds, bismuth selenide (Bi_2Se_3) and telluride (Bi_2Te_3) alloyed with antimony telluride (Sb_2Te_3). Thermocouples used for sensitive temperature measurement and heat engines utilizing the Seebeck effect also use these types of compound. The working temperature ranges from 273 to 573 K for Bi_2Te_3 (p-type conduction) and $Bi_2Te_{2.4}Se_{0.6}$ (n-type conduction). Assemblies, in which all the p $\rightarrow$ n junctions are arranged on one side of a module and the n $\rightarrow$ p junctions on the reverse side, form the basis of commercial thermoelectric devices for refrigeration or generation of direct current. Such bismuth-based thermoelements are used for converting solar energy into electric power in space ships.

Powdered Mn–Bi alloy compressed at a temperature of 300°C and under a pressure of 100–200 psi in a magnetic field has been reported to produce a permanent magnet of an energy product of 0.034 μT.A/m (4.3×10^{-6} GOe), a coercive force of 271 kA/m (3400 Oe), and a residual flux of 0.42 T (4200 G). Its high magnetostriction finds application in ultrasonic generators and receivers in the 10–100 kHz band as well as thermomagnetic writing on thin films (ca. 1000 Å) of the Mn–Bi alloy for memory applications, including holographic storage.

The next generation electrical contact materials based on bismuth-silver compositions are suitable for automobile sectors, since they do not weld or arc under heavy loads. The unique conduction properties of bismuth have led to the production of plastics with a built-in shielding protection against electromagnetic or radio-frequency interferences and electrostatic discharge which create serious disturbance to the electronic equipment around us.

Some bismuth compounds are known to exhibit superconductivity after they have been rendered metallic by lead doping. The T_c values found up to now are ~30 K for $Ba_{0.7}K_{0.3}BiO_3$, 90 K for $Bi_2Sr_2CaCu_2O_8$ and $Bi_2Sr_2Ca_3Cu_4O_{12}$, and 120 K for $Bi_2Sr_2Ca_2Cu_3O_{10}$ [88PR(B)3745, 91JMAR1595]. The lead-doped bismuth compounds with a high degree of texture have found magnet and power applications with liquid nitrogen cooling (77 K).

Ceramics represent a vast outlet for bismuth compounds, particularly bismuth oxide. Small levels of inclusions have a remarkable impact on the performances of the materials used for electronics. Typical examples of electroceramics are those derived from sintered $Bi_2O_3{\cdot}3SnO_2$, $ZnO{\cdot}Bi_2O_3$, or $2Bi_2O_3{\cdot}3TiO_2$. Many of these ceramics can work as sensors for monitoring fluids and gases. Originally applied to high voltage ceramics, the use is now expanding to low voltage and DC current devices. A more substantial inclusion of bismuth oxide lowers the processing temperature of ceramics from a typical 1400 to 1000°C or less. Single crystals of $6Bi_2O_3{\cdot}GeO_2$ are used for acoustic surface wave devices. Other uses of bismuth-based ceramics include glaze for tableware, dental fillers, human bone implants, and catheters for X-ray examination.

Bismuth also improves the optical property of glasses; the addition of major amounts of bismuth oxide imparts higher specific gravity or greater refractive indices. Bismuth phosphate is used for the manufacture of flint glass.

In the plastic industry, bismuth compounds are used as stabilizers, plasticizers and additives of some sort. They also impart a flame-retardant or smoke-inhibiting property to plastics and artificial fibers.

1.6.3. Use in chemicals and pigments

Bismuth catalysts are widely used in industrial organic synthesis. The major outlet is for the manufacture of acrylonitrile and acrolein involving a Co–Bi–Mo or Pb–Bi–Mo based catalyst. For instance, bismuth phosphomolybdate $BiPMo_{12}O_{40}$ is a highly effective catalyst for the ammo-oxidation of propylene to acrylonitrile. Other catalytic systems include ZnO–Bi_2O_5–ZrO_2 for the conversion of 2-isopropanol to acetone, Bi_2O_3–P_2O_5 for the dehydrogenative dimerization of propene to benzene, and MoO_3–Bi_2O_3 for acrylonitrile synthesis. Bismuth oleate is used as a catalyst for the manufacture of aldehydes and alcohols by the oxo process.

Bismuth compounds are added more and more to special polymers for bone implants and dental prothetic devices in order to make them detectable to X-rays. Diagnostic devices for medical purpose have been using the bismuth–germanium oxide crystals to neutralize lethal gamma rays and improve overall imaging. Basic bismuth salicylate is used to impart a pearly surface to polystyrene and phenol–formaldehyde resins.

Bismuth vanadates are a relatively new class of pigment, which have high opacity, excellent light-fastness and good chemical resistance. Due to their harmless nature, these pigments are of high industrial importance as lead- and chromium-free substitutes for currently used yellow and orange paints, chrome yellow and cadmium yellow. Relative color performances of bismuth vanadate, lead sulfochromate and cadmium sulfide are 100:82:75 for color strength and 22:16:17 for opacity [95BBI(68), 97BBI(70)]. Bismuth oxychloride is used in pigments as artificial pearl. Bismuth oxychloride as well as bismuth subnitrate are sold in the market under the names Flake White, Bismuth White or Pearl White. Some bismuth dialkyldithiocarbamates provide excellent extreme pressure performance as lubricant [98BBI(71)].

1.6.4. Pharmaceutical and cosmetic use

Bismuth is one of those rare elements considered to be safe as it is *non-toxic* and *non-carcinogenic* notwithstanding its heavy metal status. It readily forms stable complexes with polyhydrated carboxylic acids and phenols through the bismuth–oxygen bond and complex salts of this type are used as soothing agents for the treatment of digestive disorders, for outlining the alimentary tract during X-ray examination, and treating skin injuries or infections [78KOE(3)912, 80B-O, 85UE(A4)171, 99CRV2601].

The advantage of bismuth in treating intestinal disorders was noted as early as 1600. Various preparations of low bismuth doses have been employed as remedies against gastric disorders, i.e. colitis, diarrhea and peptic ulcers, because of their astringent, antiphlogistic, bacteriostatic and disinfection actions. Bismuth subcarbonate, bismuth subnitrate and complex bismuth citrate were, and still are, used in the treatment of skin injuries, alimentary diseases such as diarrhea and ulcers, and venereal diseases, though the use of these simple bismuth salts tends to decrease in developed countries. Basic bismuth salicylate (subsalicylate) is employed for anti-diarrhea, syphilis, rheumatoid arthritis and veruccae. It is taken orally in combination with other ingredients for protective antiacid action as well as antiseptic effects. Bismuth subcitrate and related salts are the only known active ingredient eradicating *Helicobacter pylori*, a bacteria inflicting ulcers, and is used widely for treatment of the relevant diseases in association with antibiotics such as amoxicillin and metronidazole [98B-R]. A colloidal bismuth preparation, tripotassium dicitratobismuthate, can accelerate the healing of gastric and duodenal ulcers. Micronized bismuth subnitrate has a similar effect, although the mechanism of action is not clear. Disodium pentaiodobismuthate tetrahydrate ($Na_2BiI_5{\cdot}4H_2O$) and cacodylate ($Bi[Me_2As(=O)O]_3{\cdot}8H_2O$) are old anti-syphilitic drugs. Typical bismuth salts used for medical purposes are listed in Table 1.7.

The use of bismuth as a beauty care product dates back to antiquity and is still used in common cosmetic products. Bismuth oxychloride is used to impart a pearlescent effect to lipstick, nail varnish, eye shadow, and make-up powder. The resulting brilliance and luster surpasses that of a natural product, guanine, which is extracted from fish scales and is more costly.

TABLE 1.7
Bismuth-based medicines[a]

Bismuth salt	Medical use
Inorganic salts	
Carbonate Phosphate Silicate Subcarbonate Subnitrate	Reacts slowly with HCl of stomach and inhibits secretion by lining the mucous membrane. Treatment for gastric and duodenal ulcers, gastritis, dyspepsia, and functional disorders of colon and peptic ulcers. In ointment used as a protective or sedative agent on inflamed skin
Aliphatic carboxylates	
Butylthiolaurate Camphorate Citrate Dipropylacetate Ethylcamphorate Potassio(sodio)tartarate Succinate	Therapeutic agent capable of acting against spirilla. Treatment for tonsillitis, infectious angina, syphilis and yaws. Preoperative treatment of appendicitis
Phenolic salts	
Iodosubgallate Salicylate Subgallate Tribromophenate Tannate	Internal astringent and absorbent for the protection of the gastrointestinal membrane. Used externally as an antiseptic. Treatment for gastroenteritis, toxic diarrhea, dermatitis, and hemorrhoids

[a] Source: [90EDM(D), 96MI].

The incorporation of bismuth to hair dyes is said to improve color and deodorize them. Some insoluble salts of bismuth are used as an ingredient for face powders and dusting powders to cover or protect skin and wounds.

1.7. TOXICOLOGY OF BISMUTH AND BISMUTH COMPOUNDS

The uniqueness of bismuth is most characterized by its low toxic nature in spite of its heavy metal status in the nitrogen family, since it is a general trend of elements that the toxicity increases as you go down a group. Hence traditionally, it has been used widely in medicine and veterinary practice. Closely related arsenic and antimony compounds are generally highly toxic. LD_{50} (mg/kg) values of bismuth oxychloride (BiOCl) and bismuth oxide are reported to be 22 000 (rat, oral) and 10 000 (mouse, oral), respectively. Accidental poisoning by bismuth and bismuth compounds has been reported to occur from the use of large doses during medical therapy rather than by exposure in the workplace. The effects of acute intoxication of bismuth include gastrointestinal disturbance, anorexia, headache and discoloration of the mucous membrane. Although no bismuth poisoning has been reported yet, neither in the laboratory nor in industry, care must be taken of the careless inhalation of dusts and fumes of bismuth compounds.

Water-insoluble inorganic bismuth salts are hardly absorbed by oral administration, and acute poisoning is seldom [98ET(1)174]. Water-soluble bismuth compounds are absorbed quickly and cause kidney damage, although the degree of such damage is usually mild. However, the relatively high toxicity of $BiO(NO_3)$, which was observed especially in children, is most unlikely to be caused by bismuth but more so by the reduction of nitrate to nitrite and subsequent methemoglobin formation by intestinal bacteria. Bismuth poisoning caused by long-term oral administration in doses exceeding 1 g/day may provoke mental disorders characterized by loss of balance, unsteadiness or confusion. These disorders stem from an accumulation of bismuth in the nerve centers. In most cases, they disappear within a period of 10 days to 2 months, during which time the bismuth is eliminated by the urinary tract. Part of the bismuth poisoning has been attributed to the transformation of bismuth compounds of low toxicity into high toxicity by intestinal bacteria. Soluble bismuth salts are excreted via urine and less soluble ones are excreted in the feces, which may be black in color due to the presence of bismuth sulfide.

Bismuth telluride, undoped or specially doped with Se or SnTe, is used in industry as a semiconductor. The hygienic standard of permissible exposure to these substances, recommended by the American Conference of Governmental Industrial Hygienists was 10 mg/m^3 for undoped bismuth telluride and 5 mg/m^3 for doped bismuth telluride, both based on a 1-year daily inhalation study in three different species of animals. Short-term exposure limits are 20 mg/m^3 for undoped bismuth telluride and 10 mg/m^3 for doped bismuth telluride [81B-CC1558]. With bismuth telluride and bismuth/antimony based metal products, adverse effects may arise from tellurium or antimony rather than bismuth. Chronic exposure during long-term therapy can cause serious consequences for humans. Symptoms of chronic intoxication by bismuth are hypersalivation, stomatitis and a grayish black seam surrounding the gum. Inhalation of $BiMe_3$ by humans causes the irritation of the respiratory tract and conjunctive tissues. Cats and dogs that inhaled $BiMe_3$ for 10–20 min showed ataxia, restlessness, or convulsions; after 24 h, severe encephalitis

TABLE 1.8
Toxicity of some bismuth compounds[a]

Compounds	LD_{50} (g/kg)	Species and route
Bismuth oxide (Bi_2O_3)	5	Rat, oral
	10	Mouse, oral
Bismuth oxychloride (BiOCl)	22	Rat, oral
Bismuth nitrate ($Bi(NO_3)_3$)	<2.5	Mouse, intraperitoneal
Bismuth titanium oxide ($Bi_2O_3{\cdot}2TiO_2$)	12	Rat, oral
	2.2	Rat, intraperitoneal
Bismuth vanadate ($BiVO_4$)	<5	Rat, oral
Bismuth dimethyldithiocarbamate	20	Mouse, oral
Bismuth sodium thioglycollate	47.2	Child (20 kg weight), oral
Bismuth potassium sodium tartrate	<3	Mammal, intramuscular
	0.055	Rabbit, intramuscular
Trimethylbismuthine	0.484	Rabbit, oral
	0.182	Rabbit, subcutaneous
Triphenylbismuthine	12	Dog, intravenous
	180	Dog, oral

[a] Source: [83EOHS(1)291, 85UE(A4)171, 95BBI(68), 2000B-L(2)521, 2000B-L(3)3585].

was observed. Some toxicological data available for bismuth compounds are shown in Table 1.8.

In the Swiss list of toxic substances $BiONO_3$ is in class 3 and $Bi_2(TiO_4)_3$ is in class 5. In a long-term carcinogenicity test, BiOCl was fed to rats over a period of 2 years (1–5% in the food); neither carcinogenic nor other toxic effects caused by BiOCl could be found. So far, there is no evidence for carcinogenicity, mutagenicity and teratogenicity of bismuth compounds.

Bismuth and its compounds do not appear to have been responsible for poisoning associated with work, so it is looked upon as the least toxic of the heavy metals currently used in industry. Bismuth and bismuth oxide are poorly soluble in blood plasma and fairly rapidly eliminated by the urinary tract. As the reported number of cases is small and adequate follow-up studies have not yet been completed, it is difficult to generalize the toxicity of bismuth compounds. So far, no strict limits appear to have been set for bismuth in air and drainage in industrialized countries.

Chapter 2

Organobismuth(III) Compounds

2.1. ORGANOBISMUTH(III) COMPOUNDS WITH Bi–GROUP 14 ELEMENT BONDS

2.1.1. Monoorganyl and diorganylbismuthines

Methylbismuthine and dimethylbismuthine are prepared by reduction of the corresponding methylbismuth chlorides or bromides with $LiAlH_4$ in dimethyl ether at low temperatures [61CB1447].

$$MeBiCl_2 + LiAlH_4 \xrightarrow[Me_2O]{} MeBiH_2$$

$$Me_2BiCl + LiAlH_4 \xrightarrow[Me_2O]{} Me_2BiH$$

They are purified by vacuum distillation at low temperatures and their boiling points at 760 mmHg are estimated, by extrapolation, to be 72°C ($MeBiH_2$) and 103°C (Me_2BiH), respectively. Methylbismuthine and dimethylbismuthine are colorless liquids and stable up to −45 and −15°C, respectively. Both bismuthines decompose at room temperature to trimethylbismuthine and bismuth metal with the evolution of gaseous hydrogen. Attempts to prepare other primary and secondary bismuthines have failed so far; black polymeric substances are obtained from the reaction of dibromo(phenyl)bismuthine or bromodiphenylbismuthine with a reducing reagent such as $LiAlH_4$ or $NaBH_4$ at low temperatures [57ZN(B)132]. Diphenylbismuthine or dimethylbismuthine is used as a cocatalyst for polymerization of olefins in a patent, but the synthesis of these diorganylbismuthines are not mentioned [66USP3285890].

Dimethylbismuthine, generated in situ from tetramethyldibismuthine and anhydrous hydrogen chloride, showed a broadened methyl peak at δ 1.26 (in CS_2/CD_2Cl_2) at −70°C [83OM1859].

2.1.2. *Triorganylbismuthines*

2.1.2.1. *Methods of synthesis (Tables 2.1–2.3)*

This class of compounds can be obtained by a variety of methods which include metathesis, oxidative addition, reduction, transmetallation, thermolysis, disproportionation, electrochemical coupling, and others. In this section, the syntheses of symmetrical and unsymmetrical triorganylbismuthines according to these methodologies are described.

2.1.2.1.1. Organometallic routes The most frequently used methodology by which a wide variety of tertiary bismuthines R_3Bi, where R may be aliphatic or aromatic, can be prepared with ease is the reaction of a bismuth halide such as $BiCl_3$ and $BiBr_3$ with an organometallic reagent. Organomagnesium, -lithium, -sodium, -cadmium, -aluminum, -silver, -zirconium, -zinc, -mercury and -silicon compounds have been employed for this purpose. Unsymmetrically substituted triorganylbismuthines of the type $R_2R'Bi$ and $RR'R''Bi$ are also prepared from R_nBiX_{3-n} ($n = 1,2$, X = halide or anionic group) or $RR'BiX$ with a suitable organometallic reagent.

(a) With Grignard reagent (method A) A wide variety of symmetrical tertiary bismuthines containing such groups as alkyl, alkenyl and aryl have been prepared by using Grignard reagents. Bismuth chloride and bromide are usually employed as the bismuth source. The reactions are customarily carried out in diethyl ether or THF under an inert atmosphere and the product is isolated by distillation or recrystallization from an alcoholic solvent.

$$BiX_3 + 3RMgX \longrightarrow R_3Bi + 3MgX_2$$

$$R_2BiX + R'MgX \longrightarrow R_2BiR' + MgX_2$$

Unsymmetrical bismuthines with two or three different organyl groups ($R_2R'Bi$ and $RR'R''Bi$) are best prepared by the reaction between a bismuth compound of the type R_2BiX or $RR''BiX$ (X = Cl, Br) and a Grignard reagent $R'MgBr$ [73JOM(47)367, 74IZV234, 83JOM(256)1]. Diarylbismuth triflate–hexamethylphosphoramide (HMPA) complexes, $[Ar_2Bi(HMPA)_2](OTf)$, are also conveniently used owing to its easy preparation and handling [96OM1951].

Synthesis of tris(1-naphthyl)bismuthine

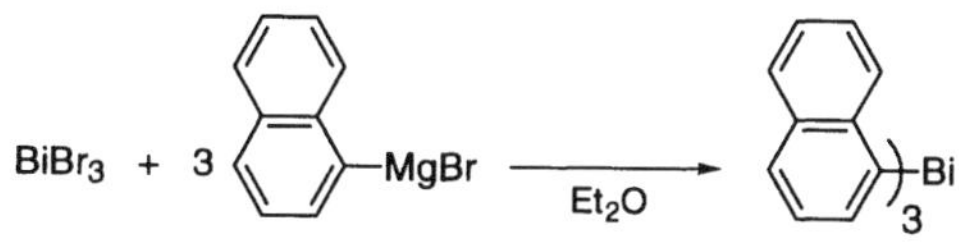

1-Naphthylmagnesium bromide, prepared from Mg (1.6 g, 66 mmol) and 1-bromonaphthalene (13.7 g, 66 mmol), was treated with anhydrous bismuth bromide (11 g, 24 mmol) dissolved in the minimum quantity of dry ether. A vigorous reaction occurred and a yellow precipitate was formed, which disappeared on shaking. Then a large quantity of solid matter gradually began to deposit. The mixture was left overnight and then heated for 1.5 h on a water bath, after which the solvent was distilled off and the residue treated with water. The insoluble solid matter was collected and dried. This product was repeatedly extracted with chloroform and treated with alcohol to give a slightly yellow crystalline deposit. On recrystallization from a mixture of chloroform and alcohol, tris(1-naphthyl)bismuthine was obtained in almost colorless needles (m.p. 235°C) [14JCS(105)2210].

Synthesis of (4-chlorophenyl)(2-methoxyphenyl)(4-methylphenyl)bismuthine

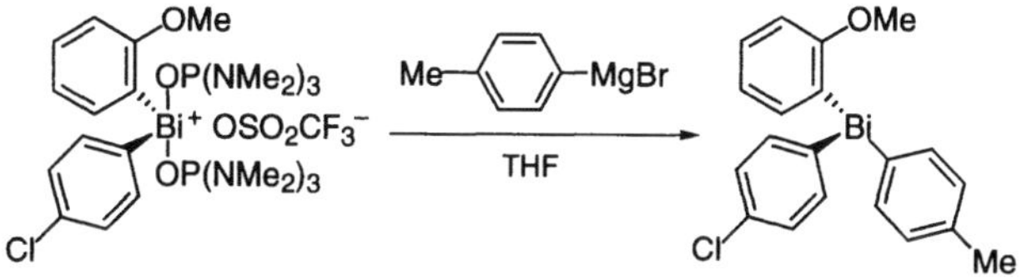

To a stirred solution of (4-chlorophenyl)(2-methoxyphenyl)bismuth triflate–HMPA complex (10 mmol) in THF (50 ml) was added, at 0°C, a solution of 4-methylphenylmagnesium bromide, generated from Mg (15 mmol) and 4-bromotoluene (15 mmol) in THF (60 ml), and the resulting mixture was stirred at the same temperature for 2 h to complete the reaction. The mixture was poured into ice-cold saturated aqueous NH_4Cl (30 ml), and the insoluble materials were filtered off through a Celite bed. The organic phase was extracted with benzene (10 ml × 5), and the combined extracts were washed with brine (10 ml × 5), dried over Na_2SO_4, and concentrated under reduced pressure to leave an oily residue, which was crystallized from benzene/MeOH (1:5) to afford the expected unsymmetrical bismuthine as colorless crystals (70%, m.p. 99–102°C) [96OM1951].

(b) With organolithium reagent (method B) The reaction between bismuth halides and organolithium compounds is an alternative convenient method for the synthesis of symmetrical tertiary bismuthines. Unsymmetrical bismuthines of the type $R_2R'Bi$ and $RR'R''Bi$ can also be obtained by the reaction between R_2BiX or $RR''BiX$ and $R'Li$ (X = Cl, Br, I). Tris(2,6-dimethoxyphenylthio)-bismuth, which is prepared from bismuth oxide and the corresponding thiol, can be converted to triarylbismuthines by treating it with an excess of aryllithiums [97JOM(548)223].

$$BiX_3 + 3\,RLi \longrightarrow R_3Bi + 3\,LiX$$

$$R_2BiX + R'Li \longrightarrow R_2BiR' + LiX$$

Synthesis of tris(4-*N,N*-dimethylaminophenyl)bismuthine

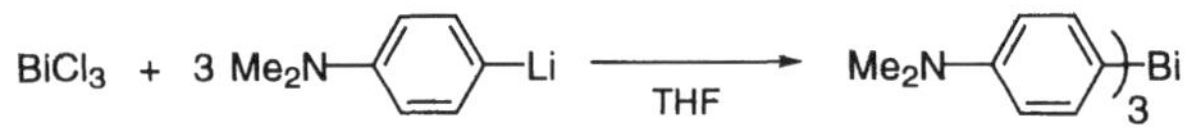

N,N-dimethyl-4-bromoaniline (1.00 g, 5 mmol) was dissolved in THF (20 ml) and cooled to −78°C. Butyllithium (3.12 ml of 1.6 M, 1.60 mmol) in hexane was added. After stirring for 30 min at −78°C, bismuth chloride (0.5 g, 1.59 mmol) dissolved in THF (10 ml) was added and the resulting mixture was stirred for 1 h at −78°C, then allowed to warm up to 23°C. After stirring for an additional 12 h at this temperature, the solvent was removed in vacuo and the residue was extracted with toluene (20 ml). The toluene extract was filtered and concentrated to about 10 ml. When the solution was allowed to stand at 0°C for a few days, the product was obtained as colorless crystals (0.96 g, 86%), m.p. 200°C (decomp.) [96OM5613].

Synthesis of methylbis(2,4,6-triisopropylphenyl)bismuthine

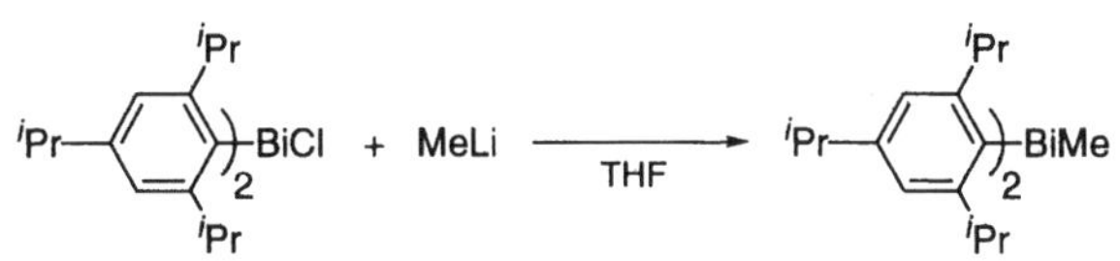

To a yellowish solution of chlorobis(2,4,6-triisopropylphenyl)bismuthine (329 mg, 0.5 mmol) in THF (10 ml) was added methyllithium (1.0 M ether solution; 0.5 ml, 0.5 mmol) at −78°C. After stirring for 10 min at −78°C, the reaction mixture was allowed to warm to room temperature gradually and the solvent was evaporated under reduced pressure. Chromatography of the solid residue on silica gel using hexane–ethyl acetate gave the expected unsymmetrical bismuthine as a colorless crystalline solid (262 mg, 77%), m.p. 53–55°C [92BCJ3504].

Tris-*ortho*-lithiation of tris[2-(diethylsulfamoyl)phenyl]bismuthine with *tert*-butyllithium followed by treatment with 3 equiv. of diaryliodobismuthine yields a branched tetrameric bismuthine, which, on similar treatment, leads to a dendrimer-type bismuthine [97CC2295].

Synthesis of a dendritic Bi_{10}-bismuthine

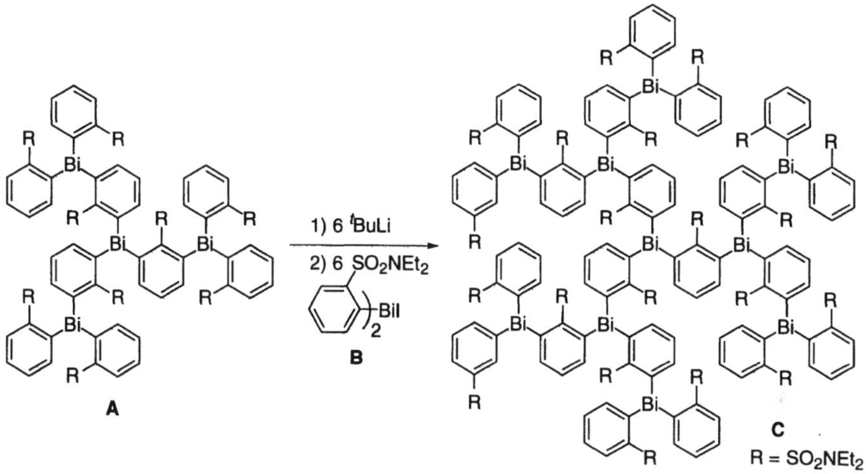

To a solution of Bi_4-bismuthine (**A**) (183 mg, 0.067 mmol) in THF (6 ml) was added *tert*-BuLi (pentane solution, 1.5 M × 0.29 ml, 0.44 mmol) at −78°C. The solution turned reddish brown within 5 min. The mixture was allowed to warm to 0°C and stirred for an additional hour at this temperature, during which time the color changed to yellowish brown. The resulting solution was cooled to −78°C and a solution of diaryliodobismuthine (**B**) (304 mg, 0.400 mmol) in THF (3 ml) was added. The resulting solution was warmed gradually to room temperature and stirred for an additional hour. Water (15 ml) was poured in and the aqueous phase was extracted with CH_2Cl_2 (20 ml × 3). The combined extracts were dried over $MgSO_4$ and the solvent was evaporated under reduced pressure to leave a product mixture, which was separated by column chromatography on silica gel with hexane–ethyl acetate as the eluent to afford crude Bi_{10}-bismuthine (**C**) (86 mg, ca. 20%) and a mixture of Bi_n-bismuthines (87 mg; $n = 6$–9). Analytically pure **C** was obtained after successive purification by gel permeation chromatography (GPC) and silica gel column chromatography (m.p. 180–185°C) [98OM4049].

(c) With organocadmium reagent (method C) Perfluoroalkyl- and perfluoroarylbismuthines have been mainly prepared by the reaction between a bismuth halide and a perfluoroalkyl- or a perfluoroarylcadmium compound [87JOM(334)323, 94JFC(69)219].

$$BiX_3 + 3/2\ R_2Cd \longrightarrow R_3Bi + 3/2\ CdX_2$$

Synthesis of tris(pentafluorophenyl)bismuthine

Bismuth bromide (1.84 g, 4.11 mmol) and $Cd(C_6F_5)_2$–diglyme solvate (3.83 g, 6.60 mmol) were dissolved in acetonitrile (10 ml) and the solution was heated at 50°C for 12 h. The solvent together with traces of diglyme and pentafluorobenzene were distilled off, and the gray solid residue was sublimed twice at 60°C/2 × 10^{-3} mmHg to give the product (2.07 g, 71%) [87JOM(334)323].

(d) With organoaluminum reagent (method D) Trialkylaluminum and lithium tetraalkylaluminate can be used for the preparation of trialkylbismuthines.

$$BiX_3 + AlR_3 \longrightarrow R_3Bi$$
$$BiX_3 + LiAlR_4 \longrightarrow R_3Bi$$

Synthesis of triethylbismuthine

$$BiCl_3 + LiAlEt_4 \xrightarrow[Et_2O]{} Et_3Bi$$

Lithium tetraethylaluminate (10.0 g, 67 mmol) in ether (50 ml) was treated with freshly sublimed bismuth chloride (15.0 g, 46 mmol) in the same solvent (50 ml). A large quantity of gray precipitate was formed. After refluxing for 1 h and subsequent hydrolysis, triethylbismuthine was extracted with ether. The bulk of the solvent was removed by fractional distillation through an efficient column and the residue was further fractionated under reduced pressure to give pure bismuthine (12.0 g, 85%), b.p. 90°C/48 mmHg [62AJC710].

(e) With organosodium reagent (method E) Sodium acetylides are used successfully for the preparation of mixed diaryl- and dialkylalkynylbismuthines [62ZAAC(317)54, 91ZN(B)1319]. Diaryl- and dialkylhalobismuthines are employed as the starting materials. Attempts to synthesize monoorganyldialkynylbismuthines from dihalo(organyl)bismuthines and 2 equiv. of sodium acetylides result in the formation of a mixture of $R_2BiC{\equiv}CR'$ and $RBi(C{\equiv}CR')_2$, which undergo a subsequent redistribution to give a mixture of $R_2BiC{\equiv}CR'$ and $Bi(C{\equiv}CR')_3$ [91ZN(B)1319]. Bismuthines bearing the cyclopentadienyl group are prepared from the corresponding sodium cyclopentadienide and bismuth chloride [60CB1417].

$$BiX_3 + 3\,RNa \longrightarrow R_3Bi + 3\,NaX$$
$$R_2BiX + R'C{\equiv}CNa \longrightarrow R_2BiC{\equiv}CR' + NaX$$

Synthesis of diphenyl(phenylethynyl)bismuthine

$$Ph_2BiCl + NaC{\equiv}CPh \xrightarrow[\text{liq } NH_3\text{–THF}]{} Ph_2BiC{\equiv}CPh$$

To a suspension of sodium phenylacetylide (2.21 g) in liquid ammonia (75 ml) placed in a 250 ml flask equipped with a stirring bar, a dropping funnel and a gas inlet apparatus, was slowly added chlorodiphenylbismuthine (6.45 g) dissolved in THF (50 ml). The reaction mixture turned yellow. After filtration, the filtrate was evaporated under reduced pressure at room temperature. Recrystallization of the residue from absolute ether gave the product as a crystalline solid (5.67 g, 75%) [62ZAAC(317)54].

(f) With other organometallic reagents (method F) Bismuthines containing the ferrocenyl or cymanthrenyl groups have been prepared from the corresponding ferrocenylsilver or cymanthrenylsilver compounds and a bismuth halide [67DOK(177)1352, 69IZV1827, 71DOK(198)590].

Triphenylbismuthine can be prepared by treating trifluoro(phenyl)silane with a mixture of bismuth hydroxide, hydrofluoric acid, and ammonium fluoride in a platinum dish [66CB1609, 69FRP1559505]. Organomercury and organozirconium compounds have been employed for the preparation of some tertiary heterocyclic bismuthines (Section 4). Dialkylzincs are also used for the preparation of trialkylbismuthines [1887CB1516, 1888CB2035].

Meta- or *para-*phenylene-bridged oligobismuthines are prepared by the reaction of chlorodiphenylbismuthine and the corresponding dicuprobenzene bis(dimethyl sulfide) complexes [98OM4049]. The products can be isolated by gel permeation chromatography (GPC) separation.

2.1.2.1.2. Onium routes (method G) Onium compounds have been used as the aryl source for the preparation of triarylbismuthines.

$$[ArN_2^+][BiCl_4^-] \xrightarrow{\text{Metal}} Ar_3Bi$$

$$[ArN_2^+][BF_4^-] + Bi \longrightarrow Ar_3Bi$$

$$3\,[ArAr'I^+][Cl^-] + 2Bi \longrightarrow Ar_3Bi + BiCl_3 + 3Ar'I$$

Complete replacement of halogen atoms in bismuth halides by aryl radicals generated from the decomposition of a diazonium compound in the presence of a metal catalyst results in the formation of triarylbismuthines. Arylazocarboxylic acid salts are also used as the aryl group source [53VMU119]. The decomposition of diaryliodonium salts in the presence of bismuth chloride or metallic bismuth also produces triarylbismuthines. Radical-induced aryl ligand exchange of diazonium salts with triarylbismuthines leads to a mixture of unsymmetrical triarylbismuthines [86T3111]. When unsymmetrical iodonium salts are used, the bismuth atom is preferentially arylated by a more electronegative aryl group [58DOK(122)1032].

Synthesis of tris(4-methylphenyl)bismuthine

$$[4\text{-}MeC_6H_4N_2^+][BiCl_4^-] \xrightarrow[Me_2CO]{Cu} (4\text{-}MeC_6H_4)_3Bi$$

A suspension of *p*-toluenediazonium salt–bismuth chloride complex (32.5 g) in cold acetone (100 ml) was treated with copper powder (52.8 g). An evolution of gas took place in several minutes and the solution became dark brown in color. After 1.5 h, an equal volume of 25% aqueous ammonia was added and the mixture was diluted to 450 ml with water. After standing for 1 h, the solid precipitate was filtered, dried, and extracted with dry chloroform. The extract was evaporated and the solid residue was crystallized from methanol to give tris(4-methylphenyl)bismuthine (3.07 g), m.p. 116–117°C [39JA3586].

2.1.2.1.3. From alkali bismuthides (method H) Alkali metal diaryl- and dialkylbismuthides (R_2BiM; M = Li, Na, K), generated in situ from the corresponding halodiorganylbismuthines and alkali metal in liquid ammonia or THF at low temperatures, readily react with alkyl and aryl halides ($R'X$) to give unsymmetrical triorganylbismuthines of the type $R_2R'Bi$ [41JA212, 85ZN(B)1476, 88ZN(B)739]. When treated with oxiranes, ethylene sulfide or propyleneimine, sodium diorganylbismuthides undergo the S_N2' type reaction to produce β-functionalized unsymmetrical triorganylbismuthines [95ZAAC(621)1403].

By treating a solution of sodium diarylbismuthide with acyl-, ethoxycarbonyl- or nitrophenyldiazonium tetrafluoroborate, triarylbismuthines bearing a substituent group incompatible with the Grignard or organolithium reagents can be obtained [97OM3565]. Treatment of lithium diphenylbismuthide with 4-bromophenol gives a small amount of (4-hydroxyphenyl)diphenylbismuthine as an unstable solid [41JA212].

Synthesis of bis(4-chlorophenyl)(1-naphthyl)bismuthine

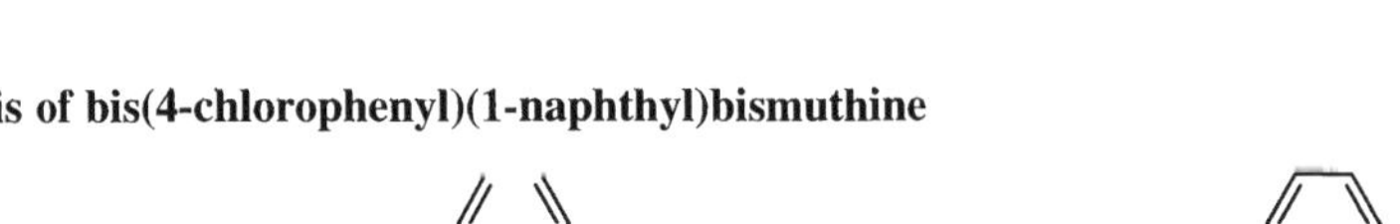

To a dark red solution of sodium bis(4-chlorophenyl)bismuthide, prepared from bis(4-chlorophenyl)iodobismuthine (2.8 g, 5 mmol), sodium (0.23 g, 10 mmol) and liquid ammonia (150 ml), was added 1-iodonaphthalene (1.27 g, 5 mmol) in ether (10 ml). Evaporation of ammonia left a residue, which was worked up as usual to give the expected triarylbismuthine (0.3 g, 10.8%), m.p. 136–137°C [41JA212].

2.1.2.1.4. Via transmetallation (method I) Metal–bismuth exchange reaction occurs between triarylbismuthine and alkyllithium or alkylsodium reagents.

$$Ar_3Bi + 3\,RM \longrightarrow 3\,ArM + R_3Bi$$
$$Ar_2Ar'Bi + Ar''Li \longrightarrow Ar_2Ar''Bi + Ar'Li$$

Although this methodology has a limited use in the preparation of triorganylbismuthines, it can be conveniently used for the selective synthesis of triarylbismuthines bearing three different aryl groups. Thus, the aryl ligand with a *tert*-butylsulfonyl group at *ortho* position, when placed in unsymmetrical triarylbismuthines, can be displaced preferentially by a given aryl group through transmetallation with aryllithium [92CC1143].

Synthesis of 4-methoxyphenylbis(4-methylphenyl)bismuthine

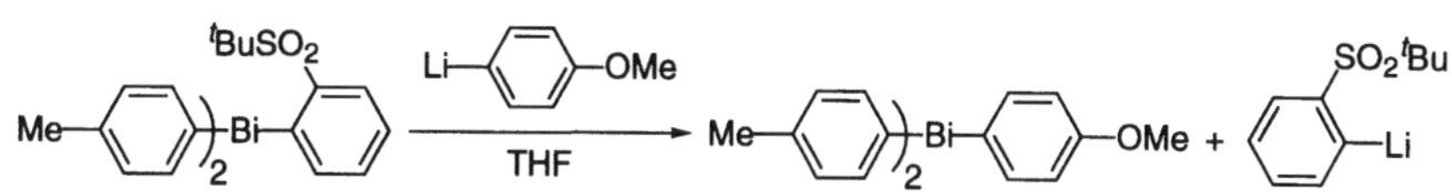

When (2-*tert*-butylsulfonylphenyl)bis(4-methylphenyl)bismuthine (588 mg, 2 mmol) was treated with 4-methoxyphenyllithium (ca. 2 mmol) in THF (10 ml) at −78°C, selective displacement occurred immediately to form 4-methoxyphenylbis(4-methylphenyl)bismuthine in 95% yield, together with *tert*-butyl phenyl sulfone [92CC1143].

2.1.2.1.5. From bismuth amides (method J) Tris(cyclopentadienyl)bismuthine is prepared by treating tris(dialkylamino)bismuthine with excess cyclopentadiene at low temperatures [95JOM(485)149]. A similar Bi–N bond cleavage reaction is employed for the synthesis of cyclopentadienyldimethylbismuthine [75JOM(88)329]. Ketene inserts into the Bi–N bonds of tris(dialkylamino)bismuthines to give tris(dialkylcarbamoylmethyl)bismuthines [87BCJ1564].

$$Bi(NR_2)_3 + 3\,CpH \longrightarrow Cp_3Bi + 3\,HNR_2$$
$$Bi(NR_2)_3 + 3\,CH_2{=}C{=}O \longrightarrow Bi(CH_2CONR_2)_2$$

Synthesis of tris(cyclopentadienyl)bismuthine

$$Bi(NMe_2)_3 + 3\,CpH \xrightarrow[THF]{} BiCp_3 + 3\,HNMe_2$$

Monomeric cyclopentadiene (10 g, 150 mmol) was added to a stirred solution of $Bi(NMe_2)_3$ (9.00 g, 26.4 mmol) in absolute THF (60 ml) at −70°C. All manipulations need to be done in the dark. After 2 h, the mixture was allowed to warm to room temperature and was stirred for an additional 2 h. The mixture turned from deep purple to red, and a dark brown solid precipitated. The solution was filtered through Celite and the red colored filtrate was concentrated to about 6 ml, cooled to −30°C, and diluted with hexane (6 ml). After drying at 0°C in vacuo, tris(cyclopentadienyl)bismuthine was obtained as deep red, transparent crystals (7.40 g, 73.2%). The filtered brown solid was dried in vacuo to give $(BiC_5H_5)_n$ as a brown, insoluble powder [95JOM(485)149].

2.1.2.1.6. Via reduction of triarylbismuth dihalides (method K) Triarylbismuth dichlorides can be reduced to triarylbismuthines by treatment with such reducing agents as hydrazine hydrate, sodium hydrosulfite, $LiAlH_4$ and organolithium reagents. This method has been used for the purification of triarylbismuthines [26JA507, 40JA665, 68LA(720)198].

$$Ar_3BiX_2 \xrightarrow{2\,e} Ar_3Bi + 2\,X^-$$

Synthesis of triphenylbismuthine

$$Ph_3BiCl_2 + H_2NNH_2 \xrightarrow[EtOH]{} Ph_3Bi$$

To a stirred suspension of triphenylbismuth dichloride (2.56 g, 5 mmol) in 95% ethanol (60 ml) was added hydrazine hydrate (2.0 g). Reduction occurred immediately. The solution was decolorized and triphenylbismuthine precipitated out of solution. Stirring was continued for 1 h and then the mixture was poured into water (300 ml). The aqueous suspension was extracted with ether, the organic layer was separated and dried, and the solvent was removed to give the product (2.1 g, 95.4%), m.p. 75–76°C [40JA665].

2.1.2.1.7. Via thermolysis of bismuth sulfinates (method L) Thermolysis of bismuth tris(arenesulfinate), which can be prepared from bismuth acetate and arenesulfinic acid in glacial acetic acid, leads to the formation of triarylbismuthines [72AJC2107].

$$Bi(O_2SAr)_3 \xrightarrow{\Delta} Ar_3Bi + 3\,SO_2$$

Synthesis of tris(4-methylphenyl)bismuthine

$$(4\text{-}MeC_6H_4SO_2)_3Bi \xrightarrow{\Delta} (4\text{-}MeC_6H_4)_3Bi$$

Bismuth tris(4-methylbenzenesulfinate) (ca. 0.1–0.2 g) was heated in a sublimation apparatus at 200°C/10^{-2}–10^{-3} mmHg. After 5 h, the crude sublimed product was collected and recrystallized from aqueous MeOH to give the expected bismuthine in 57% yield, m.p. 118–119°C [72AJC2107].

2.1.2.1.8. Via electrochemical alkylation (method M) Electrolysis of organoaluminum or organoboron compounds using a sacrificial bismuth anode gives the corresponding tertiary bismuthines, whereas the electrolysis of vinyl halides, alkyl nitriles, carboxylic acids, or acrylonitrile at a sacrificial bismuth cathode yields the corresponding functionalized trialkylbismuthines [74EL(10)1424, 62LA(652)1].

$$3\ RM\ (3\ RX) + Bi\ (electrode) \longrightarrow R_3Bi$$

Synthesis of tris(2-cyanoethyl)bismuthine

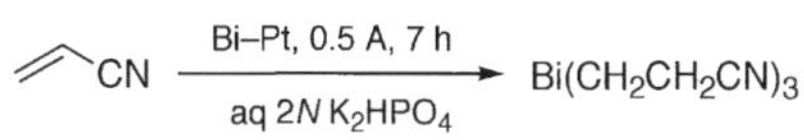

A mixture of acryrlonitrile (24.8 g) and aqueous solution of 2*N* K_2HPO_4 (145 ml) was electrolyzed (0.5 A, 7 h) at 22°C in a divided cell with a bismuth cathode and a platinum anode. During electrolysis, 4.81 g of the anode was consumed. The organic layer of the solution in the anode compartment was separated and cooled to −10°C, giving tris(2-cyanoethyl)bismuthine as white needles (8.22 g). The yields based on the current and electrode were 50 and 97%, respectively [74EL(10)1424].

2.1.2.1.9. From organic halides and bismuth metal (method N) The reaction between sodium– or potassium–bismuth alloy and alkyl or aryl halides leads to the formation of trialkyl- or triarylbismuthines, respectively [1897CB2843]. Organyl radicals generated in the gas phase react with a bismuth mirror to form organylbismuthines [1889LA(251)323, 77IC1823].

$$3\ RX + Bi\text{-}Na(K) \longrightarrow R_3Bi + 3\ NaX\ (KX)$$

$$3\ R\bullet + Bi \longrightarrow R_3Bi$$

2.1.2.1.10. Via dismutation (method O) Mixed alkyl(perfluoroalkyl)bismuthines are obtained from the dismutation reaction between trialkylbismuthines and perfluoroalkyl iodides (R_fI) [63AJC636]. Cyanodiphenylbismuthine is formed by the reaction of triphenylbismuthine and iodonitrile [15JCS(107)16].

$$R_3Bi + R_fI \longrightarrow R_2R_fBi + RI$$

$$R_3Bi + ICN \longrightarrow R_2BiCN + RI$$

Synthesis of dimethyl(trifluoromethyl)bismuthine

$$Me_3Bi + CF_3I \xrightarrow{\Delta} Me_2BiCF_3 + MeBi(CF_3)_2$$

Trimethylbismuthine (22.1 g, 86.5 mmol) and trifluoroiodomethane (21.6 g, 110 mmol) were heated together in a sealed tube at 100°C for 12 h. Fractional distillation through the traps, cooled to −46, −95, −135 and −196°C, respectively, gave a preliminary separation of the reaction products. The trap cooled to −196°C contained a trace of fluoroform, which at −135°C contained most of the unreacted trifluoroiodomethane, while that at −95°C contained most of the methyl iodide formed in the reaction. The highest boiling fraction of the reaction products all consisted of bismuth-containing materials, together with traces of methyl iodide and trifluoroiodomethane. The individual components of this fraction were isolated by vapor phase chromatography, the major constituents being dimethyl(trifluoromethyl)bismuthine (21.8 g, 82%) and methylbis(trifluoromethyl)bismuthine (5.64 g, 18%) [63AJC636].

2.1.2.1.11. From dibismuthines (method P) Diazomethane inserts into the Bi–Bi bond of tetraphenyldibismuthine to form two Bi–C bonds, resulting in the formation of bis(diphenylbismuthino)methane [84JCS(D)2365]. When diphenylbismuth phenyltelluride ($Ph_2BiTePh$) or phenylselenide ($Ph_2BiSePh$) is used instead of the dibismuthine, the corresponding bismuthine (Ph_2BiCH_2EPh; E = Te, Se) is formed [88IC3730] (Section 2.2.1.2.5). No Bi–S insertion is observed, however, with diphenylbismuth phenylsulfide (Ph_2BiSPh).

$$R_2BiBiR_2 + CH_2N_2 \longrightarrow R_2BiCH_2BiR_2$$

$$R_2BiBiR_2 + R'Li \longrightarrow R_2BiR' + R_2BiL$$

Tetramethyldibismuthine reacts with butyllithium in hexane to generate a mixture of trialkylbismuthines of the type $Me_nBu_{3-n}Bi$ ($n = 0$–3) along with lithium dimethylbismuthide [83OM1859].

Synthesis of bis(diphenylbismuthino)methane

$$Ph_2BiBiPh_2 + CH_2N_2 \xrightarrow[Et_2O]{} Ph_2BiCH_2BiPh_2$$

Tetraphenyldibismuthine (0.466 g, 0.642 mmol) dissolved in ether (75 ml) was treated with an excess of diazomethane in the same solvent, with careful exclusion of light. Within 5 min the reaction was complete as evidenced by the change in color from orange to yellow. The solution was concentrated almost to dryness under reduced pressure, hexane (25 ml) was added, and further evaporation of the more volatile solvent under reduced pressure resulted in precipitation of the colorless product. Cooling to about −30°C, filtration and drying in vacuo gave bis(diphenylbismuthino)methane (0.233 g, 49%) [84JCS(D)2365].

2.1.2.1.12. Metal coordination (method Q) Reaction of triphenylbismuthine with triamminechromium tricarbonyl in dioxane at 120°C leads to a mixture of chromium tricarbonyl π-complexes of triphenylbismuthine [81IZV1874]. Extraction and recrystallization permit separation of three types of bismuthine.

Synthesis of chromium tricarbonyl π-complexes of triphenylbismuthine

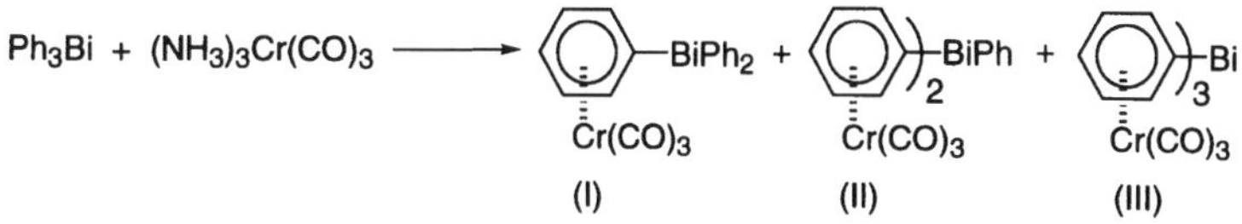

A mixture of triphenylbismuthine (4.4 g, 10 mmol), triamminechromium tricarbonyl (5.7 g, 30 mmol), and dioxane (40 ml) was heated with stirring for 7 h. After separation of the calculated amount of ammonia, the cooled mixture was filtered, the clear yellow solution was evaporated in vacuo, and the residue was treated with hot hexane. The solution was decanted and the residue was dried in vacuo. Extraction with hot cyclohexane yielded yellow crystals (0.73 g) soluble in benzene and acetone. After recrystallization from ethanol, **I** was obtained. Extraction with hot benzene of a portion of the product, after the separation of **I**, yielded a yellow compound (3.05 g), which was recrystallized from benzene to give **II**. Recrystallization of the residue from dioxane gave **III** (2.62 g) [81IZV1874].

2.1.2.1.13. Transformation of the substituted groups (method R) Transformation of the substituted groups on the organyl ligands of triorganylbismuthine yields another type of tertiary bismuthine. The carbonyl group in (formylferrocenyl)bis(4-methylphenyl)bismuthine can be converted to the imino group by treatment with excess benzylamine under reflux in benzene [96CC1693].

$$R_3Bi \longrightarrow R'_3Bi$$

$$R_2BiR' \longrightarrow R_2BiR''$$

Organogallium and organoindium compounds containing alkyldiphenylbismuthine ligands are synthesized by the reaction of triisopropylgallium or -indium with the corresponding alcohol bearing the bismuth atom in the alkyl chain [95ZAAC(621)1288].

O-tert-Butyldimethylsilyl groups on the side chains in triarylbismuthines are deprotected by tetrabutylammonium fluoride or *p*-toluenesulfonic acid to yield non-ionic water-soluble triarylbismuthines bearing neutral hydroxyl groups [98JCS(P1)2511].

Synthesis of tris{2-[*N*,*N*-bis(2-hydroxyethyl)sulfamoyl]phenyl}bismuthine

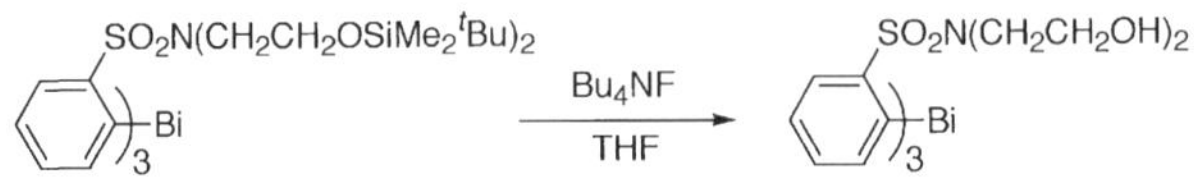

To a THF solution (15 ml) of tris{2-[*N*,*N*-bis(2-*tert*-butyldimethylsiloxy)ethyl]sulfamoyl}phenyl)bismuthine (355 mg, 0.218 mmol) was added tetrabutylammonium fluoride (THF solution; 1.35 ml, 1.35 mmol) at 0°C and the resulting mixture was stirred at room temperature for 3 h. Saturated aqueous NH_4Cl (15 ml) was added and the organic phase was removed by decantation. The aqueous phase was washed with ether (10 ml × 2) and then allowed to stand overnight. The deposit was filtered off, washed with water and recrystallized from 6% ethanol–water to yield tris{2-[*N*,*N*-bis(2-hydroxyethyl)sulfamoyl]phenyl}bismuthine (125 mg, 61%), m.p. 202–204°C [98JCS(P1)2511].

2.1.2.1.14. Disproportionation of mixed triorganylbismuthine (method S)

Mixed trialkylbismuthines of the type Me_nBiR_{3-n} ($n = 1$ or 2) are thermolabile liquids and readily undergo disproportionation to give a mixture of Me_3Bi, Me_2BiR, $MeBiR_2$ and R_3Bi [85ZN(B)1476]. A similar reaction is observed for aryldimethylbismuthines, which are converted quantitatively to trimethylbismuthine and diarylmethylbismuthines at 90°C.

$$2\ R_2BiR' \xrightleftharpoons{\Delta} R_3Bi + RBiR'_2$$

Synthesis of methylbis(4-methylphenyl)bismuthine

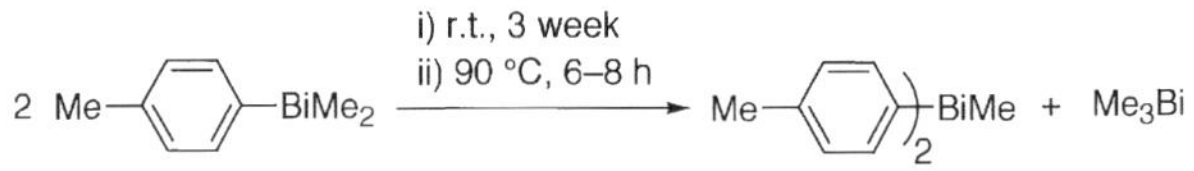

Dimethyl(4-methylphenyl)bismuthine (2 g) was allowed to stand for 3 weeks at room temperature and then heated at ca. 90°C under a stream of nitrogen (to remove trimethylbismuthine). Methylbis(4-methylphenyl)bismuthine was obtained as a colorless solid (100%) [85ZN(B)1476].

TABLE 2.1
Symmetrical triorganylbismuthines

Compound R_3Bi R	Synthetic method[a]	Yield (%)	M.p. (°C) or b.p. (°C/mmHg)	Physical data	Reference
Me	A	25	106–107/760	–	[17ZAAC(100)249]
	A	–	108/760	^{1}H-, ^{13}C-NMR	[61JINC(17)43]
	A	50–60	102–106/760	–	[66JOM(6)259]
	A	56	–	–	[97OM1655]
	F	–	110/760	–	[1887CB1516]
	–	–	–	^{1}H-NMR	[67IZV1629]
Et	A	88	123/150	–	[37JA935]
	A	46	36/1	^{1}H-NMR, MS	[86OCS(3)638]
	A	12.4	–	–	[97OM1655]
	D	86	104–105/76		[59IZV1942]
	D	85	90/48	–	[62AJC710]
	F	–	107/79	–	[1887CB1516]
	M	94	69–70/10	–	[62LA(652)1]
	S	–	–	–	[35NAT(135)1038]
	–	–	–	^{1}H-NMR	[83ZN(B)125]
Pr	A	70	127/50	–	[31BSF(49)187]
	D	70.9	89–91/10	–	[59IZV1942]
	–	–	–	^{1}H-NMR	[83ZN(B)125]
^{i}Pr	A	43	67–70/5	^{1}H-NMR, MS	[83ZN(B)125]
Bu	A	60	173/50	–	[31BSF(49)187]
	D	6	–	–	[64AG99]
	I	66	–	–	[39JA1170]
	P	–	–	^{1}H-NMR, MS	[83OM1859]
	–	–	–	^{1}H-NMR, MS	[83ZN(B)125]
^{i}Bu	D	87.2	115–116/9	–	[59IZV1942]

TABLE 2.1 *(continued)*

Compound R_3Bi R	Synthetic method[a]	Yield (%)	M.p. (°C) or b.p. (°C/mmHg)	Physical data	Reference
^{t}Bu	F	–	160–162/74	–	[1888CB2035]
	A	17.9	43–52/0.2	IR	[75JOM(87)83]
	–	–	–	^{1}H-NMR, MS	[83ZN(B)125]
	A	12	40/0.1	–	[84CB1999]
C_5H_{11}	A	60	157–158/7	–	[31BSF(49)187]
	S	–	–	–	[35NAT(135)1038]
$^{i}C_5H_{11}$	F	–	190–200/70	–	[1888CB2035]
$C_{12}H_{25}$	–	–	–	–	[39JPH(67)17]
$C_{16}H_{33}$	–	–	–	–	[39JPH(67)17]
$PhCH_2CH_2$	A	50	–	–	[88TL857]
$NCCH_2CH_2$	M	97	<70 (decomp.)	IR	[74EL(10)1424]
Me_2NCOCH_2	J	–	–	^{1}H-, ^{13}C-NMR, IR	[87BCJ1564]
Et_2NCOCH_2	J	–	–	^{1}H-, ^{13}C-NMR, IR	[87BCJ1564]
$Me_2AsCH_2CH_2CH_2$	A	15	–	^{1}H-NMR	[77JCS(D)711]
Me_3SiCH_2	A	35	140 (decomp.)	–	[58JA1336]
$(Me_3Si)_2CH$	B	90	148 (decomp.)	^{1}H-NMR, X-ray	[83IC3421]
Me_2PCH_2	B	94	Oil	^{31}P-NMR	[97JOM(529)151]
$CH_2{=}CHCH_2$	A	50	80/2	–	[63IZV1507]
$PhCH_2$	A	21	64.5–65.5	–	[57CB1176]
$2\text{-}ClC_6H_4CH_2$	A	56	54	–	[57CB1176]
$2\text{-}BrC_6H_4CH_2$	A	55	83–84	–	[57CB1176]
^{c}Hex	A	–	–	–	[14CB3257]
	A	56	–	–	[76CB1140]
CF_3	C	92.1	60/0.001	^{13}C-, ^{19}F-NMR, IR, MS, Raman	[87JOM(334)323]

	F	70	–	–	[94JFC(66)79]
	N	–	–	^{19}F-NMR, IR, MS	[77IC1823]
C_2F_5	C	65.3	85/0.001	^{19}F-NMR, MS	[87JOM(334)323]
C_3F_7	C	61.5	65/0.001 (sublim.)	^{19}F-NMR, MS	[87JOM(334)323]
C_4F_9	C	61.0	80/0.004 (sublim.)	^{19}F-NMR, MS	[87JOM(334)323]
C_6F_{13}	C	29.6	85/0.001 (sublim.)	^{19}F-NMR, MS	[87JOM(334)323]
C_8F_{17}	C	28.4	100/0.002 (sublim.)	^{19}F-NMR, MS	[87JOM(334)323]
CH_2=CH	A	23	158.1/760	–	[57JA5884]
	A	–	158.1/760	–	[57ZN(B)263]
cis-MeCH=CH	B	73.5	106/13	–	[63IZV1507]
trans-MeCH=CH	A	78	35–37/0.01	^{1}H-, ^{13}C-NMR	[83OM1859]
	B	72.5	100/9	–	[63IZV1507]
CH_2=CMe	B	62.5	86–87/12	–	[63IZV1507]
	A	41	62–64/3	^{1}H-, ^{13}C-NMR, MS	[99ZAAC(625)629]
EtCH=CH	A	–	–	–	[59BRP824944]
PhCH=CH	A	–	–	–	[27BPIP1]
Me_2C=CH	A	41	84–86/0.005	^{1}H-, ^{13}C-NMR, MS, UV	[83OM1859]
CF_2=CF	A	–	42–43/2	–	[61ZVK110]
C_5H_5	E	–	–	–	[57AG205]
	E	50–60	–	IR	[60CB1417]
	E	80	50 (decomp.)	^{1}H-NMR, IR	[70CB799]
	J	73.2	–	^{1}H-, ^{13}C-NMR, MS, X-ray	[95JOM(485)149]
MeC_5H_4	E	68	–	–	[60CB1417]
$C_5H_4ClFeC_5H_4$	F	67	203.5–204.5	–	[67DOK(177)1352]
$C_5H_5FeC_5H_3Cl$	F	46	230	–	[67DOK(177)1352]
$(C_5H_4Br)FeC_5H_4$	F	50	179.5–181	–	[69IZV1827]
$(CO)_3MnC_5H_4$	F	70	197.5–198.5	IR	[71DOK(198)590]

TABLE 2.1 *(continued)*

Compound R$_3$Bi	Synthetic method[a]	Yield (%)	M.p. (°C) or b.p. (°C/mmHg)	Physical data	Reference
$(CO)_3MnC_5H_4$	B	81	204	–	[77DOK(237)852]
	–	–	–	X-ray	[80KK805]
Ph	A	–	78	–	[04CB4620]
	A	–	77.6	–	[21ZAAC(115)253]
	A	84	76–77	–	[31JA1025]
	A	–	–	^{1}H-, ^{13}C-NMR	[75JINC(37)2347]
	A	75	71	IR, Raman	[95SA(A)2019]
	B	71	77–78	^{1}H-NMR, MS	[97JOM(548)223]
	F	100	–	–	[21JCS(119)913]
	F	50	78	–	[66CB1609]
	F	44	78	–	[69ZAAC(368)113]
	G	22	–	–	[41JA949]
	G	50.2	77.5–78	–	[46ZOB891]
	G	69	77–78	–	[58DOK(122)614]
	G	23	–	–	[58DOK(122)1032]
	G	7.3	78	–	[59DOK(125)330]
	G	61	76–77	–	[59DOK(125)1265]
	K	95.4–100	75–76	–	[40JA665]
	K	–	78	–	[20JCS(117)762]
	K	80	–	–	[26JA507]
	L	68–85	–	–	[71JOM(26)C10]
	L	79	77–78	–	[72AJC2107]
	N	–	82	–	[1887CB54]
	N	–	78	–	[1889LA(251)323]
	N	41	–	–	[13CB1675]
	–	–	–	^{1}H-NMR	[71VLU113]

2-DC$_6$H$_4$	A	69.5	73.5–74	^{1}H-NMR, IR, MS	[82JOM(240)C45]
2-MeC$_6$H$_4$	A	60	129	–	[26RZC(6)97]
	A	51	130	^{1}H-NMR, MS	[82JCC(12)53]
	A	–	130	^{1}H-, ^{13}C-NMR, IR, MS, Raman	[97SA(A)2363]
	G	6.4	–	–	[41JA949]
	G	38	130–131	–	[46ZOB891]
	G	54	130–131	–	[58DOK(122)614]
	K	100	–	–	[40JA665]
	N	–	128.5	–	[1897CB2843]
	–	–	–	^{1}H-NMR	[71VLU113]
3-MeC$_6$H$_4$	A	–	65	–	[21JCS(119)913]
	A	–	64	^{1}H-, ^{13}C-NMR, IR, MS, Raman	[97SA(A)2363]
	K	–	68	–	[26JA507]
	–	–	–	^{1}H-NMR	[71VLU113]
4-MeC$_6$H$_4$	A	60	–	–	[16JCS(109)250]
	A	68	120	–	[25RZC(5)298]
	A	40	114–115	^{1}H-NMR, MS	[82JCC(12)53]
	A	70	117	–	[96SC4569]
	B	53	117–119	^{1}H-NMR, MS	[97JOM(548)223]
	G	6	116–117	–	[39JA3586]
	G	6.7	–	–	[41JA949]
	G	31	119–120	–	[46ZOB891]
	G	42	118.5	–	[62IZV638]
	G	58	118–119	–	[58DOK(122)614]
	K	80	–	–	[26JA507]
	K	100	–	–	[40JA665]
	L	68–85	–	–	[71JOM(26)C10]
	L	57	118–119	–	[72AJC2107]
	N	–	120	–	[1889LA(251)323]
	–	–	–	^{1}H-NMR	[71VLU113]

TABLE 2.1 *(continued)*

Compound R_3Bi R	Synthetic method[a]	Yield (%)	M.p. (°C) or b.p. (°C/mmHg)	Physical data	Reference
2-$PhCH_2C_6H_4$	–	–	–	MS	[84OMS647]
2-$MeOCH_2C_6H_4$	A	–	–	–	[97JCS(P1)1609]
2-$Me_2NCH_2C_6H_4$	B	81	128–130	^1H-, ^{13}C-NMR, MS, X-ray	[96IC6179]
4-$^iPr_2NCH_2C_6H_4$	B	43	85 (decomp.)	^1H-, ^{13}C-NMR	[96OM5613]
4-(2-$PyCH_2$)C_6H_4	B	15	115–120 (decomp.)	^1H-, ^{13}C-NMR	[96OM5613]
4-$^iPrC_6H_4$	N	–	159	–	[1897CB2843]
3-CH_2=CHC_6H_4	A	–	–	–	[63USP3109851]
4-CH_2=CHC_6H_4	A	–	72–74	–	[66DOK(169)598]
4-CH_2=$CMeC_6H_4$	A	50	124	^1H-NMR	[92PM(33)1724]
4-PhC_6H_4	A	56	182–183	–	[36JA1820]
2-$CF_3C_6H_4$	A	–	119–121	–	[88JAP(K)174926]
3-$CF_3C_6H_4$	A	50	35–36	–	[88JAP(K)174926]
	A	52	–	–	[94S775]
	A	62	–	–	[96SC4569]
4-$CF_3C_6H_4$	A		–	–	[93OM1857]
4-NCC_6H_4	–	–	–	–	[84JAP(K)124347]
3-$HO_2CC_6H_4$	–	–	–	–	[86EUP173563]
4-$HO_2CC_6H_4$	K	10	162	–	[26RZC(6)97]
4-$EtO_2CC_6H_4$	G	–	–	–	[58DOK(122)614]
2-$Me_2NC_6H_4$	B	11	127–130	^1H-NMR, MS	[82JCC(12)53]
4-$Me_2NC_6H_4$	B	10	230	–	[41JA207]
	B	86	200 (decomp.)	^1H-, ^{13}C -NMR, X-ray	[96OM5613]
4-$Et_2NC_6H_4$	–	–	–	–	[70BRP1211595]
	–	–	–	–	[79JAP(K)83435]
3-$O_2NC_6H_4$	G	–	–	–	[58DOK(122)614]

4-$O_2NC_6H_4$	G	–	–	–	[58DOK(122)614]
	G	–	156	–	[62IZV638]
	G	16	232–235	^{1}H-NMR, MS	[86T3111]
	K	–	121	–	[26JA507]
	K	–	–	–	[25RZC(5)298]
3-HOC_6H_4	K	24	–	^{1}H-NMR, IR, MS	[98JCS(P1)2511]
2-$MeOC_6H_4$	A	30–65	169–170	–	[26JA507]
	A	34	168–171	–	[26RZC(6)97]
	A	20	150–152	^{1}H-NMR, MS	[82JCC(12)53]
	A	65	158	–	[96SC4569]
	A	–	159–161	^{1}H-, ^{13}C-NMR	[97JCS(P1)1609]
	A	–	165	^{1}H-, ^{13}C-NMR, IR, MS, Raman	[97SA(A)2363]
	G	6.4	–	–	[41JA949]
	G	46.5	167–169	–	[46ZOB891]
3-$MeOC_6H_4$	A	–	Oil	–	[94S775]
	A	–	70	^{1}H-, ^{13}C-NMR, IR, MS, Raman	[97SA(A)2363]
4-$MeOC_6H_4$	A	10	190	–	[26JA507]
	A	4	190	–	[26RZC(6)97]
	A	65	190	–	[96SC4569]
	G	1	–	–	[41JA949]
	G	18	185	–	[62IZV638]
	N	–	190	–	[1897CB2843]
2-$EtOC_6H_4$	A	78	121–122	–	[39JA1170]
	A	–	123–124	^{1}H-, ^{13}C-NMR	[97JCS(P1)1609]
4-$EtOC_6H_4$	A	51.8	–	–	[39JA1170]
	G	27	86–87	–	[58DOK(122)614]
	N	–	73	–	[1897CB2843]
2-$^i PrOC_6H_4$	A	–	101–102	^{1}H-, ^{13}C-NMR	[97JCS(P1)1609]
4-$^i PrOC_6H_4$	–	–	–	MS	[84OMS647]
2-$CH_2{=}CHCH_2C_6H_4$	B	61	60	^{1}H-, ^{13}C-NMR	[99T3377]

TABLE 2.1 *(continued)*

Compound R_3Bi R	Synthetic method[a]	Yield (%)	M.p. (°C) or b.p. (°C/mmHg)	Physical data	Reference
2-$MeOCH_2OC_6H_4$	B	46	95	^{1}H-NMR, IR	[98JCS(P1)2511]
3-$MeOCH_2OC_6H_4$	A	78	–	^{1}H-NMR, MS	[98JCS(P1)2511]
4-$MeOCH_2OC_6H_4$	A	86	60–61	^{1}H-NMR, MS	[98JCS(P1)2511]
3-FC_6H_4	A	40.9	69	^{1}H-, ^{19}F-NMR	[71JOM(30)365]
	A	–	–	^{1}H-NMR	[73JOM(47)367]
	A	–	–	–	[94S775]
	A	–	69	^{1}H-, ^{13}C-NMR, IR, MS, Raman	[97SA(A)2363]
4-FC_6H_4	A	66.1	93–94	–	[41JA207]
	A	37.2	92.4	^{1}H-, ^{19}F-NMR	[71JOM(30)365]
	A	–	–	^{1}H-NMR	[73JOM(47)367]
	A	76	92.4	–	[96SC4569]
	A	–	92	^{1}H-, ^{13}C-NMR, IR, MS, Raman	[97SA(A)2363]
2-ClC_6H_4	A	14.4	141	–	[41JA207]
	B	63	140–141	IR	[85JOM(288)145]
3-ClC_6H_4	A	46.3	61.5	^{1}H-NMR	[71JOM(30)365]
	A	–	–	^{1}H-NMR	[73JOM(47)367]
	A	51	53–57	–	[88JAP(K)174926]
	A	–	Oil	–	[94S775]
4-ClC_6H_4	A	–	116	–	[22JCS(121)104]
	A	65	–	–	[39JA1170]
	A	45.9	115.4	^{1}H-NMR	[71JOM(30)365]
	A	84	116	–	[96SC4569]
	G	8.4	–	–	[41JA949]
	G	–	115	–	[46ZOB897]
	G	31	115–116	–	[58DOK(122)614]
	G	19	–	–	[58DOK(122)1032]
	G	45	115	–	[62IZV638]

	K	78	–	–	[40JA665]
	K	62	–	–	[68LA(720)198]
	L	44	110–112	–	[72AJC2107]
4-BrC_6H_4	A	–	149	–	[22JCS(21)104]
	G	7	144.5–145	–	[41JA949]
	G	39.6	148–149	–	[46ZOB897]
	G	40	147–148	–	[58DOK(122)614]
	G	16	–	–	[58DOK(122)1032]
	G	65	148–149	–	[62IZV638]
	–	–	–	^{1}H-NMR	[71VLUI13]
4-IC_6H_4	G	21	207-208	–	[46ZOB897]
2-$MeSC_6H_4$	B	40	117–118	^{1}H-NMR, MS	[82JCC(12)53]
3-$NaO_3SC_6H_4$	–	–	–	–	[86EUP173563]
4-$H_2NSO_2C_6H_4$	G	50	233	–	[45BAU522]
	–	–	–	^{1}H-NMR	[71VLU113]
2-$Et_2NSO_2C_6H_4$	B	63	122–124	^{1}H-, ^{13}C-NMR, IR, MS	[98OM4049]
Ar (**1**)^b	B	63	106–108	^{1}H-NMR, IR, MS	[98JCS(P1)2511]
Ar (**2**)^b	R	61	202–204	^{1}H-NMR, IR, MS	[98JCS(P1)2511]
2-$Me_2AsC_6H_4$	B	37	Oil	^{1}H-NMR	[77JCS(D)711]
4-$Ph_2C^+C_6H_4$	–	–	–	UV	[83PS(18)121]
4-$Ph_2C(OH)C_6H_4$	–	–	–	UV	[83PS(18)121]
2,4-$Me_2C_6H_3$	A	43	159–161	–	[26RZC(6)97]
	N	–	175	–	[1889LA(251)323]
	–	–	–	^{1}H-NMR	[71VLUl13]
2,5-$Me_2C_6H_3$	N	–	194.5	–	[1897CB2843]
3,4-$Me_2C_6H_3$	N	–	175	–	[1889LA(251)323]
3,5-$Me_2C_6H_3$	–	–	–	–	[90EUP353461]
2-Me-5-^{i}PrC_6H_3	A	36.1	87	–	[41JA207]
5-Me-2-$MeOC_6H_3$	A	–	151–153	^{1}H-, ^{13}C-NMR	[97JCS(P1)1609]
2,4-$(MeO)_2C_6H_3$	A, B	–	–	–	[94JCS(P1)3479]
2,6-$(MeO)_2C_6H_3$	B	43	128	–	[83PS(14)253]

TABLE 2.1 *(continued)*

Compound R_3Bi R	Synthetic method[a]	Yield (%)	M.p. (°C) or b.p. (°C/mmHg)	Physical data	Reference
	B	–	–	^{17}O-NMR, X-ray	[94JCS(P1)3479]
2,6-$(EtO)_2C_6H_3$	B	45	134	–	[83PS(14)253]
2,6-$F_2C_6H_3$	A	81.3	133	^{1}H-, ^{13}C, ^{19}F-NMR, IR, MS, X-ray	[96ZAAC(622)2009]
3,4-$Cl_2C_6H_3$	A	90	131–131.5	–	[77JOM(136)185]
3,5-$Cl_2C_6H_3$	A	50	140–142	–	[77JOM(136)185]
4-Me-2-$O_2NC_6H_4$	K	–	125	–	[25RZC(5)298]
	K	–	126	–	[26JA507]
Ar (**3**)[b]	B	70	115	^{1}H-NMR, IR	[98JCS(P1)2511]
Ar (**4**)[b]	R	43	177–180	^{1}H-, ^{13}C-NMR, IR, MS	[98JCS(P1)2511]
2,4,6-$Me_3C_6H_2$	A, K	40	136–137	–	[38ZOB1839]
	A	40.6	134–135	–	[41JA207]
	A	70–90	136–138	^{1}H-NMR, IR, MS	[92BCJ3504]
	A, B	–	–	X-ray	[94JCS(P1)3479]
	A	–	136	^{1}H-, ^{13}C-NMR, IR, MS, Raman	[97SA(A)2363]
	B	75	139–140	^{1}H-, ^{13}C-NMR, IR, MS	[97ZAAC(623)941]
	–	–	–	X-ray	[80KK(6)945]
2,4,6-$Et_3C_6H_2$	A	70–90	82–84	^{1}H-NMR, IR, MS	[92BCJ3504]
2,4,6-$(CF_3)_3C_6H_2$	B	30	132–134	^{1}H-, ^{19}F-NMR, IR, MS, X-ray	[91JOM(402)55]
2,4,6-$Ph_3C_6H_2$	B	57	235	^{1}H-, ^{13}C-NMR, IR, MS, X-ray	[95JOM(485)141]
2,3,4-$(MeO)_3C_6H_2$	A, B	–	–	–	[94JCS(P1)3473]

2,4,6-$F_3C_6H_2$	A	74.7	109	^{1}H-, ^{13}C-, ^{19}F-NMR, IR, MS	[96ZAAC(622)2009]
3,4,5-$Cl_3C_6H_2$	A	52	241–243	–	[77JOM(136)185]
C_6D_5	A	65	71–72	^{2}H-, ^{13}C-NMR, MS, IR, Raman	[95SA(A)2019]
C_6F_5	A	70	95	IR	[69RACE119]
	A	30	94–96	^{19}F-NMR, IR, MS	[72INCL(8)271]
	A	45	97.5	^{19}F-NMR	[89CB803]
	C	71	60/0.002 (sublim.)	^{19}F-NMR, MS	[87JOM(334)323]
	–	–	–	X-ray	[90CB761]
C_6Cl_5	B	–	–	IR	[79JOM(171)333]
1-$C_{10}H_7$	A	–	235	–	[14JCS(105)2210]
	A	63	235	–	[96SC4569]
	G	15	235–236	–	[46ZOB891]
4-Me-1-$C_{10}H_6$	–	–	–	–	[88CL2021]
2-C_4H_3S	A	40	137.5	–	[29CB1710]
Ph·$Cr(CO)_3$	Q	31	240 (decomp.)	^{1}H-NMR, IR	[81IZV1874]
R (**5**)[b]	B	60	275–277	–	[71SAG77]
Ph(^{210}Bi)	A	–	–	–	[64JOM(2)59]
2-MeC_6H_4(^{210}Bi)	A	–	–	–	[72R436]
3-MeC_6H_4(^{210}Bi)	A	–	–	–	[66R98]
4-MeC_6H_4(^{210}Bi)	A	–	–	–	[66R98]
4-BrC_6H_4(^{210}Bi)	G	–	–	–	[72R436]
4-ClC_6H_4(^{210}Bi)	A	–	–	–	[72R436]
4-$MeOC_6H_4$(^{210}Bi)	A	–	–	–	[66R98]
4-$H_2NSO_2C_6H_4$(^{210}Bi)	G	–	–	–	[67R738]
2,4,6-$Me_3C_6H_2$(^{210}Bi)	A	–	–	–	[66R376]
$[Me_4N^+][Bi(CF_3)_4^-]$	–	quant.	–	^{13}C-, ^{19}F-NMR	[99JCS(D)657]

[a] For notation, see Section 2.1.2.1.
[b] For structures, see Scheme 2.1.

TABLE 2.2
Mixed triorganylbismuthines

Compound			Synthetic method[a]	Yield (%)	M.p. (°C) or b.p. (°C/mmHg)	Physical data	Reference
R^1	R^2	R^3					
$R^1_2R^2Bi$							
Me	CF_3		O	82	121/760	IR	[63AJC636]
	CCl_3		P	–	–	^{1}H-NMR, MS	[83OM1859]
	C_2F_5		O	84	130/760	IR	[63AJC636]
	C_3F_7		O	92	137/760	IR	[63AJC636]
	Et		H	79	32/19	^{1}H-NMR, MS	[85ZN(B)1476]
	Pr		H	85	27–28/3–4	^{1}H-NMR, MS	[85ZN(B)1476]
	^{i}Pr		H	72	43/20	^{1}H-NMR, MS	[85ZN(B)1476]
	Bu		H	70	55–57/8	^{1}H-NMR, MS	[85ZN(B)1476]
			P	48.1	–	^{1}H-NMR, MS	[83OM1859]
	$CH_2{=}CHCH_2$		H	90	28–29/5	^{1}H-NMR, MS	[85ZN(B)1476]
	$PhCH_2$		A	52	64–65/0.1	^{1}H-NMR, MS	[83OM1859]
	$MeCH(OH)CH_2$		H	66	–	^{1}H-, ^{13}C-NMR	[95ZAAC(621)1403]
	Me_3SiCH_2		H	33	–	^{1}H-NMR, MS	[88ZN(B)739]
	Me_3GeCH_2		H	50	–	^{1}H-NMR	[88ZN(B)739]
	Me_3SnCH_2		H	33	–	^{1}H-NMR	[88ZN(B)739]
	$EtO_2CC(N_2)$		J	93.2	210 (decomp.)	^{1}H-, ^{13}C-NMR, IR	[75JOM(93)339] [77JOM(129)55]
	$MeC{\equiv}C$		E	65	–	^{1}H-, ^{13}C-NMR	[91ZN(B)1319]
	^{t}BuC≡C		E	64	–	^{1}H-, ^{13}C-NMR	[91ZN(B)1319]
	$PhC{\equiv}C$		E	50	–	^{1}H-, ^{13}C-NMR	[91ZN(B)1319]
	C_5H_5		J	49.6	–	^{1}H-NMR, IR, MS	[75JOM(88)329]
	$4\text{-}MeC_6H_4$		H	50	40–65/0.01	^{1}H-NMR, MS	[85ZN(B)1476]
	$4\text{-}MeOC_6H_4$		H	40	50–70/0.01	^{1}H-NMR, MS	[85ZN(B)1476]
	$2,4,6\text{-}Me_3C_6H_2$		H	47	60–85/0.01	^{1}H-NMR, MS	[85ZN(B)1476]

Ph	CF_3	O	76	123/760 (decomp.)	IR	[63AJC636]
	C_5H_{11}	A	–	–	–	[35NAT(135)1038]
	MeCH(OH)CH_2	H	71	–	^{1}H-, ^{13}C-NMR	[95ZAAC(621)1403]
	PhCH(OH)CH_2	H	76	–	^{1}H-, ^{13}C-NMR	[95ZAAC(621)1403]
	Ph	H	–	–	MS	[83ZN(B)125]
Bu	Me	P	–	–	^{1}H-NMR, MS	[83OM1859]
	Ph	H	–	–	MS	[83ZN(B)125]
CF_3	Me	O	18	132/760	IR	[63AJC636]
	Et	O	14	130/760 (decomp.)	IR	[63AJC636]
	Ph	C	–	–	^{19}F-NMR, MS	[91JOM(417)C47]
	4-MeC_6H_4	C	33	–	^{19}F-NMR	[94JFC(69)219]
	3-FC_6H_4	C	57	–	^{19}F-NMR	[94JFC(69)219]
	4-FC_6H_4	C	58	–	^{19}F-NMR	[94JFC(69)219]
	4-$CF_3C_6H_4$	C	83	–	^{19}F-NMR	[94JFC(69)219]
C_2F_5	Me	O	16	147/760	IR	[63AJC636]
C_3F_7	Me	O	8	162/760	IR	[63AJC636]
Ph	Me	B	89	105–110/0.02	^{1}H-, ^{13}C-NMR, IR, MS	[85CB1031]
	Et	B	70	155–160/0.7	^{1}H-, ^{13}C-NMR, IR, MS	[85CB1031]
		H	39	–	^{1}H-NMR, MS	[83JOM(253)C21]
	^{i}Pr	–	–	–	–	[85KK1171]
	Bu	B	91	128–130/0.02	^{1}H-, ^{13}C-NMR, IR, MS	[85CB1031]
	^{t}Bu	–	–	–	–	[85KK1171]
	C_5H_{11}	B	92	150–155/0.07	^{1}H-, ^{13}C-NMR, IR, MS	[85CB1031]
	C_6H_{13}	B	91	135–140/0.02	^{1}H-, ^{13}C-NMR, IR, MS	[85CB1031]
	c-C_5H_9	B	61	165–170/0.05	^{1}H-, ^{13}C-NMR, IR, MS	[85CB1031]
	PhC_2CH_2	A	76	–	–	[88TL857]
	$Ph_2C(OH)CH_2$	R	48	118–120	^{1}H-NMR, MS	[85CB1039]
	CF_3	C	58	–	^{19}F-NMR, MS	[91JOM(417)C47]
	Ph_3C	A	–	59–60	–	[41JA207]
	$PhSeCH_2$	P	–	–	^{1}H-NMR	[88IC3730]
	$PhTeCH_2$	P	–	–	^{1}H-NMR, IR, MS	[88IC3730]
	MeCH(NH_2)CH_2	H	68	72	^{1}H-, ^{13}C-NMR	[95ZAAC(621)1403]

TABLE 2.2 *(continued)*

Compound R^1	R^2	R^3	Synthetic method[a]	Yield (%)	M.p. (°C) or b.p. (°C/mmHg)	Physical data	Reference
Ph	MeCH(OH)CH_2		H	72	–	^{1}H-, ^{13}C-NMR	[95ZAAC(621)1403]
	PhCH(OH)CH_2		H	68	–	^{1}H-, ^{13}C-NMR	[95ZAAC(621)1403]
	2-HO-^{c}Hex		H	53	–	^{1}H-, ^{13}C-NMR	[95ZAAC(621)1403]
	$HSCH_2CH_2$		H	49	73	^{1}H-, ^{13}C-NMR	[95ZAAC(621)1403]
	$^iPr_2$GaOCH(Me)CH_2		R	>90	42 (decomp.)	^{1}H-, ^{13}C-NMR	[95ZAAC(621)1288]
	$^iPr_2$InOCH(Me)CH_2		R	>90	–	^{1}H-, ^{13}C-NMR	[95ZAAC(621)1288]
	$(Me_3Si)_2$CH		A	84	–	^{1}H-NMR, MS	[99OM328]
	PhCH=CH		A	50	–	–	[90JAPS(40)835]
	MeC≡C		E	45	69	^{1}H-, ^{13}C-NMR	[91ZN(B)1319]
	^{t}BuC≡C		E	65	94	^{1}H-, ^{13}C-NMR	[91ZN(B)1319]
	PhC≡C		E	75.3	101	–	[62ZAAC(317)54]
	4-MeC_6H_4		A	85	43–45	^{1}H-NMR, MS	[86T3111]
	2-$HCOC_6H_4$		B	–	–	–	[95OM3848]
	2-$MeCOC_6H_4$		B	23	96–98	^{1}H-, ^{13}C-NMR, IR	[95OM3848]
	4-$HO_2CC_6H_4$		–	–	–	–	[86EUP173563]
	4-CH_2=CHC_6H_4		A	50	–	^{1}H-NMR, MS	[92PM(33)1724]
			A	51.7	42–43.5	–	[63MAC(62)183]
	4-CH_2=C(Me)C_6H_4		A	60	–	^{1}H-NMR, MS	[92PM(33)1724]
	(CH_2=CH)$SiMe_2C_6H_4$		A	60	–	^{1}H-NMR, MS	[92PM(33)1724]
	2-$Me_2NC_6H_4$		B	15	72–73	^{1}H-NMR, MS	[82JCC(12)53]
	4-$Me_2NC_6H_4$		H	–	187	–	[41JA212]
	2-$Ph_2PC_6H_4$		B	30	168–170	MS, IR	[76JOM(122)351]
	2-$Ph_2AsC_6H_4$		B	35	155–157	MS, IR	[76JOM(122)351]
	2-$Ph_2SbC_6H_4$		B	57	59–61	^{1}H-NMR, MS	[82JCC(12)53]
	4-HOC_6H_4		H	6.6	179–180	–	[41JA212]
	2-$MeOC_6H_4$		B	82	74	^{1}H-NMR, MS	[82JCC(12)53]
	2-$MeSC_6H_4$		B	61	–	^{1}H-NMR, MS	[82JCC(12)53]

Ph	3-FC_6H_4	A	78	56–57	^{19}F-NMR	[73JOM(47)367]
	4-FC_6H_4	A	47	53–54	^{19}F-NMR	[73JOM(47)367]
	2-ClC_6H_4	A	22	68	MS	[82JCC(12)53]
	4-ClC_6H_4	A	33	83–83.5	–	[41JA207]
	C_6T_5	–	22.2	–	–	[89R69]
	C_6F_5	–	–	–	–	[85KK1171]
	1-$C_{10}H_7$	A	–	189–119	–	[14JCS(105)2210]
		A	–	115–116	–	[21JCS(119)913]
		A	57	114–115	–	[41JA207]
		H	40.8	–	–	[41JA212]
		I	–	118	–	[22JCS(121)104]
		K	96.2	–	–	[40JA665]
	Ph·$Cr(CO)_3$	Q	13	110	^{1}H-NMR, IR	[81IZV1874]
2-MeC_6H_4	4-ClC_6H_4	A	78	80–83	–	[83JOM(256)C1]
		A	85	88–91	^{1}H-NMR, IR, MS	[96OM1951]
	4-BrC_6H_4	–	–	–	MS	[84OMS647]
	1-$C_{10}H_7$	A	59.5	112–114	–	[41JA207]
4-MeC_6H_4	Me	S	100	44	^{1}H-NMR, MS	[85ZN(B)1476]
	$MeCH(NH_2)CH_2$	H	74	117	^{1}H-, ^{13}C-NMR	[95ZAAC(621)1403]
	$MeCH(OH)CH_2$	H	54	–	^{1}H-, ^{13}C-NMR	[95ZAAC(621)1403]
	CF_3	C	88	–	^{19}F-NMR	[94JFC(69)219]
	HC≡C	E	43.2	90–91	–	[62ZAAC(317)54]
	MeC≡C	E	50	100	^{1}H-, ^{13}C-NMR	[91ZN(B)1319]
	^{t}BuC≡C	E	84	84	^{1}H-, ^{13}C-NMR	[91ZN(B)1319]
	PhC≡C	E	83.2	91	–	[62ZAAC(317)54]
	Ph	A	85	67–69	^{1}H-NMR, MS	[86T3111]
	4-$EtO_2CC_6H_4$	H	10	–	^{1}H-NMR, IR, MS	[97OM3565]
	2-$HCOC_6H_4$	B	–	–	–	[95OM3848]
	2-$MeCOC_6H_4$	B	23	114–116	^{1}H-, ^{13}C-NMR, IR	[95OM3848]
	4-$MeCOC_6H_4$	H	11	101–104	^{1}H-NMR, IR, MS	[97OM3565]

TABLE 2.2 *(continued)*

Compound			Synthetic method[a]	Yield (%)	M.p. (°C) or b.p. (°C/mmHg)	Physical data	Reference
R^1	R^2	R^3					
4-MeC_6H_4	2-OxC_6H_4 (**1**)[b]		A	63	112–113	^{1}H-, ^{13}C-NMR, IR	[98OM1013]
	4-$O_2NC_6H_4$		G	12	129–132	^{1}H-NMR, MS	[86T3111]
			H	15	92–94	^{1}H-NMR, MS	[97OM3565]
	4-$MeOC_6H_4$		A	65	109–112	^{1}H-NMR, MS	[86T3111]
			I	95	–	–	[92CC1143]
	2-^{t}$BuSO_2C_6H_4$		B	62	147–149	^{1}H-NMR, IR, MS, X-ray	[93JCS(P1)1169]
	2-^{t}$BuSO_2$-3-DC_6H_3		B	88	148–150	^{1}H-NMR	[93JCS(P1)1169]
	Ar (**2**)[b]		B	62	159–161	^{1}H-NMR, IR, MS	[93JCS(P1)1169]
	2-$Et_2NSO_2C_6H_4$		B	62	–	^{1}H-NMR, IR, MS	[98OM4049]
	Ar (**3**)[b]		B	54	112–114	^{1}H-NMR, IR, MS	[98JCS(P1)2511]
	4-FC_6H_4		A	73	70–73	^{19}F-NMR	[77JOM(136)185]
	4-ClC_6H_4		A	55	96–97	–	[41JA207]
			I	87	–	–	[92CC1143]
	1-$C_{10}H_7$		A	42	129–130	–	[41JA207]
			B	80	133–134	^{1}H-NMR, IR, MS	[97OM3565]
			H	19.3	129–130	–	[41JA212]
			H	46	133–134	^{1}H-NMR, IR, MS	[97OM3565]
	9-Anthryl		B	51	148–151	^{1}H-NMR, IR, MS	[97OM3565]
	9-Phenanthryl		B	51	165–170	^{1}H-NMR, IR, MS	[97OM3565]
			H	50	165–170	^{1}H-NMR, IR, MS	[97OM3565]
	Ar (**4**)[b]		B	23	149–151	^{1}H-NMR, IR, UV	[96CC1693]
	Ar (**5**)[b]		B	20	102–104	^{1}H-NMR, IR, UV	[96CC1693]
	Ar (**6**)[b]		R	93	–	–	[96CC1693]
4-$HO_2CC_6H_4$	Ph		–	–	–	–	[86EUP173563]
4-$Me_2NC_6H_4$	4-FC_6H_4		A	50	116–118	^{19}F-NMR	[77JOM(136)185]
4-$O_2NC_6H_4$	4-MeC_6H_4		G	15	145–149	^{1}H-NMR, MS	[86T3111]

		H	5	150–153	^{1}H-NMR, MS	[97OM3565]
2-$MeOC_6H_4$	4-ClC_6H_4	A	86	93–95	^{1}H-NMR, IR, MS	[96OM1951]
4-$MeOC_6H_4$	Me	S	100	36	^{1}H-NMR, MS	[85ZN(B)1476]
	4-MeC_6H_4	A	16	130–134	^{1}H-NMR, MS	[86T3111]
	2-$HCOC_6H_4$	B	–	–	–	[95OM3848]
	2-$MeCOC_6H_4$	B	–	Oil	–	[95OM3848]
	4-FC_6H_4	A	30	137–139	^{19}F-NMR	[77JOM(136)185]
	1-$C_{10}H_7$	A	10.9	135–136	–	[41JA207]
4-$EtOC_6H_4$	1-$C_{10}H_7$	A	13.9	131–132	–	[41JA207]
2-$Et_2NSO_2C_6H_4$	4-MeC_6H_4	B	64	–	^{1}H-NMR, IR, MS	[98OM4049]
Ar (**3**)[b]	4-MeC_6H_4	B	69	Oil	^{1}H-NMR, IR, MS	[98JCS(P1)2511]
3-FC_6H_4	CF_3	C	56	–	^{19}F-NMR	[94JFC(69)219]
4-FC_6H_4	CF_3	C	76	–	^{19}F-NMR	[94JFC(69)219]
3-ClC_6H_4	4-FC_6H_4	A	–	–	^{19}F-NMR	[77JOM(136)185]
	PhC≡C	E	82.2	140 (decomp.)	–	[62ZAAC(317)54]
4-ClC_6H_4	2-MeC_6H_4	A	45.2	104–104.5	–	[41JA207]
	4-FC_6H_4	A	87	95–96	^{19}F-NMR	[77JOM(136)185]
	2-$CHOC_6H_4$	B	–	–	–	[95OM3848]
	2-$MeCOC_6H_4$	B	38	129–131	^{1}H-, ^{13}C-NMR, IR	[95OM3848]
	1-$C_{10}H_7$	A	46.8	138–139	–	[41JA207]
		H	10.8	136–137	–	[41JA212]
	Ar (**7**)[b]	B	36	135–137	^{1}H-NMR, IR	[98OM1711]
	Ar (**8**)[b]	B	42	115–116	^{1}H-NMR, IR	[98OM1711]
4-$CF_3C_6H_4$	CF_3	C	87	–	^{19}F-NMR	[94JFC(69)219]
3,4-$Cl_2C_6H_3$	4-FC_6H_4	–	–	–	^{19}F-NMR	[77JOM(136)185]
2,4,6-$Me_3C_6H_2$	Me	B	84	44–45	^{1}H-NMR, IR, MS	[92BCJ3504]
		S	100	51	^{1}H-NMR, MS	[85ZN(B)1476]
	PhC≡C	B	84	112–114	^{1}H-NMR, MS	[92BCJ3504]
	Ph	–	–	–	MS	[84OMS647]
	4-ClC_6H_4	A	50	130–131	–	[83JOM(256)C1]
	1-$C_{10}H_7$	A	31.4	151–151.5	–	[41JA207]
2,4,6-$^iPr_3C_6H_2$	Me	B	77	53–55	^{1}H-NMR, IR, MS	[92BCJ3504]
	PhCH=CH	B	62	139–141	^{1}H-NMR, IR, MS	[92BCJ3504]

TABLE 2.2 *(continued)*

Compound R^1	R^2	R^3	Synthetic method[a]	Yield (%)	M.p. (°C) or b.p. (°C/mmHg)	Physical data	Reference
2,4,6-$^iPr_3C_6H_2$	PhC≡C		B	–	139–141	^{1}H-NMR, IR, MS	[92BCJ3504]
	4-MeC_6H_4C≡C		B	–	127–129	^{1}H-NMR, IR, MS	[92BCJ3504]
	4-ClC_6H_4C≡C		B	–	105–107	^{1}H-NMR, IR, MS	[92BCJ3504]
3,4,5-$Cl_3C_6H_2$	4-FC_6H_4		–	–	–	^{19}F-NMR	[77JOM(136)185]
1-$C_{10}H_7$	4-BrC_6H_4		A	67	133–136	–	[83JOM(256)C1]
	4-ClC_6H_4		A	63	135	–	[83JOM(256)C1]
	2-$HCOC_6H_4$		B	17	170-172	^{1}H-, ^{13}C-NMR, IR	[95OM3848]
	2-$MeCOC_6H_4$		B	14	235–237 (decomp.)	^{1}H-, ^{13}C-NMR, IR	[95OM3848]
Ph·$Mn(CO)_3$	Ph		B	85	120	IR	[87IZV187]
Ph·$Cr(CO)_3$	Ph		Q	43	196	^{1}H-NMR, IR, MS	[92BCJ3504]
Ph	CN		E	–	210–212 (decomp.)	–	[88JAP(K)174926]
			O	–	210 (decomp.)	–	[15JCS(107)16]
			O	52	210	–	[22JCS(121)91]
			O	87	210 (decomp.)	–	[78IJC(16A)778]
2-MeC_6H_4	CN		E	–	>149 (decomp.)	–	[88JAP(K)174926]
3-MeC_6H_4	CN		E	–	140 (decomp.)	–	[88JAP(K)174926]
4-MeC_6H_4	CN		E	–	>195 (decomp.)	–	[88JAP(K)174926]
3-$CF_3C_6H_4$	CN		E	–	>125 (decomp.)	–	[88JAP(K)174926]
2-$MeOC_6H_4$	CN		E	–	93–96	–	[88JAP(K)174926]

4-$MeOC_6H_4$	CN		E	–	>175 (decomp.)	–	[88JAP(K)174926]
3-FC_6H_4	CN		E	–	>147 (decomp.)	–	[88JAP(K)174926]
4-FC_6H_4	CN		E	31	>155 (decomp.)	–	[88JAP(K)174926]
3-ClC_6H_4	CN		F	–	136–140 (decomp.)	–	[88JAP(K)174926]
	CN		F	–	145–150 (decomp.)	–	[88JAP(K)174926]
$R^1R^2R^3Bi$							
Ph	4-BrC_6H_4	1-$C_{10}H_7$	A	44	129–130	–	[83JOM(256)C1]
2-MeC_6H_4	4-$MeOC_6H_4$	4-ClC_6H_4	A	80	120–123	^{1}H-NMR, IR, MS	[96OM1951]
	2-^{i}$PrOC_6H_4$	4-ClC_6H_4	A	10	70–70.5	^{1}H-NMR, MS	[83JOM(256)C1] [84OMS647]
	2-$MeOC_6H_4$	4-ClC_6H_4	A	70	99–102	^{1}H-NMR, IR, MS	[96OM1951]
	4-$MeOC_6H_4$	4-ClC_6H_4	I	70	118–120	–	[92CC1143]
	4-$MeOC_6H_4$	2-^{t}$BuSO_2C_6H_4$	A	88	155–157	^{1}H-NMR, IR, MS	[93JC5(P1)1169]
	4-$MeOC_6H_4$	1-$C_{10}H_7$	I	88	108–110	–	[92CC1143]
	4-FC_6H_4	2-^{t}$BuSO_2C_6H_4$	A	90	167–169	^{1}H-NMR, IR, MS	[93JCS(P1)1169]
	4-ClC_6H_4	2-^{t}$BuSO_2C_6H_4$	A	95	157–159	–	[92CC1143]
	2-^{t}$BuSO_2C_6H_4$	1-$C_{10}H_7$	A	98	182–184	^{1}H-NMR, IR, MS	[93JCS(P1)1169]
4-$CF_3C_6H_4$	4-$MeOC_6H_4$	2-OxC_6H_4 (**1**)[b]	A	60	148–150	^{1}H-NMR, IR, MS	[99OM5668]
	4-$MeOC_6H_4$	2-C_4H_3S	–	–	–	^{1}H-NMR, MS	[99OM5668]
4-ClC_6H_4	2-^{i}$PrOC_6H_4$	2,4,6-$Me_3C_6H_2$	A	45	110–112	^{1}H-NMR	[83JOM(256)C1]
	2-^{i}$PrOC_6H_4$	1-$C_{10}H_7$	A	39	130–133	^{1}H-NMR, MS	[83JOM(256)C1] [84OMS647]
	2-$PhCH_2OC_6H_4$	2,4,6-$Me_3C_6H_2$	A	55	103–104	^{1}H-NMR	[83JOM(256)C1]

[a] For notation, see Section 2.1.2.1.
[b] For structures, see Scheme 2.2.

TABLE 2.3
Organylbismuthines with multiple bismuth atoms

Compound	Synthetic method[a]	Yield (%)	M.p. (°C)	Physical data	Reference
Bi_2-bismuthines					
$Me_2BiC(N_2)BiMe_2$	J	–	–	^{1}H-NMR	[75JOM(93)339]
^{c}$Hex_2BiCH_2CH_2Bi^{c}Hex_2$	–	–	–	–	[85KK1171]
$Ph_2BiCH_2BiPh_2$	H	53	91	^{1}H-, ^{13}C-NMR, IR, MS	[85CB1039]
$Ph_2BiCH_2CH_2CH_2BiPh_2$	H	57	–	^{1}H-NMR	[88ZN(B)739]
$Ph_2BiCH_2CH_2CH_2CH_2BiPh_2$	H	58	–	^{1}H-NMR	[88ZN(B)739]
$Ph_2BiCH_2CH_2CH_2CH_2CH_2BiPh_2$	H	46	–	^{1}H-NMR	[88ZN(B)739]
$Ph_2BiC{\equiv}CBiPh_2$	A, E	–	145 (decomp.)	–	[55AG233]
	E	–	133 (decomp.)	–	[62ZAAC(317)54]
$(4\text{-}MeC_6H_4)_2BiC{\equiv}CBi(4\text{-}MeC_6H_4)_2$	E	56.2	125 (decomp.)	–	[62ZAAC(317)54]
$(4\text{-}ClC_6H_4)_2BiC{\equiv}CBi(4\text{-}ClC_6H_4)_2$	B	52.6	135 (decomp.)	–	[62ZAAC(317)54]
Bi_2(A) (**1**)[b]	B	14–20	160–162	IR, UV	[67M(98)731]
	B	33	–	^{1}H-, ^{13}C-NMR, X-ray	[99JOM(584)179]
	F	27	158–160	^{1}H-, ^{13}C-NMR, MS, UV	[98OM4049]
Bi_2(B) (**2**)[b]	F	26	114–115	^{1}H-NMR, MS, UV	[98OM4049]
Bi_2(C) (**3**)[b]	F	37	143	^{1}H-, ^{13}C-NMR, IR, MS, UV	[98OM4049]
Bi_2(D) (**4**)[b]	F	2	95–100	^{1}H-NMR, IR, MS	[98OM4049]
Bi_2(E) (**5**)[b]	B	7	42–46	^{1}H-NMR, IR, MS	[98JCS(P1)2511]
Bi_2(F) (**6**)[b]	R	19	137–141	^{1}H-NMR, IR, MS	[98JCS(P1)2511]
Bi_2(G) (**7**)[b]	B	20	–	^{1}H-, ^{13}C-NMR, X-ray	[96OM887]
Bi_2(H) (**8**)[b]	B	21	120–122	^{1}H-NMR, IR, UV	[96CC1693]
Bi_2(I) (**9**)[b]	B	40	218–220	^{1}H-NMR, IR, MS, X-ray	[98OM4049]

Bi_3-bismuthines					
Bi_3(A) (**10**)[b]	F	13	55–58	^{1}H-NMR, MS, UV	[98OM4049]
Bi_3(B) (**11**)[b]	F	8	39–43	^{1}H-NMR, MS, UV	[98OM4049]
Bi_3(C) (**12**)[b]	B	18	132	^{1}H-NMR, IR, MS, UV	[98OM4049]
Bi_3(D) (**13**)[b]	B	6	102–108	^{1}H-NMR, IR, MS	[98OM4049]
Bi_3(E) (**14**)[b]	B	37	57–63	^{1}H-NMR, IR, MS	[98JCS(P1)2511]
Bi_3(F) (**15**)[b]	R	16	148–152	^{1}H-NMR, IR, MS	[98JCS(P1)2511]
Bi_4-bismuthines					
Bi_4(A) (**16**)[b]	F	9	75–79	^{1}H-NMR, MS, UV	[98OM4049]
Bi_4(B) (**17**)[b]	F	2	59–62	^{1}H-NMR, MS, UV	[98OM4049]
Bi_4(C) (**18**)[b]	B	49	162–164	^{1}H-, ^{13}C-NMR, IR, MS, UV	[97CC2295] [98OM4049]
Bi_4(D) (**19**)[b]	B	6	102–108	^{1}H-, ^{13}C-NMR, IR, MS	[97CC2295] [98OM4049]
Bi_{10}-bismuthines					
Bi_{10}(C) (**20**)[b]	B	20	180–185	^{1}H-, ^{13}C-NMR, IR, MS, UV	[97CC2295] [98OM4049]

[a] For notation, see Section 2.1.2.1.
[b] For structures, see Scheme 2.3.

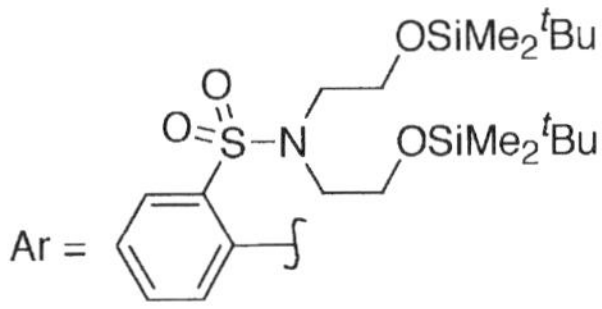

(1)

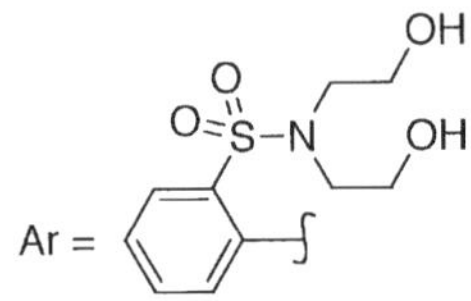

(2)

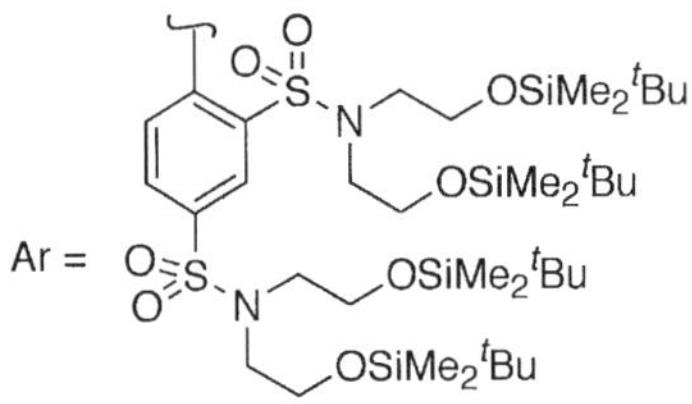

(3)

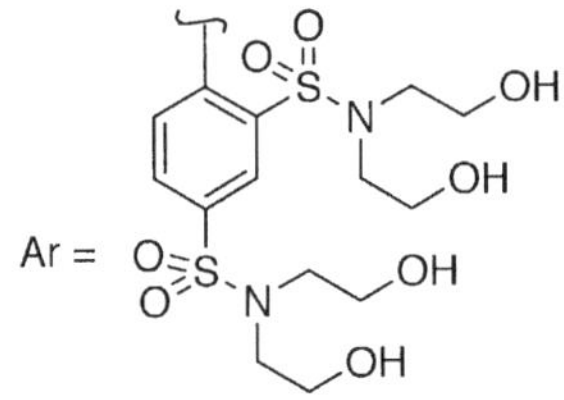

(4)

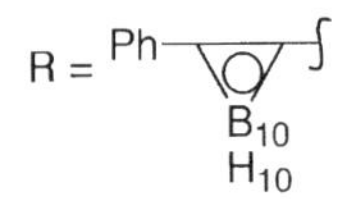

(5)

Scheme 2.1. Structures of compounds listed in Table 2.1.

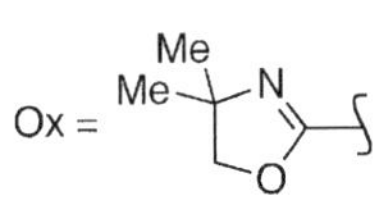

(1)

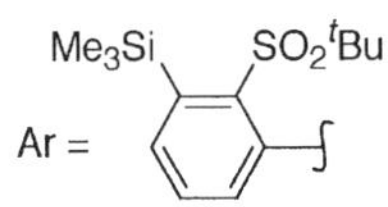

(2)

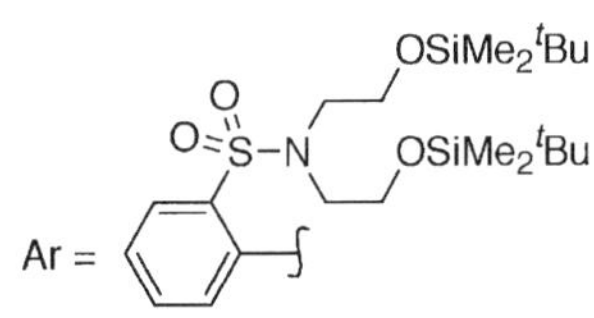

(3)

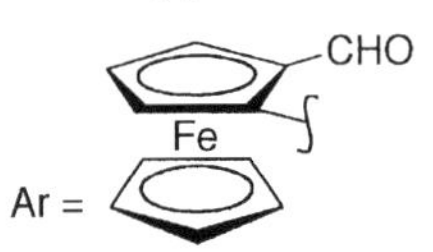

(4)

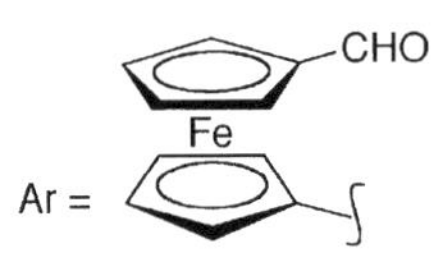

(5)

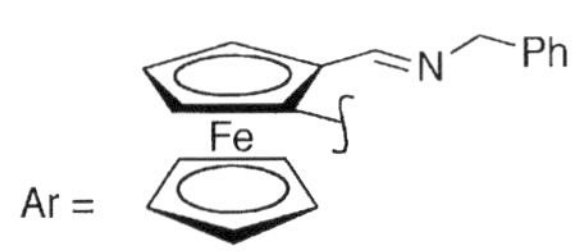

(6)

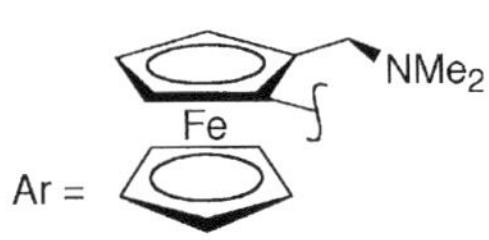

(7)

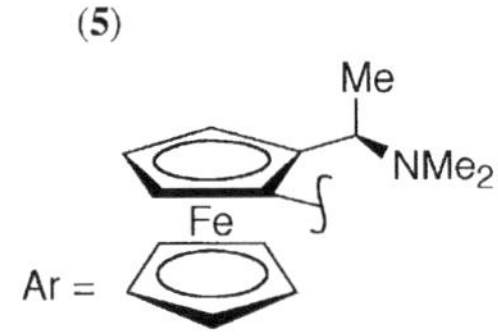

(8)

Scheme 2.2. Structures of compounds listed in Table 2.2.

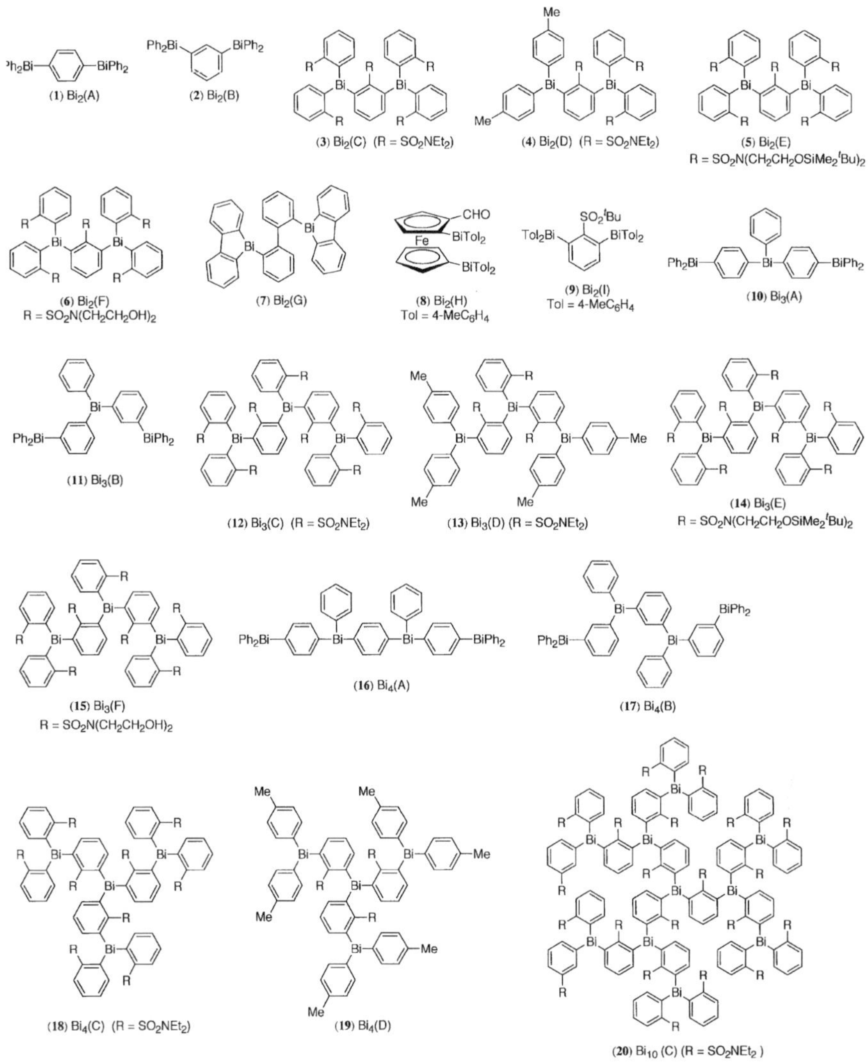

Scheme 2.3. Structures of compounds listed in Table 2.3.

2.1.2.2. Properties

Trialkylbismuthines are mainly liquids at room temperature and can be distilled under atmospheric or reduced pressure. Reported boiling points of trimethyl-, triethyl-, and tripropylbismuthines are 102–106°C/760 mmHg, 104–105°C/76 mmHg, and 89–91°C/10 mmHg, respectively [66JOM(6)259, 59IZV1942]. Trialkylbismuthines of higher molecular weight and most triarylbismuthines are solids at room temperature. Melting points of tribenzylbismuthine and triphenylbismuthine are 64.5–65.5°C [57CB1176] and 77–78°C, respectively. The physical form of mixed alkyldiaryl- and dialkylarylbismuthines depend on the structure of organyl ligands; methyldiphenylbismuthine is a liquid, while methylbis(2,4,6-trimethylphenyl)bismuthine is a solid [85CB1031, 85ZN(B)1476]. Lower members of trialkylbismuthines are spontaneously ignited in air, suggesting the ease of oxidative cleavage of the Bi–C_{alkyl} bond by molecular oxygen (Section 2.1.2.3). Although the Bi–C_{alkyl} bond in alkyldiphenylbismuthines is readily cleaved by air and light [85CB1031], coordinative or bulky aryl groups can stabilize this labile Bi–C_{alkyl} bond to such an extent that the compounds can be handled without decomposition in open air. Thus, compounds of the Ar_2BiMe type bearing 2,4,6-trimethylphenyl (mesityl) or 2,4,6-triisopropylphenyl group can be handled safely in air and purified by column chromatography [92BCJ3504].

Tertiary alkylarylbismuthines ($R_nAr_{3-n}Bi$; $n = 1$ or 2) tend to undergo dismutation on storage to produce a mixture of all possible trialkyl-, dialkylaryl-, alkyldiaryl- and triarylbismuthines [85ZN(B)1476]. Mixed trialkylbismuthines also undergo a similar reaction. In contrast to the alkyl counterparts, triarylbismuthines are generally stable and can be purified by recrystallization from an alcoholic solvent. Most of the organobismuthines are soluble in benzene, dichloromethane, chloroform, THF, diethyl ether, DMSO and EtOAc, slightly soluble in hexane, and nearly insoluble in MeOH. However, tris(3-hydroxyphenyl)bismuthine is soluble in MeOH, and multi-branched triarylbismuthines bearing the bis(2-hydroxyethyl)sulfamoyl moiety on the side chains are considerably soluble in water under neutral conditions [98JCS(P1)2511].

It has been reported that highly hindered tris[2,4,6-tris(trifluoromethyl)phenyl]bismuthine is unstable in solution even under an inert atmosphere and, hence, found in the solid state when exposed to air, probably owing to the steric crowding around the bismuth atom [91JOM(402)55]. In contrast, tris(2,4,6-triphenylphenyl)bismuthine is a thermally stable solid with a definite melting point at 235°C [95JOM(485)141]. Based on the molecular weights measured by a cryoscopic method, it has been clarified that tertiary bismuthines are not associated. For instance, triarylbismuthines and alkynylbismuthines of the type $Ar_2BiC{\equiv}CR$ give normal molecular weights cryoscopically in benzene solutions [20JCS(117)762, 62ZAAC(317)54].

Many values have been reported for the bond dissociation energy of trialkylbismuthines. They are apparently smaller in value when compared with those of lighter pnictogen counterparts. The mean gas-phase Bi–C bond dissociation energy of triphenylbismuthine, estimated by means of a rotating bomb calorimeter, is 46 kcal/mol [79JCT(11)187]. The polarographic half-wave potential of triphenylbismuthine, determined in a glyme solution with Ag/$AgClO_4$ as the reference electrode, has been found to be 3.1 V [66JA467]. The value is smaller than those of the corresponding phosphorus, arsenic and antimony compounds.

The protonic basicity (pK_b) of triphenylbismuthine has been estimated to be 8.81 by a potentiometric method in AcOH [70IC408]. The value is higher than those of phosphorus and antimony analogs obtained by the same measurement. A similar measurement in fluorosulfonic acid has shown that the K_b values of triphenylarsine and triphenylstibine are higher than that of the bismuthine [72JINC(34)2523]. The lower basicity of tertiary bismuthines is readily seen in their coordination chemistry; trialkyl- and triarylbismuthines have little use as a donor in the transition metal complex formation. The lone pair in tertiary bismuthines has inherently an *s* character due to low efficiency of the hybridization between the 6*s* and 6*p* orbitals. As a result, the number of known coordination compounds, in which triorganylbismuthines are used as donor ligands for transition metals, is very limited when compared with other nitrogen family members. The comparison of the bond angles and bond distances observed for a series of compounds of the type $[Cr(CO)_5EPh_3]$ ($E = P$, As, Sb, and Bi) by X-ray crystallographic analysis has shown that the Bi–M bond is essentially *s* in nature and the ligand strength increases in the order $Ph_3Bi < Ph_3As < Ph_3P$ [79CC639] (Section 2.4.2).

Electron diffraction analysis of gaseous trimethylbismuthine as well as X-ray crystallographic analysis of several triorganylbismuthines have established that the geometry of the bismuth atom in these compounds is essentially trigonal pyramidal (Fig. 2.1). From these analyses, the Bi–C bond distance and C–Bi–C bond angle of trimethylbismuthine are 2.267 Å and 96.7°, respectively [73JMOS(17)429]. Triphenylbismuthine has a mean Bi–C bond distance of 2.24 Å and a mean C–Bi–C bond angle of 94°, indicating a very minor contribution of 6*s*–6*p* orbital hybridization at bismuth [68JCS(A)2059].

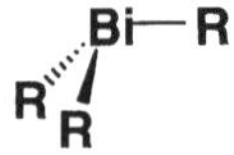

Fig. 2.1. A representative structure of triorganylbismuthines.

The average Bi–C bond distances and C–Bi–C bond angles in sterically hindered triarylbismuthines, such as tris[bis(trimethylsilyl)methyl]bismuthine (2.328 Å and 102.9°) [83IC3421], tris(2,4,6-trimethylphenyl)bismuthine (2.29, 2.31 Å and 102, 103°) [80KK945, 94JCS(P1)3479], tris[2,4,6-tris(trifluoromethyl)phenyl]bismuthine (2.367 Å and 106.0°) [91JOM(402)55], tris(2,6-dimethoxyphenyl)bismuthine (2.262 Å and 99.4°) [94JCS(P1)3479] and tris(2,4,6-triphenylphenyl)bismuthine (2.357 Å and 105.8°) [95JOM(485)141], are considerably lengthened and widened, when compared with those of parent bismuthines. The unusual bond parameters observed for the last compound are partly ascribed to the molecular packing effect, induced by THF solvate molecules. The crystal structure of the deep-red colored tris(cyclopentadienyl)bismuthine has been determined at −70°C, where the bismuth atom possesses a distorted trigonal pyramidal geometry directly bonded to three carbon atoms of the cyclopentadienyl ligands in a monohapto fashion; the mean Bi–C bond length and C–Bi–C bond angle are 2.38 Å and 94.7°, respectively [95JOM(485)149]. The bismuth center is also found to be weakly coordinated to the cyclopentadienyl ring of a neighboring molecule with a Bi–Cp(centroid) distance of 3.17 Å. A heterocyclic bismuthine bearing an alkynyl group, 10-(4-chlorophenylethynyl)phenothiabismine-5,5-dioxide, possesses a pyramidal configuration around the bismuth atom [92JCS(P1)1593]. The shorter length of the Bi–C_{alkyne} bond (2.211(8) Å) when compared with the Bi–C_{Ar} bonds (2.283(3) and 2.272(7) Å) reflects the difference in the electronic nature of the relevant carbon atoms attached to the bismuth atom. In a chromium complex [$Cr(CO)_5BiPh_3$], the average C–Bi–C bond angle is 98.7°, which is slightly wider than that of Ph_3Bi [79CC639]. This indicates that the contribution of the *s* character to the lone pair decreases slightly when the bismuth center is coordinated to a transition metal.

The dipole moment of triphenylbismuthine is reported to be essentially zero in benzene, cyclohexane and octane, suggesting that either each Ph–Bi group is effectively non-polar or that the resultant moments of three Ph–Bi bonds are compensated by the moment of the lone pair [32ZPCB401, 63JCS1739, 41JOC421]. The dipole moments observed for tris(4-methylphenyl)bismuthine (0.66 D), tris(4-fluorophenyl)bismuthine (2.59 D), and tris(4-chlorophenyl)bismuthine (2.37 D) are in good agreement with the theoretical values (0.60, 2.60 and 2.40 D, respectively) calculated by assuming a C–Bi–C bond angle of 93° [77JMOS(40)89]. The dipole moment of 1,4-bis(diphenylbismuthino)benzene is also reported to be zero in benzene solution [67M(98)731]. An average diamagnetic susceptibility of the trivalent bismuth atom has been determined to be -36.98×10^{-6} cgs units by measuring three different kinds of triarylbismuthines [58JIC(35)573]. By comparing with the obtained values for organoantimony compounds, it has been concluded that the Bi–C bond is more ionic than the Sb–C bond.

There have been many studies on the IR, Raman, MS, NMR and UV–Vis spectroscopies of tertiary bismuthines. The Raman spectrum of Me_3Bi shows two very strong broad lines at 171 and 460 cm^{-1}, which are assigned as the degenerated two stretching modes and two bending modes derived from the pyramidal structure of this compound with bismuth at the apex of the pyramid [40JCP366]. The IR and Raman spectra of Et_3Bi and $^{t}Bu_3Bi$, in which the Bi–C bending and stretching modes are shown, have also been reported [75JOM(87)83, 64JMSP(14)320]. Based on the IR spectra of a black and an orange-red form of tris(cyclopentadienyl)bismuthine, it has been concluded that the former modification consists of the Bi–C π-bonds and the latter exists as a rapid equilibrium mixture of a π-bonded and a σ-bonded form [70CB799]. This interpretation may be supported by the recent X-ray crystallographic analysis of the red Cp_3Bi [95JOM(485)149]. The vibrational spectrum of Ph_3Bi shows several characteristic heavy element sensitive bands attributable to the bending and stretching modes of the phenyl–Bi bond. In one study, the principal Bi–Ph stretching vibration is assigned to two strong bands at 237 and 220 cm^{-1} in the IR spectrum, and to a polarized band at 237 cm^{-1} and a depolarized band at 219 cm^{-1} in the Raman spectrum in benzene solutions [69AS(23)12]. The Bi–Ph bending mode is assigned to a band at 157 cm^{-1} in the IR and a band at 155 cm^{-1} in the Raman spectra. The values for the Bi–Ph bands were calculated by assuming a C_{3v} structure for Ph_3Bi, and since they are in excellent agreement with the observed ones, it is believed that triphenylbismuthine possesses a C_{3v} symmetry with bismuth at the apex.

Trimethylbismuthine exhibits broad absorptions at λ_{max} 211.5, 222.5 and 260.0 nm in the far-UV region [71JA822]. Triphenylbismuthine shows characteristic absorptions at λ_{max} 248 and 280 nm. There have been several interpretations for its UV spectrum; the interpretation most likely to be accepted is that the shoulder at 280 nm is due to the π–π^* transition of benzene, and the peak at 248 nm due to the dissociation process [61CJC171]. The initial assignment for the band at 280 nm as the n–π* transition has been negated both by theoretical work and the study of the temperature dependence of UV spectra for a series of triaryl compounds of element groups 13, 14 and 15 [72ZSK(13)701, 76IZV338, 77IZV1035]. The UV spectra of 1,4-bis(diphenylbismuthino)benzene and higher oligomeric polyphenylbismuth compounds bridged by a 1,3- or 1,4-phenylene unit are very similar to that of Ph_3Bi, suggesting that the phenylene bridge does not contribute to the delocalization of π-electrons over the molecules of these oligomers [67M(98)731, 98OM4049].

The ^{1}H-NMR spectrum of trimethylbismuthine in CCl_4 shows a single peak at δ 1.08 (versus Me_4Si) [67IZV1629]. In a series of trimethyl compounds of the type Me_3E (E = N, P, As, Sb, Bi), a linear relationship, $\chi = 1.03\delta + 0.797$, is observed between the δ_{Me} value and the electronegativity (χ) of the central atom. The methyl peak of methyldiphenylbismuthine appears at δ 1.45 in $CDCl_3$ (versus Me_4Si) [85CB1031]. The vinyl proton in 10-(2,2-diphenylvinyl)phenothiabismine is observed at δ 9.02 in $CDCl_3$ (versus Me_4Si) [92JCS(P1)1593]. The ^{1}H-NMR spectrum of triphenylbismuthine in CCl_4 or $CDCl_3$ exhibits the *ortho*-, *meta*- and *para*-protons at δ 7.7, 7.4 and 7.39 (versus Me_4Si) [71VLU113; 97JOM(548)223].

The NQR spectrum of triphenylbismuthine has also been reported [53PR(89)1305, 69VLU167, 74JMOS(23)329]. As bismuth has a spin of 9/2, bismuthine exhibits four lines, $\nu_1 = 29.785$, $\nu_2 = 55.214$, $\nu_3 = 83.516$ and $\nu_4 = 111.438$. The quadrupole coupling constant eQq_{zz} is 669.06 ± 0.13 MHz, indicating that the bismuth bonds have approximately 8–9% *s* character.

In the mass spectra of tertiary bismuthines, the parent peak is difficult to detect [68OMS127, 69TL111, 72OMS1183, 76OMS1019, 84OMS647]. This is probably due to the low bond dissociation energy of the Bi–C bonds. Instead, strong fragment peaks due to R_2Bi^+, RBi^+ and Bi^+ (209) are observed for most tertiary organobismuth compounds.

2.1.2.3. Reactions

Almost all reactions of trialkylbismuthines begin with the cleavage of the carbon–bismuth bond. They are rapidly oxidized in air, affording R_2BiOR, RBiO and polymeric bismuth oxides containing alkyl groups [42JA392]. Thermal decomposition of triethylbismuthine at 192–224°C gives ethane, ethylene, butane and butylenes as major gaseous products [87ZOB2034]. This result indicates that ethyl radicals are formed by homolysis of the

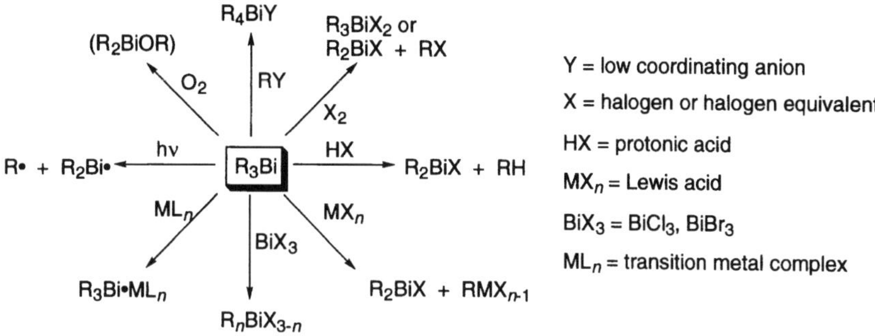

Scheme 2.4. Typical reactions of trialkylbismuthines.

Bi–C bonds of the bismuthine. Most trialkylbismuthines readily react with halogens at room temperature to form the corresponding dialkylhalobismuthines and halogenoalkanes. Thus, treatment of trimethylbismuthine with molecular chlorine results in complete cleavage of the Bi–C bonds even at low temperatures. Trimethylbismuthine is oxidized by sulfuryl chloride to trimethylbismuth dichloride at low temperatures, but it readily undergoes reductive coupling to give chlorodimethylbismuthine and methyl chloride at room temperature [94AG(E)976]. A few trialkenylbismuthines are similarly brominated by bromine to give trialkenylbismuth dibromides [63IZV1507]. The Bi–C_{alkyl} bond in the mixed alkyldiarylbismuthines is readily cleaved by halogens at room temperature to give diarylhalobismuthines as the predominant bismuth product [85CB1031, 92BCJ3504]. This type of Bi–C_{alkyl} bond is also sensitive to the action of air and light (Scheme 2.4).

Prolonged heating of a C_6D_6 solution of triphenylbismuthine gives PhC_6D_5, Ph_2 and $(C_6D_5)_2$, suggesting that phenyl radicals are formed by homolytic cleavage of the Bi–C bonds [65ZOB481]. When heated under a H_2 atmosphere, triphenylbismuthine is reduced to the metallic bismuth with evolution of benzene [30CB1110]. In contrast to alkyl derivatives, most triarylbismuthines do not react with molecular oxygen but add halogens onto the bismuth atom to yield the corresponding triarylbismuth dihalides. Such oxidative halogenation can be easily accomplished by using a variety of halogenating agents, which include molecular bromine, chlorine and fluorine, xenon difluoride, sulfur monochloride, sulfuryl chloride, thionyl chloride, iodine trichloride and phenyliodine dichloride (Section 3.1.1). However, the reaction between triarylbismuthines and iodine usually results in iodolysis to give the corresponding diaryliodobismuthines and iodoarenes. Hydrogen peroxide, dinitrogen trioxide as well as a mixture of dinitrogen trioxide and nitrous oxide also cleave oxidatively the Bi–C bonds of triarylbismuthines. Ozonized oxygen reacts with triphenylbismuthine in toluene, acetone, methyl ethyl ketone or ethyl trifluoroacetate to afford triphenylbismuth dicarboxylates [93JCS(P1)2411]. When the reaction is carried out in dichloromethane, triphenylbismuth dichloride is obtained. Compounds of the general formula Ar_3BiX_2, where X is an anionic group other than halogen, have been obtained by the oxidative addition of an appropriate electrophile to triarylbismuthines (Section 3.1.1). Thus, triphenylbismuth dibenzoate is prepared from the reaction of triphenylbismuthine with benzoyl peroxide [85ZOB73]. The addition of carbene, nitrene or some oxygen units to triarylbismuthine produces the corresponding triarylbismuth ylide, imide and oxide, respectively (Sections 3.5.1, 3.6.1 and 3.8.1).

The bismuth–carbon bond is readily cleaved by acid reagents such as hydrochloric acid, sulfuric acid, nitric acid, hydrazoic acid, carboxylic acids, sulfonic acids, hydrogen sulfide, thiophenol, selenols, acyl halides, perfluoroalkyl iodides and others. This type of Bi–C bond cleavage reaction has been widely employed for the preparation of compounds of the $R_{3-n}BiX_n$ type (n = 1–3) (Sections 2.2 and 2.3). For example, all Bi–C bonds in triphenylbismuthine are readily cleaved by three equiv. of triflic acid (CF_3SO_3H) to give bismuth triflate [99TL285], which shows extremely efficient catalytic activity in the Friedel-Crafts acylation of various aromatic compounds (Section 5.4.2). Dicarboxylic acids such as oxalic, malonic, succinic and maleic acids react with triphenylbismuthine to form the five-, six- or seven-membered ring bismuth compounds together with 2 equiv. of benzene (Section 4.2.4). Similar cleavage of the Bi–C bond of triphenyl- or trimethylbismuthine with a variety of mercaptocarboxylic acids has been used for the preparation of the corresponding heterocyclic bismuth compounds bearing both Bi–S and Bi–O bonds. Triethylbismuthine can be used for the detection of the SH group [37JA935]. The reaction between triphenylbismuthine and benzenethiol takes place by a homolytic chain mechanism to afford phenylbismuth bis(benzenethiolate) [70JCS(B)735]. A variety of metal salts can cleave the Bi–C bond of triorganylbismuthines, and these reactions are employed as a convenient method for the preparation of bismuth compounds of the type R_2BiX and $RBiX_2$, where X denotes halogen atoms or other anionic groups (Sections 2.2 and 2.3.1). This type of reaction is especially useful when

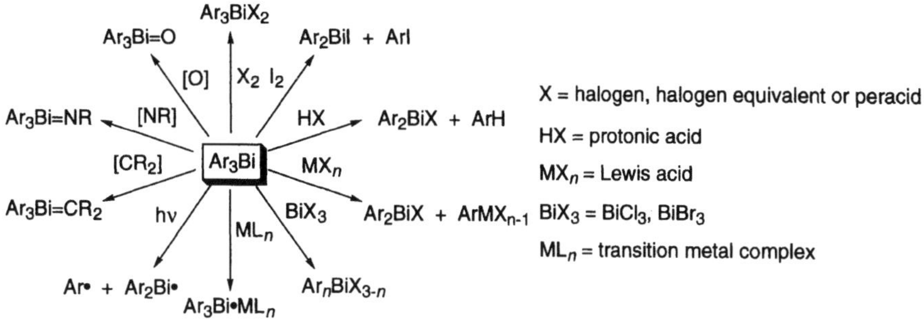

Scheme 2.5. Typical reactions of triarylbismuthines.

BiX_3 (X = Cl, Br) is used as a Lewis acid; this has been used now for a long time in the preparation of halobismuthines (Section 2.3.1). Carbon acids such as 1,3-diketones can cleave all Bi–C bonds in triphenylbismuthine to yield the corresponding bismuth tris(enolate) complexes, which may be used as the material for Bi-based superconducting films through gas phase deposition [93ZNK(38)1205, 98ICA(275/276)340]. When treated with perfluoroalkyl iodide, one or two alkyl groups in trialkylbismuthines are cleaved to form the mixed bismuthines of the formula $R_nR_{f(3-n)}Bi$ (n = 1,2; R_f denotes a perfluoroalkyl moiety) [63AJC636]. Benzoyl chloride cleaves the Bi–C bond of a sterically crowded alkynyldiarylbismuthine to give the corresponding diarylchlorobismuthine and α,β-alkynyl ketone [92BCSJ3504]. On treatment with carbon tetrachloride, tributylbismuthine is converted to butyldichlorobismuthine, probably with dibutyl(trichloromethyl)bismuthine as an intermediate [85ZOB586] (Scheme 2.5).

Action of sodium or potassium metal on triorganylbismuthines in liquid ammonia has been reported to cleave all of the Bi–C bonds and liberate metallic bismuth [86OCS(3)636, 74JA112]. However, it has also been reported that potassium diphenylbismuthide can be generated as a highly reactive species by the reaction between potassium metal and triphenylbismuthine in THF [83JOM(253)C21]. Triphenylbismuthine reacts with metallic mercury at high temperatures to form diphenylmercury and biphenyl with recovery of the original bismuthine [13CB1675].

Organometallic hydrides, derived from silicon, germanium or tin, cleave the Bi–C bond of triethylbismuthine to give compounds bearing the respective Bi–Si, Bi–Ge or Bi–Sn bond(s) of the type $Et_{3-n}Bi(MEt_3)_n$ (n = 1,2,3 for M = Ge; n = 3 for Si, Sn) [65ZOB394, 66DOK(166)99, 66JOM(6)474]. Transition metal hydride complexes also react with trimethylbismuthine; heating a mixture of $CpMH(CO)_3$ (M = Mo, W) with Me_3Bi at 90–100°C for 100–150 h yields $Cp(CO)_3MBiMe_2$ and methane [77JOM(134)C32]. Selenium dioxide does not add oxygen to the bismuth but cleaves the Bi–C bond of triphenylbismuthine [77JCS(D)641]. This contrasts with the observed reactivities of the phosphorus, arsenic and antimony counterparts, which are all converted to the corresponding oxides by this reagent.

Irradiation of triphenylbismuthine with ultraviolet light in organic solvent causes the homolytic cleavage of the Bi–C bond; benzene, biphenyl and phenylation products of the solvent as well as metallic bismuth are obtained [63JCS5612]. The Bi–C bond of alkyldiphenylbismuthines is cleaved even by ambient light [85CB1031]. When triphenylbismuthine is heated at 160°C with 9,10-phenanthrenequinone in the cavity of an ESR spectrometer, the ESR signal due to a semiquinone radical complexed with the diphenylbismuth unit is observed [79JOM(182)C49]. This indicates that the single electron transfer from the bismuthine to the quinone occurs and yields the out-of-cage radical. Anodic oxidation of triphenylbismuthine in acetonitrile affords acetanilide and biphenyl [97EA(42)1979]. Such electrochemical behavior of the bismuthine is quite different from that of triphenylstibine, which yields pentavalent compounds of the type Ph_3SbX_2. Triphenylbismuthine can quench the excited states of several carbonyl compounds [73JP(2)471, 75TL4361].

Although the reported examples are limited, triarylbismuthine can transfer its aryl moiety to a transition metal complex. This type of transmetallation produces a reactive arylmetal species, which can react with olefins, alcohols, amines and carbon monooxide, or undergoes homocoupling to give biaryl via reductive elimination. Synthetic applications of tertiary bismuthines to oxidation, arylation and C–C bond formation are described in Sections 5.2.4, 5.4.3 and 5.5.2. Although transmetallation occurs between triarylbismuthines and organolithium or organosodium reagent, this reaction has little significance for the preparation of organylbismuthines [50JA8]. However, when a coordinating, electron-deficient aryl group is attached to the bismuth center, this type of transmetallation can be employed for the preparation of triarylbismuthines. Based on this methodology, some bismuthines bearing three different aryl ligands are successfully synthesized (Section 2.1.2.1.4). Transmetallation between $(Ph_2Bi)_2CH_2$ and PhLi generates Ph_2BiCH_2Li, which adds to the carbonyl carbon of benzaldehyde, benzophenone and heptanal [80AG(E)723, 85CB1039]. On quenching with perchloric acid at room temperature, styrene, 1,1-diphenylethene and 1-octene, respectively, are obtained in moderate yields.

When a sulfonyl group is attached to the aromatic ring of triarylbismuthines, the *ortho*-proton (i.e. *meta* to the bismuth) can be deprotonated by such lithium reagents as ^{t}BuLi and *N*-lithium 2,2,6,6-tetramethylpiperidide [98OM4049]. The resulting (lithioaryl)bismuthines are trapped by diarylhalobismuthines of the type Ar_2BiX to afford polyarylbismuthines bearing two bismuth atoms. When treated successively with 3 equiv. of ^{t}BuLi and 3 equiv. of Ar_2BiX, tris[2-(*N,N*-diethylsulfamoyl)phenyl]bismuthine is readily converted to a dendritic tetrabismuth compound and related oligomers (Section 2.1.2.1.2). Based on this method, several highly branched as well as dendritic bismuthine oligomers have been constructed. Another type of bismuth-containing polymer is synthesized by homo- and co-polymerization of triarylbismuthines bearing a vinyl substituent at the *para*-position; diphenyl(4-vinylphenyl)bismuthine, (4-isopropenylphenyl)diphenylbismuthine and tris(4-isopropenylphenyl)bismuthine are polymerized or copolymerized radically or anionically [92PM(33)1724, 97CM1335]. Glass transition temperatures of the resulting polymers are determined as a function of the bismuth comonomer content. Radiopacities in millimeters of aluminum per millimeter of polymer are proportional to the molar bismuth content of the transparent copolymer specimens.

Due to the low nucleophilicity of the lone pair, tertiary bismuthines have found little use in the construction of onium salts and transition metal complexes. Trimethylbismuthine reacts with methyl triflate in acetonitrile to yield tetramethylbismuthonium triflate, which has so far been the only known example of tetraalkylbismuthonium salts [94AG(E)976]. Some transition metal complexes coordinated by tertiary bismuthine have been reported. They are described in Section 2.4.

Representative reactions of triphenylbismuthine are summarized in Table 2.4.

TABLE 2.4
Reactions of triphenylbismuthine

Reagent[a]	Condition	Product[a]	Reference
–	260–270°C, 75 h/C_6D_6	Ph–C_6D_5, Ph_2, C_6D_5–C_6D_5	[65ZOB481]
hν	ArH[b]	Bi, $(PhBi)_n$, Ph_2, Ph–Ar	[63JCS5612]
	40–50°C, 150 h/C_6D_6	Ph-C_6D_5, Ph_2, C_6D_5–C_6D_5	[65ZOB481]
	200 h/$^{14}C_6H_6$	$^{14}C_6H_5Ph$, Ph_2, $(^{14}C_6H_5)_2$	[65ZOB481]
hν, TiO_2, O_2	MeOH	Ph_3Bi=O	[96JOC2895]
γ-Radiation (^{60}Co)	PhH	$(PhBi)_n$, Ph_2, $C_{12}H_{12}$	[65JPC2880]
H_2 (60 atom)	225°C/xylene	Bi, PhH	[30CB1110]
H_2, Rh–Al_2O_3	r.t.	–	[55JA4262]
W-7 Raney-Ni	Reflux/MeOH	PhH	[62JCS3746]
N_2O_3	–	$Bi(NO_3)_3$, $[PhN_2][NO_3]$	[39ZOB771]
N_2O_3–NO	–	$Bi(NO_3)_3$, $[PhN_2][NO_3]$	[39ZOB771]
SO_2	r.t., 5 min	$Ph_3BiS_2O_4$	[68CC1474]
SO_3	CH_2ClCH_2Cl	Ph_3BiSO_3	[61ZAAC(308)33]
	−10°C/CH_2ClCH_3Cl	Ph_2BiOSO_2Ph	[86ICA(113)43]
H_2O_2	–	Infusible solid	[62JOC3851]
$Me_3SiOOSMe_3$	110°C, 4 h/pet. ether	–	[74JOM(73)217]
O_3	r.t./toluene	$Ph_3Bi(OCOH)_2$	[93JCS(P1)2411]
	r.t./RCOR′[c]	$Ph_3Bi(OCOR)_2$	[93JCS(P1)2411]
	r.t./CH_2Cl_2	Ph_3BiCl_2	[93JCS(P1)2411]
SeO_2	Reflux, 10 h/ PhH-EtOH	$Bi_2(SeO_3)_3$, $PhSeO_2H$	[77JCS(D)641]
R_3NO[d]	–	PhH	[62CJC181]
PhI=O	r.t./CH_2Cl_2	Ph_3Bi=O	[94TL8197]
$MeCO_2Me$	r.t./MeCOMe	Ph_3BiCO_3, $Ph_3Bi(OAc)_2$	[93JCS(P1)2411]
$(PhCOO)_2$	68°C	$Bi(OH)_3$, Ph–Ph	[27JCS209]
	Reflux/petrol. ether	$Ph_3Bi(OCOPh)_2$	[27JCS209]
$(RCO)_2O$-$^tBuO_2H$[e]	r.t./Et_2O	$Ph_3Bi(OCOR)_2$	[85ZOB73] [93IZV2043]
Ac_2O-$^tBuO_2H$	70°C, 35 h/toluene	$Ph_3Bi(OAc)_2$	[95IZV964]
$NaBO_3$	r.t., 1 h/AcOH	$Ph_3Bi(OAc)_2$	[96SC4569]
$Pb(OAc)_4$	CCl_4	$Ph_3Bi(OAc)_2$	[38ZOB1839]
	AcOH	$Ph_3Bi(OAc)_2$	[64CB789]
$PhI(OAc)_2$	r.t., 7 h/CH_2Cl_2	$Ph_3Bi(OAc)_2$	[98TL4313]
$PhCONO_3$	−15°C, 16 h/CCl_4	$Ph_3Bi(NO_3)_2$	[24JCS(125)854]
$NaClNSO_2Tol$	Reflux, 3 h/MeCN	Ph_3Bi=NSO_2Tol	[64CB789] [91CL105]
	aq. HX	(Bi–C cleavage)	[74PS(4)25]
PhI=NSO_2Tol	r.t./CH_2Cl_2	Ph_3Bi=NSO_2Tol	[96JCR(S)24]
$N_2CC_4Ph_4$[f]	140°C	Ph_3Bi=CC_4Ph_4	[67CC10421]
N_2CR_2[g]	80°C, 2–2.5 h/PhH	Ph_3Bi=CR_2	[88S319]
HNO_3 (fum.)	–	m-$(O_2N)_2C_6H_4$	[24JCS(125)854]
H_2SO_4	–	PhH	[16JCS(109)250]
HNO_3, H_2SO_4	−15°C, 2 h	$Bi(NO_3)_3$, $Bi_2(SO_4)_3$, m-$(O_2N)_2C_6H_4$	[24JCS(125)854]

TABLE 2.4 *(continued)*

Reagent[a]	Condition	Product[a]	Reference
$ClSO_3H$	$CHCl_3$	PhH	[16JCS(109)250]
$(Cl_2CSO_2)_3$	Et_2O	$[Ph_3RBi^+]Cl^-$	[63JPR(22)1]
HCl	–95°C	Disproportionation	[61JCS1238]
HBr	–	$BiBr_3$, PhH	[77JINC(39)565]
Me_3NHCl	130°C/EtOH	$BiCl_3$, PhH	[48ZOB936]
$(MeSO_2)_2NH$	20°C/CH_2Cl_2	$Ph_2BiN(SO_2Me)_2$, PhH	[95ZAAC(621)1746]
HCO_2H	50°C, 1 h	$Bi(OCOH)_3$, $Bi(OH)(OCOH)_2$, PhH	[39ZOB2283]
AcOH	50°C, 1 h	$Bi(OH)_2(OAc)$, BiO(OAc), PhH	[39ZOB2283]
	–	Ph_2BiOAc, PhH	[95IZV964]
RCO_2H[h]	50–100°C	$Bi(OCOR)_3$, PhH	[41ZOB379]
$HOCH_2CO_2H$	Reflux, 2 h/toluene	$PhBi(OCOCH_2OH)_2$, PhH	[64NEP6405308]
$ClCH_2CO_2H$	45 min/$CHCl_3$	$Bi(OCOCH_2Cl)_3$, PhH	[79ZOB1185]
CF_3CO_2H	Reflux, 1 h/ether	$Ph_2BiOCOCF_3$, PhH	[84HCA586]
$PhCO_2H$	100°C, 2 h	$PhBi(OCOPh)_2$, PhH	[51JA2880]
$PhCH{=}CHCO_2H$	150°C, 1.5 h	$PhBi(OCOCH{=}CHPh)_2$, PhH	[51JA2880]
$4\text{-}HOC_6H_4CO_2H$	Reflux, 2 h/toluene	$PhBi(OCOC_6H_4OH)_2$, PhH	[64NEP6405308]
$(CO_2H)_2$	–	BiX_3, PhH	[20CB2267]
$HO_2C\text{–}Z\text{–}CO_2H$[i]	60°C/MeCOMe	$PhBi(O_2C\text{–}Z\text{–}CO_2)$, PhH	[80ZN(B)1000]
$HO_2CH{=}CHCO_2H$	–	$Ph_2BiO_2CCH{=}CHCO_2BiPh_2$	[93AOMC137]
CF_3SO_3H	–	$Ph_2BiOSO_2CF_3$, PhH	[95JCS(P1)2543]
	−78°C to r.t.	$Ph_nBi(OSO_2CF_3)_{3-n}$ ($n = 0$–2), PhH	[99TL285]
Ph_4AsCl, HCl	Reflux/EtOH, MeOH	$[Ph_4As][Ph_nBiCl_{4-n}]$ ($n = 1, 2$), PhH	[69JOM(20)99]
Ph_4AsCl	Reflux/EtOH	No reaction	[69JOM(20)99]
$4\text{-}O_2NC_6H_4OH$	75°C, 1 h	$BiO(OC_6H_4O_2N)$, PhH	[49ZOB1675]
$4\text{-}MeC_6H_4OH$	100°C, 3 h	$Bi_2O_3{\cdot}(4\text{-}MeC_6H_4OH){\cdot}(H_2O)$, PhH	[49ZOB1675]
$2,4,6\text{-}Cl_3C_6H_2OH$	150°C, 3 h	$Bi(OC_6H_2Cl_3)_3$, PhH	[49ZOB1675]
C_6F_5OH	Reflux/toluene	$Bi(OC_6F_5)_3$, PhH	[93IC5136]
$1\text{-}C_{10}H_7OH$	130°C, 3 h	Bi, PhH	[47ZOB1307]
H_2S	EtOH–NH_3	Bi_2S_3, PhH	[1889LA(251)323]
PhSH	Reflux, 2 h/$CHCl_3$	$PhBi(SPh)_2$, PhH	[51JA2880]
	35°C, 5 h/PhH	$PhBi(SPh)_2$, PhH	[70JCS(B)735]
	Reflux, 15 h/xylene	$Bi(SPh)_3$, PhH	[51JA2880]
	75–130°C, 1–3 h	$Bi(SPh)_3$, PhH	[50ZOB2093]
$2\text{-}HSC_6H_4CO_2Me$	100°C, 2 h	$PhBi(SC_6H_4CO_2Me)_2$, PhH	[51JA2880]
	Reflux, 15 h/xylene	$Bi(SC_6H_4CO_2Me)_3$, PhH	[51JA2880]
$HSCH_2CO_2H$	Reflux/PhH	$C_6H_9O_6S_3Bi$	[51JA2880]
$HSCH_2CH_2OH$	Reflux, 2 h/toluene	$PhBi(SCH_2CH_2OH)_2$	[64NEP6405308]
$2\text{-}HSC_6H_4CO_2H$	150°C, 1 h	$PhBi(\text{–}SC_6H_4CO_2\text{–})$, PhH	[51JA2880]
	100°C, 3 h	$\text{–}OC_6H_4SBi(OH)OC\text{–}$	[52ZOB643]
HSCN	Et_2O	Ph_2BiSCN, $PhBi(SCN)_2$, PhH	[23JCS(123)1046]

TABLE 2.4 *(continued)*

Reagent[a]	Condition	Product[a]	Reference
$PhSO_2H$	20°C/Et_2O	$Ph_nBi(O_2SPh)_{3-n}$ ($n = 1, 2$), PhH	[71JOM(26)C10]
$Hg(O_2SPh)_2$	r.t., 5 h/$CHCl_3$	$Bi(O_2SPh)_3$, $PhHgO_2SPh$	[72AJC2107]
$Hg(O_2STol)_2$	r.t./MeOH	$Bi(O_2STol)_3$, $PhHgO_2STol$	[72AJC2107]
$Se(SeCN)_2$	r.t., 30 h/$CHCl_3$	$Ph_2BiSeCN$, Se, PhSeCN	[26JCS1648]
$Te(CN)_2$	Ether	Ph_2BiCN, PhTeCN	[26JCS1648]
F_2	$CFCl_3$	Ph_3BiF_2	[78AG(E)214]
Cl_2	–5°C/$CHCl_3$	Ph_3BiCl_2	[52LA(578)136]
	Ligroin	Ph_3BiCl_2	[1887CB54]
	0°C/pet. ether	Ph_3BiCl_2	[43ZAAC(250)257]
Br_2	Ligroin	Ph_3BiBr_2	[1887CB54]
	MeCN	Ph_3BiBr_2	[66JCS(A)520]
I_2	MeCN	Ph_2BiI, $PhBiI_2$, BiI_3, PhI	[66JCS(A)520]
	Et_2O	$PhBiI_2$	[64CB789]
	Et_2O	$PhBiI_2$, Ph_2BiI, BiI_3, PhI	[24JCS(125)854]
	Et_2O	Ph_2BiI, PhI, I_2	[15JCS(107)16]
	r.t., 15 min/Et_2O; THF	$PhBiI_2\cdot$(THF)	[92JCS(D)1967]
ICl	Et_2O	Ph_2BiCl, PhI	[15JCS(107)16]
	Et_2O	Ph_3BiCl_2, PhI	[20JCS(117)762]
	100°C, 4.5 h/$CHCl_3$	Ph_2BiCl, $BiCl_3$, PhI	[60TTFI(2)317]
IBr	Et_2O	Ph_2BiBr, PhI	[15JCS(107)16]
	MeCN	Ph_2BiBr, $PhBiBr_2$, PhI	[66JCS(A)520]
ICN	Reflux/PhH	Ph_2BiCN, PhI	[15JCS(107)16]
BrCN	Heat, 4 h/PhH	No reaction	[15JCS(107)16]
	120°C, 2.5 h	Ph_2BiBr, BiOBr, PhCN	[24JCS(125)854]
ClCN	80°C, 1 h	Ph_2BiCl, PhCN	[24JCS(125)854]
$(CN)_2$	Reflux/PhH	No reaction	[15JCS(107)16]
H_2NCN	Heat/PhH	Ph_3Bi (recovered), $(NCNH)_2$	[15JCS(107)16]
$(SCN)_2$	PhH	Ph_2BiSCN, PhSCN	[69JOM(21)211]
	0°C, 4.5 h/Et_2O	Ph_2BiSCN, $Bi(SCN)_3$, PhSCN	[22JCS(121)91]
SO_2Cl_2	Pet. ether	Ph_3BiCl_2	[16JCS(109)250]
$SOCl_2$	–	Ph_3BiCl_2	[16JCS(109)250]
S_2Cl_2	–	Ph_3BiCl_2	[16JCS(109)250]
NBS	–5°C, 24 h/MeCN	$Ph_3BiBr(NR_2)$	[79JPR(321)370]
$AlCl_3$	Heat, 6 h/$CHCl_3$	$BiCl_3$, $PhAlCl_2$, PhH	[51ZOB93]
$TlCl_3$	r.t./Et_2O	Ph_2BiCl, Ph_2TlCl	[22JCS(121)104]
	Reflux, 5 h/$CHCl_3$	Ph_2BiCl, $BiCl_3$, Ph_2TlCl, PhH	[57TTFI(1)321]
$(C_6F_5)_2TlBr$	Reflux, 2 h/Et_2O	Ph_2TlBr, $(C_6F_5)_3Bi$	[65JCS3411]
$SiCl_4$	Et_2O	Ph_2BiCl	[24JCS(125)864]
	Reflux, 5 h/$CHCl_3$	Ph_2BiCl, $BiCl_3$, PhH	[60TTFI(2)317]
$SnCl_4$	PhH; H_2O	BiOCl, Ph_2SnCl_2, $Ph_2SnCl(OH)$	[24JCS(125)864]
	Reflux, 5 h/$CHCl_3$	Ph_2BiCl, Ph_2SnCl_2, PhH, Ph_2	[60TTFI(2)317]
$GeCl_2$	r.t., 12 h/toluene	Bi, Ph_2GeCl_2	[94ZOB1051]
PCl_3	Et_2O	Ph_2BiCl, $PhPCl_2$, Ph_2PCl	[24JCS(125)864]
	Et_2O	Ph_2BiCl, $PhPCl_2$	[22JCS(121)104]

TABLE 2.4 *(continued)*

Reagent[a]	Condition	Product[a]	Reference
$AsCl_3$	Et_2O	Ph_2BiCl, $PhAsCl_2$, Ph_2AsCl	[24JCS(125)864]
	10 min/Et_2O	Ph_2BiCl, $PhAsCl_2$	[22JC5(121)104]
$SbCl_3$	Et_2O	Ph_2BiCl, Ph_3SbCl_2	[24JCS(125)864]
	Et_2O	Ph_2BiCl, BiOCl, Ph_3SbCl_2	[22JCS(121)104]
	Reflux, 5 h/$CHCl_3$	$BiCl_3$, Ph_2SbCl, PhH	[57TTFI(1)321]
$BiCl_3$	Et_2O	Ph_2BiCl	[15JCS(107)16]
	0°C/Et_2O	Ph_2BiCl	[86T3111]
	Acetone	$PhBiCl_2$	[72JOM(44)317]
	Et_2O; THF	$PhBiCl_2$·(THF)	[93JCS(D)637]
	Reflux, 4 h/$CHCl_3$; HCl	Ph_2BiCl, $BiCl_3$, PhH, Ph_2	[53SSOK1308]
$BiBr_3$	Et_2O	Ph_2BiBr	[41JA207]
	r.t., 15 min/Et_2O; THF	Ph_2BiBr·(THF)	[92JCS(D)1967]
	Et_2O	$PhBiBr_2$	[21JCS(119)913]
	r.t., 30 min/Et_2O; THF	$PhBiBr_2$·(THF)	[92JCS(D)1967]
$TiCl_4$	0°C, 15 h/Et_2O	Ph_2BiCl	[24JCS(125)864]
	Reflux, 5 h/$CHCl_3$	$BiCl_3$	[60TTFI(2)317]
$FeCl_3$	Heat, 6 h/$CHCl_3$	$BiCl_3$, PhH	[51ZOB93]
$CuCl_2·H_2O$	15 min/EtOH	Ph_2BiCl, CuCl	[22JCS(121)104]
$CuCl_2$	Et_2O	Ph_2BiCl, CuCl	[34JCS405]
	Reflux, 5 h/$CHCl_3$	Ph_2BiCl, $BiCl_3$, Ph_2	[57TTFI(1)321]
$CuBr_2$	Et_2O	Ph_2BiCl, CuBr	[34JCS405]
$ZnCl_2$	Reflux, 5 h/$CHCl_3$	Ph_2BiCl, $BiCl_3$, Ph_2, PhH	[57TTFI(1)321]
$HgCl_2$	Et_2O	Ph_2BiCl, PhHgCl	[22JCS(121)104]
	0°C/Et_2O	Ph_2BiCl, PhHgCl	[69JCS(A)2138]
	Reflux, 5 h/$CHCl_3$	$BiCl_3$, Ph_2Hg, PhHgCl, PhH	[57TTFI(1)321]
	H_2O	BiOCl, PhHgCl, PhH	[1897CB2843]
$PhCH_2Cl$	r.t., 8 weeks/Et_2O	No reaction	[22JCS(121)104]
	r.t.	BiOCl, HCl, Ph_2CH_2	[22JCS(121)104]
MeI	–	No reaction	[22JCS(121)104]
MeCOCl	Reflux, 4 h/CCl_4	Ph_2BiCl, MeCOPh	[62CJC181]
PhCOCl	Reflux, 6 h/CCl_4	Ph_2BiCl, PhCOPh	[22JCS(121)104]
$AgNO_3$	Reflux, 5 h/$CHCl_3$	$BiCl_3$, $(PhAg)_2AgNO_3$	[22JCS(121)104]
	r.t./Et_2O	AgPh	[60TTFI(2)317]
Li	r.t./Et_2O	Bi, PhLi	[38ZOB1831]
Na	liq. NH_3	–	[41JA212]
K	liq. NH_3	Bi, PhH	[74JA112]
K	25°C/THF	Ph_2BiK	[83JOM(253)C21]
Sb	300°C, 5 min	Ph_3Sb	[65JOM(3)420]
Se	Reflux/mesitylene	Ph_2Se, Ph_2Se_2	[99JOC3722]
Te	Reflux/mesitylene	Ph_2Te, Ph_2Te_2	[99JOC3722]
Hg	250°C, 10 min	Bi, Ph_2Hg, Ph_2	[13CB1675]
	Reflux, 5 h/$CHCl_3$	Bi, Ph_2Hg	[57TTFI(1)321]
Hg (labeled)	55°C, 2.3–5 h/PhH	Bi, Ph_2Hg (labeled)	[71JOM(27)149]
Et_3SnH	140°C, 75 h	Bi, Et_3SnPh, H_2	[64ZOB2787]

TABLE 2.4 *(continued)*

Reagent[a]	Condition	Product[a]	Reference
$H_2NCH_2CH_2NH_2$	PhH	No reaction	[50JCS617]
KNH_2	liq. NH_3	BiN	[67ZAAC(349)328]
BuLi	Et_2O	Ph_nBiBu_{3-n}, PhLi (n=0–3)	[50JA8]
TolLi	Et_2O	Ph_nBiTol_{3-n}, PhLi (n=0–3)	[67JOM(8)491]
$^{14}C_6H_5Li$	Pentane–Et_2O	$(^{14}C_6H_5)_3Bi$, PhLi	[88R537]
C_6Me_6	–	No complex formation	[68LA(713)30]
Np_3Bi	190°C, 2 h	Ph_2NpBi, NpH	[22JCS(121)104]
$Me_2S{=}CR_2$[j]	200–230°C, 30 min	$MeSR'$	[78IZV1643]
Ph_2Se_2	Reflux/mesitylene	Ph_2Se	[99JOC3722]
Ph_2Te_2	Reflux/mesitylene	Ph_2Te	[99JOC3722]
$C_6T_5^+$	β-Decay from C_6T_6	$[Ph_3BiC_6T_5^+]$	[89R69]
9,10-PQ[k]	160°C/diglyme	9,10-PQ=BiPh$_2$	[79JOM(182)C49]
4,5-TQ[l]	160°C/diglyme	4,5-TQ=BiPh$_2$	[79JOM(182)C49]

[a] For details, see references.
[b] ArH = pyridine, toluene, *tert*-butylbenzene.
[c] $RCOR'$ = MeCOMe, EtCOMe, CF_3CO_2Et.
[d] R_3NO = 2,4,4-trimethyl-1-pyrroline-1-oxide.
[e] R = H, Me, Et, Bu, Ph, CH=CH_2.
[f] CC_4Ph_4 = C(–CPh)=CPh–CPh=CPh–).
[g] CR_2 = C(–$COCH_2CH_2CH_2CO$–), $C(SO_2Ph)_2$.
[h] R = H, Me.
[i] Z = $(CH_2)_n$ (n=0, 1, 2), CH=CH.
[j] CR_2 = C(–$COCH_2CMe_2CH_2CO$–).
[k] 9,10-PQ = 9,10-phenanthrenequinone.
[l] 4,5-TQ = benzodithiophene-4,5-dione.

2.2. ORGANOBISMUTH(III) COMPOUNDS WITH Bi–GROUP 15 OR Bi–GROUP 16 ELEMENT BONDS

2.2.1. Diorganylbismuth compounds

2.2.1.1. Methods of synthesis (Tables 2.5–2.9)

The most general method for the synthesis of diorganylbismuth compounds is the nucleophilic substitution of halo- or alkoxydiorganylbismuthines. Diorganylbismuth compounds are obtained by the action of an equimolar amount of an acidic compound on triorganylbismuthine. However, they are often accompanied by a monoorganylbismuth or inorganic bismuth compound as a by-product.

(a) Via nucleophilic substitution of diorganylbismuth halides (method A)

Synthesis of dimethylbismuth thiolates

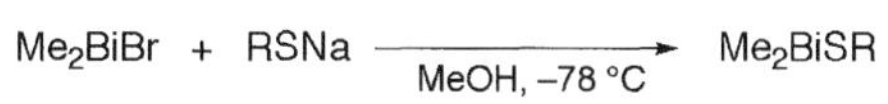

Under a nitrogen atmosphere, a solution of bromodimethylbismuthine (3.19 g, 10 mmol) in absolute methanol was added to a solution of sodium thiolate prepared from thiol (10 mmol) and sodium (0.23 g, 10 mmol) in absolute methanol at −78°C, giving dimethylbismuth thiolate as yellow crystals in 10–58% yields [83ZAAC(505)147].

Synthesis of diphenylbismuth 3-fluorobenzenethiolate

To a solution of sodium 3-fluorobenzenethiolate (0.6 g, 4 mmol) in absolute ethanol (20 ml) was added bromodiphenylbismuthine (1.7 g, 4 mmol). The reaction mixture was heated to reflux until a yellow clear solution resulted. The solvent was removed under reduced pressure and the residue was washed with water and dried. After recrystallization from petroleum ether, diphenylbismuth 3-fluorobenzenethiolate was obtained as yellow needles, 2.3 g (99%) [75JOM(86)383].

(b) Via nucleophilic substitution of diorganylbismuth alkoxides (method B)

Synthesis of diphenylbismuth thiolates

Under a nitrogen atmosphere, a solution of sodium ethoxide from sodium (0.23 g, 10 mmol) in absolute ethanol was added dropwise to a solution of bromodiphenylbismuthine (4.43 g, 10 mmol) in absolute ethanol (40 ml) with stirring. The reaction mixture was heated to boiling point and the precipitated sodium bromide was filtered off. Without isolating diphenylbismuth ethoxide, a solution of a given thiol (10 mmol) in ethanol was added to the filtrate to obtain a yellow solution. The corresponding diphenylbismuth thiolate immediately precipitated as yellow crystals, which were filtered off and recrystallized from an appropriate solvent [78ZAAC(439)134].

(c) Via cleavage of Bi–C or Bi–N bonds with acidic compounds (method C)

Synthesis of diphenylbismuth trifluoroacetate

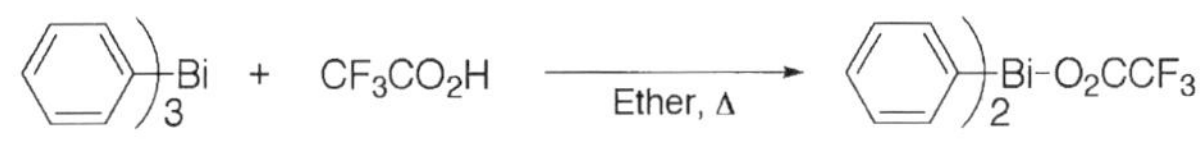

A solution of triphenylbismuthine (0.44 g, 1 mmol) and trifluoroacetic acid (76.5 μl, 1 mmol) in anhydrous ether (4 ml) was heated to reflux under argon for 1 h, until the substrate had been consumed (TLC). The mixture was cooled to 0°C in an ice bath, and the resulting white precipitate was filtered off, washed with cold ether, and dried in vacuo to give diphenylbismuth trifluoroacetate (0.32 g, 67%) [84HCA586].

Synthesis of dimethylbismuth azide

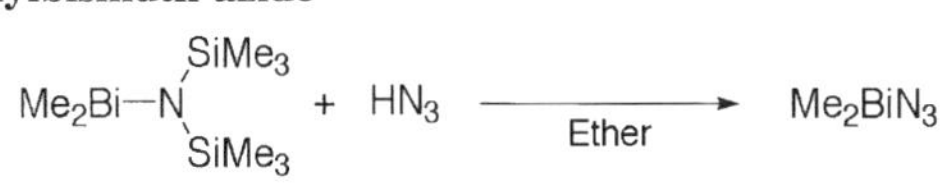

Hydrogen azide (highly toxic) in ether (2.3 M, 20 ml) was added to an ice-cooled solution of dimethylbismuth *N,N*-bis(trimethylsilyl)amide (18.3 g, 45 mmol) in the same solvent (50 ml) in the dark, and the resulting mixture was stirred for 1 day. The white separated precipitate was washed with small portions of ether and dried in vacuo to give dimethylbismuth azide as a microcrystalline solid (11.0 g, 86%) [85ZN(B)1320].

(d) Via cleavage of the Bi–C bond with neutral compounds (method D)

Synthesis of diphenylbismuth benzenesulfonate

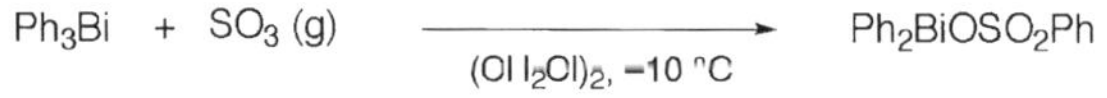

Dry gaseous sulfur trioxide was bubbled through a solution of triphenylbismuthine (10 g, 22.7 mmol) in 1,2-dichloroethane (40 ml) for 2 h at −10°C. On addition of ether (400 ml), a white precipitate was formed. This was filtered off and extracted with dry ethanol (3 ml × 50). Dilution of the combined ethanol extracts with petroleum ether (b.p. ca. 40°C) gave a white solid, which was filtered off, washed with ether, and dried to give diphenylbismuth benzenesulfonate (3.2 g, 27%) [86ICA(113)43].

(e) Via dismutation between dibismuthine and dichalcogenide (method E)

Synthesis of bis(4-methylphenyl)bismuth benzenethiolate

$$(4\text{-}MeC_6H_4)_2Bi\text{–}Bi(C_6H_4\text{-}4\text{-}Me)_2 + PhSSPh \xrightarrow[CH_2Cl_2,\ 0\ °C]{} (4\text{-}MeC_6H_4)_2Bi\text{–}SPh$$

Diphenyl disulfide (0.14 g, 0.64 mmol) was added with vigorous stirring to a solution of tetrakis(4-methylphenyl)dibismuthine (0.50 g, 0.64 mmol) in dichloromethane (15 ml) at 0°C under an argon atmosphere. The reaction mixture turned pale yellow within several minutes. After 15 min the reaction mixture was concentrated to a half volume in vacuo, diluted with pentane (3 ml), and poured onto crushed dry ice. After decantation, the yellow residue was washed with pentane and dried in vacuo to obtain the product (0.57 g, 89%) [87ZN(B)695].

(f) Via insertion of a free element into the Bi–Bi bond (method F)

Synthesis of μ-selenobis(diphenylbismuth)

$$Ph_2Bi\text{–}BiPh_2 + Se \xrightarrow[PhMe,\ r.t.]{} Ph_2BiSeBiPh_2$$

To a solution of tetraphenyldibismuthine (0.24 g, 0.32 mmol) in toluene (25 ml) was added powdered selenium (0.03 g, 0.38 mmol) and the resulting mixture was stirred at room temperature in the dark. After 1 h the selenium disappeared and the solution turned yellow. Evaporation to a small volume, dilution with heptane, filtration of a solid precipitate, and drying in vacuo gave the product in 33% yield [88IC3730].

(g) Via reductive elimination of the Bi(V) compound by thermolysis (method G)

Synthesis of diphenylbismuth thiocyanate

(a)

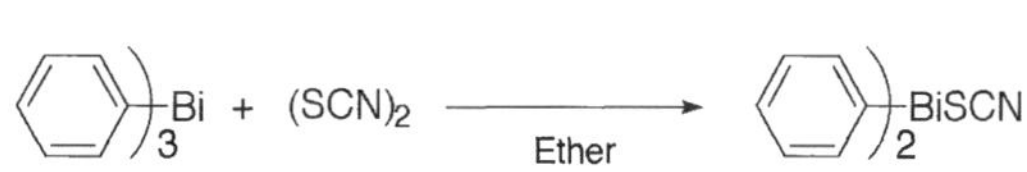

Lead thiocyanate (7.5 g) was added to a solution of bromine (1 ml) in absolute ether (40 ml) and the mixture was shaken for 33 min under ice-cooling. The mixture was filtered and the residue was washed with absolute ether (60 ml). A solution of thiocyanogen in ether (34 ml) thus obtained was added to an ice-cold solution of triphenylbismuthine (2.93 g) in the same solvent. Yellow crystals began to separate on the wall of the reaction vessel, but became oily after having been shaken for 4.5 h. The product was extracted with cold alcohol to give diphenylbismuth thiocyanate (1.7 g) [22JCS(121)91].

(b)

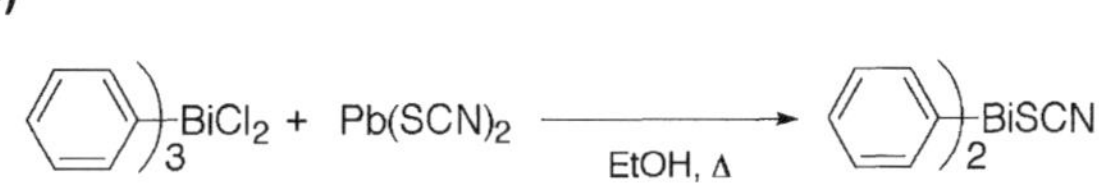

Triphenylbismuth dichloride (5.1 g) and lead thiocyanate (6 g) were boiled in ethanol (200 ml)

for 30 min. The resulting bright yellow solution was decanted and the solid residue was heated in ethanol for an additional 30 min. The combined solution was concentrated to deposit a yellow amorphous powder, which was filtered off. On standing, the filtrate deposited crystals of diphenylbismuth thiocyanate, which have recrystallized from a mixture of alcohol and acetone [22JCS(121)91].

Synthesis of diphenylbismuth azide

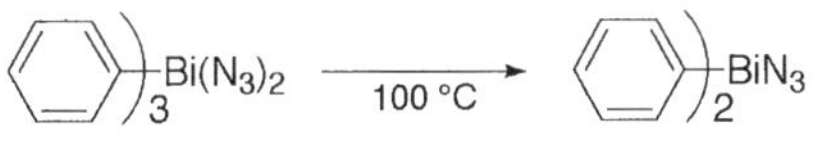

Triphenylbismuth diazide (2 g) was gradually warmed to 100°C to give a pasty solid with a characteristic pungent odor. Extraction with cold light petroleum to remove triphenylbismuthine (0.8 g), followed by crystallization of the residue from hot alcohol gave diphenylbismuth azide as a microcrystalline solid [34JCS405].

(h) Via metathesis reaction (method H)

Synthesis of diphenylbismuth hexafluorophosphate-bis(hexamethylphosphoramide)-complex

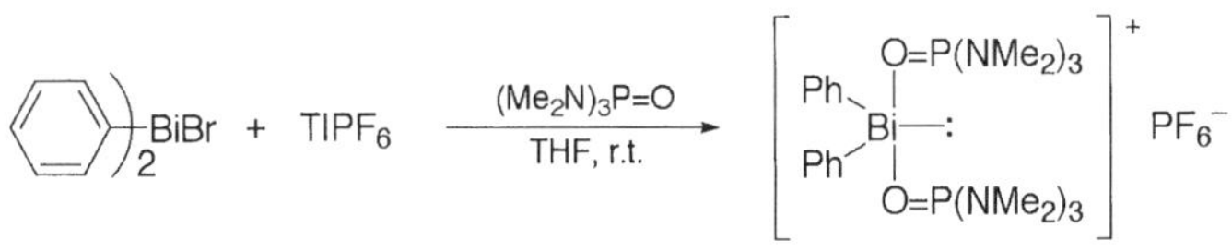

Hexamethylphosphoramide (HMPA; 0.1 ml, 0.575 mmol) was added to a stirred solution of bromodiphenylbismuthine (0.127 g, 0.287 mmol) in THF (5 ml) at room temperature to give a colorless solution. A solution of thallium hexafluorophosphate (0.100 g, 0.287 mmol) in the same solvent (6 ml) was added and led to the immediate formation of a pale yellow precipitate. The mixture was stirred for 30 min and then filtered through Celite to give a slightly cloudy filtrate which was reduced to about 5 ml in volume under vacuum. Hexane (20 ml) was added as an overlayer, and solvent diffusion, over a period of several days at −22°C, afforded the complex as colorless crystals in 76% yield [96JCS(D)443].

2.2.1.2. Properties and reactions

2.2.1.2.1. Diorganylbismuth compounds with Bi–N bonds (Table 2.5)

(a) Diorganylbismuth amides The chemistry of organobismuth

compounds with the N–Bi bond has been studied very little to date. Dimethylbismuth *N*-methyl-*N*-(trimethylsilyl)amide **1** has been prepared by the reaction of bromodimethylbismuthine with the corresponding lithium amide [66JOM(6)259]. The bismuth amide is sensitive to oxygen and moisture, and reacts with cyclopentadiene [75JOM(88)329] or ethyl diazoacetate [75JOM(93)339] to give (cyclopentadienyl)dimethylbismuthine **2** or ethyl diazo(dimethylbismuthino)acetate **3**, respectively.

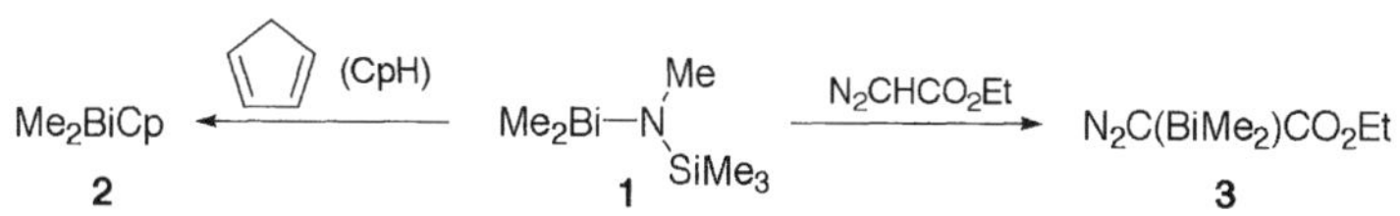

Bis(4-methylphenyl)bismuth *N*-(4-methylphenyl)sulfonylamide **4** is obtained in moderate yield by the reaction of tris(4-methylphenyl)bismuth [*N*-(4-methylphenyl)sulfonyl]imide with anise alcohol or cinnamyl alcohol [96JCR(S)24]. This amide is insoluble in benzene, but soluble in chloroform, and not so sensitive to oxygen and moisture in the solid state. However, it slowly decomposes to tris(4-methylphenyl)bismuthine and a white powdery deposit when the amide is stood in $CDCl_3$.

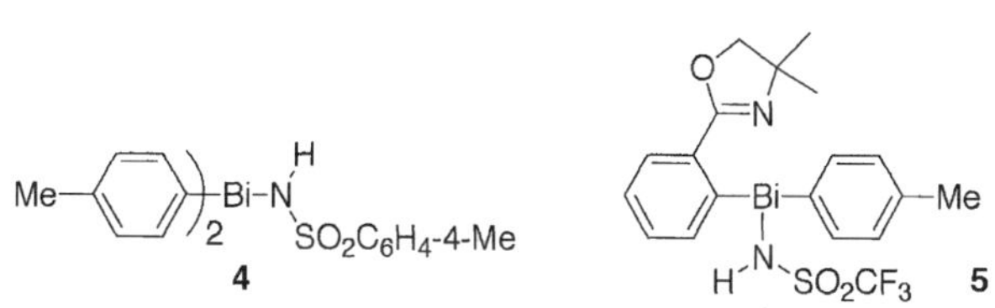

By contrast, (4-methylphenyl)[2-(4,4-dimethyl-2-oxazolin-2-yl)phenyl]-bismuth *N*-(trifluoromethanesulfonyl)amide (**5**) is prepared in quantitative yield by the action of a 1:1 molar mixture of potassium *tert*-butoxide and trifluoromethanesulfonyl amide ($TfNH_2$) on chloro(4-methylphenyl)[2-(4,4-dimethyl-2-oxazolin-2-yl)phenyl]bismuthine in THF at room temperature [98OM1013]. In contrast to amide **4**, the amide **5** is thermally stable and does not disproportionate when stood in a solution.

Diphenylbismuth bis(methanesulfonyl)amide (**6**) is prepared by an acidolysis of Ph_3Bi with $HN(SO_2Me)_2$ in 85% yield. Compound **6** can be prepared by a reaction of Ph_2BiCl with $AgN(SO_2Me)_2$ in 26% yield

[95ZAAC(621)1746]. X-Ray crystallography of **6** has revealed that the compound exists as a cyclic dimer, in which two bismuth centers form a 12-membered $(BiOSNSO)_2$ ring. The Bi–O bond lengths are 2.397(6) and 2.466(6) Å and the geometry of bismuth center is described as a trigonal bipyramidal structure.

(b) Diorganylbismuth azides Dimethylbismuth azide Me_2BiN_3 is obtained as slightly moisture-sensitive fine crystals by acidolysis of trimethylbismuthine with hydrogen azide. It is insoluble in ether, benzene, pentane, THF, ethanol, pyridine and chloroform. The azide is not explosive and, on heating up to 150°C, only undergoes slow decomposition [71ZAAC(381)103]. Diphenylbismuth azide Ph_2BiN_3 is prepared by the reaction of chlorodiphenylbismuthine and sodium azide in pyridine at ambient temperature. It is soluble in benzene, but an efficient crystallization procedure is not yet available [68JOM(13)529]. On standing at 25°C, the azide undergoes disproportionation to give a mixture of products, from which Ph_3Bi is obtained. In contrast to diphenylantiomony azide, this azide has been reported not to be reactive toward triphenylphosphine [68JOM(13)529]. Thermal decomposition of triphenylbismuth diazide $Ph_3Bi(N_3)_2$ at 100°C leads to a mixture of PhN_3, Ph_3Bi, and Ph_2BiN_3 [34JCS405, 73JOM(50)129]. Goel et al. have claimed that Ph_2BiN_3 prepared by metathesis is not pure enough analytically and the thermal decomposition of $Ph_3Bi(N_3)_2$ is a better choice in order to obtain a purer azide [79SA(A)339]. The IR spectrum of Ph_2BiN_3 exhibits peaks at 2050, 1315, 1265, 1260 and 655 cm^{-1} due to $\nu_a(N_3)$, $2\delta(N_3)$, $\nu_s(N_3)$ and $\delta(N_3)$ vibrational modes, respectively, and a strong absorption at 305 cm^{-1} due to $\nu(Bi–N_3)$ vibration while its Raman spectrum shows peaks at 2048 and 652 cm^{-1} due to $\nu_a(N_3)$ and $\delta(N_3)$ vibrational modes, respectively [79SA(A)339]. Müller et al. have prepared dimethylbismuth azide in good yield by treatment of dimethylbismuth *N,N*-bis(trimethylsilyl)amide with hydrogen azide. An X-ray crystallographic study of this azide revealed that bismuth atoms are linked together via the α-N atom of the azido group to form an endless zig-zag chain [85ZN(B)1320].

By metathesis between bromo(di-*tert*-butyl)bismuthine $^tBu_2BiBr$ and dipotassium sulfurdiimide K_2SN_2, bis(di-*tert*-butylbismuthino)sulfurdiimide $^tBu_2BiN{=}S{=}NBi^tBu_2$ is formed as yellow crystals in 10% yield. This compound is labile and decomposes slowly at room temperature. It shows reversible thermochromism on cooling. Under an inert atmosphere at −40°C in the dark, it can be kept for a long period. Its IR spectrum shows peaks at 1125 and 1000 cm^{-1} due to the vibration of the N=S=N bond [84CB1999].

2.2.1.2.2. Diorganylbismuth compounds with Bi–As or Bi–Sb bonds (Table 2.5)

Compounds containing the Bi–As or Bi–Sb bond were reported by Ashe et al. as the product from the exchange reaction of tetramethyldi-bismuthine $Me_2BiBiMe_2$ with tetramethyldiarsine $Me_2AsAsMe_2$ or tetramethyldistibine $Me_2SbSbMe_2$. These compounds are not obtainable in pure form [86JOM(303)197]. $Pr_2BiSbMe_2$ is formed by a similar reaction; the equilibrium between $Pr_2BiBiPr_2$ and $Me_2SbSbMe_2$ in benzene has been studied by ^{1}H-NMR and the estimated equilibrium constant is 0.7 [86ZN(B)1129].

$$Pr_2BiBiPr_2 \quad + \quad Me_2SbSbMe_2 \quad \rightleftharpoons \quad 2\ Pr_2BiSbMe_2$$

TABLE 2.5
Diorganylbismuth and related compounds with Bi–N, Bi–As or Bi–Sb bonds

Compound		Synthetic method[a]	Yield (%)	M.p. (°C) or b.p. (°C/mmHg)	Physical data	Reference
Bi–N compounds						
R_2BiX						
R = Me	$X = N_3$	C	75	150> (decomp.)	IR, MS, Raman	[71ZAAC(381)103]
		–	–	–	IR	[75JOM(93)339]
		B	86	150–152 (decomp.)	X-ray	[85ZN(B)1320]
	$X = Me(Me_3Si)N$	A	37.6	31–32/0.1	^1H-NMR	[66JOM(6)259]
		A	–	–	–	[75JOM(93)339]
R = ^tBu	$X = N{=}S{=}NBi^tBu_2$	A	10	–	^1H-, ^{13}C-NMR, MS, IR	[84CB1999]
R = Ph	$X = N_3$	G	–	168	–	[34JCS405]
		A	76.5	160.5–162	IR	[68JOM(13)529]
		G	–	–	IR	[73JOM(50)129]
		A	–	168 [EtOH]	IR, Raman	[79SA(A)339]
	$X = (MeSO_2)_2N$	A/C	26/85	228–231 (decomp.)	^1H-NMR, X-ray	[95ZAAC(621)1746]
R = 4-MeC_6H_4	X = 4-$MeC_6H_4SO_2NH$	G	39	163–165	^1H-NMR, MS, IR	[96JCR(S)24]
R^1R^2BiX						
R^1 = 4-MeC_6H_4	$X = CF_3SO_2NH$	A	100	153–155	^1H-, ^{13}C-NMR, IR	[98OM1013]

R^2 =

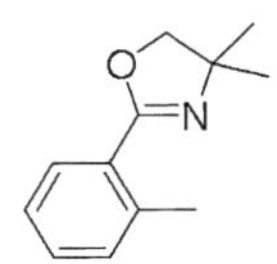

Bi–As compounds					
$Me_2BiAsMe_2$	E	–	–	^{1}H-NMR, Raman	[86JOM(303)197]
Bi–Sb compounds					
$Me_2BiSbMe_2$	E	–	–	^{1}H-NMR, Raman	[86JOM(303)197]
$Pr_2BiSbMe_2$	E	–	–	^{1}H-NMR, MS	[86ZN(B)1129]

[a] For notation, see Section 2.2.1.1.

2.2.1.2.3. Diorganylbismuth compounds with Bi–O bonds (Table 2.6)

(a) Diorganylbismuth alkoxides The chemistry of diorganylbismuth alkoxides has received little attention much like their structures and reactivities. Diethylbismuth ethoxide is obtained by the controlled oxidation of Et_3Bi [42JA392]. Dimethylbismuth and diphenylbismuth trimethylsiloxide are prepared by the reaction of the corresponding diorganylbismuth bromides with sodium trimethylsiloxide. Dimethylbismuth siloxide is the first spectrally characterized alkoxide. It decomposes at 60°C but can be purified by sublimation [68ZAAC(363)84]. Wieber et al. have investigated the preparation and reactions of diphenylbismuth alkoxides (methoxide, ethoxide, isopropoxide, phenoxide) with various thiols. Alkoxides are moisture sensitive and *irritate the skin severely*! They are insoluble in hexane and carbon tetrachloride, but soluble in benzene. Molecular weight determination of the alkoxides in benzene has revealed that they exist as a monomer in solution. Diphenylbismuth alkoxides react with thiols at ambient temperature to give the corresponding thiolates in good yield [78ZAAC(439)134]. $Ph_2BiBiPh_2$ reacts with *p*-benzoquinone to give 1,4-$C_6H_4(OBiPh_2)_2$ as a yellow solid in 83% yield. This compound is only sparingly soluble in common organic solvents [84JCS(D)2365]. Phenol and 2,4,6-trinitrophenol react with Et_3Bi slowly in boiling dioxane; however the products have not been identified [37JA935]. The reaction of Et_3Bi with phenols in boiling toluene gives diethylbismuth phenoxides. Both Et_2BiOPh and $Et_2BiOC_6F_5$ are too sensitive toward moisture and air to obtain satisfactory elemental analysis. High solubility of these compounds in organic solvents as well as the ^{1}H-NMR spectrum of $Et_2BiOC_6F_5$ in solution indicate that they exist either in a monomeric form or as rapidly exchanging oligomers in solution. Mass spectra of both phenoxides show peaks corresponding to M^+, Et_2Bi^+ and $EtBi^+$ ions. A crystallographic study of Et_2BiOPh and $Et_2BiOC_6F_5$ has shown that both compounds are isomorphous, forming a chiral helical chain through the intermolecular coordination between the bismuth and oxygen atoms. The Bi–O bond lengths are 2.4105(7) for $Et_2BiOC_6F_5$ and 2.382(7) Å for Et_2BiOPh [92CC1021].

(b) Cyclic diorganylbismuth alkoxides Many diarylbismuth alkoxides with the bismuth atom(s) incorporated in a cyclic structure are known (Section 4.2.1). By the action of aryllithiums on 1-acetoxy-3,3-bis(trifluoromethyl)-3*H*-2,1-benzoxabismole **1**, the acetoxy group is

replaced by the aryl group to give the corresponding benzoxabismole derivatives. Known compounds include **2–9** [92JA7906, 95JA3922].

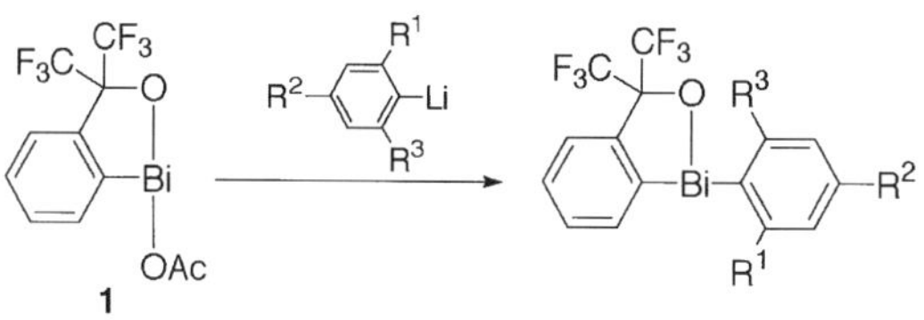

2; $R^1 = R^3 = H$, $R^2 = CF_3$
3; $R^1 = R^3 = H$, $R^2 = OMe$
4; $R^1 = Me_2NCH_2$, $R^2 = R^3 = H$
5; $R^1 = MeOC(Me)_2$, $R^2 = R^3 = H$
6; $R^1 = MeOC(CF_3)_2$, $R^2 = R^3 = H$
7; $R^1 = MeHNCH_2$, $R^2 = R^3 = H$
8; $R^1 = R^3 = Me_2NCH_2$, $R^2 = H$
9; $R^1 = R^3 = Me_2NCH_2$, $R^2 = {}^tBu$

An alternative approach to 1-aryl-3,3-bis(trifluoromethyl)-3*H*-2,1-benzoxabismole derivatives involves the reaction between the bismuthate complexes **10** with water or MeI. By this method, compounds **5** [95HAC293], **6** [92TL6653], **11** [95JA3922], and **12–14** are prepared [92TL6653, 95HAC293]. Transformations of this type of compound are described in detail in Section 4.2.1.

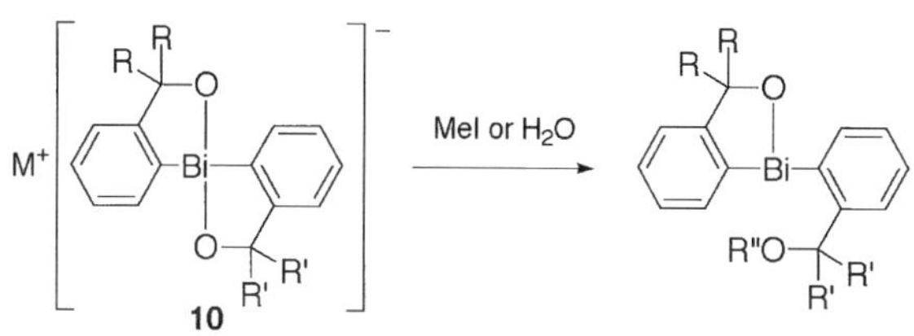

5; $R = CF_3$, $R' = Me$, $R'' = Me$
6; $R = R' = CF_3$, $R'' = Me$
11; $R = CF_3$, $R' = H$, $R'' = Me$
12; $R = R' = CF_3$, $R'' = H$
13; $R = Me$, $R' = CF_3$, $R'' = H$
14; $R = R' = Me$, $R'' = Me$

(c) μ-Oxobis(diorganylbismuth) By the action of molecular oxygen at −30°C, tetramethyldibismuthine is converted to μ-oxobis(dimethylbismuth) in 95% yield, which exothermally decomposes above 16°C to give Me_3Bi and bismuth oxide Bi_2O_3 [84ZN(B)887]. Similarly, tetrapropyldibismuthine reacts with molecular oxygen at −30°C to yield μ-oxobis(dipropylbismuth) as a colorless solid, which decomposes around 40°C to Pr_3Bi and Bi_2O_3. It is soluble in benzene and chloroform [86ZN(B)1129]. Tetradibisumuthine is similarly oxidized in air to give μ-oxobis[tris(2,4,6-trimethylphenyl)] as colorless crystals. This compound can also be prepared by the action of sodium hydroxide on bromobis(2,4,6-trimethylphenyl)bismuthine [95ZN(B)735]. It crystallizes from EtOH in the solvated form with 0.5 EtOH. The Bi–O bond lengths and the Bi–O–Bi bond angle determined by X-ray crystallographic analysis are 2.095(18) and 2.117(19) Å, and 117.1(8)°, respectively. Its UV–Vis spectrum shows absorption maxima at λ_{max} 390 and

240 nm [95ZN(B)735]. Mass spectra of these three μ-oxo compounds exhibit characteristic peaks due to M^+, $R_3Bi_2O^+$, $R_2Bi_2O^+$ and RBi_2O^+ ions.

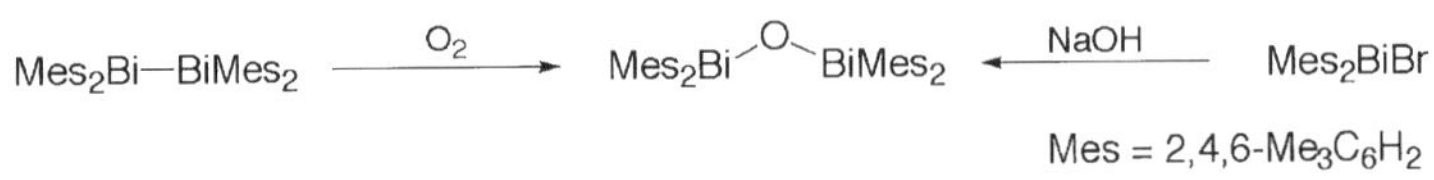

(d) Diorganylbismuth carboxylates Diorganylbismuth carboxylates are prepared by two different methods; one is the substitution reaction of diorganylbismuth halides or ethoxides with carboxylate anions or carboxylic acids, and the other is the cleavage of the Bi–C bond in triorganylbismuthines with carboxylic acids. The former method is generally preferred, since the desired products are obtainable in better yield. Diarylbismuth ethoxides are a good precursor to this type of compound [82ZN(B)815]. Carboxylic acids such as acetic, chloroacetic, dichloroacetic, trichloroacetic, butyric, maleic and benzoic acids all react with Et_3Bi in boiling dioxane, although the products have not been identified [37JA935]. Phenylbismuth dicinnamate and dibenzoate are obtained as a slightly soluble white powder in moderate yield by the reaction between Ph_3Bi and the corresponding acids [51JA2880]. By similar acidolysis, diphenylbismuth trifluoroacetate, bis(4-methylphenyl)bismuth trifluoroacetate, bis(4-methoxylphenyl)bismuth trifluoroacetate, bis(4-ethoxylphenyl)bismuth trifluoroacetate and bis(pentafluorophenyl)bismuth trifluoroacetate are prepared. However, this procedure often gives a complex mixture of diarylbismuth carboxylate $Ar_2Bi(O_2CR)$ and arylbismuth dicarboxylate $ArBi(O_2CR)_2$ depending on the solubility of the relevant compounds. Thus, the acidolysis of triphenylbismuthine with acetic acid gives phenylbismuth diacetate as the sole product, whereas the corresponding reaction of trifluoroacetic acid (molar ratio 1:1) gives $Ph_2Bi(O_2CCF_3)$ as a solid precipitate.

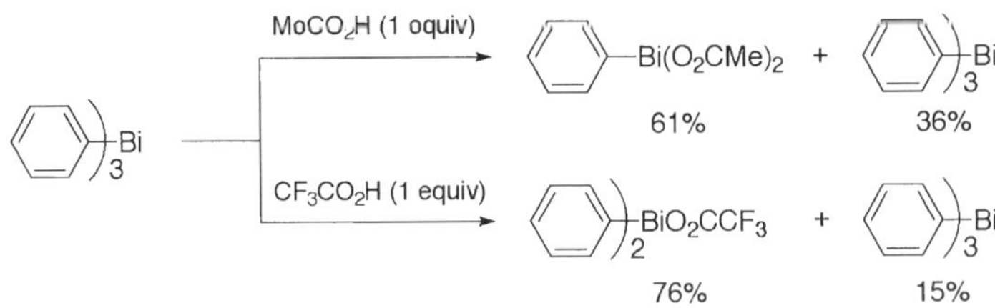

Acidolysis of tris(4-ethoxylphenyl)bismuthine with an equimolar trifluoroacetic acid gives an intimate mixture of the corresponding mono- and dicarboxylates $Ar_2Bi(O_2CR)$ and $ArBi(O_2CR)_2$ in a ratio of 9:1 [84AJC527]. Diarylbismuth carboxylates are readily hydrolyzed and no efficient method for purification is available at present. Diarylbismuth carboxylates undergo a redistribution reaction in DMSO to give a mixture of triarylbismuthine and arylbismuth dicarboxylate [84AJC527]. From the reaction of Ph_3Bi and maleic acid, bis(diphenylbismuth) maleate is obtained. In its IR spectrum, $\nu(CO_2)$ absorptions appear at 1588, 1483, 1395 and 1253 cm^{-1} [93AOMC137].

(e) Diorganylbismuth sulfinates Diphenylbismuth benzenesulfinate is obtained by the reaction of triphenylbismuthine and benzenesulfinic acid in benzene–$CHCl_3$ under reflux conditions, or in ether at ambient temperature. The sulfinate is very poorly soluble in common organic solvents, suggesting that the compound exists in the polymeric form. On heating this sulfinate at 216°C for 4 h under vacuum (10^{-2}–10^{-3} mmHg), gaseous SO_2 is evolved to produce triphenylbismuthine in good yield [71JOM(26)C10].

$$Ph_2BiO_2SPh \xrightarrow[216\ ^\circ C,\ 10^{-2} \sim 10^{-3}\ mmHg]{} Ph_3Bi + SO_2$$

(f) Diorganylbismuth sulfonates Treatment of triarylbismuthines with arenesulfonic acid gives the corresponding diarylbismuth sulfonates Ar_2BiOSO_2R in good yield [86ICA(113)43]. The sulfonates are also obtained in moderate yield by the insertion reaction between triarylbismuthine and gaseous dry sulfur trioxide. On prolonged storage, they are slowly hydrolyzed by atmospheric moisture to afford a basic bismuth sulfonate and trace amounts of triarylbismuthine. Disproportionation of the sulfonates yields triarylbismuthine and arylbismuth disulfonate. Several diarylbismuth triflate–HMPA complexes are prepared by the action of trimethylsilyl triflate and methanol on triarylbismuthines in the presence of HMPA [96OM1951]. These complexes are stable colorless crystals or oils, readily soluble in CH_2Cl_2, $CHCl_3$, benzene, THF and methanol without decomposition. The HMPA ligand is effective in protecting the triflates from hydrolysis by atmospheric moisture; without HMPA, these triflates become less stable in open air. ^1H-

NMR spectra of diarylbismuth triflate–HMPA complexes show a characteristic signal due to the *ortho* protons of aromatic rings in a low field (δ 8.17–8.37). In IR spectra, broad and well-defined bands appear in the 1350–1050 cm^{-1} region due to the triflate group. FAB mass spectra exhibit strong peaks due to the $[M–OTf]^+$ and/or $[M–OTf, HMPA]^+$ fragments. In a similar manner, unsymmetrical diarylbismuth triflate–HMPA complexes are prepared easily from mixed triarylbismuthines; more electron-rich aryl groups are removed preferentially from mixed triarylbismuthines.

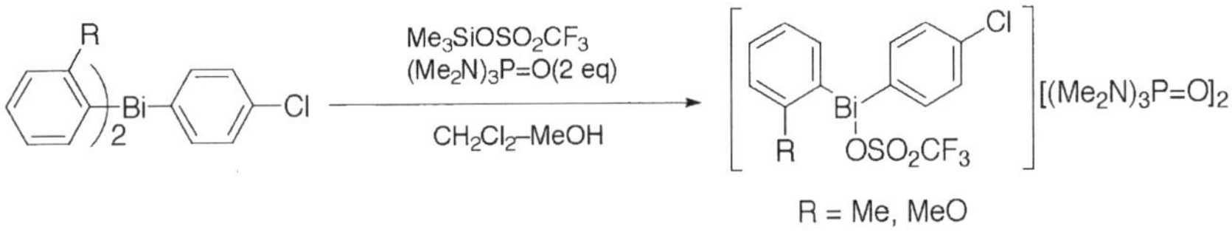

Treatment of diarylbismuth triflate–HMPA complexes with aryl Grignard reagents gives the corresponding triarylbismuthines. This reaction is applied to the synthesis of triarylbismuthines with three different aryl groups [96OM1951]. Treatment of the bis(4-methylphenyl)bismuth trifluoromethanesulfonate–HMPA complex with a solution of sodium naphthalenide in THF at −78°C generates sodium bis(4-methylphenyl)bismuthide in situ, which is reacted with iodoarenes to give the corresponding arylbis(4-methylphenyl)-bismuthine [97OM3565].

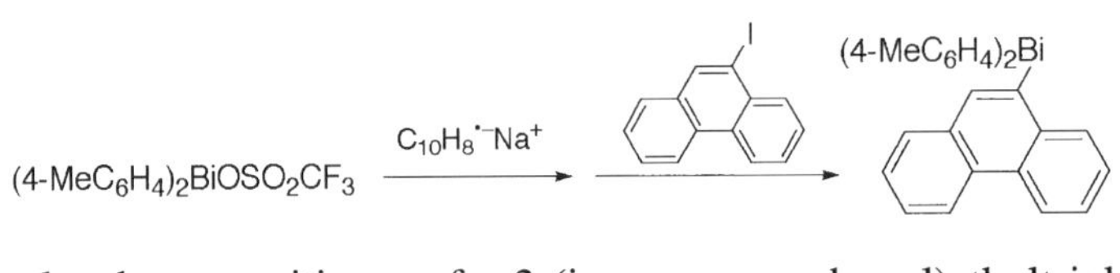

Thermal decomposition of 2-(isopropoxycarbonyl)ethyltriphenylbismuthonium trifluoromethanesulfonate leads to the formation of diphenylbismuth trifluoromethanesulfonate in 16–84% yield [95JCS(P1)2543]. Recently, a preparation of diphenylbismuth trifluoromethanesulfonate has reported, however, there is only limited information on its properties [99TL285].

TABLE 2.6
Diorganylbismuth and related compounds with Bi–O bonds

Compound R	R′	Synthetic method[a]	Yield (%)	M.p. (°C) or b.p. (°C/mmHg)	Physical data	Reference
R_2BiOH						
Me		A	–	–	–	[1887CB(20)1516]
		C	–	–	–	[1887CB(21)2038]
		G	–	–	–	[71JCS(C)1660]
Et		G	–	–	–	[71JCS(C)1660]
Ph		A	–	–	–	[27BTI3]
		D	–	–	–	[62ZAAC(317)54]
4-MeC_6H_4		D	–	–	–	[62ZAAC(317)54]
4-ClC_6H_4		D	–	–	–	[62ZAAC(317)54]
R_2BiOR'						
Me	Me_3Si	A	10	42–44	^{1}H-NMR, IR	[68ZAAC(363)84]
Et	Et	G	49	–	–	[42JA392]
	C_6F_5	C	–	166–168	^{1}H-, ^{13}C-, ^{19}F-NMR, X-ray	[92CC1021]
	Ph	C	–	174	X-ray	[92CC1021]
Ph	Me	A	ca.70	191 (decomp.) [MeOH]	^{1}H-NMR	[78ZAAC(439)134]
	Et	A	ca.70	143–146 (decomp.) [EtOH]	^{1}H-NMR	[78ZAAC(439)134]
	^{i}Pr	A	ca.70	123–127 (decomp.) [^{i}PrOH]	^{1}H-NMR	[78ZAAC(439)134]

TABLE 2.6 *(continued)*

Bi–O compounds R	R′	Synthetic method[a]	Yield (%)	M.p. (°C) or b.p. (°C/mmHg)	Physical data	Reference
Ph	Ph	A	ca.70	179–182 (decomp.) [EtOH]	^{1}H-NMR	[78ZAAC(439)134]
	Me_3Si	A	50	70–74/1	^{1}H-NMR	[68ZAAC(363)84]
$R_2BiOBiR_2$						
Me		E	95	16 (decomp.)	^{1}H-NMR	[75JOM(93)339]
		F	95	16 (decomp.)	^{1}H-NMR, MS	[84ZN(B)887]
Pr		F	Quant.	40 (decomp.)	^{1}H-NMR, MS	[86ZN(B)1129]
2,4,6-$Me_3C_6H_2$		A, F	55/51.6	131–144 [EtOH]	^{1}H-NMR, MS, UV, X–ray	[95ZN(B)735]
1,4-$(R_2BiO)_2C_6H_4$						
Me		F	89	–	^{1}H-NMR, MS	[95ZN(B)735]
Ph		F	83	–	IR	[84JCS(D)2365]
4-MeC_6H_4		F	94	100 (decomp.) [CH_2Cl_2–C_5H_{12}]	^{1}H-NMR, MS	[87ZN(B)695]
R_2BiO_2CR'						
Ph	Me	A, B	59–77/13	149	IR	[82ZN(B)815]
	Et	A	90	122	IR	[82ZN(B)815]
	Pr	A	64–76	99	IR	[82ZN(B)815]

	^{i}Pr	A	51–78	119	IR	[82ZN(B)815]
	^{t}Bu	A	71–69	230	IR	[82ZN(B)815]
	$MeCO(CH_2)_2$	A	67	104	IR	[82ZN(B)815]
	$PhCONHCH_2$	B	–	–	X-ray	[88AX(C)828]
	$CH_2NHC(O)Ph$	C	–	282–284	IR	[93AOMC137]
	Ph	A	59–79	129	IR	[82ZN(B)815]
	CF_3	–	–	–	^{1}H-NMR	[94JCS(D)2545]
		C	67	>350	IR, MS	[84HCA586]
		C	76	213–215	^{1}H-NMR, IR, MS	[84AJC527]
$4\text{-}MeC_6H_4$	CF_3	C	66	>250	^{1}H-NMR, IR, MS	[84AJC527]
$4\text{-}MeOC_6H_4$	CF_3	C	–	–	^{1}H-, ^{19}F-NMR, IR	[84AJC527]
$4\text{-}EtOC_6H_4$	CF_3	C	–	–	^{1}H-, ^{19}F-NMR, IR	[84AJC527]
Ph_2BiO_2SPh		C	–	–	IR	[71JOM(26)C10]
R_2BiO_3SR'						
Ph	Ph	C/D	84/25	250 (decomp.) [Et_2O]	^{1}H-NMR, IR	[86ICA(113)43]
	$4\text{-}MeC_6H_4$	C	79	196 (decomp.) [Et_2O]	^{1}H-NMR, IR	[86ICA(113)43]
		G	16–84	–	^{1}H-NMR	[95JCS(P1)2543]
		C	–	–	^{1}H-, ^{13}C-, ^{19}F-NMR	[99TL285]
$4\text{-}MeC_6H_4$	Ph	C	98	>250 (decomp.) [Et_2O]	^{1}H-NMR, IR	[86ICA(113)43]
	$4\text{-}MeC_6H_4$	C, D	80	199 (decomp.) [Et_2O]	^{1}H-NMR, IR	[86ICA(113)43]

TABLE 2.6 *(continued)*

Bi–O compounds R	R′	Synthetic method[a]	Yield (%)	M.p. (°C) or b.p. (°C/mmHg)	Physical data	Reference
$R_2BiO_3SCF_3(HMPA)_2$						
Ph		C	100	–	^{1}H-NMR, IR, MS	[96OM1951]
4-MeC_6H_4		C	100	–	^{1}H-NMR, IR, MS	[96OM1951]
2-MeC_6H_4		C	100	–	^{1}H-NMR, IR, MS	[96OM1951]
2-$MeOC_6H_4$		C	100	93–100 (decomp.)	^{1}H-NMR, IR, MS	[96OM1951]
$RR'BiO_3SCF_3(HMPA)_2$						
2-MeC_6H_4	4-ClC_6H_4	C	100	–	^{1}H-NMR, IR, MS	[96OM1951]
2-$MeOC_6H_4$	4-ClC_6H_4	C	100	–	^{1}H-NMR, IR, MS	[96OM1951]

[a] For notation, see Section 2.2.1.1.

2.2.1.2.4. Diorganylbismuth compounds with Bi–S bonds (Table 2.7)

(a) Diorganylbismuth thiocyanates Treatment of Ph_2BiBr with lead thiocyanate in alcoholic solution gives Ph_3Bi and diphenylbismuth thiocyanate Ph_2BiSCN [22JCS(121)91]. The thiocyanate can also be prepared by the reaction of Ph_3BiCl_2 with lead thiocyanate [22JCS(121)91], acidolysis of Ph_3Bi with hydrogen thiocyanate [23JCS(123)1046], and treatment of Ph_3Bi with thiocyanogen $(SCN)_2$ in cold ether [22JCS(121)91]. The reaction of Ph_2BiCl and sodium thiocyanate in aqueous solution also gives Ph_2BiSCN [79SA(A)339]. Ph_2BiSCN forms pale yellow crystals, which are soluble in most organic solvents except for petroleum ether. It is indefinitely stable when dry and decomposes slowly in solution, giving Ph_3Bi and a yellow precipitate, possibly composed of bismuth thiocyanate and polymerized thiocyanogen $(SCN)_n$ [22JCS(121)91]. The reaction of Ph_3Bi and $(SCN)_2$ in CH_2Cl_2 at 0°C gives a yellow oil, which is recrystallized from EtOH–petroleum to give Ph_2BiSCN in 48% yield. Recrystallization from $CHCl_3$–hexane gives a chloroform solvate $[Ph_2Bi(SCN)]_2 \cdot CHCl_3$ as crystals, of which the crystallographic structure has been elucidated by X-ray crystallographic; the thiocyanate group bridges intermolecularly with another bismuth atom to form a rectangular spiral chain, in which the alternate bismuth atoms are coordinated by two nitrogen or two sulfur atoms. The bond lengths and angles around the bismuth atoms are as follows: Bi–N bond lengths, 2.53(6) and 2.52(5) Å; Bi–S bond lengths, 2.87(1) and 2.88(1) Å; N–Bi–N bond angle, 166.2(16)°; S–Bi–S bond angle, 177.0(4)° [95JCS(D)383]. The IR spectrum of Ph_2BiSCN in the solid state shows peaks at 2102 ν(CN), 2084 ν(CN), 762 ν(CS), 462 δ(SCN) and 238 ν(Bi–SCN) cm^{-1} [79SA(A)339], while in Nujol mull it exhibits peaks at 2085 ν(CN), 767 ν(CS) and 470 ν(CS) cm^{-1} [95JCS(D)383]. The Raman spectrum in the solid state shows peaks at 2100 ν(CN), 2085 ν(CN), 760 ν(CS), 460 δ(SCN), 238 ν(Bi–SCN) and 215 ν(Bi–Ph) cm^{-1} [79SA(A)339]; in acetone solution, it shows a strong band at 2125 cm^{-1}, accompanied by minor bands at 724 and 438 cm^{-1} [95JCS(D)383]. The mass spectrum of Ph_2BiSCN does not show the parent ion peak, instead prominent peaks due to Ph_2Bi^+, $PhBi^+$ and Bi^+ fragments are observed [95JCS(D)383]. Dimethylbismuth thiocyanate, Me_2BiSCN, is prepared by the action of thiocyanogen on trimethylbismuthine. This compound has not been studied in detail, but its IR spectrum shows peaks at 2110 ν(CN), 795 ν(CS), 450 and 440 δ(SCN) cm^{-1} [69JOM(20)211].

(b) Diorganylbismuth thiolates The reactivity of Et_3Bi toward various thiols has been examined. These include 1-butyl, 1-heptyl, benzyl, phenyl, 4-methylphenyl and 2-naphthyl thiols, 1-mercaptobenzothiazole and thioacetic acid, although the products have not been identified definitely [37JA935]. Diphenylbismuth benzenethiolate is prepared by the reaction of chlorodiphenylbismuthine and thiophenol [51JA2880]. Diphenylbismuth 3- and 4-fluorobenzenethiolates were prepared and the electron donating effect of the Ph_2BiS group has been examined in comparison with the antimony congeners; the Ph_2BiS group is shown to be more electron-releasing than the Ph_2SbS group [75JOM(86)383]. Wieber et al. have prepared dimethylbismuth [83ZAAC(505)147] and diphenylbismuth thiolates [77ZAAC(436)101, 78ZAAC(439)134] by the reaction of diorganylbismuth halides with thiolate anions, and the reaction of diorganylbismuth alkoxides with thiols. The thiolates are colored yellow and soluble in such polar solvents as acetone, acetonitrile, DMSO and DMF. Diarylbismuth thiolates are not sensitive to moisture but thermolabile, and decompose relatively rapidly in solution (especially in the case of dimethyl compounds) to give the corresponding triorganylbismuthine and organylbismuth dithiolate [78ZAAC(439)134, 83ZAAC(505)147]. For example, Me_2BiSPh decomposes at 57°C endothermally [84ZN(B)1668]. In addition, the thiolates are light sensitive and decompose slowly under diffused light to form a dark colored substance [77ZAAC(436)101]. The reaction between dibismuthines and diphenyl disulfide also gives diorganylbismuth benzenethiolates [84ZN(B)1668, 88IC3730]. This type of reaction usually proceeds quickly and is completed within a few hours, even at 0°C. In contrast to Ph_2BiEPh (E = Se and Te), Ph_2BiSPh does not react with diazomethane, probably because of the stronger nature of the Bi–S bond [88IC3730].

(c) μ-Thiobis(diorganylbismuth) Tetramethyldibismuthine reacts with elemental sulfur at −30°C to give μ-thiobis(dimethylbismuth) as a yellow powder in 93% yield, which endothermally decomposes at 28°C to Me_3Bi and Bi_2S_3 [84ZN(B)887]. μ-Thiobis(dipropylbismuth) can be prepared similarly. It is a light orange solid and decomposes to Pr_3Bi and a reddish brown solid above 25°C. The decomposition is accelerated by light [86ZN(B)1129]. Mass spectra of these μ-thiobis(dialkylbismuth) show characteristic peaks due to M^+ and $R_3Bi_2S^+$ ions. μ-Thiobis(diphenylbismuth) decomposes slowly through disproportionation even under an inert atmosphere [84JCS(D)2365].

μ-Thiobis[bis(2,4,6-trimethylphenyl)bismuth] is obtained as yellow crystals by the action of sodium sulfide on bromobis(2,4,6-trimethylphenyl)-bismuthine in a mixture of ether and water. X-Ray crystallographic analysis of this compound has shown the Bi–S bond lengths to be 2.520(7) and 2.545(6) Å, and the Bi–S–Bi bond angle to be 98.7(3)° (Table 6.5, compound **33**). Its mass spectrum shows characteristic peaks due to M^+, $Ar_3Bi_2^+$, Ar_3BiS^+, Ar_3Bi^+, Ar_2Bi^+ and $ArBi^+$ ions, and the UV–Vis spectrum shows absorption maxima at λ_{max} 450, 350 and 250 nm [95ZN(B)735].

(d) Diorganylbismuth alkylthiocarbonates Diphenylbismuth [80ZAAC (470)171] and dimethylbismuth alkylthiocarbonates [83ZAAC(505)150] are obtained as yellow colored substances by the reaction of R_2BiBr and sodium alkylthiocarbonates, which undergo disproportionation at ambient temperature to give a mixture of the corresponding triorganylbismuthine and organobismuth bis(alkylthiocarbonates). No further information is available on the chemistry of this type of compound.

TABLE 2.7
Diorganylbismuth and related compounds with Bi–S bonds

Compound R	R′	Synthetic method[a]	Yield (%)	M.p. (°C)	Physical data	Reference
R_2BiSCN						
Me		G	–	66	IR	[69JOM(20)211]
Ph		A, G	–	122–122.5	–	[22JCS(121)91]
		C	–	123.5	–	[23JCS(123)1046]
		G	–	122	IR	[69JOM(20)211]
		A	–	125 [EtOH]	IR, Raman	[79SA(A)339]
		G	48	123 [EtOH-petroleum; $CHCl_3$-C_6H_{14}]	IR, MS, Raman, X-ray	[95JCS(D)383]
R_2BiSR'						
Me	Ph	E	80	57 (decomp.)	^{1}H-, ^{13}C-NMR, MS	[84ZN(B)1668]
	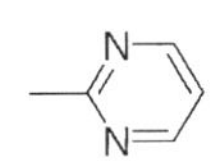	A	57	127 [MeOH]	^{1}H-NMR, MS	[83ZAAC(505)147]
	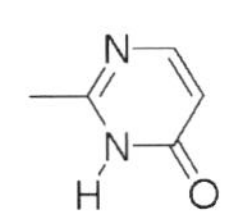	A	58	82 [MeOH]	^{1}H-NMR, MS	[83ZAAC(505)147]
	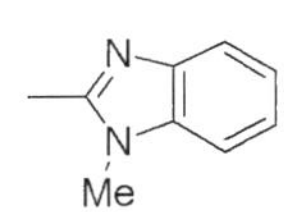	A	57	125 [MeOH]	^{1}H-NMR, MS	[83ZAAC(505)147]

	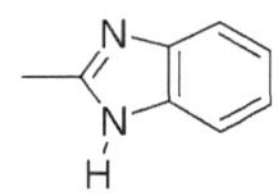	A	10	152 [MeOH]	^{1}H-NMR, MS	[83ZAAC(505)147]
	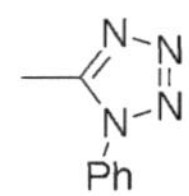	A	27	137 [MeOH]	^{1}H-NMR, MS	[83ZAAC(505)147]
R_2BiSR'						
Ph	Et	A	50	53 [EtOH]	^{1}H-NMR, MS	[77ZAAC(436)101]
	^{i}Pr	B	50	56 [EtOH]	^{1}H-NMR, MS	[78ZAAC(439)134]
	^{t}Bu	B	70	97 [EtOH]	^{1}H-NMR, MS	[78ZAAC(439)134]
		A	66	97 [EtOH]	^{1}H-NMR, MS	[77ZAAC(436)101]
	^{c}Hex	B	73	116 [EtOH]	^{1}H-NMR, MS	[78ZAAC(439)134]
	CH_2Ph	A	53	76 (decomp.) [EtOH]	^{1}H-NMR, MS	[77ZAAC(436)101]
	$(CH_2)_2SBiPh_2$	A	68	124 (decomp.) [MeCN]	^{1}H-NMR	[77ZAAC(436)101]
	$(CH_2)_3SBiPh_2$	A	64	105 (decomp.) [MeCN]	^{1}H-NMR	[77ZAAC(436)101]
	$(CH_2)_5SBiPh_2$	B	65	84 (decomp.) [EtOH]	^{1}H-NMR	[78ZAAC(439)134]
	Ph	A	80	160 (decomp.) [PhH]	–	[51JA2880]
		E	46	–	^{13}C-NMR, IR	[88IC3730]
	$4\text{-}MeC_6H_4$	B	68	72–75 [EtOH]	^{1}H-NMR, MS	[78ZAAC(439)134]
	$2\text{-}H_2NC_6H_4$	C	65	105–110	IR, MS	[87POL1593]
	$3\text{-}FC_6H_4$	A	99	79–80	^{19}F-NMR	[75JOM(86)383]
	$4\text{-}FC_6H_4$	A	99	89–90	–	[74IZV927]
		A	99	89–90	^{19}F-NMR	[75JOM(86)383]
	$4\text{-}ClC_6H_4$	A	42	90 [EtOH]	^{1}H-NMR, MS	[77ZAAC(436)101]

TABLE 2.7 *(continued)*

Compound R	R′	Synthetic method[a]	Yield (%)	M.p. (°C)	Physical data	Reference
4-MeC_6H_4	Ph	E	89	104 (decomp.) [CH_2Cl_2-C_5H_{12}]	^{1}H-NMR, MS	[87ZN(B)695]
$R_2BiSBiR_2$						
Me		C	–	–	–	[1888CB(21)2035]
		F	93	28 (decomp.)	^{1}H-NMR, MS	[84ZN(B)887]
Pr		F	Quant.	25 (decomp.)	^{1}H-NMR, MS	[86ZN(B)1129]
Ph		A	–	–	–	[1889LA(251)323]
		F	47.8	–	IR	[84JCS(D)2365]
4-MeC_6H_4		F	82	89 (decomp.)	^{1}H-NMR, MS	[87ZN(B)695]
2,4,6-$Me_3C_6H_2$		A	52	163	^{1}H-NMR, MS, UV, X-ray	[95ZN(B)735]
$Ph_2BiSCOMe$		A	41	80 (decomp.) [^{i}PrOH]	^{1}H-NMR, MS	[77ZAAC(436)101]
R_2BiS_2COR'						
Me	Me	A	47	56 [MeOH]	^{1}H-NMR, MS	[83ZAAC(505)150]
	Et	A	34	99 [MeOH–EtOH]	^{1}H-NMR, MS	[83ZAAC(505)150]
	Pr	A	31	84 [MeOH–PrOH]	^{1}H-NMR, MS	[83ZAAC(505)150]
	^{i}Pr	A	26	107 [MeOH–^{i}PrOH]	^{1}H-NMR, MS	[83ZAAC(505)150]
	Bu	A	13	68 [MeOH–BuOH]	^{1}H-NMR, MS	[83ZAAC(505)150]
	^{i}Bu	A	9	78 [MeOH–^{i}BuOH]	^{1}H-NMR, MS	[83ZAAC(505)150]
Ph	Me	A	70	96 [MeCN]	^{1}H-NMR	[80ZAAC(470)171]

Et	A	66	77 [EtOH]	^{1}H-NMR	[80ZAAC(470)171]
Pr	A	68	86 [PrOH]	^{1}H-NMR	[80ZAAC(470)171]
^{i}Pr	A	72	135 [^{i}PrOH]	^{1}H-NMR, X-ray	[80ZAAC(470)171]
Bu	A	64	72 [BuOH]	^{1}H-NMR	[80ZAAC(470)171]
^{i}Bu	A	61	90 [^{i}BuOH]	^{1}H-NMR	[80ZAAC(470)171]

[a] For notation, see Section 2.2.1.1.

2.2.1.2.5. Diorganylbismuth compounds with Bi–Se or Bi–Te bonds (Table 2.8)

(a) Organylbismuth compounds with a Bi–Se bond Diphenylbismuth selenocyanate $Ph_2BiSeCN$ and bis(4-methylphenyl)bismuth selenocyanate $(4\text{-}MeC_6H_4)_2BiSeCN$ were prepared initially by the action of selenocyanogen $(SeCN)_2$ on the corresponding triarylbismuthines [26JCS1648]. This reaction was reinvestigated later in detail by Goel et al. [79SA(A)339]. The IR spectrum of $Ph_2BiSeCN$ shows peaks at 2116 ν(CN), 2088 ν(CN), 560 ν(CSe), 405 δ(SeCN), 395 δ(SeCN), 236 ν(Bi–SeCN) and 216 ν(Bi–Ph) cm^{-1}, respectively.

$$Ar_3Bi + (SeCN)_2 \longrightarrow Ar_2BiSeCN \quad (Ar = Ph, 4\text{-}MeC_6H_4)$$

By the reaction of tetraorganyldibismuthine R_2BiBiR_2 and diphenyl diselenide PhSeSePh, compounds of the type $R_2BiSePh$ are obtained [84ZN(B)1668, 88IC3730]. Dimethylbismuth benzeneselenolate $Me_2BiSePh$ is obtained as a yellow solid in 79% yield; it is light-sensitive and thermolabile but can be kept safely at −30°C for long periods. It is soluble in CH_2Cl_2, $CHCl_3$, DMSO and acetone, but less soluble in EtOH and non-polar solvents. At room temperature and within 3 h, it undergoes complete disproportionation to Me_3Bi and $MeBi(SePh)_2$. The rate of disproportionation is faster in CH_2Cl_2 than in DMSO [84ZN(B)1668]. Similarly, $Ph_2BiSePh$ is obtained in 54% yield and its X-ray crystallographic study shows that the Bi–Se bond length is 2.704(3) Å and the Se–Bi–C bond angles are 97.4(3) and 87.4(3)° [88IC3730]. $Pr_2BiBiPr_2$ reacts with elemental selenium to give a μ-selenobis(dipropylbismuth) $Pr_2BiSeBiPr_2$ as an orange-yellow liquid, which is soluble in benzene and $CHCl_3$, and decomposes into Pr_3Bi and a black solid above 25°C [86ZN(B)1129]. Similarly, tetramethyldibismuthine is converted to the corresponding μ-seleno compound as yellow crystals, which are not so sensitive to oxygen. This μ-seleno compound is thermolabile and decomposes within several minutes at room temperature; however, it can be kept for a fortnight below −30°C [95ZN(B)735].

Mass spectra of these $R_2BiSeBiR_2$ compounds show characteristic peaks due to M^+, $R_3Bi_2Se^+$, $R_2Bi_2Se^+$ and RBi_2Se^+ ions. The insertion reaction between $Ph_2BiBiPh_2$ and elemental selenium to $Ph_2BiSeBiPh_2$ has been mentioned briefly [88IC3730]. μ-Selenobis[bis(2,4,6-trimethylphenyl)bismuth] is obtained as yellow crystals by the action of sodium selenide on bromobis(2,4,6-trimethylphenyl)bismuthine in liquid ammonia. An X-ray crystallographic study of this μ-seleno compound has shown that the Bi–Se bond length is 2.651(1) Å and the Bi–Se–Bi bond angle is 91.2(1)° (Table 6.5, compound **41**). Its mass spectrum (DCI^+; NH_3) shows peaks due to $[M + H]^+$ and $Ar_3Bi_2SeH^+$ ions, whereas the mass (EI) spectrum exhibits peaks due to Ar_3BiSe^+, Ar_2BiSe^+, Ar_2Bi^+ and $ArBi^+$ fragments. The UV–Vis spectrum shows absorption maxima at λ_{max} 460, 370 and 230 nm [95ZN(B)735].

(b) Organylbismuth compounds with Bi–Te bonds By the reaction of $Me_2BiBiMe_2$ and PhTeTePh, $Me_2BiTePh$ is obtained as a yellowish fine powder. Its half-life in CH_2Cl_2 at room temperature is 5–6 h. It decomposes to give a mixture of Me_3Bi and $MeBi(TePh)_2$ above 16°C. The compound can also be prepared by the action of Me_3Bi on Ph_2Te_2 [84ZN(B)1668]. In a similar manner, dipropylbismuth 4-methylbenzenetellurolate is obtained as a reddish brown liquid, which is soluble in benzene, toluene and pentane, and can be kept for several months at −20°C. However, it decomposes rapidly at 25°C and disproportionates to ditelluride under atmospheric conditions [85ZN(B)848]. Similarly, $Ph_2BiBiPh_2$ and Ph_2Te_2 give $Ph_2BiTePh$ as an orange solid in 69% yield. This compound reacts smoothly with diazomethane in THF at room temperature to yield Ph_2BiCH_2TePh as a yellow solid. This mode of insertion reaction becomes increasingly difficult as the atomic number of chalcogen element decreases [88IC3730].

$$Ph_2BiTePh + CH_2N_2 \xrightarrow[THF]{} Ph_2BiCH_2TePh$$

The mass spectra of dialkylbismuth aryltellurides show peaks due to M^+, $[M–R]^+$, $[M–2R]^+$ and $BiTe^+$ ions, whereas the mass spectrum of $Ph_2BiTePh$ shows peaks due to M^+, $[M–TePh]^+$, $[M–TePh_2]^+$ and $[M–BiPh]^+$ ions. The formation of $Me_2BiTeMe$ is confirmed by ^1H-NMR, though the compound has not been isolated yet [86JOM(308)289].

Tetrapropyldibismuthine reacts with elemental tellurium to give a μ-telluro compound $Pr_2BiTeBiPr_2$ as a red liquid, which is soluble in common organic solvents. By warming or under diffused light, it decomposes to Pr_3Bi and a black solid [86ZN(B)1129]. Similarly, tetramethyldibismuthine reacts with tellurium to give the corresponding μ-telluro compound, $Me_2BiTeBiMe_2$, as red crystals. This compound is not so sensitive to oxygen but thermolabile, and decomposes within 15 min even at −30°C. At −78°C, however, it can be kept for several weeks [95ZN(B)735]. In contrast to alkyl derivatives, Ph_2Bi-$BiPh_2$ does not react with elemental tellurium [88IC3730]. μ-Tellurobis-[bis(2,4,6-trimethylphenyl)bismuth] is obtained as a reddish brown powder by the action of sodium telluride on bromodimesitylbismuthine. This compound can also be prepared by the reaction of tetrakis(2,4,6-trimethylphenyl)dibismuthine and tellurium [95ZN(B)735]. Its CI mass spectrum (DCI^+; NH_3) shows peaks due to $[M{\cdot}2NH_3]^+$ and M^+ ions, whereas its EI mass spectrum exhibits peaks due to Ar_3BiTe^+, Ar_2Bi^+ and Ar_2Te^+ fragments. The UV–Vis spectrum shows absorption maxima at λ_{max} 480, 380, 305 and 260 nm [95ZN(B)735].

TABLE 2.8
Diorganylbismuth and related compounds with Bi–Se and Bi–Te bonds

R	R′	Synthetic method[a]	Yield (%)	M.p. (°C)	Physical data	Reference
Bi–Se compounds						
$R_2BiSeCN$						
Ph		G	–	–	–	[26JCS1648]
		A	–	122 (decomp.) [EtOH]	IR, Raman	[79SA(A)339]
4-MeC_6H_4		G	–	–	–	[26JCS1648]
R_2BiSeR'						
Me	Me	E	–	–	^1H-, ^{77}Se-NMR, Raman	[86JOM(308)289]
	Ph	E	79	40 (decomp.)	^1H-, ^{13}C-NMR, MS	[84ZN(B)1668]
Ph	Ph	E	54	–	^{13}C-NMR, IR, X-ray	[88IC3730]
4-MeC_6H_4	Ph	E	92	29 (decomp.) [CH_2Cl_2-C_5H_{12}]	^1H-NMR, MS	[87ZN(B)695]
	4-MeC_6H_4	F	74	106 (decomp.)	^1H-NMR, MS	[87ZN(B)695]
$R_2BiSeBiR_2$						
Me		F	85	–	^1H-NMR, MS	[95ZN(B)735]
Pr		F	Quant.	25 (decomp.)	^1H-NMR, MS	[86ZN(B)1129]
Ph		F	33	–	IR	[88IC3730]
2,4,6-$Me_3C_6H_2$		A	64	63	^1H-NMR, MS, UV, X-ray	[95ZN(B)735]

TABLE 2.8 (*continued*)

R	R′	Synthetic method[a]	Yield (%)	M.p. (°C)	Physical data	Reference
Bi–Te compounds						
R_2BiTeR'						
Me	Me	E	–	–	^{1}H-, ^{125}Te-NMR, Raman	[86JOM(308)289]
	Ph	E, G	72	16 (decomp.)	^{1}H-, ^{13}C-NMR, MS	[84ZN(B)1668]
Pr	4-MeC_6H_4	E	Quant.	25 (decomp.)	^{1}H-, ^{125}Te-NMR, MS	[85ZN(B)848]
		E	–	–	^{1}H-, ^{125}Te-NMR, Raman	[86JOM(308)289]
Ph	Ph	E	69	–	^{13}C-NMR, IR, MS	[88IC3730]
4-MeC_6H_4	Ph	E	85	37 (decomp.)	^{1}H-NMR, MS	[87ZN(B)695]
$R_2BiTeBiR_2$						
Me		F	63	–	^{1}H-NMR, MS	[95ZN(B)735]
Pr		F	Quant.	–	^{1}H-NMR, MS	[86ZN(B)1129]
2,4,6-$Me_3C_6H_2$		A	54.6	58 (decomp.)	^{1}H-NMR, MS, UV	[95ZN(B)735]

[a] For notation, see Section 2.2.1.1.

2.2.1.2.6. Cationic and anionic diorganylbismuth compounds (Table 2.9)

The Lewis acid behavior of organometallic compounds derived from Group 15 elements has not been studied in detail. In this subsection, diorganylbismuth compounds with a positive $[R_2BiL_2]^+$ or negative $[R_2BiL_2]^-$ bismuth species are described. In the case of cationic diorganylbismuth compounds, ligand L represents a neutral two-electron donor, i.e. a Lewis base. In the case of anionic diorganylbismuth compounds, the ligand L is a pseudohalide anion such as azide, thiocyanate or a carboxylate.

(a) Cationic diorganylbismuth compounds Known cationic diorganylbismuth compounds include compounds **1–9**. All these compounds are prepared in moderate yield by metathesis reaction between Ar_2BiX (X = Cl, Br) and $AgBF_4$, $AgClO_4$ or $TlPF_6$ in the presence of the corresponding Lewis base [79JOM(166)365, 93JOM(460)C22, 96JCS(D)443]. Compound **9** is air- and moisture-stable, soluble in polar solvents such as acetone and ethanol, but insoluble in non-polar organic solvents [79JOM(166)365].

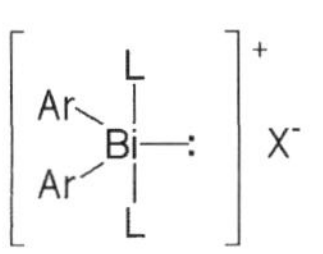

1; Ar = Ph, L = Py, X = BF_4
2; Ar = Ph, L = Py, X = PF_6
3; Ar = Ph, L = HMPA, X = BF_4
4; Ar = Ph, L = HMPA, X = PF_6
5; Ar = ^{p}Tol, L = HMPA, X = PF_6
6; Ar = Mes, L = HMPA, X = PF_6
7; Ar = Ph, L = $Ph_3P{=}O$, X = BF_4
8; Ar = Mes, L = $Ph_3P{=}O$, X = PF_6
9; Ar = Ph, L = $Ph_3As{=}O$, X = ClO_4

In addition, an intramolecularly coordinated cationic diorganylbismuth compound, bis[2-(*N*,*N*-dimethylaminomethyl)phenyl]bismuth hexafluorophosphate **10**, has been prepared in 65% yield by the action of $TlPF_6$ on the corresponding chlorobismuthine [97OM3597].

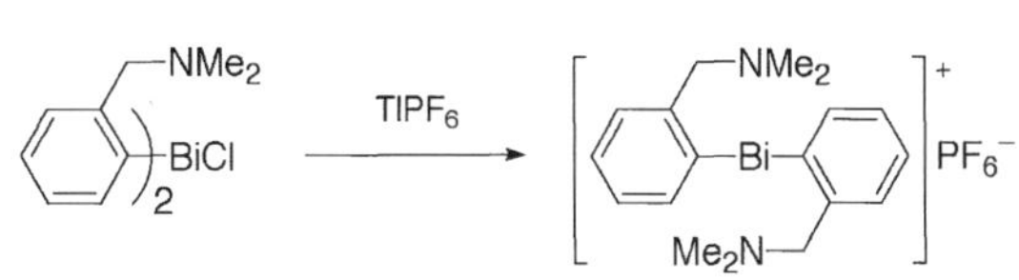

10

^1H, ^{13}C and ^{19}F-NMR spectral data are available for almost all cationic diorganylbismuth compounds. The most characteristic are ^1H-NMR signals due to the deshielded protons; generally, *o*-, *m*-, and *p*-protons of compounds **1–5** and **7** are observed at δ 8.2, 7.7 and 7.4, respectively [96JCS(D)443]. Compound **9** exhibits IR absorptions at 1090 and 625 cm^{-1} due to the ClO_4^- ion, 835 cm^{-1} due to AsO stretching vibration, and 378 cm^{-1} due to Bi–O stretching vibration, which suggests that the cationic center has a C_{2v} geometry in the solid state [79JOM(166)365]. X-Ray structure analysis of compounds **3**, **4**, **6** and **7** have elucidated that these four compounds possess a four coordinated disphenoidal geometry, in which two aryl groups occupy equatorial positions and two Lewis base ligands occupy axial positions [96JCS(D)443]. For example, the average Bi–C bond length and C–Bi–C bond angle are 2.211 Å and 94.3°, respectively. The HMPA ligands coordinate with an average Bi–O distance of 2.336 Å and O–Bi–O angle of 172.4(2)° [96JCS(D)443].

(b) Anionic diorganylbismuth compounds (bismuthate complexes) Known anionic diorganylbismuth compounds which contain the Bi–pnictogen or Bi–chalcogen bond include compounds **1–5**. Compounds **1** and **2** are prepared by metathesis reaction between $Ph_2BiCl_2^-$ with the corresponding sodium pseudohalide, NaN_3 or NaSCN. [79JOM(166)365]

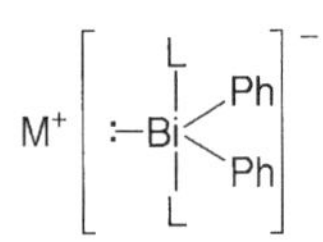

1; M = Et_4N, L = N_3
2; M = Et_4N, L = SCN
3; M = $Me_2NH(CH_2)_2NMe_2$, L = CF_3CO_2
4; M = $Me_2NH(CH_2)_3NMe_2$, L = CF_3CO_2
5; M = Ph_4Bi, L = CF_3CO_2

Compound **5** can be obtained by three different procedures; (i) reaction of pentaphenylbismuth Ph_5Bi and phenylbismuth bis(trifluoroacetate) $PhBi(O_2CCF_3)_2$, (ii) reaction of tetraphenylbismuth trifluoroacetate $Ph_4BiO_2CCF_3$ and diphenylbismuth trifluoroacetate $Ph_2BiO_2CCF_3$, and (iii) tetraphenylbismuth trifluoroacetate $Ph_4BiO_2CCF_3$ and triphenylbismuth bis(trifluoroacetate) $Ph_3Bi(O_2CCF_3)_2$ in 86, 68 and 50% yield, respectively [84HCA586].

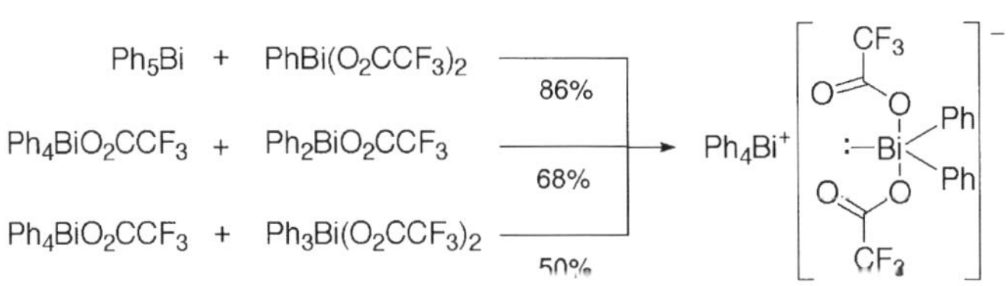

Compounds **3** and **4** are generated by disproportionation of the corresponding phenyltetrakis(trifluoroacetato)bismuthate $[PhBi(O_2CCF_3)_4]^{2-}$ derivatives; however, attempts to isolate compounds **3** and **4** have failed probably because of their subsequent disproportionation reactions [94JCS(D)2545].

The IR spectra of compounds **1**, **2** and **5** are recorded; **1** shows a very strong

IR absorption at 2010 cm^{-1} due to the antisymmetric N_3 stretching, while **2** shows IR absorptions at 2030 (CN stretching) and 470 cm^{-1} (NSC bending vibration), indicating the presence of *N*-bonded NCS groups [79JOM(166)365]. The IR spectrum of compound **5** exhibits absorptions at 1680 and 1180–1120 cm^{-1} [84HCA586].

$$2[Hdam][Ph_2Bi(O_2CCF_3)_2] \rightleftharpoons 2[Ph_2Bi(O_2CCF_3)] + [H_2dam][O_2CCF_3]_2 + dam$$

$$2[Hdam][O_2CCF_3] \rightleftharpoons [H_2dam][O_2CCF_3]_2 + dam$$

$$dam = Me_2N(CH_2)_nNMe_2 \ (n = 2, 3)$$

Compounds **1** and **2** show Raman bands in acetone at 280 and 235 cm^{-1}, respectively, due to the Bi–X symmetric stretching mode. The mass spectrum of compound **5** contains characteristic signals due to $Ph_2Bi(O_2CCF_3)_2^+$, Ph_4Bi, Bi_2^+, $PhBiO_2CCF_3^+$ and Ph_2Bi^+ ions.

X-Ray structure analysis of compound **5** revealed that the geometry of the bismuthate unit is described best as a trigonal bipyramidal structure with a stereochemically active lone pair of electrons. Two phenyl groups occupy the equatorial sites in a propeller fashion, and two trifluoroacetoxy groups occupy the apical sites. The average Bi–C bond length and C–Bi–C angle are 2.25 Å and 95°, respectively, and the average Bi–O bond length and O–Bi–O bond angle are 2.39 Å and 172°, respectively. The carbonyl oxygen atom of one trifluoroacetate group is oriented toward the counter cation Ph_4Bi^+, with the Bi–O distance of 3.11 Å [84HCA586].

TABLE 2.9
Anionic and cationic diorganylbismuth compounds

Compound	Synthetic method[a]	Yield (%)	M.p. (°C)	Physical data	Reference
Anionic compounds					
$[Me_2NH(CH_2)_2NMe_2][Fh_2Bi(O_2CCF_3)_2]$	–	–	–	^{1}H-NMR	[94JCS(D)2545]
$[Me_2NH(CH_2)_3NMe_2][Fh_2Bi(O_2CCF_3)_2]$	–	–	–	^{1}H-NMR	[94JCS(D)2545]
$[Ph_4Bi][Ph_2Bi(O_2CCF_3)_2]$	H	86	142–147 [PhH-Et_2O]	^{13}C-NMR, IR, MS, X-ray	[84HCA586]
$[Et_4N][Ph_2Bi(SCN)_2]$	H	ca. 70	81–83 (decomp.)	IR, Raman	[79JOM(166)365]
$[Et_4N][Ph_2Bi(N_3)_2]$	H	ca. 70	83–85	IR, Raman	[79JOM(166)365]
Cationic compounds					
$[Ph_2Bi(HMPA)_2][BF_4]$	H	70	–	^{1}H-NMR, X-ray	[93JOM(460)C22]
$[Ph_2Bi(HMPA)_2][PF_6]$	H	76	–	^{1}H-, ^{13}C-, ^{31}P-NMR, X-ray	[96JCS(D)443]
$[(4\text{-}MeC_6H_4)_2Bi(HMPA)_2][PF_6]$	H	64	–	^{1}H-, ^{13}C-, ^{31}P-NMR	[96JCS(D)443]
$[(2,4,6\text{-}Me_3C_6H_2)_2Bi(HMPA)_2][PF_6]$	H	60	–	^{1}H-, ^{13}C-, ^{31}P-NMR, X-ray	[96JCS(D)443]
$[Ph_2Bi(OPPh_3)_2][BF_4]$	H	52	–	^{11}B-, ^{31}P-NMR, X-ray	[96JCS(D)443]
$[(2,4,6\text{-}Me_3C_6H_2)_2Bi(OFPh_3)_2][PF_6]$	H	60	–	^{1}H-, ^{31}P-NMR	[96JCS(D)443]
$[Ph_2Bi(OAsPh_3)_2][ClO_4]$	H	80	190	IR	[79JOM(166)365]
$[Ph_2Bi(PyH)_2][BF_4]$	H	30	–	^{1}H-NMR	[96JCS(D)443]
$[Ph_2Bi(PyH)_2][PF_6]$	H	98	–	^{1}H-, ^{13}C-, ^{31}P-NMR	[96JCS(D)443]
$\{[2\text{-}(Me_2NCH_2)C_6H_4]_2B\}[PF_6]$	H	65	–	^{1}H-NMR, MS, X-ray	[97OM3597]

[a] For notation, see Section 2.2.1.1.

2.2.2. Dibismuthines and dibismuthenes

Dibismuthines R_2BiBiR_2 are a type of compound which contain a bismuth–bismuth bond. They have characteristic physical and chemical properties due to the Bi–Bi bond. The most interesting is thermochromism: when melted or dissolved in organic solvents, some dibismuthines show drastic color changes. Recent studies have revealed that the intermolecular Bi···Bi interaction of dibismuthines in the solid state has a significant influence on this phenomenon. The Bi–Bi bond is cleaved easily by the attack of various reagents. Recently, oligomeric organylbismuth (I) compounds, $(RBi)_3$ and $(RBi)_4$ were reported to exist. They form small ring structures with Bi–Bi bondings, and there is an equilibrium between these two compounds in a solution.

In contrast to dibismuthines, the chemistry of dibismuthenes RBi=BiR remained almost untouched until very recently. In this section, the latest study of a dibismuthene is also discussed.

2.2.2.1. Methods of synthesis of dibismuthines (Table 2.10)

In 1922, Denham examined the reaction between bismuth suboxide BiO and methyl iodide at 260–262°C and claimed the isolation of ''bismuth dimethyl, $BiMe_2$'' as a distillate of the reaction mixture [21JA2367]. In 1935, Paneth et al. reported the first synthesis of thermochromic ''bis(dimethylbismuth)'' by the reaction of methyl radical with bismuth metal [35JCS366]. Later, Ashe et al. investigated this compound and isolated tetramethyldibismuthine as a thermolabile compound [82OM1408]. Since then, several methods have been reported for the preparation of this type of compound; all manipulations need to be performed under nitrogen or argon, since dibismuthines are so air-sensitive and react rapidly with oxygen. Tetraorganyldibismuthines are prepared mainly in the same manner as the corresponding distibines. Thus, the reaction of triorganylbismuthines with sodium in liquid ammonia gives sodium diorganylbismuthides as a red solution, which on treatment with 1,2-dichloroethane, affords tetraorganyldibismuthines (methods A and B). Treatment of diorganylbismuth halides with sodium in liquid ammonia also affords dibismuthines (method C). The improved preparation of a dibismuthine involves the reduction of diphenylbismuth iodide with dicyclopentadienyl cobalt(II), where the reaction can be carried out in THF (method D). Dimerization of bismabenzene forms a bicyclic dibismuthine which, however, cannot be isolated (method E).

(a) Via reaction of sodium bismuthides and dihaloethane (method A)

Synthesis of tetramethyldibismuthine

$$Me_3Bi + Na \xrightarrow[\text{liq. } NH_3,\ -33°C]{} Me_2BiNa \xrightarrow{(CH_2Cl)_2} Me_2BiBiMe_2$$

Action of sodium metal (0.5 g, 22 mmol) on trimethylbismuthine (2.9 g, 11 mmol) in liquid ammonia (100 ml) at −33°C gave sodium dimethylbismuthide as a red solution. After the addition of 1,2-dichloroethane (1.1 g, 11 mmol), ammonia was allowed to evaporate. The residue was extracted with pentane and filtered off. Three recrystallizations of the product from pentane at −20°C gave pure tetramethyldibismuthine (2.6 g, 70%) [82OM1408].

(b) Via reaction of lithium bismuthides and dihaloethane (method B)

Synthesis of tetrakis(trimethylsilyl)dibismuthine

$$(Me_3Si)_2BiLi\ (THF)_2 \xrightarrow[\text{pentane, } -30\ °C]{(CH_2Br)_2} (Me_3Si)_2BiBi(SiMe_3)_2$$

A solution of lithium bis(trimethylsilyl)bismuthide·2 THF (25.3 g, 50 mmol) in pentane (100 ml) was added slowly to a cooled (−30°C) solution of 1,2-dibromoethane (4.7 g, 25 mmol) in the same solvent (100 ml). As the reaction proceeded, lithium bromide precipitated from a yellowish orange reaction mixture. Liberated ethylene gas (24.6 mmol) was collected into an inverted stand cylinder above water. After completion of the reaction, the precipitate was filtered off and washed successively with pentane. The volatiles were distilled off in vacuo at 20°C to leave a solid residue, which was recrystallized from pentane. The yield was 13.5 g (76%) [82ZN(B)91].

(c) Via reaction of halogenodiorganylbismuthines and sodium metal (method C)

Synthesis of tetraphenyldibismuthine

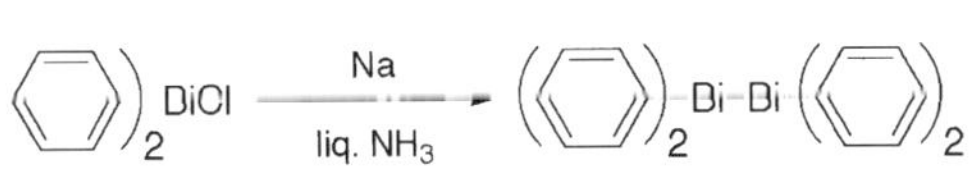

Sodium metal (0.23 g, 10 mmol) was added piecewise to a well stirred mixture of chlorodiphenylbismuthine (3.58 g, 10 mmol) in liquid ammonia (100 ml). By warming to 0°C, ammonia was evaporated from the resulting green-black solution. The residue was taken up in toluene (25 ml) and filtered off. On addition of methanol (10 ml), the dibismuthine separated from the solution as air-sensitive red crystals (1.55 g, 45%) [83OM1859].

(d) Via reaction of halogenodiorganylbismuthines and dicyclopentadienyl-cobalt(II) (method D)

Synthesis of tetraphenyldibismuthine

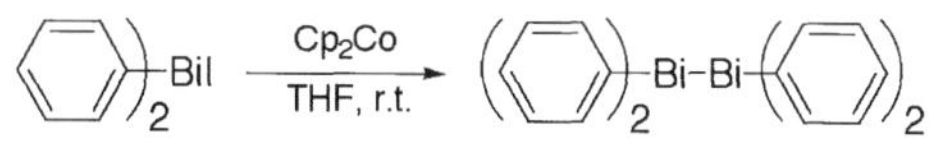

To a solution of iododiphenylbismuthine (8.806 g, 17.97 mmol) in THF (250 ml), dicyclopentadienylcobalt(II) (3.368 g, 17.83 mmol) dissolved in the same solvent was added dropwise at room temperature over a period of 5 min. After stirring for 10 min, the precipitated dicyclopentadienyl iodocobalt(III) was filtered off and the orange solution was evaporated to dryness under reduced pressure at room temperature. The residual solid was recrystallized from dry hot ethanol (600 ml). The dibismuthine obtained as orange needles was collected by filtration and dried in vacuo. Yield 80% [88IC3730].

(e) Via dimerization of bismabenzenes (method E)

Synthesis of bismabenzene dimer

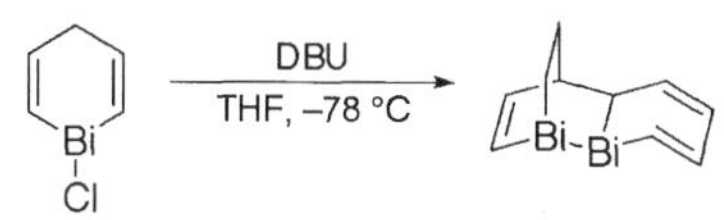

To a solution of 1-chlorobismacyclohexa-2,5-diene (0.800 g, 2.6 mmol) in THF (100 ml) was added DBU (0.288 g, 2.3 mmol) at −78°C in THF (50 ml). The base precipitated as hydrochloride. The resulting organylbismuth compound was identified as a [4 + 2]-dimer of bismabenzene by ^{1}H-NMR THF-d_8 at −78°C. On warming the reaction mixture to −10°C, however, the dibismuthine turned into monomeric bismabenzene [82JA5693].

(f) Via reaction of dichloromethylbismuthine and lithium phosphide complex (method F)

Synthesis of *cis*-Mo(CO)$_4$(PMe$_2$-BiMeBiMe-PMe$_2$) complex

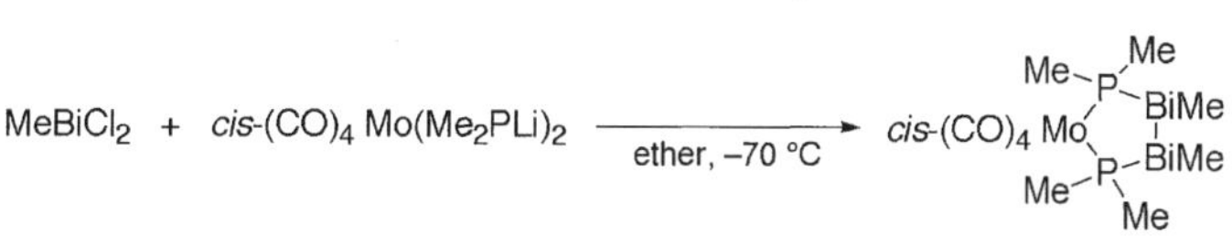

cis-Tetracarbonyl bis(lithium dimethylphosphide) molybdenum (10.0 mmol), prepared by deprotonation of *cis*-tetracarbonyl bis(dimethylphosphine) molybdenum with butyllithium or methyllithium, was treated with an equivalent amount of dichloromethylbismuthine (3.15 g, 10.7 mmol) in diethyl ether (15 ml) at −70°C. The reaction mixture was stirred at −70°C for 2 h, then allowed to warm to room temperature for 2–3 h, and stirred subsequently at this temperature for 4–5 h. The volatiles were removed in vacuo to leave a residue, which was suspended in dichloromethane, and the insoluble part was separated by means of a centrifuge. The residue after the operation was subjected to thin-layered chromatography using a mixture of pentane and dichloromethane (1:3) as an eluent. The yield of the dibismuthine was 0.225 g (3%) [77CB3430].

(g) Via reduction of an alkyldichlorobismuthine (method G)

Synthesis of bis(trimethylsilyl)methyldibismuthine trimer and tetramer

$$(Me_3Si)_2CHBiCl_2 \xrightarrow[THF]{Mg} [(Me_3Si)_2CHBi]_n \quad (n = 3, 4)$$

At -80°C a solution of $(Me_3Si)_2CHBiCl_2{\cdot}0.5Et_2O$ (1.48 g, 3.11 mmol) in THF (20 ml) was added dropwise with stirring to magnesium filings (0.25 g, 10.28 mmol) activated with 1,2-dibromoethane in THF (10 ml). Thereafter the mixture was warmed to -35°C, stirred for 6 h at -35°C and 1 h at -15°C. The solvent was removed in vacuo and the residue was extracted with petroleum ether (250 ml). The dark brown solution was filtered at -10°C through a frit and concentrated to dryness. After removal of the solvent a dark brown solid consisting of a trimer and tetramer of the title compound remained. Yield 0.78 g, 68% [98AG(E)3175].

TABLE 2.10
Tetraorganyldibismuthines and diorganyldibismuthenes

Compound R	Synthetic method[a]	Yield (%)	M.p. (°C)	Thermochromism	Physical data	Reference
R_2Bi–BiR_2						
Me	–	–	–	–	–	[21JA2367]
	–	–	–	–	–	[34TFS(30)179]
	–	–	–	Positive	–	[35JCS366]
	A	70	−12.5	Positive	^{1}H-, ^{13}C-NMR, MS, Raman, UV	[82OM1408]
	A	70	−12.5	Positive	^{1}H-, ^{13}C-NMR, MS, UV	[83OM1859]
	C	78	−17	Positive	^{1}H-NMR, Reaction	[84ZN(B)887]
	–	–	–	–	–	[84ZN(B)1668]
	–	–	–	–	–	[86JOM(308)289]
	–	–	–	–	X-ray	[88ZN(B)952]
	D	–	–	–	^{1}H-NMR	[91G(121)93]
Et	–	–	–	–	–	[35JCS366]
	A	64	40 (decomp.)	Positive	^{1}H-NMR, IR, MS, UV	[82AG(E)439]
	A	63	–	–	^{1}H-NMR, MS	[83ZN(B)125]
Pr	–	–	–	–	^{1}H-NMR, Raman	[86ZN(B)1129]
	–	–	–	–	–	[85ZN(B)848]
	A	75	–	–	^{1}H-NMR, MS	[83ZN(B)125]
^{i}Pr	A	53	–	–	^{1}H-NMR, MS	[83ZN(B)125]
Bu	A	81	–	–	^{1}H-NMR, MS	[83ZN(B)125]
H_2C=CMe	A	67	13.5	Positive	^{1}H-, ^{13}C-NMR, IR, MS, UV	[83OM1859]
Me_2C=CH	A	82	29–29.5	Negative	^{1}H-, ^{13}C-NMR, IR, MS, UV	[83OM1859]
Ph	C	–	–	–	–	[41JA212]
	D	–	–	–	–	[57ZN(B)132]
	C	–	–	–	–	[66JA467]
	–	–	–	–	–	[66JA5117]
	–	–	–	–	–	[66JA5121]
	C	45	124–126 (decomp.)	Negative	^{1}H-, ^{13}C-NMR, MS, UV	[83OM1859]

TABLE 2.10 *(continued)*

Compound R	Synthetic method[a]	Yield (%)	M.p. (°C)	Thermochromism	Physical data	Reference
	C	45	–	Negative	MS, X-ray	[83CC507]
	C	60.4	–	–	IR, MS, X-ray	[84JCS(D)2365]
	C/D	54/80	–	–	–	[88IC3730]
	D	43–52	–	–	–	[91G(121)93]
	–	–	–	–	–	[92CC453]
	A	–	–	–	X-ray	[92AX(C)917]
4-MeC_6H_4	D	45	24 (decomp.)	–	^{1}H-NMR, MS	[87ZN(B)695]
	D	50	–	–	^{1}H-NMR	[91G(121)93]
2,4,6-$Me_3C_6H_2$	C/D	40.7/52.6	170 (decomp.)	Positive	^{1}H-NMR, MS, UV	[95ZN(B)735]
Me_3Si	B	76	148	Positive	^{1}H-, ^{13}C-NMR, IR, MS	[82ZN(B)91]
	B	–	–	Positive	X-ray	[83ZAAC(506)42]
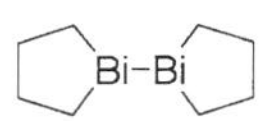	A	58	46–47	Positive	^{1}H-, ^{13}C-NMR, MS, UV	[83OM1859]
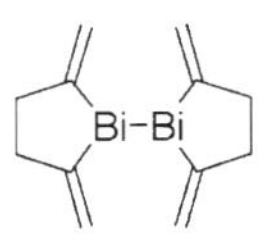	C	–	>0 (decomp.)	Positive	^{1}H-NMR, MS, UV	[87OM1185]
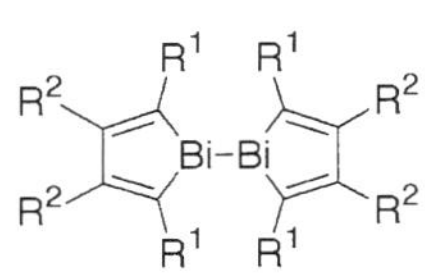						

$R^1 = Me, R^2 = H$	A	–	>95 (decomp.)	Positive	^1H-, ^{13}C-NMR, MS	[84OM495]
	–	–	–	–	UV	[86JOM(303)197]
	A	61	101–102 (decomp.)	Positive	^1H-, ^{13}C-NMR, MS, UV, X-ray	[92OM2743]
$R^1 = Me, R^2 = Me$	A	61	128–129	Positive	–	[92OM3492]
$R^1 = Me_3Si, R^2 = Me$	A	64	121–122	Positive	^1H-, ^{13}C-NMR, IR	[92OM3492]
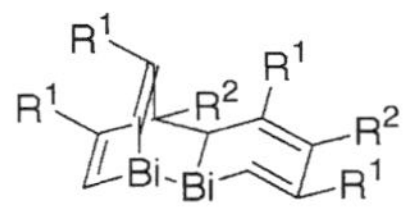						
$R^1 = H, R^2 = H$	E	–	–	–	^1H-NMR	[76TL415]
	E	–	–	–	^1H-NMR	[82JA5693]
$R^1 = D, R^2 = H$	E	–	–	–	^1H-NMR	[76TL415]
$R^1 = H, R^2 = Me$	E	–	–	–	^1H-NMR	[82JA5693]
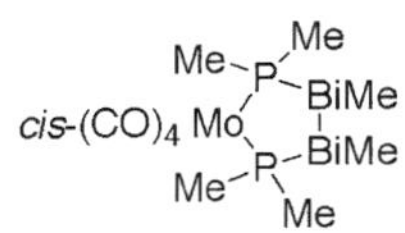	F	3	–	–	^1H-, ^{31}P-NMR, IR, MS	[77CB3430]
Polybismuthines						
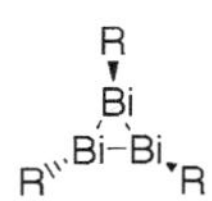	G	–	–	–	^1H-, ^{13}C-NMR	[98AG(E)3175]
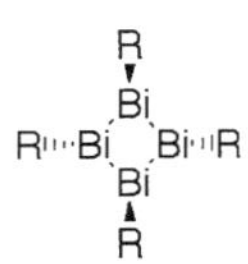	G	68	89–91 [petroleum ether]	–	^1H-, ^{13}C-NMR, X-ray	[98AG(E)3175]
RBi=BiR[b,c]						

TABLE 2.10 *(continued)*

Compound R	Synthetic method[a]	Yield (%)	M.p. (°C)	Thermochromism	Physical data	Reference
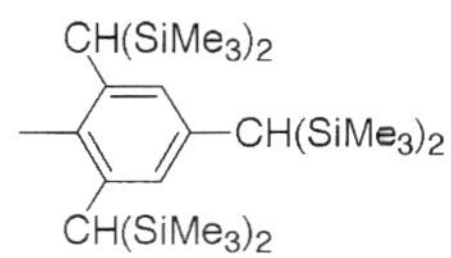	A	–		–	Raman, UV, X-ray	[97SCI(277)78]
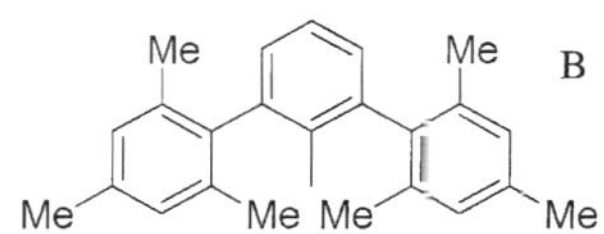	B	30	215–219 (decomp.)	–	^{1}H-, ^{13}C-NMR, UV, X-ray	[99JA3357]
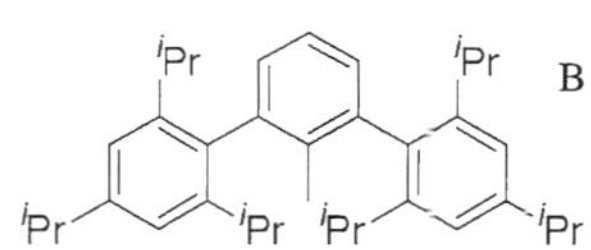	B	83	>250 (decomp.)	–	^{1}H-, ^{13}C-NMR, UV	[99JA3357]

[a] For notation, see Section 2.2.2.1.
[b] For notation, see Section 2.2.2.4.
[c] A stibabismuthene of a structure RBi=SbR is obtained by the reaction of $RBiBr_2$ and $RSbH_2$ in the presence of DBU at low temperature [UR].

2.2.2.2. Physical properties of dibismuthines

There is a leading review on the thermochromism and other physical properties of dibismuthines [90AOC(30)77]. Suitably substituted binuclear compounds of the heavier main group elements often associate in the solid state. These compounds exhibit short intramolecular contacts which indicate the formation of secondary intermolecular bonding. One of the most interesting characteristics of dibismuthines is the dramatic change in color on melting the solid (thermochromism). However, the intermolecular bonding is rather sensitive to substitution pattern, so the thermochromic effect is not always observed for all dibismuthines (see Table 2.10).

Most known dibismuthines are red in color, both in liquid form and in solution in organic solvents. A series of tetravinyldibismuthines show absorption maxima in the range 270–330 nm, but spectral assignments have not been made yet. On crystallization, non-thermochromic dibismuthines show little visual color change. The solid colors of two dibismuthines, tetrakis(2-methyl-1-propenyl)dibismuthine and tetraphenyldibismuthine, have been characterized by diffuse reflectance [83OM1859, 83CC507]. In both cases, only modest changes in the absorption maxima are observed between the solid and a solution. On the other hand, the intense colors shown by the solid phases of the thermochromic dibismuthines are red, shifted by 300–370 nm from their solution phase maxima [83OM1859, 84OM495, 87OM1185]. Thus, tetrakis(2,4,6-trimethylphenyl)dibismuthine exists as a reddish brown solid, which reversibly becomes pink-colored on cooling with liquid nitrogen. The UV–Vis spectrum of this dibismuthine shows two absorption maxima at 350 and 530 nm. Even when cooled with liquid nitrogen, it does not show any shift of the absorption maxima [95ZN(B)735].

Structural data are now available for four dibismuthines: thermochromic tetrakis(trimethylsilyl)dibismuthine [83ZAAC(506)42], tetramethyldibismuthine [88ZN(B)952] and 2,2′,5,5′-tetramethylbibismole [92OM2743], and non-thermochromic tetraphenyldibismuthine [83CC507, 84JCS(D)2365, 92AX(C)917]). As for thermochromic dibismuthines, there is a Bi–Bi···Bi–Bi chain structure; the intramolecular Bi–Bi bond is in the range 2.99–3.12 Å, while the intermolecular Bi···Bi separation is in the range 3.58–3.80 Å. Non-thermochromic tetraphenyldibismuthine does not show any close intermolecular Bi···Bi contacts. Electronic structure calculations have been carried out using both effective core potential and relativistically parameterized semi-empirical molecular orbital methods for diorganylbismuthines and tetraorganyldibismuthines [93OM343].

Raman spectroscopy is extremely useful for the characterizing of dibismuthines, since the Bi–Bi stretching vibrations give rise to intense, easily recognizable peaks. The $\nu_{\mathrm{Bi-Bi}}$ stretching is close to 110 cm^{-1} for all dibismuthines irrespective of the thermochromic or non-thermochromic nature, and of the solid or liquid state [83OM1859]. This insensitivity to the substitution patterns suggests that these bands are largely due to the localized Bi–Bi vibrations.

2.2.2.3. Chemical properties of dibismuthines

Dibismuthines are very labile; they react readily with free radicals, electrophilic and nucleophilic reagents, which all cleave the Bi–Bi bond. Some dibismuthines are thermolabile; tetramethyldibismuthine decomposes at 25°C quantitatively into trimethylbismuthine and bismuth metal. The half-life of this dibismuthine is approximately 6 h in a dilute benzene solution [82OM1408]. Tetraphenyldibismuthine [83CC507] and 2,2′,5,5′-tetramethylbibismole are stable up to 100°C [92OM2743].

Thermal decomposition occurs more rapidly under ultraviolet irradiation or in a halogenated hydrocarbon such as chloroform or carbon tetrachloride. Even the thermostable tetraphenyldibismuthine does decompose in chloroform at 25°C [83OM1859]. These interesting findings are explained best by a free radical mechanism.

$$\mathrm{Me_2BiBiMe_2} + \mathrm{BrCCl_3} \longrightarrow \mathrm{Me_2BiBr} + \mathrm{Me_2BiCCl_3} + \mathrm{Me_3Bi} + \mathrm{C_2Cl_6} + \mathrm{Cl_2C{=}CCl_2}$$

Thus, the thermal decomposition of tetramethyldibismuthine in the presence of bromotrichloromethane may take place via the following chain mechanism. (i) Tetramethyldibismuthine decomposes first into two dimethylbismuth radicals. (ii) These radicals react with bromotrichloromethane to form bromodimethylbismuthine bromide and a trichloromethyl radical. (iii) The trichloromethyl radical reacts with the dibismuthine to form dimethyl(trichloromethyl)bismuthine and dimethylbismuth radical. The termination step is the self-combination of the trichloromethyl radicals. In fact, the major product of the reaction is bromodimethylbismuthine bromide and the second product is dimethyl(trichloromethyl)bismuthine. Trimethylbismuthine, hexachloroethane and tetrachloroethylene are also formed in smaller quantities [83OM1859].

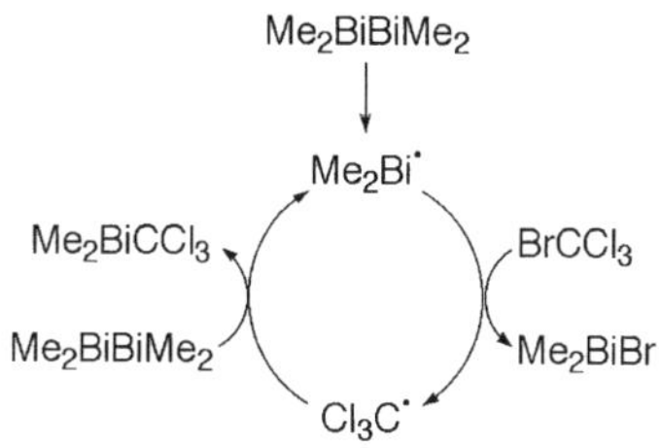

The reaction of tetramethyldibismuthine with hydrogen chloride affords hydrogen gas, bismuth metal, trimethylbismuthine and chlorodimethylbismuthine. These products are likely to be formed by the decomposition of thermolabile dimethylbismuthine, which can be detected at −78°C by ^{1}H-NMR measurement [83OM1859].

$$Me_2BiBiMe_2 + HCl \longrightarrow Me_2BiCl + \left[Me_2BiH\right] \longrightarrow Me_3Bi + H_2 + Bi$$

Tetramethyldibismuthine is cleaved with butyllithium to give lithium dimethylbismuthide and butyldimethylbismuthine. The former product reacts subsequently with butyllithium, eventually leading to all possible butylmethylbismuthines [83OM1859]. The reaction of tetramethyldibismuthine and phenyllithium has been reported to give dimethylphenylbismuthine [84ZN(B)887].

The Bi–Bi bond is cleaved easily by the action of compounds containing the dichalcogen bond. Thus, bis(dipropylbismuth) oxide, sulfide, selenide and telluride are formed in good yields by the reaction of tetrapropyldibismuthine with elemental chalcogens [84JCS(D)2365, 86ZN(B)1129]. In a similar manner, tetraorganyldibismuthines react with diorganyl disulfides, diselenides and ditellurides to give the corresponding organylchalchogenobismuthines in good yields [84ZN(B)1668, 85ZN(B)848, 86JOM(308)289, 88IC3730]. Tetraphenyldibismuthine reacts with benzoquinone and diazomethane to yield the respective addition insertion products, 1,4-bis(diphenylbismuthoxy)benzene and bis(diphenylbismuthyl)methane [84JCS(D)2365]. In contrast to the diorganyl dichalchogenides, tetraorganyldibismuthine and tetraorganyldipnictogens undergo a mutual element exchange reaction at 25°C both in a solution and as a neat liquid mixture. The formation of (dimethylarsino)dimethylbismuthine $Me_2AsBiMe_2$ and (dimethylstibino)dimethylbismuthine $Me_2SbBiMe_2$ is observed by ^{1}H-NMR from a solution of equimolar tetramethyldibismuthine and tetramethyldiarsine or -distibine in C_6D_6, and their equilibrium constants are estimated as 9×10^{-3} and 1.2, respectively [86JOM(303)197], although the isolation of these mixed dipnictogens has not been attempted.

The reaction of tetramethyldibismuthine with iodine gives all possible iodo(methyl)bismuthines as well as elemental bismuth [83OM1859]. Similarly, tetraphenyldibismuthine reacts with iodine to give iododiphenylbismuthine as the single product [84JCS(D)2365].

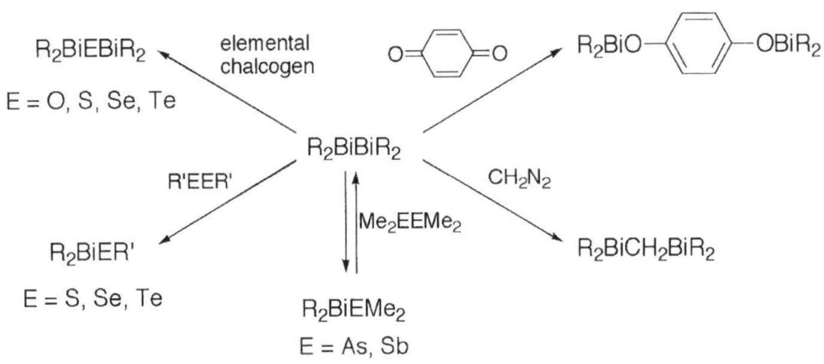

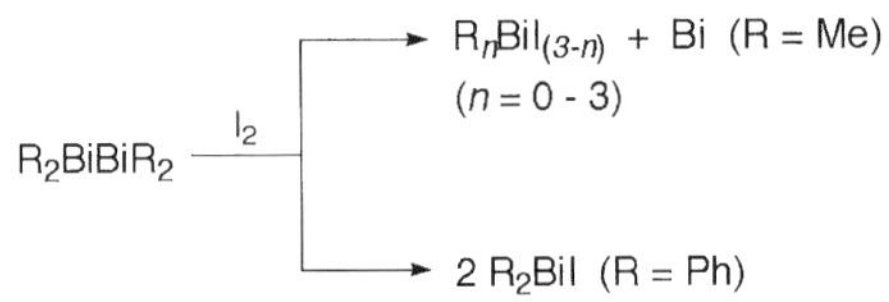

The existence of oligomeric or polymeric monoorganylbismuthines, $(RBi)_n$ has been discussed for a number of years. Reduction of $PhBiBr_2$ with $LiBH_4$ has reported to generate $(PhBi)_n$ as an oxygen-sensitive compound [57ZN(B)132]. Thermal decomposition of tetraethyldibismuthine leads to the formation of triethylbismuthine and poly(ethylbismuth) $(EtBi)_n$ as a black solid [82AG(E)439]. These poly(organylbismuth) compounds cannot be isolated. An attempt to isolate a poly(organylbismuth) using 2,4,6-triphenylphenyl group as an organyl substituent has failed [95JOM(485)141].

Very recently, trimer and tetramer of methylbis(trimethylsilyl)bismuthines $[(Me_3Si)_2CHBi]_n$ (n = 3 and 4) were prepared by the reaction between the corresponding alkyldichlorobismuthine with magnesium in THF at −35°C [98AG(E)3175].

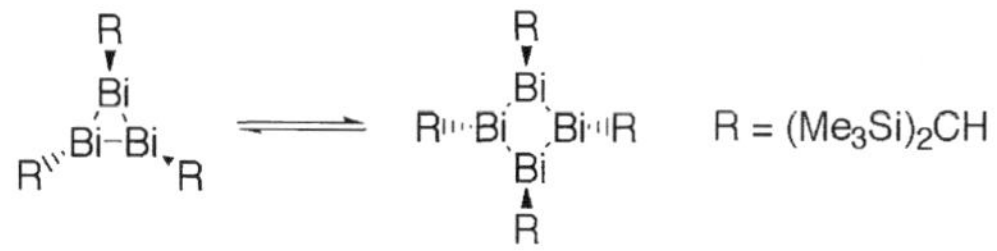

According to ^{1}H-NMR measurements, there is an equilibrium between the trimer and tetramer with the equilibrium constant of 40 ± 5 mol/l in C_6D_6 at 23°C. On cooling the solution, black crystals of tetramer are deposited. Both compounds are air-sensitive and thermolabile; they can be kept for several months at −28°C, but they easily decompose at ambient temperature. In a solution, a *cis–trans* and an all-*trans* structure is suggested by ^{1}H-NMR for the trimer and tetramer, respectively. The structure of the tetramer was elucidated by an X-ray crystallographic analysis to show that each molecule contains a folded four-membered bismuth ring. The Bi–Bi bond lengths are in the range 2.970(5)–3.044(2) Å, and the Bi–Bi–Bi bond angles are very small which fall in the range 78.97(8)–79.93(6)°. In addition, the molecule has a transannular interaction with Bi–Bi distance of 3.834 and 3.859 Å.

2.2.2.4. Dibismuthenes

Dibismuthenes are compounds possessing the Bi=Bi double bond. The possible existence of this type of compound had been predicted earlier by theoretical calculations [90CC1724]. Tokitoh et al. were successful in isolating the first example of a stable dibismuthene, TbtBi=BiTbt **1** where Tbt denotes the 2,4,6-tris[bis(trimethylsilyl)methyl]phenyl group. The dibismuthene is obtained as poorly soluble deep purple single crystals by the deselenation of a triselenatribismane with hexamethylphosphorous triamide in toluene at 100°C in a sealed tube [97SCI(277)78].

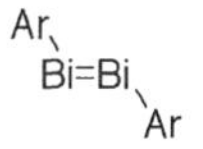

1; Ar = 2,4,6-tris[bis(trimethylsilyl)methyl]phenyl
2; Ar = 2,6-bis(2,4,6-trimethylphenyl)phenyl
3; Ar = 2,6-bis(2,4,6-triisopropylphenyl)phenyl

Very recently, two novel dibismuthenes **2** and **3** stabilized by 2,6-bis(2,4,6-trimethylphenyl)phenyl and 2,6-bis(2,4,6-triisopropylphenyl)phenyl groups, respectively, have been prepared by the reduction of the corresponding aryldichlorobismuthines with potassium and fully characterized [99JA3357].

(a) Via deselenation of a triselenatribismane (method A)

Synthesis of 1,2-bis{2,4,6-tris[bis(trimethylsilyl)methyl]phenyl}dibismuthene

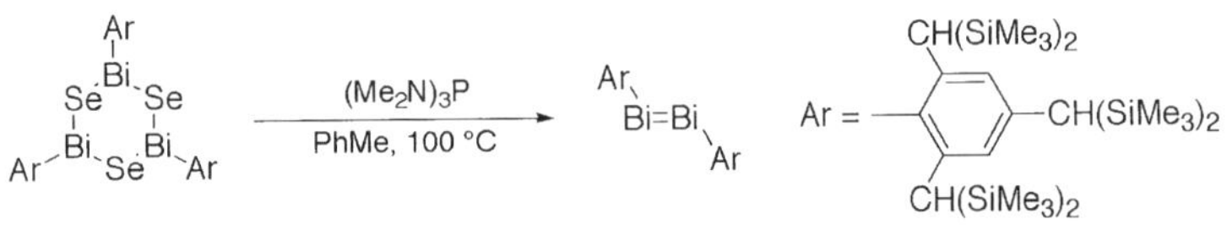

A mixture of 2,4,6-tris{2,4,6-tris[bis(trimethylsilyl)methyl]phenyl}-1,3,5-triselena-2,4,6-tribismane, hexamethylphosphorous triamide and toluene were heated in a sealed tube at 100°C to obtain the title dibismuthene as poorly soluble deep purple crystals [97SCI(277)78].

(b) Via reduction of aryldichlorobismuthines (method B)

Synthesis of 1,2-bis[2,6-bis(2,4,6-triisopropylphenyl)phenyl]dibismuthene

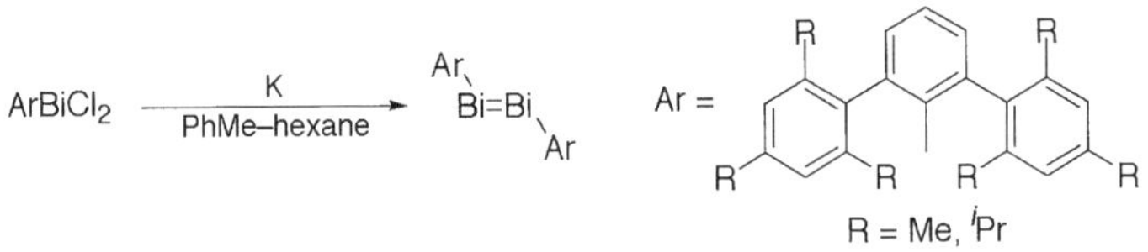

A solution of 2,6-bis(2,4,6-triisopropylphenyl)phenyldichlorobismuthine (1.00 g, 1.31 mmol) in hexane/toluene (3:1, 80 ml) was added to a stirred suspension of finely divided potassium (0.103 g, 2.63 mmol), and the mixture was stirred overnight. The solution developed a deep purple color, and all the potassium was consumed. The solution was filtered under a nitrogen atmosphere and concentrated to ca. 40 ml under reduced pressure. Cooling in a freezer at −20°C overnight afforded deep purple crystals (0.90 g, 83%) [99JA3357].

Compound **1** is stable in the solid state; its purple color is kept for several hours under atmospheric conditions. It is slowly oxidized by atmospheric oxygen to give 1,3,2,4-dioxadibismuthene $(TbtBiO)_2$ as colorless crystals in quantitative yield; the oxidation proceeds rapidly in solution [97SCI(277)78]. High thermal stability of compounds **2** and **3** has been reported, though there is no information on their reactivity [99JA3357].

The dibismuthene **1** shows two absorption maxima at λ_{max} 660 and 525 nm in hexane, suggesting the double bonding nature of the Bi–Bi bond even in the solution. UV–Vis spectra of dibismuthenes show absorption maxima at λ_{max} 970 and 505 nm for **2** and 1025, 913 and 518 nm for **3** in hexane, respectively [99JA3357]. The Fourier transformation Raman spectrum of **1** exhibits a strong band due to the Bi=Bi stretching at 134 cm^{-1}

[97SCI(277)78], which is 31 cm^{-1} shorter than that of the Bi–Bi bond (103 cm^{-1}) in $Ph_2BiBiPh_2$ [83OM1859].

Molecular structures of two dibismuthenes have been determined by X-ray crystallography. Dibismuthene **1** exists as the *trans* form with a center of symmetry, and the sterically protecting Tbt groups surround the Bi=Bi bond effectively in order to prevent its oligomerization. The bond length and angles around the Bi=Bi bond in **1** offer important structural information; the Bi=Bi bond length of 2.8206(8) Å is 6% shorter than that of the Bi–Bi single bond (2.990(2) Å) in $Ph_2BiBiPh_2$ [84JCS(D)2365], while the Bi=Bi–C bond angle of 100.5(2)° is considerably larger than that of $Ph_2BiBiPh_2$ (90.9(5) and 91.6(5)°) [97SCI(277)78]. The basic structure of dibismuthene (**2**) is similar to that of **1**: the structure of **2** possesses a center of symmetry in the middle of the Bi=Bi bond. The Bi=Bi bond length and the Bi=Bi–C bond angle are estimated at 2.8327(14) Å and 92.5(4)°, respectively [99JA3357].

2.2.3. Monoorganylbismuth compounds

Compared with the chemistry of diorganylbismuth compounds R_2BiX dealt with in Section 2.2.1, the chemistry of monoorganylbismuth compounds $RBiX_2$ has not been studied in detail except for compounds which contain Bi–S bonds.

2.2.3.1. Methods of synthesis (Tables 2.11–2.15)

(a) Via nucleophilic substitution of dihalogenoorganylbismuthines (method A)

Synthesis of phenylbismuth diethoxides

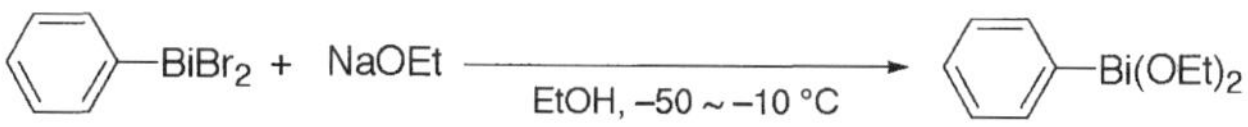

To a solution of dibromophenylbismuthine (8.92 g, 20 mmol) in ethanol (50 ml) cooled to −50°C was added dropwise a solution of sodium ethoxide (prepared from sodium 0.92 g in ethanol), and the reaction mixture was allowed to warm to −10°C gradually. As the reaction proceeded, sodium bromide and phenylbismuth diethoxide precipitated as a white solid. The reaction mixture was allowed to stand at room temperature for a short time and then filtered to remove sodium bromide. The filtrate was rapidly cooled to −20°C to give phenylbismuth diethoxide as a colorless microcrystalline solid, which was filtered at low temperature and dried in vacuo. The reaction was quantitative, but the yield of the product was 2.1 g (28%) due to thermal decomposition during isolation [76ZAAC(423)47].

(b) Via nucleophilic substitution of dialkoxyorganylbismuthines (method B)

Synthesis of methylbismuth dithiolates

$$MeBiBr_2 + NaOMe \xrightarrow{50\ °C} MeBi(OMe)_2 \xrightarrow[r.t.]{RSH} MeBi(SR)_2$$

To a solution of dibromomethylbismuthine (3.84 g, 10 mmol) in absolute methanol was added dropwise a solution of sodium methoxide (prepared from sodium 0.46 g, 20 mmol in methanol), and the resulting mixture was warmed gradually to 50°C. The precipitated sodium bromide was filtered off and a methanol solution of an appropriate thiol (20 mmol) was added dropwise to the filtrate at room temperature. The volatiles were removed off in vacuo and the residue was cooled to precipitate a yellow solid, which was filtered and recrystallized from ethanol. The yields of methylbismuth dithiolate from 2-propanethiol, 2-methyl-2-propanethiol, cyclohexanethiol and 4-methylbenzenethiol were in the range 57–75% [78ZAAC(439)139].

(c) Cleavage of the Bi–C bond with protic acid (method C)

Synthesis of phenylbismuth oxalate

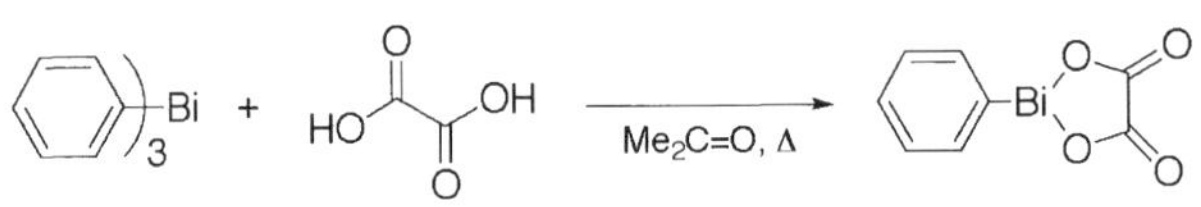

To a solution of oxalic acid (0.61 g, 6.8 mmol) in acetone (50 ml) was added a solution of triphenylbismuthine (3.0 g, 6.8 mmol) in the same solvent (150 ml). The reaction mixture was heated under reflux for 2 h. The white precipitate was filtered off and washed with acetone several times. The combined filtrate and washings were concentrated to give phenylbismuth oxalate in 82% yield [80ZN(B)1000].

(d) Via disproportionation of diorganylbismuth compounds (method D)

Synthesis of phenylbismuth bis(dimethyldithiocarbamate)

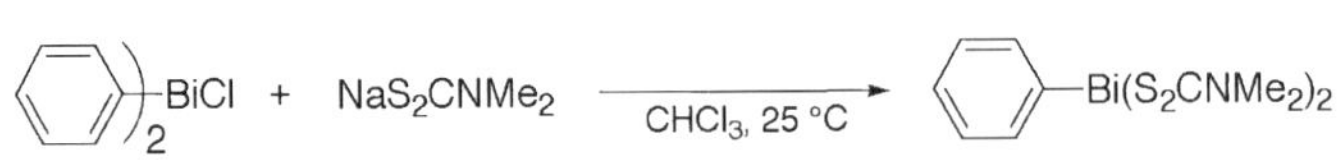

Chlorodiphenylbismuthine (10 mmol) was treated with sodium dimethyldithiocarbamate dihydrate (10 mmol) in chloroform (50 ml) for 1 h at 25°C to give phenylbismuth bis(dimethyldithiocarbamate) in 84% yield; the other product was triphenylbismuthine (64%) [68JOM(11)627].

Synthesis of phenylbismuth bis(pyridine-1-oxide-2-thiolate)

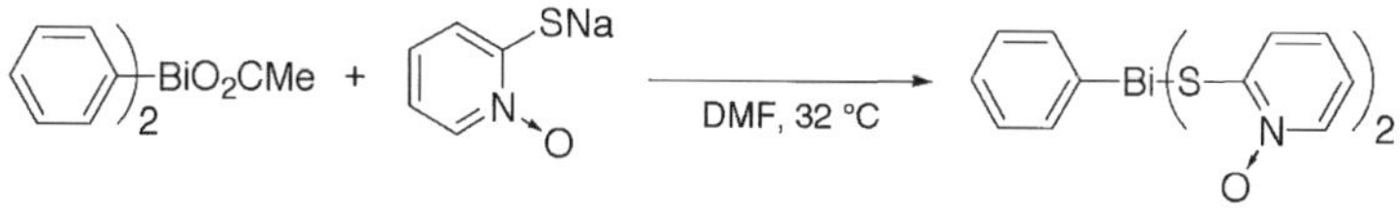

Diphenylbismuth acetate (164.7 g, 390 mmol) and sodium pyridine-1-oxide-2-thiolate (59.7 g, 390 mmol, 97.3% active) were stirred in DMF (3900 ml) for 15 min at 32°C, and then heated with stirring at 65°C for 1.25 h. The resulting yellow solution was filtered and the filtrate was mixed with dichloromethane and water (1:1:2). The bottom layer was separated and cooled in an ice bath. Yellow crystals of the crude product were washed successively with ether and water, and then dissolved in warm *N,N*-dimethylacetamide (ca. 60°C). When the warm solution was filtered normally, small amounts of a solid were observed and shown to be largely bismuth tris(pyridine-1-oxide-2-thiolate). By adding water, phenylbismuth bis(pyridine-1-oxide-2-thiolate) was precipitated, collected by filtration and washed three times with water and ether. The yield of the product was 42% [72JCS(D)1120].

(e) Via insertion of a hetero element component into the Bi–heteroatom bond (method E)

Synthesis of methylbismuth bis(*O*-alkyldithiocarbonate)

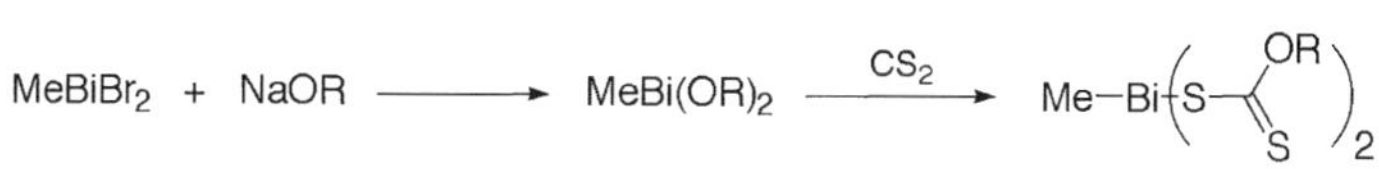

Dibromomethylbismuthine (3.84 g, 10 mmol) was treated with sodium alkoxide (methoxide, ethoxide or isopropoxide; 20 mmol) in the corresponding alcohol to give methylbismuth dialkoxide and sodium bromide. The latter was removed by filtration and freshly distilled carbon disulfide (ca. 10 ml) was added to the filtrate. The mixture was heated under gentle reflux and turned dark brown to black. After 10 min, the mixture was filtered, decolorized, and evaporated to remove the solvent. The residue was carefully recrystallized from an appropriate solvent, avoiding too long and vigorous heating. The yield of methylbismuth bis(*O*-alkyldithiocarbonate) was 38–43% [78ZAAC(439)139].

(f) Alkylation of inorganic bismuth compounds (method F)

Synthesis of (2,2-diphenylethyl)bismuth bis(benzenethiolate)

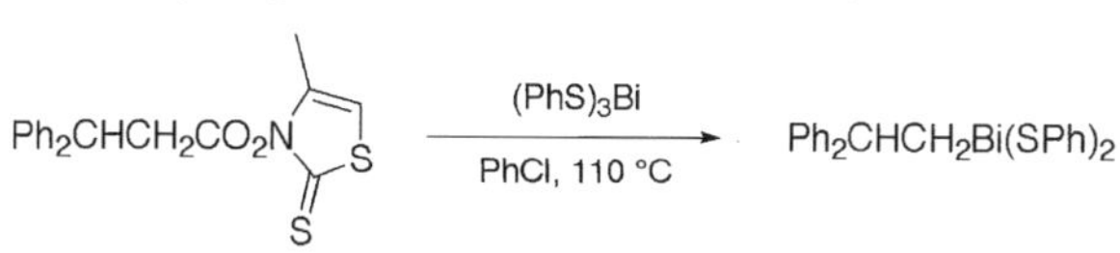

A degassed solution of 2,2-diphenylethyl thiohydroxamate (0.153 g, 0.43 mmol) and bismuth tris(benzenethiolate) (0.463 g, 0.86 mmol) in dry chlorobenzene (20 ml) was heated at 110°C for 2 h under an inert atmosphere. The solvent was then removed in vacuo, and the residue purified by flash chromatography on silica gel, first eluting with pentane–dichloromethane (4:1) to remove diphenyl disulfide, then with pentane–dichloromethane (1:9) to obtain (2,2-diphenylethyl)bismuth bis(phenylthiolate) as a yellow crystalline solid (0.210 g, 80%) [89T2615].

(g) Via oxygenation of a dibismuthene (method G)

Synthesis of 2,4-bis{2,4,6-tris[bis(trimethylsilyl)methyl]phenyl}-1,3,2,4-dioxadibismuthene

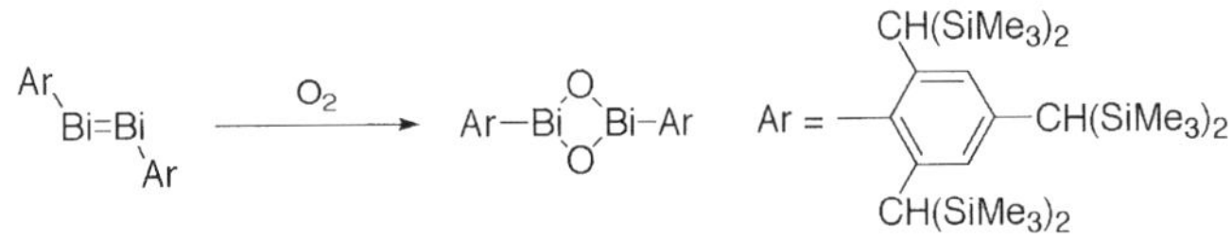

Crystals of 1,2-bis{2,4,6-tris[bis(trimethylsilyl)methyl]phenyl}dibismuthene were exposed to air at room temperature for several days to give the title 1,3,2,4-dioxadibismuthene as crystals in good yield. The reaction proceeded more rapidly in solution [97SCI(277)78].

(h) Via metathesis reaction (method H)

Synthesis of phenyltetrakis(hexamethylphosphoramido)bismuth bis(hexafluorophosphate)

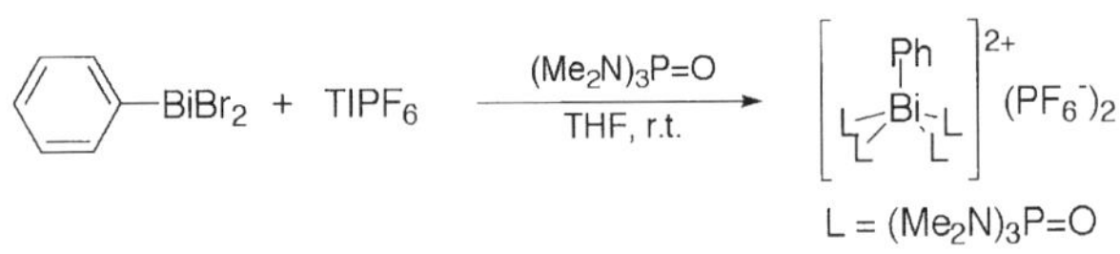

Hexamethylphosphoramide (HMPA; 0.16 ml, 0.924 mmol) was added dropwise to a stirred solution of dibromophenylbismuthine (0.206 g, 0.464 mmol) in THF (5 ml) at room temperature, which resulted in a colorless solution. Subsequent addition of thallium hexafluorophosphate (0.162 g, 0.464 mmol) in THF (6 ml) immediately led to the formation of a pale yellow precipitate. The mixture was stirred for 30 min and then filtered through a Celite bed, resulting in a slightly cloudy filtrate, which was concentrated under vacuum to about 5 ml. Hexane (20 ml) was added as an overlayer and the solvent diffusion over a period of days at −30°C gave the complex as colorless crystals in 30% yield [96JCS(D)443].

2.2.3.2. Properties

2.2.3.2.1. Monoorganylbismuth compounds with Bi–N bonds (Table 2.11)

By the reaction of Me_2BiBr containing 20% of $MeBiBr_2$ with lithium (trimethylsilyl)methylamide in hexane at 0°C, methylbismuth bis[(trimethylsilyl)methylamide] is obtained in 43% yield. The accompanying dimethylbismuth (trimethylsilyl)methylamide (38%) is separated by fractional distillation; the former compound boils at 70–71°C/0.1 mmHg and the latter at 31–32°C/0.1 mmHg. These compounds are not sensitive toward oxygen and moisture [66JOM(6)259]. By treatment of $PhBiBr_2$ with silver bis(methanesulfonyl)amide $AgN(SO_2Me)_2$, phenylbismuth bis(methanesulfonyl)amide $PhBi[N(SO_2Me)_2]_2$ is prepared in 39% yield [95ZAAC(621)1746].

TABLE 2.11
Monoorganylbismuth compounds with Bi–N bonds

Bi–N compound	Synthetic method[a]	Yield (%)	M.p. (°C) or b.p. (°C/mmHg)	Physical data	Reference
$MeBi[NMe(SiMe_3)]_2$	A	43	70–71/0.1	^{1}H-NMR	[66JOM(6)259]
$PhBi[N(SO_2Me)_2]_2$	A	39	123 (decomp.) [$MeCN/Et_2O$]	^{1}H-NMR	[95ZAAC(621)1746]

[a] For notation, see Section 2.2.3.1.

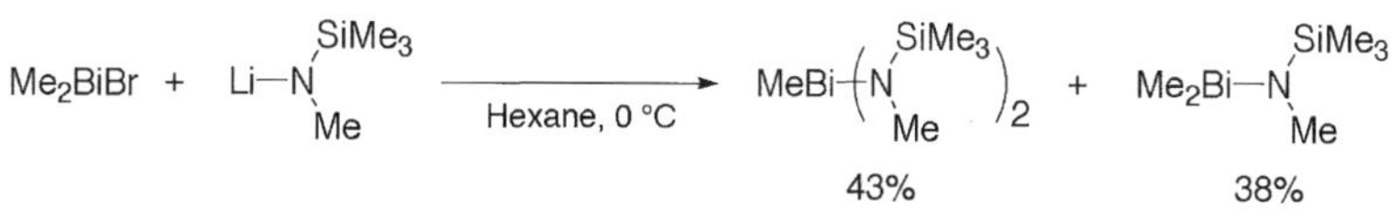

2.2.3.2.2. Monoorganylbismuth compounds with Bi–O bonds (Table 2.12)

(a) Organylbismuth oxides Polymeric phenylbismuthine $(PhBi)_n$ prepared from $PhBiBr_2$ and $LiBH_4$, is reported to react with atmospheric oxygen to give polymeric phenylbismuth oxide $(PhBiO)_n$ as a white solid. However, these compounds are not well characterized [57ZN(B)132]. By atmospheric oxidation of dibismuthene TbtBi=BiTbt, where Tbt denotes 2,4,6-tris[bis(trimethylsilyl)methyl]phenyl group, a dimeric arylbismuth oxide, 1,3,2,4-dioxadibismuthene $(TbtBiO)_2$, is obtained as colorless crystals in quantitative yield [97SCI(277)78].

(b) Organylbismuth dialkoxides The chemistry of organobismuth dialkoxides $RBi(OR')_2$ has been studied little. Wieber et al. have prepared methylbismuth diethoxide $MeBi(OEt)_2$ in 87% yield by the action of sodium ethoxide on $MeBiBr_2$ [76ZAAC(423)40]. Similarly, methylbismuth dimethoxide and diisopropoxide were prepared [78ZAAC(439)139]. Phenylbismuth diethoxide $PhBi(OEt)_2$ is also prepared by this method,

though its yield is low (28%) [76ZAAC(423)47]. These alkoxides are considerably air- and moisture-sensitive and decompose easily at room temperature. By thermal analysis, the ethoxide has been found to decompose at 64°C. Thermal decomposition of $PhBi(OEt)_2$ produces Ph_3Bi and bismuth triethoxide $Bi(OEt)_3$, and on treatment with thiols, the latter gives bismuth trithiolates $Bi(SR)_3$ [76ZAAC(423)47]. These organobismuth dialkoxides are poorly soluble in aprotic solvents, but soluble in alcohols. The mass spectrum of $MeBi(OEt)_2$ shows characteristic peaks due to M^+, $MeBiOEt^+$, $MeBi^+$ and $EtOBi^+$ ions [76ZAAC(423)40]. The diethoxide reacts with various thiols and 1,2-diols to give the corresponding methylbismuth dithiolates (vide infra) and 2-methyl-1,3,2-dioxabismolanes in moderate to good yields. By this method, the following cyclic dialkoxides have been prepared; bismolane derivatives **1–5** [76ZAAC(423)40]. Carbon disulfide converts the dialkoxides to the corresponding bis(dithiocarbonate)s (vide infra) in 30–40% yield [78ZAAC(439)139]. By metathesis reaction between $MeBiBr_2$ and $NaOSiMe_3$, methylbismuth bis(trimethylsilyloxide) $MeBi(OSiMe_3)_2$ is obtained as colorless needles after rapid sublimation. By gentle heating, it decomposes to bismuth oxide Bi_2O_3 and hexamethyldisiloxane $(Me_3Si)_2O$. The disiloxide is soluble in benzene and ether [68ZAAC(363)84].

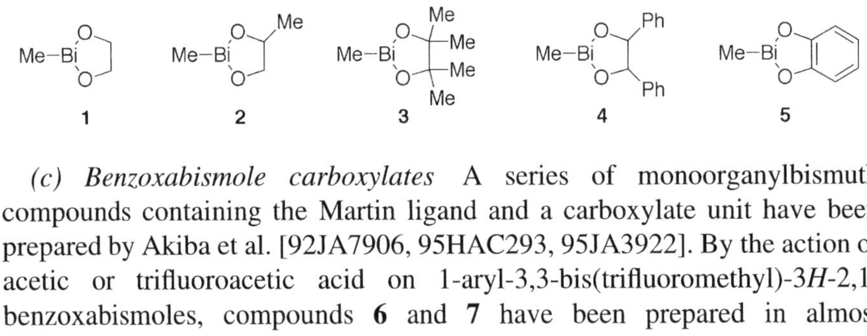

1 **2** **3** **4** **5**

(c) Benzoxabismole carboxylates A series of monoorganylbismuth compounds containing the Martin ligand and a carboxylate unit have been prepared by Akiba et al. [92JA7906, 95HAC293, 95JA3922]. By the action of acetic or trifluoroacetic acid on 1-aryl-3,3-bis(trifluoromethyl)-3*H*-2,1-benzoxabismoles, compounds **6** and **7** have been prepared in almost quantitative yield via cleavage of the Bi–C bond. These and related compounds are tabulated in Table 4.2 and their chemistry is described in Section 4.2.1.

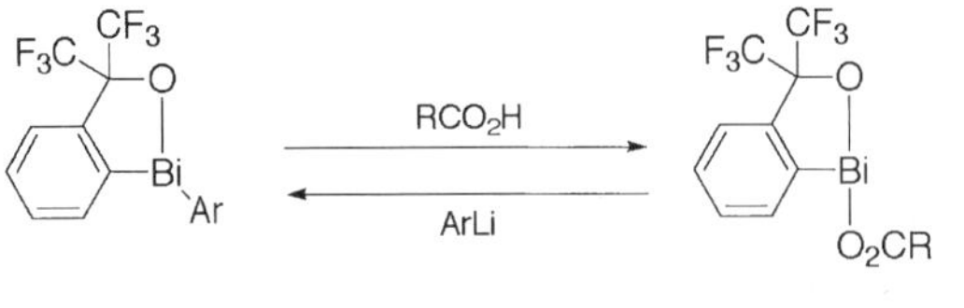

6; R = Me
7; R = CF_3

(d) Organylbismuth dicarboxylates By substitution reaction between $MeBiBr_2$ and sodium carboxylates, methylbismuth dicarboxylate $MeBi(O_2CR)_2$ is obtained. By this methodology, methylbismuth diacetate and dipivalate have been prepared as white solids. These compounds are also accessible by the reaction between $MeBi(OEt)_2$ and free carboxylic acids [83ZAAC(505)138]. $MeBi(O_2CMe)_2$ is poorly soluble in common organic solvents; only DMSO can dissolve this compound well. It begins to decompose above 153°C endothermally with a wide temperature range. In contrast, the corresponding dipivalate is soluble in polar solvents and melts at exactly 89°C. This difference may be attributed to intermolecular coordination; in the case of diacetate, the oxygen atom of the acetoxy group interacts with the bismuth atom of a neighboring molecule. However, the steric bulkiness of the pivaloyl group rules out such an interaction. In both carboxylates, the C=O stretching band appears at 1555 cm^{-1} [83ZAAC(505)138].

The use of triethylbismuthine Et_3Bi for qualitative and quantitative determination of active and acidic hydrogens has been examined. The bismuthine reacts with carboxylic acids such as acetic, chloroacetic, dichloroacetic, trichloroacetic, butyric, maleic and benzoic acids in an expected way. However, the identification of the respective products has not been performed [37JA935]. Similar cleavage of the Bi–C bonds in triarylbismuthines with carboxylic acids to arylbismuth dicarboxylates has been reported by Deacon et al. By this method, $PhBi(O_2CMe)_2$, $PhBi(O_2CCF_3)_2$, 4-MeC_6H_4-$Bi(O_2CMe)_2$, 4-$MeC_6H_4Bi(O_2CCF_3)_2$, 4-$MeOC_6H_4Bi(O_2CCF_3)_2$, 4-$EtOC_6H_4$-$Bi(O_2CCF_3)_2$ and $MeBi(O_2CMe)_2$ are prepared. Unfortunately, there is no successful procedure as yet available for their purification, due to the high hydrolytic instability of aryl derivatives as well as high oxygen sensitivity of the methyl derivative [84AJC527]. These dicarboxylates undergo a subsequent reversible ligand redistribution reaction to give an equilibrium mixture of $R_2Bi(O_2CR')$ and $Bi(O_2CR')_3$; for example, the equilibrium constant of $PhBi(O_2CMe)_2$ in DMSO has been estimated to be 0.90 at 37°C [84AJC527]. Therefore, no analytically pure dicarboxylates can be obtained by this acidolysis method. The solubility and yield of organylbismuth dicarboxylates change depending on their substituent groups. In the case of *p*-alkoxyphenyl derivatives, an inseparable mixture of Ar_3Bi, $Ar_2Bi(O_2CCF_3)$ and $ArBi(O_2CCF_3)_2$ is obtained. Phenylbismuth diacetate and dimaleate are prepared by a similar procedure from the corresponding acids in boiling ether. The latter product $PhBi(O_2CCH{=}CHCO_2H)_2$ separates out from the reaction mixture as a high melting solid in 26% yield [90JCS(D)899]; the C=O stretching bands are observed at 1585 and 1533 cm^{-1} [90JCS(D)899; 93AOMC137].

With the reaction between Ph_3Bi and aliphatic dicarboxylic acids such as oxalic, malonic, succinic and maleic acids in boiling acetone, cyclic organobismuth dicarboxylates have been prepared as analytical pure samples. Depending on dicarboxylic acids, five-, six- and seven-membered cyclic compounds **8–11** are obtained in 15–99% yields. These products are white solids insoluble in common organic solvents, except for phenylbismuth maleate which is soluble in DMSO. In formic and acetic acids as well as in mineral acids, these compounds readily dissolve with decomposition to give the corresponding bismuth salts. The mass spectra of these cyclic compounds show peaks due to $PhBi^+$, Bi^+ and $Ph\text{-}Ph^+$ fragment ions. Their IR spectra exhibit absorption bands at 1600 (C=O), 1000, 1050, 730 and 690 (Bi–Ph) cm^{-1} [80ZN(B)1000]. Known cyclic bismuth dicarboxylates are tabulated in Table 4.2.

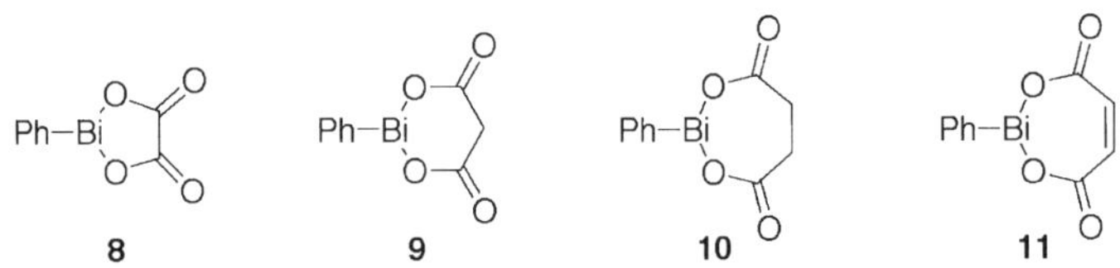

Similarly, mercaptocarboxylic acids (H_2Mc) react with bismuthines R_3Bi to give the corresponding organobismuth mercaptocarboxylates RBiMc. The reaction between Me_3Bi and H_2Mc in acetone always gives MeBiMc at all concentrations employed. However, the corresponding reaction with Ph_3Bi depends on the reaction conditions: in $CHCl_3$ and $CHCl_3$–MeOH, all Bi–C bonds are cleaved to form Bi(Mc)(HMc), whereas in dilute acetone solution, PhBiMc is obtained [81ZN(B)70].

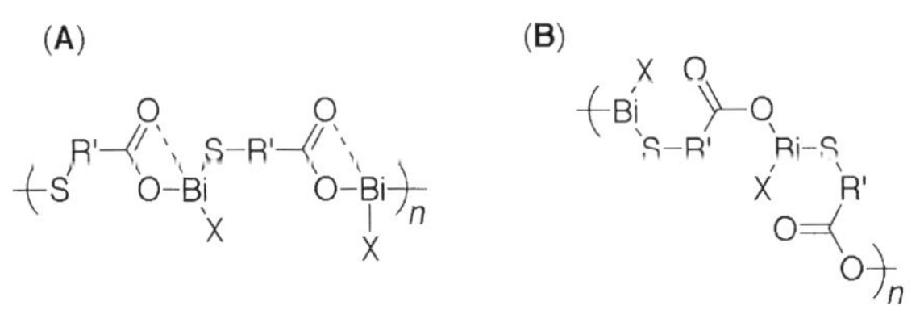

12; X = Ph, R' = $-CH_2CH_2-$
13; X = Ph, R' = $-CH(CH_3)-$
14; X = Me, R' = $-CH_2-$
15; X = Me, R' = $-CH_2CH_2-$
16; X = Me, R' = $-CH(CH_3)-$
17; X = Me, R' = $-C_6H_4\text{-}2-$

From the reactions of Me_3Bi or Ph_3Bi with mercaptoacetic, 2- or 3-mercaptopropionic and 2-mercaptobenzoic acids, compounds **12–17** are obtained as analytically pure samples in 43–77% yields. These compounds are insoluble in common organic solvents, probably because of their polymeric nature. Their IR spectra exhibit absorption bands at 1520–1540 (ν_{as}; CO_2), 1340–1385 (ν_{sy}; CO_2), 455–465 (ν; Bi–C) and 305–325 (ν; Bi–S) cm^{-1}, respectively [81ZN(B)70].

(e) Organylbismuth disulfinates Information is quite limited on the property of this type of compound. A dithionite structure has been ruled out [71JOM(32)C11]. When heated at 190°C for 5 h under 10^{-2}–10^{-3} mmHg, phenylbismuth bis(benzenesulfinate) $PhBi(O_2SPh)_2$ gives Ph_3Bi under evolution of SO_2. $Bi(O_2SPh)_3$ and $Ph_2Bi(O_2SPh)$ also give Ph_3Bi under similar conditions [71JOM(26)C10]. Phenylbismuth bis(benzenesulfinate) can be prepared from $PhBiCl_2$ and $NaSO_2Ph$ or from Ph_3Bi and $PhSO_2H$. The reaction of an equimolar Ph_3Bi and mercury bis(4-methylbenzenesulfinate) $Hg(O_2SC_6H_4\text{-}4\text{-}Me)_2$ in MeOH or $CHCl_3$ has been reported to give $PhBi(O_2SC_6H_4\text{-}4\text{-}Me)_2$ and Ph_2Hg, although the experimental detail is not available. The IR spectrum of $PhBi(O_2SC_6H_4\text{-}4\text{-}Me)_2$ shows a SO_2 stretching frequency at 925 cm^{-1} as a broad band [72AJC2107].

$$Ph_3Bi \xrightarrow[\text{MeOH}]{2\ Hg(O_2SC_6H_4\text{-}4\text{-}Me)_2} PhBi(O_2SC_6H_4\text{-}4\text{-}Me)_2 + 2\ PhHgO_2SC_6H_4\text{-}4\text{-}Me$$

(f) Organylbismuth disulfonates The literature contains a few examples of organylbismuth disulfonates. Treatment of $(4\text{-}MeC_6H_4)_3Bi$ with benzenesulfonic acid in ether affords 4-methylphenylbismuth bis(phenylsulfonate), $4\text{-}MeC_6H_4Bi(O_3SPh)_2$ in 92% yield. $4\text{-}MeC_6H_4Bi(O_3SC_6H_4\text{-}4\text{-}Me)_2$ is formed as a minor product by the reaction between $(4\text{-}MeC_6H_4)_3Bi$ and gaseous SO_3 in 1,2-dichloroethane at −10°C. The IR spectrum of $4\text{-}MeC_6H_4Bi(O_3SPh)_2$ exhibits peaks at 1270 (ν_{as}; SO_3) and 1065 (ν_{sy}; SO_3) cm^{-1} [86ICA(113)43]. Disproportionation of $PhBi(O_3SPh)_2$ to Ph_3Bi and $Ph_2Bi(O_3SPh)$ in DMSO-d_6 has also been reported. Recently, a preparation of phenylbismuth bis(trifluoromethanesulfonate) was reported, however, there is only limited information on its property [99TL285].

TABLE 2.12
Monoorganylbismuth compounds with Bi–O bonds

Compounds R	R′	Synthetic methods[a]	Yield (%)	M.p. (°C) or b.p. (°C/mmHg)	Physical data	Reference
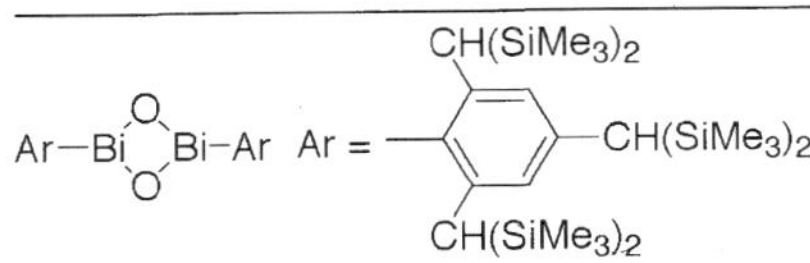		G	–	–	–	[97SCI(277)78]
$RBi(OR')_2$						
Me	Me	A	70	135–140 (decomp.) [MeOH]	^{1}H-NMR	[78ZAAC(439)139]
	Et	A	87	64 (decomp.)	^{1}H-NMR, MS	[76ZAAC(423)40]
		A	46	154 (decomp.) [EtOH]	^{1}H-NMR	[78ZAAC(439)139]
	^{i}Pr	A	68	135–140 (decomp.) [^{i}PrOH]	^{1}H-NMR	[78ZAAC(439)139]
	Me_3S	A	5	63–65, 70/1	^{1}H-NMR, IR	[68ZAAC(363)84]
Ph	Et	A	28	65 (decomp.)	^{1}H-NMR	[76ZAAC(423)47]
	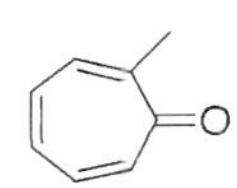	A	78	>330	^{1}H-, ^{13}C-NMR, IR, MS	[95CB335]
$RBi(O_2CR')_2$						
Me	Me	B	80	153 (decomp.) [EtOH]	^{1}H-NMR, IR	[83ZAAC(505)138]
		C	83	141–143 (decomp.)	^{1}H-NMR, IR, MS	[84AJC527]

	^{i}Bu	A	54	89 (decomp.) [AcOH]	^{1}H-NMR, IR	[83ZAAC(505)138]
Ph	Me	C	61	184	^{1}H-NMR, IR, MS	[84AJC527]
	C_7H_{15}	A	–	–	–	[93AOMC137]
	PhCH=CH	C	20	>230	–	[51JA2880]
	HO_2CCH=CH	C	26	>300	^{13}C-NMR, IR	[90JCS(D)899]
		A	–	–	IR	[93AOMC137]
	CF_3	C	78	–	^{1}H-, ^{19}F-NMR, IR	[84AJC527]
$RBi(O_2CR')_2$						
Ph	Ph	C	44	215–216	–	[51JA2880]
4-MeC_6H_4	Me	C	82	–	^{1}H-NMR, IR	[84AJC527]
	CF_3	C	55	–	^{1}H-, ^{19}F-NMR, IR	[84AJC527]
4-$MeOC_6H_4$	CF_3	C	–	–	^{1}H-, ^{19}F-NMR	[84AJC527]
4-$EtOC_6H_4$	CF_3	C	–	–	^{1}H-, ^{19}F-NMR	[84AJC527]
$EtBi(ONO_2)_2$		A	–	–	–	[1854LA(92)371]
$RBi(O_2SR')_2$						
Ph	Ph	C	–	–	IR	[71JOM(26)C10]
		C	–	–	^{1}H-NMR	[71JOM(32)C11]
	4-MeC_6H_4	–	–	–	IR	[72AJC2107]
$RBi(O_3SR')_2$						
Ph	CF_3	C	–	–	^{1}H-, ^{13}C-, ^{19}F-NMR	[99TL285]
4-MeC_6H_4	Ph	C	92	228 (decomp.) [Et_2O]	^{1}H-NMR, IR	[86ICA(113)43]
	4-MeC_6H_4	C	–	–	^{1}H-NMR	[86ICA(113)43]

[a] For notation, see Section 2.2.3.1.

2.2.3.2.3. Monoorganylbismuth compounds with Bi–S bonds (Table 2.13)

(a) Organylbismuth dithiolates Organylbismuth dithiolate has not been studied much. Although the literature contains many examples of this type of compound, their chemical nature (especially their reactivity) has not been examined much. The simplest organylbismuth dithiolate, methylbismuth di(methylthiolate) $MeBi(SMe)_2$, has been prepared by treatment of $MeBiBr_2$ with lithium methylthiolate in 76% yield [88ICA(152)49, 88M1317, 88BM(1)69]. The ^{1}H-NMR spectrum of $MeBi(SMe)_2$ in DMSO-d_6 shows signals at δ 1.29 (Bi–Me) and 2.79 (S–Me), while its MS spectrum exhibits peaks at 418 Bi_2^+, 318 M^+, 303 $[M–Me]^+$, 288 $[M–2Me]^+$, 271 $[M–SMe]^+$, 256 $[M–SMe_2]^+$, 241 BiS^+, 224 $[M–2SMe]^+$, 209 Bi^+ and 94 $Me_2S_2^+$ [88M1317]. Wieber et al. have prepared this type of compound by the reaction between methylbismuth diethoxide $MeBi(OEt)_2$ and aliphatic or aromatic thiols and 1,2-dithiols. The advantage of this method is that highly pure $MeBi(SR)_2$ can be obtained in good yield [76ZAAC(423)40, 78ZAAC(439)139]. Phenyl-bismuth diethoxide $PhBi(OEt)_2$ can also be used for the preparation of $PhBi(SR)_2$ [76ZAAC(423)47]. Using this methodology, the following compounds are prepared: methylbismuth bis(benzenethiolate), bis(ethanethiolate), bis(phenylmethanethiolate), bis(2-hydroxyethanethiolate), di(2-benzothiazolethiolate), and dithiabismolane derivatives **1** and **2** [76ZAAC(423)40], methylbismuth bis(2-propanethiolate), bis(2-methyl-2-propanethiolate), bis(cyclohexanethiolate), bis(4-methylbenzenethiolate) [78ZAAC(439)139], phenylbismuth bis(ethanethiolate), bis(phenylmethanethiolate), bis(benzenethiolate), bis(2-hydroxyethanethiolate) and compound **3** [76ZAAC(423)47].

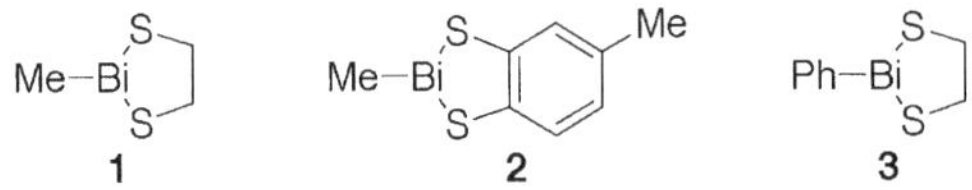

1 **2** **3**

In contrast to the bismuthines with the Bi–O bond, these organylbismuth dithiolates are soluble in common organic solvents and insensitive to moisture. However, they are thermally unstable and decompose easily above melting point. They also decompose slowly in daylight [76ZAAC(423)47].

Eight-membered heterocyclic compounds are prepared by the reaction of $MeBi(OEt)_2$ or $PhBi(OEt)_2$ with bis(2-mercaptoethyl) sulfide and bis(2-mercaptoethyl) ether in EtOH at low temperatures. Using this method, bismocanes **4–7** are obtained as yellow crystals in good yields

[85JOM(290)133]. These cyclic bismuth compounds are also tabulated in Table 4.2.

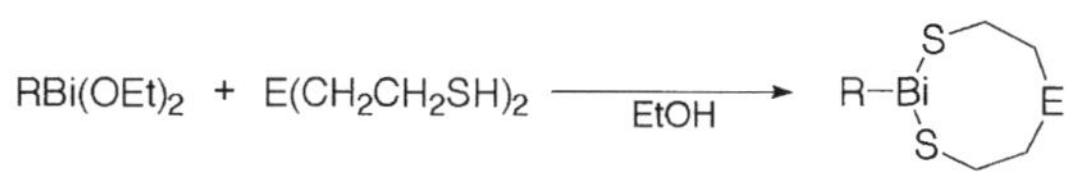

4; R = Me, E = O
5; R = Me, E = S
6; R = Ph, E = O
7; R = Ph, E = S

Their MS spectra show peaks due to M^+, $[M–R]^+$, BiR^+ and BiS^+ fragment ions. The IR and Raman spectra exhibit absorption bands at 457–472 (ν; Bi–Me) and 240–302 cm^{-1} (ν; BiS_2). The X-ray crystal structure analysis of compound **6** has shown that its ring possesses the chair–chair conformation with a weak transannular Bi···O interaction of the Bi–O distance, 2.97(1) Å. The Bi–S bond lengths are 2.560(2) and 2.602(3) Å. There are two intermolecular Bi···S contacts with distances of 3.440(3) and 3.509(3) Å, which assign the bismuth center a ψ-monocapped octahedral geometry [85JOM(290)133]. Wieber et al. have prepared several organylbismuth dithiolates by the reaction of dibromoorganylbismuthine and sodium thiolates, which include compounds **8**–**12**. These organylbismuth dithiolates possess relatively high melting points (134–256°C) [85ZAAC(520)59]. The metathesis reaction of diphenylbismuth acetate Ph_2BiO_2CMe with sodium 1-oxopyridine-2-thiolate results in disproportionation to give thiolate **13** in 42% yield. This product can also be obtained by the reaction between diiodophenylbismuthine $PhBiI_2$ and sodium 1-oxopyridine-2-thiolate [72JCS(D)1120]. The X-ray crystallographic study has shown that the bismuth center has a distorted octahedral structure with the *cis*-configuration of sulfur and oxygen atoms. The Bi–S bond lengths are 2.635(9) and 2.712(6) Å, and the Bi–O bond lengths are 2.51(2) and 2.54(1) Å, respectively [72JCS(D)1120].

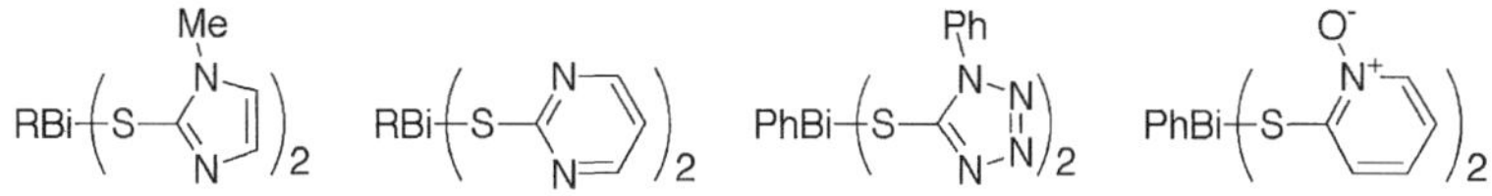

8; R = Me
10; R = Ph

9; R = Me
11; R = Ph

12

13

Klapötke et al. have prepared several organylbismuth dithiolates of biological activity. By the reaction of lithium (2,6-dichlorobenzenethiolate) with $MeBiBr_2$ at room temperature, methylbismuth bis(2,6-dichlorobenzenethiolate) is obtained as yellow crystals in 71% yield. The antibiological activity of this dithiolate against bacteria, yeasts and molds has been examined; this compound possesses the ability to inhibit growth of bacteria [87ZN(B)940].

By similar metathesis reaction, phenylbismuth bis(4-chlorobenzenethiolate) and methylbismuth bis(4-aminobenzenethiolate) are obtained as yellow to orange crystals. The latter thiolate has been functionalized by the action of methyl iodide to give methylbismuth bis(4-methylammoniobenzenethiolate) diiodide as a deliquescent red solid, the antibacterial activity of which is 10^3 times more effective than methylarsenic bis(4-aminobenzenethiolate). Treatment of this diiodide with $AgNO_3$ or $TlNO_3$ gives the corresponding dinitrate. A hydrophilic ammonium benzenethiolate has been shown to possess some potential as a novel bactericide [87JOM(331)299].

The acidic cleavage of Bi–C bonds in triorganylbismuthines with thiols also produces organylbismuth dithiolates. The use of Et_3Bi for the qualitative and quantitative determination of active and acidic hydrogens has been examined. This bismuthine can react with a variety of thiols such as butane-, heptane-, phenylmethane-, benzene-, 4-methylbenzene and 2-naphthalene-thiols, 1-mercaptobenzothiazole, and thioacetic acid. However, identification of the resulting products has not been made [37JA935]. By the action of benzenethiol, Ph_3Bi, $(4\text{-}MeC_6H_4)_3Bi$ and $(4\text{-}ClC_6H_4)_3Bi$ are transformed in moderate yields into $PhBi(SPh)_2$, $4\text{-}MeC_6H_4Bi(SPh)_2$ and $4\text{-}ClC_6H_4Bi(SPh)_2$, respectively [51JA2880]. Phenylbismuth di(2-carbomethoxybenzenethiolate) and phenylbismuth thiosalicylate are prepared similarly. Thioglycolic acid can also cleave the Bi–C bonds of bismuthines, though the products are not obtained in pure form. Prolonged heating with an excess amount of thiol in boiling xylene causes cleavage of all three Bi–C bonds in Ph_3Bi to afford the corresponding $Bi(SR)_3$. Tris(1-naphthyl)bismuthine does not react with thiophenol even in boiling xylene, but diphenyl(1-naphthyl)bismuthine treated with thiophenol in boiling $CHCl_3$ for 3 h gives 1-naphthylbismuth di(benzenethiolate). The solubilities of these dithiolates are dependent on their substituent: phenylbismuth bis(2-carbomethoxybenzenethiolate) can be recrystallized from petroleum ether, whereas phenylbismuth thiosalicylate is not soluble in any organic solvent [51JA2880]. Treatment of Ph_3Bi with an excess of 2-mercaptoaniline gives phenylbismuth bis(2-aminobenzenethiolate) as a bright yellow solid. The mass spectrum of this dithiolate shows characteristic peaks due to M^+, $PhBi^+$ and $[M\text{–}C_6H_4(S)NH_2]^+$ ions. Its IR spectrum exhibits absorptions due to the Bi–S bond at 320 (ν_{as}) and 305 (ν_s) cm^{-1} [83ICA(73)141].

By the reaction between $Bi(SPh)_3$ and a carbon-centered radical $R^{\cdot}$, generated by decarboxylation of thiohydroxamate esters (the Barton reaction), an organylbismuth dithiolate can be prepared; a 1:2 mixture of 3-(3,3-diphenylpropionoxy)-4-methylthiazole-2(3*H*)-thione and $Bi(SPh)_3$ is heated at 110°C for 2 h in chlorobenzene under an inert atmosphere, and the resulting mixture is passed on a silica gel column to afford 2,2-diphenylethylbismuth di(benzenethiolate) as yellow crystals [89T2615].

The biological activity is reviewed on five organylbismuth dithiolates, $MeBi(SMe)_2$, $MeBi(SC_6H_3\text{-}2,6\text{-}Cl_2)_2$, $PhBi(SC_6H_4\text{-}4\text{-}Cl)_2$, $MeBi(SC_6H_4\text{-}4\text{-}NH_2)_2$ and $[MeBi(SC_6H_4\text{-}4\text{-}NH_2Me)_2]^{2+}[I^-]_2$. These dithiolates possess higher microbiological activity compared with Ph_3Bi and Ph_3BiCl_2 [88BM(1)69].

(b) Organylbismuth dithiocarbonates (xanthates) Treatment of methylbismuth dialkoxides $MeBi(OR)_2$ with carbon disulfide gives the corresponding dithiocarbonates $MeBi(S_2COR)_2$ in moderate yield.

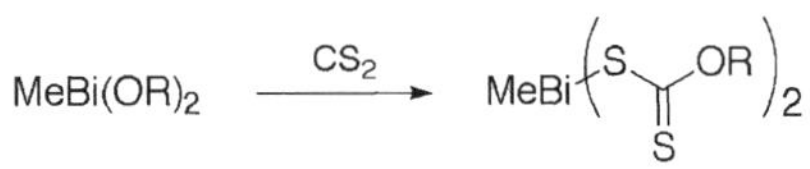

By this procedure, methylbismuth bis(*O*-methyldithiocarbonate), bis(*O*-ethyldithiocarbonate) and bis(*O*-isopropyldithiocarbonate) are obtained as colorless crystals. These compounds are soluble in common organic solvents and melt around 120°C; they decompose exothermally above the melting points. Since they decompose slowly in daylight, they should be kept under cooling in the dark [78ZAAC(439)139]. The X-ray crystal structure analysis of phenylbismuth bis(*O*-methyldithiocarbonate) has shown that the bismuth center possesses a distorted octahedral structure with two pairs of short and long Bi–S bonds, 2.649(3) and 2.670(3) Å, and 2.961(2) and 3.079(3) Å [82ZAAC(485)217].

(c) Organylbismuth dithiocarbamates The metathesis reaction between chlorodiphenylbismuthine Ph_2BiCl with sodium *N,N*-dimethyldithiocarbamate NaS_2CNMe_2 in $CHCl_3$ for 1 h at 25°C gives the corresponding dithiocarbamate $PhBi(S_2CNMe_2)_2$ (84%) and Ph_3Bi (64%), instead of the expected $Ph_2BiS_2CNMe_2$. A similar reaction with NaS_2CNEt_2 gives a mixture of $PhBi(S_2CNEt_2)_2$ (38%), Ph_3Bi (46%) and $Bi(S_2CNEt_2)_3$ (11%). The IR spectra of these dithiocarbamates show absorptions due to the CN stretching at 1502 (Me) and 1490 (Et) cm^{-1}, respectively. The UV–Vis spectra exhibit absorption maxima at λ_{max} 259 (Me) and 262 (Et) nm [68JOM(11)627]. Several organylbismuth

bis(diorganyldithiocarbamate)s are prepared by metathesis reaction between $RBiBr_2$ and $NaS_2CNR'_2$ as yellow crystals, in moderate to good yields; this includes methylbismuth bis(dimethyldithiocarbamate), bis(diethyldithiocarbamate) and bis(piperidyl-dithiocarbamate), phenylbismuth bis(dimethyldithiocarbamate), bis(diethyl-dithiocarbamate) and bis(piperidyldithiocarbamate) [79ZAAC(448)89]. All these dithiocarbamates are indefinitely stable and soluble in many organic solvents; however, phenyl derivatives are generally less soluble in MeOH and EtOH when compared with methyl derivatives. The X-ray crystal structure of methylbismuth bis(diethyldithiocarbamate) has been elucidated; there are intermolecular Bi···S interactions within the distance of 2.67(1)–3.36(1) Å, which cause the bismuth center to take a distorted pentagonal-pyramidal geometry, i.e. a dimeric structure. In contrast to the solid state, the compound exists as a monomer in benzene solution [79ZN(B)1037]. [2-(2-Pyridyl)phenyl]bismuth bis(diethyldithiocarbamate) and phenylbismuth bis(diethyldithiocarbamate) are obtained by similar metathesis reaction. The X-ray crystallographic study of these thiocarbamates has revealed that the former compound possesses a pseudopentagonal pyramidal structure of the monomeric form and the latter a pseudosquare pyramidal structure of the dimeric form with an interaction through the Bi···S contact. The Bi–S bond lengths are in the range 2.671(2)–2.942(2) Å and both S_2CNR_2 moieties are nearly planar [90JCS(D)899].

(d) Organylbismuth bis(diphenyldithiophosphinate) This type of compound has been studied less. The first and only known example is 2,4,6-trimethylphenylbismuth bis(diphenyldithiophosphinate), which has been prepared by the action of $NH_4Ph_2PS_2$ on $(2,4,6\text{-}Me_3C_6H_2)_2BiBr$. The reaction gives not the expected monodithiophosphinate but the bis(dithiophosphinate) in quantitative yield. Its IR spectrum shows two absorptions at 636 and 544 cm^{-1} due to the asymmetric and symmetric PS_2 vibrations. The mass spectrum exhibits characteristic peaks due to MH^+ and $[M\text{–}Mes]^+$ ions. The X-ray crystallographic study of $2,4,6\text{-}Me_3C_6H_2Bi(S_2PPh_2)_2$ has shown that its crystals consist of discrete monomeric molecules, which possess a symmetry plane containing the 2,4,6-trimethylphenyl (mesityl) group and bismuth atom. Due to the through-space interaction between the bismuthand S atoms, the bismuth center possesses a square-pyramidal geometry [94JOM(470)93].

TABLE 2.13
Monoorganylbismuth compounds with Bi–S bonds

Compound R	R′	Synthetic method[a]	Yield (%)	M.p. (°C)	Physical data	Reference
RBi(SR′)₂						
Me	Me	–	–	–	Bioactive data	[88ICA(152)49]
		A	75.7	–	Bioactive data, ^{1}H-NMR, MS	[88M1317]
		–	–	–	Bioactive data	[88BM(1)69]
	Et	B	67	91	^{1}H-NMR, MS	[76ZAAC(423)40]
	^{i}Pr	B	72	80–82 [MeOH]	^{1}H-NMR, MS	[78ZAAC(439)139]
	^{t}Bu	B	75	150 (decomp.) [MeOH]	^{1}H-NMR, MS	[78ZAAC(439)139]
	^{c}Hex	B	68	131 [MeOH]	^{1}H-NMR, MS	[78ZAAC(439)139]
	$HOCH_2CH_2$	B	88	86	^{1}H-NMR, MS	[76ZAAC(423)40]
	$PhCH_2$	B	83	90–92	^{1}H-NMR, MS	[76ZAAC(423)40]
	Ph	B	80	119 (decomp.)	^{1}H-NMR, MS	[76ZAAC(423)40]
		C	92	–	^{1}H-NMR, MS	[70JCS(B)735]
		D	Quant.	127 (decomp.)	^{1}H-NMR, MS	[84ZN(B)1668]
	$4\text{-}MeC_6H_4$	B	57	180–182	^{1}H-NMR, MS	[78ZAAC(439)139]
	$4\text{-}H_2NC_6H_4$	B	86.1	220 (decomp.)	Bioactive data, ^{1}H-NMR, IR, MS	[87JOM(331)299]
		–	–	–	Bioactive data	[88ICA(152)49]
		–	–	–	Bioactive data	[88BM(1)69]
	$4\text{-}(MeH_2N^+)(I^-)C_6H_4$	A	97.9	40 (deliquescence)	Bioactive data, ^{1}H-NMR, IR, MS	[87JOM(331)299]
		–	–	–	Bioactive data	[88ICA(152)49]
		–	–	–	Bioactive data	[88BM(1)69]
RBi(SR′)₂						
Me	$4\text{-}(MeH_2N^+)(NO_3^-)C_6H_4$	A	51.2	45 (deliquescence)	^{1}H-NMR	[87JOM(331)299]
	$2,6\text{-}Cl_2C_6H_3$	A	71	202 (decomp.)	Bioactive data, ^{1}H-NMR, IR, MS	[87ZN(B)940]

TABLE 2.13 (*continued*)

Compounds R	R′	Synthetic method[a]	Yield (%)	M.p. (°C)	Physical data	Reference
		–	–	–	Bioactive data	[88BM(1)69]
Me	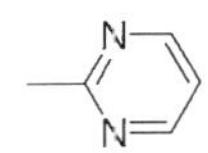	A	74	256 [EtOH]	^{1}H-NMR	[85ZAAC(520)59]
	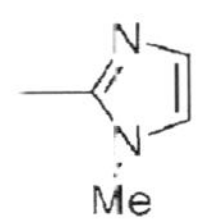	A	89	227 [EtOH]	^{1}H-NMR	[85ZAAC(520)59]
	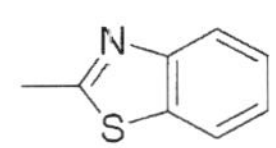	B	34	156 (decomp.)	MS	[76ZAAC(423)40]
Me-^{13}C	Me	A	–	–	MS	[88M1317]
Et	Ph	C	85	98–99	–	[70JCS(B)735]
Ph_2CHCH_2	Ph	F	80	–	^{1}H-NMR, MS	[89T2615]
Ph	Et	B	60	74 (decomp.)	^{1}H-NMR, MS	[76ZAAC(423)47]
	$HOCH_2CH_2$	B	75	57 (decomp.)	^{1}H-NMR, MS	[76ZAAC(423)47]
	$PhCH_2$	B	72	90 (decomp.)	^{1}H-NMR, MS	[76ZAAC(423)47]
	Ph	C	26	170 (decomp.) [$CHCl_3$]	–	[51JA2880]
		C	41	158 (decomp.)	–	[70JCS(B)735]
		B	81	170	^{1}H-NMR, MS	[76ZAAC(423)47]
	4-ClC_6H_4	A	75.1	165 (decomp.)	Bioactive data, ^{1}H-NMR, IR, MS	[87JOM(331)299]
		–	–	–	Bioactive data	[88BM(1)69]
		–	–	–	Bioactive data	[87ZN(B)940]

$RBi(SR')_2$						
Ph	2-$MeO_2CC_6H_4$	C	5	107–108	–	[51JA2880]
	3-$NH_2C_6H_4$	C	–	198–200	IR, MS	[83ICA(73)141]
	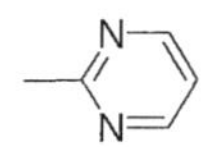	A	51	211 [EtOH]	^{1}H-NMR	[85ZAAC(520)59]
	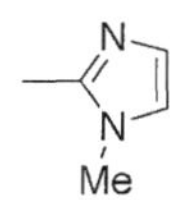	A	70	250 [EtOH]	^{1}H-NMR	[85ZAAC(520)59]
	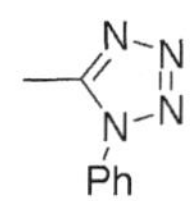	A	78	134 [EtOH]	^{1}H-NMR	[85ZAAC(520)59]
	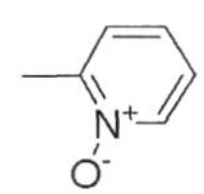	A, D	42	215–218 [DMF-H_2O]	IR, X-ray	[72JCS(D)1120]
4-MeC_6H_4	Ph	C	39	155 (decomp.)	–	[51JA2880]
1-$C_{10}H_7$	Ph	C	Low	>240	–	[51JA2880]
4-ClC_6H_4	Ph	C	39	170 (decomp.) [PhH]	–	[51JA2880]
$RBi(S_2CNR'_2)_2$						
Me	Me	A	55	185 [CCl_4]	^{1}H-NMR	[79ZAAC(448)89]
	Et	A	76.3	141 [EtOH]	^{1}H-NMR	[79ZAAC(448)89]
	–	–	–	–	X-ray	[79ZN(B)1037]
	R'_2 = -$(CH_2)_5$-	A	59	192 [EtOH]	^{1}H-NMR	[79ZAAC(448)89]

TABLE 2.13 *(continued)*

Compounds R	R′	Synthetic method[a]	Yield (%)	M.p. (°C)	Physical data	Reference
Ph	Me	D	98	210–211 [MeCN]	IR, UV	[68JOM(11)627]
		A	58	219 [EtOH]	^{1}H-NMR	[79ZAAC(448)89]
Ph	Et	D	21	131–133 [MeCN]	IR, UV	[68JOM(11)627]
		A	65	137 [EtOH]	^{1}H-NMR	[79ZAAC(448)89]
		A	–	–	^{13}C-NMR, X-ray	[90JCS(D)899]
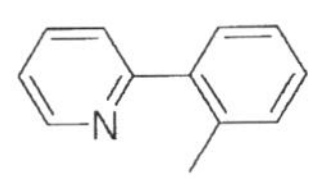	R'_2 = -$(CH_2)_5$-	A	46	228 [EtOH–$CHCl_3$]	^{1}H-NMR	[79ZAAC(448)89]
	Et	A	88	222–224 [MeCN]	^{13}C-NMR, IR	[90JCS(D)899]
RBi(S_2COR')$_2$						
Me	Me	E	41	127 [MeOH]	^{1}H-NMR	[78ZAAC(439)139]
	Et	E	43	113 [EtOH]	^{1}H-NMR	[78ZAAC(439)139]
	^{i}Pr	E	38	126 [^{i}PrOH]	^{1}H-NMR	[78ZAAC(439)139]
Ph	Me	–	–	–	X-ray	[82ZAAC(485)217]
2,4,6-$Me_3C_6H_2$-Bi(S_2PPh_2)$_2$		A	100	180 (decomp.) [PhH]	^{1}H-NMR, IR, MS, X-ray	[94JOM(470)93]

[a] For notation, see Section 2.2.3.1.

2.2.3.2.4. Monoorganylbismuth compounds with Bi–Se or Bi–Te bonds (Table 2.14)

The first preparation of this type of compound was reported by Wieber et al. in 1984. By thermolysis at 80°C for 10 h under a dry nitrogen stream, Me_2Bi-SePh gives an equimolar mixture of Me_3Bi and methylbismuth bis(benzene-selenolate) $MeBi(SePh)_2$ in quantitative yield. This bisselenolate has remarkably high thermal stability; when heated in DMSO or DMF at 130°C, it decomposes to diphenyl diselenide PhSeSePh and an insoluble polymeric substance. The resulted insoluble matter is identified as polymeric methylbismuth, $(MeBi)_n$ [84ZN(B)1668]. There is a significant contrast of thermal stability between $MeBi(SePh)_2$ and $Me_2BiSePh$, the latter decomposing at room temperature. The MS spectrum of $MeBi(SePh)_2$ exhibits characteristic peaks due to M^+, $[M–Me]^+$, $Me_2BiSePh^+$, $[M–PhSe]^+$, $PhSeBi^+$ and $Ph_2Se_2^+$ ions.

In the case of the tellurium counterpart, a similar disproportionation of $Me_2BiTePh$ to Me_3Bi and methylbismuth bis(benzenetellurolate) MeBi-$(TePh)_2$ is observed. The formation of $MeBi(TePh)_2$ has been confirmed by elemental analysis; however, it decomposes rapidly at room temperature.

The first cyclic bismuth diselenolate has been prepared by Klapötke, who treated $MeBiBr_2$ with dilithium benzene-1,2-diselenolate in ether under high

TABLE 2.14
Monoorganylbismuth compounds with Bi–Se bonds

Compound	Synthetic method[a]	Yield (%)	M.p. (°C)	Physical data	Reference
$MeBi(SePh)_2$	E	Quant.	120 (decomp.)	^{1}H-NMR, MS	[84ZN(B)1668]
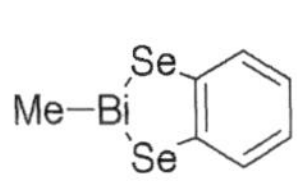	A	62	142–147 (decomp.)	^{1}H-NMR, IR, MS	[87POL1593]
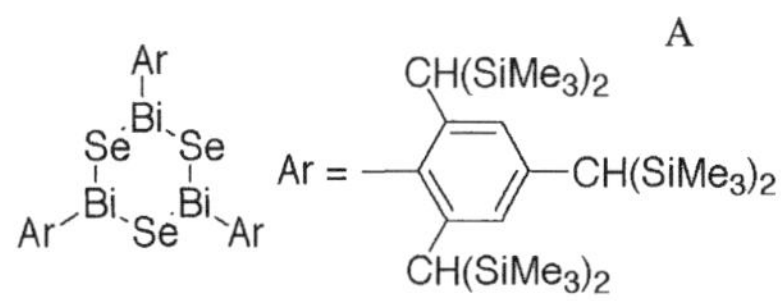	A	–	–	–	[97SCI(277)78]

dilution conditions and obtained 2-methyl-1,3,2-benzodiselenabismole in 62% yield.

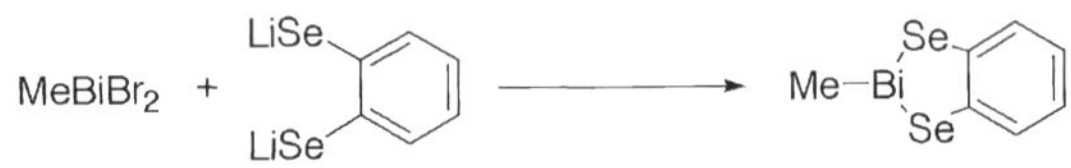

The mass spectrum exhibits characteristic peaks due to M^+, $[M–Me]^+$, Bi_2^+, $[M–Se]^+$ and $BiSe_2^+$ ions. The ^{1}H-NMR spectrum shows two methyl peaks at δ 1.59 and 1.96 due to two conformers with an *endo* or *exo* methyl group. The *endo* methyl group appears at a higher field due to the anisotropic effect of the aromatic ring. The VT-NMR measurement between −60 and +60°C gives no significant change of the spectrum pattern and, therefore, the coalescence temperature should lie at a temperature above 60°C [87POL1593]. On treatment with Li_2Se in THF, a highly hindered aryldichlorobismuthine, $TbtBiCl_2$ [Tbt = 2,4,6-tris(bis(trimethylsilyl)methyl)phenyl] affords a six-membered heterocyclic compound, 2,4,6-triaryl-1,3,5-triselena-2,4,6-tribismane, as a stable solid [97SCI(277)78].

2.2.3.2.5. Cationic and anionic monoorganylbismuth compounds (Table 2.15)

(a) Cationic monoorganylbismuth compounds The cationic monoorganylbismuth compounds $[RBiL_4]^{2+}$ have been studied little; the only known example is phenyltetrakis(hexamethylphosphoramido)bismuth bis-(hexafluorophosphate) $[PhBi(HMPA)_4][PF_6]_2$, prepared via metathesis of $PhBiBr_2$ with $TlPF_6$ in the presence of HMPA [96JCS(D)443]. Its ^{1}H-NMR spectrum exhibits *o*-, *m*- and *p*-proton signals at δ 8.96, 8.10 and 7.46, respectively. The X-ray structure analysis elucidated that the bismuth center possesses a square pyramidal geometry, in which the phenyl group occupies the apical site. The Bi–C bond length is 2.27(1) Å and the Bi···O distances are in the range 2.32(1)–2.36(1) Å. Additionally, there is a close interaction between the bismuth center and one fluorine atom of the PF_6 anion [Bi···F, 3.28(3) Å] [96JCS(D)443].

There is no information available on the thermal and chemical stabilities of $[PhBi(HMPA)_4][PF_6]_2$, though a possible contamination by $[Ph_2Bi(HMPA)_2][PF_6]$ has been confirmed by ^{1}H-NMR measurement.

(b) Anionic monoorganylbismuth compounds There has been a series of studies on the $[RBiL_4]^{2-}$ anion containing trifluoroacetate as the ligand and

ammonium as the counter cation. Known complexes include compounds **1–7** [93AG(E)589, 94JCS(D)2545], which are obtained as colorless crystals via two steps: (i) acidolysis of Ph_3Bi with CF_3CO_2H in a 1:4 ratio to obtain $PhBi(O_2CCF_3)_2$, (ii) neutralization of the remaining acid with the corresponding amines. In contrast with $PhBi(O_2CCF_3)_2$, the anionic part $[PhBi(O_2CCF_3)_4]^{2-}$ in compounds **1–7** is air-stable in the solid sate. These salts are soluble in polar solvents such as THF and acetone, but insoluble in CH_2Cl_2 and $CHCl_3$. Their stability depends on the counter cations: compounds **1** and **2** are stable for days in solution, but compound **4** decomposes easily when dissolved [93AG(E)589]. Aliphatic diammonium cations cause disproportionation of $[PhBi(O_2CCF_3)_4]$ anion in solution, while in the case of phenylenediammonium and pyridinium cations, the anion is stable in solution [94JCS(D)2545].

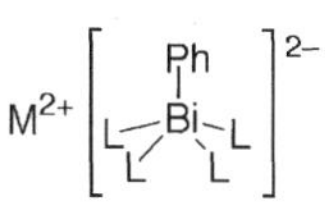

L = CF_3CO_2

1; M = $Me_2NH(CH_2)_2NHMe_2$
2; M = $Me_2NH(CH_2)_3NHMe_2$
3; M = $Et_2NH(CH_2)_3NHEt_2$
4; M = $(Me_2NHCH_2)_2CHOH$
5; M = $Me_2NH(CH_2)_4NHMe_2$
6; M = 1,4-$(Me_2NH)_2C_6H_4$
7; M = $(C_5H_4NH)_2$

Their ^1H-NMR spectra show *o*-, *m*- and *p*-proton signals due to the phenyl group at δ 8.8, 7.9 and 7.3, respectively, and these chemical shifts are not changed significantly by the difference of counter cations. They show IR absorptions in the range 1692–1678, 1482–1429, 1430–1384, 1205–1186 and 1133–1129 cm^{-1}. Except for compound **3**, they have been characterized by X-ray crystallography. There are three types of structure **I–III** for the bismuth center of the $[PhBi(O_2CCF_3)_4]$ anion, depending on their counter cations, due to the hydrogen-bonding between the ammonium group and oxygen atoms of the carboxylate group. In all three structures, the phenyl group occupies the apical site. In structure **I**, the bismuth center possesses a square-pyramidal geometry and four carboxylate groups form a cup shape. In structure **II**, one of the carboxylate group chelates to the bismuth center to form a distorted pentagonal-pyramidal geometry. Structure **III** can be described as an intermediate between **I** and **II**; its bismuth center is five coordinated, but four carboxylate groups do not form the cup structure. The Bi–C and Bi–O bond lengths are in the range 2.19(2)–2.29(3) and 2.31(2)–2.62(4) Å, respectively [94JCS(D)2545].

TABLE 2.15
Anionic and cationic monoorganylbismuth compounds

Compound	Synthetic method[a]	Yield (%)	M.p. (°C)	Physical data	Reference
Anionic compounds					
$[Me_2NH(CH_2)_2NHMe_2][PhBi(O_2CCF_3)_4]$	C	–	–	^{1}H-NMR, X-ray	[93AG(E)589]
	C	Good	151	^{1}H-NMR, IR, X-ray	[94JCS(D)2545]
$[Me_2NH(CH_2)_3NHMe_2][PhBi(O_2CCF_3)_4]$	C	63	–	^{1}H-NMR, X-ray	[93AG(E)589]
	C	63	160	^{1}H-NMR, IR, X-ray	[94JCS(D)2545]
$[Et_2NH(CH_2)_2NHEt_2][PhBi(O_2CCF_3)_4]$	C	Good	103–105	^{1}H-NMR, IR	[94JCS(D)2545]
$[(Me_2NHCH_2)_2CHOH][PhBi(O_2CCF_3)_4]$	C	–	–	^{1}H-NMR, X-ray	[93AG(E)589]
	C	64	151	^{1}H-NMR, IR, X-ray	[94JCS(D)2545]
$[Me_2NH(CH_2)_4NHMe_2][PhBi(O_2CCF_3)_4]$	C	Good	138	^{1}H-NMR, IR, X-ray	[94JCS(D)2545]
$[1,4\text{-}(Me_2NH)_2C_6H_4][PhBi(O_2CCF_3)_4]$	C	33	102	^{1}H-NMR, IR, X-ray	[94JCS(D)2545]
$[PyH]_2[PhBi(O_2CCF_3)_4]$	C	68	123	^{1}H-NMR, IR, X-ray	[94JCS(D)2545]
Cationic compounds					
$[PhBi(HMPA)_4][PF_6]_2$	H	30	–	^{1}H-NMR, X-ray	[96JCS(D)443]

[a] For notation, see Section 2.2.3.1.

2.2.4. Bismuth(III) compounds with three Bi-Group 15 or 16 element bonds

2.2.4.1. Methods of synthesis (Tables 2.16, 2.19, 2.22 and 2.23)

(a) Via nucleophilic substitution of Bi–X bonds (method A)

Synthesis of bismuth tris(dimethylamide)

$$BiCl_3 \quad + \quad Li[NMe_2] \xrightarrow[\text{THF, -78 °C}]{} Bi(NMe_2)_3$$

A solution of lithium dimethylamide (1.20 g, 24.0 mmol) in THF (15 ml) was added to a stirred solution of bismuth chloride (2.47 g, 7.8 mmol) in THF (15 ml) at −78°C (dry ice/ethanol bath). The resulting yellow solution was stirred for 1 h and then allowed to warm to room temperature. All volatiles were removed under vacuum to leave a yellow-green oily residue, which was extracted with hexane (20 ml) and filtered through a Celite bed to afford a yellow solution. Removal of all volatiles from this filtrate under vacuum, followed by sublimation onto a cold finger, kept at dry ice temperature, afforded the titled amide $Bi(NMe_2)_3$ as a yellow crystalline solid (1.66 g, 62%). The optimum conditions for sublimation are warming the flask to 30°C under a pressure of about 10^{-2} mmHg [91IC4680].

Synthesis of bismuth triethoxide

$$BiCl_3 \quad + \quad NaOEt \xrightarrow[\text{EtOH}]{} Bi(OEt)_3$$

Bismuth chloride (6.3 g, 20 mmol) was dissolved in absolute ethanol, and sodium ethoxide (4.08 g, 60 mmol) was added dropwise in the same solvent. After exothermic reaction, sodium chloride precipitated as a fine white solid. This was filtered off and the filtrate was concentrated in vacuo at room temperature to give $Bi(OEt)_3$ as colorless crystals with a fruit-like odor. The compound is very sensitive to moisture and decomposes slowly on prolonged storage. It is poorly soluble in organic solvents. Yield, 6.0 g (86%) [76ZAAC(423)47].

Synthesis of bismuth malate

$$Bi(NO_3)_3{\cdot}5H_2O + HO_2CCH_2CH(OH)CO_2H \xrightarrow[H_2O]{} Bi[O_2CCH_2CH(O)CO_2]{\cdot}H_2O$$

To a hot aqueous solution of L-malic acid (1.34 g, 10.0 mmol) was added slowly bismuth nitrate pentahydrate (4.85 g, 10.0 mmol). The homogeneous reaction mixture gradually turned into a suspension. The mixture was stirred for 15 min at 60°C and the insoluble materials were removed by filtration. The clear filtrate was kept hot in a Dewar bath for crystallization. During the course of 4 weeks, colorless cubic crystals deposited gradually. These were separated, washed with acetone and dried [93CB51].

Synthesis of bismuth tris(tropolonate)

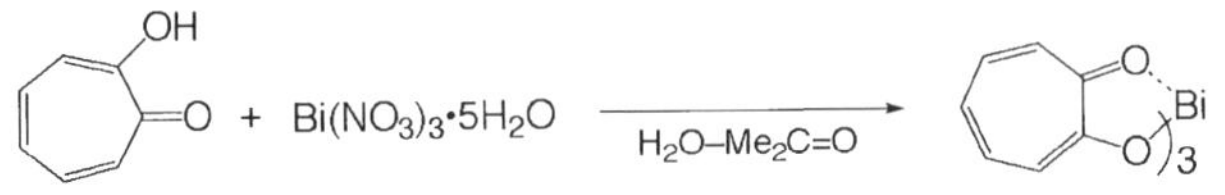

A solution of tropolone (1.22 g, 10 mmol) in distilled water (250 ml) and acetone (150 ml) was heated under reflux and to this solution was added finely powdered $Bi(NO_3)_3 \cdot 5H_2O$ (1.45 g, 3.0 mmol) in small portions. The mixture was stirred vigorously until a clear solution was obtained. The hot solution was filtered and the filtrate was concentrated until the solution became turbid. The reaction mixture was allowed to cool slowly to room temperature. The resulting yellow precipitate was filtered off, washed twice with acetone (30 ml), and dried in vacuo over P_4O_{10}. Yield, 1.55 g (90%) [95CB335].

(b) Via cleavage of Bi–C bond(s) (method B)

Synthesis of bismuth tris(chloroacetate)

$$Ph_3Bi + ClCH_2CO_2H \xrightarrow[CHCl_3]{} Bi(O_2CCH_2Cl)_3 + PhH$$

A solution of Ph_3Bi (1 g) in dry $CHCl_3$ (6 ml) was added to a solution of chloroacetic acid (0.644 g) in the same solvent (2 ml), and the resulting mixture was stirred for 45 min. The precipitate formed was filtered off and vacuum dried to obtain bismuth tris(chloroacetate) (0.97 g, 90%) [79ZOB1185].

(c) Via cleavage of Bi–N or Bi–O bond(s) (method C)

Synthesis of bismuth tris(2,4,6-trimethylbenzenethiolate)

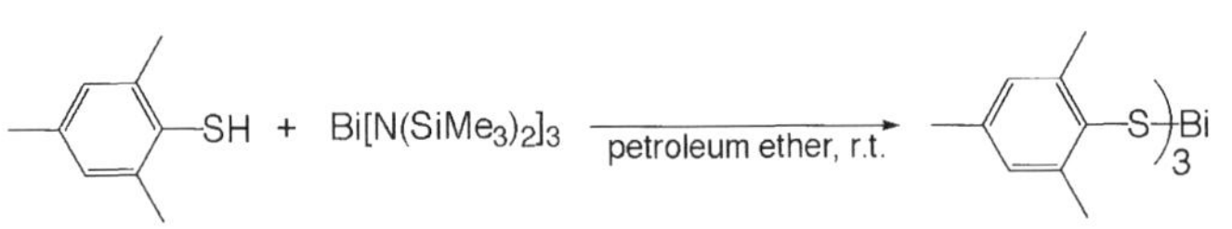

A solution of bismuth tris[bis(trimethylsilyl)amide] $Bi[N(SiMe_3)_2]_3$ (0.98 g, 1.42 mmol) in light petroleum (10 ml) was added to 2,4,6-trimethylbenzenethiol (1.0 g, 4.25 mmol) in the same solvent (40 ml) at room temperature. The resulting red solution was stirred for 1 h and then the solvent was removed in vacuo. The residue was recrystallized from MeCN–ethyl acetate (3:1 v/v) to give bismuth tris(2,4,6-trimethylbenzenethiolate) as orange crystals (1.1 g, 86%) [95JCS(D)1649].

Synthesis of bismuth tris(benzenesulfinate)

$$PhSO_2H + Bi(OAc)_3 \xrightarrow[AcOH]{} Bi(O_2SPh)_3$$

Solutions of benzenesulfinic acid (ca. 4 mmol) and bismuth acetate (1.0 mmol) in glacial acetic acid (50 and 100 ml, respectively) were mixed and evaporated to dryness under vacuum. The residue was washed with chloroform and dried under a stream of nitrogen to give bismuth tris(benzenesulfinate) in 82% yield [72AJC2107].

Synthesis of bismuth pyridine-2,6-dicarboxylate

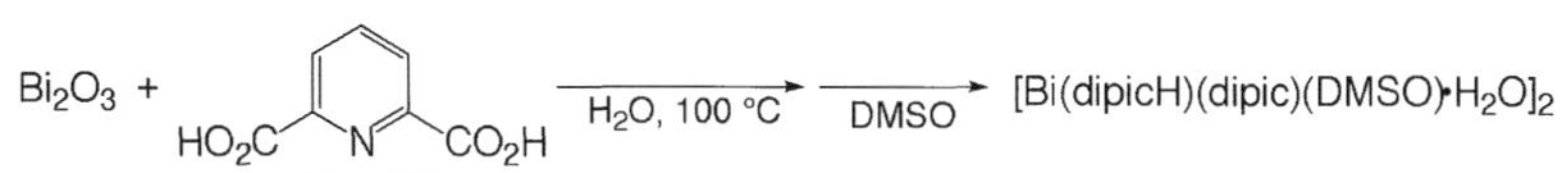

(dipicH$_2$)

A suspension of Bi_2O_3 (500 mg, 1.07 mmol) in water (40 ml) was heated under reflux in the presence of pyridine-2,6-dicarboxylic acid (dipicolinic acid; 720 mg, 4.3 mmol) for 2 h. The resulting mixture was stirred overnight at 110°C, then the solution was concentrated to a 25-ml volume to precipitate $[Bi(dipicH)(dipic)(DMSO)\cdot H_2O]_2$ quantitatively [92MGMC217].

(d) Via insertion of a molecule into the Bi–X bond (method D)

Synthesis of bismuth tris(dimethylcarbamate)

$$Bi(NMe_2)_3 + CO_2 \xrightarrow[THF]{} Bi(O_2CNMe_2)_3$$

An excess amount of CO_2 was passed to a solution of $Bi(NMe_2)_3$ in THF (10 ml), under ice-cooling. White slurry of fine solid resulted. Replacement of THF by petroleum ether gave a filterable precipitate, which was recrystallized from $CHCl_3$–petroleum ether to give bismuth tris(dimethylcarbamate) (4.12 g, 87%) [75JINC(37)2011].

Synthesis of bismuth tris(dimethyldithiocarbamate)

$$Bi(NMe_2)_3 + CS_2 \xrightarrow[THF]{} Bi(S_2CNMe_2)_3$$

When $Bi(NMe_2)_3$ (1.70 g, 5 mmol) in THF was mixed with a slight excess of CS_2 (18 mmol), an exothermic reaction took place to give a yellow slurry. By adding petroleum ether to this slurry, bismuth tris(dimethyldithiocarbamate) was obtained as a yellow solid (2.84 g, 95%) [75JINC(37)2011].

Synthesis of bismuth tris(*N*-phenyl-*N'*,*N'*-dimethylureide)

When $Bi(NMe_2)_3$ (1.71 g, 5 mmol) was mixed with phenyl isocyanate (1.79 g, 15 mmol) in

THF, a mild exothermic reaction took place with the disappearance of the yellow colour of the bismuth amide. After removal of THF in a rotatory evaporator, the remaining slurry was triturated with petroleum ether to give bismuth tris(*N*-phenyl-*N*′,*N*′-dimethylureide) (2.92 g, 87%) [75JINC(37)2011].

(e) Via disproportionation (method E)

Synthesis of bismuth tris(diethyldithiocarbamate)

$$Ph_2BiCl + NaS_2CNMe_2 \xrightarrow[CHCl_3]{} Bi(S_2CNMe_2)_3$$

The reaction of Ph_2BiCl (0.01 mole) and sodium diethylcarbamate trihydrate (0.01 mol) in $CHCl_3$ (50 ml) at 25°C for 1 h gave $PhBi(S_2CNMe_2)_2$ (38%), Ph_3Bi (46%) and $Bi(S_2CNMe_2)_3$ (11%). In the preparation of $PhBi(S_2CNMe_2)_2$, Ph_3Bi and $Bi(S_2CNMe_2)_3$ were always obtained as important side products [68JOM(11)627].

2.2.4.2. Bismuth(III) compounds with three Bi–N bonds (Table 2.16)

(a) Bismuth amides (aminobismuthines)

(i) Synthesis: The chemistry of bismuth amides has not been studied quite so much as that of the other 15 group family congeners. In recent years, however, this type of compound has attracted attention and they could be potential precursors for superconducting materials of high critical temperature.

Tris[methyl(trimethylsilyl)amino]bismuthine $Bi[N(Me)SiMe_3]_3$ is prepared in 54% yield by the reaction of $BiBr_3$ with the corresponding lithium amide in hexane. This compound can be distilled under reduced pressure [66JOM(6) 259]. Similarly, tris[bis(trimethylsilyl)amino]bismuthine $Bi[N(SiMe_3)_2]_3$ is obtained as a pale yellow powder by treating $BiCl_3$ with lithium hexamethyldisilylamide [89POL1579, 80JCS(D)2428, 95JCS(D)1649].

Simple bismuth amides $Bi(NR_2)_3$ (R = Me, Et and Pr) are obtained as yellow liquid or crystals by treating $BiCl_3$ with the corresponding lithium amides in THF-petroleum ether. In this preparation, the use of THF as solvent is indispensable; all attempts in hexane, cyclohexane, petroleum ether and ether failed [75JINC(37)2011]. These amides are thermolabile and gradually turn jet black at room temperature; however, they can be stored indefinitely at dry ice temperature. Norman et al. have prepared tris(diphenylamino)bismuthine $Bi(NPh_2)_3$ as a moisture-sensitive solid [89POL1579]. The same authors reported that $Bi(NMe_2)_3$ is rather light sensitive, bright sunlight making it turn black; it is better to store it in a freezer in the dark [91IC4680].

Bismuth amides bearing a sterically crowded substituent are also known. Treatment of LiNHAr with $BiCl_3$, where Ar = 2,4,6-tri-*tert*-butylphenyl group, does not afford the expected bismazene ArN(H)–Bi=NAr, but the corresponding bismuth amide, $Bi[N(H)(2,4,6\text{-}^tBu_3C_6H_2)]_3$ **1**, is obtained in low yield [96IC4013]. This is quite different from the reaction of ArNHLi with PCl_3 or $AsCl_3$, where the formation of the corresponding azene (ArN(H)–E=NAr; E = P and As) is predominant. The hindered bismuth amide can be recrystallized from hexane; however, it decomposes in ether and CH_2Cl_2 over a period of days. When decomposed thermally, it releases elemental bismuth [96IC4013]. Changing the aryl substituent on the imino nitrogen to the 2,6-diisopropylphenyl group results in the formation of a cyclic bismuth amide **2** irrespective of the reaction conditions employed, i.e. the molar ratio of the reactants and temperatures [94IC4607]. The addition of $^tBuNH_2$ to $Bi(NMe_2)_3$ generates $[(Me_2N)Bi(\mu\text{-}N^tBu)]_2$ **3** in situ, which is subsequently treated with ^tBuNHLi to give a complex $[Bi_2(N^tBu)_4]Li_2{\cdot}2$ THF **4** of a bimetallic cubane structure. The complex is thermally stable and can be stored indefinitely under argon [96ICA(248)9].

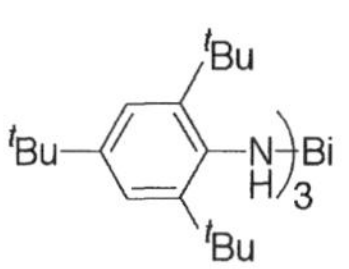

1

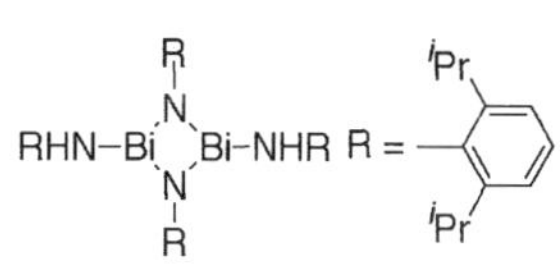

2

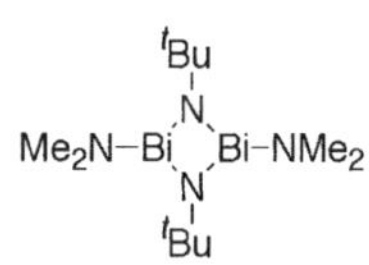

3

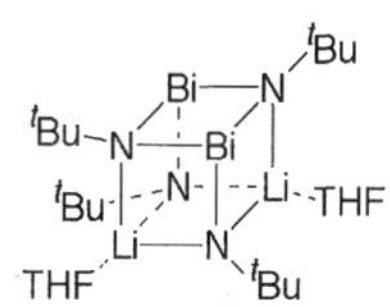

4

(ii) Spectral properties: $Bi[N(Me)SiMe_3]_3$ shows ^{1}H-NMR peaks in CCl_4 at δ − 4.5 (Me_3Si) and −213.0 (NMe) Hz [66JOM(6)259], while $Bi[N(SiMe_3)_2]_3$ exhibits a ^{1}H-NMR signal at δ 0.42 in C_6H_6 [80JCS(D)2428] or δ 0.37 in C_6D_6 [95JCS(D)1649]. The ^{1}H-NMR spectra of bismuth amides exhibit peaks at δ 3.45 (s) for $Bi(NMe_2)_3$, 0.29 (t) and 3.62 (q) for $Bi(NEt_2)_3$, and 0.85 (t), 1.45 (m) and 3.50 (q) for $Bi(NPr_2)_3$, while the Raman spectra show absorptions at 550 and 270 cm^{-1} for $Bi(NEt_2)_3$ and 408 and 322 cm^{-1} for $Bi(NPr_2)_3$ [75JINC(38)2011]. Upon Raman irradiation, $Bi(NMe_2)_3$ suffers extensive decomposition and no peaks are observed. Reinvestigation of ^{1}H-NMR of $Bi(NMe_2)_3$ in toluene-d_8 at 295 K has shown a signal at δ 3.72, which becomes significantly broader at 183 K and shifts to δ 3.85. The ^{13}C-NMR of $Bi(NMe_2)_3$ in toluene-d_8 at 295 K exhibits a peak at δ 48.4 which becomes broader at 203 K and disappears at 183 K [91IC4680]. From the VT-NMR measurement of $Bi[N(SiMe_3)_2]_3$, the activation energy for the Bi–N bond rotation has been estimated as 41 kJ/mol [91IC4680].

(iii) Structures: The first X-ray structure of a homoleptic bismuth amide was solved by Norman et al. in 1989 for $Bi(NPh_2)_3$, which shows that each bismuth center has a trigonal pyramidal geometry. The Bi–N bond lengths are in the range 2.12(2)–2.28(2) Å, while the N–Bi–N bond angles are in the range 96(1)–101(1)° [89POL1579]. In the case of $Bi(NMe_2)_3$, the Bi–N bond lengths and the N–Bi–N bond angles are in the range 2.180(21)–2.189(18) Å and 96.2(9)–98.3(5)°, respectively [91IC4680]. The X-ray structure of amide **1** has revealed that a sterically hindered substituent has little influence on both the bond length and bond angles between the bismuthand N atoms. The Bi–N bond lengths and N–Bi–N bond angles being in the range 2.14(2)–2.214(13) Å and 82.7(6)–106.7(6)°, respectively [96IC4013]. Cyclic bismuth amide **2** has both cyclic and acyclic Bi–N bonds; however, there is no significant difference between them, the bond lengths being in the range 2.158(4)–2.174(5) Å and bond angles in the range 78.5(2)–97.4(2)°, respectively [94IC4607]. The X-ray structure of the cubic bismuth amide complex **4** shows Bi–N bond lengths of 2.101(7)–2.237(7) Å and an average N–Bi–N bond angle of 81.3° for a Bi_2N_2 ring and 87.9° for a BiN_2Li ring [96ICA(248)9].

(iv) Reactions: Bismuth amides are known as good precursors for bismuth alkoxides [90IC358, 91POL437] and bismuth thiolates [95JCS(D)1649]; treatment of bismuth amides with alcohols or thiols gives the corresponding substituted products and secondary amines.

$Bi(NR_2)_3$ (R = Me, Et, Pr) reacts with CO_2, CS_2, phenyl isocyanate and phenyl isothiocyanate to produce the corresponding bismuth tris(*N*,*N*-dialkylcarbamate), tris(dialkyldithiocarbamate) $Bi(S_2CNR_2)_3$, tris(*N*-phenyl-*N*′,*N*′-dialkylureide) $Bi[NPhCONR_2]_3$ and tris(*N*-phenyl-*N*′,*N*′-dialkylthioureide) $Bi[NPhCSNR_2]_3$, respectively, in good to excellent yields. Treatment of $Bi(NR_2)_3$ (R = Me, Et, Pr) with acetic anhydride forms $Bi(O_2CMe)_3$ in more than 90% yield [75JINC(37)2011]. Two bismuth amides, $Bi(NMe_2)_3$ and $Bi(NEt_2)_3$, react with ketene to give the corresponding insertion products, $Bi(CH_2CONMe_2)_3$ and $Bi(CH_2CONEt_2)_3$, in almost quantitative yields (Scheme 2.6). The reactivity of MX_3, where M = As, Sb or Bi, X = SR, OR or NR_2, toward ketene increases in the order As < Sb < Bi for M, and SR < OR < NR_2 for X. The highly polar nature as well as the weakness of the Bi–N bond are considered to be responsible for the enhanced nucleophilicity of the nitrogen atom [87BCJ1564].

(b) Bismuth ureides and thioureides There is only one report on bismuth

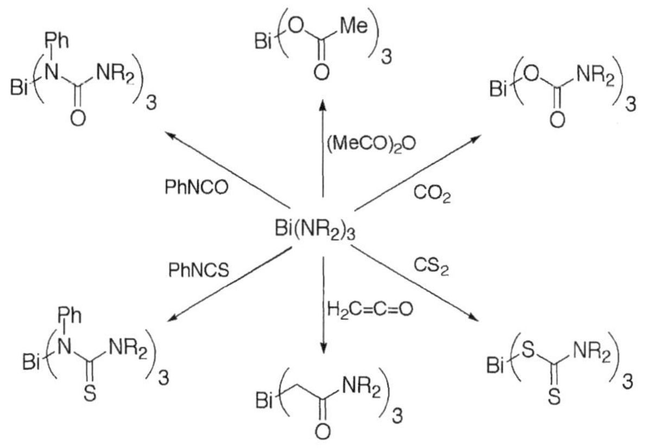

Scheme 2.6.

ureides and thioureides, Bi[NPhCONR$_2$]$_3$ and Bi[NPhCSNR$_2$]$_3$ (R = Me, Et, Pr), which are obtained in good to excellent yields by the insertion of phenyl isocyanate or phenyl isothiocyanate into the Bi–N bond of bismuth amides [75JINC(37)2011]. These compounds are not so stable and decompose when stored in air, giving the corresponding urea and thiourea derivatives. The IR spectrum of bismuth ureides exhibits two absorption bands due to the NCO linkage at 1530 and 1632–1650 cm^{-1}. On the other hand, bismuth thioureides showed two IR absorption bands due to the NCS linkage at 1302–1310 and 1520–1535 cm^{-1} [75JINC(37)2011].

TABLE 2.16
Bismuth(III) compounds with three amino or amido ligands

Compound		Synthetic method[a]	Yield (%)	M.p. (°C) or b.p. (°C/mmHg)	Physical data	Reference
R						
Bi(NR$_2$)$_3$						
Me		A	20	40–40.5 46–49/0.16	^{1}H-NMR	[75JINC(37)2011]
		–	–	–	–	[87BCJ1564]
		A	62	–	^{1}H-, ^{13}C-NMR, X-ray	[91IC4680]
Et		A	52	80–82/0.09	^{1}H-NMR, Raman	[75JINC(37)2011]
Pr		A	41	110–112/0.02	^{1}H-NMR, Raman	[75JINC(37)2011]
Me_3Si		A	–	–	^{1}H-, ^{13}C-NMR	[89POL1579]
		A	30	90 (decomp.)	^{1}H-NMR	[80JCS(D)2428]
		A	42.5	–	^{1}H-, ^{13}C-NMR	[95JCS(D)1649]
		A	30	–	^{1}H-, ^{13}C-NMR, X-ray	[89POL1579]
Bi(NR^{1}R^2)$_3$						
R^1 = Me	R^2 = Me_3Si	A	54	27–29, 90–92/0.1	–	[66JOM(6)259]
R^1 = H	R^2 = 2,4,6-^{t}$Bu_3C_6H_2$	A	11	144–147 (decomp.) [C_6H_{14}]	^{1}H-, ^{13}C-NMR, IR, Raman, X-ray	[96IC4013]

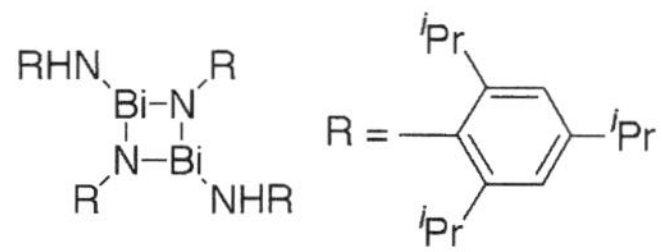	A	57	163	^{1}H-, ^{13}C-NMR, IR, MS, X-ray	[94IC4607]
Bi(NPhCONR$_2$)$_3$					
Me	D	87	212–215 (decomp.)	^{1}H-NMR, IR	[75JINC(37)2011]
Et	D	92	184–185 (decomp.)	^{1}H-NMR, IR	[75JINC(37)2011]
Pr	D	89	146–150 (decomp.)	^{1}H-NMR, IR	[75JINC(37)2011]
Bi(NPhCSNR$_2$)$_3$					
Me	D	94	117–120 (decomp.) [$CHCl_3$–CCl_4]	^{1}H-NMR, IR	[75JINC(37)2011]
Et	D	42	235 (decomp.)	^{1}H-NMR, IR	[75JINC(37)2011]
Pr	D	76	74–75 (decomp.)	^{1}H-NMR, IR	[75JINC(37)2011]
[$^{t}Bu_4Bi_2$]Li_2·2THF	C	33	108–110	^{1}H-NMR, IR, X-ray	[96ICA(248)9]

[a] For notation, see Section 2.2.4.1.

2.2.4.3. *Bismuth(III) compounds with three Bi–O bonds (Table 2.19)*

(a) Bismuth alkoxides

(i) Synthesis: The chemistry of bismuth alkoxides was almost neglected until the 1990s and, therefore, only in recent years, has various information on the nature of bismuth alkoxides become available. A variety of alkoxides are derived not only from aliphatic and aromatic alcohols but also from enol compounds. In particular, bismuth alkoxides has attracted considerable attention as a potential precursor for bismuth-containing superconducting materials in sol-gel or chemical vapor deposition (CVD) processes [95AG(E)2187].

The most common procedure for preparing bismuth alkoxides is the substitution reaction of bismuth halides with alkali metal alkoxides. The alternative is treatment of bismuth amides with alcohols. Bismuth alkoxides such as trimethoxide, triethoxide and triisopropoxide were first prepared in 80–93% yields by the reaction between $BiCl_3$ and alkali metal alkoxides by Mehrotra et al. in 1966. Bismuth alkoxides are less soluble in organic solvents and the yields after purification by sublimation are generally very low (10–20%) [66IJC537]. Known aliphatic alkoxides include $Bi(OEt)_3$ [76ZAAC(423)47], $Bi(O^iPr)_3$ [90IC358], $Bi(O^tBu)_3$ [91POL437], $Bi(OCMe_2Et)_3$, $Bi(OC_2H_4OMe)_3$, $Bi(OC_2H_4NMe_2)_3$ and $Bi[OCH(Me)CH_2NMe_2]_3$ [90IC358]. In the case of $Bi[OCH(CF_3)_2]_3$, several cluster structures have been reported [92AG(E)451]. Wieber et al. have prepared $Bi(OEt)_3$ as colorless crystals, which have a fruity odor and are significantly moisture-sensitive [76ZAAC(423)47]. This alkoxide has been found to be useful as a precursor for bismuth thiolates. Bismuth oxide Bi_2O_3 heated in boiling ethylene glycol gives bismuth ethylene glycolate $[Bi_4(\mu\text{-}\eta^1\text{-}OCH_2CH_2OH)_2(\mu_4\text{-}\eta^2\text{-}OCH_2CH_2O)(\mu_2\text{-}\eta^2\text{-}OCH_2CH_2O)_4]_\infty$ as a gray precipitate in 98% yield [92SRI(22)289]. Bismuth siloxide $Bi(OSiMe_3)_3$ is obtained as a powdery compound by the reaction between $BiCl_3$ and $NaOSiMe_3$. This siloxide is labile and turns rapidly dark at room temperature due to decomposition [68ZAAC(363)84]. $Bi(OSiPh_3)_3$ is also known [91POL437].

Bismuth phenoxide $Bi(OC_6F_5)_3$ has been found to form several cluster structures which incorporate such solvent molecules as THF or toluene [93IC5136]. This phenoxide is accessible by acid cleavage of the Bi–C bond in Ph_3Bi with pentafluorophenol in boiling toluene [92AG(E)451, 93IC5136]. Bismuth phenoxides such as $Bi(OC_6H_3\text{-}2,6\text{-}Me_2)_3$ [89CC1628] and bismuth catecholates (vide infra) [90HAC233] are also known.

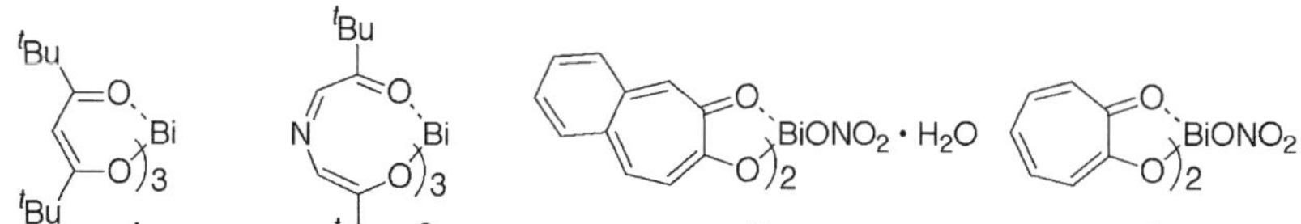

1 **2** **3** **4**

Bismuth enolates such as bismuth tris(1,3-di-*tert*-butylpropane-1,3-dionate) $Bi(dpm)_3$ **1** [91POL437] and bismuth tris(4-aza-1,7-dioxa-2,6-di-*tert*-butylhepta-2,5-dien-1,4,7-triylate) **2** are reported [85JA3397, 87JA627]. TGA analysis has shown that the thermal stability of bismuth alkoxides decreases in the order of $Bi(O^tBu)_3 > Bi(OEt)_3 > Bi(dpm)_3$ [91POL437]. Many bismuth tropolonates have been prepared [66JA4856, 95CB335] and some of these are found to possess biological (antimicrobial) activities [95CB335, 97CCR345].

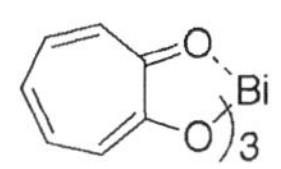

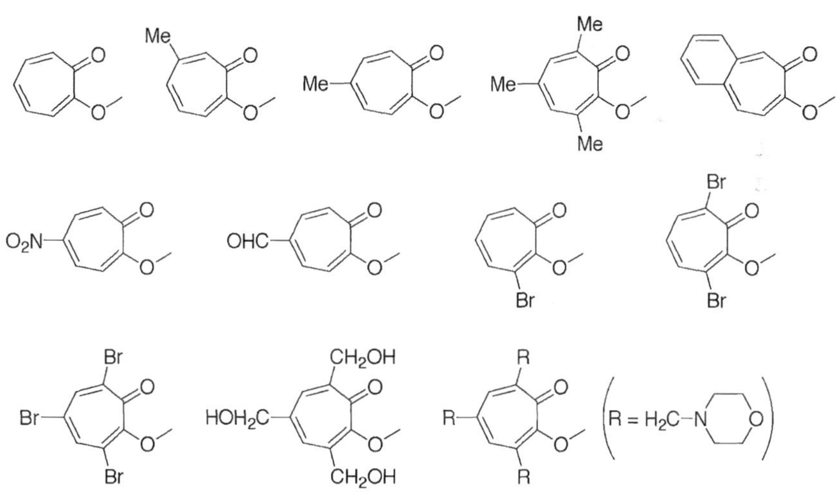

(ii) Spectral properties: The ^{1}H-NMR spectrum of $Bi(OEt)_3$ shows peaks at δ 1.23 (Me) and 4.06 (CH_2O) in CCl_4 [76ZAAC(423)47]. $Bi(O^tBu)_3$ exhibits a ^{1}H-NMR peak at δ 1.42 [89CC1628], while $Bi(OC_2H_4OMe)_3$ shows peaks at δ 3.22 (MeO), 3.51 (CH_2) and 5.08 (CH_2OBi) in C_6D_6 [90IC358]; the latter three signals appear at δ 3.37, 3.39, 3.41, 3.50–3.54 and 3.7–3.77 in

$CDCl_3$ [91POL437]. The ^{1}H-NMR spectrum of $Bi(OC_2H_4NMe_2)_3$ exhibits peaks at δ 2.12 (Me_2N) 2.40 (CH_2), and 5.12 (CH_2OBi) in C_6D_6 [90IC358]. Bismuth tropolonate shows peaks at δ 6.98 (5-H), 7.03 (3,7-H) and 7.70 (4,6-H) in DMSO-d_6 [95CB335]. The ^{1}H-NMR spectrum of $Bi(OSiMe_3)_3$ shows a peak at δ −0.092 in CCl_4 [68ZAAC(363)84].

More important spectral information on bismuth alkoxides is obtained from IR spectra; the absorptions due to the Bi–O bonds appear at 520 cm^{-1} for $Bi(OEt)_3$ [91POL437], 459 cm^{-1} for $Bi(O^tBu)_3$ [91POL437], 524 cm^{-1} for $Bi(OC_2H_4OMe)_3$ [91POL437] and 543 cm^{-1} for $Bi(OSiMe_3)_3$ [68ZAAC(363)84]. The bismuth ethylene glycolate shows IR absorption due to ν(Bi–OR) at 586 cm^{-1} [92SRI(22)289]. In the case of bismuth tropolonates, the ν(Bi–O) stretching falls in the range 466–611 cm^{-1} and mostly around 500 cm^{-1} [95CB335]. In contrast, the ν(Bi–O) value of $Bi(dpm)_3$ shifts dramatically to 274 cm^{-1} [91POL437].

The mass spectrum of $Bi(O^tBu)_3$ exhibits characteristic signals due to $[M-Me]^+$, $[M-O^tBu]^+$, $BiOH(O^tBu)^+$, $BiC_4H_6^+$, $BiOC_3H_2^+$, $BiOMe^+$, $BiOH^+$ and Bi^+ fragment ions, while that of $Bi(dpm)_3$ shows signals due to $Bi(dpm)_2^+$, $Bi(dpm)(C_3H_5)^+$, $Bi(dpm)^+$, $Bi(C_7H_{11}O_2)^+$ and Bi^+ fragment ions [91POL437]. In the case of bismuth tropolonates, the mass spectra show a prominent peak due to the M^+ ion, but sometimes the $[M-Tp]^+$ ion (Tp denotes a tropolonato ligand) is also observed [95CB335].

(iii) Structures: Many structural studies of bismuth alkoxides have been made in these two decades, and the first report on their X-ray crystallographic study appeared in 1989. Structurally, bismuth alkoxides can be divided into three categories, monomeric, oligomeric and polymeric, and the monomeric structure is not the major one. Evans et al. reported in 1989 the first X-ray structure analysis of a $Bi(OC_6H_3\text{-}2,6\text{-}Me_2)_3$, which shows that the compound has a monomeric structure with an average Bi–O distance of 2.091(5) Å and O–Bi–O angles of 92(2)°, with the three ArO groups not arrayed in any regular pattern [89CC1628]. Similarly, aquabis(4,5-benzotropolonato)bismuth nitrate **3** has a monomeric form with the Bi–O bond lengths in the range 2.130–2.323 Å; there is no suitable regular polyhedron to describe the coordination sphere [95CB335]. Bismuth siloxide $Bi(OSiPh_3)_3$ forms a monomeric structure with three THF molecules within a capped octahedral geometry. The average Bi–OSi distance 2.04(1) Å is shorter than the length of the average coordination bond, 2.95(1) Å, between the bismuth center and the oxygen atom of the THF molecules [91POL437]. The nine-coordinate 20 electron bismuth enolate complex **2** also forms a monomeric structure, in which six oxygen atoms

and three nitrogen atoms surround the bismuth center. The geometry around the bismuth center has been described as a face-capped twisted trigonal prism. The average Bi–O and Bi–N bond lengths are 2.51(3) and 2.54(3) Å, respectively [85JA3397]. Interestingly, this bismuth enolate contains two unique molecules in the crystallographic unit. Since the trio of the attached ligands forms a propeller arrangement, each system becomes chiral [85JA3397]. The X-ray structure of bismuth bis(tropolonate) nitrate **4** has been shown to be dimeric. Two bismuth centers are surrounded by six and seven oxygen atoms, respectively, and there is no appropriate polyhedron to describe their respective geometries [95CB335].

Both $Bi[OCH(CF_3)_2]_3$ and $Bi(OC_6F_5)_3$ are reported to form the dimeric structures with appropriate solvents. For the former, a THF-solvated compound, $\{Bi[OCH(CF_3)_2]_3(THF)\}_2$ **5**, has been reported [92AG(E)451, 93IC5136]. Its bismuth center possesses a distorted octahedral geometry, in which two sites are occupied asymmetrically by bridging alkoxide ligands. For the latter, the THF and toluene-solvated structures are known: $[Bi(OC_6F_5)_2(\mu\text{-}OC_6F_5)(PhMe)]_2\cdot 2$ PhMe **6**, $[Bi(OC_6F_5)_2(\mu\text{-}OC_6F_5)(PhMe)]_2$ **7**, $[Bi(OC_6F_5)_2(\mu\text{-}OC_6F_5)(THF)_2]_2\cdot 2(C_6H_{14})$ **8** and $[Bi(OC_6F_5)_2(\mu\text{-}OC_6F_5)(THF)_2]_2$ **9** [92AG(E)451, 93IC5136]. The structures solvated by toluene have the four-coordinated bismuth center with a distorted octahedral geometry with two bridging phenoxide ligands. The bismuth–oxygen bond lengths are summarized in Table 2.17 [93IC5136].

The cluster structures of $Bi(OC_6F_5)_3$ have been characterized by X-ray crystallography. Both $Bi_6(\mu_3\text{-}O)_7(\mu_3\text{-}OC_6F_5)[Bi(OC_6F_5)_4]_3\cdot(THF)_2$ and $Bi_6(\mu_3\text{-}O)_7(\mu_3\text{-}OC_6F_5)[Bi(OC_6F_5)_4]_3\cdot(PhMe)_2$ contain the octahedral $[Bi_6(\mu_3\text{-}O)_7(\mu_3\text{-}OC_6F_5)]^{3+}$ cores and three bismuthate cores $[Bi(OC_6F_5)_4]^-$. These bismuth aggregates contain nine bismuth atoms, seven oxide ligands, thirteen C_6F_5O ligands and two THF molecules. They are the largest bismuth aggregates which have been structurally characterized to date [92CC1638].

TABLE 2.17
Bismuth–oxygen bond lengths (Å) of solvated dimeric bismuth alkoxides **5–9**

	5	**6**	**7**	**8**	**9**
$Bi\text{–}(\mu\text{-}OR)_{short}$	2.188(7)	2.210(8)	2.168(7)	2.198(9)	2.245(4)
$Bi\text{–}(\mu\text{-}OR)_{long}$	2.688(7)	2.571(7)	2.5553(9)	2.75(1)	2.661(4)
$Bi\text{–}OR_{terminal}$	2.116(7)	2.147(8)	2.136(8)	2.21(1)	2.151(4)
	2.064(8)	2.088(9)	2.089(8)	2.132(9)	2.111(3)
$Bi\text{–}O_{THF}$	2.575(7)	–	–	2.58(1)	2.615(4)
		–	–	2.62(1)	2.562(4)

The X-ray structure of $Bi(OC_2H_4OMe)_3$ has shown that this alkoxide forms the polymer chain, which can be represented by the formula $[Bi(\mu\text{-}\eta^1\text{-}OC_2H_4OMe)_2(\eta^1\text{-}OC_2H_4OMe)]_\infty$. The structure is a one-dimensional zig-zag chain with bridging alkoxide ligands. Interestingly, this alkoxide has good solubility in organic solvents, and the cryoscopic molecular weight determination suggests the possibility of a dimeric structure in benzene solution, the chelation of ether oxygen atoms to the bismuth occurring in the solution [90CC301, 90IC358]. The structure of the same compound is discussed later [91POL437].

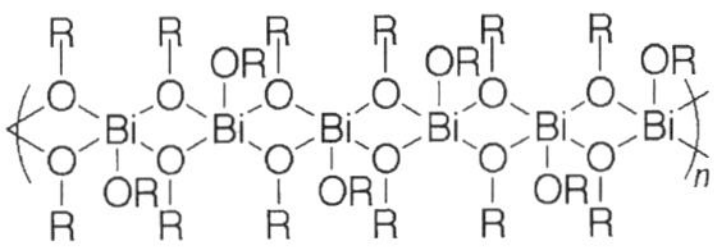

The geometry around the bismuth center is a basal-edge-shaped square pyramid with Bi–O bond distances of 2.07(2)–2.11(1), 2.20(1)–2.21(1) and 2.53(1)–2.58(1) Å. The terminal ligands are bound to the bismuth more tightly than the bridging ligands [90IC358]. The interaction between the bismuth center and the MeO oxygen has been pointed out; the estimated distance is 2.90–3.10 Å, the values being significantly shorter than the sum of the van der Waals radii of the bismuth and oxygen atoms (3.47 Å) [91POL437].

(iv) Reactions: The chemistry of some volatile bismuth alkoxides has been developed recently in connection with the production of high temperature superconductive materials based on the MOCVD techniques. However, there is only limited information available on the reaction modes of bismuth alkoxides in general. Treatment of $Bi(OEt)_3$ with PhSH, EtSH, $PhCH_2SH$ and HOC_2H_4SH leads to the formation of the corresponding bismuth thiolates $Bi(SR)_3$ in 70–84% yield [76ZAAC(432)47]. Partial hydrolysis of $Bi(O^tBu)_3$ in aqueous THF gives a sol, which is transformed to a gel after 12 h. In addition, complete hydrolysis or thermal treatment at 550°C for 15 h of an amorphous powdery $Bi(O^tBu)_3$ yields cubic bismuth(III) oxide Bi_2O_3. On the other hand, complete hydrolysis of $Bi(OSiPh_3)_3(THF)_3$ gives an amorphous powder, which is heated in air at 475°C to afford bismuth oxohydroxocarbonate $(BiO)_4CO_3(OH)_2$ as a black material [91POL437].

(b) Bismuth alkoxide nitrates

Treatment of $Bi(NO_3)_3{\cdot}5H_2O$ in a mixture of MeCN and MeOH with 12-

crown-4, 18-crown-6, triethylene glycol (EO3), tetraethylene glycol (EO4), pentaethylene glycol (EO5) or hexaethylene glycol (EO6) gave the corresponding complexes, $[Bi(NO_3)_2(EO3^-)]_2$, $[Bi(NO_3)_2(EO4^-)]_2 \cdot 2$ MeOH, $[Bi(NO_3)_2(EO5)]$ $[Bi(NO_3)_2(EO5^{2-})\text{-}Bi(NO_3)_3] \cdot 2\ H_2O$, $[Bi(NO_3)_2(EO6^-)]_2$, $[Bi(NO_3)_3$(12-crown-4)] and $[Bi(NO_3)_3(OH_2)_3] \cdot$18-crown-6, the structures of which are characterized by X-ray crystallography [92JA2960]. The first X-ray structure analysis of the bismuth tropolonato complex has been carried out for nitratobis (tropolonato)bismuth **4**, which shows that the intermolecular Bi–O interaction with distances of 2.666 and 2.688 Å leads to the formation of a dimeric structure [95CB335]. This nitrate exhibits a relatively strong antimicrobial activity against *Helicobacter pylori.*

(c) Bismuth catecholates

Bismuth catecholates may be classified better as bismuth alkoxides, but the known compound examples are not ''pure'' catecholates. These compounds are discussed in Section 4.2.4.1.

(d) Bismuth carboxylates

(i) Synthesis: Historically, bismuth carboxylates have long been important as drugs for chemotherapy; bismuth subgallate, subcitrate, tartarate and subsalicylate are used to treat syphilis, hypertension, infection as well as gastrointestinal disorders, and also used as cosmetics to control skin conditions [97CB669, 99CRV2601]. Since the 1970s, bismuth subsalicylate (BSS) and colloidal bismuth subcitrate (CBS) have been used worldwide for the treatment of diarrhea, dyspepsia and peptic ulcers [97CB669]. In spite of this however, their chemical nature has not been understood very well yet. This is probably due to their poor solubility in common organic solvents. In this section, recent progress on the study of bismuth carboxylates is presented, but the studies on their biological aspects are not included.

For the preparation of bismuth carboxylates, four methods are available: (i) a substitution reaction of Bi–X bonds with carboxylate anions, (ii) the cleavage of Bi–C bonds in Ar_3Bi with carboxylic acids, (iii) a ligand-exchange reaction of bismuth carboxylates with different carboxylic acids, (iv) the treatment of Bi_2O_3 with carboxylic acids. These four methods (i)–(iv) are denoted in Table 2.19: (i) as method A, (ii) as method B, (iii) and (iv) as method C, respectively.

Thus, $Bi(O_2CMe)_3$ is prepared by method A which involves the reaction between $BiBr_3$ and TlO_2CMe in acetic acid [79ZOB1185]. The reaction of

i) $BiX_3 + M(O_2CR) \xrightarrow{-MX} Bi(O_2CR)_3$ (X = Cl, Br) (M = Ag, Tl)

ii) $Ph_3Bi + RCO_2H \xrightarrow{-PhH} Bi(O_2CR)_3$

iii) $Bi(O_2CR)_3 + R'CO_2H \xrightarrow{-RCO_2H} Bi(O_2CR')_3$

$Bi(NO_3)_3 \cdot 5H_2O + RCO_2H \xrightarrow{-HNO_3} Bi(O_2CR)_3$

iv) $Bi_2O_3 + RCO_2H \xrightarrow{-H_2O} Bi(O_2CR)_3$

$BiCl_3$ with AgO_2CCF_3 gives $Bi(O_2CCF_3)_3$ in good yield as a very moisture-sensitive compound [75IC1722]. The formation of $Bi(O_2CC_7H_{15})_3$ from $PhBiCl_2$ and $NaO_2CC_7H_{15}$ has been reported, but its characterization is not sufficient [93AOMC137]. Method B has only limited uses; treatment of Ph_3Bi with boiling chloroacetic acid precipitates bismuth tris(chloroacetate) $Bi(O_2CCH_2Cl)_3$ in 90% yield [79ZOB1185]. By method C which involves the ligand exchange between $Bi(O_2CMe)_3$ and isobutyric acid, bismuth triisobutylate $Bi(O_2C^iPr)_3$ is obtained quantitatively; this method needs a large excess of carboxylic acid [79ZOB1185].

Bismuth nitrate can be converted into bismuth malate $Bi[O_2CCH_2CH(O)CO_2]$ and bismuth tartarate trihydrate $Bi[O_2CCH(OH)CH(OH)CO_2][O_2CCH(OH)CH(OH)CO_2H]$ simply by mixing with the corresponding carboxylic acids [93CB51]. The reaction of $Bi(NMe_2)_3$ or $Bi(NEt_2)_3$ with acetic anhydride gives $Bi(O_2CMe)_3$ in almost quantitative yield [75JINC(37)2011]. Method C which involves acidolysis is the best choice for practical purposes. $Bi(O_2CMe)_3$ is obtained quantitatively by treating Bi_2O_3 with boiling acetic acid [79ZOB1185]. In a similar way, bismuth triformate [69ACS1525, 79ZOB1185], tripivalate [93CC335], trihexanoate [79ZOB1185], tris(trifluoroacetate) [72JINC(34)3913], tris(2-pyridinecarboxylate) and tris(2-pyrazinecarboxylate) [91NJC927] are prepared. This type of reaction sometimes gives products of a more complicated nature: treatment of Bi_2O_3 with L-mandelic acid ($mandH_2$) in boiling water leads to an oxygen-bridged complex $[(mandH)_2Bi(\mu\text{-}O)Bi(mandH)_2]$ in 93% yield, whereas the same reaction with glycolic acid ($glcH_2$) gives a polymeric complex $[(\mu\text{-}\eta^2\text{-}glc)Bi(\mu_4\text{-}O)Bi(\mu\text{-}\eta^2\text{-}glc)]_\infty$ in 96% yield [92SRI(22)289].

In DMSO, pyridine-2,6-dicarboxylic acid (dipicolinic acid, $dipH_2$) reacts with Bi_2O_3 to form a dimeric complex $[Bi(dipH)(dip)(DMSO)\cdot H_2O]_2$, which is soluble in hot DMSO [92MGMC217].

The chemical nature of bismuth carboxylates can be changed dramatically by the modification of carboxylate ligands. Such simple bismuth carboxylates as triformate [69ACS1525] and tris(trifluoroacetate) [75IC1722] are moisture-sensitive and decompose easily under atmospheric conditions. Among the seven known bismuth pyridine- and pyrazine-carboxylates, only bismuth dipicolinate (2,6-pyridinedicalboxylate) is soluble in DMSO and DMF [91NJC927, 92MGMC217]. Bismuth tris(2,3-hydrogenpyrazinedicarboxylate) is soluble in DMSO, too [98NJC973]. A bismuth salt of 3,5-pyrazoledicarboxylic acid is insoluble in common organic solvents [98NJC973]. Bismuth mandelate is soluble in DMSO, but insoluble in water and acetone [92SRI(22)289]. Two of these soluble bismuth carboxylates possess the intra- and intermolecularly coordinating ligands and are not affected by moisture; they are usually prepared in aqueous media [92MGMC217, 92SRI(22)289].

(ii) Spectral properties: Due to the poor solubility of bismuth carboxylates in common organic solvents, information on their NMR spectra is poor. However, the ^1H-NMR spectrum of bismuth mandelate has been determined in DMSO-d_6 [92SRI(22)289]. ^1H-NMR spectra of water-soluble bismuth lactate [92CB43] and bismuth citrate [91IC4210, 93IC5322, 96JCS(D)2417] are also reported.

The IR spectra of bismuth carboxylates are more important; they generally show peaks at 1630–1654 (CO_2 asymmetric), 1333–1452 (CO_2 symmetric), and 396–470 (Bi–O) cm^{-1}. Representative data are summarized in Table 2.18. In the case of two μ-oxo compounds characteristic peaks due to ν(Bi–O–Bi) appear at 532 and 511 cm^{-1} [92SRI(22)289].

Mass spectra of bismuth carboxylates have not been studied in detail. The mass spectrum of $Bi(O_2CCF_3)_3$ contains characteristic peaks due to $Bi_2O_3^+$, $Bi_2O_2^+$, $BiO_2CCF_3^+$, BiO_2^+ and BiO^+ fragments [75IC1722].

(iii) Structures: Many X-ray crystallographic studies have been made on bismuth carboxylates. The first report has the elucidation of the crystal structure of $Bi(O_2CH)_3$ by Stålhandske [69ACS1525], where the bismuth centers are surrounded by nine oxygen atoms forming a deformed octahedron geometry. This means that the significant intermolecular interaction leads to the formation of chain polymer structure. The Bi–O bond lengths fall into

TABLE 2.18
IR data of some bismuth carboxylates (in cm^{-1})

Compound[a]	$\nu_{as}(CO_2H)$	$\nu_{as}(CO_2)$	$\nu_s(CO_2)$	ν(Bi–O)	Reference
$Bi(O_2CCF_3)_3$	–	1632	1444	–	[72JINC(34)3913]
$Bi(O_2CCF_3)_3$	–	1641	1452	–	[75IC1722]
$Bi[2-(O_2C)Py]_3$	–	1653	1337	–	[91NJC927]
$Bi[2-(O_2C)Pz]_3$	–	1654	1333	–	[91NJC927]
$Bi[2-(O_2C)-6-(CO_2H)Py]_2(OH)$	1701	1630	1358	419	[91NJC927]
$Bi[2-(O_2C)-3-(CO_2H)Py]_2(OH)$	1701	1630	1360	447	[91NJC927]
$Bi[2-(O_2C)-5-(CO_2H)Py]_2(OH)$	1699	1641	1362	440	[91NJC927]
$Bi[2-(O_2C)-3-(CO_2H)Py]_2(OH)$	–	1637	1366	396 470	[91NJC927]
$[(mandH)_2Bi(\mu-O)Bi(mandH)_2]$	–	1605	1323	–	[92SRI(22)289]
$[(\mu-\eta^2-glc)Bi(\mu_4-O)Bi(\mu-\eta^2-glc)]_\infty$	–	1563	1404	571	[92SRI(22)289]

[a] Abbreviations Py and Pz refer to pyridine and pyrazine, respectively.

three ranges, 2.34–2.40, 2.49–2.56 and 2.75–2.79 Å, showing the existence of three different types of formate ligand [69ACS1525]. The crystallographic structure of $Bi(O_2CMe)_3$ revealed that the bismuth center is surrounded by nine oxygen atoms to form an irregular nine-vertex polyhedron [91KK909]. The Bi–O bond lengths are classified into three categories: short bonds of 2.25(2), 2.34(2) and 2.38(2) Å, longer bonds of 2.52(2), 2.55(2) and 2.62(2) Å, and intermolecular Bi–O contacts of 2.81(2), 2.83(2) and 2.94(1) Å [91KK909]. In the case of $Bi(O_2C^tBu)_3$, the isolated tetrameric units are found in the crystal lattice, where every bismuth center is surrounded by nine oxygen atoms from three different types of pivalate group; the Bi–O distances are 2.24 and 2.50 Å, 2.87 and 3.16 Å, and 2.50 and 3.27 Å, respectively [93CC335]. The crystal structure of bismuth lactate, $Bi[O_2C\text{-}CH(OH)Me]_3$, shows that the intermolecular interaction forms the three-dimensional network of bismuth carboxylates. In the asymmetric unit, there are three different types of lactate ligand, where nine oxygen atoms surround the bismuth center [92CB43].

Among the known bismuth compounds with multidentate carboxylic acids, the crystal structure of $[Bi(dipicH)(dipic)(DMSO)\cdot H_2O]_2$ has been elucidated. The asymmetric unit contains two bismuth centers of pentagonal bipyramid geometry. These bismuth centers are surrounded by four oxygen atoms and two nitrogen atoms of the dipicolinato groups, and one oxygen atom of a DMSO molecule [92MGMC217]. Bismuth citrates have been subjected to many studies by X-ray crystallography. The ratio of bismuth atom to citrate anions, the coordination numbers of the bismuth centers, and the degree of polymerization of the asymmetric units change considerably depending on the counter cations of the citrate anions [97CB669]. Reported crystallographic studies include compounds $K_{4.75}(NH_4)_{0.25}[Bi_2(cit)_2(Hcit)]\cdot 13H_2O$ [91IC4210], $K\{Bi(cit)\}\cdot 3H_2O$ [91IC2579], $Na_2[Bi_2(cit)_2]\cdot 7H_2O$ [96JCS(D)2417], $K(NH_4)[Bi_2(cit)_2]\cdot 4H_2O$ [93IC5322], $(NH_4)_6[Bi_6O_4(OH)_4(Hcit)_4]\cdot 2H_2O$ [92CL1967] and $(NH_4)_{12}[Bi_{12}O_8(cit)_8]\cdot 10H_2O$ [95IC2447]. The bismuth centers in bismuth L-(−)-malate and bismuth L-(+)-tartarate are surrounded by nine oxygen atoms [93CB51]. Bismuth complexes of triethylenetetraaminehexaacetic acid and *trans*-cyclohexane-1,2-diaminetetraacetic acid have also been studied by X-ray crystallography [96JCS(D)2023].

(iv) Reactions: Several bismuth carboxylates find industrial applications as the catalysts for vapor phase conversion of acetylene to vinyl acetate, high temperature preparation of phthalic esters, high-pressure air oxidation of aliphatic hydrocarbons, and oxo processes.

Prolonged heating of $Bi(O_2CMe)_3$ with amines or alcohols around 150°C affords the respective acetylated derivatives in moderate yields [73JOC764]. $Bi(O_2CMe)_3$ can be used successfully for the Prévost type reaction of olefins [89CC407]. The reaction is highly stereospecific; no *trans* isomer is detectable under wet conditions, while no *cis* isomer is formed under dry conditions.

Bismuth mandelate has been found to be an effective catalyst for the oxidative C–C bond cleavage of epoxides to carboxylic acids in dry DMSO medium [93TL2601]. The reaction of several styrene oxides in a DMSO-O_2 system has been studied kinetically to demonstrate that an electron-donating group on the phenyl ring accelerates the reaction [94JOM(482)119]. Bismuth mandelate works also as a catalyst for the oxidative C–C bond cleavage by molecular oxygen of a series of α-keto alcohols to form the corresponding carboxylic acids [95T4991]. DMSO is the choice solvent for this purpose. Under a nitrogen atmosphere, little or none of the expected acid is obtained (Section 5.2.3).

(e) Bismuth carbamates

There are three known examples of bismuth carbamates, two of which, $Bi(O_2CNMe_2)_3$ and $Bi(O_2CNEt_2)_3$, are prepared by the reaction of the corresponding $Bi(NR_2)_3$ and CO_2 [75JINC(37)2011]. Dipropylcarbamate is prepared similarly, though it does not give satisfactory elemental analysis.

The dimethylcarbamate shows IR peaks at 1400 and 1350 cm^{-1} due to $\nu(CO_2)$, while the $Bi(O_2CNEt_2)_3$ shows them at 1510 and 1430 cm^{-1}. The ^1H-NMR data of these compounds are also available.

(f) Bismuth sulfinates

Only limited information on the chemistry of bismuth sulfinates is available. Two articles were published by Deacon et al. for this type of compound in 1971–1972. The sulfinates are accessible by three methods: (i) reaction of Ph_3Bi and mercury arenesulfinate $Hg(O_2SAr)_2$ (Ar=Ph, 4-MeC_6H_4) in $CHCl_3$ or MeOH at room temperature, (ii) a substitution reaction between $BiCl_3$ and silver arenesulfinate AgO_2SPh, and (iii) treatment of $Bi(O_2CMe)_3$ with a slight excess of arenesulfinic acid in glacial acetic acid [71JOM(26)C10, 72AJC2107]. These methods are referred to as methods B, A and C, respectively, in Table 2.19. The most common is the ligand exchange reaction using arenesulfinic acid (method C). By this method, $Bi(O_2SPh)_3$ **1**, $Bi(O_2SC_6H_4$-4-$Me)_3$ **2**, $Bi(O_2SC_6H_4$-4-$Cl)_3$ **3**, $Bi[O_2SC_6H_4$-4-(NHCOMe)$]_3$ **4** and $Bi(O_2SC_6H_2$-2,4,6-$^iPr_3)_3$ **5** are prepared in 64–86% yields. The mercury salt method has disadvantages in that it uses poisonous mercuric compounds and it is difficult to separate the product from the accompanying arylmercuric sulfinates.

$$\text{i)}\quad Ph_3Bi + 3\,Hg(O_2SAr)_2 \xrightarrow[\text{MeOH}]{} Bi(O_2Ar)_3 + 3\,Ph_2Hg$$

$$\text{ii)}\quad BiCl_3 + 3\,AgO_2SPh \longrightarrow Bi(O_2SPh)_3 + 3\,AgCl$$

$$\text{iii)}\quad Bi(O_2CMe)_3 + 3\,ArSO_2H \xrightarrow[\text{MeCO}_2\text{H}]{} Bi(O_2SAr)_3 + 3\,MeCO_2H$$

The IR spectra of $Bi(O_2SAr)_3$ show strong bands, due to $\nu_{as}(SO_2)$ and $\nu_s(SO_2)$, around 943–880 cm^{-1}, suggesting that the sulfinate ligands bind to the bismuth center as *O*-sulfinates and, thus, ruling out the possibility of *S*-sulfinates or ionic structure. Low solubility of $Bi(O_2SAr)_3$ supports their polymeric nature in the solid state [72AJC2107].

When heated at around 180–210°C in vacuum, $Bi(O_2SAr)_3$ decomposes to give the corresponding Ar_3Bi, the reported yields of which are Ph (79%), 4-MeC_6H_4 (57%) and 4-ClC_6H_4 (44%). By thermal decomposition of sulfinate **3**, di(4-chlorophenyl)sulfone and *S*-(4-chlorophenyl)-4-chlorobenzenethiosulfonate are obtained as side products. Thermal decomposition of sulfinate **4** gives acetanilide, whereas similar treatment of sulfinate **5** results in extensive decomposition to unidentified products. Synthetically, a pyrolysis reaction is not considered to be useful for obtaining Ar_3Bi.

(g) Bismuth sulfonates

Rather surprisingly, the reported chemistry of bismuth sulfonate is limited to bismuth trifluoromethanesulfonate (triflate), which has been prepared in almost quantitative yield by treating $Bi(O_2CCF_3)_3$ with trifluoromethanesulfonic acid CF_3SO_3H in CF_3CO_2H [83IJC(A)814]. However, the use of excess CF_3SO_3H is required to complete the reaction. $Bi(O_3SCF_3)_3$ is a white hygroscopic solid and stable up to 100°C under atmospheric conditions. The triflate decomposes at 100–400°C to form $Bi_2O_2SO_4$, which, on further heating at 400–600°C, gives Bi_2O_3. The triflate is insoluble in common organic solvents except for DMSO and DMF. $Bi(O_3SCF_3)_3$ can also be prepared by the reaction of triflic anhydride and Bi_2O_3, however the reaction requires prolonged heating [90JFC(48)421]. Very recently, a more efficient method for the preparation of $Bi(O_3SCF_3)_3$ has been reported: the direct reaction of CF_3SO_3H on triphenylbismuthine [99TL285]. In this method, there is no need to use an excess triflic reagent.

The IR spectrum of the triflate shows peaks at 1380 ν_{as}(SO), 1200 ν_s(SO) and 840 ν(SO) cm^{-1} due to the S–O stretching of the terminal triflate ligand. In addition, absorptions around 1305, 1130 and 1000 cm^{-1} are assigned to the ν_{as}(SO), ν_s(SO) and ν(SO) vibration modes of the bridging or chelating triflate groups. These findings suggest that $Bi(O_3SCF_3)_3$ forms the oligomeric or polymeric chain structure in the solid state. The ^{19}F-NMR spectrum of $Bi(O_3SCF_3)_3$ shows a peak at δ 78.3; the terminal and bridging triflate groups are indistinguishable on NMR time scale [83IJC(A)814].

Bismuth triflate forms complexes with pyridine, bipyridyl and $Ph_3P{=}O$. The IR spectra of these compounds also exhibit absorption bands due to the terminal and bridging triflate groups, which is in agreement with the presence of two types of triflate ligand [83IJC(A)814].

The X-ray structure of the hydrated triflate shows that the bismuth center is surrounded by nine molecules of water, $[(CF_3SO_3)_3Bi(H_2O)_9]$ [95AG(E)2416]. This unique geometry of the bismuth center may be described as a tricapped trigonal prism, the bismuth atom possessing a C_{3h} symmetry. The Bi–O(H_2O) bond distances therein are 2.448(2) and 2.577(2) Å.

(h) Bismuth tetraphenylimidodiphosphinate

The title compound $Bi[OPPh_2NPPh_2O]_3$ has been prepared as an air-stable solid in 50% yield by the reaction of $BiCl_3$ and potassium tetraphenylimidodiphosphinate $K[OPPh_2NPPh_2O]$.

Its ^{31}P-NMR spectrum shows a peak at δ -17.3, which is close to the δ -17.0 value of $Pb[OPPh_2NPPh_2O]_2$ [80INCL189]. The FAB-mass spectrum of this compound contains characteristic peaks due to M^+, $[M–Ph]^+$ and $[M–OPPh_2NPPh_2O]^+$ fragments [96IC3948]. X-Ray crystallography has shown that six oxygen atoms surround the bismuth center to form a C_{3v} geometry. The ligands coordinate to the bismuth atom asymmetrically through two types of Bi–O bond; the short bonds are in the range 2.256(7)–2.281(7) Å, while the longer ones fall in the range 2.337(8)–2.404(8) Å. The bond lengths and bond angles are comparable with those of other metal complexes containing the same $[OPPh_2NPPh_2O]$ ligand [96IC3948].

2.2.4.4. Bismuth(III) compounds with three Bi–S bonds (Tables 2.22 and 2.23)

(a) Bismuth thiolates

Bismuth thiolates represent one of the most studied inorganic bismuth compounds, though publications on this type of compound are few. In recent years, increasing attention has been given to the potential usefulness of bismuth thiolates as fungicides, antitumor agents, vulcanization catalysts and as well as precursors to Bi_2S_3 for solution or vapor phase deposition purposes. Many spectroscopic data are available, but their reactivity has not been studied in detail. In this section, bismuth thiolates are divided into two categories: aliphatic and aromatic thiolates.

(i) Bismuth alkanethiolates Although the study of aliphatic bismuth thiolates began as early as the last century [1876BSF(25)183, 1877JPR(15)209], the first spectral identification of these compounds was made by Wieber and Baudis only two decades ago [76ZAAC(423)47]. Three aliphatic thiolates, $Bi(SEt)_3$ **1**, $Bi(SC_2H_4OH)_3$ **2**, and $Bi(SCH_2Ph)_3$ **3**, are obtained by treating $Bi(OEt)_3$ with the corresponding thiols in EtOH; this produces yellow to reddish-orange colored crystals, soluble in common organic solvents. These thiolates decompose when heated at high temperatures or exposed to daylight for a prolonged time [76ZAAC(423)47]. A mixture of BiF_3 and Me_3SiS^tBu,

TABLE 2.19
Bismuth compounds with three organyl oxygen ligands

Compound R	Synthetic method[a]	Yield (%)	M.p. (°C) or b.p. (°C/mmHg)	Physical data	Reference
Bi(OR)₃					
Me	A	93	190/0.08 (sublim.)	–	[661JC537]
Et	A	90	123–127/1.0 (sublim.)	–	[661JC537]
	A	86	105 (decomp.)	^{1}H-NMR	[76ZAAC(423)47]
	C	Quant.	–	IR	[91POL437]
^{i}Pr	A	80	139–141/0.5 (sublim.)	–	[66IJC537]
	C	80	–	–	[901C358]
^{t}Bu	A	80	–	^{1}H-, ^{13}C-NMR, IR	[89CC1628]
	A	ca. 50	80–100/10^{-2} (sublim.)	^{1}H-NMR, IR, MS	[90CC301]
	A	80	154	^{1}H-NMR, IR, MS	[91POL437]
$EtMe_2C$	C	80	–	^{1}H-NMR	[901C358]
$MeOC_2H_4$	C	89	–	^{1}H-NMR, X-ray	[901C358]
	A	ca. 50	–	^{1}H-, ^{13}C-NMR, IR, X-ray	[90CC301]
	A/C	51/60	–	^{1}H-, ^{13}C-NMR, IR, X-ray	[91POL437]
$Me_2NC_2H_4$	C	79	–	^{1}H-NMR	[90IC358]
$Me_2NCH_2CH(Me)$	C	50	–	^{1}H-NMR	[90IC358]
2,6-$Me_2C_6H_3$	A	80	–	^{1}H-, ^{13}C-NMR, IR, X-ray	[89CC1628]
Me_3Si	A	35	(decomp. at r.t.)	^{1}H-NMR, IR	[68ZAAC(363)84]
Ph_3Si	C	62	–	^{1}H-, ^{29}Si-NMR, IR, X-ray	[91POL437]
Bridged bismuth alkoxides[b]					
$[Bi_4(\mu\text{-}\eta^1\text{-}OCH_2CH_2OH)_2(\mu_4\text{-}\eta^2\text{-}OCH_2CH_2O)(\mu_2\text{-}\eta^2\text{-}OCH_2CH_2O)_4]_\infty$	C	98	–	IR	[92SRI(22)289]
$[Bi(OCH(CF_3)_2)_2(\mu\text{-}OCH(CF_3)_2)(THF)]_2$	A	39	–	^{1}H-, ^{19}F-NMR, IR, X-ray	[93IC5136]
$[Bi(OC_6F_5)_2(\mu\text{-}OC_6F_5)(PhMe)]_2\cdot 2PhMe$	B	51	–	^{19}F-NMR, IR, X-ray	[93IC5136]

TABLE 2.19 *(continued)*

Compound R	Synthetic method[a]	Yield (%)	M.p. (°C) or b.p. (°C/mmHg)	Physical data	Reference
$[Bi(OC_6F_5)_2(\mu\text{-}OC_6F_5)(PhMe)]_2$	B	–	–	X-ray	[93IC5136]
$[Bi(OC_6F_5)_2(\mu\text{-}OC_6F_5)(THF)_2]_2$	B	–	–	IR, X-ray	[93IC5136]
$[Bi(OC_6F_5)_2(\mu\text{-}OC_6F_5)(THF)_2]_2{\cdot}2C_6H_{14}$	B	–	–	X-ray	[93IC5136]
$[Bi_8(tsal)_{12}(DMF)_6](DMF)_6$[b]	A	28	–	X-ray	[94CL2123]
$Bi_6(\mu_3\text{-}O_7)(\mu_3\text{-}OC_6F_5)[Bi(OC_6F_5)_4]_3{\cdot}2THF_2$	B	–	–	X-ray	[92CC1638]
$Bi_6(\mu_3\text{-}O_7)(\mu_3\text{-}OC_6F_5)[Bi(OC_6F_5)_4]_3{\cdot}2PhMe$	B	–	–	X-ray	[92CC1638]
$\{Bi[OCH(CF_3)_2]_3(THF)\}_2$	A	–	–	^1H-, ^{19}F-NMR, IR, X-ray	[92AG(E)451]
$[Bi(OC_6F_5)_3(PhMe)]_2{\cdot}2PhMe$	B	–	–	^{19}F-NMR, IR, X-ray	[92AG(E)451]
$Bi(OSiPh_3)_3(THF)_2$	–	–	–	^1H-, ^{29}Si-NMR, IR, X-ray	[91POL437]
Bismuth enolates					
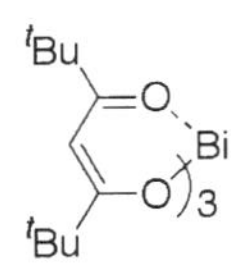	A	41	78	^1H-NMR, IR, MS	[91POL437]
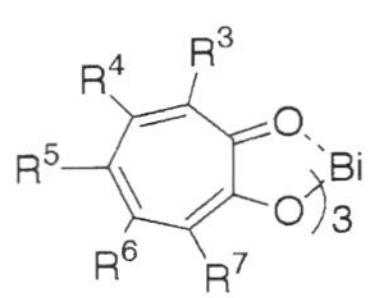					
$R^3 = R^4 = R^5 = R^6 = R^7 = H$	A	–	323–326 (decomp.) $[CHCl_3\text{–}MeCN]$	–	[66JA4856]
	A	90	>330	^1H-, ^{13}C-NMR, IR, MS, Bioactive data	[95CB335]

$R^4 = Me, R^3 = R^5 = R^6 = R^7 = H$	A	41	273	^{1}H-NMR, IR, Bioactive data	[95CB335]
$R^5 = Me, R^3 = R^4 = R^6 = R^7 = H$	A	95	302 (decomp.)	^{1}H-, ^{13}C-NMR, IR	[95CB335]
$R^3 = R^5 = R^7 = Me, R^4 = R^6 = H$	A	72	275	^{1}H-, ^{13}C-NMR, IR	[95CB335]
$R^5 = O_2N, R^3 = R^4 = R^6 = R^7 = H$	A	79	330 (decomp.)	^{1}H-NMR, IR, Bioactive data	[95CB335]
$R^5 = CHO, R^3 = R^4 = R^6 = R^7 = H$	A	91	290 (decomp.)	^{1}H-, ^{13}C-NMR, IR, MS	[95CB335]
$R^3 = Br, R^4 = R^5 = R^6 = R^7 = H$	A	74	215 (decomp.)	^{1}H-, ^{13}C-NMR, IR	[95CB335]
$R^3 = R^7 = Br, R^4 = R^5 = R^6 = H$	A	89	233	^{1}H-NMR, IR	[95CB335]
$R^3 = R^5 = R^7 = Br, R^4 = R^6 = H$	A	96	313	^{1}H-NMR, IR	[95CB335]
$R^3 = R^5 = R^7 = CH_2OH, R^4 = R^6 = H$	A	63	275 (decomp.)	^{1}H-NMR, IR	[95CB335]
$R^3 = R^5 = R^7 = CH_2N(C_2H_4)_2O, R^4 = R^6 = H$	A	98	183	^{1}H-, ^{13}C-NMR, IR	[95CB335]
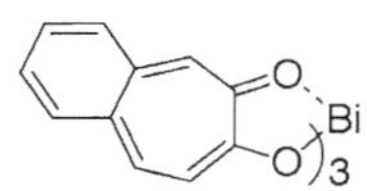	A	89	265–266	^{1}H-NMR, IR, Bioactive data	[95CB335]
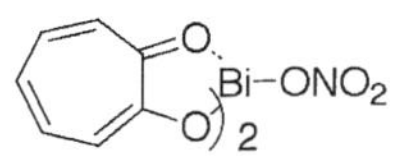	A	55	271–275	^{1}H-, ^{13}C-NMR, IR, MS, X-ray	[95CB335]
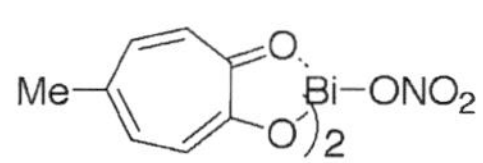	A	81	276–279	^{1}H-, ^{13}C-NMR, IR	[95CB335]
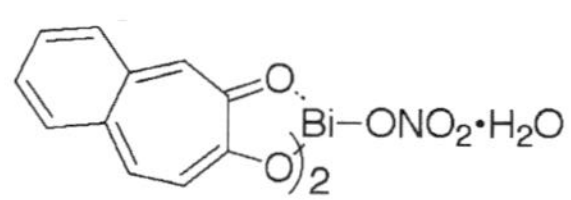	A	57	180	^{1}H-, ^{13}C-NMR, IR, MS, X-ray	[95CB335]
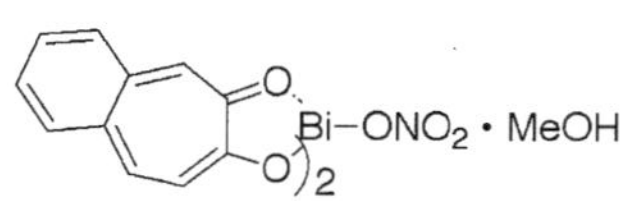	A	–	–	–	[95CB335]

TABLE 2.19 *(continued)*

Compound R	Synthetic method[a]	Yield (%)	M.p. (°C) or b.p. (°C/mmHg)	Physical data	Reference
					[85JA3397]
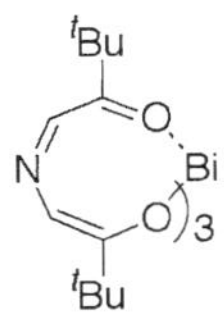	A	>90	202 (decomp.) [C_5H_{12}]	^{1}H-, ^{13}C-NMR, X-ray	[87JA627]
		30	202 (decomp.) [^{i}PrOH]	^{1}H-, ^{13}C-, ^{15}N-, ^{17}O-NMR, UV	
Bi(O$_2$CR)$_3$					
H	C	Quant.	–	–	[79ZOB1185]
	C	–	–	X-ray	[69ACS1525]
Me	A/C	–/Quant	>250	–	[79ZOB1185]
	C	96	–	–	[75JINC(37)2011]
	C	–	–	X-ray	[91KK909]
^{i}Pr	C	Quant.	–	–	[79ZOB1185]
^{i}Bu	C		164	X-ray	[93CC335]
C_5H_{11}	C	Quant.	–	–	[79ZOB1185]
C_7H_{15}	A	–	–	–	[93AOMC137]
$CH_3CH(OH)$	C	–	188 (decomp.)	^{1}H-NMR, X-ray	[92CB43]
CF_3	C	Quant.	–	IR	[72JINC(34)3913]
	A	–	186–189	IR, MS	[751C1722]
CH_2Cl	B/C	90/Quant.	195 (decomp.)	–	[79ZOB1185]
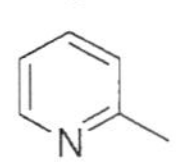	C	98	–	IR	[91NJC927]

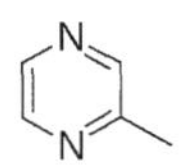	C	88	–	IR	[91NJC927]
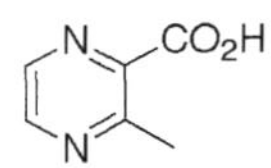	A	80	–	^{1}H-, ^{13}C-NMR, IR	[98NJC973]
Unsymmetrical bismuth carboxylates					
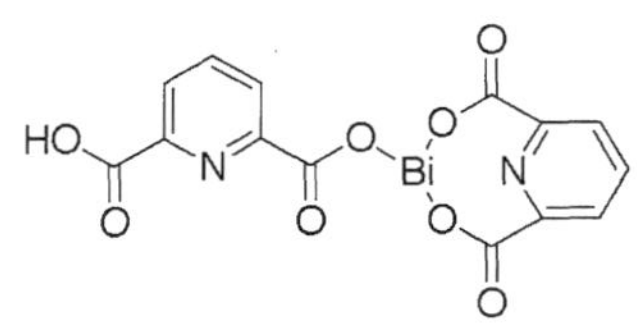	C	Quant.	–	IR, X-ray	[92MGMC217]
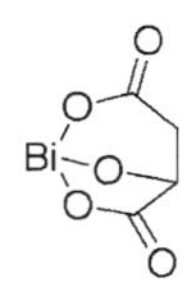	A	–	–	X-ray	[93CB51]
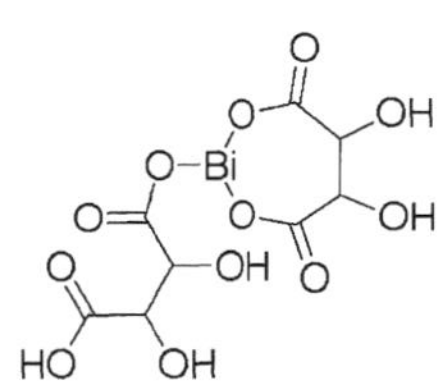	A	–	–	X-ray	[93CB51]

TABLE 2.19 *(continued)*

Compound R	Synthetic method[a]	Yield (%)	M.p. (°C) or b.p. (°C/mmHg)	Physical data	Reference
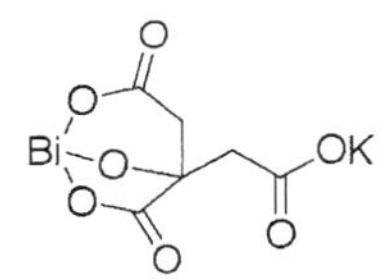	–	–	–	X-ray	[91IC2579]
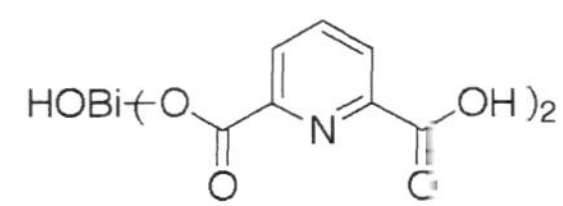	C	98	–	^{1}H NMR, IR	[91NJC927]
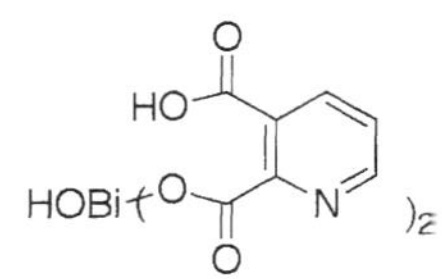	C	96	–	IR	[91NJC927]
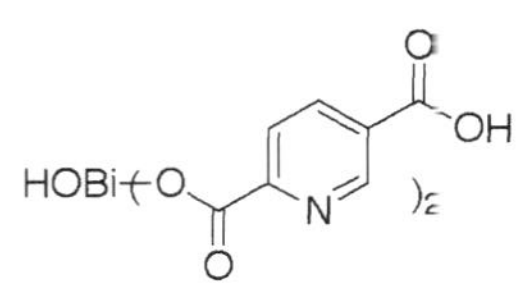	C	97	–	IR	[91NJC927]
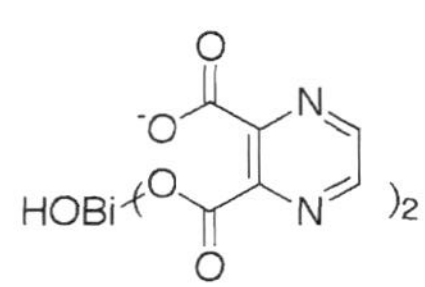	C	96	–	IR	[91NJC927]
	A	83	–	IR	[98NJC973]

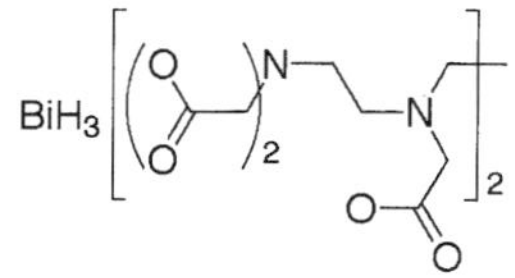	C	–	–	IR, X-ray	[96JCS(D)2023]
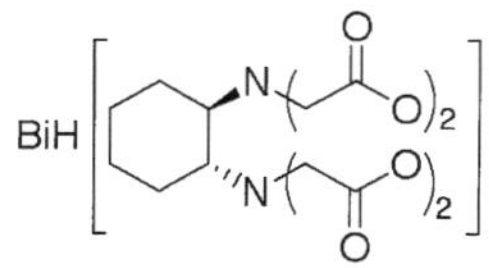	C	–	–	IR, X-ray	[96JCS(D)2023]
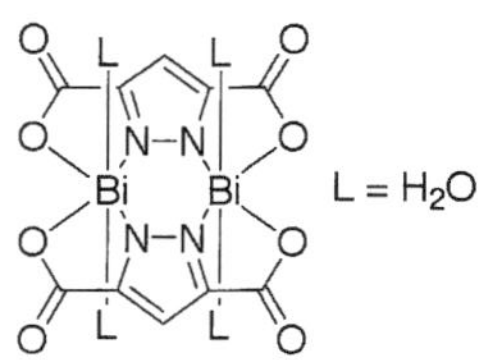	A	89	–	IR	[98NJC973]
Polymeric bismuth carboxylates[b]					
[(μ-η^2-glc)Bi(μ_4-O)Bi(μ-η^2-glc)]$_\infty$	C	96	–	IR	[92SRI(22)289]
[(mandH)$_2$Bi(μ-O)Bi(mandH)$_2$]	C	93	–	^{1}H-NMR, IR	[92SRI(22)289]
Bismuth carbamates					
Bi(O_2CNMe$_2$)$_3$	D	87	217–219 (decomp.) [$CHCl_3$–petroleum ether]	^{1}H-NMR, IR	[75JINC(37)2011
Bi(O_2CNEt$_2$)$_3$	D	62	125–128 (decomp.)	^{1}H-NMR, IR	[75JINC(37)2011

TABLE 2.19 *(continued)*

Compound R	Synthetic method[a]	Yield (%)	M.p. (°C) or b.p. (°C/mmHg)	Physical data	Reference
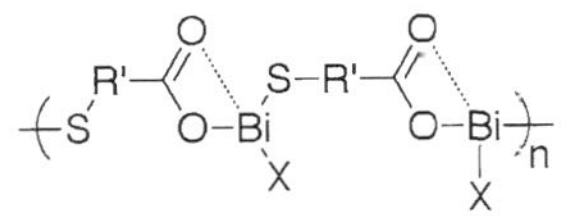					
X = SCH_2CO_2H	B/C	80	118 (decomp.)	^1H-NMR, IR	[81ZN(B)70]
X = $SCH_2CH_2CO_2H$	B/C	65	155 (decomp.)	^1H-NMR, IR	[81ZN(B)70]
X = $O_2CCH(Me)SH$	B/C	80	126 (decomp.)	^1H-NMR, IR	[81ZN(B)70]
X = $O_2CC_6H_4$-2-SH	B/C	77	274 (decomp.)	^1H-NMR, IR	[81ZN(B)70]
Bismuth nitrato glycolates[b]					
$[Bi(NO_3)_2(EO3^-)]_2$	A	–	>360	X-ray	[92JA2960]
$[Bi(NO_3)_2(EO4^-)]_2 \cdot 2MeOH$	A	–	129.5–140	X-ray	[92JA2960]
$[Bi(NO_3)_2(EO5^-)]$	A	–	189–200 (decomp.)	X-ray	[92JA2960]
$[Bi(NO_3)_2(EO5^{2-})Bi(NO_3)_3] \cdot 2H_2O$					
$[Bi(NO_3)_2(EO6^-)]_2$	A	–	113–117.5	X-ray	[92JA2960]
$[Bi(NO_3)_3(12\text{-crown-}4)]$	–	–	235–260 (decomp.)	X-ray	[92JA2960]
$[Bi(NO_3)_3(H_2O)_3] \cdot 18\text{-crown-}6$	–	–	105 (decomp.)	X-ray	[92JA2960]
$Bi(O_2SR)_3$					
Ph	B/C	–	–	IR	[71JOM(26)C10]
	A/B/C	–/66/82	–	IR	[72AJC2107]
4-MeC_6H_4	B/C	–	–	IR	[71JOM(26)C10]
	B/C	79/64	–	IR	[72AJC2107]
4-ClC_6H_4	C	80	–	IR	[72AJC2107]
4-$(MeCONH)C_6H_4$	C	67	–	IR	[72AJC2107]
2,4,6-$^iPr_3C_6H_2$	C	86	–	IR	[72AJC2107]

Bismuth sulfonate					
$Bi(O_3SCF_3)_3$	C	Quant.	>100	^{19}F-NMR, IR	[83IJC(A)814]
	C	98	–	IR	[90JFC(48)421]
	C	89	–	^{13}C-, ^{19}F-NMR, IR	[99TL285]
$[(CF_3SO_3)_3Bi(H_2O)_9]$	C	81.0	–	IR, X-ray	[95AG(E)2416]
Bismuth imidodiphosphate					
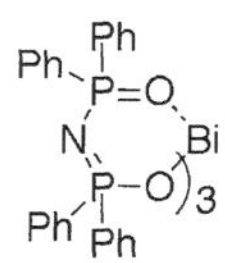	A	50	198–201 [CH_2Cl_2–C_7H_{16}] 271–273 [PhH–C_6H_{14}]	^{31}P-NMR	[80INCL189]
	A	–	–	MS, X-ray	[96IC3948]

[a] For notation, see pp. 155 or 159 in Section 2.2.4.1.

[b] Abbreviations: tsalH_2, thiosalicylic acid 2-$HSC_6H_4CO_2H$; glcH, glycolic acid $HOCH_2CO_2H$; mandH, mandelic acid $C_6H_5CH(OH)CO_2H$; EO*n* (n = 3–6), ethylene glycols $HO(CH_2CH_2O)_{n-1}CH_2CH_2OH$.

when shaken in a sealed tube for 3 days at 25°C, gives $Bi(S^tBu)_3$ **4** in quantitative yield [81JINC(43)1469]. This thiolate can be stored in an evacuated tube, however, it decomposes in open air or on dissolution in CH_2Cl_2. Recently, bismuth thiolates **5–9** are prepared in good to excellent yields by treating cyclic *Bi*-chloro-1,3-dithia-2-bismacycloalkanes with dithiols in the presence of aqueous $NaNO_3$ [96JA3225] (Table 4.2). These compounds **5–9** are poorly soluble in common organic solvents.

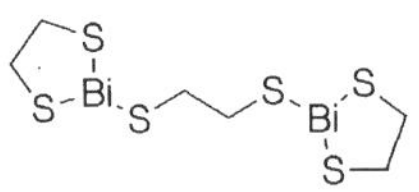

5

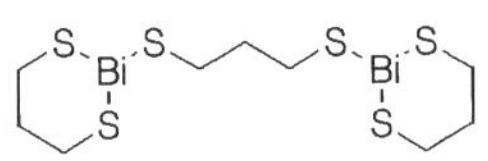

6

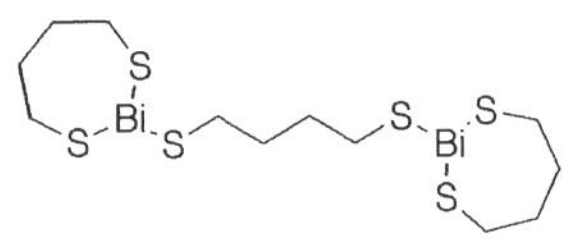

7

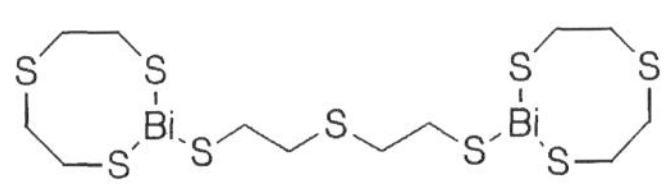

8

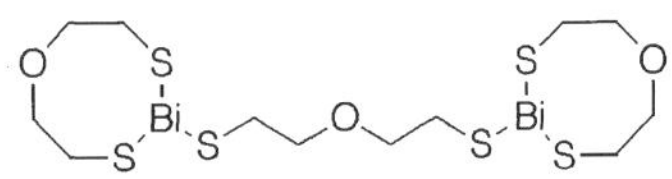

9

The formation of bismuth glutathionate $Bi(GS)_3$ **10** from the reaction of ranitidine bismuth citrate, $[Bi(edta)]^-$, and glutathione has been confirmed by ^1H- and ^{13}C-NMR. This complex is stable over a range of pH 2–10. The reaction of bismuth citrate with glutathione in red blood cells has been examined by means of ^1H-NMR [96CEJ701]. Bismuth dithizonate (3-mercapto-1,5-diphenylformazan complex) **11** is obtained by shaking a $CHCl_3$ solution of dithizone and aqueous Bi^{3+}, prepared from metallic bismuth, concentrated H_2SO_4, NH_3 (to adjust pH 5) and KCN [82AX(B)2140].

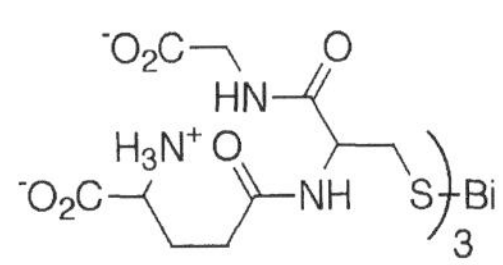

10

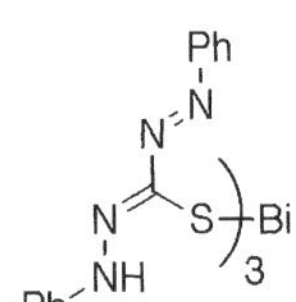

11

Mass spectral data are available for compounds **1** and **5–9**; only an M^+ signal is reported for compound **1** [76ZAAC(423)47], while little fragmented ion peaks are observed in their EI mass spectra for compounds **5–9**. APCI mass spectral data of compounds **5**, **6**, **8** and **9** in MeCN with 1% HCl exhibit a similar trend; monocyclic bismuthenium cations provide predominant signals due to the cleavage of one Bi–S bond [96JA3225].

IR and Raman spectra of compound **4** are reported as follows; IR: 578 and 443 ν(CS), 420 and 377 ν(BiS) and 348 ν(BiSC) cm^{-1}, and Raman: 580 ν(CS), 424 and 383 ν(BiS), 351 and 227 ν(BiSC), and 120 ν(SBiS) cm^{-1} [81JINC(43)1469]. In IR spectra of compounds **5–9**, strong absorption bands are observed at 660–677, 422–433, 318–335 and 240–294 cm^{-1}, while in their Raman spectra strong absorptions appear at 633–682, 294–313, 230–278, 159–169, and 68–92 cm^{-1} [96JA3225].

X-Ray crystallographic structures of compounds **5** and **8** have been characterized. In the asymmetric unit, two identical molecules are observed in both compounds. The endocyclic Bi–S bond lengths in compound **5** are 2.552(6), 2.553(7), 2.565(7) and 2.591(7) Å, and the exocyclic Bi–S bond lengths are 2.633(7) and 2.635(7) Å. In addition, there are intra- and intermolecular Bi–S interactions in the range 3.134(7)–3.902(7) Å, forming a six and a seven coordinated bismuth center, respectively [96JA3225]. The Bi–S bond lengths in compound **8** are 2.559(4), 2.576(4), 2.597(5) and 2.599(4) Å for endocyclic linkages, and 2.566(5) and 2.615(4) Å for exocyclic ones. Distances of the transannular Bi–S interactions are 3.071(5) and 3.197(5) Å. Other intra- and intermolecular Bi–S contacts found are 3.192(4)–3.346(5) Å [96JA3225].

The X-ray structure analysis of compound **11** has shown that the bismuth center forms a distorted octahedral geometry. The Bi–S bond lengths are 2.606(3), 2.607(3) and 2.613(3) Å, and the Bi–N intramolecular distances are 2.678(9), 2.706(8) and 2.746(10) Å [82AX(B)2140].

(ii) Bismuth arenethiolates Many bismuth thiolates have been prepared from aromatic thiols and are characterized in detail. Known compounds include compounds **12** [17CB21, 51JA2880, 79JINC(41)1390], **13** [95JCS(D)2129], **14** [95JCS(D)2129], **15** [95JCS(D)2129], **16** [95JCS(D)1649], **17** [95JCS(D)1649], **18** [93IC2972, 95JCS(D)1649], **19** [65JINC(27)1431, 83ICA(73)141, 86POL1513], **20** [86POL1513], **21** [51JA2880], **22** [88JFC(38)367] and **23** [68CJC2699, 95JCS(D)2129, 95POL311].

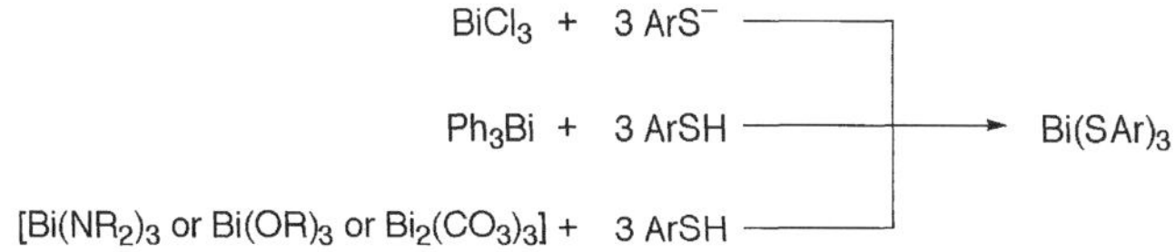

They can be prepared by the following three procedures; (i) a substitution reaction between $BiCl_3$ and thiolate anions [86POL1513, 93IC2972,

95JCS(D)2129], (ii) cleavage of three Bi–C bonds in Ph_3Bi by thiols [51JA2880, 95JCS(D)2129], and (iii) treatment of $Bi(NR_2)_3$, $Bi(OR)_3$ or $Bi_2(CO_3)_3$ with thiols [76ZAAC(423)47, 83ICA(73)141, 95JCS(D)1649]. Though all these methods give good yield of the desired bismuth thiolates, the first one is the most simple and hence the most convenient. There is an interesting example of the reaction between $BiCl_3$ and three equivalents of $NaSC_6F_5$ in THF, where an unexpected complex $[Na_2(THF)_4][Bi(SC_6F_5)_5]$ is formed instead of the expected thiolate **23** [95POL311].

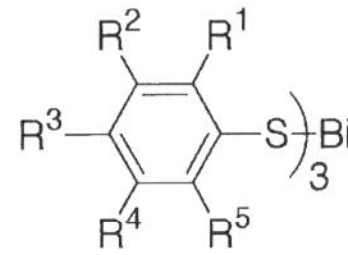

12; $R^1 = R^2 = R^3 = R^4 = R^5 = H$
13; $R^3 = Me$, $R^1 = R^2 = R^4 = R^5 = H$
14; $R^1 = R^5 = Me$, $R^2 = R^3 = R^4 = H$
15; $R^2 = R^4 = Me$, $R^1 = R^3 = R^5 = H$
16; $R^1 = R^3 = R^5 = Me$, $R^2 = R^4 = H$
17; $R^1 = R^3 = R^5 = {}^iPr$, $R^2 = R^4 = H$
18; $R^1 = R^3 = R^5 = {}^tBu$, $R^2 = R^4 = H$
19; $R^1 = NH_2$, $R^2 = R^3 = R^4 = R^5 = H$
20; $R^1 = NH_3Cl$, $R^2 = R^3 = R^4 = R^5 = H$
21; $R^1 = CO_2Me$, $R^2 = R^3 = R^4 = R^5 = H$
22; $R^3 = F$, $R^1 = R^2 = R^4 = R^5 = H$
23; $R^1 = R^2 = R^3 = R^4 = R^5 = F$

In most cases, the reaction and subsequent manipulation for product isolation are carried out under an inert, oxygen-free atmosphere to avoid oxidative degradation [51JA2880, 93IC2972, 95JCS(D)2129]. Sterically crowded bismuth thiolates **16–18** are reported to be air-stable in the solid state. In solution, however, even these compounds are readily oxidized to form diaryl disulfides [95JCS(D)1649]. Compound **19** is air- and moisture-insensitive over a period of 1 week in the solid state [83ICA(73)141]. The solubility of these thiolates depends greatly on the structure of the thiolate ligands attached; thiolate **12** is soluble in Et_2O, $CHCl_3$ and C_6H_6 [79JINC(41)1390], thiolates **16–18** are soluble in common organic solvents [95JCS(D)1649]; thiolate **23** is soluble in MeOH, acetone, Et_2O, $CHCl_3$ and C_6H_6 [68CJC2699]. Thiolates **19** and **20** are soluble in pyridine, DMSO and DMF, but insoluble in $CHCl_3$, C_6H_6, CCl_4, EtOH and CS_2 [86POL1513].

^1H and ^{13}C-NMR data of thiolates **12** [76ZAAC(423)47], **13–15** [95JCS(D)2129] and **16–18** [93IC2972, 95JCS(D)1649] are available, though they do not provide the diagnostic features of bismuth thiolates. ^{19}F-NMR of thiolate **23** shows signals at δ 130.3, 158.2 and 165.2, which are not so different from those of the corresponding thiolates of Zn, Cd, Hg and Pb [68CJC2699]. Only limited data are available for mass spectrometry; **12** shows an M^+ ion peak [76ZAAC(423)47], while **18** exhibits not a molecular ion peak M^+ but a peak due to $ArSSAr^+$ [93IC2972]. The mass spectrum of **19** contains signals due to M^+, $BiSC_6H_4\text{-}2\text{-}NH_2^+$, $HSC_6H_4\text{-}2\text{-}NH_2^+$ and $SC_6H_4\text{-}2\text{-}NH_2^+$ ions [83ICA(73)141]. Compounds **19**, **20** and **23** show IR peaks at 675–627 ν(C–S) and 313–250 cm^{-1} ν(Bi–S) [68CJC2699, 83ICA(73)141, 86POL1513], while thiolate **12** exhibits a peak at 313 cm^{-1} ν(Bi–S) [79JINC(41)1390]. Thiolate **19** shows the UV–Vis absorption maxima at 400 and 301 nm in pyridine, 400, 277.3 and 267.5 nm in DMF, 400 and 295 nm in DMSO, and 400, 280, 226 and 215.4 nm in dioxane, while thiolate **20** shows the absorption maxima at 400 and 305 nm in pyridine, 387.5, 300 and 267.5 nm in DMF, and 400 and 290 nm in DMSO [86POL1513].

The first X-ray structure analysis of a bismuth thiolate without inter- and intramolecular contacts has been performed on compound **18** by Cowley et al., which shows that the crystal contains two independent molecules. Each molecule has the pyramidal bismuth center with two sets of the Bi–S bond length; 2.556(7), 2.558(7) and 2.562(7) Å for one molecule and 2.554(7), 2.555(7) and 2.569(8) Å for the other. The S–Bi–S bond angles are 99.1(2), 96.2(2) and 97.6(2)° for the former molecule and 90.3(2), 100.0(2) and 104.5(2)° for the latter [93IC2972]. On the other hand, the X-ray crystallographic structure of thiolate **23** reveals that two molecules form a centrosymmetric dimer, in which each bismuth center has a four-coordinated disphenoidal geometry [95POL311]. In each molecule, the Bi–S bond lengths are 2.532(2), 2.540(2) and 2.584(2) Å and the intermolecular Bi–S interaction distance is 3.323(2) Å. The S–Bi–S bond angles are 87.57(5), 92.76(6) and 96.51(6)° [95POL311]. In addition, molecular complexes of thiolate **23** with eight Lewis bases are known and include $[(F_5C_6S)_3Bi(S{=}PPh_3)]$, $[K(18\text{-crown-}6)]^+[(C_6F_5S)_3Bi(NCS)]^-$, $[(C_6F_5S)_3Bi(HMPA)_2]$, $[(C_6F_5S)_3Bi(O{=}PPh_3)]\cdot CH_2Cl_2$, $[(C_6F_5S)_3Bi(dmpu)_2]$, $\{(C_6F_5S)_3Bi[S{=}C(NHMe)_2]_3\}$ $[(C_6F_5S)_3Bi(OSPh_2)_2]$, and $\{(C_6F_5S)_3Bi[S{=}C(NH_2)_2]\}$ [95JCS(D)1163]. These complexes are obtained by crystallizing thiolate **23** in the presence of 3–4 equiv. of the corresponding ligands.

Bismuth thiolates from heteroaromatic thiols are also known. Bismuth tris(1-oxopyridine-2-thiolate) **24** is obtained in 89% yield from the reaction of $BiCl_3$ and sodium 1-oxopyridine-2-thiolate. Compound **24** can be washed safely with water, EtOH, Et_2O, hot DMSO and hot *N,N*-dimethylacetamide owing to its insolubility in common organic solvents [72JCS(D)1120]. Treatment of $Bi(NO_3)_3\cdot 5H_2O$ with 3-(trimethylsilyl)pyridine-2-thiol in MeOH in the presence of Et_3N gives compound **25** in 55% yield as bright yellow crystals, which is soluble in common organic solvents such as CH_2Cl_2, MeCN and DMF [91IC4784]. The IR spectrum of thiolate **25** shows absorption bands due to ν(Si–C) and ν(Bi–S) in the range 850–700 and 350–250 cm^{-1}, respectively.

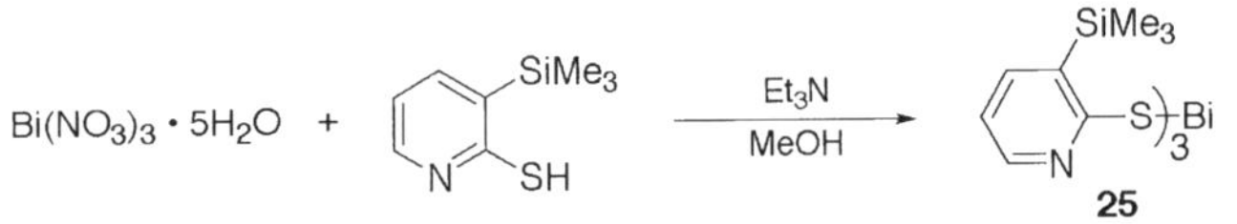

25

The X-ray crystal structure analysis of thiolate **25** has shown that the polyhedron of the bismuth center is considered to be intermediate between the capped-octahedral and pentagonal-bipyramidal geometry. The Bi–S bond lengths are 2.544(6), 2.604(6) and 2.647(7) Å, and the Bi–N intramolecular distances are 2.69(2) and 2.83(2) Å [91IC4784].

The reaction of bismuth arenethiolates has been studied very little; there is only one example of the reaction between thiolate **23** and 2 equiv. of {Bi[M(CO)$_3$(η-C_5H_5)]$_3$} (M=Mo and W) to form a complex [Bi(SC_6F_5){M(CO)$_3$(η-C_5H_5)}$_2$] [95JCS(D)2129].

(b) Bismuth mercaptocarboxylates

Bismuth mercaptocarboxylates are compounds derived from mercaptocarboxylic acids, in which the bismuth atom is bound to different heteroatoms via covalent Bi–OCO and Bi–S bonds. There are four known examples of this type of compound **1–4**, as shown below [81ZN(B)70]. Two polymeric structures are proposed for these compounds; one is a linear chain structure (A), and the other is a polymer containing a pseudo five-membered ring structure (B).

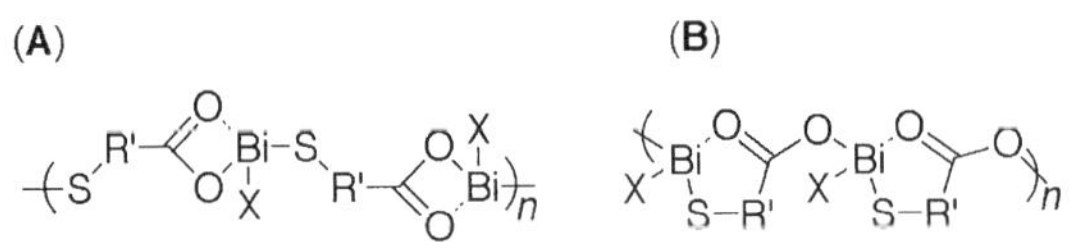

1; X = SCH_2CO_2H
2; X = $SCH_2CH_2CO_2H$
3; X = $O_2CCH(CH_3)SH$
4; X = $O_2CC_6H_4$-2-SH

The IR spectra of these compounds contain characteristic absorption bands at 1540–1510 cm^{-1} due to ν_{as}(COO), 1370–1380 cm^{-1} due to ν_{sy}(COO), and 290–340 cm^{-1} due to ν(Bi–S). In addition, compounds **1** and **2** show signals at 1710 and 1690 cm^{-1} due to ν(C=O), and compounds **3** and **4** exhibit signals at 2560 and 2580 cm^{-1} due to ν(S–H) [81ZN(B)70].

The X-ray structure analysis of bismuth thiosalicylate has revealed that the compound forms an octanuclear cluster core $[Bi_8(tsal)_{12}(DMF)_6](DMF)_6$, where $tsalH_2$ denotes a thiosalicylic acid unit. The presence of stereochemically active lone pairs of electrons is suggested for each bismuth center [94CL2123].

(c) Bismuth thioenolates

This type of compound $Bi[PhC(S)CHC(O)Ar]_3$, where Ar=Ph **1**, 4-$MeOC_6H_4$ **2** and 4-ClC_6H_4 **3**, is prepared by treating $BiCl_3$ with benzoyl(thiobenzoyl)methane derivatives in the presence of sodium acetate [92POL1219]. They are orange to red colored crystals, poorly soluble in benzene but can be recrystallized from CH_2Cl_2–MeOH. The IR spectra of these compounds contain common absorption bands in the range 1580–1586 cm^{-1} due to ν(C=O), 1230–1249 and 840–842 cm^{-1} due to ν(C–S), 485–495 cm^{-1} due to ν(Bi–O), and 350–356 cm^{-1} due to ν(Bi-S). ^1H- and ^{13}C-NMR data are available but do not provide any diagnostic features.

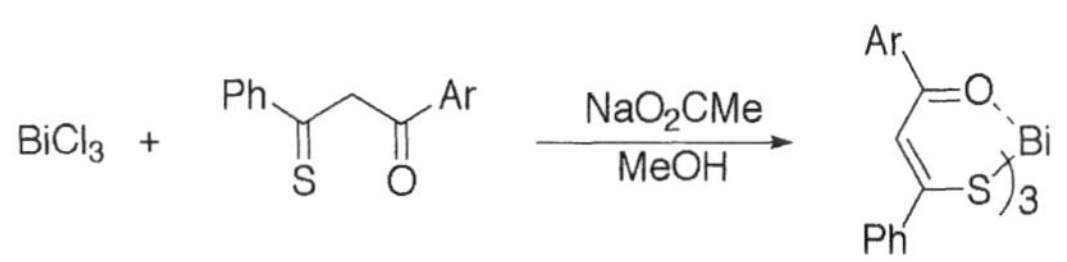

1; Ar = Ph
2; Ar = C_6H_4-4-OMe
3; Ar = C_6H_4-4-Cl

The X-ray crystallographic study of compound **1** has shown that the bismuth center is described best as a distorted pentagonal bipyramidal. The Bi–S bond lengths are 2.581(4), 2.627(5) and 2.731(3) Å, and the Bi–O bond lengths are 2.536(9), 2.575(11) and 2.650(9) Å. In addition, there is some intermolecular interaction between the bismuth and sulfur atoms (3.494(5) and 3.551(5) Å), which causes a dimeric structure in the solid state [92POL1219].

(d) Bismuth dithiocarboxylates

Bismuth tris(2-amino-1-cyclopentene-1-dithiocarboxylate) **1** and 2-(ethylamino)cyclopentene-1-dithiocar-boxylate **2** are obtained as red crystals by treating $BiCl_3$ or BiOCl with the corresponding dithiocarboxylic acids in EtOH. Compound **1** is insoluble in common organic solvents except for DMF and DMSO, in which it is sparingly soluble. Compound **2** is moderately soluble in common organic solvents [87IC1453].

1; R = H
2; R = Et

The X-ray structure analysis of compound **2** has shown that the bismuth center is surrounded by six sulfur atoms to form a trigonal antiprismatic geometry. All three dithiocarboxylate ligands are non-equivalent and the Bi–S bonds can be divided into two types: short (2.617(2), 2.626(1) and 2.647(1) Å) and long (2.963(2), 3.040(1) and 3.108(2) Å). One of the sulfur atom coordinates weakly to the neighboring bismuth center (3.689 Å) to form a dimeric structure in the solid state [87IC1453].

(e) Bismuth dithiocarbamates

Four bismuth dithiocarbamates $Bi(S_2CNR_2)_3$ are prepared and characterized which include **1** (R=Me), **2** (R=Et), **3** ($R=CH_2CH_2OH$) and **4** (R=Pr). For the preparation of compounds **1**, **2** and **4**, the insertion reaction of CS_2 to the corresponding $Bi(NR_2)_3$ has been employed. The reaction undergoes exothermally to give the product in good yield [75JINC(37)2011].

1; R = Me
2; R = Et
4; R = Pr

Rather unexpectedly, the substitution reaction of BiX_3 with sodium dithiocarbamates has not been employed except for the preparation of compound **2** [68JOM(11)627]. Direct mixing of $BiCl_3$ with Et_2NH and CS_2 in petroleum ether gives compound **2** in 74% yield [70INCL71], while treatment of $BiONO_3$ with the corresponding carbamic acid in the presence of HCl affords compound **3** [97POL1211]. By treating Ph_2BiCl with Et_2NCS_2Na, compound **2** is obtained as a side product, along with Ph_3Bi and $PhBi(S_2CNEt_2)_2$ [68JOM(11)627].

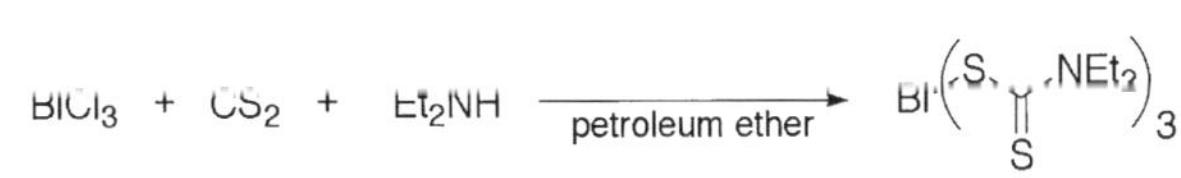

Except for compound **3**, ^{1}H-NMR, IR and UV spectral data are available and some of them are summarized in Table 2.20. Little difference has been observed between the spectra recorded in mull and solution, suggesting that compound **2** exists as a monomer in solution [70INCL71]. Values of λ_{max} and

TABLE 2.20
IR and UV spectra of bismuth dithiocarbamates

Compound	IR (cm^{-1})[a]				UV		References
	N–C_{vib}	C=S_{st}	C–N_{de}	C=S_{vib}	λ_{max} (nm)	ε ($mol^{-1}cm^{-1}$)	
$Bi(Me_2NCS_2)_3$	1495	1245	–	1130	261	5.2×10^4	[75JINC(37)2011]
	1502	–	–	–	259	5.93×10^4	[68JOM(11)627]
$Bi(Et_2NCS_2)_3$	1490	–	–	–	262	6.21×10^4	[68JOM(11)627]
	1500	1270	1430	1145, 990	–	–	[70INCL71]
	1480	1270	–	1140	263	4.1×10^4	[75JINC(37)2011]
$Bi(Pr_2NCS_2)_3$	1475	1270	–	1145	263	4.1×10^4	[75JINC(37)2011]

[a] Abbreviations: vib, vibration; st, stretching; de, deforming.

ε of these carbamates are very similar [68JOM(11)627, 75JINC(37)2011]. The electrochemical behavior of compound **3** is also reported [97POL1211].

The X-ray crystallographic structures of compounds **2** [76JCS(D)791] and **3** [97POL1211] have been characterized. The bismuth center of compound **2** is surrounded by eight sulfur atoms to form a C_3 symmetry, where some intermolecular interaction has been observed. The Bi–S bonds can be divided into two types: short (2.595(5)–2.775(5) Å) and long (2.956(5)–2.964(4) Å) [76JCS(D)791]. The bismuth center of compound **3** is an eight-coordinated polyhedron of a distorted square antiprism. The compound forms a dimeric structure in the solid state and two of the six carbamate ligands bridge. The Bi–S bonds can be classified into three types according to their lengths: 2.669(7)–2.729(7) Å, 2.819(7)–2.885(7) Å, and 3.076(9)–3.179(9) Å [97POL1211]. The carbamate groups of both compounds are non-equivalent, forming stereochemically active lone pairs of electrons.

(f) Bismuth dithiocarbonates (xanthates)

Five examples of bismuth tri(dithiocarbonate)s are known: *O*-methyldithiocarbonate **1**, *O*-ethyldi-thiocarbonate **2**, *O*-isopropyldithiocarbonate **3**, *O*-cyclohexyldithio-carbonate **4**, and *O*-benzyldithiocarbonate **5**. All these compounds are prepared by a simple substitution reaction of $BiCl_3$ with potassium dithiocarbonate [76AJC559, 85ICA(99)177, 92JCSR(22)231].

1; R = Me **2**; R = Et **3**; R = ^{i}Pr **4**; R = ^{c}Hex **5**; R = CH_2Ph

The IR spectra of these dithiocarbonates are similar to those of common

TABLE 2.21
IR spectra of bismuth dithiocarbonates (cm^{-1})

Compound	C–O	C–S	Reference
$Bi(S_2COEt)_3$	1210	1035	[76AJC559]
$Bi(S_2CO^iP_2)_3$	1210	1015	[85ICA(99)177]
$Bi(S_2CO^cHex)_3$	1250, 1225, 1207	1040, 1022	[92JCSR(22)231]
$Bi(S_2COCH_2Ph)_3$	1264, 1250	1052, 1038, 1030	[92JCSR(22)231]

metal dithiocarbonates and their absorption bands due to the C–O and C–S bonds are summarized in Table 2.21. The UV spectrum of *O*-ethyldithiocarbonate **2** is reported; λ_{max} and ε values are 365 (n–π^*, ε, 2000), 330 (charge transfer, ε, 7500), 285 (π–π^*, ε, 17000), 260 (charge transfer, ε, 22000), and 229 nm (n–σ^*, ε, 62000 [76AJC559].

The X-ray crystallographic structures of bismuth dithiocarbonates **1–5** have been studied. All these structures are dimeric or polymeric due to the intermolecular interaction among the bidentate xanthate ligands. Compounds **1** [87AJC743], **2** [94MGMC727], **4** and **5** [92JCSR(22)231] all exist as centrosymmetric dimers, while compound **3** [84ICA(81)L33, 85ICA(99)177] forms a ladder-like polymeric structure. The coordination modes of the bismuth center in the former compounds are similar: the bismuth centers are surrounded by seven sulfur atoms, forming a distorted pentagonal bipyramidal geometry. The coordination polyhedron of compound **3** is described as a distorted bicapped trigonal prism [85ICA(99)177]. Three dithiocarbonate ligands are non-equivalent, and the Bi–S bonds in compound **1** can be classified into two types: short (2.596(3), 2.720(3) and 2.747(3) Å) and long (2.933(4), 2.961(4) and 2.998(4) Å] [87AJC743].

(g) Bismuth dithiophosphinates

The dithiophosphinato complexes of main group and transition metal elements have been studied extensively. In the case of bismuth, $Bi(S_2PR_2)_3$ **1** (R=Me) [94POL547], **2** (R=Et) [64CB2306, 87JCS(D)1257, 89ACS511], **3** (R=Pr) [64CB2306], and **4** (R=Ph) [71CB1892, 87JCS(D)145] are known. These compounds are obtained as yellow solids by the substitution reaction of $BiCl_3$ and NaS_2PR_2 [64CB2306, 94POL547], or by the reaction of $Bi(NO_3)_3{\cdot}5H_2O$ and NaS_2PR_2 in the presence of nitric acid [71CB1892]. They appear to be insensitive towards moisture and oxygen.

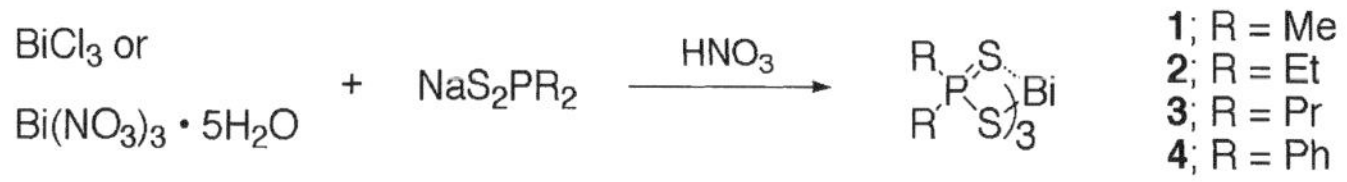

Spectral data of these dithiophosphinates are currently limited. ^{1}H, ^{13}C and ^{31}P-NMR spectra are reported for compound **1**. Its ^{31}P-NMR spectrum shows a signal at δ 60.4 [94POL547]. The mass spectrum of compound **1** does not show the parent peak M^+, but the daughter peaks due to $[M–Me_2PS]^+$, $[M–Me_2PS_2]^+$, $[M–2Me_2PS_2]^+$, $BiSPMe_2^+$, BiS^+ and bismuthions are observed [94POL547]. Compound **4** exhibits IR absorption bands at 688, 626 ν_{as}(PS), 602, 551 ν_s(PS), 485, 589 and 251 ν(BiS) cm^{-1} [71CB1892], while **1** shows IR peaks at 580 and 492 (PS_2) and 260 ν(BiS) cm^{-1} [94POL547]. The UV–Vis spectrum of compound **4** shows three absorption maxima at λ_{max} 235 (ε=3.4×10^3 (mol^{-1} cm^{-1}), 291 (ε=1.7×10^4 mol^{-1} cm^{-1}) and 314 (shoulder) nm [71CB1892]. The molar conductivity of compound **3** is estimated to be 0.182 Ω^{-1} cm^2 mol^{-1} [64CB2306].

The X-ray crystallographic structures of compounds **1** [94POL547], **2** [87JCS(D)1257, 89ACS511] and **4** [87JCS(D)145] have been elucidated. Compound **2** exists as a monomer in the solid state; its bismuth center is surrounded by six sulfur atoms to form a distorted octahedral geometry. The Bi–S bond lengths are not equal and fall in the range 2.753(2)–2.858(2) Å [89ACS511]. The crystal structure of $Bi(S_2PEt_2)_3{\cdot}C_6H_6$ is also elucidated. Its bismuth center is monomeric and the lone pair of electrons is not sterically active. In the crystal, three dithiophosphinato ligands are almost equivalent; the Bi–S distances are 2.782(5) and 2.794(5) Å [87JCS(D)1257] (Table 6.5, compound **95**). In contrast to compound **2**, the crystal structures of compounds **1** and **4**, $Bi(S_2PMe_2)_3$ and $Bi(S_2PPh_2)_3$, are dimeric. Compound **1** exists as a centrosymmetric dimer which is bridged by one of the dithiophosphinato ligands. The geometry of the bismuth center is described as a distorted pentagonal bipyramid and three ligands are not equivalent. The Bi–S bonds can be divided into two types: short (2.641(3), 2.782(3) and 2.714(3) Å) and long (2.891(3), 3.025(3) and 3.165(3) Å). The longest Bi–S distance is that of the bridging dithiophosphinato group: 3.183(3) Å [94POL547]. The structure of compound **4** is very similar to that of compound **1**; two molecules form a dimeric structure, in which all dithiophosphinato ligands are not equivalent. Its Bi–S bond lengths are similar to those of compound **1** [87JCS(D)145].

Both structures may possess a stereochemically active lone pair of electrons, though this is only speculative at present.

(h) Bismuth dithioarsinates

Bismuth tris(dimethyldithioarsinate) is the only reported example of this type of compound [97HAC317]. It is obtained as light yellow crystals by treatment of $Bi(NO_3)_3{\cdot}5H_2O$ with sodium dimethyldithioarsinate NaS_2AsMe_2 in MeCN. Its mass spectrum contains signals due to M^+, $[M{-}S]^+$, $Me_4As_2S_5{}^+$, $Me_4As_2S_3Bi^+$, $Me_4As_2S_2Bi^+$ and Me_2AsS^+ ions. The X-ray crystal structure analysis has revealed that the compound exists as a dimer which contains the centrosymmetric monomeric unit with a pentagonal bipyramidal geometry similar to $Bi(S_2PMe_2)_3$. The bismuth center is surrounded by seven sulfur atoms with short (2.658(4), 2.735(4) and 2.802(4) Å) and long (2.881(4), 3.048(4) and 3.177(4) Å) bond distances; the intermolecular Bi–S distance is 3.141(4) Å [97HAC317]. No further information is available on its spectra and reactivity.

(i) Bismuth dithiophosphates

The chemistry of bismuth dithiophosphate, $Bi[S_2P(OR)_2]_3$, has not been studied in detail though the following six compounds are known: diethyl- **1** (R=Et) and diisopropyldithiophosphate **2** (R=^{i}Pr), and cyclic dithiophosphates **3–6**. These compounds are prepared as yellow crystalline solids by treating $BiCl_3$ with ammonium dithiophosphate $NH_4^+[S_2P(OR)_2]^-$ in acetone [74IC135, 84POL1337, 89POL483], or by the reaction of $BiCl_3$ and dithiophosphoric acid in boiling benzene [84POL1337].

1; R = Et
2; R = ^{i}Pr

They appear to be insensitive towards moisture and oxygen. Their solubility depends on the ligands employed; compounds **1** and **2** are soluble even in alkanes [74IC135, 89POL483], and compounds **3** and **5** are soluble in benzene, $CHCl_3$ and CCl_4. However, compound **4** is not soluble in benzene, and compound **6** is slightly soluble in benzene and insoluble in acetone [84POL1337].

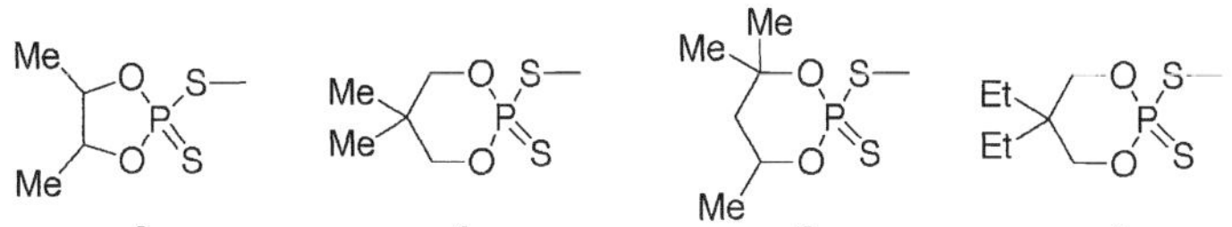

3 **4** **5** **6**

There is only limited information available on the spectral data of bismuth dithiophosphates. ^{1}H-NMR spectra of compounds **3**, **5** and **6** are reported. Compounds **3** and **6** exhibit ^{31}P-NMR signals at δ 115.29 and 97.90, respectively [84POL1337].

Compounds **1** and **2** have been characterized by X-ray crystallography. Both compounds exist as monomers in the solid state; six sulfur atoms of non-equivalent three dithiophosphato ligands surround the bismuth center to form a distorted octahedral geometry. The Bi–S bonds of compound **1** are divided into two types: short (2.750(1), 2.747(14) and 2.778(12) Å) and long (2.781(11), 2.795(13) and 2.820(16) Å) [89POL483]. In the case of compound **2**, the corresponding values are 2.690(6), 2.702(6) and 2.714(6) Å for the short distances and 2.858(6), 2.878(6) and 2.886(6) Å for the long ones [74IC135].

(j) Bismuth imidodithiodiphosphinate

Two imidodithiodiphosphinato bismuth complexes are known, *P,P,P′,P′*-tetramethyl derivative **1** [87JCC(16)315] and *P,P,P′,P′*-tetraphenyl derivative **2** [80INCL189, 82IC2097]. Metal complexes bearing these ligands are quite rare and compound **1** is the first example of this class of metal complex. Both compounds are prepared as air-stable solids by treating $Bi(NO_3)_3{\cdot}5H_2O$ with $K[SPR_2NPR_2S]$ in MeOH. Compound **1** is a lustrous yellow-green solid [87JCC(16)315] and compound **2** is a bright yellow solid [80INCL189]. Compound **2** exhibits a ^{31}P-NMR signal at $\delta -34.9$, while the corresponding dioxodiphosphinate shows a ^{31}P-NMR signal at $\delta -17.3$ [80INCL189].

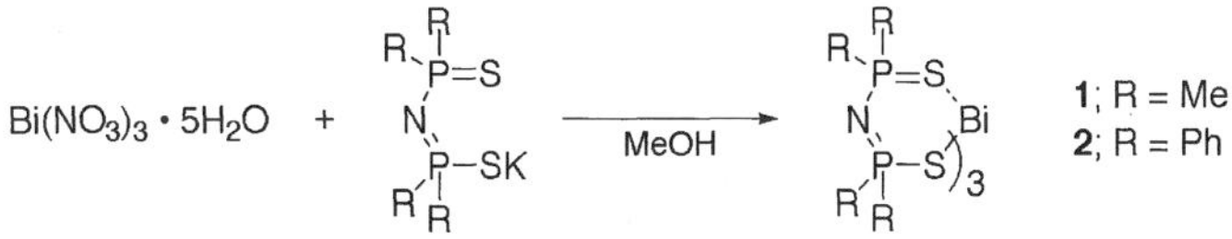

The X-ray structure analysis of compound **2** has shown that the bismuth center possesses a distorted octahedral geometry, formed by six sulfur atoms of three non-equivalent imidodithiodiphosphinato ligands. The Bi–S bonds are divided into two types: short (2.770(3), 2.728(3) and 2.772(3) Å) and long (2.832(3), 2.851(3) and 2.986(3) Å). The molecule exists as a monomer in the solid state [82IC2097] and its crystallographic structure has been compared with that of In[$SPPh_2NPPh_2S$] [92M391].

TABLE 2.22
Bismuth compounds with three organyl sulfur ligands

Compound R	Synthetic method[a]	Yield (%)	M.p. (°C)	Physical data	Reference
$Bi(SR)_3$					
Bismuth aliphatic thiolates					
Et	C	74	80–82 (decomp.) [EtOH]	^{1}H-NMR, MS	[76ZAAC(423)47]
^{t}Bu	A	Quant.	150 (decomp.)	^{1}H-NMR, IR, Raman	[81JINC(43)1469]
HOC_2H_4	C	70	77 (decomp.) [EtOH]	^{1}H-NMR	[76ZAAC(423)47]
$PhCH_2$	C	84	88–90	^{1}H-NMR	[76ZAAC(423)47]
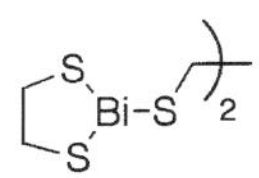	A	97	145 (decomp.)	IR, MS, Raman, X-ray	[96JA3225]
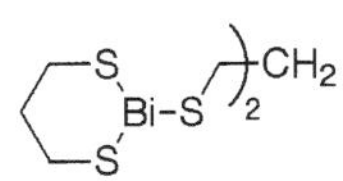	A	92	170 (decomp.)	IR, MS, Raman	[96JA3225]
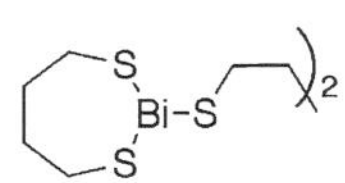	A	98	140 (decomp.)	IR, Raman	[96JA3225]
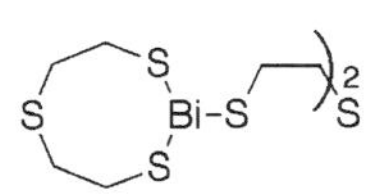	A	67	180 (decomp.)	IR, MS, Raman, X-ray	[96JA3225]
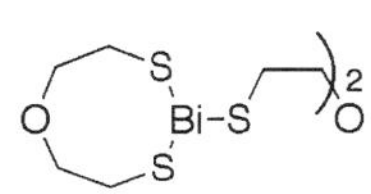	A	99	177–178	IR, MS, Raman	[96JA3225]

TABLE 2.22 *(continued)*

Compound R	Synthetic method[a]	Yield (%)	M.p. (°C)	Physical data	Reference
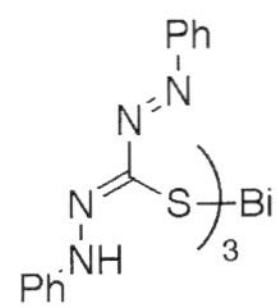	A	–	–	X-ray	[82AX(B)2140]
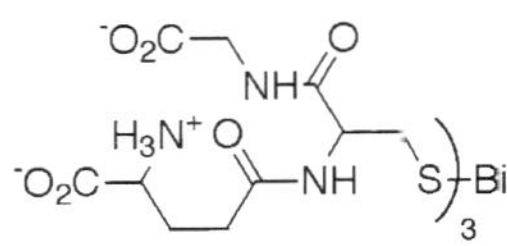	C	–	–	^{1}H-, ^{13}C-NMR	[96CEJ701]
Bismuth aromatic thiolates					
Ph	A	–	98–99 [EtOH]	–	[17CB21]
	B	100	90–91	–	[51JA2880]
	C	78	98 (decomp.) [EtOH]	^{1}H-NMR, MS	[76ZAAC(423)47]
	A	79	95–97 (decomp.)	IR	[79JINC(41)1390]
4-MeC_6H_4	A	62	–	^{1}H-NMR	[95JCS(D)2129]
2,6-$Me_3C_6H_2$	A	78	–	^{1}H-, ^{13}C- NMR	[95JCS(D)2129]
3,5-$Me_3C_6H_2$	A	49	–	^{1}H-, ^{13}C- NMR	[95JCS(D)2129]
2,4,6-$Me_3C_6H_2$	C	86	142–144	^{1}H-, ^{13}C-NMR	[95JCS(D)1649]
2,4,6-$^{i}Pr_3C_6H_2$	C	85	102–104	^{1}H-, ^{13}C-NMR	[95JCS(D)1649]
2,4,6-$^{t}Bu_3C_6H_2$	A	77	105–110 (decomp.)	^{1}H-NMR, MS, X-ray	[93IC2972]
	C	86	166–168	^{1}H-, ^{13}C- NMR	[95JCS(D)1649]
2-$H_2NC_6H_4$	A	86	79–80	-	[65JINC(27)1431]
	C		165–167	IR, MS	[83ICA(73)141]
	A	–	–	IR, UV	[86POL1513]
2-($Cl^{-}H_3N^{+}$)C_6H_4	A	–	–	IR, UV	[86POL1513]

2-(MeO_2C)C_6H_4	B	78	141–143 [petroleum ether]	–	[illegible]
4-FC_6H_4	A	80	140 (decomp.)	–	[88JFC(38)367]
C_6F_5	A	93	290 (decomp.)	^{19}F-NMR, IR	[68CJC2699]
	–	–	–	X-ray	[95POL311]
	B	70	–	^{19}F-NMR	[95JCS(D)2129]
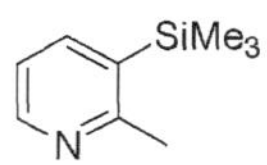	A	55	–	IR, X-ray	[91IC4784]
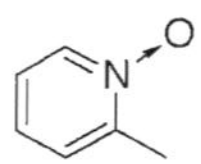	A	89	266–267.5 (decomp.)	–	[72JCS(D)1120]
Solvated bismuth thiolates					
[(C_6F_5S)$_3$Bi(S=PPh_3)]	–	ca. 50	–	X-ray	[95JCS(D)1163]
[K(18-crown-6)]$^+$[(C_6F_5S)$_3$Bi(NCS)]$^-$	–	ca. 50	–	X-ray	[9SJCS(D)1163]
[(C_6F_5S)$_3$Bi(O=PPh_3)]·CH_2Cl_2	–	ca. 50	–	X-ray	[95JCS(D)1163]
[(C_6F_5S)$_3$Bi(HMPA)$_2$]	–	ca. 50	–	X-ray	[95JCS(D)1163]
[(C_6F_5S)$_3$Bi(Dmpu)$_2$]	–	ca. 50	–	X-ray	[95JCS(D)1163]
[(C_6F_5S)$_3$Bi{S=C(NHMe)$_2$}$_3$]	–	ca. 50	–	X-ray	[95JCS(D)1163]
[(C_6F_5S)$_3$Bi{S=C(NH_2)$_2$}]	–	ca. 50	–	–	[95JCS(D)1163]
[(C_6F_5S)$_3$Bi($OSPh_2$)$_2$]	–	ca. 50	–	–	[95JCS(D)1163]
Bismuth thioenolates					
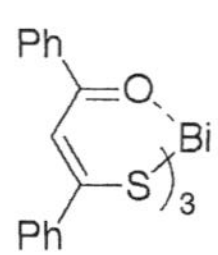	A	85–90	–	^{1}H-, ^{13}C-NMR, IR, X-ray	[92POL1219]
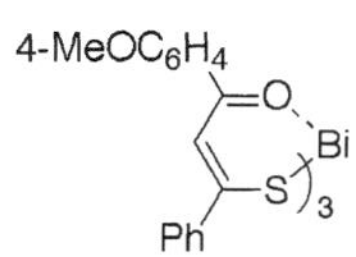	A	85–90	–	^{1}H-, ^{13}C -NMR, IR	[92POL1219]

TABLE 2.22 *(continued)*

Compound R	Synthetic method[a]	Yield (%)	M.p. (°C)	Physical data	Reference
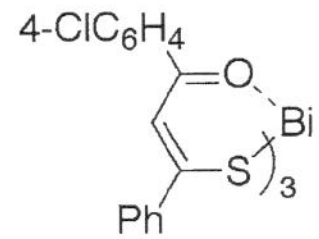	A	85–90	–	^{1}H-, ^{13}C-NMR, IR	[92POL1219]
Bismuth dithiocarboxylates					
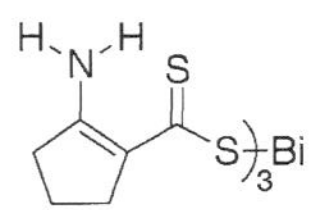	A	99	–	–	[87IC1453]
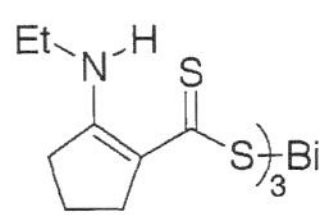	A	98	–	X-ray	[87IC1453]
$Bi(S_2CX)_3$					
Bismuth dithiocarbamates					
X = NMe_2	D	95	210 [CCl_4]	^{1}H-NMR, IR, UV	[75JINC(37)2011]
X = NEt_2	E	60	185–186	IR, UV	[68JOM(11)627]
	A	63	193–195	IR	[70INCL71]
	D	92	170.5–171.5	^{1}H-NMR, IR, UV	[75JINC(37)2011]
	–	–	–	X-ray	[76JCS(D)791]
X = NPr_2	D	88	170.5–171.5	IR, UV	[75JINC(37)2011]
X = $N(CH_2CH_2OH)_2$	A	–	–	X-ray	[97POL1211]

$Bi(S_2CX)_3$					
Bismuth dithiocarbonates					
X = OMe	–	–	–	X-ray	[87AJC743]
X = OEt	A	–	–	IR, UV	[76AJC559]
	–	–	–	X-ray	[94MGMC727]
X = O^iPr	–	–	–	X-ray	[84ICA(81)L33]
	A		169 (decomp.)	^1H-, ^{13}C-NMR, IR, X-ray	[85ICA(99)177]
X = O^cHex	–	–	–	IR, X-ray	[92JCSR(22)231]
X = OCH_2Ph	–	–	–	IR, X-ray	[92JCSR(22)231]
$Bi(S_2PX_2)_3$					
Bismuth dithiophosphinates					
X = Me	A	62	242 [PhH]	^1H-, ^{13}C-, ^{31}P-NMR, IR, MS, X-ray	[94POL547]
X = Et	A	94.3	94.5 [^iPrOH]	–	[64CB2306]
				X-ray	[87JCS(D)1257]
	–	–	–	X-ray	[89ACS511]
X = Pr	A	94.5	201–202	–	[64CB2306]
X = Ph	A	75	–	IR, ES	[71CB1892]
	–	–	–	X-ray	[87JCS(D)145]
Bismuth dithiophosphates					
X = OEt	A	–	–	X-ray	[89POL483]
X = O^iPr	A	–	67–68 [$C_{10}H_{22}$]	X-ray	[74IC135]
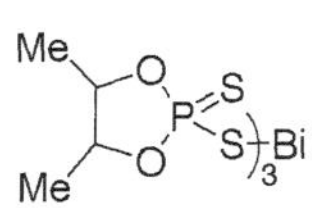	A	81	260	^1H-, ^{31}P-NMR	[84POL1337]
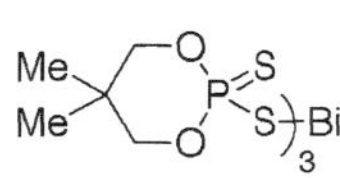	A	45	290 (decomp.)	–	[84POL1337]

TABLE 2.22 *(continued)*

Compound R	Synthetic method[a]	Yield (%)	M.p. (°C)	Physical data	Reference
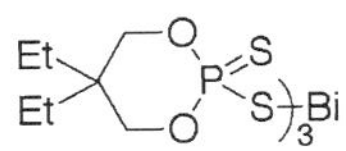	A	59	255	^{1}H-, ^{31}P-NMR	[84POL1337]
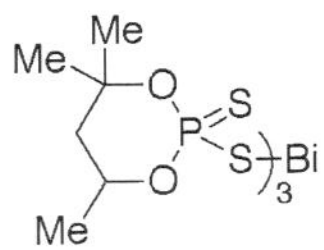	A	93	80 (decomp.)	^{1}H-NMR	[84POL1337]
Bismuth imidodithiodiphosphinates					
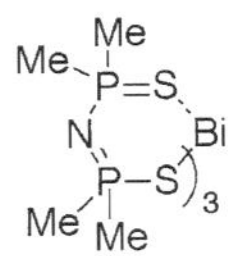	A	47	185–190 (decomp.) [CH_2Cl_2–EtOH]	^{1}H-NMR	[87JCC(16)315]
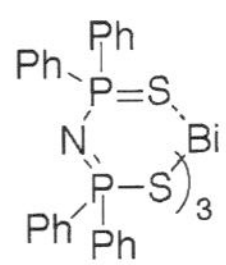	A – –	57 – –	235–236 [MeOH] – –	^{31}P-NMR X-ray –	[80INCL189] [82IC2097] [92M391]
Bismuth dithioarsinates					
$Bi(S_2AsMe_2)_3$	A	–	127	MS, X-ray	[97HAC317]

[a] For notation, see Section 2.2.4.1.

2.2.4.5. Bismuth(III) compounds with three Bi–Se bonds (Table 2.23)

(a) Bismuth selenolates

Examples of homoleptic bismuth selenolates $Bi(SeR)_3$ are few. Three compounds **1–3** are known, which are prepared by treating $Bi[N(SiMe_3)_2]_3$ with the corresponding areneselenols in light petroleum. They are colored yellow to red, air-stable in the solid state, and can be sublimed at moderate temperatures. They are soluble in common organic solvents and a mixture of MeCN and ethyl acetate is the solvent of choice for recrystallization [95JCS(D)1649]. ^{1}H- and ^{13}C-NMR data of these compounds are available; chemical shift values are similar to those of the corresponding thiolates.

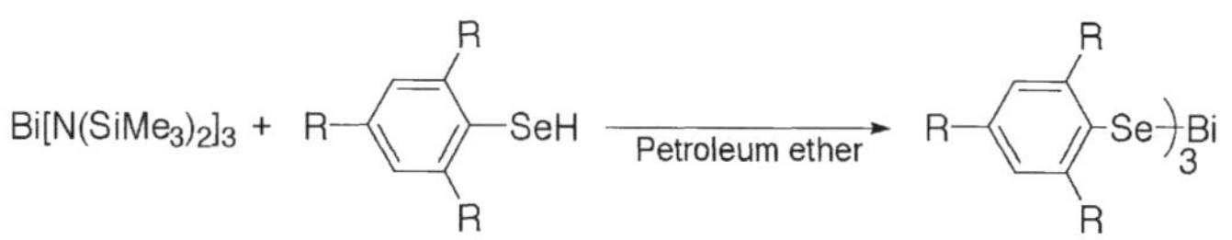

1; R = Me
2; R = ^{i}Pr
3; R = ^{t}Bu

Bismuth selenides and tellurides are used for laser-sensitive optical recording media. The thermolytic behavior has been investigated on compounds **1** and **3**; the thermolysis of compound **1** at 200°C under 5.0×10^{-6} mbar gives elemental bismuth, with ArSeSeAr as a by-product. On the other hand, the thermolysis of compound **3** at 200°C under 1.6×10^{-6} mbar results in the formation of a mixture of elemental bismuth, Bi_2Se_3, and Ar_2Se [95JCS(D)1649].

(b) Bismuth diselenocarbamates

The first preparation of this type of compounds **1–3**, has been performed by Manoussakis et al., who treated $BiCl_3$ with CSe_2 in the presence of the corresponding secondary amines [73IC3015]. By a similar procedure, $Bi[Se_2CN(CH_2)_5]_3$ **4** was obtained as a red solid [75ZAAC(417)235]. These compounds are air-stable red solids, soluble in CCl_4, $CHCl_3$, CH_2Cl_2, acetone and benzene, but insoluble in MeOH, EtOH, petroleum ether and water.

1; R = Et
2; R = Bu
3; R = CH_2Ph
4; $R_2 = (CH_2)_5$

^{1}H-NMR spectra are available for compounds **1–3**; their chemical shifts are close to those of arsenic and antimony counterparts with the same ligands [73IC3015]. The IR spectra of these compounds contain absorption bands at

1500–1410 cm^{-1} due to the C–N double and single bonds, and two sets of bands at 1200–1100 and 950–800 cm^{-1} due to the CSe(=Se) function [73IC3015]. The IR spectrum of compound **4** shows absorptions at 1478, 1428 and 818 cm^{-1} [75ZAAC(417)235]. The electronic spectra of compounds **1**–**4** exhibit three absorption bands at 33.3–35.6 (π–π*, logε=4.7–4.9), 31–32 (shoulder, logε=4.5–4.6) and 23–24 kK (n–π*, logε=3.5–4.1) [73IC3015, 75ZAAC(417)235]. The third band appears only in the electronic spectra of bismuth diselenocarbamates and is not detectable in the spectra of the arsenic and antimony counterparts.

TABLE 2.23
Bismuth compounds with three organyl selenium ligands

Compounds	Synthetic method[a]	Yield (%)	M.p. (°C)	Physical data	Reference
$Bi(SeC_6H_2\text{-}2,4,6\text{-}Me_3)_3$	C	68	116 (decomp.)	^1H-, ^{13}C-NMR	[95JCS(D)1649]
$Bi(SeC_6H_2\text{-}2,4,6\text{-}{}^iPr_3)_3$	C	87	122–124	^1H-, ^{13}C-NMR	[95JCS(D)1649J
$Bi(SeC_6H_2\text{-}2,4,6\text{-}{}^tBu_3)_3$	C	87	150 (decomp.)	^1H-, ^{13}C-NMR	[95JCS(D)1649}
$Bi(Se_2CNEt_2)_3$	A	85	227 (dccomp.)	^1H-NMR, IR, UV	[73IC3015]
$Bi(Se_2CNBu_2)_3$	A	68	133–135	^1H-NMR, IR, UV	[73IC3015]
$Bi[Se_2CN(CH_2Ph)_2]_3$	A	75	152–154	^1H-NMR, IR, UV	[73IC3015]
$Bi[Se_2CN(CH_2)_5]_3$	A	90	251 (decomp.)	IR, UV	[75ZAAC(417)235]

[a] For notation, see Section 2.2.4.1.

2.3. ORGANOBISMUTH(III) COMPOUNDS WITH Bi–GROUP 17 ELEMENT BONDS

2.3.1. Methods of synthesis (Tables 2.24–2.29)

2.3.1.1. Redistribution route (method A)

Mono- and dihalobismuthines of the type R_2BiX and $RBiX_2$ (X=Cl, Br, I) are prepared best by a redistribution reaction between a triorganylbismuthine (R_3Bi) and a bismuth halide (BiX_3).

$$n\,R_3Bi + (3-n)\,BiX_3 \longrightarrow 3\,R_nBiX_{3-n} \qquad (n = 1, 2)$$

Both alkyl and aryl derivatives are accessible by this method with similar ease. Monohalobismuthine R_2BiX is the main product when 1 mol of BiX_3 is reacted with 2 mol of R_3Bi. If the molar ratio is reversed, dihalobismuthine $RBiX_2$ is obtained as the major product. Diethyl ether, tetrahydrofuran, benzene, dichloromethane and chloroform are the solvents of choice for this reaction. This reaction is, however, essentially based on the equilibration among four different bismuth compounds, R_nBiX_{3-n} (n=0–3), in a given solution. Thus, recrystallization from a suitable solvent is necessary to obtain an analytically pure compound.

Synthesis of chlorodiphenylbismuthine

$$2\,Ph_3Bi + BiCl_3 \xrightarrow[Et_2O]{} 3\,Ph_2BiCl$$

An ethereal solution of triphenylbismuthine (2.5 g, 5.7 mmol) was mixed with a solution of bismuth chloride (1.3 g, 4.1 mmol) in the same solvent. Deposited colorless crystals (3.0 g) were collected and recrystallized from benzene to obtain pure chlorodiphenylbismuthine, m.p. 184–185°C [15JCS(107)16].

2.3.1.2. Metathesis route (method B)

Mono- and diiodobismuthines are prepared by treating chloro-, bromo-, dichloro- or dibromobismuthines with an excess of sodium or potassium iodide. Though not isolated, diarylfluorobismuthines have also been used as the precursor for the synthesis of diarylchloro-, bromo- and iodobismuthines [93JCS(P1)2969, 95OM3843].

$$R_nBiX_{3-n} + (3-n)\,MY \longrightarrow R_nBiY_{3-n} + (3-n)\,MX \qquad (n = 1, 2)$$

Synthesis of iododiphenylbismuthine

$$Ph_2BiCl + NaI \xrightarrow[EtOH]{} Ph_2BiI$$

A mixture of chlorodiphenylbismuthine (23 g, 58 mmol), sodium iodide (9 g, 60 mmol) and absolute ethanol (200 ml) was stirred for 12 h at room temperature, and then poured into boiling water (600 ml). Iododiphenylbismuthine separated out in the form of yellow flakes which were filtered off, washed with water and dried in a vacuum desiccator. The crude product was treated with hot ethyl acetate to extract the iodobismuthine. The extract was cooled to room temperature and the resulting red by-product was removed by filtration. The solvent was distilled from the filtrate under diminished pressure in a steam of nitrogen. The residue was powdered, triturated with petroleum ether (20 ml) in order to remove triphenylbismuthine, and the petroleum ether solution decanted. This washing was repeated several times to obtain pure product, m.p. 132–134°C [31JA1025].

2.3.1.3. Organometallic route (method C)

The reaction of an organometallic compound and a bismuth halide in an appropriate molar ratio provides a convenient method for the preparation of mono- and dihalobismuthines, as long as the redistribution among R_nBiX_{3-n} species ($n = 0$–3) does not take place during the reaction.

$$BiX_3 + n\,RM \longrightarrow R_nBiX_{3-n} + (3-n)\,MX \quad (n = 1, 2)$$

$$RBiX_2 + R'M \longrightarrow RR'BiX + MX$$

A sterically crowded aryl Grignard or aryllithium reagent allows the selective introduction of one or two aryl group(s) onto the bismuth center [92BCJ3504, 95JOM(485)141, 99JA3357]. Unsymmetrical diarylhalobismuthines are synthesized by the reaction of aryldihalobismuthines with 1 equiv. of Grignard or lithium reagent [81JOM(212)C7]. Aryllithium reagents with a coordinating group, such as the amino moiety at *ortho* positions of the aromatic ring, couple with a half equiv. of bismuth halide to afford the corresponding diarylhalobismuthines [92ICA(198-200)271, 97IC2770].

Synthesis of chlorobis(2,4,6-triisopropylphenyl)bismuthine

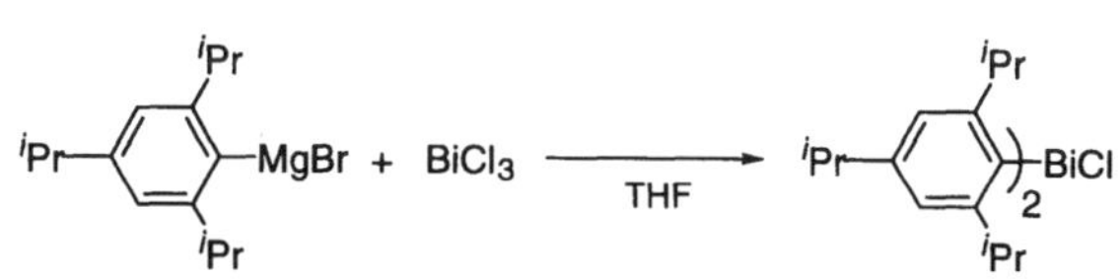

To a THF (30 ml) solution of freshly prepared 2,4,6-triisopropylphenylmagnesium bromide (31 mmol) was added an ethereal solution (10 ml) of bismuth chloride (4.20 g, 13.3 mmol) at 0°C.

After stirring for 6 h at room temperature, the reaction mixture was filtered through a Celite bed. Removal of the solvent from the filtrate, extraction of the solid residue with benzene, and evaporation of the solvent left a yellow sticky oil, which was chromatographed over a silica gel column using hexane/benzene as the eluent to give chlorobis(2,4,6-triisopropylphenyl)bismuthine as yellow crystals (7.8 g, 90%), m.p. 130–132°C [92BCJ3504].

Organometallic compounds other than the Grignard and lithium reagents have also been used for synthetic purposes. Bismuth chloride induces the cleavage of the C–C bond in a siloxycyclopropane, or the Si–C bond in trimethylsilylcyclopentadiene, to give the corresponding alkyldichlorobismuthines and/or dialkylchlorobismuthines [85OM641, 90JOM(386)177]. The reaction between tetraphenyl- or diethyldiphenylstannane and bismuth bromide leads to a mixture of dibromo(phenyl)bismuthine and halostannanes [79ZOB1185]. Heterocyclic tin and cadmium compounds are utilized for the synthesis of heterocyclic bismuthines (Sections 4.1.1 and 4.1.2). Dibromo(methyl)bismuthine is prepared by the reaction of dimethylzinc and bismuth bromide [887CB1516]. The interaction between organolead compounds and bismuth halides can be used for the preparation of halobismuthines bearing the phenyl, vinyl or ethynyl groups [22JCS(121)978, 39JOC162].

Synthesis of dichloro[2-(isopropoxycarbonyl)ethyl]bismuthine

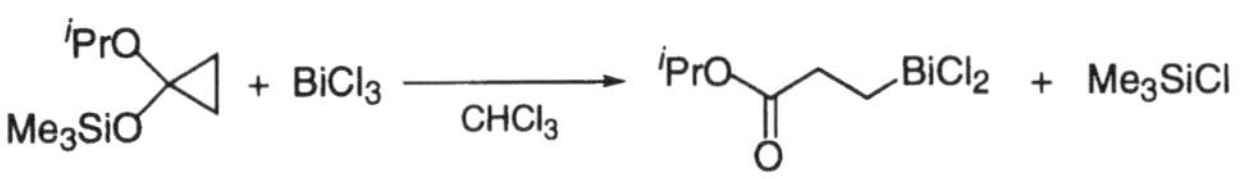

To a suspension of bismuth chloride (276 mg, 0.874 mmol), dried at ca. 300°C for 10 min, in chloroform (1 ml) was added 1-isopropoxy-1-trimethylsiloxycyclopropane (0.180 ml, 1.874 mmol) at −60°C with stirring. The mixture was agitated under ultrasonic irradiation for 1.5 h at room temperature. The solvent was removed, and the residual white solid was recrystallized from benzene to give the expected alkyldichlorobismuthine (89%), m.p. 144–145°C (decomp.) [85OM641].

2.3.1.4. *Via halogenolysis (method D)*

When treated with a halogen or its equivalent at ambient temperature, trialkylbismuthines are converted to the corresponding dialkylhalo- or alkyldihalobismuthines. Treatment of triphenylbismuthine with bromine affords triphenylbismuth dibromide, which undergoes reductive coupling in boiling benzene to yield bromodiphenylbismuthine and bromobenzene [14JCS(105)2210] (Sections 3.1.1.1 and 3.1.3). A similar reaction between triarylbismuthines and iodine directly affords diaryliodobismuthines and/or aryldiiodobismuthines. Action of bromine on bromodiphenylbismuthine gives dibromo(phenyl)bismuthine [14JCS(105)2210].

$$R_3Bi + X_2 \longrightarrow R_2BiX + RX$$

$$R_2BiX + X_2 \longrightarrow RBiX_2 + RX$$

When the alkyl and aryl groups are both bound to the bismuth atom as mixed bismuthines, the alkyl ligand is replaced preferentially by halogen [85CB1031, 92BCJ3504].

When one of the aryl ligands has a coordinating substituent at a suitable position of the aromatic ring or *ortho*-side chain, a mixed triarylbismuthine undergoes a selective Bi–C bond cleavage. This gives the corresponding halodiarylbismuthine that possesses the hypervalent structure around the bismuth center through the coordination by the relevant substituent group [93JCS(P1)1169, 98OM4049, 98JCS(P1)2511]. In the case where two different aryl ligands are attached to the bismuth center in mixed triarylbismuthines, a more electron-rich aryl group is preferentially cleaved by halogen to give unsymmetrical diarylhalobismuthines.

Synthesis of (2-*tert*-butylsulfonylphenyl)iodo(4-methylphenyl)bismuthine

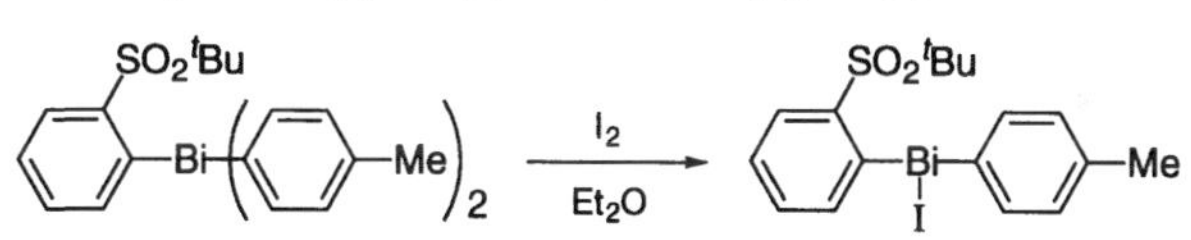

To a well stirred suspension of [(2-*tert*-butylsulfonyl)phenyl]bis(4-methylphenyl)bismuthine (588 mg, 1 mmol) in diethyl ether (10 ml) was added dropwise a solution of iodine (254 mg, 1 mmol) in the same solvent (10 ml) under ambient conditions and until the starting material was completely consumed. The resulting pale yellow solution was concentrated under reduced pressure to leave an oily residue, which was recrystallized from methanol–benzene (5:1) to afford the iodobismuthine as pale yellow crystals (624 mg, 100%), m.p. 142–144°C [93JCS(P1)1169].

2.3.1.5. Via acidolysis (method E)

Lewis acids such as $AlCl_3$, $TlCl_3$, $SiCl_4$, $SnCl_4$, PCl_3, $AsCl_3$, $SbCl_3$, $TiCl_4$, $FeCl_3$, $HgCl_2$, $CuCl_2$ and $ZnCl_2$, as well as protonic acids such as HCl and HBr, can all cleave the Bi–C bond of triorganylbismuthines to afford the corresponding mono- and/or dihalobismuthines. Boron trifluoride also cleaves a Bi–C bond of triarylbismuthines to generate the diarylfluorobismuthines, but they have not been isolated due to the low stability [93JCS(P1)1169].

However, on subsequent treatment with the respective halide anions, the fluorobismuthines can be converted to diarylchloro-, bromo- and iodobismuthines (Section 2.3.1.2).

$$R_3Bi + MX_m \longrightarrow R_{3-n}BiX_n + R_nMX_{m-n} \quad (n = 1, 2)$$

$$R_3Bi + n\,HX \longrightarrow R_{3-n}BiX_n + n\,HR \quad (n = 1, 2)$$

Among different organyl ligands attached to the bismuth atom in unsymmetrical triarylbismuthines, the most electron-rich ligand is preferentially removed [83JOM(256)C1]. When one of the aryl ligands has a coordinating substituent at a suitable position in the molecule, a mixed triarylbismuthine suffers a selective Bi–C bond cleavage to give a halodiarylbismuthine, in which the hypervalent structure is formed at the bismuth center through the coordination from the donating substituent [93JCS(P1)2969, 95OM3848]. Using this methodology, an optically active diarylhalobismuthine bearing the chirality at the bismuth center has been prepared [98OM1711].

Synthesis of chloro[2-(*N,N*-dimethylaminomethyl)phenyl](4-methoxyphenyl)bismuthine

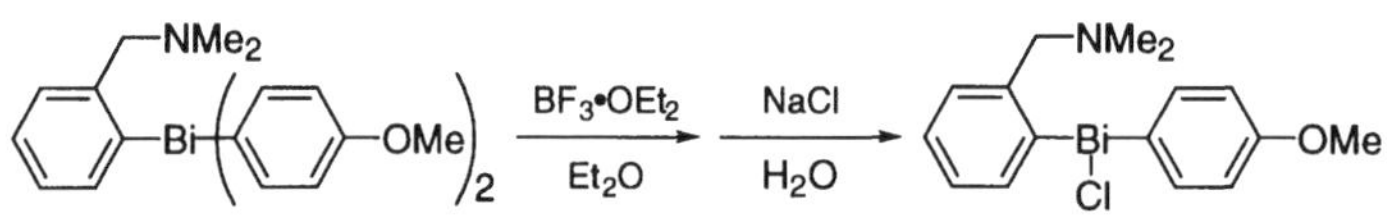

To a well stirred suspension of chlorobis(4-methoxyphenyl)bismuthine (ca. 10 mmol), generated from tris(4-methoxyphenyl)bismuthine (3.53 g, 6.66 mmol) and bismuth chloride (1.05 g, 3.33 mmol) in ether (15 ml), was added dropwise at −50°C a suspension of lithiated *N,N*-dimethylbenzylamine (ca. 10 mmol) in the same solvent (20 ml). The resulting mixture was stirred for 3 h, during which the temperature was gradually raised to the ambient temperature. After the addition of boron trifluoride diethyl etherate (ca. 15 mmol) at room temperature, the mixture was quenched with brine (20 ml) and extracted with ethyl acetate (50×4 ml). The combined extracts were evaporated under reduced pressure to leave a brown oily residue, which was chromatographed on silica gel using hexane–ethyl acetate (3:1) as the eluent to afford the crude chlorobismuthine. Recrystallization from hexane–benzene (5:1) gave white crystals as a pure compound (1.46 g, 30%), m.p. 145–147°C [93JCS(P1)2969].

Treatment of triarylbismuthines with acyl chloride gives diarylchlorobismuthine and/or aryldichlorobismuthines along with the corresponding ketones [22JCS(121)104]. Benzoyl chloride cleaves the Bi–$C_{alkynyl}$ bond of a sterically crowded alkynyldiarylbismuthine to give diarylchlorobismuthine and an alkynyl ketone [92BCJ3504]. The C–Cl bond in carbon tetrachloride is cleaved by tributylbismuthine to generate dibutyl(trichloromethyl)bismuthine, which is subsequently converted to butyldichlorobismuthine via the elimination of a carbene [85ZOB586]. The Bi–C bond can also be cleaved by chloroform on silica gel, resulting in the formation of the Bi–Cl bond [98OM1013].

$$R_3Bi + R'COX \longrightarrow R_2BiX + R'COR$$

$$R_3Bi + R'X \longrightarrow R_2BiX + R'R$$

Synthesis of chlorobis(4-methylphenyl)bismuthine

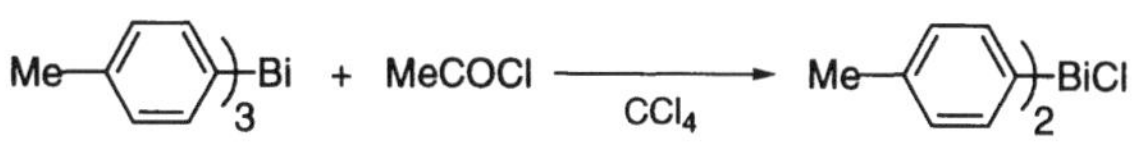

When a mixture of tris(4-methylphenyl)bismuthine (2 g, 4.1 mmol), acetyl chloride (0.5 g, 6.4 mmol) and carbon tetrachloride was heated for 4 h on a water bath, the corresponding diarylchlorobismuthine was obtained as a white solid (1.4 g), which was recrystallized from cold dry chloroform–light petroleum, m.p. 180°C. The mother liquors and washings were united and heated for 5 h to afford chlorobis(4-methylphenyl)bismuthine as yellowish crystals, m.p. 206°C [22JCS(121)104].

2.3.1.6. From alkyl halides and bismuth metal (method F)

Activated bismuth metal reacts with alkyl halides or trifluoromethyl iodide in the presence or absence of metallic copper at high temperatures to afford a mixture of mono- and dihalobismuthines [61JINC213, 63AJC636, 97JA724].

$$RX + Bi \longrightarrow R_nBiX_{3-n} \quad (n = 1, 2)$$

Dibromo(pentamethylcyclopentadienyl)bismuthine has been prepared by this method [89CB287]. It has been suggested that metallic bismuth may insert into the carbon–halogen bond of allylic halides to generate a reactive allylic bismuth(III) halide, though this has not been confirmed directly [90BCJ1738] (Table 5.7). Electrolysis of alkyl halides, using a sacrificial bismuth cathode, forms alkylhalobismuthines [74EL(10)1424].

Synthesis of dibromo(methyl)bismuthine

$$MeBr + Bi \xrightarrow{Cu} MeBiBr_2$$

A mixture of pulverized bismuth (40 g) and copper powder (8 g) was heated at 350°C under an atmosphere of hydrogen for 3 h. Methyl bromide was then introduced at a rate of 5 ml/min at 250°C. A yellow crystalline solid was formed beyond the heating zone. Over an interval of 16 h, yellow crystals (2 g) were collected [61JINC(16)213].

2.3.1.7. *From germylbismuthines and hydrogen chloride (method G)*

The Bi–Ge bond is readily cleaved by hydrogen chloride to form the Bi–Cl bond [78JOM(162)289].

$$(Ar_3Ge)_nBiEt_{3-n} + n\,HCl \longrightarrow Et_{3-n}BiCl_n + n\,Ar_3GeH \quad (n = 1, 2)$$

Synthesis of dichloro(ethyl)bismuthine

$$[(C_6F_5)_3Ge]_2BiEt + 2\,HCl \xrightarrow[THF]{} EtBiCl_2$$

Dry hydrogen chloride gas (0.03 g, 0.82 mmol) was passed to a solution of ethylbis[tris(-pentafluorophenyl)germyl]bismuthine (0.72 g, 0.52 mmol) in THF (5 ml) at room temperature. After 20 min the solvent was evaporated and the residue was extracted twice with hot hexane to give dichloro(ethyl)bismuthine as a yellow solid (0.14 g, 64%), m.p. 154–158°C [78JOM(162)289].

2.3.1.8. *From dibismuthines (method H)*

Iodine readily cleaves the Bi–Bi bond of tetraphenyldibismuthine to give iododiphenylbismuthine [84JCS(D)2365, 91G(121)93] (Section 2.2.2.3).

$$R_2BiBiR_2 + I_2 \longrightarrow 2\,R_2BiI$$

Synthesis of iododiphenylbismuthine

$$Ph_2BiBiPh_2 + I_2 \xrightarrow[Me_2CO]{} 2\,Ph_2BiI$$

To a suspension of tetraphenyldibismuthine (0.300 g, 0.413 mmol) in acetone (25 ml) was slowly added a solution of iodine (0.107 g, 0.422 mmol) in the same solvent (10 ml). The color of the solution turned from orange to yellow. The reaction mixture was filtered and the filtrate was concentrated to a small volume. Dilution with water (20 ml) precipitated analytically pure iododiphenylbismuthine, which was filtered off and dried in vacuo over phosphorus pentoxide. Yield, 0.102 g (25.2%) [84JCS(D)2365].

2.3.1.9. *Via disproportionation (method I)*

Disproportionation reaction of halodiorganylbismuthines leads to a mixture of dihaloorganylbismuthines and triorganylbismuthines [62ZAAC(317)54, 22JCS(121)104].

$$2\,R_2BiX \longrightarrow R_3Bi + RBiX_2$$

2.3.1.10. From diazonium salt–bismuth halide complexes (method J)

In the reaction between an aryldiazonium salt and bismuth chloride, diarylchlorobismuthine is formed together with the desired triarylbismuthine [41JA949].

$$[ArN_2^+][BiCl_4^-] \longrightarrow Ar_2BiCl + N_2 \ (+ \ Ar_3Bi)$$

2.3.1.11. Transformation of the substituted group (method K)

The formyl group on the side chain in mixed halodiarylbismuthines can be converted to the imino group by treatment with excess benzylamine under reflux in benzene [95OM3848, 98OM1711].

$$ArAr'BiX \longrightarrow ArAr''BiX$$

2.3.1.12. Halobismuthine complex salts and related compounds

The bismuth center in diarylhalobismuthines and aryldihalobismuthines is positively charged and possesses the Lewis acidity. Pyridine, 2,2′-bipyridine, 1,10-phenanthroline, hexamethylphosphoric triamide (HMPA), triphenylphosphine oxide, diphenyl sulfoxide and *N,N′*-dimethylpropylene urea (DMPU) can coordinate to the bismuth center to give the corresponding halobismuthine complexes $Ar_nBiX_{3-n}{\cdot}L_m$ ($n = 1{,}2$; $m = 1{,}2$). A THF molecule can also coordinate to the bismuth atom in bromodiphenylbimuthine and dihalophenylbismuthines [92JCS(D)1967, 93JCS(D)637]. When the metathesis between a diarylhalobismuthine (Ar_2BiX) and a metal salt MY (M=Ag, Tl; Y=BF_4, PF_6, ClO_4) is carried out in the presence of these donors, the corresponding cationic complexes of the type $[Ar_2Bi{\cdot}L_2^+][Y^-]$ are formed [79JOM(166)365, 93JOM(460)C22, 96JCS(D)443].

$$Ar_2BiX + n\,L \longrightarrow [Ar_2BiL_n]X$$

$$ArBiX_2 + n\,L \longrightarrow [ArBiL_n]X_2$$

Synthesis of diphenylbismuth tetrafluoroborate–bis(hexamethylphosphoramide) complex

$$Ph_2BiBr + AgBF_4 \xrightarrow[THF]{2\,(Me_2N)_3P{=}O} [Ph_2Bi(HMPA)_2][BF_4]$$

HMPA (0.058 g, 0.325 mmol) was added dropwise to a stirred solution of bromodiphenylbismuthine (0.072 g, 0.172 mmol) in THF (5 ml) at room temperature. This was followed by the addition of a solution of $AgBF_4$ (0.032 g, 0.162 mmol) in THF (5 ml), which led to the formation of a pale yellow precipitate that turned dark grey after a few minutes. The mixture was filtered through a Celite bed and the colorless filtrate was concentrated under vacuum to about 5 ml. Hexane (20 ml) was added as an overlayer, and solvent diffusion over a period of days at −20°C afforded diphenylbismuth tetrafluoroborate–HMPA complex as colorless crystals (70%) [93JOM(460)C22].

Reaction of diarylhalobismuthines with ammonium, arsonium or phosphonium halides yields the corresponding bismuthate complexes of the type $[ER_4^+][Ar_2BiX_2^-]$ (X=Cl, Br, I; E=N, As, P) [69JOM(20)99, 79JOM(166)365, 88JAP(K)174926, 92JCS(D)1967, 93JCS(D)637]. Similar treatment of dihalo(phenyl)bismuthines gives the ate complexes of the type $[ER_4^+][PhBiX_3^-]$. These compounds and related complexes are listed in Table 2.28.

$$Ar_2BiX + R_4EY \longrightarrow [R_4E^+][Ar_2BiXY^-]$$

$$ArBiX_2 + R_4EX \longrightarrow [R_4E^+][ArBiX_3^-]$$

Synthesis of tetraphenylarsonium dibromodiphenylbismuthate

$$Ph_2BiBr + Ph_4AsBr \xrightarrow[Me_2CO]{} [Ph_4As^+][Ph_2BiBr_2^-]$$

An equimolar amount of bromodiphenylbismuthine and tetraphenylarsonium bromide were dissolved in boiling acetone. By gradually evaporating the solution, the arsonium salt separated out as pale yellow needles, m.p. 179–182°C [69JOM(20)99].

2.3.1.13. Arene–bismuth halide π-complexes

Treatment of a bismuth trihalide with electron-rich arenes affords arene–BiX_3 π-complexes. When the reaction is carried out in the presence of $AlCl_3$, the corresponding cationic π-complexes are formed [87AG(E)74]. The arenes employed so far are benzene, toluene, mesitylene, hexamethylbenzene, perylene, pyrene, acenaphthene, phenanthrene, fluorene, and fluoranthene–*p*-xylene (1/0.5). The isolated compounds are summarized in Table 2.29.

$$n\,BiX_3 + ArH \longrightarrow (BiX_3)_n\text{–}ArH \quad (n = 1, 2)$$

Synthesis of mesitylene–bismuth chloride π-complex

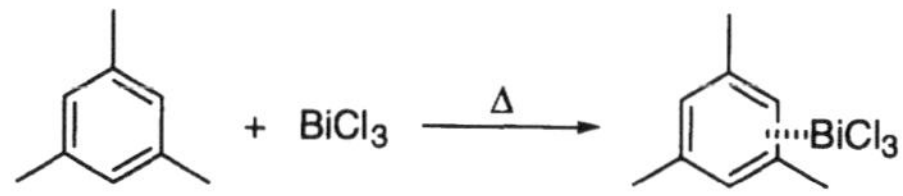

A suspension of bismuth chloride (0.25 g, 0.79 mmol) in mesitylene (25 ml) was heated to 60°C and then filtered. A colorless filtrate was obtained, from which $BiCl_3 \cdot 1,3,5\text{-}Me_3C_6H_3$ precipitated out as colorless plates (0.3 g, 87%), m.p. 157°C (decomp.) [86AG(E)757].

TABLE 2.24
Halodiorganylbismuthines and related complex salts

Compound R	Synthetic method[a]	Yield (%)	M.p. (°C) or b.p. (°C/mmHg)	Physical data	Reference
Chlorides (R_2BiCl)					
Me	D	–	116	–	[1887CB1516]
Et	C				[39JOC162]
	G	96	–	–	[78JOM(162)289]
$^iPrO_2CCH_2CH_2$	C	44	–	^{1}H-NMR, IR	[85OM641]
$(Me_3Si)_2CH$	C	51	108–110/0.0001	–	[85JA1062]
	C	93	71–73	^{1}H-, ^{13}C-NMR, MS, X-ray	[99OM328]
C_5H_5	E	98	–	IR	[60CB1417]
1,2,4-$^tBu_3C_5H_2$	C	54	145 (decomp.)	^{1}H-, ^{13}C-NMR, X-ray	[94CB1335]
C_5Me_5	C	30	74 (decomp.)	^{1}H-, ^{13}C-NMR	[98AST(11)42]
Ph	A, D	79	184–185	–	[15JCS(107)16]
	A	99.4	–	–	[41JA207]
	A	80	190	–	[86T3111]
	D	–	182	–	[21JCS(119)913]
	D	11	183–184	–	[24JCS(125)854]
	D	–	–	X-ray	[97ZN(B)149]
	E	88	185–186	–	[22JCS(121)36]
	E	–	183	–	[22JCS(121)104]
	E	68	184–185	–	[24JCS(125)864]
	E	72	184	–	[34JCS405]
	E	–	–	–	[69JCS(A)2138]
Ph(^{210}Bi)	A	–	–	–	[64JC(15)70]
2-MeC_6H_4	E	–	–	–	[69JCS(A)2138]
4-MeC_6H_4	A	–	181.5	–	[16JCS(109)250]
	A	–	181	–	[21JCS(119)913]
	A	93.3	181.5	–	[41JA207]
	A	–	183–185	MS	[76JOM(122)351]
	A	92	185	–	[86T3111]

	E	–	180	–	[22JCS(121)104]
	E	–	–	–	[22JCS(121)256]
	E	–	–	–	[69JCS(A)2138]
2-$Me_2NCH_2C_6H_4$	C	50	150–152	^{1}H-, ^{13}C-NMR, MS, X-ray	[97IC2770]
4-PhC_6H_4	A	–	–	–	[36JA1820]
2-$MeOCH_2C_6H_4$	E	22	140–142	^{1}H-NMR, IR	[97JCS(P1)1609]
2-$CHOC_6H_4$	B	12	208–210 (decomp.)	^{1}H-NMR, IR	[95OM3848]
2-$MeO_2CC_6H_4$	J	10.3	180–181	–	[41JA949]
2-$EtO_2CC_6H_4$	J	6.5	147–148	–	[41JA949]
3-$EtO_2CC_6H_4$	J	36	130–135 (decomp.)	–	[62IZV638]
4-$Me_2NC_6H_4$	A	–	140 (decomp.)	–	[62ZAAC(317)54]
4-$MeOC_6H_4$	A	–	158–159 (decomp.)	–	[62ZAAC(317)54]
2-$^t BuSO_2C_6H_4$	E–B	–	218–221	^{1}H-NMR, IR, MS	[93JCS(P1)1169]
2-$Et_2NSO_2C_6H_4$	D	78	127–128	1H-NMR, IR, MS	[98OM4049]
3-FC_6H_4	A	100	118–119	^{1}H-, ^{19}F-NMR	[94JFC(69)219]
4-FC_6H_4	A	100	129–130	^{1}H-, ^{19}F-NMR	[94JFC(69)219]
4-ClC_6H_4	A	89.3	158–160	–	[41JA207]
4-BrC_6H_4	A	–	165	–	[62ZAAC(317)54]
4-$CF_3C_6H_4$	A	100	185–187	^{1}H-, ^{19}F-NMR	[94JFC(69)219]
4-(CH_2=CH)-2-$MeOC_6H_3$	C	–	–	–	[63USP3109851]
2,4,6-$Me_3C_6H_2$	A	41	124–126 (decomp.)	^{1}H-, ^{13}C-NMR, IR	[97ZAAC(623)941]
	A	86	–	^{1}H-, ^{13}C-NMR, IR	[98JOM(560)211]
2,4,6-$(CF_3)_3C_6H_2$	B	10	142–143	^{1}H-, ^{19}F-NMR, MS, X-ray	[91JOM(402)55]
2,4,6-$^i Pr_3C_6H_2$	C	90	130–132	^{1}H-NMR, IR, MS	[92BCJ3504]
2,4,6-$Ph_3C_6H_2$	C	59	158–163	^{1}H-, ^{13}C-NMR, IR, MS	[95JOM(485)141]
1-$C_{10}H_7$	A	–	168	–	[21JCS(119)913]
	D	–	240 (decomp.)	–	[15JCS(107)16]
	E	–	–	–	[22JCS(121)256]
	E	–	–	–	[48ZOB936]
8-Me_2N-1-$C_{10}H_6$	C	50	254	^{1}H-, ^{13}C-NMR, MS	[97IC2770]

TABLE 2.24 *(continued)*

Compound R	Synthetic method[a]	Yield (%)	M.p. (°C) or b.p. (°C/mmHg)	Physical data	Reference
Bromides (R_2BiBr)					
Me	D	–	–	–	[1887CB1516]
	F	85.7	–	–	[74EL(10)1424]
Et	D	–	–	–	[1887CB1516]
	F	75.8	–	–	[74EL(10)1424]
^{i}Bu	D	–	–	–	[1888CB2035]
^{t}Bu	D	40	–	^{1}H-NMR, IR, MS	[84CB1999]
$^{i}C_5H_{11}$	D	–	–	–	[1888CB2035]
CH_2=CMe	D	–	–	^{1}H-, ^{13}C-NMR, MS, X-ray	[99ZAAC(625)629]
Ph	A	–	157–158	–	[1889LA(251)323]
	A	85	–	–	[41JA207]
	A	88	155–157	–	[52LA(578)136]
	C	75.4	157–158	–	[22JCS(121)978]
	D	–	154–156	–	[14JCS(105)2210]
	D	92	157	–	[15JCS(107)16]
	D	36	157	–	[24JCS(125)854]
	E	–	158	–	[34JCS405]
4-MeC_6H_4	–	–	–	^{1}H-NMR	[71JOM(32)C11]
4-$Me_2NC_6H_4$	A	–	–	–	[77JOM(136)185]
4-FC_6H_4	A	–	–	–	[77JOM(136)185]
4-ClC_6H_4	A	–	159	–	[22JCS(121)104]
2,4,6-$Me_3C_6H_2$	A	81	160 (decomp.)	^{1}H-NMR, MS, X-ray	[94JOM(470)93]
1-$C_{10}H_7$	D	–	–	–	[15JCS(107)16]
Iodides (R_2BiI)					
CF_3	D, F	–	–	–	[63AJC636]
$NCCH_2CH_2$	F	84.5	–	IR	[74EL(10)1424]
Ph	A	–	–	^{13}C-NMR	[91G(121)93]
	B	–	133	–	[1897CB2843]

	B	–	132–134	–	[31JA1025]
	B	64.8	132–134	–	[41JA207]
	B	–	134–136	–	[88JAP(K)174926]
	D	–	133–134	–	[15JCS(107)16]
	D	–	133	–	[24JCS(125)854]
	D	–	–	^{13}C-NMR	[91G(121)93]
	H	25.2	–	–	[84JCS(D)2365]
	H	–	–	^{13}C-NMR	[91G(121)93]
3-MeC_6H_4	B	–	108–111 (decomp.)	–	[88JAP(K)174926]
4-MeC_6H_4	B	45.5	147–148	–	[41JA207]
	B	–	151–154 (decomp.)	–	[88JAP(K)174926]
2-$Et_2NSO_2C_6H_4$	D	78	142–145	^{1}H-NMR, IR, MS	[98OM4049]
Ar[b]	D	93	Oil	^{1}H-NMR, IR, MS	[98JCS(P1)2511]
4-FC_6H_4	B	–	>137 (decomp.)	–	[88JAP(K)174926]
4-ClC_6H_4	B	34.7	139–140	–	[41JA207]
	D	49	139	–	[22JCS(121)104]
	B	69	142–145 (decomp.)	–	[88JAP(K)174926]
2,4,6-$Me_3C_6H_2$	B	–	135–137 (decomp.)	^{1}H-NMR, IR, MS	[92BCJ3504]
1-$C_{10}H_7$	D	–	–	–	[15JCS(107)16]
Hexafluorophosphate ($[R_2Bi^+][PF_6^-]$)					
2$Me_2NCH_2C_6H_4$	B	65	232–234	^{1}H-, ^{13}C-NMR, MS, X-ray	[97OM3597]
Tetrachloroaluminate ($[R_2Bi^+][AlCl_4^-]$)					
1,2,4-$^{t}Bu_3C_5H_2$	B	72	–	^{1}H-, ^{13}C-NMR, X-ray	[97ZN(B)398]

[a] For notation, see Section 2.3.1.

[b]

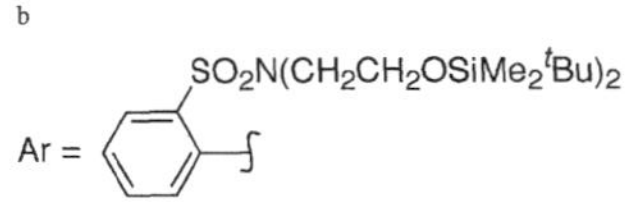

TABLE 2.25
Dihaloorganylbismuthines and related complex salts

Compound R	Synthetic method[a]	Yield (%)	M.p. (°C)	Physical data	Reference
Dichlorides ($RBiCl_2$)					
Me	A	–	242	–	[1887CB1516]
	A	96	–	MS	[97OM1655]
Et	A	87	–	–	[97OM1655]
	D	–	–	–	[1887CB1516]
	E	–	–	–	[1854LA(92)371]
	G	64	154–158	–	[78JOM(162)289]
Bu	E	96	–	–	[85ZOB586]
C_5H_{11}	–	–	–	–	[35NAT(135)1038]
$ClCH_2$	E	–	–	–	[52ZOB1528]
$PhCH_2$	–	–	–	–	[70USP3504005]
$^iPrO_2CCH_2CH_2$	C	89	144–145 (decomp.)	^{1}H-, ^{13}C-NMR, IR	[85OM641]
$(Me_3Si)_2CH$	E	92	132–134	^{1}H-, ^{13}C-NMR, MS, X-ray	[99OM328]
$CH_2{=}CH$	C	–	–	–	[59TL1]
C_5H_5	C	76.9	–	^{1}H-NMR, X-ray	[90JOM(386)177]
$^iPr_4C_5H$	C	84	158 (decomp.)	^{1}H-, ^{13}C- NMR, X-ray	[94CB1335] [96ZAAZ(622)923]
1,2,4-$^tBu_3C_5H_2$	C	82	148 (decomp.)	^{1}H-NMR	[94CB1335]
C_5Me_5	I	–	–	–	[98AST(11)42]
2-Py$(Me_3Si)_2C$[b]	C	44	194	^{1}H-, ^{13}C-NMR, MS	[91CC1560]
Ph	A, E	–	–	–	[72JOM(44)317]
	–	–	–	^{1}H-NMR	[71JOM(32)C11]
4-MeC_6H_4	A	–	206–207	–	[21JCS(119)913]
	A	60	209	–	[86T3111]
	E	–	206	–	[22JCS(121)104]
2-$Me_2NCH_2C_6H_4$	A	93	228–230	MS	[97IC2770]
2-$MeO_2CC_6H_4$	J	2	220–221	–	[41JA949]
4-$Me_2NC_6H_4$	I	–	–	–	[62ZAAC(317)54]
4-$MeOC_6H_4$	I	–	–	–	[62ZAAC(317)54]

2-$Et_2NSO_2C_6H_4$	A	85	177–181	^{1}H-NMR, IR, MS	[98OM4049]
4-BrC_6H_4	I	–	–	–	[62ZAAC(317)54]
2,6-$(Me_2NCH_2)_2C_6H_3$	C	39	250 (decomp.)	^{1}H-, ^{13}C-NMR, MS	[92ICA(198-200)271]
2,6-$Mes_2C_6H_3$[c]	C	65	220–224	^{1}H-, ^{13}C-NMR, IR, X-ray	[99JA3357]
2,6-$Trip_2C_6H_3$[d]	C	91	255–257	^{1}H-, ^{13}C-NMR, IR	[99JA3357]
2,4,6-$Me_3C_6H_2$	A	71	212	^{1}H-, ^{13}C-NMR, IR, MS	[97ZAAC(623)941]
2,4,6-$Ph_3C_6H_2$	A	63	–	IR, MS	[95JOM(485)141]
	C	20	–	X-ray	[97JOM(530)71]
1-$C_{10}H_7$	E	–	–	–	[22JCS(121)104]
Dibromides ($RBiBr_2$)					
Me	C	–	214	–	[1887CB1516]
	C	–	214	–	[22BSF(31)545]
	D	–	–	–	[66JOM(6)259]
	F	15–20	–	–	[61JINC(16)213]
Et	C	–	–	–	[15CB1473]
	D	–	–	–	[1887CB1516]
^{i}Bu	A	–	124	–	[1888CB2035]
$^iC_5H_{11}$	A	–	134	–	[1888CB2035]
C_5Me_5	F	69	230 (decomp.)	^{1}H-, ^{13}C-NMR, MS	[89CB287]
CH_2=CMe	D	70	248–250	–	[63IZV1507]
	D	–	–	^{1}H-, ^{13}C-NMR	[99ZAAC(625)629]
$C_{16}H_{33}$	–	–	–	–	[39JPH(67)17]
Ph	A	–	202	–	[21JCS(119)913]
	A, D	–	–	IR, UV	[72JOM(44)317]
	A	–	–	X-ray	[94JMAC(4)891]
	C	96	201	–	[79ZOB1185]
	D	–	205–206	–	[14JCS(105)2210]
	D	–	201–202	–	[15JCS(107)16]

TABLE 2.25 (*continued*)

Compound R	Synthetic method[a]	Yield (%)	M.p.(°C)	Physical data	Reference
2-MeC_6H_4	A	95.5	181	–	[41JA207]
	–	–	–	^{1}H-NMR	[71JOM(32)C11]
3-MeC_6H_4	–	–	–	^{1}H-NMR	[71JOM(32)C11]
4-MeC_6H_4	–	–	–	^{1}H-NMR	[71JOM(32)C11]
2-$MeOC_6H_4$	–	–	171	–	[42CRV(30)281]
3-FC_6H_4	A	100	203–205 (decomp.)	^{1}H-, ^{19}F-NMR	[94JFC(69)219]
4-FC_6H_4	A	–	–	–	[77JOM(136)185]
	A	100	230 (decomp.)	^{1}H-, ^{19}F-NMR	[94JFC(69)219]
4-ClC_6H_4	I	–	244	–	[22JCS(121)104]
4-$CF_3C_6H_4$	A	100	192–195 (decomp.)	^{1}H-, ^{19}F-NMR	[94JFC(69)219]
2-(2′-Py)C_6H_4	C	76	248–250	^{13}C-NMR	[90JCS(D)899]
1-$C_{10}H_7$	A	–	208	–	[21JCS(119)913]
	A	89	–	–	[41JA207]
2,4,6-$Me_3C_6H_2$	A	66	173	^{1}H-, ^{13}C-NMR, IR, MS, X-ray	[97ZAAC(623)941]
Diiodides ($RBiI_2$)					
Me	E	–	225 (decomp.)	–	[1887CB1516]
	F	40	–	X-ray	[97JA724]
Et	B	–	–	–	[1887CB1516]
	B	–	–	–	[1854LA(92)371]
CF_3	D, F	–	–	–	[63AJC636]
$^iPr_4C_5H$	C	90	178–180 (decomp.)	^{1}H-, ^{13}C-NMR, X-ray	[94CB1335]
Ph	A	–	–	^{13}C-NMR	[91G(121)93]
	D	–	193–194	–	[24JCS(125)854]
	D	48.5	180	–	[64CB789]
	D	–	–	IR, UV	[72JOM(44)317]
2-$Me_2NCH_2C_6H_4$	B	71	–	^{1}H-, ^{13}C-NMR, MS, X-ray	[97IC2770]

Tetrachlorobismuthate (($[^tRBi^+][BiCl_4^-]_2)_2$)					
$^iPr_4C_5H$	C	66	60 (decomp.)	^1H-NMR, X-ray	[94CB1335]
Mixed halides (*RBiXY*)					
R = , X = , Y =					
Bu, Cl, OPh	I	–	–	–	[70USP3504005]
$PhCH_2$, Cl, CN	I	–	–	–	[70USP3504005]
Ph, Cl, Br	D	–	–	–	[15JCS(107)16]
Ph, Br, $N(SO_2Me)_2$	B	79	141 (decomp.)	^1H-NMR	[95ZAAC(621)1746]
Ph, Br, SPh	I	–	–	–	[70USP3505005]
$(\eta^5\text{-}Cp_2^*Al)(\eta^5\text{-}Cp^*Bi)\ (\mu\text{-}AlI_4)(AlI_4)_2$	C	–	–	X-ray	[99ZACC(625)1354]

[a] For notation, see Section 2.3.1.

[b]

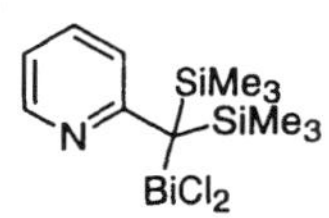

[c] Mes = 2,4,6-$Me_3C_6H_2$.

[d] Trip = 2,4,6-$^iPr_3C_6H_2$.

TABLE 2.26
Mixed halodiorganylbismuthines

Compound R[1]	R[2]	Synthetic method[a]	Yield (%)	M.p. (°C)	Physical data	Reference
Chlorides (R^1R^2BiCl)						
$^iPr_4C_5H$	1,2,3-$^iPr_3C_5H_2$	C	68	112 (decomp.)	^{1}H-, ^{13}C-NMR	[94CB1335]
R (**1**)[b]	R′ (**1**)[b]	C	–	–	MS, X-ray	[91CC1560]
Ph	2-$Me_2NCH_2C_6H_4$	E-B	47	145–147	^{1}H-NMR, IR	[93JCS(P1)2969]
	Ar (**2**)[b]	E-B	48	136–138	^{1}H-NMR, MS, X-ray	[93JCS(P1)2969] [95OM1542]
4-MeC_6H_4	2-$Me_2NCH_2C_6H_4$	E-B	42	145–147	^{1}H-NMR, IR, X-ray	[93JCS(P1)2969]
	2-$CHOC_6H_4$	E-B	–	–	–	[95OM3848]
	2-$MeCOC_6H_4$	E-B	90	175–177	^{1}H-, ^{13}C-NMR, IR	[95OM3848]
	2-(PhN=CH)C_6H_4	K	69	210–212	^{1}H-NMR, IR	[95OM3848]
	2-OxC_6H_4 (**3**)[b]	E	80	184–186	^{1}H-, ^{13}C-NMR, IR	[98OM1013]
	2-$^tBuSO_2C_6H_4$	E-B	90	170–172	^{1}H-NMR, IR, MS, X-ray	[93JCS(P1)1169]
	2-$^tBuSO_2$-3-DC_6H_3	E-B	85	167–169	^{1}H-NMR	[93JCS(P1)1169]
	8-Me_2N-1-$C_{10}H_6$	C	40	193–195	^{1}H-NMR, IR, MS	[93JCS(P1)2969]
	Ar (**4**)[b]	E-B	59	185–190 (decomp.)	^{1}H-NMR, IR, MS, X-ray	[98OM4049]
4-$MeOC_6H_4$	2-$Me_2NCH_2C_6H_4$	E-B	30	145–147	^{1}H-NMR, IR, MS	[93JCS(P1)2969]
4-FC_6H_4	2-$^tBuSO_2C_6H_4$	E-B	70	147–149	^{1}H-NMR, IR, MS	[93JCS(P1)1169]
4-ClC_6H_4	2-$Me_2NCH_2C_6H_4$	E-B	60	174–176 (decomp.)	^{1}H-NMR, IR	[93JCS(P1)2969]
1-$C_{10}H_7$	2-$Me_2NCH_2C_6H_4$	E-B	40	200–203 (decomp.)	^{1}H-NMR, IR	[93JCS(P1)2969]
	Ar (**2**)[b]	E-B	45	180–184	^{1}H-NMR, MS	[93JCS(P1)2969]
Bromides (R^1R^2BiBr)						
Ph	2-$Me_2NCH_2C_6H_4$	B	100	163–165	^{1}H-NMR, IR, MS	[93JCS(P1)2969]

	2-$HCOC_6H_4$	E-B	20	138–140	^{1}H-, ^{13}C-NMR, IR	[95OM3848]
	2-$MeCOC_6H_4$	E-B	87	146–148	^{1}H-, ^{13}C-NMR, IR	[95OM3848]
4-MeC_6H_4	2-$Me_2NCH_2C_6H_4$	C	70	126	^{1}H-, ^{13}C-NMR	[81JOM(212)C7]
	2-$MeOCH_2C_6H_4$	C	31	123–124	^{1}H-NMR	[81JOM(212)C7]
	2-$HCOC_6H_4$	E-B	24	139–141	^{1}H-, ^{13}C-NMR, IR	[95OM3848]
4-MeC_6H_4	2-$MeCOC_6H_4$	E-B	83	174–176 (decomp.)	^{1}H-, ^{13}C-NMR, IR, X-ray	[95OM3848]
	2-(PhN=CH)C_6H_4	K	82	218–220	^{1}H-NMR, IR	[95OM3848]
	2-^{t}Bu$SO_2C_6H_4$	E-B	90	159–161	^{1}H-NMR, IR, MS	[93JCS(P1)1169]
	Ar (**5**)[b]	E-K-B	20	190–192 (decomp.)	^{1}H-NMR, IR	[98OM1711]
4-$MeOC_6H_4$	2-$HCOC_6H_4$	E-B	16	125–127	^{1}H-, ^{13}C-NMR, IR	[95OM3848]
	2-$MeCOC_6H_4$	E-B	12	153–155	^{1}H-, ^{13}C-NMR, IR	[95OM3848]
4-ClC_6H_4	2-$HCOC_6H_4$	E-B	27	155–157	^{1}H-, ^{13}C-NMR, IR	[95OM3848]
	2-$MeCOC_6H_4$	E-B	86	157–159 (decomp.)	^{1}H-, ^{13}C-NMR, IR	[95OM3848]
1-$C_{10}H_7$	2-$HCOC_6H_4$	E-B	14	193–195 (decomp.)	^{1}H-, ^{13}C-NMR, IR	[95OM3848]
	2-$MeCOC_6H_4$	E-B	82	225–227 (decomp.)	^{1}H-, ^{13}C-NMR, IR	[95OM3848]
Iodides (R^1R^2BiI)						
4-MeC_6H_4	2-$HCOC_6H_4$	E-B	88	141–143	^{1}H-, ^{13}C-NMR, IR	[95OM3848]
	2-$MeCOC_6H_4$	E-B	89	148–150	^{1}H-, ^{13}C-NMR, IR	[95OM3848]
	2-^{t}Bu$SO_2C_6H_4$	D	100	142–144	^{1}H-NMR, IR, MS	[93JCS(P1)1169]
	Ar (**5**) (*exo*)[b]	E-K-B	16	212–214	^{1}H-NMR, IR	[98OM1711]
4-ClC_6H_4	Ar (**6**) (*exo*)[b]	E-B	83	153–157 (decomp.)	^{1}H-NMR, IR	[98OM1711]
	Ar (**7**) (*exo*)[b]	E-B	83	176–179 (decomp.)	^{1}H-, ^{13}C-NMR, IR, X-ray	[98OM1711]

[a] For notation, see Section 2.3.1.
[b] For structures, see Scheme 2.7.

TABLE 2.27
Halobismuthine complexes

Compound R	L	n	Yield (%)	M.p. (°C)	Physical data	Reference
Bromides ($R_2BiBr{\cdot}L$)						
Ph	THF		–	–	X-ray	[92JCS(D)1967]
	Ph_3PO		87	–	–	[95JOM(496)59]
2,4,6-$Me_3C_6H_2$	Ph_2SO		68	–	X-ray	[95JOM(496)59]
	$(Me_2N)_3PO$		77	–	X-ray	[95JOM(496)59]
Iodides ($R_2BiI{\cdot}L$)						
Ph	4-Me-pyridine		–	–	X-ray	[99JCS(D)2837]
Tetrafluoroborates ($[R_2BiL_2][BF_4]$)						
Ph	Pyridine		30	–	^{1}H-NMR	[96JCS(D)443]
	Ph_3PO		52	–	^{1}H-, ^{11}B-, ^{31}P-NMR, X-ray	[96JCS(D)443]
	$(Me_2N)_3PO$		70	–	^{1}H-, ^{13}C-, ^{11}B-, ^{19}F-, ^{31}P-NMR, X-ray	[93JOM(460)C22]
Hexafluorophosphate ($[R_2BiL_2][PF_6]$)						
Ph	Pyridine		98	–	^{1}H-, ^{13}C-, ^{31}P-NMR	[96JCS(D)443]
	$(Me_2N)_3PO$		76	–	^{1}H-, ^{13}C-, ^{31}P-NMR, X-ray	[96JCS(D)443]
4-MeC_6H_4	$(Me_2N)_3PO$		64	–	^{1}H-, ^{13}C-, ^{31}P-NMR	[96JCS(D)443]
2,4,6-$Me_3C_6H_2$	Ph_3PO		60	–	^{1}H-, ^{31}P-NMR	[96JCS(D)443]
2,4,6-$Me_3C_6H_2$	$(Me_2N)_3PO$		60	–	^{1}H-, ^{13}C-, ^{31}P-NMR, X-ray	[96JCS(D)443]
Perchlorate ($[R_2BiL_2][ClO_4]$)						
Ph	Ph_3AsO		80	190	IR, EC	[79JOM(166)365]
Dichlorides ($RBiCl_2{\cdot}L_n$)						
Ph	THF	1	89	–	X-ray	[93JCS(D)637]
	Pyridine	2	77	136–137	–	[66BCJ1823]
	4-Me-pyridine	2	–	–	X-ray	[99JCS(D)2837]
	2,2′-Bipyridine	1	–	244–247	IR, EC	[72JOM(44)317] [64IC500]
	1,10-Phenanthroline	1	–	>275	IR, EC	[72JOM(44)317] [64IC500]

Dibromides ($RBiBr_2 \cdot L_n$)						
Ph	THF	1	89	–	X-ray	[92JCS(D)1967]
	Pyridine	2	100	145–146	–	[66BCJ1823]
		2	–	147	–	[88JAP(K)174926]
		2	–	–	X-ray	[99JCS(D)2837]
	4-Me-pyridine	2	–	–	X-ray	[99JCS(D)2837]
	4-^{t}Bu-pyridine	2	–	–	X-ray	[99JCS(D)2837]
	2,2′-Bipyridine	1	–	238–240	IR, UV, EC	[72JOM(44)317] [64IC500]
		1	–	234	–	[88JAP(K)174926]
	1, 10-Phenanthroline	1	–	>275	IR, UV, EC	[72JOM(44)317] [64IC500]
	Ph_3PO[a]	1	>49	–	X-ray	[95JOM(496)59]
	Dmpu[a, b]	1	81	–	X-ray	[95JOM(496)59]
		2	45	–	X-ray	[95JOM(496)59]
$2-MeC_6H_4$	2,2′-Bipyridine	1	–	>154 (decomp.)	–	[88JAP(K)174926]
$3-MeC_6H_4$		1	–	>165 (decomp.)	–	[88JAP(K)174926]
$4-MeC_6H_4$		1	–	>195 (decomp.)	–	[88JAP(K)174926]
$3-CF_3C_6H_4$	Pyridine	2	-	>150 (decomp.)	-	[88JAP(K)174926]
	2,2′-Bipyridine	1	–	>155 (decomp.)	–	[88JAP(K)174926]
$3-FC_6H_4$	Pyridine	2	31	>140 (decomp.)	–	[88JAP(K)174926]
	2,2′-Bipyridine	1	–	>160 (decomp.)	–	[88JAP(K)174926]
$4-FC_6H_4$		1	–	>140 (decomp.)	–	[88JAP(K)174926]
$3-ClC_6H_4$	Pyridine	2	–	>173 (decomp.)	–	[88JAP(K)174926]
	2,2′-Bipyridine	1	–	>165 (decomp.)	–	[88JAP(K)174926]
$4-ClC_6H_4$	Pyridine	2	–	>175 (decomp.)	–	[88JAP(K)174926]
	2,2′-Bipyridine	1	–	180 (decomp.)	–	[88JAP(K)174926]
Diiodides ($RBiI_2 \cdot L$)						
Ph	THF		91	–	X-ray	[92JCS(D)1967]
	4-Me-pyridine	2	–	–	–	[99JCS(D)2837]
	4-^{t}Bu-pyridine	2	–	–	–	[99JCS(D)2837]
	2,2′-Bipyridine		–	212–213	IR, UV	[72JOM(44)317] [64IC500]
	1,10-Phenanthroline			>275	IR, UV	[72JOM(44)317] [64IC500]
Bis(hexafluorophosphate) ($[RBiL_4][(PF_6)_2]$)						
Ph	$(Me_2N)_3PO$		30	–	^{1}H-NMR, X-ray	[96JCS(D)443]

[a] Exists as a dimer.
[b] Dmpu = *N,N′*-dimethylpropylene urea.

TABLE 2.28
Halobismuthine ate complex salts and related compounds

Compound			Yield (%)	M.p. (°C)	Physical data	Reference
$M^+[R_3BiX^-]$						
R=	X=	M=				
CF_3	F	Me_4N	Quant.	–	^{13}C-, ^{19}F-NMR	[99JCS(D)657]
$M^+[R_2BiX_2^-]$						
R=	X_2=	M=				
Ph	Cl_2	Me_4N	–	155–156	UV, EC	[69JOM(20)99]
		Et_4N	80–90	96	IR, Raman, EC	[79JOM(166)365]
		Ph_4As	–	165–167	UV, EC	[69JOM(20)99]
	Br_2	Me_4N	–	112–115	–	[88JAP(K)174926]
		Et_4N	80–90	105–108	IR, Raman, EC	[79JOM(166)365]
			–	118–120	–	[88JAP(K)174926]
		Me_3PhN	–	135–138	–	[88JAP(K)174926]
		Ph_4As	–	179–182	UV, EC	[69JOM(20)99]
		Ph_4P	–	–	X-ray	[92JCS(D)1967]
3-MeC_6H_4	Br_2	Et_4N	–	106–108	–	[88JAP(K)174926]
4-MeC_6H_4	Br_2	Me_4N	–	>140 (decomp.)	–	[88JAP(K)174926]
		Et_4N	–	158–161	–	[88JAP(K)174926]
		Me_3PhN	–	165–167	–	[88JAP(K)174926]
Ph	I_2	Et_4N	55	–	X-ray	[93JCS(D)637]
		Ph_4As	–	201–203	UV, EC	[69JOM(20)99]
	Br, Cl	Me_4N	–	168–172 (decomp.)	–	[88JAP(K)174926]
		Et_4N	–	98–101	–	[88JAP(K)174926]

		Me_3PhN	–	136–139	–	[88JAP(K)174926]
3-MeC_6H_4	Br, Cl	Et_4N	–	100–102	–	[88JAP(K)174926]
4-MeC_6H_4	Br, Cl	Me_4N	–	>155 (decomp.)	–	[88JAP(K)174926]
		Et_4N	–	145–149	–	[88JAP(K)174926]
		Me_3PhN	–	158–160	–	[88JAP(K)174926]
3-ClC_6H_4	Br, Cl	Me_4N	40	>155 (decomp.)	–	[88JAP(K)174926]
		Et_4N	–	76–79	–	[88JAP(K)174926]
4-ClC_6H_4	Br, Cl	Me_4N	40	>160 (decomp.)	–	[88JAP(K)174926]
		Et_4N	–	118–121	–	[88JAP(K)174926]
Ph	$(CN)_2$	Et_4N	70	97–100	IR, EC	[79JOM(166)365]
	$(SCN)_2$	Et_4N	70	81–83 (decomp.)	IR, Raman, EC	[79JOM(166)365]
		Ph_4As	–	115–117	UV, EC	[69JOM(20)99]
	$(N_3)_2$	Et_4N	70	83–85	Raman, EC	[79JOM(166)365]
$M^+[RBiX_3^-]$						
R=	X_3=	M=				
Ph	Cl_3	Ph_4As	–	229–231	UV, EC	[69JOM(20)99]
		4-^{t}BupyH[b]	–	–	X-ray	[99JCS(D)2837]
	Br_3	Bu_4N[a]	70	–	X-ray	[93JCS(D)637]
		Ph_4As	–	228–230	UV, EC	[69JOM(20)99]
	I_3	Et_4N[a]	–	–	X-ray	[92JCS(D)1967]
		Ph_4As	–	225–228	UV, EC	[69JOM(20)99]
2-(2′-Py)C_6H_4	Br_2, I	Et_4N	85	155–157	IR	[90JCS(D)899]
	Br_2, O_2CEt	Na	84	>250	IR	[90JCS(D)899]

[a] Exists as a dimer.
[b] 4-*tert*-Butylpyridine adduct.

TABLE 2.29
Bismuth halide–arene π-complexes

Compound	Yield (%)	M.p. (°C)	Physical data	Reference
$BiCl_3$–arene complexes				
$BiCl_3 \cdot C_6H_6$	–	140 (decomp.)	X-ray	[93CC799]
$BiCl_3 \cdot$1,2-$Me_2C_6H_4$	86	159 (decomp.)	X-ray	[93ZAAC(619)1073]
$BiCl_3 \cdot$1,3-$Me_2C_6H_4$	75	163 (decomp.)	X-ray	[93ZAAC(619)1073]
$BiCl_3 \cdot$1,4-$Me_2C_6H_4$	60	169 (decomp.)	X-ray	[93ZAAC(619)1073]
$BiCl_3 \cdot$1,3,5-$Me_3C_6H_3$	92	157 (decomp.)	X-ray	[87CB1837]
$(BiCl_3)_2 \cdot C_6Me_6$	97	212–215 (decomp.)	X-ray	[87CB1829]
$[BiCl_2 \cdot C_6MeH_5^+][AlCl_4^-]$	77	83	–	[87AG(E)74]
$[BiCl_2 \cdot C_6Me_6^+][AlCl_4^-]$	83	180	X-ray	[87AG(E)74]
$BiCl_3 \cdot$pyrene		–	X-ray	[88JCS(D)191]
$BiBr_3$–arene complexes				
$BiBr_3 \cdot$1,3,5-$Me_3C_6H_3$	89	118	–	[87CB1837]
$(BiBr_3)_2 \cdot C_6Me_6$	95	162 (decomp.)	–	[87CB1829]
$(BiBr_3)_2 \cdot$perylene	–	–	UV	[90JCS(D)1675]
$(BiBr_3)_2 \cdot$pyrene	–	–	UV	[90JCS(D)1675]
$(BiBr_3)_2 \cdot$acenaphthene	–	–	UV	[90JCS(D)1675]
$(BiBr_3)_2 \cdot$phenanthrene	–	–	UV	[90JCS(D)1675]
$(BiBr_3)_2 \cdot$fluorene	–	–	UV	[90JCS(D)1675]
$(BiBr_3)_4 \cdot$(fluoranthene)·(*p*-xylene)	–	145	IR, UV, X-ray	[90JCS(D)1675]
$(BiBr_3)_2 \cdot$(pyrene)([2, 2']paracyclophane)	–	–	IR, X-ray	[95ICA(237)169]

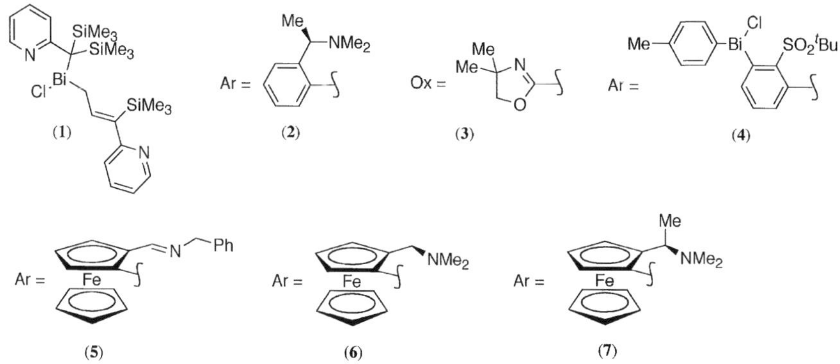

Scheme 2.7. Structures of compounds listed in Table 2.26.

2.3.2. Properties

Lower members of dialkylhalobismuthines are extremely reactive and spontaneously flammable in air. Crystalline mono- and dihalobismuthines bearing no specific ligand are poorly soluble in non-polar organic solvents, probably due to the intermolecular polymeric association between the bismuth and halogen atoms in the solid state (vide infra). Halobismuthines other than iodides are generally moisture-sensitive and gradually hydrolyzed under atmospheric conditions. However, through complexation with an appropriate donor molecule, halobismuthines become soluble in common organic solvents and their Bi–X bond becomes less sensitive towards moisture. Triphenylphosphine oxide, triphenylarsine oxide, HMPA and DMSO work well for this purpose. X-Ray crystallographic analyses have shown that a THF molecule can often coordinate to the cationic bismuth center of chlorodiphenylbismuthine and a series of dihalo(phenyl)bismuthines in the solid state (vide infra). The presence of a coordinating functional group or a sterically demanding group in an organyl ligand moiety stabilizes the labile Bi–Cl and Bi–Br bonds considerably. For example, arylchloro[2-(dimethylaminomethyl)phenyl]bismuthine and chlorobis(2,4,6-triisopropylphenyl)bismuthine can be handled safely in open air and treated on a silica gel column without decomposition [93JCS(P1)2969, 92BCJ3504]. The solubility of these halobismuthines in organic solvents is much higher than the corresponding phenyl derivatives. Diaryliodobismuthines are usually less sensitive towards moisture and their preparation can be carried out in alcohols [31JA1025]. In solution, however, many halodiorganylbismuthines undergo slow disproportionation and give a mixture of triorganylbismuthine, dihaloorganylbismuthine and original halodiorganylbismuthine. The arene–BiX_3 π-complexes are usually unstable in vacuo to liberate the arene ligand.

Molecular weight and electrical conductance measurements have shown that this class of compound exists as the monomeric molecular species in acetone and dichloromethane [79SA(A)339, 69JOM(20)99], while vibrational and electronic spectroscopies indicate that they exist as the polymeric form in the solid state via the intermolecular halogen bridging [72JOM(44)317, 79SA(A)339].

X-Ray crystallographic analyses have been performed for a variety of halobismuthines. They possess several types of coordination geometry, which are strongly dependent on the ligand bound or coordinated to the bismuth atom. Chlorodiphenylbismuthine forms polymeric chains via the bent μ_2-chlorine bridges and almost linear Cl–Bi–Cl units (175.2(2)°) [97ZN(B)149]. The bismuth center thereby possesses a distorted trigonal bipyramidal geometry with two phenyl groups and a pair of unshared electrons at equatorial sites and two chlorine atoms at apical sites. The Bi–C and Bi–Cl bond lengths are 2.124(6)–2.144(6) and 2.746(4)–2.763(3) Å, respectively. The bismuth atom in bromobis(2,4,6-trimethylphenyl)bismuthine has a trigonal pyramidal configuration with the bond angles of 89.4(3)–106.4(3)° around the bismuth atom [94JOM(470)93]. The Bi–Br bond length of 2.696(2) Å is much shorter than the intermolecular Bi··Br distance of 3.795(3) Å. The related bond parameters found in dibromo(2,4,6-trimethylphenyl)bismuthine, which consists of two crystallographycally independent molecules, are 88.2(7)–106.8(7)°, 2.71 Å (average) and 3.02 Å (average), respectively [97ZAAC(623)941]. This suggests that the intermolecular Bi–Br interaction of dibromo(2,4,6-trimethylphenyl)bismuthine is stronger than that of bromobis(2,4,6-trimethylphenyl)-bismuthine.

Chlorobis[2,4,6-tris(trifluoromethyl)phenyl]bismuthine also possesses a trigonal pyramidal bismuth center with the bond angles around the bismuth atom of 87.8(2)–106.9(3)°, a Bi–Cl bond length of 2.463(3) Å, and relatively longer Bi–C bond lengths of 2.338(7)–2.356(8) Å [91JOM(402)55]. On the other hand, the bismuth atom in diiodo(methyl)bismuthine has a square pyramidal local geometry, where four iodine atoms form the base and one methyl group occupies an apical site [97JA724]. The mean Bi–I and Bi–C bond lengths are 3.11 and 2.28 Å, respectively, and the mean C–Bi–I bond angle is 88.1°. Each basal square shares *trans* edges with its two nearest neighbors to form a one-dimensional inorganic BiI_2 chain along the *b*-axis, and all methyl groups are aligned on one side of the BiI_2 basal plane. The crystal structure of dibromo(phenyl)bismuthine comprises one-dimensional zig-zag chains in which the bismuth atoms are linked through bromine bridges [94JMAC(4)891]. The essential unit of the structure is a dimeric $Bi_2Ph_2Br_6$ fragment, and each bismuth center adopts a square-based pyramidal geometry with the phenyl group at the apical position. The Bi–Br bond lengths (2.8801(9)–2.9253(11) Å) differ by only 0.045 Å.

When such a coordinative group as carbonyl, sulfonyl or amino is placed at a spatially suitable position in the molecule, the bismuth center tends to adopt a disphenoidal or pseudotrigonal bipyramidal geometry. The coordinating oxygen or nitrogen atom and the halogen atom occupy the axial positions, and the two *ipso* carbons of the aryl groups and a pair of unshared electrons are located at the equatorial positions. For example, in chloro[2-(dimethylaminomethyl)phenyl](4-methylphenyl)bismuthine, the Bi–N bond distance (2.525(6) Å) is much shorter than the sum of their van der Waals radii 3.66 Å [93JCS(P1)2969]. The relatively long Bi–Cl bond length of 2.700(2) Å as well as the wide N–Bi–Cl bond angle of 162.9(1)° support the view that this molecule has a pseudotrigonal bipyramidal structure with the chlorine and nitrogen atoms at the apical sites. The intramolecular coordination is also observed in bis[2-(dimethylaminomethyl)phenyl]bismuth hexafluorophosphate, in which the bismuth center possesses a pseudotrigonal bipyramidal structure with two nitrogen atoms at the apical positions and two *ipso* carbon atoms at the equatorial positions [97OM3597]. The Bi–N bond lengths of 2.491(1) and 2.509(1) Å, as well as the N–Bi–N bond angle of 164.8(4)° support the hypervalent character of the cationic bismuth center. An optically active diaryliodobismuthine (ArAr′BiI; Ar=2-[1-(*R*)-(dimethylamino)ethyl]-ferrocenyl, Ar′=4-chlorophenyl) has been structurally characterized; the 4-chlorophenyl group occupies the *exo* position with respect to the iron atom and the bismuth center possesses a pseudotrigonal bipyramidal geometry with the Bi–C bond lengths of 2.19(1)–2.24(1) Å, a Bi–I bond length of 2.969(1) Å and a Bi–N distance of 2.71(1) Å [98OM1711]. The chiral bismuth center adopts the (*S*)-configuration, in which the apical imido nitrogen atom occupies the position *trans* to the iodine atom (N–Bi–I bond angle, 161.4(2)°) and coordinates intramolecularly to the bismuth center, with two *ipso* carbon atoms located at the equatorial sites.

Recrystallization of bromodiphenylbismuthine from hexane–tetrahydrofuran (THF) affords a THF complex $Ph_2BiBr{\cdot}(THF)$ as yellow crystals [92JCS(D)1967]. This complex exists as a monomeric form and possesses an equatorially vacant trigonalbipyramidal geometry around the bismuth center, in which the phenyl groups occupy the equatorial sites and the bromine and THF ligand the axial sites. Neutral Lewis base adducts of bromobis(2,4,6-trimethylphenyl)bismuthine have also been characterized; $Mes_2BiBr{\cdot}(OSPh_2)$ and $Mes_2BiBr{\cdot}(HMPA)$ are both monomeric, where each bismuth center has a disphenoidal geometry with an axial bromine atom, an axial ligand donor and two equatorial aryl groups [95JOM(496)59]. The Bi–Br bond lengths in these complexes (2.741(1)–2.769(2) Å) are larger than the corresponding value (2.696 Å) in bromobis(2,4,6-trimethylphenyl)bismuthine, indicating that the lone pair of the donor molecule coordinates to the Bi–Br σ^* orbital. This is supported by the data of the Bi–O bond distances (2.589(7)–2.64(1) Å) and the O–Bi–Br bond angles (171.7(2)–176.1(3)°). THF complexes of the formula $PhBiX_2{\cdot}(THF)$, where X denotes chlorine, bromine or iodine, form the halogen-bridged polymeric chains, in which each bismuth center is located in a square-based pyramidal coordination environment [92JCS(D)1967, 93JCS(D)637]. In these molecules, the phenyl group occupies the apical site, whilst two *cis*-halides, a bridging halide from an adjacent monomer unit and the THF ligand occupy the four basal positions. The halogen bridges in the chlorine derivative are quite asymmetric, where three Bi–Cl bonds and three C–Bi–Cl bond angles are found to be 2.543(3), 2.654(3) and 2.934(3) Å, and 92.6(2), 88.9(2) and 82.8(2)°, respectively [93JCS(D)637]. The Bi–O bond length of the chlorine derivative (2.608(7) Å) is significantly shorter than the corresponding values of the bromide (2.671(8) Å) and iodide (2.808(17)–2.813(18) Å), indicating that the secondary bonding is strongest for the iodide and weakest for the chloride. This trend reflects the electronegativity of halogen atoms, which increases from iodine to chlorine. So, on going from the iodide to chloride, the bismuth center develops an increasingly greater partial positive charge, which makes the THF ligand more strongly bound and results in a shorter Bi–O bond length. The average difference of 0.280 Å between the primary and secondary Bi–Cl distances in $PhBiCl_2{\cdot}(THF)$ is longer than the corresponding values of 0.213 Å in $PhBiBr_2{\cdot}(THF)$ and 0.152 Å in $PhBiI_2{\cdot}(THF)$. It is most likely that, on going from chloride to iodide, the orbitals become larger and more diffuse leading to better overlap between the lone pair and the bismuth σ^* orbital. Some donor adducts of the type $PhBiBr_2{\cdot}L_n$ (n=1,2) take on a bromine-bridged dimeric structure with a similar square-based pyramidal geometry around each bismuth center [95JOM(496)59, 99JCS(D)2837]. In both known cases, L is located *trans* to X.

Four-coordinate, ten-electron bismuthate complexes of the type $[ER'^{+}_{4}]$

Fig. 2.2. Representative structures of diorganylhalobismuthines.

$[R_2BiX_2^-]$ (E=N, As, P) have a pseudotrigonal bipyramidal geometry at the bismuth center with two halogen atoms at axial sites and two organyl ligands and a pair of unshared electrons at equatorial sites [92JCS(D)1967, 93JCS(D)637]. For example, in an orange colored anion complex, $[NEt_4^+]$ $[Ph_2BiI_2^-]$, the Bi–I bond lengths are 3.1415(8) and 3.0517(8) Å, and the I–Bi–I and C–Bi–C bond angles are 177.87(2) and 93.6(3)°, respectively. On the other hand, a yellow colored complex $[NEt_4^+][PhBiI_3^-]$ exsists as a dimeric form and each bismuth atom possesses a square-based pyramidal coordination geometry with apical phenyl groups and iodines in the basal plane. The bismuth atoms are 0.162(2) and 0.098(2) Å above their respective basal planes (Fig. 2.2).

In general, the R_2BiX type compounds coordinate one two-electron donor (L) to afford a four-coordinate disphenoidal species $R_2BiX{\cdot}L$, whilst the $RBiX_2$ type compounds coordinate two two-electron-donors, giving a five-coordinate, square-based pyramidal species $RBiX_2{\cdot}L_2$ (Figs. 2.2 and 2.3).

The X-ray analysis of a red thermolabile, light-sensitive $C_5H_5BiCl_2$ has shown that the cyclopentadienyl ring (Cp) is π-bonded to the bismuth atom and a pseudorotational disorder due to packing requirements results in a mixture of η^2- (Bi–C bond lengths; 2.47 and 2.48 Å) and η^3-bonded states (Bi–C bond lengths; 2.38, 2.61 and 2.64 Å) [90JOM(386)177]. A similar coordination environment is found in the substituted derivatives, $^iPr_4C_5HBiX_2$ (X=Cl, I), where the Cp ring coordinates to the bismuth center with a η^3-bonding mode [94CB1335]. The Bi–C bond lengths of the chloride (2.46(1)–2.67(2) Å) are almost the same as those of the iodide (2.48(2)–2.63(2) Å). On the other hand, the aromatic ring in a cationic complex, $[(1,2,4\text{-}^tBu_3C_5H_2)_2$

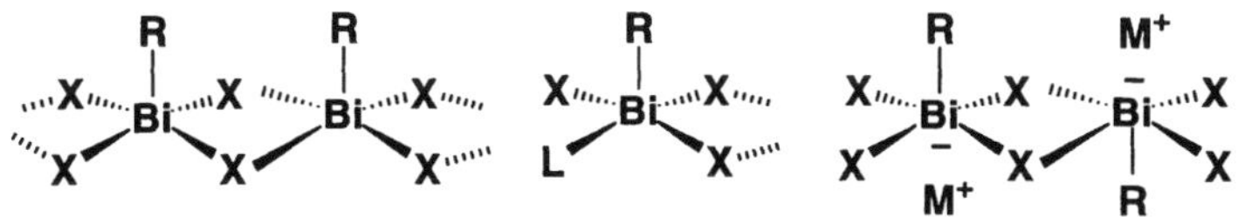

Fig. 2.3. Representative structures of organyldihalobismuthines.

Bi^+][$AlCl_4^-$], has a η^5-bonding mode with the Bi–C bond distances of 2.648(3)–2.726(3) Å [97ZN(B)398]. Only a slight bending of the Cp_{center}–Bi–Cp_{center} axis (172.9°) is observed.

Several BiX_3–arene π-complexes have been structurally characterized. Single crystals of the 1:1 compound $BiCl_3 \cdot C_6H_6$ are interpreted best as a layer coordination polymer, built up from quasi-molecular $BiCl_3$ units via the weak chlorine bridges [93CC799]. The bismuth atom has a distorted pentagonal bipyramidal geometry with the arene molecule at an apical position. The mean distance of the bismuth atom from the plane is 3.14 Å and the mean Bi–C bond distance is 3.43 Å, indicating that the benzene coordinates to the bismuth center with the η^6-mode. The crystals of the related mesitylene complex $BiCl_3 \cdot 1,3,5\text{-}Me_3C_6H_3$ consist of independent sheets of $BiCl_3$ networks [87AG(E)74, 87CB1837]. The bismuth center of one sheet possesses a distorted pentagonal bipyramidal geometry like $BiCl_3 \cdot C_6H_6$, while the other exhibits a distorted octahedral coordination. In both sheets, the arene molecule coordinates to the bismuth center with the Bi–C_{center} bond distances of 2.99–3.11 Å, which are shorter than those of $BiCl_3 \cdot C_6H_6$. This seems to reflect the difference in the π-donating ability of the arene ligands. The Bi–C_{center} bond distance (2.72 Å) of a cationic complex, [$BiCl_2 \cdot C_6H_6^+$] [$AlCl_4^-$], is much shorter than that (3.07 Å) found in a neutral sandwich complex, $(BiCl_3)_2 \cdot C_6Me_6$, suggesting that the cationic bismuth center interacts more strongly with the arene [86AG(E)757, 87CB1829].

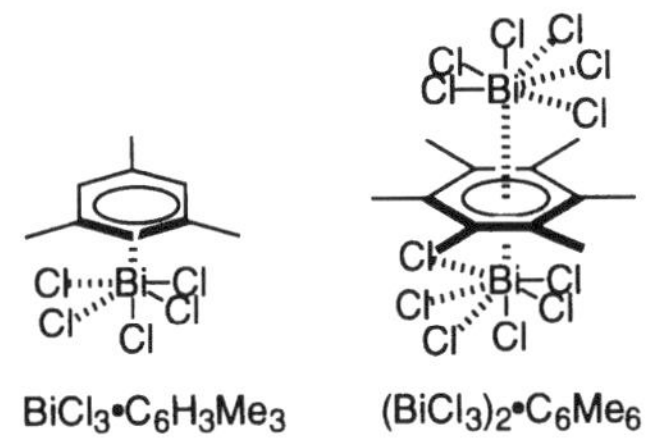

The IR and Raman spectra of chlorodiphenylbismuthine and bromodiphenylbismuthine have been reported [79SA(A)339]. The Raman spectra of these compounds in acetone solutions show a strong, polarized band at 204 cm^{-1} and a weak, depolarized band at 215 cm^{-1}. These two characteristic bands are assigned as the symmetric and asymmetric Bi–Ph stretching frequencies. The additional bands observed at 270 cm^{-1} (Ph_2BiCl) and 170 cm^{-1} (Ph_2BiBr) are attributed to the Bi–X stretching frequency. These data agree with the assumption that both bismuth compounds possess the structure of C_{3v} symmetry. By contrast, the vibration spectra of both compounds in the solid state exhibit no Bi–X stretching band, suggesting intermolecular association. The UV spectra have been determined in dichloromethane solutions for chlorodiphenylbismuthine (λ_{max} 316 nm), bromodiphenylbismuthine (λ_{max} 325 nm) and dibromo(phenyl)bismuthine (λ_{max} 341 nm) [69JOM(20)99]. The electronic spectrum of bismuth chloride in benzene shows a long-wavelength absorption at λ_{max} 291 nm which is assigned to a metal-localized *sp* transition [96JOM(515)245]. The photoredox chemistry of bismuth chloride in benzene is also studied.

2.3.3. Reactions

Mono- and dihalobismuthines of the type R_2BiX and $RBiX_2$ (X=Cl, Br) bearing no specific ligand are generally reactive and readily decomposed by moisture, alcohols and ammonia. Dialkylhalobismuthines are extremely sensitive and spontaneously flammable in air. By placing a coordinating group in a suitable position of the molecule, the Bi–X bond can be made less sensitive. Diarylhalobismuthines and aryldihalobismuthines are also reactive and gradually hydrolyzed by atmospheric moisture. However, there are known several *ortho*-substituted derivatives which can be handled without decomposition in air. Such stabilization arises from either steric protection of or intramolecular coordination to the bismuth center. By contrast, iodobismuthines are much less reactive than the corresponding chloro and bromo compounds (Scheme 2.8).

Halodiorganylbismuthines and dihalomonoorganylbismuthines undergo metathetical reaction with a wide variety of inorganic salts. This reaction is

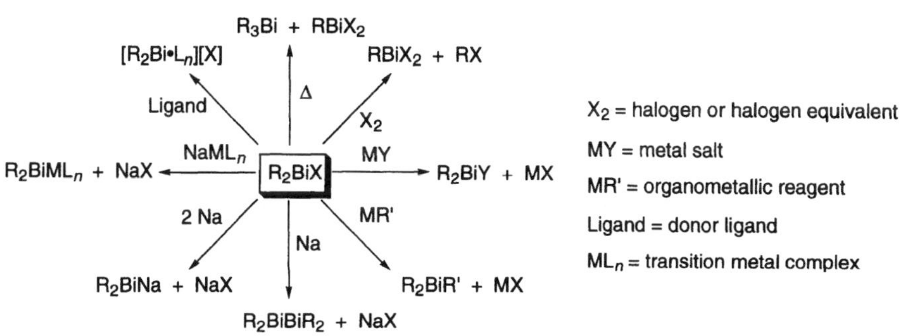

Scheme 2.8. Typical reactions of halodiorganylbismuthines.

commonly used for obtaining iodo- and diiodobismuthines as well as compounds of the general formula R_2BiY and $RBiY_2$, where Y denotes an anionic group such as OR, OCOR, ONO_2, $OSiMe_3$, NR_2, SR, S_2CNR_2, S_2COR, CN, N_3, SCN and SeCN (Section 2.2). Several compounds containing one or two bismuth–transition metal covalent bonds have also been prepared by this method (Section 2.5). When treated with NaOH, bromobis(2,4,6-trimethylphenyl)bismuthine undergoes dehydrative coupling to afford bis[bis(2,4,6-trimethylphenyl)bismuthyl] oxide instead of the expected bis(2,4,6-trimethylphenyl)bismuth hydroxide [95ZN(B)735].

In liquid ammonia or in THF at low temperatures, halodiorganylbismuthines are reduced by 2 equiv. of alkali or alkaline earth metal, such as Li, Na, K, Ca and Ba, to give a highly reactive species of the formula R_2BiM or $(R_2Bi)_2M$ as a deep red solution [41JA212]. These bismuthides are thermally unstable and dissociate slowly to the corresponding tertiary bismuthines and inorganic bismuth compounds at room temperature. However, they can react with several haloalkanes and haloarenes at low temperatures to yield the corresponding mixed tertiary bismuthines (Section 2.1.2.1.3). When an equivalent amount of sodium metal or cobaltocene (Cp_2Co) is used for the reduction, tetraorganyldibismuthines R_2BiBiR_2 are obtained (Section 2.2.2). Although all attempts to obtain a diaryldibismuthene starting from 2,4,6-$Ph_3C_6H_2BiCl_2$ have resulted in the formation of metallic bismuth and organic products [95JOM(485)141], reductive coupling of aryldichlorobismuthines bearing more bulky *ortho*-substituents (2,6-$R_2C_6H_3BiCl_2$; R = 2,4,6-$Me_3C_6H_2$ or 2,4,6-$^iPr_3C_6H_2$) yielded the expected dibismuthenes (ArBi=BiAr) [99JA3357]. The reduction of dichloro(methyl)bismuthine and chlorodimethylbismuthine with $LiAlH_4$ at low temperatures gives methylbismuthine ($MeBiH_2$) and dimethylbismuthine (Me_2BiH), respectively, while a similar reduction of chlorodiphenylbismuthine and dibromo(phenyl)bismuthine yields a polymeric substance of empirical formula $(PhBi)_n$ (Section 2.1.1). The electrolytic reduction of chlorodiphenylbismuthine has also been attempted; however, the resulting anion is unstable and rapidly liberates bismuth metal [66JA467, 66JA5117]. Attempted reduction of diarylhalobismuthines and aryldihalobismuthines with hydrazine hydrate leads to the formation of triarylbismuthines [41JA949, 41JA212].

As has been mentioned in Section 2.3.1.12, many oxygen and nitrogen donor adducts are prepared from mono- and dihalobismuthines and structurally characterized. The oxygen donors include THF, triphenylphosphine oxide, triphenylarsine oxide, diphenyl sulfoxide, HMPA and *N,N'*-dimethylpropylene urea. Cationic species $[Ph_2Bi{\cdot}L_2^+]$, where L is triphenylphosphine oxide, triphenylarsine oxide or HMPA, have been prepared by the halide abstraction of the corresponding halobismuthines with AgX or TlX (X=BF_4, PF_6, ClO_4) and subsequent complexation with donor molecules [79JOM(166)365, 96JCS(D)443, 97OM3597]. The nitrogen donors include pyridine, 2,2′-bipyridine and 1,10-phenanthroline. When treated with pyridine, bromodiphenylbismuthine is converted to a pyridine adduct $PhBiBr_2{\cdot}$(pyridine)$_2$ [72JOM(44)317, 64IC500]. This adduct is also obtained from the reaction of bromodiphenylbismuthine and 2 equiv. of pyridine, suggesting a possible intervention from the redistribution of Ph_2BiBr.

Halodiphenylbismuthines and dihalo(phenyl)bismuthines react with tetraorganylarsonium salts to form the bismuthinate complexes of the formula $[R_4As^+][Ph_2BiX_2^-]$ and $[R_4As^+][PhBiX_3^-]$, where the bismuth moiety exists as the anionic species. Similar types of ammonium and phosphonium salts have also been prepared by this methodology (Section 2.3.1.12).

2.4. TRIORGANYLBISMUTHINE–TRANSITION METAL COMPLEXES

2.4.1. Methods of synthesis (Table 2.30)

Due to the weak Bi–C bond as well as the strong *s* character of a lone pair of electrons in tertiary bismuthines, relatively few coordination compounds have been reported for the bismuthines in comparison with the congeners of lighter elements of the nitrogen family. Trialkylbismuthines bearing methyl, ethyl and cyclohexyl groups react with $Cr(CO)_5(THF)$ to yield $Cr(CO)_5(BiR_3)$ (R = Me, Et or ^{c}Hex) [79CC639]. Of interest, a carbene complex $(CO)_5Cr{=}C(OMe)Me$ also produces the same adducts by treatment with R_3Bi [76CB1140]. Photoinduced CO-extraction from $[Et_4N][V(CO)_6]$ and $VCp(CO)_4$, in the presence of trialkylbismuthines (R_3Bi) yields $[Et_4N][V(CO)_5(BiR_3)]$ and $VCp(CO)_3(BiR_3)$, respectively [78CB1978]. Tris(*tert*-butyl)bismuthine can replace one of the CO groups in $Ni(CO)_4$, $Cr(CO)_6$, $Mo(CO)_6$ and $W(CO)_6$ to give the corresponding bismuthine-coordinated complexes of the type $M(CO)_n(Bi{}^tBu_3)$ [75JOM(87)83]. A weakly coordinating THF ligand in a cationic iron complex, $[FeCp(CO)_2(THF)][BF_4]$, is readily displaced by triorganylbismuthines [91JOM(403)183].

Transition metal complexes coordinated by a triarylbismuthine have also been reported. A unique ligand, tris[2-(dimethylarsino)phenyl]bismuthine, reacts with nickel halides NiX_2 (X = Cl, Br or I) in the presence of sodium tetraphenylborate to afford $[NiX{\cdot}Bi(2\text{-}Me_2AsC_6H_4)_3][BPh_4]$, where the nickel center has a square-pyramidal geometry [75CC164]. A few related nickel and cobalt complexes bearing a similar type of bismuthine are reported [77JCS(D)711]. Triphenylbismuthine coordinates to $[Ir(CO)(PPh_3)_2][ClO_4]$ to give a cationic complex $[Ir(CO)(PPh_3)_2(BiPh_3)][ClO_4]$, although this bismuthine behaves as the phenyl donor in the rhodium complex-catalyzed reactions [71SK(B)317, 95BCJ950] (Section 5.4.3.2). Treatment of triarylbismuthines with palladium(II) salts, such as $PdCl_2$ and $Pd(OAc)_2$, results in the cleavage of thc Bi–C bond and the liberation of metallic bismuth [71JOM(33)C81, 76JOM(111)C1, 88T5661]. This type of transmetallation has been employed for the Pd-catalyzed arylation of some olefins [77BCJ2021] (Section 5.4.3.2). Crystalline complexes $M(CO)_5(BiPh_3)$, where M = Cr, Mo or W, are obtained by the UV irradiation of $M(CO)_6$ in the presence of triphenylbismuthine in benzene [71JINC(33)892]. These complexes are stable in air but decompose rapidly in solution. Triphenylbismuthine can also coordinate to uranium and niobium chlorides

[65CJC1879, 70JINC(32)3249]. Mono- and di-metal complexes of chromium and tungsten coordinated by 1,4-bis(diphenylbismuthino)benzene are known [99JOM(584)179]

Triphenylbismuthine adsorbs from solution onto gold, but it tends to form oxygen-containing oligomeric or polymeric species on gold [92L90]. Thermodynamics of complex formation of triphenylbismuthine with silver(I), copper(II) and mercury(II) salts have been studied, though no bismuthine-complex has been isolated [77ACS(A)775].

Synthesis of [$C_5H_5Fe(CO)_2(BiR_3)$][BF_4]

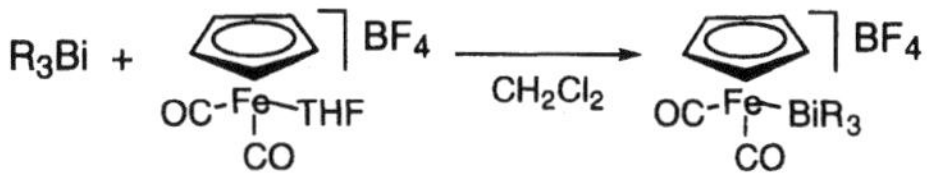

To an ethereal solution of BiR_3 (15 mmol, R=Me, ^{i}Pr or Ph) was added [$C_5H_5Fe(CO)_2(THF)$][BF_4] (2.0 g, 6.0 mmol) and the resulting suspension was stirred for 12–16 h at room temperature. After evaporation of the solvent, the product was extracted with dichloromethane, precipitated with ether, and recrystallized from dichloromethane/ether (1:1–1:4). Yield, 25–51% [91JOM(403)183].

Synthesis of [$NiX(BiAr_3)$][BPh_4]

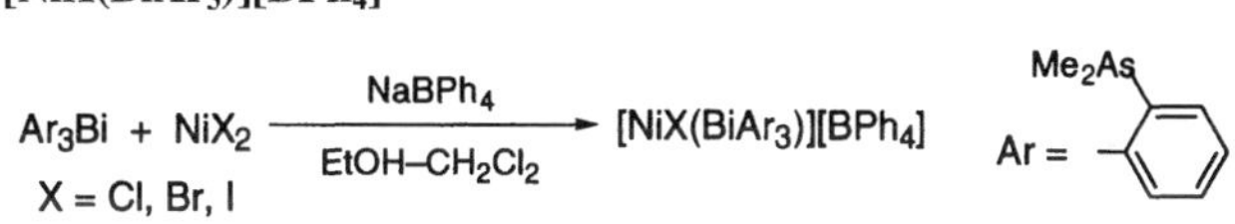

An appropriate nickel(II) halide hydrate (1 mmol, X = Cl, Br or I) was dissolved in dry ethanol (10 ml), and tris(2-dimethylarsinophenyl)bismuthine (1 mmol) was added in dichloromethane. The resulting dark red solution was stirred briefly and then $NaBPh_4$ (0.35 g, 1 mmol) was added in ethanol (10 ml). The product [$NiX(BiAr_3)$][BPh_4] was precipitated by adding an excess of ether and recrystallized from dichloromethane–ethanol. Yield 60% [77JCS(D)711].

2.4.2. Properties and reactions

The stability of this class of compound depends on the oxidation state, coordination number and the ligands attached to the relevant transition metal center as well as on the substituent of the coordinating bismuthine. For example, molybdenum(0) and tungsten(0) complexes of the type $M(CO)_5(BiPh_3)$ (M = Mo, W) are air stable and possess definite melting points [71JINC(33)892], while molybdenum(II) and tungsten(II) complexes such as $MI_2(CO)_3(MeCN)(BiPh_3)$ and $MI_2(CO)_3(PPh_3)(BiPh_3)$ (M = Mo, W)

TABLE 2.30
R_3Bi coordinated transition metal complexes

Complex	Yield (%)	M.p. (°C)	Physical data	Reference
Vanadium				
$VCp(CO)_3(BiEt_3)$	30	–	^{51}V-NMR, IR	[78CB1978]
$VCp(CO)_2(BiEt_3)_2$	–	–	^{51}V-NMR, IR	[78CB1978]
$VCP(CO)_2(BiPh_2Ar)^a$	–	–	^{51}V-NMR, IR	[81ZN(B)451]
$[NEt_4][V(CO)_5(BiEt_3)]$	–	–	^{51}V-NMR, IR	[78CB1978]
$[NEt_4][V(CO)_5(BiPh_2Ar)]^a$	–	–	^{51}V-NMR, IR	[81ZN(B)451]
Niobium				
$NbCl_5(BiPh_3)$	–	–	–	[65CJC1879]
$(NbCl_5)_2(BiPh_3)$	–	–	–	[65CJC1879]
Chromium				
$Cr(CO)_5(BiMe_3)$	79	68	^1H-NMR, IR, MS	[76CB1140]
$Cr(CO)_5(BiEt_3)$	72	19	^1H-NMR, IR, MS	[76CB1140]
$Cr(CO)_5(BiBu_3)$	35	–	^1H-NMR, IR, MS	[84ZAAC(510)104]
$Cr(CO)_5(Bi^iBu_3)$	62	138 (decomp.)	^1H-NMR, IR, Raman	[75JOM(87)83]
$Cr(CO)_5(Bi^cHex_3)$	76	120 (decomp.)	^1H-NMR, IR, MS	[76CB1140]
$Cr(CO)_5(BiPh_3)$	–	115 (decomp.)	IR, MS	[71JINC(33)892]
			X-ray	[79CC639]
	10	–	^1H-, ^{13}C-NMR, IR, MS	[97JOM(545)111]
$Cr(CO)_5(p\text{-}Ph_2BiC_6H_4BiPh_2)$	–	–	^1H-, ^{13}C-NMR, IR	[99JOM(584)179]
$[Cr(CO)_5]_2(p\text{-}Ph_2BiC_6H_4BiPh_2)$	–	–	^1H-, ^{13}C-NMR, IR	[99JOM(584)179]
Molybdenum				
$Mo(CO)_5(BiEt_3)$	–	–	IR	[63BSF1583]
$Mo(CO)_5(BiBu_3)$	36	–	^1H-NMR, IR, MS	[84ZAAC(510)104]
$Mo(CO)_5(Bi^iBu_3)$	53	158 (decomp.)	^1H-NMR, IR, Raman	[75JOM(87)83]
$Mo(CO)_5(BiPh_3)$	–	–	IR	[63BSF1583]
	–	–	IR	[68JCS(A)1195]

	–	–	^{1}H-NMR	[70CC1621]
	–	105 (decomp.)	IR, MS	[71JINC(33)892]
	10	–	^{1}H-, ^{13}C-NMR, IR, MS, X-ray	[97JOM(545)111]
$Mo(CO)_3$(1,10-phenanthroline)($BiPh_3$)	–	–	IR	[66IC2119]
$MoI_2(CO)_3(NCMe)(BiPh_3)$	43	–	^{1}H-NMR, IR	[88ICA(150)217]
$MoI_2(CO)_3(PPh_3)(BiPh_3)$	24	–	IR	[88ICA(150)217]
$MoI_2(CO)_3(PPh_2{}^c Hex)(BiPh_3)\cdot CH_2Cl_2$[b]	43	–	^{1}H-NMR, IR	[88ICA(150)217]
$MoI_2(CO)_3[P(OPh)_3](BiPh_3)$	68	–	IR	[88ICA(150)217]
$MoI_2(CO)_3(BiPh_3)_2$	54	–	IR	[88ICA(150)217]
$MoI_2(CO)_3(AsPh_3)(BiPh_3)$	63	–	IR	[88ICA(150)217]
$MoI_2(CO)_3(SbPh_3)(BiPh_3)$	33	–	IR	[88ICA(150)217]
$[MoI(CO)_3(bipy)(BiPh_3)]I$	67	–	IR	[88ICA(150)217]
$[Mo(\mu\text{-}I)I(CO)_3(BiPh_3)]_2$	47	–	IR	[88ICA(150)217]
$[MoCp(CO)_2(BiPh_3)_2][BF_4]$	–	–	^{13}C-NMR, IR	[91POL665]
$MoCP(CO)(Me)(BiPh_2Ar)$[a]	60	–	–	[81ZN(B)451]
Tungsten				
$W(CO)_5(BiBu_3)$	48	–	^{1}H-NMR, IR, MS	[84ZAAC(510)104]
$W(CO)_5(Bi^tBu_3)$	49	132 (decomp.)	^{1}H-NMR, IR, Raman	[75JOM(87)83]
$W(CO)_5(BiPh_3)$	–	–	–	[69IC83]
	–	132 (decomp.)	IR, MS	[71JINC(33)892]
	10	–	^{1}H-, ^{13}C-NMR, IR, MS, X-ray	[97JOM(545)111]
$WI_2(CO)_3(NCMe)(BiPh_3)$	31	–	^{1}H-NMR, IR	[88ICA(150)217]
$WI_2(CO)_3(PPh_3)(BiPh_3)\cdot MeCN$	57	–	^{1}H-NMR, IR	[88ICA(150)217]
$WI_2(CO)_3[P(OPh)_3](BiPh_3)$	62	–	IR	[88ICA(150)217]
$WI_2(CO)_3(BiPh_3)_2$	53	–	^{13}C-NMR, IR	[88ICA(150)217]
$WI_2(CO)_3(SbPh_3)(BiPh_3)\cdot CH_2Cl_2$	32	–	^{1}H-NMR, IR	[88ICA(150)217]
$W(CO)_5(p\text{-}Ph_2BiC_6H_4BiPh_2)$	–	–	^{1}H-, ^{13}C-NMR, IR	[99JOM(584)179]
$[W(CO)_5]_2(p\text{-}Ph_2BiC_6H_4BiPh_2)$	–	–	^{1}H-, ^{13}C-NMR, IR	[99JOM(584)179]
Manganese				
$MnCp(CO)_2(BiPh_3)$	44	–	^{13}C-NMR, IR	[67CJC161]
$MnCp(CO)_2(BiPh_2Ar)$[a]	–	–	^{31}P-NMR, IR	[81ZN(B)451]

TABLE 2.30 (*continued*)

Complex	Yield (%)	M.p. (°C)	Physical data	Reference
Iron				
$Fe(CO)_4(BiEt_3)$	47	–	^{1}H-NMR, IR, MS	[84ZAAC(510)104]
$Fe(CO)_4(BiPr_3)$	44	–	^{1}H-NMR, IR, MS	[84ZAAC(510)104]
$Fe(CO)_4(BiBu_3)$	84	–	^{1}H-NMR, IR, MS	[84ZAAC(510)104]
$Fe(CO)_4(BiPh_3)_2$	–	–	–	[41ZAAC(248)84]
$[FeCp(CO)_2(BiMe_3)][BF_4]$	42	–	^{1}H-, ^{13}C-NMR, IR, MS	[91JOM(403)183]
$[FeCp(CO)_2(Bi^iPr_3)][BF_4]$	25	–	^{1}H-, ^{13}C-NMR, IR,	[91JOM(403)183]
$[FeCp(CO)_2(BiPh_3)][BF_4]$	51	–	^{1}H-, ^{13}C-NMR, IR, X-ray	[91JOM(403)183]
Cobalt				
$Co[CH(CSMe)_2]_2(BiPh_3)$	–	–	UV	[71IC1080]
$[CoBr(BiAr_3)][BPh_4]^c$	40	–	^{1}H-NMR, UV	[77JCS(D)711]
Nickel				
$Ni(CO)_3(BiEt_3)$	–	–	IR	[63BSF1583]
$Ni(CO)_3(BiBu_3)$	–	–	^{13}C-NMR	[83JOM(243)305]
$Ni(CO)_3(Bi^tBu_3)$	76	65 (decomp.)	^{1}H-NMR, IR	[75JOM(87)83]
$[Ni(OH_2)(BiR_3)][ClO_4]_2^d$	65	–	^{1}H-NMR, IR, UV	[77JCS(D)711]
$[NiCl(BiAr_3)][BPh_4]^c$	60	–	^{1}H-NMR, UV	[77JCS(D)711]
$[NiBr(BiAr_3)][BPh_4]^c$	60	–	^{1}H-NMR, UV	[77JCS(D)711]
$[NiI(BiAr_3)][BPh_4]^c$	60	–	^{1}H-NMR, UV	[77JCS(D)711]
$[Ni_2(BiAr_3)_3][ClO_4]_4^c$	55	–	^{1}H-NMR, IR, UV	[77JCS(D)711]
Others				
$Rh_2(OAc)_4[P(OMe)_3](BiPh_3)$	–	–	^{31}P-NMR	[85JCS(D)629]
$[Rh(COD)(BiPh_3)_2][ClO_4]$	–	–	–	[78JMOC(4)231]
$[Ir(CO)(PPh_3)_2(BiPh_3)][ClO_4]$	–	–	–	[71SK(B)317]
$[Ag(BiPh_3)][ClO_4]$	–	–	IR	[62JCS2854]
	–	–	–	[77ACS(A)775]
$[Cu(BiPh_3)][ClO_4]$	–	–	–	[78ACS(A)933]

$[Hg(BiPh_3)][ClO_4]_2$	–	–	–	[78ACS(A)933]
$[U_2Cl_9(BiPh_3)]Cl$	–	190 (decomp.)	EC	[70JINC(32)3249]
$Tl(CF_3COCHCOCF_3)(BiPh_3)$	–	–	MS	[74CL733]

[a] Ar = 2-$Ph_2PC_6H_4$.
[b] ^{c}Hex = cyclohexyl.
[c] Ar = 2-$Me_2AsC_6H_4$.
[d] R = $Me_2AsCH_2CH_2CH_2$.

are moderately air-sensitive (particularly in solution) and thermally unstable (much less stable than their PPh_3, $AsPh_3$ and $SbPh_3$ counterparts), though they can be stored for several days under nitrogen at 0°C [88ICA(150)217]. However, the first pentacarbonyl complexes decompose rapidly in solution.

Some transition metal complexes coordinated by Ph_3Bi have been characterized by X-ray analyses. An iron complex, $[FeCp(CO)_2(BiPh_3)][BF_4]$, is composed of the discrete ions $[FeCp(CO)_2(BiPh_3)]^+$ and BF_4^- [91JOM(403)183]. Coordination at the iron atom is pseudooctahedral (three-legged piano-stool geometry) with a Fe–Bi bond distance of 2.570(1) Å, which is slightly longer than the sum of the covalent radii (2.53 Å). The Bi–C bond lengths of 2.218(6)–2.226(6) Å are in agreement with those of the uncoordinated ligand Ph_3Bi, while the mean C–Bi–C bond angle of 101.1° is larger than that of Ph_3Bi (94°). The structures of $Cr(CO)_5(EPh_3)$, where E is P, As, Sb or Bi, have also been characterized [79CC639]. Based on the data of a series of the Cr–E–C bond angles, and both Cr–E and E–C bond lengths, it has been concluded that the Bi–Cr bond has more *s* character than the corresponding E–Cr bond, where E = P, As or Sb. The corresponding values of the bismuth derivative is 114.6(2)–123.6(2)° (Cr–Bi–C), 2.705(1) Å (Cr–Bi) and 2.212(9)–2.222(9) Å (Bi–C), respectively. The $BiPh_3$ geometry is not significantly different from the Cr, Mo or W complexes of the type $M(CO)_5(BiPh_3)$ [97JOM(545)111].

IR and NMR spectra of these complexes have sometimes been used to examine the structure as well as the nature of the ligand attached to a transition metal center. For instance, from ^{51}V-NMR data of vanadium complexes of the type $VCp(CO)_3(ER_3)$, where E is P, As, Sb or Bi, the coordinating ability has been shown to follow the order: $Et_3Bi < R_3As < R_3P < R_3Sb$ [78CB1978].

2.5. ORGANOBISMUTH(III) COMPOUNDS WITH Bi–TRANSITION METAL LINKAGE

2.5.1. Methods of synthesis (Table 2.31)

Several organobismuth compounds are known in which one or two bismuth–transition metal covalent bonds are contained. Metathesis between R_nBiX_{3-n} (n=1, 2; X=Cl, Br or I) and $[ML_n]^-$ has been employed for the preparation of Cr, Mo, W, Mn, Fe, Co and Ni complexes bearing the Bi–M bond (method A). In some cases, this method accompanies such side reactions as disproportionation of the desired compound. For instance, $(Me_3Si)_2CHBiCl_2$ reacts with $Na_2[W(CO)_5]$ in THF to yield two kinds of clusters, $W_2(CO)_8(\mu\text{-}\eta^2\text{-}Bi_2)[\mu\text{-}BiMeW(CO)_5]$ and $[W(CO)_5]_3(\mu_3\text{-}\eta^2\text{-}Bi_2)$, instead of the expected alkylbismuth complex [85JA1062, 86IC4836, 90JOM(399)C21]. The Bi–C bond cleavage of trimethylbismuthine with a metal hydride such as $Cp(CO)_3MH$ (M=Mo, W) is also applicable to the Bi–M bond construction, where 1 equiv. of methane is liberated (method B). When the relatively weak Bi–Bi bond of dibismuthine is cleaved by heat or light in the presence of a binuclear metal complex, the Bi–M bond can be also formed (method C). For instance, a chromium complex, $Cr_2Cp_2(CO)_6$, reacts thermally with tetraphenyldibismuthine at room temperature to give the crossover product $CrCp(CO)_3$–$BiPh_2$, whereas the molybdenum and tungsten analogs react with the same dibismuthine under exposure to diffused daylight to yield similar types of complexes $MCp(CO)_3$–$BiPh_2$ (M=Mo, W). This crossover reaction using tetraphenyldibismuthine has been employed for the preparation of $Co(BiPh_2)(CO)_4$ which is readily converted to $Co(BiPh_2)(CO)_3(PPh_3)$ by treatment with triphenylphosphine. A few manganese complexes have been prepared by the reaction between dimethylbismuthide anion and $MnBrL_n$ type compounds (method D). Trimethylphosphine and triphenylphosphine can replace a CO ligand of $MCp(CO)_n(BiR_2)$ type complexes (M=Mo, W or Fe) (method E). Alkyl halides (RX) can react with an anionic iron complex, $[Et_4N]_3[Bi\{Fe(CO)_4\}_3]$, to give either $[Et_4N]_2[\{Fe(CO)_4\}_3BiR]$ or $[Fe(BiR)(CO)_4]_2$ depending on the alkyl group R as well as the leaving group X (method F). The former anionic products can be converted to the latter cyclic complexes in acetic acid. Tetrakis(methyldiphenylphosphine)palladium reacts with bromodiphenylbismuthane to give $[Bi_4\{Pd(PPh_2Me)_2\}_4][Ph_2BiBr_2]_2$ [99JA4409].

$$R_nBiX + ML_n^- \longrightarrow R_nBiML_n + X^- \quad \text{(Method A)}$$

$$Me_3Bi + HML_n \longrightarrow Me_2BiML_n + MeH \quad \text{(Method B)}$$

$$Ph_2BiBiPh_2 + L_nM\text{–}ML_n \longrightarrow 2\ Ph_2BiML_n \quad \text{(Method C)}$$

$$R_nBi^- + XML_n \longrightarrow R_nBiML_n + X^- \quad \text{(Method D)}$$

$$R_nBiML \xrightarrow{L'} R_nBiML' + L \quad \text{(Method E)}$$

$$[Et_4N]_3[Bi(FeL_n)_3] + RX \longrightarrow [Et_4N]_2[RBi(FeL_n)_3] + [RBiFeL_n]_2 \quad \text{(Method F)}$$

Synthesis of $[Ph_3P{=}N{=}PPh_3][Ph_2BiFe(CO)_4]$

$$Ph_2BiCl + Na_2Fe(CO)_4 \xrightarrow[\text{THF}]{[Ph_3P{=}N{=}PPh_3]Cl} [PPN][Ph_2BiFe(CO)_4]$$

A suspension of chlorodiphenylbismuthine (0.460 g, 1.15 mmol) in THF (15 ml) was added dropwise to a stirred solution of $Na_2Fe(CO)_4$ (0.246 g, 1.15 mmol) in the same solvent (10 ml). The resulting brown solution was stirred for 0.5 h, filtered, and $[Ph_3P{=}N{=}PPh_3]Cl$ (0.66 g) was added to the filtrate. The solution was stirred for a short time and then filtered to remove a gray solid. The solvent was removed and toluene (40 ml) was added to the red oily residue. On standing at room temperature, $[Ph_3P{=}N{=}PPh_3][Ph_2BiFe(CO)_4]$ separated out as a yellow crystalline solid (66%) [91IC2788].

Synthesis of $CrCp(BiPh_2)(CO)_3$

$$Ph_2BiBiPh_2 + Cr_2Cp_2(CO)_6 \xrightarrow[\text{PhMe}]{} 2\ CrCp(BiPh_2)(CO)_3$$

Tetraphenyldibismuthine (0.79 g, 1.09 mmol) was added to a toluene solution of $Cr_2Cp_2(CO)_6$ (0.35 g, 0.87 mmol) at room temperature and brought an immediate color change from emerald green to orange brown. The solution was evaporated to dryness under reduced pressure, and the residue was recrystallized from hot heptane (25 ml) to give the chromium(II) compound, $CrCp(BiPh_2)(CO)_3$, as red crystals (68%) [91IC1274].

2.5.2. Properties and reactions

Properties of this class of compounds are considerably dependent on both the metal and ligands involved. For instance, $[Ph_3P{=}N{=}PPh_3][Ph_2Bi$-

TABLE 2.31
Organobismuth compounds with Bi–transition metal bonds

Complex	Synthetic method[a]	Yield (%)	M.p. (°C)	Physical data	Reference
Chromium					
$CrCp(BiMe_2)(CO)_3$	A	–	49–51 (decomp.)	^{1}H-NMR, IR	[77JOM(134)C32]
$CrCp(BiPh_2)(CO)_3$	C	68	–	IR	[91IC1274]
$[CrCp(CO)_3]_2(BiMe)$	A	–	125 (decomp.)	^{1}H-NMR, IR	[77JOM(134)C32]
Molybdenum					
$MoCp(BiMe_2)(CO)_3$	B	–	44–46 (decomp.)	^{1}H-NMR, IR	[77JOM(134)C32]
$MoCp(BiPh_2)(CO)_3$	A, C	19	–	^{13}C-NMR, IR	[91IC1274]
$MoCp(BiPh_2)(CO)_2(PPh_3)$	E	26	–	^{1}H-NMR, IR	[91IC1274]
$[MoCp(CO)_3]_2(BiMe)$	A	–	126–128 (decomp.)	^{1}H-NMR, IR	[77JOM(134)C32]
$[Mo(CO)_4][FeCp(PMe_3)(CO)(BiMe_2)]_2$	E	60	101–103	^{1}H-, ^{31}P-NMR, IR	[83JOM(252)C23]
Tungsten					
$WCp(BiMe_2)(CO)_3$	B	57	61–63 (decomp.)	^{1}H-NMR, IR	[77JOM(134)C32]
$WCp(BiPh_2)(CO)_3$	A, C	24	–	^{13}C-NMR, IR	[91IC1274]
$WCp(BiPh_2)(CO)_2(PPh_3)$	E	–	–	IR	[91IC1274]
$[WCp(CO)_3]_2(BiMe)$	A	37.1	134–136 (decomp.)	^{1}H-NMR, IR	[77JOM(134)C32]
$[WCp(CO)_3][Cr(CO_5)](BiMe_2)$	–	–	107–109 (decomp.)	^{1}H-NMR, IR	[77JOM(134)C32]
$W_2(CO)_8(\mu\text{-}\eta^2\text{-}Bi_2)[\mu\text{-}BiMeW(CO)_5]$	A	12	160	IR, UV, X-ray	[86IC4836]
Manganese					
$Mn(BiPh_2)(CO)_5$	C	40	–	IR	[91IC1274]
	A	33	75–78	^{1}H-NMR, IR, MS, X-ray	[91IC2788]
$Mn(Bi\text{-}\eta^1\text{-}C_4Me_2H_2)(CO)_5$	D	54	124–125	^{1}H-, ^{13}C-NMR, IR, MS, X-ray	[93JOM(447)197]
$Mn(Bi\text{-}\eta^5\text{-}C_4Me_2H_2)(CO)_3$	D	90	78–80	^{1}H-, ^{13}C-NMR, IR, MS	[93JOM(447)197]

TABLE 2.31 *(continued)*

Complex	Synthetic method[a]	Yield (%)	M.p. (°C)	Physical data	Reference
Rhenium					
$Re(BiPh_2)(CO)_5$	A	66	–	–	[64DOK(156)383]
Iron					
$FeCp(BiMe_2)(CO)_2$	A	51	50–52/10^{-4} (sublim.)	^{1}H-NMR, IR	[82JOM(228)C29]
$FeCp(BiMe_2)(CO)(PMe_3)$	E	87	–	^{1}H-, ^{31}P-NMR, IR	[82JOM(228)C29]
$[Fe(BiMe)(CO)_4]_2$	F	34	142–143 (decomp.)	^{1}H-NMR, IR, MS, X-ray	[89IC3164]
$[FeCp(CO)_2]BiMe$	A	10	115–117	^{1}H-NMR, IR	[82JOM(228)C29]
$Fe(BiPh_2)_2(CO)_4$	A	72	–	^{1}H-NMR, IR, X-ray	[91IC2788]
$Na[Fe(BiPh_2)(CO)_4]$	A	–	–	–	[91IC2788]
$[Ph_3P{=}N{=}PPh_3][Fe(BiPh_2)(CO)_4]$	A	66	Oil	^{1}H-NMR, IR, X-ray	[91IC2788]
$[Fe(BiMe)(CO)_4]_2$	F	40	–	^{1}H-NMR, IR, MS	[93OM4926]
$[Fe(BiEt)(CO)_4]_2$	F	54	120	^{1}H-NMR, IR	[93OM4926]
$[Fe(Bi^iPr)(CO)_4]_2$	F	11	255	^{1}H-NMR, IR, MS	[93OM4926]
$[Fe(BiBu)(CO)_4]_2$	F	32	154	^{1}H-NMR, IR, MS	[93OM4926]
$[Fe(Bi^tBu)(CO)_4]_2$	F	20	140	^{1}H-NMR, IR, MS, X-ray	[93IC2212]
$[Fe(BiCH_2Ph)(CO)_4]_2$	F	21.6	117	^{1}H-NMR, IR, MS	[93OM4926]
$[Fe(BiCH_2CH_2CH_2Cl)(CO)_4]_2$	F	22	91	^{1}H-NMR, IR, MS	[93OM4926]
$[Fe(BiCH_2CH_2CH_2CH_2Br)(CO)_4]_2$	F	8	310	^{1}H-NMR, IR, MS	[93OM4926]
$[Fe(BiPh)(CO)_4]_2$	A	50	–	^{1}H-NMR, IR, X-ray	[91IC2788]
$[NEt_4]_2[\{Fe(CO)_4\}_3Bi^iPr]$	F	85	–	^{1}H-NMR, IR	[93OM4926]
$[NEt_4]_2[\{Fe(CO)_4\}_3BiBu]$	F	87	–	^{1}H-NMR, IR	[93OM4926]
$[NEt_4]_2[\{Fe(CO)_4\}_3Bi^tBu]$	F	98	–	^{1}H-NMR, IR, X-ray	[93IC2212]
$[NEt_4]_2[\{Fe(CO)_4\}_3BiCH_2Ph]$	F	79	–	^{1}H-NMR, IR	[93OM4926]
$[NEt_4]_2[\{Fe(CO)_4\}_3BiCH_2CH_2CH_2Cl]$	F	92	–	^{1}H-NMR, IR	[93OM4926]
$[NEt_4]_2[\{Fe(CO)_4\}_3BiCH_2CH_2CH_2CH_2Br]$	F	–	–	–	[93OM4926]

Cobalt					
$Co(BiPh_2)(CO)_4$	C	–	–	IR	[84JCS(D)2535]
$Co(BiPh_2)(CO)_3(PPh_3)$	E	65	–	IR, X-ray	[84JCS(D)2535]
$Co(BiPh_2)(pyridine)L^b$	A	–	225–230 (decomp.)	–	[65AG130]
$CoCp_2(BiPh_2Cl_2)$	A	77	–	IR	[91G(121)93]
Nickel					
$[Ni_{10}(BiMe)_2(CO)_{18}][NMe_4]_2$	A	45	–	^1H-, ^{13}C-NMR, IR, MS, X-ray	[97OM1655]
$[Ni_{10}(BiEt)_2(CO)_{18}][NMe_4]_2$	A	–	–	IR, X-ray	[97OM1655]
$[Bi_4\{Pd(PPh_2Me)_2\}_4][Ph_2BiBr_2]$	–	16	–	^1H-, ^{31}P-NMR, X-ray	[99JA4409]

[a] For notation, see Section 2.5.1.
[b] L=dimethylglyoximate.

$Fe(CO)_4$] is yellow and moderately air-sensitive, whereas $(Ph_2Bi)_2Fe(CO)_4$ is red and very air-sensitive in solution [91IC2788]. The latter undergoes disproportionation slowly in solution to produce $[PhBiFe(CO)_4]_2$.

Several complexes bearing the Bi–M σ bond have been characterized by X-ray crystallographic methods. The crystal structure of red-brown $Co(BiPh_2)(CO)_3(PPh_3)$ consists of well separated monomers [84JCS(D)2535]. The cobalt atom possesses a distorted trigonal bipyramidal geometry surrounded by three equatorial carbonyl groups with the mean Co–C bond length of 1.80 Å and the mean C–Co–C bond angle of 119.3°. The phosphorus and bismuth atoms occupy the axial sites with the Bi–Co–P bond angle of 169.3(3)°. The geometry of the bismuth atom is essentially pyramidal with the Bi–Co bond length of 2.692(5) Å. The bond angles around the bismuth are 95.7(8)–100.9(5)°, and are considerably larger than those of Ph_3Bi and $Ph_2BiBiPh_2$. This may be attributed to steric repulsion between the carbonyl and phenyl groups. The normal Bi–C bond distances of 2.25(2)–2.28(2) Å, however, suggest that the residual lone pair on the bismuth atom has a considerable *s* character.

The structures of several iron complexes have also been determined. The iron atom in $[Ph_3P{=}N{=}PPh_3][Ph_2BiFe(CO)_4]$ has a trigonal bipyramidal coordination with the Ph_2Bi group occupying an axial position [91IC2788]. Two Ph_2Bi groups are bound to the octahedral iron center in $(Ph_2Bi)_2Fe(CO)_4$. The Bi–Fe bond distances in these complexes are 2.676(4) and 2.823(2)–2.832(2) Å, respectively. The difference observed may be due partly to the difference in steric interaction between the phenyl and carbonyl groups, the more crowded $(Ph_2Bi)_2Fe(CO)_4$ showing the longer Bi–Fe bond. The bismuth center in these complexes exhibits a pyramidal geometry with a stereochemically active lone pair of electrons. The average Bi–C distances and C–Bi–C angles are 2.30 Å and 94.9° for $[Ph_3P{=}N{=}PPh_3][Ph_2BiFe(CO)_4]$, and 2.23 Å and 96.2° for $(Ph_2Bi)_2Fe(CO)_4$. These values are almost identical to those of Ph_3Bi and $Ph_2BiBiPh_2$, suggesting that the bonding to the bismuth atom is primarily *p* in character, with little contribution from sp^3 hybridization. In an anionic iron complex, $[Et_4N]_2[{}^iBuBi\{Fe(CO)_4\}_3]_2$, the bismuth center possesses an extremely distorted tetrahedral geometry; the Fe–Bi–C and Fe–Bi–Fe bond angles are 96.7(2)–107.7(2)° and 112.7(3)–116.4(3)°, respectively [93IC2212]. The distortion about bismuth is due to the steric hindrance of the $Fe(CO)_4$ groups. A dimeric complex $[PhBiFe(CO)_4]_2$ is known and can be envisioned as the coupling product of a bismuthene–iron complex

($PhBiFe(CO)_4$) [91IC2788]. The structure around the bismuth center in $Ph_2BiMn(CO)_5$ is found to be similar to that of the above-mentioned iron complexes [91IC2788]. The Bi–Mn bond distance of 2.842(2) Å is shorter than the one in the more sterically crowded $Bi[Mn(CO)_5]_3$.

Chapter 3

Organobismuth(V) Compounds

3.1. TRIORGANYLBISMUTH DIHALIDES AND RELATED COMPOUNDS

3.1.1. Methods of synthesis (Tables 3.1–3.4)

3.1.1.1. Oxidative addition to triorganylbismuthines

Oxidation of triarylbismuthines to pentavalent triarylbismuth dihalides is accomplished using a variety of halogenating agents such as molecular halogens, sulfur dichloride, disulfur dichloride, thionyl chloride, sulfuryl chloride, phenyliodine dichloride, xenon difluoride, iodine pentafluoride and iodine trichloride. Lead tetraacetate, benzoyl peroxide and benzoyl nitrate can convert triarylbismuthines to the corresponding triarylbismuth(V) dicarboxylates or dinitrates. Oxidative addition of *N*-bromosuccinimide to triphenylbismuthine gives Ph_3BiBr(succinimide).

(a) Via halogenation with molecular halogens (method A) Triarylbismuth difluorides, dichlorides and dibromides are usually prepared by the stoichiometric reaction between triarylbismuthines and the corresponding halogens.

$$R_3Bi + X_2 \longrightarrow R_3BiX_2$$

Iodine usually cleaves one of the Bi–C bonds in triarylbismuthines to give diaryliodobismuthines. Although trialkylbismuthines are readily cleaved by chlorine and bromine, tri(1-propenyl)bismuthine can be brominated by molecular bromine at low temperatures to give tri(1-propenyl)bismuth dibromide [63IZV1507].

Synthesis of tris(2,4,6-trimethylphenyl)bismuth dichloride

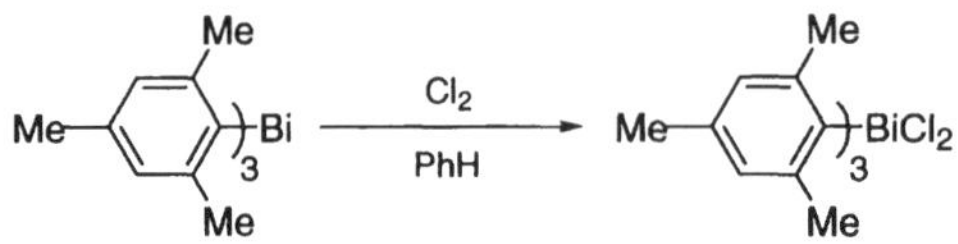

Tris(2,4,6-trimethylphenyl)bismuthine (1 g, 1.8 mmol) dissolved in dry benzene was treated with dry gaseous chlorine until an excess was evident. Partial evaporation of the solvent, followed by addition of methanol, gave tris(2,4,6-trimethylphenyl)bismuth dichloride as a pale green solid (0.85 g, 76%), m.p. 149–150°C [41JA207].

(b) Via halogenation with sulfuryl halides and other halogen equivalents (method B) Triarylbismuthines are converted to the corresponding dichlorides by means of sulfuryl chloride. Sulfur dichloride, disulfur dichloride, thionyl chloride, phenyliodine dichloride and iodine trichloride can be used similarly for this purpose. Xenon difluoride is a versatile fluorine source, directly oxidizing triarylbismuthines to the corresponding difluorides.

$$R_3Bi + EX_2 \longrightarrow R_3BiX_2 + E$$

This method is convenient because the reagents are easier to handle compared with molecular chlorine or fluorine. Trimethylbismuth dichloride is obtainable by using sulfuryl chloride at low temperatures [94AG(E)976]. However, an attempt to synthesize trimethylbismuth difluoride by using xenon difluoride has failed.

Synthesis of trimethylbismuth dichloride

$$Me_3Bi + SO_2Cl_2 \xrightarrow[Et_2O]{} Me_3BiCl_2$$

Freshly degassed sulfuryl chloride (1.22 g, 9.0 mmol) was condensed into a dry ethereal solution (10 ml) of trimethylbismuthine (2.3 g, 9.1 mmol) at −196°C. The mixture was stirred at −100°C, then warmed to −78°C over 2 h, and the solvent together with sulfur dioxide was pumped off at 10^{-3} mbar to give trimethylbismuth dichloride as a colorless solid (2.9 g, 95%) [94AG(E)976].

(c) Via oxidation with ozone, peracids and their equivalents (method C) Ozonized oxygen converts triarylbismuthines to the corresponding dicarboxylates in a suitable solvent such as toluene, ethyl acetate or ethyl propionate [93JCS(P1)2411].

$$Ar_3Bi + O_3/Solvent \longrightarrow Ar_3Bi(OCOR)_2$$

The carboxyl ligands introduced are derived from the solvent employed. When this reaction is performed in dichloromethane, triarylbismuth dichloride is isolated in good yield.

Synthesis of triphenylbismuth diformate

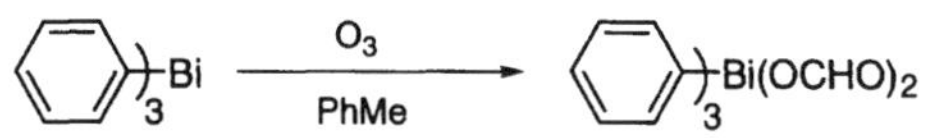

Ozonized oxygen (10 mmol/h) was passed into toluene (60 ml) for 1 h at −78°C to give a dark-green solution, to which was added triphenylbismuthine (440 mg, 1 mmol) in toluene (2 ml) in one portion. The resulting pale yellow suspension was allowed to warm to ambient temperature while being flushed with argon to give a pale yellow solution; this was evaporated under reduced pressure to leave a light brown residue, which was recrystallized from benzene–hexane (1:3) to give triphenylbismuth diformate (510 mg, 96%) as crystals (m.p. 140–150°C, decomp.) [93JCS(P1)2411].

Triarylbismuthines are oxidized to the corresponding dicarboxylates by treatment with a mixture of a carboxylic acid and an organic peroxide. Sodium perborate can be used instead of organic peroxides [96SC4569]. Peracids convert triarylbismuthines directly to the corresponding dicarboxylates.

$$Ar_3Bi + RCO_2H + R'OOH \longrightarrow Ar_3Bi(OCOR)_2$$

$$Ar_3Bi + RCO_2OH \longrightarrow Ar_3Bi(OCOR)_2$$

Synthesis of triphenylbismuth dibenzoate

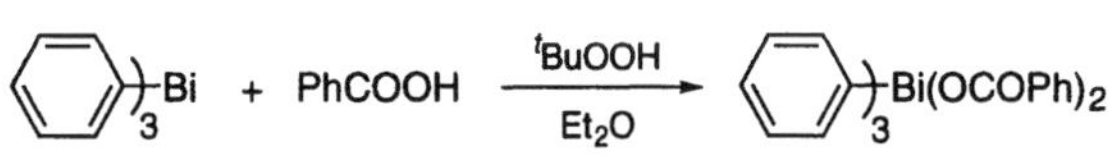

A mixture of triphenylbismuthine (5 mmol), *tert*-butyl hydroperoxide (5 mmol), benzoic acid (10 mmol) and anhydrous diethyl ether (50 ml) was stirred at room temperature. After a day, the liquid part was distilled off at a reduced pressure and the residue was crystallized from ethyl acetate to give triphenylbismuth dibenzoate. Yield 87%, m.p. 169°C [85ZOB73].

3.1.1.2. Via metathetical reaction of triarylbismuth dihalides and related compounds

Metathetical reaction of triarylbismuth dihalides and their derivatives is widely used for the preparation of triarylbismuth(V) compounds of the type Ar_3BiX_2, where X denotes an anionic group.

(a) With metal salts (method D) Triarylbismuth dichlorides, dibromides and diformates undergo metathetical reaction with a variety of sodium, potassium, lead and silver salts to give triarylbismuth(V) compounds of the type Ar_3BiX_2.

$$Ar_3BiX_2 + 2\,MY \longrightarrow Ar_3BiY_2 + 2\,MX$$

Triphenylbismuth(V) derivatives such as difluoride, dicyanate, diazide, dicyanide, dinitrate, diacetate, bis(haloacetate), bis(dihaloacetate), bis(trihaloacetate), dibenzoate, carbonate, oxalate, bis(succinimide), oxide, bis(pentachlorophenoxide), bis(pentafluorophenoxide) and many other dicarboxylates are prepared using this methodology. Attempts to obtain the chiral helical extended structures by the reaction between Ar_3BiCl_2 and AgOCOR*, where R* (racemic) contains one chiral center and one hydroxyl group, have resulted in the exclusive formation of enantiomeric (*R*,*R*)- and (*S*,*S*)-derivatives [97JCS(D)2009]. Triphenylbismuth molybdate and chromate are also prepared by the metathesis of the corresponding dibromide with ammonium molybdate or chromate [80IC3863, 71CJC2529]. By contrast, attempts to prepare triphenylbismuth diperchlorate and bis(tetrafluoroborate) by the reaction between triphenylbismuth dichloride and the corresponding silver salts have resulted in the formation of oxobismuth(V) compounds of the type $[Ph_3BiOBiPh_3]X_2$ and/or tetraphenylbismuthonium salts.

Synthesis of triphenylbismuth diacetate

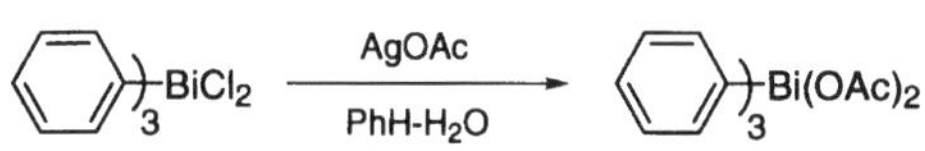

A benzene solution of triphenylbismuth dichloride and an aqueous solution of silver acetate were mixed with vigorous stirring. The benzene layer was then separated and concentrated under reduced pressure. Dilution with light petroleum caused triphenylbismuth diacetate to precipitate. This was recrystallized from a mixture of benzene and light petroleum, m.p. 173°C [70CJC2488].

Triphenylbismuth dichloride and bis(trifluoroacetate) react with one equiv. of sodium salt of *p*-nitrophenol to give the corresponding triphenylbismuth(V) compounds of the type $Ph_3BiX(OAr)$ [85JCS(P1)2657]. Sodium or guanidium salts of 3,5-di-*tert*-butylphenol and 8-hydroxyquinoline are also transferrable by this method.

Although Challenger and co-workers reported that triphenylbismuth(V) hydroxides of the structure $Ph_3Bi(OH)X$ (X=OH, Cl, Br, N_3 or CN) are formed by this methodology [20JCS(117)762, 34JCS405], the existence of these compounds is now questionable. For instance, Goel and Prasad found that the product formed by the reaction between triphenylbismuth dichloride and alcoholic sodium or potassium hydroxide is oxybis(triphenylbismuth) dichloride [72JOM(36)323].

Hexafluoropropene oxide is also a good reagent for the nucleophilic formation of the Bi–F bond in triphenylbismuth(V) compounds [99JFC(93)103].

(b) Via reactions with carboxylic acids, phenols and others (method E) A number of triphenylbismuth dicarboxylates are prepared from an interaction of triarylbismuth dihalides with carboxylic acids. Triphenylbismuth carbonate reacts with various aliphatic carboxylic acids to yield the corresponding dicarboxylates. Reaction between triarylbismuth diacetate and tropolones gives the corresponding triarylbismuth ditropolonates [96AG(E)67].

Treatment of triphenylbismuth carbonate with sulfonic acids affords triphenylbismuth disulfonates, which are converted to triphenylbismuth dichloride by metathesis [86ZAAC(539)110].

$$Ar_3BiX_2 + 2\,RCO_2H \xrightarrow[-2\,HX]{Base} Ar_3Bi(OCOR)_2$$

$$Ar_3Bi(OCOR)_2 + 2\,R'CO_2H \longrightarrow Ar_3Bi(OCOR')_2 + 2\,RCO_2H$$

$$Ar_3BiCO_3 + 2\,R'SO_3H \longrightarrow Ar_3Bi(OSO_2R')_2 + H_2O$$

Synthesis of triphenylbismuth bis(ethylthioacetate)

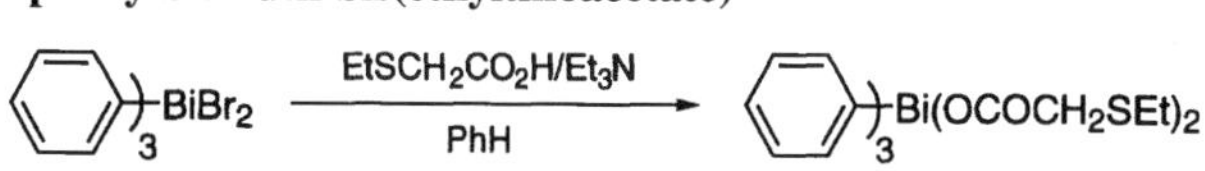

Triethylamine (1 g, 9.9 mmol) was added to a benzene solution of triphenylbismuth dibromide (1.2 g, 2 mmol) and ethylthioacetic acid (1.2 g, 5 mmol). After stirring for 1 h, the precipitated triethylamine hydrobromide was filtered off, and the filtrate was evaporated. The residue was recrystallized from acetonitrile to obtain the corresponding dicarboxylate (m.p. 119°C) [75JINC(37)2559].

3.1.1.3. Via miscellaneous approaches (method F)

Triphenylbismuthine is oxidized by *N*-bromosuccinimide, lead tetraacetate or benzoyl nitrate to give triphenylbismuth bromo(succinimide), diacetate and dinitrate, respectively. When treated with iodosobenzene in dichloromethane, tris(2,4,6-trimethylphenyl)bismuthine is converted to tris(2,4,6-trimethylphenyl)bismuth dichloride in a moderate yield [97JCS(P1)1609]. Triphenylbismuthine reacts irreversibly with sulfur dioxide within 5 min to give crystalline precipitates of composition $Ph_3BiS_2O_4$ [68CC1474].

Synthesis of triphenylbismuth dinitrate

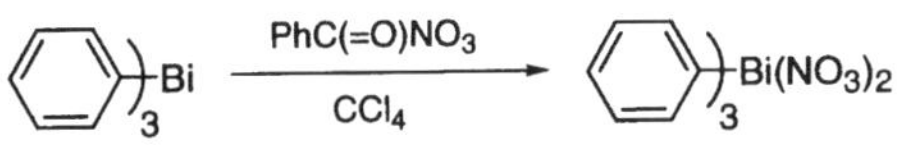

A solution of triphenylbismuthine (22 g) in dry carbon tetrachloride (50 ml) was cooled to −15°C and treated with a similarly cooled solution of benzoyl nitrate prepared in situ from benzoyl chloride (37 g) and silver nitrate (50 g). The mixture was well shaken, left to stand for 16 h and filtered. The residue was extracted with chloroform repeatedly and the combined filtrate and extracts were diluted with light petroleum to precipitate the dinitrate as a solid (27.5 g). This, on further extraction with cold ether, left insoluble triphenylbismuth dinitrate (12.5 g) [24JCS(125)854].

Nitration of the benzene ring of triphenylbismuth dinitrate is a convenient route to obtain the nitro-substituted triarylbismuth(V) derivatives [24JCS(125)854, 26JA507, 34JCS1258]. Oxidation of tris(4-methylphenyl)-bismuth dichloride by potassium permanganate or chromic acid gives the corresponding tris(carboxyphenyl)bismuth dichloride, which can be converted to the methyl ester in low yield by treatment with methanol in the presence of HCl [26JA507]. Sulfonation of tris(2-methylphenyl)bismuth dichloride results in the formation of a tris(hydroxy-2-methylphenyl)bismuth(V) compound after hydrolysis [26JA507].

$$Ar_3BiX_2 + HZ \longrightarrow Ar'_3BiZ_2$$

$$Ar_3BiX_2 \xrightarrow{[O]} Ar'_3BiX_2$$

Deprotection of tris(methoxymethoxyphenyl)bismuth dichlorides affords the corresponding tris(hydroxyphenyl)bismuth dichlorides, the *para*-isomer of which is acetylated by acetic anhydride/pyridine at the hydroxyl group to yield tris(4-acetoxyphenyl)bismuth dichloride [98JCS(P1)2511].

Triphenylbismuth difluoride has been synthesized by thermal decomposition of a cationic triphenylbismuth(V) complex $[Ph_3Bi(OAsPh_3)_2][BF_4]_2$ in dichloromethane [73IC944]. Tetraphenylbismuth tribromide is transformed thermally to triphenylbismuth dibromide [52LA(578)136], while tetraphenylbismuth triphenylsiloxide is readily cleaved by HCl to give triphenylbismuth dichloride [75DOK(225)581].

Synthesis of tris(4-nitrophenyl)bismuth dinitrate

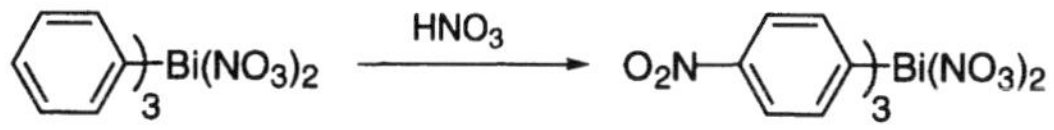

Triphenylbismuth dinitrate (10 g) was dissolved in fuming nitric acid at 0°C, and the resulting mixture was allowed to stand at 0°C for 12 h. The reaction mixture was poured into ice water (500 ml) and the deposit was filtered off, dried and recrystallized from ethyl acetate to give the dinitrate as a pale yellow solid, which exploded on heating at about 147°C [26JA507].

TABLE 3.1
Triorganylbismuth dihalides and nitrates

Compound	Synthetic method[a]	Yield (%)	M.p. (°C)	Physical data	Reference
Fluorides (R_3BiF_2)					
Ph	A	78	158	^{13}C-, ^{19}F-NMR	[78AG(E)214]
	D	–	158.5–159	–	[22JCS(121)91]
	D	–	127.5	IR, EC	[70CJC2488]
	D	90	158–161	^{1}H-NMR	[93JCS(P1)2411]
	D	45	148–153	^{19}F-NMR	[99JFC(93)103]
2-MeC_6H_4	B	100	165–166 (decomp.)	^{1}H-NMR, MS	[99OM5668]
	D	–	204–207 (decomp.)	–	[88JAP(K)174926]
3-MeC_6H_4	D	–	135–138	–	[88JAP(K)174926]
4-MeC_6H_4	D	–	120–123	–	[88JAP(K)174926]
	D	–	123–125	^{1}H-NMR	[94JCS(P1)1739]
3-$CF_3C_6H_4$	D	–	146–149	–	[88JAP(K)174926]
2-$MeOC_6H_4$	D	–	210–213 (decomp.)	–	[88JAP(K)174926]
4-$MeOC_6H_4$	B	100	–	^{1}H-NMR	[98OM4332]
3-FC_6H_4	D	–	149–153	–	[88JAP(K)174926]
3-ClC_6H_4	D	44	194–197	–	[88JAP(K)174926]
4-ClC_6H_4	B	100	–	^{1}H-NMR	[98OM4332]
	D	–	97–103	–	[88JAP(K)174926]
2,6-$F_2C_6H_3$	B	78.6	207 (decomp.)	^{1}H-, ^{13}C-, ^{19}F-NMR, IR, MS	[97ZAAC(623)122]
C_6F_5	B	–	155–162	^{19}F-NMR, MS	[89CB803]
	B	100	–	^{13}C-, ^{19}F-NMR, MS	[89CJC1949]
$C_6F_5(Ar_3Bi)$ (**1**)[b]	B	–	–	^{19}F-NMR, X-ray	[90CB761]
Chlorides (R_3BiCl_2)					
Me	B	95	–	^{1}H-NMR, X-ray	[94AG(E)976]
Ph	A	–	140	–	[1887CB54]
	A	–	141.5	–	[1889LA(251)323]
	A	–	141	–	[38ZOB1839]
	A	–	146–147	–	[43ZAAC(250)257]
	A	–	–	–	[64JOM(2)59]

	A	–	159–160	IR, EC	[70CJC2488]
	B	–	–	–	[16JCS(109)250]
	B	96	156–160	–	[86T3111]
	B, E	–	120	–	[20JCS(117)762]
	C	42	156–158	^{1}H-NMR	[93JCS(P1)2411]
	–	–	–	X-ray	[68JCS(A)2539]
Ph (^{210}Bi)	A	–	–	–	[63JINC(25)318]
2-MeC_6H_4	A	–	160	–	[1897CB2843]
	A	–	160	–	[26RZC(6)97]
	A	–	169–171	–	[88JAP(K)174926]
	D	87	172	^{1}H-, ^{13}C-NMR	[99T1341]
3-MeC_6H_4	A	–	132–133	–	[26JA507]
	A	37	151–153	–	[58DOK(122)614]
	A	–	126–129	–	[88JAP(K)174926]
4-MeC_6H_4	A	–	147	–	[1889LA(251)323]
	A	–	147	–	[25RZC(5)298]
	A	–	142–145	–	[88JAP(K)174926]
	B	–	–	–	[21JCS(119)913]
	B	98	147–149	–	[86T31111]
2-$CF_3C_6H_4$	A	–	135–137	–	[88JAP(K)174926]
3-$CF_3C_6H_4$	A	–	158–160	–	[88JAP(K)174926]
4-$CF_3C_6H_4$	B	–	140–143	–	[93OM1857]
4-^{i}PrC_6H_4	A		208	–	[1897CB2843]
4-PhC_6H_4	A	100	198–200	–	[36JA1820]
4-(2-Py)$_2$$NCH_2C_6H_4$	B	74	–	–	[96OM5613]
2-$HO_2CC_6H_4$	F	15	157–159	–	[26JA507]
	F	9	157–159	–	[26RZC(6)97]
4-$HO_2CC_6H_4$	F	–	130	–	[26JA507]
	F	–	130	–	[26RZC(6)97]
2-$MeO_2CC_6H_4$	F	–	135–137	–	[26RZC(6)97]
4-$MeO_2CC_6H_4$	F	–	103	–	[26JA507]
	F	–	130	–	[25RZC(5)298]
	F	–	107	–	[26RZC(6)97]
4-$EtO_2CC_6H_4$	A	20	138–139	–	[58DOK(122)614]

TABLE 3.1 *(continued)*

Compound	Synthetic method[a]	Yield (%)	M.p. (°C)	Physical data	Reference
4-$Me_2NC_6H_4$	B	83	145	^{1}H-NMR	[72ZN(B)591]
	B	88	–	^{1}H-NMR	[96OM5613]
3-$O_2NC_6H_4$	A	26	131–132	–	[58DOK(122)614]
4-$O_2NC_6H_4$	A	5.5	160–161	–	[58DOK(122)614]
	B	95	165	–	[86T3111]
	F	50	132–134	–	[26JA507]
	F	–	132	–	[25RZC(5)298]
2-HOC_6H_4	F	100	158–160	^{1}H-NMR, IR, MS	[98JCS(P1)2511]
3-HOC_6H_4	F	100	–	^{1}H-NMR, IR, MS	[98JCS(P1)2511]
4-HOC_6H_4	F	100	–	^{1}H-, ^{13}C-NMR, IR	[98JCS(P1)2511]
2-$MeOC_6H_4$	A	–	185–189	–	[88JAP(K)174926]
	C	23	196–197	^{1}H-, ^{13}C-NMR, IR	[97JCS(P1)1609]
4-$MeOC_6H_4$	A	–	133	–	[26RZC(6)97]
	B	85	135	–	[86T3111]
2-$EtOC_6H_4$	C	–	190 (decomp.)		[97JCS(P1)1609]
2-$MeOCH_2OC_6H_4$	B	–	–	^{1}H-NMR, MS	[98JCS(P1)2511]
3-$MeOCH_2OC_6H_4$	B	–	–	^{1}H-NMR	[98JCS(P1)2511]
4-$MeOCH_2OC_6H_4$	B	49	74–79	^{1}H-, ^{13}C-NMR, IR, MS	[98JCS(P1)2511]
4-$MeCO_2C_6H_4$	F	61	161	^{1}H-, ^{13}C-NMR, MS	[98JCS(P1)2511]
3-FC_6H_4	A	–	91–93	–	[88JAP(K)174926]
4-FC_6H_4	A	68	137.5	^{1}H-, ^{13}C-, ^{19}F-NMR, IR, MS	[89CB803]
	A	–	136–139	–	[88JAP(K)174926]
	B	–	134–136	–	[93OM1857]
3-ClC_6H_4	A	58	83–85	–	[88JAP(K)174926]
4-ClC_6H_4	A	100	141	–	[22JCS(121)104]
	A	17	170	–	[58DOK(122)614]
	A	–	187–190	–	[88JAP(K)174926]
4-BrC_6H_4	A	–	155	–	[22JCS(121)104]
2,4-$Me_2C_6H_3$	A	–	161	–	[1889LA(251)323]

	A	80	161	–	[26RZC(6)97]
	D	65	169	^{1}H-, ^{13}C-NMR	[99T1341]
2,5-$Me_2C_6H_3$	A	–	167.5	–	[1897CB2843]
2-Me-5-$^{i}PrC_6H_3$	A	89.4	163–164	–	[41JA207]
2-Me-4-$HO_2CC_6H_3$	F	13	142–157	–	[26RZC(6)97]
2-Me-3-$O_2NC_6H_3$	F	–	147–149	–	[26RZC(6)97]
	F	–	149–150	–	[26JA507]
4-Me-2-$O_2NC_6H_3$	F	–	156	–	[26JA507]
	F	–	156	–	[25RZC(5)298]
2-O_2N-4-$HO_2CC_6H_3$	F	–	250–260 (decomp.)	–	[26JA507]
	F	–	260	–	[26RZC(6)97]
2-O_2N-4-$MeO_2CC_6H_3$	F	–	250–260	–	[26JA507]
2,6-$F_2C_6H_3$	A	94.2	131 (decomp.)	^{1}H-, ^{13}C-, ^{19}F-NMR, IR, MS	[97ZAAC(623)122]
Ar (**2**)[b]	F	–	105–110	–	[24JCS(125)854]
Ar (**3**)[b]	F	–	136	–	[24JCS(125)854]
Ar (**4**)[b]	F	–	146	–	[24JCS(125)854]
Ar (**5**)[b]	F	–	148–149	–	[24JCS(125)854]
2,4,6-$Me_3C_6H_2$	A	–	150 (decomp.)	–	[38ZOB1839]
	C	80	–	–	[97JCS(P1)1609]
C_6F_5	D	100	42 (decomp.)	^{13}C-, ^{19}F-NMR, MS	[97JFC(84)69]
1-$C_{10}H_7$	A	–	167–169	–	[20JCS(117)762]
	B	–	–	–	[21JCS(119)913]
	B	–	–	–	[99OM3016]
2-C_4H_3S	A	–	–	–	[32CB777]
Bromides (R_3BiBr_2)					
cis-MeCH=CH	A	62.5	65–66	–	[63IZV1507]
trans-MeC=CH	A	71.3	142	–	[63IZV1507]

TABLE 3.1 (*continued*)

Compound	Synthetic method[a]	Yield (%)	M.p. (°C)	Physical data	Reference
$CH_2{=}CMe$	A	70	248–250	–	[63IZV1507]
Ph	A	–	119	–	[1887CB54]
	A	–	122	–	[1889LA(251)323]
	A	–	118	–	[14JCS(105)2210]
	A	–	119	–	[25RZC(5)298]
	A	79	144–146	–	[52LA(578)136]
	A	–	–	–	[66JCS(A)520]
	A	–	139	IR	[70CJC2488]
	A	–	–	^1H-, ^{13}C-NMR	[75JINC(37)2347]
	D	91	139–141	^1H-NMR	[93JCS(P1)2411]
$2\text{-}MeC_6H_4$	A	–	125	–	[1897CB2843]
$3\text{-}MeC_6H_4$	A	–	92	–	[24JCS(125)864]
	A	–	52	–	[26JA507]
$4\text{-}MeC_6H_4$	A	–	111	–	[25RZC(5)298]
	A	–	111–112	–	[1889LA(251)323]
$4\text{-}^iPrC_6H_4$	A	–	150	–	[1897CB2843]
$4\text{-}PhC_6H_4$	A	–	–	–	[36JA1820]
$4\text{-}MeOC_6H_4$	A	–	103	–	[1897CB2843]
	A	–	–	–	[26JA507]
$2\text{-}EtOC_6H_4$	A	100	127–128	–	[41JA207]
$4\text{-}ClC_6H_4$	A	–	105	–	[22JCS(121)104]
$4\text{-}BrC_6H_4$	A	–	107 (decomp.)	–	[22JCS(121)104]
$2,4\text{-}Me_2C_6H_3$	A	–	117	–	[1889LA(251)323]
$2,5\text{-}Me_2C_6H_3$	A	–	130	–	[1897CB2843]
$2\text{-}Me\text{-}5\text{-}^iPrC_6H_3$	A	84	101–113	–	[41JA207]
$2,4,6\text{-}Me_3C_6H_2$	A	90	250 (decomp.)	–	[38ZOB1839]
	A	56.5	91–93	–	[41JA207]
$1\text{-}C_{10}H_7$	A	–	119–121	–	[14JCS(105)2210]

Nitrates ($R_3Bi(ONO_2)_2$)					
Ph	F	50	–	–	[24JCS(125)854]
	F	84	159–161	–	[88T6387]
	D	–	–	–	[1889LA(251)323]
	F	96	145–147	–	[88T6387]
	D	–	–	–	[25RZC(5)298]
	D	–	157	–	[43ZAAC(250)257]
	D	–	148 (explodes)	IR, EC	[70CJC2488]
	D	38	127–130 (decomp.)	–	[88JAP(K)174926]
2-MeC_6H_4	D	–	–	–	[1897CB2843]
	D	–	–	–	[26RZC(6)97]
	D	–	–	–	[26JA507]
	D	–	150–153 (decomp.)	–	[88JAP(K)174926]
3-MeC_6H_4	D	–	150–160 (explodes)	–	[26JA507]
4-MeC_6H_4	D	–	–	–	[1889LA(251)323]
	D	–	–	–	[25RZC(5)298]
	D	–	157–160 (decomp.)	–	[88JAP(K)174926]
4-PhC_6H_4	D	–	162	–	[36JA1820]
2-$MeO_2CC_6H_4$	F	–	97	–	[26RZC(6)97]
4-$MeO_2CC_6H_4$	F	–	–	–	[26RZC(6)97]
3-$O_2NC_6H_4$	F	–	145	–	[25CB1893]
	F	–	–	–	[34JCS1258]
4-$O_2NC_6H_4$	F	–	147	–	[25RZC(5)298]
	F	–	147	–	[26JA507]
4-ClC_6H_4	F	–	163–165	–	[24JCS(125)854]
4-Me-2-$O_2NC_6H_3$	F	–	160	–	[25RZC(5)298]
	F	–	160 (explodes)	–	[26JA507]
O_2N-4-$MeO_2CC_6H_3$	F	–	150–160 (explodes)	–	[26JA507]
Ar (**2**)[b]	F	83	–	–	[24JCS(125)854]
Ar (**3**)[b]	F	–	–	–	[1897CB2843]
	F	–	–	–	[24JCS(125)854]
Ar (**4**)[b]	F	72	–	–	[24JCS(125)854]
Ar (**5**)[b]	F	100	–	–	[24JCS(125)854]

[a] For notation, see Section 3.1.1.
[b] (**1**) $(C_6F_5)_3BiF_2 \cdot 2Bi(C_6F_5)_3$. For other structures, see Scheme 3.1.

TABLE 3.2
Triorganylbismuth(V) compounds

Compound R^1	R^2	Synthetic method[a]	Yield (%)	M.p. (°C)	Physical data	Reference
Carboxylates ($R^1_3Bi(OCOR^2)_2$)						
Ph	H	C	86	–	–	[93IZV2043]
		C	96	140–150 (decomp.)	^{1}H-, ^{13}C-NMR, IR, MS, X-ray	[93JCS(P1)2411]
4-MeC_6H_4	H	C	98	126–128	^{1}H-, ^{13}C-NMR, IR, MS	[93JCS(P1)2411]
Ph	Me	C	34	187–189 (decomp.)	–	[64CB789]
		C	100	186	–	[85ZOB73]
		C	65	184	–	[88ZOB1270]
		C	90	187	–	[93IZV2043]
		C	59	182	IR	[95IZV964]
		C	90	189	^{1}H-NMR	[96SC4569]
		C	77	192–194	^{1}H-, ^{13}C-NMR, IR	[97T4137]
		C	80	189	–	[98T4313]
		D	–	173	IR, EC	[70CJC2488]
		D	–	168–172	–	[88JAP(K)174926]
		D	74	160–164	^{1}H-NMR	[93JCS(P1)2411]
		E	99	162	–	[20JCS(117)762]
		E	–	148–150	–	[16JCS(109)250]
		D	–	152–153	–	[26JA507]
Ph	Me (**1**)[b]	E	–	136–137	–	[20JCS(117)762]
2-MeC_6H_4	Me	C	78	129	^{1}H-NMR	[98T4313]
3-MeC_6H_4	Me	D	–	148–149	–	[26JA507]
		D	–	155–158	–	[88JAP(K)174926]
4-MeC_6H_4	Me	C	–	196–198 (decomp.)	^{1}H-, ^{13}C-NMR, IR, MS	[93JCS(P1)2411]
		C	72	172	^{1}H-NMR	[96SC4569]
		C	65	172	–	[98T4313]
		D	–	162	–	[26JA507]
		D	–	187–189	–	[88JAP(K)174926]
Ar (**2**)[b]	Me	D	80	95–97	^{1}H-, ^{13}C-NMR, X-ray	[96OM5613]

3-$CF_3C_6H_4$	Me	C	61	94	^{1}H-NMR	[96SC4569]
		C	–	–	–	[98T4313]
		D	–	89–92	–	[88JAP(K)174926]
2-$MeOC_6H_4$	Me	C	71	147	^{1}H-NMR	[96SC4569]
		C	65	147	–	[98T4313]
4-$MeOC_6H_4$	Me	C	67	170–173	^{1}H-NMR	[96SC4569]
		C	65	170–173	–	[98T4313]
Ar (**3**)[b]	Me	C	63	121–124	^{1}H-, ^{13}C-NMR	[99T3377]
3-FC_6H_4	Me	D	–	>165 (decomp.)	–	[88JAP(K)174926]
4-FC_6H_4	Me	C	85	162	^{1}H-NMR	[96SC4569]
		C	68	162	–	[98T4313]
		D	–	>193 (decomp.)	–	[88JAP(K)174926]
3-ClC_6H_4	Me	D	52	158–160	–	[88JAP(K)174926]
4-ClC_6H_4	Me	C	81	154–156	^{1}H-NMR	[96SC4569]
		C	67	154–156	–	[98T4313]
		D	–	158–162	–	[88JAP(K)174926]
1-$C_{10}H_7$	Me	C	65	95 (decomp.)	^{1}H-NMR	[96SC4569]
		C	–	–	–	[98T4313]
2,4-$Me_2C_6H_3$	Me	C	72	135–136	^{1}H-, ^{13}C-NMR	[99T1341]
Ph	Et	C	52	–	–	[88ZOB1270]
		C	80	160	–	[93IZV2043]
		C	4	157–159	^{1}H-, ^{13}C-NMR	[93JCS(P1)2411]
		E	–	–	^{13}C-NMR	[75JINC(37)2347]
	Bu	C	90	91	–	[85ZOB73]
		C	71	91	–	[93IZV2043]
	^{t}Bu	C	–	146–148	^{1}H-, ^{13}C-NMR, IR, MS	[93JCS(P1)2411]
		E	76	130	^{1}H-, ^{13}C-NMR, IR	[97T4137]
	C_7H_{15}	C	–	–	–	[85ZOB73]
	$C_{11}H_{23}$	C	–	–	–	[85ZOB73]
	FCH_2	D	–	165–166	IR, EC	[70CJC2488]
	F_2CH	D	–	146	IR, EC	[70CJC2488]
	CF_3	C	–	147–149	^{1}H-NMR	[93JCS(P1)2411]
		D	–	133	IR, EC	[70CJC2488]
		D	–	–	X-ray	[91JOM(419)283]

TABLE 3.2 *(continued)*

Compound R^1	R^2	Synthetic method[a]	Yield (%)	M.p. (°C)	Physical data	Reference
		E	95	143–144	^{1}H-NMR, IR	[94SC989]
		E	87	–	–	[97TL365]
4-MeC_6H_4	CF_3	E	81.5	178 (decomp.)	^{1}H-NMR, IR	[94SC989]
4-$Me_2NC_6H_4$	CF_3	D	71	120	^{1}H-NMR, X-ray	[96OM5613]
2,6-$F_2C_6H_3$	CF_3	D	98.2	151 (decomp.)	^{1}H-, ^{13}C-, ^{19}F-NMR, IR, MS	[97ZAAC(623)122]
C_6F_5	CF_3	D	100	83 (decomp.)	^{13}C-, ^{19}F-NMR, MS	[97JFC(84)69]
Ph	$ClCH_2$	D	–	166	IR, EC	[70CJC2488]
		E	82	169 (decomp.)	^{1}H-NMR, IR	[94SC989]
		E	96.5	155–156 (decomp.)	–	[51JA4470]
	Cl_2CH	D	–	118–119	IR, EC	[70CJC2488]
		E	82	119–119.5	^{1}H-NMR IR	[94SC989]
	CCl_3	D	–	158	IR, EC	[70CJC2488]
		E	91	160	^{1}H-NMR, IR	[94SC989]
	$BrCF_2$	D	–	171–172	IR, EC	[70CJC2488]
		E	81	167–169	^{1}H-NMR, IR	[94SC989]
4-MeC_6H_4	$BrCF_2$	E	80.7	146–147	^{1}H-NMR, IR	[94SC989]
Ph	$NCCH_2$	D	–	123–124	IR, EC	[70CJC2488]
		E	80	126–127	^{1}H-NMR, IR	[94SC989]
	$MeSCH_2$	E	–	115	^{1}H-NMR, IR	[75JINC(37)2559]
	$EtSCH_2$	E	–	119	^{1}H-NMR, IR	[75JINC(37)2559]
	$PrSCH_2$	E	–	93	^{1}H-NMR, IR	[75JINC(37)2559]
	^{i}$PrSCH_2$	E	–	105	^{1}H-NMR, IR	[75JINC(37)2559]
	$BuSCH_2$	E	–	78	^{1}H-NMR, IR	[75JINC(37)2559]
	$PhSCH_2$	E	–	106	^{1}H-NMR, IR	[75JINC(37)2559]
	$PhCH_2SCH_2$	E	–	106	^{1}H-NMR, IR	[75JINC(37)2559]
	$PhOCH_2$	E	–	143	^{1}H-NMR, IR	[75JINC(37)2559]
	$HO(CH_2)_3$	D	84	–	^{1}H-NMR, X-ray	[97JCS(D)2009]
4-$Me_2NC_6H_4$	$HO(CH_2)_3$	D	75	–	^{1}H-NMR, X-ray	[97JCS(D)2009]

Ph	4-$HOC_6H_4(CH_2)_2$	D	70	–	^{1}H-NMR, X-ray	[97JCS(D)2009]
Ph	Ph($HOCH_2$)CH	D	95	–	^{1}H-, ^{13}C-NMR, X-ray	[97JCS(D)2009]
4-$Me_2NC_6H_4$	HOCH(Et)	D	70	–	^{1}H-, ^{13}C-NMR, X-ray	[97JCS(D)2009]
	HOCH(Me)CH_2	D	70	–	^{1}H-, ^{13}C-NMR, X-ray	[97JCS(D)2009]
	Ph(C_5H_9)CH	D	70	–	^{13}C-NMR	[97JCS(D)2009]
	Ph($HOCH_2$)CH	D	89	–	^{1}H-, ^{13}C-NMR. X-ray	[97JCS(D)2009]
Ph	CH_2=CH	C	99	151	–	[85ZOB73]
		C	89	151	–	[93IZV2043]
	Ch_2=CMe	E	80	170	–	[64NEP6405309]
	PhCH=CH	D	50	176–178 (decomp.)	–	[51JA4470]
	Ph	C, D	–	172–173	–	[27JCS209]
		C	87	169	–	[85ZOB73]
		C	59	–	–	[88ZOB1270]
		C	84	169	–	[93IZV2043]
		E	–	–	^{1}H-, ^{13}C-NMR	[75JINC(37)2347]
		E	75	167–169	–	[97TL365]
4-MeC_6H_4	Ph	D	69	168–169 (decomp.)	–	[51JA4470]
Ph	4-CH_2=CHC_6H_4	E	–	170–180	–	[64NEP6405309]
	2-$HO_2CC_6H_4$	E	63.3	168–169 (decomp.)	–	[51JA4470]
	2-HOC_6H_4	D	52	184–185 (decomp.)	–	[51JA4470]
	4-HOC_6H_4	D	–	–	–	[97JCS(D)2009]
		E	81.6	250 (decomp.)	–	[51JA4470]
	4-$H_2NC_6H_4$	E	69.8	148 (decomp.)	–	[51JA4470]
	2-$H_2NC_6H_4$	E	90.2	95–96 (decomp.)	–	[51JA4470]
	4-$Me_2NC_6H_4$	E	68	171	^{1}H-, ^{13}C-NMR, IR	[97T4137]
	2-C_4H_3O	–	–	–	X-ray	[88AX(C)830]
4-MeC_6H_4	2-HOC_6H_4	D	65.5	164–165 (decomp.)	–	[51JA4470]
2-MeC_6H_4	2-HOC_6H_4	D	51	150–151 (decomp.)	–	[51JA4470]
4-ClC_6H_4	2-HOC_6H_4	D	85.7	187 (decomp.)	–	[51JA4470]
Ph	1,2-$C_6H_4(CO_2)_2$ (**4**)^b	D	58.3	155–165 (decomp.)	–	[51JA4470]
	$(CO_2)_2$ (**5**)^b	F	100	–	–	[85ZOB73]
		D	40	177 (decomp.)	IR	[71CJC2529]
Ph	$CO_2(CH_2)_2CO_2$ (**6**)^b	E	–	–	–	[16JCS(109)250]
	CO_3	C	40	150–155 (decomp.)	IR	[93JCS(P1)2411]

TABLE 3.2 *(continued)*

Compound R^1	R^2	Synthetic method[a]	Yield (%)	M.p. (°C)	Physical data	Reference
		D	–	>220	–	[20JCS(117)762]
		D	–	164–165 (decomp.)	IR	[71CJC2529]
		D	100	155 (decomp.)	–	[85JCS(P1)2657]
4-MeC_6H_4	CO_3	D	100	139	–	[85JCS(P1)2667]
4-$MeOC_6H_4$	CO_3	D	99	140–145 (decomp.)	–	[86T3111]
4-$O_2NC_6H_4$	CO_3	D	85	170–176	–	[86T3111]
Sulfonates ($R^1_3Bi(OSO_2R^2)_2$)						
Ph	F	E	98	–	^1H-, ^{19}F-NMR, IR	[95JFC(70)237]
	Me	E	82	172 (decomp.)	IR	[86ZAAC(539)110]
		E	82	166–170	^1H-, ^{13}C-NMR, IR	[97TL365]
	CF_3	E	71	107 (decomp.)	IR	[86ZAAC(539)110]
		E	98	>155 (decomp.)	^1H-, ^{19}F-NMR, IR	[95JFC(70)237]
	$HOCH_2CH_2$	E	70	130–132 (decomp.)	IR	[86ZAAC(539)110]
	Ph	E	87	176 (decomp.)	IR, X-ray	[86ZAAC(539)110]
	4-MeC_6H_4	E	77	190–194	^1H-NMR, IR	[85JCS(P1)2657]
		E	87	178 (decomp.)	IR	[86ZAAC(539)110]
		E	75	–	–	[97TL365]
	2,4-$(O_2N)_2C_6H_3$	E	80	198 (decomp.)	IR	[86ZAAC(539)110]
2,6-$F_2C_6H_3$	CF_3	D	82.7	119 (decomp.)	^1H-, ^{13}C-, ^{19}F-NMR, IR, MS	[97ZAAC(623)122]
C_6F_5	CF_3	D	100	−15 (decomp.)	^{13}C-, ^{19}F-NMR	[97JFC(84)69]
Oximates (R_3BiX_2)						
R	X					
Ph	Me_2C=NO	DE	56	66	^1H-NMR, IR	[87SRI(17)559]
	MeEtC=NO	DE	60	92	^1H-NMR, IR	[87SRI(17)559]
	MePrC=NO	DE	52	88	^1H-NMR, IR	[87SRI(17)559]
	MePhC=NO	DE	62	114	^1H-NMR, IR	[87SRI(17)559]
	Ph_2C=NO	DE	66	120	^1H-NMR, IR	[87SRI(17)559]

	$(CH_2)_5C{=}NO$	DE	54	58	^{1}H-NMR, IR	[87SRI(17)559]
Miscellaneous						
R_3BiX_2 type						
R	X					
Ph	Succinimide (**7**)[b]	D	–	–	^{1}H-NMR	[79JPR(321)370]
	N_3	D	–	94–95	–	[34JCS405]
		D	–	95 (decomp.)	IR, Raman	[73JOM(50)129]
		D	40	95	^{1}H-NMR, IR, MS	[92JCR(S)34]
4-MeC_6H_4	N_3	D	70	110	^{1}H-NMR, IR, MS	[92JCR(S)34]
Ph	CN	D	–	134 (decomp.)	IR, Raman	[73JOM(50)129]
	OH	D	–	–	–	[34JCS405]
Ar (**8**)[b]	OH	D	–	108 (decomp.)	–	[24JCS(125)854]
Ph	OCN	D	–	125–126	IR, EC	[70CJC2488]
		D	–	128.5–129	–	[27JCS209]
	CF_3CH_2O	D	>80	–	^{1}H-NMR, IR, EC	[94IJC(A)687]
	^{t}BuOO	C	–	–	–	[88ZOB712]
	$Br_3C_6H_2O$	D	–	–	IR, EC	[94IJC(A)687]
	C_6F_5O	D	47	131	^{1}H-, ^{19}F-NMR, IR, X-ray	[98OM1347]
	C_6Cl_5O	D	>80	–	IR, EC	[94IJC(A)687]
		D	88	155	^{1}H-NMR, IR, X-ray	[98OM1347]
	2,4,6-$(O_2N)_3C_6H_2O$	E	–	–	–	[16JCS(109)250]
	O_2N-tropolonate (**9**)[b]	E	73.8	–	^{1}H-, ^{13}C-NMR, IR, X-ray	[96AG(E)67]
	PhS	D	35.0	44	–	[51JA4470]
	C_6Cl_5S	E	>80	–	IR, EC	[94IJC(A)687]
4-MeC_6H_4	Tropolonate (**10**)[b]	E	69	–	^{1}H-, ^{13}C -NMR, IR, X-ray	[96AG(E)67]
R	X_2					
Ph	S_2O_4	E	–	126–128 (decomp.)	IR	[68CC1474]
	SO_4	E	–	–	–	[16JCS(109)250]
		E	–	> 280	–	[20JCS(117)762]
		E	70	200 (decomp.)	IR	[71CJC2529]

TABLE 3.2 (*continued*)

Compound			Synthetic method[a]	Yield (%)	M.p. (°C)	Physical data	Reference
	CrO_4		D	40	111 (explodes)	IR	[71CJC2529]
	SeO_4		E	100	180 (decomp.)	IR	[71CJC2529]
	MoO_4		D	87	–	^{1}H-, ^{13}C-, ^{17}O-NMR	[80IC3863]
$[Bu_4N]_2[Ph_3Bi(MoO_4)_2]^{2-}$			D	91	–	^{1}H-, ^{13}C-, ^{17}O-NMR	[80IC3863]
R_3BiXY type							
R	X	Y					
Ph	Cl	3,5-$^t Bu_2C_6H_3O$	E	45	98–101	^{1}H-NMR, MS	[85JCS(P1)2657]
		8-QO (**11**)[b]	D	80	115–117	IR, EC	[72JOM(38)91]
			D	80	119–124	X-ray	[84HCA586]
		2-Me-8-QO (**12**)[b]	D	70	130	^{1}H-NMR, IR, UV, X-ray	[83JOM(253)317]
		OH	F	–	160–161	–	[20JCS(117)762]
		$4-O_2NC_6H_4O$	D	86	157 (decomp.)	^{1}H-NMR, IR, MS	[85JCS(P1)2657]
		$N(SO_2Me)_2$	B	78	50 (decomp.)	^{1}H-NMR	[95ZAAC(621)1746]
	Br	Succinimide (**7**)[b]	B	64.8	130–132 (decomp.)	^{1}H-NMR, IR	[79JPR(321)370]
		8QO (**11**)[b]	D	80	98–101	IR, EC	[72JOM(38)91]
		OH	F	–	147–148	–	[20JCS(117)762]
		C_6F_5O	D	59	124	^{1}H-, ^{19}F-NMR, IR, X-ray	[98OM1347]
		C_6Cl_5O	D	85	125	^{1}H-NMR, IR, X-ray	[98OM1347]
		2-Me-8-QO (**12**)[b]	D	40	115	^{1}H-NMR, IR, UV	[83JOM(253)317]
	I	AsF_6	D	85	47	^{1}H-NMR, IR, Raman	[89JOM(379)251]
	CF_3CO_2	8-QO (**11**)[b]	D	86	129–130 (decomp.)	^{1}H-NMR, IR, MS	[85JCS(P1)2657]
		$4-O_2NC_6H_4O$	D	91	139–145 (decomp.)	^{1}H-NMR, IR	[85JCS(P1)2657]
	CN	OH	D	–	136	–	[34JCS405]
			F	–	137–158 (decomp.)	–	[88JAP(K)174926]
$2-MeC_6H_4$	CN	OH	F	–	133–136 (decomp.)	–	[88JAP(K)174926]
$3-MeC_6H_4$	CN	OH	F	–	154–156 (decomp.)	–	[88JAP(K)174926]
$4-MeC_6H_4$	CN	OH	F	–	150 (decomp.)	–	[88JAP(K)174926]

4-FC_6H_4	CN	OH	F	–	150–153 (decomp.)	–	[88JAP(K)174926]
3-ClC_6H_4	CN	OH	F	70	139–143 (decomp.)	–	[88JAP(K)174926]
4-ClC_6H_4	CN	OH	F	–	138–142 (decomp.)	–	[88JAP(K)174926]
Complex ([Ph_3BiL_2]X_2)							
L	X						
Ph_3AsO	NO_3			–	–	IR, EC	[73IC944]
Me_2SO	ClO_4			–	116 (decomp.)	IR, EC	[73IC944]
$(CD_3)_2SO$	ClO_4			–	120 (decomp.)	IR, EC	[73IC944]
C_5H_5NO	ClO_4			–	Explodes	IR, EC	[73IC944]
Ph_3PO	ClO_4			–	169 (decomp.)	IR, EC	[73IC944]
Ph_3AsO	ClO_4			–	Explodes	IR, EC	[73IC944]
	BF_4			–	166 (decomp.)	IR, EC	[73IC944]
	BF_4 (**13**)[b]			–	154	–	[73IC944]
Ph_3PO	PF_6			–	167 (decomp.)	IR, EC	[73IC944]
Ph_3AsO	PF_6			–	161–162 (decomp.)	IR, EC	[73IC944]
	PF_6 (**13**)[b]			–	163–164 (decomp.)	–	[73IC944]
	PF_6 (**14**)[b]			–	150 (decomp.)	–	[73IC944]
	PF_6 (**15**)[b]			–	135 (decomp.)	–	[73IC944]

[a] For notation, see Section 3.1.1.

[b] (**1**) 1/2 AcOH adduct; (**2**) Ar = 4-(2-Py)$_2$NC$H_2C_6H_4$; (**3**) Ar = 2-(CH_2=CHCH_2)C_6H_4; (**13**) acetone (Me_2CO) adduct; (**14**) diethylketone (Et_2CO) adduct; (**15**) acetylacetone ((MeCO)$_2CH_2$) adduct. For other structures, see Scheme 3.2.

TABLE 3.3
Mixed triorganylbismuth dihalides and related compounds

Compound R[1]	R[2]	R[3]	Synthetic method[a]	Yield (%)	M.p. (°C)	Physical data	Reference
$R^1_2R^2BiX_2$ type							
Fluoride ($R^1_2R^2BiF_2$)							
$4\text{-}MeC_6H_4$	$2\text{-}OxC_6H_4$[b]		B	98	184–187	^{1}H-NMR, IR, MS	[99OM5668]
Chlorides ($R^1_2R^2BiCl_2$)							
Ph	$4\text{-}MeC_6H_4$[b]		A	46.8	109–110	–	[41JA207]
			B	85	121–123	–	[86T3111]
	$1\text{-}C_{10}H_7$		A	–	141–142	–	[22JCS(121)91]
$2\text{-}MeC_6H_4$	$1\text{-}C_{10}H_7$		A	59.5	140	–	[41JA207]
$4\text{-}MeC_6H_4$	Ph		B	95	132–135	MS	[86T3111]
	$4\text{-}MeOC_6H_4$		B	89	120–122	^{1}H-NMR, MS	[86T3111]
	$4\text{-}O_2NC_6H_4$		B	85	–	–	[86T3111]
	$2\text{-}OxC_6H_4$[b]		B	80	187–189 (decomp.)	^{1}H-, ^{13}C-NMR, IR	[98OM1013]
	$1\text{-}C_{10}H_7$		A	95.0	147	–	[41JA207]
	9-Anthryl		B	88	170–173	^{1}H-NMR, IR, MS	[97OM3565]
	9-Phenanthryl		B	90	145–147	^{1}H-NMR, MS	[97OM3565]
$4\text{-}ClC_6H_4$	$2\text{-}MeC_6H_4$		A	86.5	132–133	–	[41JA207]
	$1\text{-}C_{10}H_7$		A	85	132	–	[41JA207]
$4\text{-}MeOC_6H_4$	$4\text{-}MeC_6H_4$		B	90	130–134	^{1}H-NMR	[86T3111]
$O_2NC_6H_4$	Ph		E	–	136	–	[1897CB2843]
$4\text{-}O_2NC_6H_4$	$4\text{-}MeC_6H_4$		B	80	–	–	[86T3111]
Bromides ($R^1_2R^2BiBr_2$)							
Ph	$1\text{-}C_{10}H_7$		A	–	140	–	[14JCS(105)2210]
			A	76	–	–	[20JCS(117)762]
$2\text{-}MeC_6H_4$	$1\text{-}C_{10}H_7$		A	76.5	122	–	[41JA207]
$4\text{-}MeC_6H_4$	$1\text{-}C_{10}H_7$		A	90.0	126–127	–	[41JA207]
$4\text{-}ClC_6H_4$	$2\text{-}MeC_6H_4$		A	92	109–110	–	[41JA207]
$4\text{-}ClC_6H_4$	$1\text{-}C_{10}H_7$		A	88.6	102–103	–	[41JA207]

Nitrates ($R^1_2R^2Bi(ONO_2)_2$)							
$O_2NC_6H_4$	Ph		F	–	–	–	[1897CB2843]
Benzoates ($R^1_2R^2Bi(OCOPh)_2$)							
Ph	$4\text{-}PhC_6H_4$		D	–	145–147	–	[41JA212]
	$2\text{-}C_{10}H_7$		D	–	138–140	–	[41JA212]
Carbonates ($R^1_2R^2BiCO_3$)							
Ph	$4\text{-}MeC_6H_4$		D	96	142–146	–	[86T3111]
$4\text{-}MeC_6H_4$	Ph		D	95	134–136	–	[86T3111]
	$4\text{-}MeOC_6H_4$		D	95	110–115	–	[86T3111]
	$4\text{-}O_2NC_6H_4$		D	80	108–112	–	[86T3111]
	$4\text{-}MeC_6H_4$		D	80	118–121	–	
$4\text{-}O_2NC_6H_4$	$4\text{-}MeC_6H_4$		D	85	126–134	–	[86T3111]
Sulfates ($R^1_2R^2BiSO_4$)							
$2\text{-}MeC_6H_4$	$2\text{-}Me\text{-}4\text{-}(HO_3S)C_6H_3$		F	–	166–167	–	[26RZC(6)97]
$2\text{-}Me\text{-}4\text{-}(HO_3S)C_6H_3$	$2\text{-}MeC_6H_4$		F	–	174	–	[26RZC(6)97]
$R^1R^2R^3BiX_2$ type							
Fluorides ($R^1R^2R^3BiF_2$)							
$2\text{-}MeC_6H_4$	$4\text{-}ClC_6H_4$	$4\text{-}MeOC_6H_4$	D	84	119–120 (decomp.)	^{1}H-NMR, IR, MS	[99OM5668]
$4\text{-}MeC_6H_4$	$4\text{-}ClC_6H_4$	$2\text{-}MeOC_6H_4$	D	66	167–170 (decomp.)	^{1}H-NMR, IR, MS	[99OM5668]
	$4\text{-}ClC_6H_4$	$2\text{-}^iPrOC_6H_4$	D	80	135–147 (decomp.)	^{1}H-NMR, IR, MS	[99OM5668]
	$4\text{-}CF_3C_6H_4$	$2\text{-}OxC_6H_4$[b]	B	78	199–201 (decomp.)	^{1}H-NMR, IR, MS	[99OM5668]
Chlorides ($R^1R^2R^3BiCl_2$)							
$2\text{-}MeC_6H_4$	$4\text{-}ClC_6H_4$	$4\text{-}MeOC_6H_4$	B	100	152–156	^{1}H-NMR, IR, MS	[99OM5668]
$4\text{-}MeC_6H_4$	$4\text{-}ClC_6H_4$	$2\text{-}MeOC_6H_4$	B	100	175–178	^{1}H-NMR, IR, MS	[99OM5668]
$2\text{-}MeC_6H_4$	$4\text{-}ClC_6H_4$	$2\text{-}^iPrOC_6H_4$	B	–	–	–	[99OM5668]

[a] For notation, see Section 3.1.1.

[b] Ox =

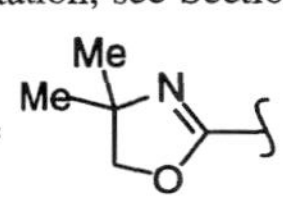

TABLE 3.4
Triorganylbismuth dihalides with multiple bismuth atoms

Compound	Synthetic method[a]	Yield (%)	M.p. (°C)	Physical data	Reference
Bi_2(A) (**1**)[b]	B	100	180–185 (decomp.)	^{1}H-NMR, MS, UV	[98OM4049]
Bi_2(B) (**2**)[b]	B	100	148–154 (decomp.)	^{1}H-NMR, MS, UV	[67M(98)731]
Bi_2(C) (**3**)[b]	A	–	135–137	–	[98OM4049]
Bi_3(A) (**4**)[b]	B	100	122–130 (decomp.)	^{1}H-NMR, MS, UV	[98OM4049]
Bi_3(B) (**5**)[b]	B	100	87 (decomp.)	^{1}H-NMR, MS, UV	[98OM4049]
Bi_4(A) (**6**)[b]	B	100	152–160 (decomp.)	^{1}H-NMR, MS, UV	[98OM4049]
Bi_4(B) (**7**)[b]	B	100	103–106 (decomp.)	^{1}H-NMR, MS, UV	[98OM4049]
Bi_4(D) (**8**)[b]	B	55	156–157	^{1}H-NMR, IR, MS	[97CC2295] [98OM4049]

[a] For notation, see Section 3.1.1.
[b] For structures, see Scheme 3.3.

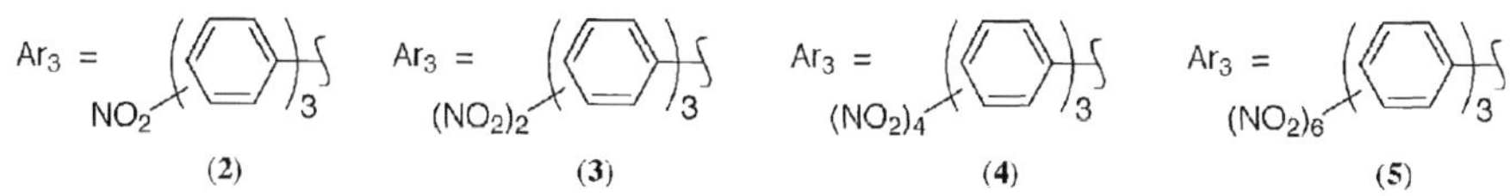

Scheme 3.1. Structures of compounds listed in Table 3.1.

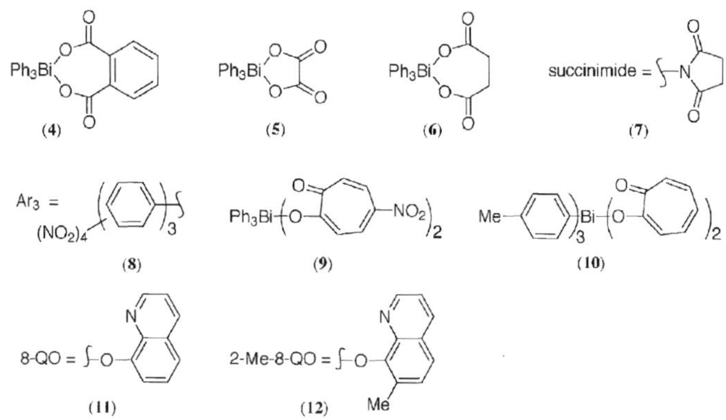

Scheme 3.2. Structures of compounds listed in Table 3.2.

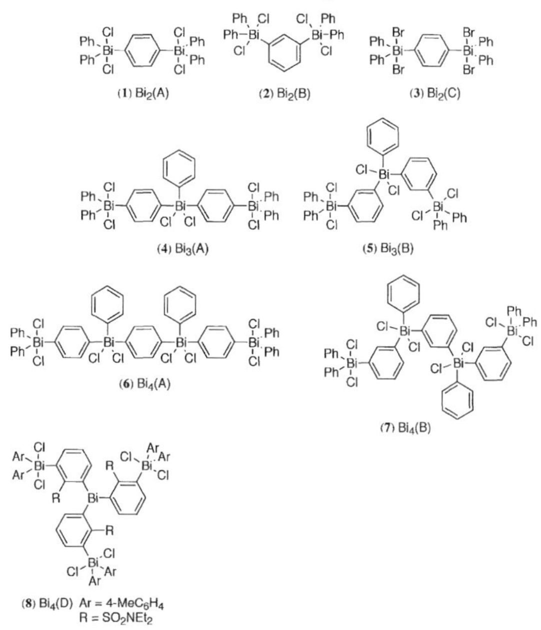

Scheme 3.3. Structures of compounds listed in Table 3.4.

3.1.2. Properties

Most triarylbismuth(V) compounds of the type Ar_3BiX_2 are crystalline, non-hygroscopic solids, which are soluble in benzene, chloroform, dichloromethane, dioxane, THF and acetone, slightly soluble in alcohols and diethyl ether and insoluble in hexane. Tris(hydroxyphenyl)bismuth dichlorides are soluble in methanol but hardly soluble in water [98JCS(P1)2511]. Triphenylbismuth carbonate is insoluble in most organic solvents, suggesting that the polymeric structure of the Ph_3Bi units is linked by the carbonate anions. The thermal stability of these compounds is highly dependent on the nature of the anionic groups attached. For example, the stability of triphenylbismuth dihalides decreases in the order fluoride>chloride>bromide≫iodide: the diiodide has not been isolated so far. The pure difluoride and dichloride are stable up to 200 and 150°C, respectively [22JCS(121)91]. Triphenylbismuth dicarboxylates, disulfonates and dinitrate are mostly stable, whereas the diazide is relatively unstable and decomposes rapidly at 100°C to yield diphenylbismuth azide and phenyl azide [34JCS405, 73JOM(50)129]. Triphenylbismuth dicyanide undergoes a similar type of decomposition on heating. Both diazide and dicyanide are rather unstable in solution and readily degrade. Trimethylbismuth dichloride is thermally unstable and decomposes rapidly at room temperature, although its crystal structure has been characterized by low-temperature X-ray chrystallographic analysis at −143°C [94AG(E)976].

Most triorganylbismuth dihalides and related compounds have a non-ionic, five-coordinate structure, in which the bismuth atom possesses a trigonal bipyramidal geometry (Fig. 3.1). The X-ray analyses of trimethylbismuth dichloride (mean Bi–C and Bi–Cl bond lengths, 2.20 and 2.61 Å; mean C–Bi–C and Cl–Bi–Cl bond angles, 120 and 178°, respectively) and triphenylbismuth dichloride (mean Bi–C and Bi–Cl bond lengths, 2.12 and 2.57 Å; mean C–Bi–C and Cl–Bi–Cl bond angles, 120 and 176°, respectively) have clearly shown that three carbon atoms adjacent to the bismuth occupy the

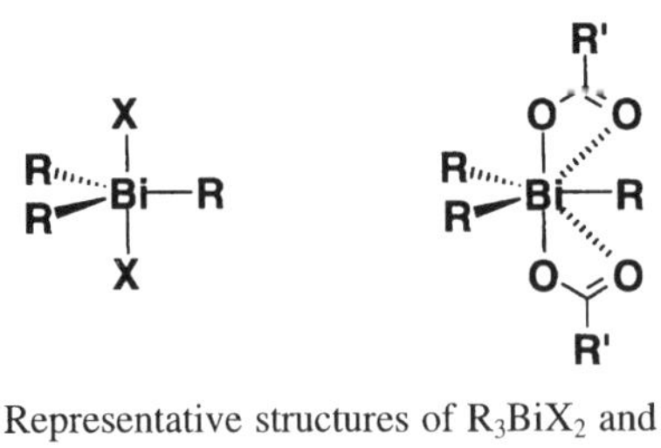

Fig. 3.1. Representative structures of R_3BiX_2 and $R_3Bi(OCOR')$.

equatorial sites and two chlorine atoms occupy the axial sites [94AG(E)976, 68JCS(A)2539]. Other triarylbismuth dichlorides show a similar geometry. Triphenylbismuth(V) alkoxide compounds, $Ph_3Bi(OAr)_2$ and $Ph_3BiBr(OAr)$ ($Ar{=}C_6F_5$ and C_6Cl_5), also possess a distorted trigonal bipyramidal coordination geometry, in which the axial Bi–O bonds are 2.22–2.25 Å [98OM1347]. The X-ray investigations of triphenylbismuth dicarboxylates revealed that the bismuth center possesses a distorted trigonal (or pentagonal, if both of the carboxylate oxygens are considered) bipyramidal structure, weakly coordinated by the neighboring carbonyl oxygen atoms [91JOM(419)283, 93JCS(P1)2411]. For example, in crystalline triphenylbismuth bis(trifluoroacetate), the non-bonded bismuth–oxygen distances are only ca. 30% longer than the bonded ones (2.308(7)–2.309(7) Å). Triphenylbismuth di(benzenesulfonate) also adopts a distorted trigonal bipyramidal geometry with $Bi–O_{axial}$ bond lengths of 2.276(5)–2.280(4) Å, which are shorter than the non-bonded Bi–O distances (3.129(5)–3.431(6) Å) [86ZAAC(539)110]. Various types of triphenylbismuth(V) dicarboxylate bearing a hydroxyl group at the alkyl chain have been structurally characterized [97JCS(D)2009]. In these molecules, unusual hydrogen-bonded extended structures such as double strands and double layers have been observed. Intramolecular coordination is also observed in the crystal structure of chloro(2-methyl-8-quinolinolato)triphenylbismuth, in which the bismuth atom has a distorted octahedral geometry and the nitrogen atom coordinates to the bismuth center with a relatively weak Bi–N bond of 2.71(2) Å [83JOM(253)317].

Molecular weight, electrical conductance, IR and Raman spectra have been examined for a number of compounds of the type Ph_3BiY_2 (Y = F, Cl, Br, NO_3, N_3, CN, NCO, OAc and other carboxylates) [66JCS(A)520, 70CJC2488, 73JOM(50)129]. They have a normal molecular weight in benzene solutions and behave as a molecular species in polar solvents. The dipole moment of triphenylbismuth dichloride is essentially zero in benzene [42JA173, 43ZAAC(250)257]. This result indicates that the dichloride has a trigonal bipyramidal geometry in solution as well as in the solid state. An osmometric measurement of triphenylbismuth molybdate in chloroform showed the polymeric structure in which the Ph_3Bi units are linked together by the bridging MoO_4 groups [80IC3863]. In the IR spectrum of the difluoride, a strong band attributable to the Bi–F stretching has been observed at 412 cm^{-1}, suggesting the linear nature of the F–Bi–F linkage. The IR spectra of the dinitrate, diazide and dicyanide show strong bands of the covalently bound anionic groups at 1510–1520 (asym), 2034–2063 (asym) and 2128–2140 cm^{-1}, respectively. In the spectra of the dicarboxylates, both symmetric and asymmetric carboxylate stretching frequencies are observed with a difference of their frequencies, (asym–sym), of 215–340 cm^{-1}, indicating that the anionic groups are present as the unidentate ligand. All of the available data clearly show that the anionic groupings (Y) are covalently bound to the bismuth atom and the molecule has a non-ionic, monomeric five-coordinated structure in accordance with the trigonal bipyramidal skeleton. The IR spectra of cationic complexes of the type $(Ph_3BiL_2)Y_2$ (L=Me_2SO, PyO, Ph_3PO, Ph_3AsO; Y=ClO_4, BF_4, PF_6) exhibit the presence of anionic perchlorate, tetrafluoroborate or hexafluorophosphate groups as well as the cationic Ph_3BiL_2 moiety. Based on the electrical conductance measurements in nitromethane, it has been suggested that these complexes behave as a 1:2 electrolyte [73IC944].

3.1.3. Reactions

Trimethylbismuth dichloride is thermally unstable and decomposes to chlorodimethylbismuthine and methyl chloride at room temperature [94AG(E)976]. This type of reductive coupling is not observed for triphenylbismuth dichloride at room temperature, but prolonged heating or photoirradiation of a solution of the phenyl derivative results in decomposition [22JCS(121)91]. Triarylbismuth(V) compounds of the Ar_3BiX_2 type also undergo ligand coupling to give Ar_2BiX and ArX at the temperature depending on the anionic groups.

Reduction of triarylbismuth dihalides to the parent triarylbismuthines can be performed by using a variety of reducing agents, which include hydrazine hydrate, sodium hydrosulfite, liquid ammonia, $LiAlH_4$, $NaBH_4$, sodium sulfide and sodium dialkyldithiocarbamate. This type of reduction has been used for the purification of tris(3-methylphenyl)bismuthine which is purified with difficulty in the trivalent state [26JA507]. The electrolytic reduction of triphenylbismuth dibromide has been found to be a one-step, two-electron process where the bromine atoms are released as bromide ions [66JA467].

Metathesis of triarylbismuth dichloride with a variety of metal salts such as fluoride, azide, cyanide, carboxylates and sulfonates has been used frequently for the synthesis of triarylbismuth(V) compounds of the type Ar_3BiY_2, where Y is the corresponding anionic group (Section 3.1.1). The reaction of triphenylbismuth dichloride with mercuric chloride in an alkaline medium leads to triphenylbismuthine, while treatment of triphenylbismuth dicyanide with mercuric oxide results in the formation of triphenylbismuthine oxide (Section

3.8.1). Pentacoordinate cationic complexes of the type $(Ph_3BiL_2)Y_2$ can be prepared by the reaction of triphenylbismuth dichloride with a silver salt AgY in the presence of a ligand molecule L [73IC944]. These complexes are stable in the solid state, but decompose in solution. Triphenylbismuth difluoride is formed from the thermal decomposition of $[Ph_3Bi(OAsPh_3)_2][BF_4]_2$.

It has been reported that $Ph_3BiBr(OC_6X_5)$ (X = F, Cl) does not exist as such in solution due to a very rapid redistribution equilibrium with $Ph_3Bi(OC_6X_5)_2$ and Ph_3BiBr_2 [98OM1347]. A similar type of equilibration is also observed between Ph_3BiX_2 and Ph_3BiY_2 (X, Y = F, Cl, Br) in solution, where the equilibrium constants correlate well with the difference in electronegativity between the X and Y anionic species.

Treatment of triphenylbismuth dichloride with 2 equiv. of phenyllithium gives pentaphenylbismuth. This method has been widely used for the preparation of unsymmetrical pentaarylbismuth compounds bearing two different kinds of aryl ligand (Section 3.4.1). Attempts to prepare quaternary bismuth halides by treatment of triarylbismuth dihalides with one equiv. of arylmagnesium bromide have been unsuccessful; the compounds obtained from the reaction between triphenylbismuth dibromide and phenylmagnesium bromide are triphenylbismuthine, bromodiphenylbismuthine, dibromo(phenyl)bismuthine and bromobenzene [20JCS(117)762, 14JCS(105)2210]. By contrast, the Lewis acid-promoted reaction between triarylbismuth difluorides and organosilanes, organotins or organoborons can afford a variety of bismuthonium salts (Section 3.3.1).

The reaction of triphenylbismuth carbonate with cyclic 1,3-dicarbonyl compounds, such as dimedone and Meldrum's acid, gives highly stabilized triphenylbismuthonium ylides (Section 3.5.1). Moderately stabilized acyclic bismuthonium ylides can be prepared from triphenylbismuth dichloride and a 1,3-diketone or a 1,3-disulfone in the presence of a suitable base. A similar base-promoted reaction of triphenylbismuth dihalides with *N*-sulfonylamines or amides has been used for the preparation of stabilized bismuthine imides bearing the *N*-sulfonyl or *N*-acyl moiety (Section 3.6.1).

Barton and co-workers have investigated the use of triarylbismuth dihalides and related compounds for synthetic purposes. These compounds can be used for the oxidation or arylation of a variety of alcohols, enols, amines, phenols, thiols, hydrazine, nitroalkanes and others (Sections 5.2.4, 5.4.3 and 5.5.2).

Typical reaction modes of triarylbismuth dihalides and derivatives are shown in Scheme 3.4. Representative reactions of triphenylbismuth dichloride are listed in Table 3.5.

TABLE 3.5
Reactions of triphenylbismuth dichloride

Reagent[a]	Conditions	Product[a]	Reference
–	100°C (sealed tube)	No reaction	[15JCS(107)16]
–	Reflux/PhH	Ph_2BiCl	[15JCS(107)16]
– (Cu)	Reflux/MeCOMe	Ph_3Bi, Ph_2BiCl, PhCl	[20JCS(117)762]
H_2O	ROH	No reaction	[20JCS(117)762]
	Steam distillation	$Ph_3Bi(OH)Cl$	[60TTFI(2)317]
H_2S	NH_3–ROH	Ph_3Bi, S_x, NH_4Cl	[50JCS617]
Electric reduction	DMF	Ph_3Bi	[66JA467]
Na	Reflux, 9 h/Et_2O	$Ph_3Bi(OH)Cl$	[20JCS(117)762]
	Liquid NH_3	Ph_3Bi	[41JA212]
$N_2H_4{\cdot}H_2O$	EtOH	Ph_3Bi	[40JA665]
$Na_2S_2O_4$	0°C/MeCOMe–H_2O	Ph_3Bi	[26JA507]
NH_3	Liquid NH_3	Ph_3Bi	[41JA212]
NH_3 (gas)	$CHCl_3$	$Ph_3Bi(OH)Cl$	[20JCS(117)762]
$LiBH_4$	−60°C/Et_2O	Ph_3Bi	[56ZN(B)755]
$LiAlH_4$	−95°C/Et_2O	Ph_3Bi	[56ZN(B)755]
NaS_2CNR_2	25°C, 2 h/$CHCl_3$	Ph_3Bi	[68JOM(11)627]
PhLi	−75°C/Et_2O	Ph_5Bi	[52LA(578)136]
4-ClC_6H_4Li	−70°C, 1 h/Et_2O	$Ph_3(4\text{-}ClC_6H_4)_2Bi$	[68LA(720)198]
ArLi[b]	−78°C/Et_2O	Ph_3Ar_2Bi	[89CB803] [90AG(E)213]
PhMgBr	–	Bi(III) compounds	[15JCS(107)16]
4-MeC_6H_4MgBr	Et_2O	Ph_3Bi	[20JCS(117)762]
NaOH	r.t., 2 h/MeCOMe–MeOH	$Ph_3Bi(Cl)OBi(Cl)Ph_3$	[72JOM(36)323]
KOH	Reflux, 3 h/EtOH	Ph_3Bi	[20JCS(117)762]
	Cold, 5.5 h/H_2O	Ph_3Bi	[20JCS(117)762]
H_2SO_4	10°C	Ph_3BiSO_4, HCl	[20JCS(117)762]
$ClSO_3H$	Cold $CHCl_3$	HCl + unknown compounds	[16JCS(109)250]
NaN_3	60–65°C, 10 min/H_2O	Ph_3Bi, PhN_3	[34JCS405]
	5°C, 8 h/H_2O	$Ph_3Bi(N_3)_2$	[73JOM(50)129]
	0°C/MeCN–H_2O	$Ph_3Bi(N_3)_2$	[92JCR(S)34]
NaOQ[c]	CH_2Cl_2–MeOH	$Ph_3Bi(OQ)Cl$	[72JOM(38)91]
NaOAr[d]	0.5–1 h/THF	$Ph_3Bi(OAr)Cl$	[85JCS(P1)2657]
NaON=CRR′[e]	r.t., 3 h/PhH	$Ph_3Bi(ON{=}CRR')_2$	[87SRI(17)559]
NaOCOR[f]	r.t., 36 h/dioxane	$Ph_3Bi(OCOR)_2$	[51JA4470]
NaSPh	r.t., 36 h/dioxane	$Ph_3Bi(SPh)_2$	[51JA4470]
KCN	5 h/H_2O	$Ph_3Bi(OH)CN$	[34JCS405]
	5°C, 8 h/H_2O	$Ph_3Bi(CN)_2$	[73JOM(50)129]
	Heat, 1 h/EtOH–H_2O	Ph_2BiCN	[22JCS(121)91]
	8.5 h/PhH	No reaction	[22JCS(121)91]
KF	Reflux/EtOH–H_2O	Ph_3BiF_2	[22JCS(121)91]
KHF_2	Reflux, 3.5 h/EtOH–H_2O	Ph_3Bi + unknown compounds	[22JCS(121)91]
KI	EtOH	Ph_2BiI	[27BTI3]
K_2CO_3	8 h/H_2O	Ph_3BiCO_3	[71CJC2529]
	5 min/MeCOMe–H_2O	Ph_3BiCO_3	[85JCS(P1)2657]
$K_2C_2O_4$	8 h/H_2O	$Ph_3BiC_2O_4$	[71CJC2529]

TABLE 3.5 (*continued*)

Reagent[a]	Conditions	Product[a]	Reference
$K_2Cr_2O_4$	8 h/H_2O	$Ph_3BiCr_2O_4$	[71CJC2529]
KSeCN	15 h/pet. ether	Ph_3Bi, $Ph_2BiSeCN$, PhSeCN	[26JCS1648]
AgF	PhH	Ph_3BiF_2	[70CJC2488]
Ag_2O	60°C, 1 h/MeCOMe–H_2O	Ph_3Bi	[20JCS(117)762]
	H_2O	$Ph_3Bi(OH)Cl$, Ph_3Bi, $Bi(OH)_3$	[34JCS405]
	35–40°C/H_2O	$Ph_3Bi(OH)_2$	[34JCS405]
	r.t., 5 h/PhH–H_2O	$Ph_3Bi{=}O$	[72JOM(36)323]
	2 h/PhH	$Ph_3Bi{=}O$	[72JOM(36)323]
$AgNO_3$	PhH	$Ph_3Bi(NO_3)_2$	[70CJC2488]
	r.t., 3–4 h/MeCOMe–H_2O	$[Ph_3BiOBiPh_3][NO_3]_2$	[72JOM(36)323]
	EtOH	$Ph_3Bi(NO_3)_2$	[1889LA(251)323]
AgOCN	4 h/Et_2O	$Ph_3Bi(OCN)_2$	[27JCS209]
	PhH	$Ph_3Bi(OCN)_2$	[70CJC2488]
AgOCOR[g]	PhH–H_2O	$Ph_3Bi(OCOR)_2$	[70CJC2488]
AgOCOR[h]	23°C, 17 h/THF	$Ph_3Bi(OCOR)_2$	[97JCS(D)2009]
$AgClO_4$	1 h/MeCOMe	$[Ph_3BiCH_2COMe][ClO_4]$	[71JCS(A)562]
	r.t., 3 h/PhH–H_2O	$[Ph_3BiOBiPh_3{\cdot}(H_2O)_2][ClO_4]_2$	[72JOM(36)323]
	EtOH–MeCOMe	$[Ph_4Bi][ClO_4]$	[66JA2342]
	EtOH	$[Ph_3BiOBiPh_3{\cdot}(H_2O)_2][ClO_4]_2$	[73JCS(D)1394]
	30 min/MeCOEt	$[Ph_4Bi][ClO_4]$	[73JCS(D)1394]
$AgBF_4$	1 h /MeCOMe	$[Ph_3BiCH_2COMe][BF_4]$	[71JCS(A)562]
	30 min/MeCOEt	$[Ph_4Bi][BF_4]$	[73JCS(D)1394]
$Pb(SCN)_2$	Reflux, 1 h/EtOH	Ph_2BiSCN, $(SCN)_x$	[22JCS(121)91]
$Pb(OAc)_2$	AcOH	$Ph_3Bi(OAc)_2$	[26JA507]
$HgCl_2$	Reflux, 30 min/EtOH–H_2O	BiOCl, PhHgCl, Cl_2, PhH	[37ZOB143]
$CH_2(COMe)_2$, NaH	−20°C, 1 h/THF	$Ph_3Bi{=}C(COMe)_2$	[89CC1749]
$CH_2(SO_2R)_2$[i]	10 min/NaOMe–MeOH	$Ph_3Bi{=}C(SO_2R)_2$	[94JFC(66)75]
H_2NSO_2R[j]	10 min/NaOMe–MeOH	$Ph_3Bi{=}NSO_2R$	[93JFC(63)179]
HON=CRR′[e] Et$_3$N	r.t., 3 h/PhH	$Ph_3Bi(ON{=}CRR')_2$	[87SRI(17)559]

[a] For details, see references.
[b] Ar = 2,6-$F_2C_6H_3$, 2-FC_6H_4.
[c] QO = 8-quinolinate.
[d] Ar = 4-$O_2NC_6H_4$, 3,5-$^t Bu_2C_6H_3$.
[e] (R,R′) = (Me,Me), (Me,Et), (Me,Pr), (Me,Ph), (Ph,Ph), –$(CH_2)_5$–.
[f] R = 2-HOC_6H_4, PhCH=CH.
[g] R = Me, CF_3, CF_2H, CFH_2, CCl_3, CCl_2H, $CClH_2$, $CBrH_2$, CH_2CN.
[h] R = $HOCH_2CH_2CH_2$, 4-$HOC_6H_4CH_2CH_2$, Ph($HOCH_2$)CH.
[i] R = CF_3, C_4F_9, $OCH_2CF_2CF_2H$.
[j] R = CF_3, C_4F_9, 4-MeC_6H_4.

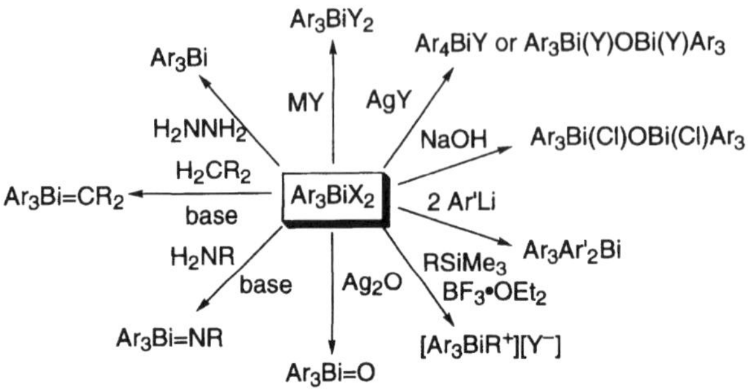

Scheme 3.4. Typical reactions of triarylbismuth dihalides and derivatives.

3.2. OXYBIS(TRIORGANYLBISMUTH) COMPOUNDS

3.2.1. Methods of synthesis (Table 3.6)

3.2.1.1. From triarylbismuth dihalides (method A)

In some limited solvent systems, a metathetical reaction between triphenylbismuth dichloride and silver perchlorate results in the formation of oxybis(triphenylbismuth) diperchlorate [72JOM(36)323]. This type of compound is only obtained when a mixture of water–benzene is used as the reaction medium. When the same reaction is carried out in absolute ethanol or in pure acetone, tetraphenylbismuthonium perchlorate and acetonyltriphenylbismuthonium perchlorate are obtained as the respective product (Section 3.3.1).

$$2\ Ph_3BiX_2 + 2\ AgY \xrightarrow{(H_2O)} Ph_3Bi(Y)OBi(Y)Ph_3 + 2\ AgX$$

Oxybis(triphenylbismuth) dinitrate is also obtainable by treating an acetone solution of triphenylbismuth dichloride with an aqueous solution of silver nitrate.

Synthesis of oxybis(triphenylbismuth) diperchlorate

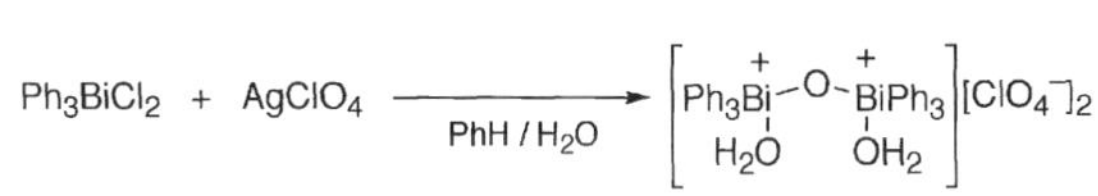

A benzene solution (25 ml) of triphenylbismuth dichloride (1.6 g, 3 mmol) and an aqueous solution (25 ml) of silver perchlorate (1.38 g, 5 mmol) were mixed together and stirred for 3 h at room temperature. The benzene layer was separated and the remaining mixture was filtered. The residue, which contained silver chloride and most of the oxybis(triphenylbismuth) diperchlorate, was treated with 95% ethanol (50 ml). The mixture was filtered and the filtrate was concentrated under vacuum. The compound was finally obtained as white crystals by adding diethyl ether slowly and cooling down to 5°C. Yield, 50%, m.p. 148°C (decomp.) [72JOM(36)323].

Triarylbismuth dihalides are converted to oxybis(triarylbismuth) dihalides by treatment with sodium hydroxide in alcoholic solutions [72JOM(36)323, 96OM5613]. Treatment of triarylbismuth difluorides with hexamethyldisiloxane in the presence of trimethylsilyl triflate gives oxybis(triarylbismuth) ditriflates [94JCS(P1)1739].

$$Ar_3BiX_2 + NaOH \longrightarrow 1/2\ Ar_3Bi(X)OBi(X)Ar_3 + NaX$$

$$Ar_3BiF_2 + 1/2\ Me_3SiOSiMe_3 \xrightarrow{Me_3SiOTf} 1/2\ Ar_3Bi(OTf)OBi(OTf)Ar_3 + 2\ Me_3SiF$$

Synthesis of oxybis[tris(4-*N,N*-dimethylaminophenyl)bismuth] dichloride

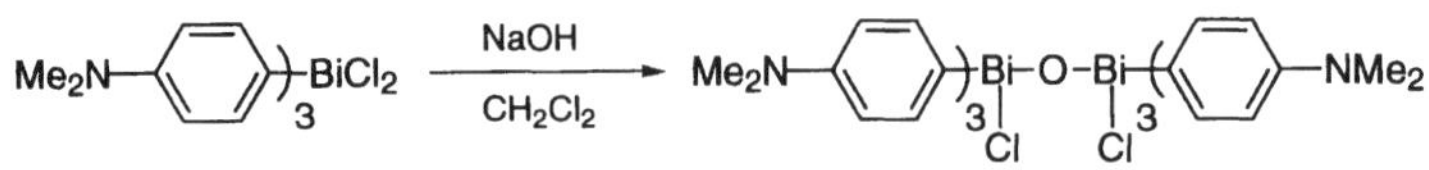

Tris(4-*N,N*-dimethylaminophenyl)bismuth dichloride (0.100 g, 0.156 mmol) was dissolved in reagent grade dichloromethane (20 ml) in the presence of 1.5 equiv. of NaOH (5 mg). The solution was stirred for 3 h and filtered. The filtrate was concentrated to about 4 ml and kept at 0°C. After a few days, oxybis[tris(4-*N,N*-dimethylaminophenyl)bismuth] dichloride was obtained as yellow crystals in 35% yield (m.p. 159°C, decomp.) [96OM5613].

Synthesis of oxybis(triphenylbismuth) ditriflate

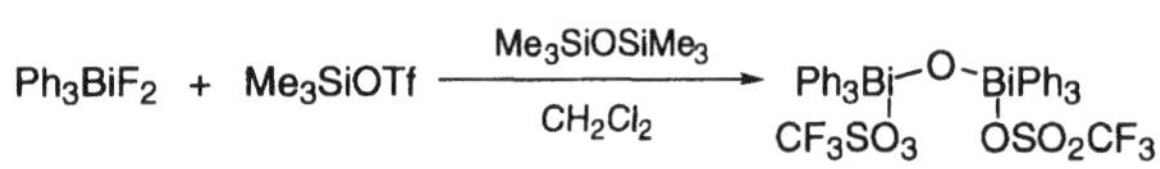

To a well stirred solution of triphenylbismuth difluoride (478 mg, 1 mmol) in dichloromethane (5 ml) at 0°C was added trimethylsilyl triflate (0.19 ml, 1 mmol). After 1 h, hexamethyldisiloxane (0.11 ml, 0.5 mmol) was added and stirring was continued for a further 24 h at ambient temperature to complete the reaction. The solvent was removed under reduced pressure to leave an oily residue, which was recrystallized from dichloromethane to give oxybis(triphenylbismuth) ditriflate as colorless crystals in 95% yield, m.p. 205–206°C [94JCS(P1)1739].

3.2.1.2. From other oxybis(triarylbismuth) compounds (method B)

Various oxybis(triphenylbismuth) compounds can be prepared by the reaction of oxybis(triphenylbismuth) dichloride and silver salts [72JOM(36)323]. A combination of silver(I) oxide and sulfonic acids can also be used instead of silver sulfonates [88JOM(342)185].

$$Ph_3Bi(X)OBi(X)Ph_3 + 2\ AgY \longrightarrow Ph_3Bi(Y)OBi(Y)Ph_3 + 2\ AgX$$

Synthesis of oxybis(triphenylbismuth) dicyanate

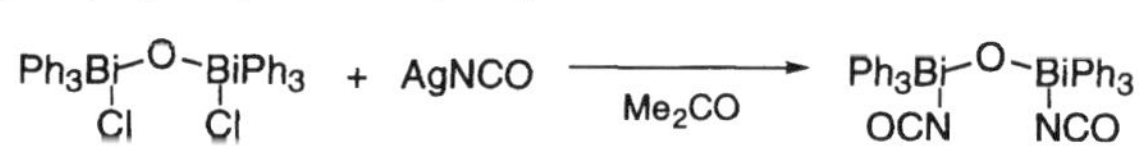

Silver cyanate (0.3 g, 2 mmol) was added to an acetone (50 ml) solution of oxybis(triphenylbismuth) dichloride (970 mg, 1 mmol) and the mixture was stirred for 5 h at around 5°C. It was then filtered and the filtrate was concentrated under vacuum. By adding water to the solution, oxybis(triphenylbismuth) dicyanate was obtained as a fine white powder. The compound was filtered and crystallized twice from dry acetone and diethyl ether; Yield 40%, m.p. 139°C (decomp.) [72JOM(36)323].

Synthesis of oxybis(triphenylbismuth) bis(benzenesulfonate)

A mixture of silver oxide (0.5 mmol), oxybis(triphenylbismuth) dichloride (0.5 mmol), chloroform (20 ml) and methanol (5 ml) was stirred for 15 min. To this mixture was added benzenesulfonic acid (1 mmol) and it was stirred for a further 3 h at room temperature. It was then filtered, washed with methanol (5 ml), and the filtrate was concentrated under vacuum. By adding ethyl acetate to the concentrate, oxybis(triphenylbismuth) bis(benzenesulfonate) was obtained as a fine white powder. The compound was filtered and crystallized from ethyl acetate. Yield, 58%, m.p. 155–157°C (decomp.) [88JOM(342)185].

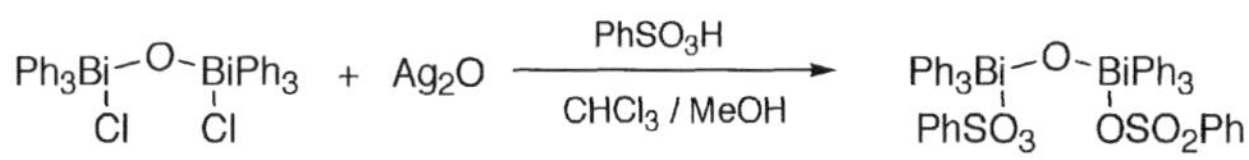

3.2.1.3. Ligand coordination to oxybis(triarylbismuth) compounds (method C)

Treatment of anhydrous oxybis(triphenylbismuth) perchlorate with oxygen-donor bases results in the formation of pentacoordinate cationic complexes [73IC944]. Dimethyl sulfoxide, pyridine *N*-oxide, triphenylphosphine oxide and triphenylarsine oxide are commonly employed ligands.

$$Ph_3Bi(X)OBi(X)Ph_3 + 2\,L \longrightarrow [Ph_3Bi(L)OBi(L)Ph_3]X_2$$

Synthesis of oxybis(triphenylbismuth) diperchlorate-bis(triphenylphosphine oxide)

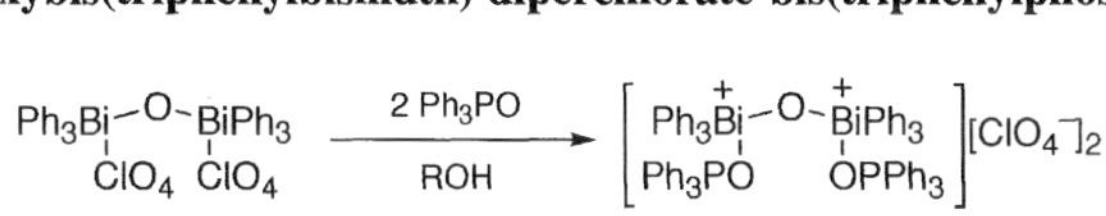

Anhydrous oxybis(triphenylbismuth) diperchlorate and triphenylphosphine oxide were mixed in a 1:2 molar ratio in absolute alcohol. The solution was allowed to stand overnight. Dropwise addition of petroleum ether (b.p. 35–60°C) until the solution became turbid, followed by refrigeration, gave the product as crystals. The complex was recrystallized from absolute alcohol containing free ligand, washed successively with benzene and petroleum ether, and dried in vacuo [73IC944].

TABLE 3.6
Oxybis(triarylbismuth) compounds

Compound		Synthetic method[a]	Yield (%)	M.p. (°C)	Physical data	Reference
$R_3BiXOBiXR_3$						
R	X					
Ph	Cl	A	60–70	147 (decomp.)	IR, EC	[72JOM(36)323]
	Br	A	60–70	135 (decomp.)	IR, EC	[72JOM(36)323]
	NO_3	A	60–70	155–156	IR, EC	[72JOM(36)323]
	NCO	B	40	139 (decomp.)	IR, EC	[72JOM(36)323]
	CF_3CO_2	B	40	156–157	IR, EC	[72JOM(36)323]
	ClO_4	A	100	144 (decomp.)	IR, EC	[72JOM(36)323]
	$MeSO_3$	B	71	158–162 (decomp.)	IR	[88JOM(342)185]
	$PhSO_3$	B	58	155–157 (decomp.)	IR	[88JOM(342)185]
	$4\text{-}MeC_6H_4SO_3$	B	62	170–173 (decomp.)	IR	[88JOM(342)185]
	$2,4\text{-}(O_2N)_2C_6H_3SO_3$	B	40	110–115 (decomp.)	IR	[88JOM(342)185]
	$HOCH_2CH_2SO_3$	B	52	150–152 (decomp.)	IR	[88JOM(342)185]
	CF_3SO_3	B	20	120–121 (decomp.)	IR	[88JOM(342)185]
		A	95	205–206	^1H-, ^{13}C-NMR, IR, MS, X-ray	[94JCS(P1)1739]
$4\text{-}Me_2NC_6H_4$	Cl	A	35	159 (decomp.)	^1H-NMR, X-ray	[96OM5613]
$4\text{-}MeC_6H_4$	CF_3SO_3	A	96	199–200	^1H-, ^{13}C-NMR, IR, MS	[94JCS(P1)1739]
$[R_3Bi(L)OBi(L)R_3][ClO_4]_2$						
R	L					
Ph	H_2O	A	50	148 (decomp.)	IR, EC	[72JOM(36)323]
	Me_2SO	C	–	–	IR, EC	[73IC944]
	$(CD_3)_2SO$	C	–	–	IR	[73IC944]
	C_5H_5NO	C	–	135 (decomp.)	IR, EC	[73IC944]
	Ph_3PO	C	–	–	IR, EC	[73IC944]
	Ph_3AsO	C	–	117 (decomp.)	IR, EC	[73IC944]

[a] For notation, see Section 3.2.1.

3.2.2. Properties and reactions

Oxybis(triphenylbismuth) compounds of the type $(Ph_3BiX)_2O$ are soluble in dichloromethane, ethanol and acetonitrile. When X is a soft anion such as perchlorate and trifluoromethanesulfonate, it is thermally stable. Oxybis(triphenylbismuth) diperchlorate and ditriflate possess a distorted trigonal bipyramidal structure around each bismuth atom with the bridging oxygen atom at one apical site and the anionic group at the other. Three phenyl carbon atoms bound to each bismuth atom are in a plane slightly below the bismuth with respect to the bridging oxygen. For example, in the molecule of $[Ph_3Bi(OSO_2CF_3)]_2O$, the respective mean O–Bi–O and C–Bi–C angles are 177.9 and 119.0°, and each bismuth atom is located above the plane, formed by three *ipso* phenyl carbon atoms, by 0.20–0.24 Å and bent slightly towards the bridging oxygen atom [93JCS(P1)1739]. The Bi–O (bridge) lengths (2.038 and 2.091 Å) agree with the value expected for the covalent Bi–O single bond, while the Bi–O (triflate) bond lengths (2.58 and 2.53 Å) are too long to be covalent, which suggests the contribution of some tetrahedral character of the bismuth center to the distortion of this molecule. The Bi–O–Bi angle of 134.0° is much larger than the expected tetrahedral value (109.8°) probably because of the steric repulsion among six attached phenyl groups (Fig. 3.2).

Vibrational spectroscopy and conductance measurements have been performed to elucidate the structural features of oxybis(triphenylbismuth) compounds of the type $(Ph_3BiX)_2O$, where X is Cl, Br, NO_3, NCO, CF_3CO_2 or ClO_4 [72JOM(36)323]. In the IR spectra, they show a strong band at ca. 630–620 cm^{-1}, which is assigned to the Bi–O–Bi asymmetric stretching vibration. Except for diperchlorate, all oxybis compounds studied are non-ionic in the solid state, the anionic group acting as the monodentate ligand. Conductance measurements demonstrate that both hydrated and anhydrous diperchlorates behave as a 1:2 electrolyte in acetonitrile, but all other oxybis compounds are not significantly ionized. In dichloromethane, however, even the diperchlorates are likely to be non-ionized.

The bismuth center in oxybis(triphenylbismuth) diperchlorate is positively

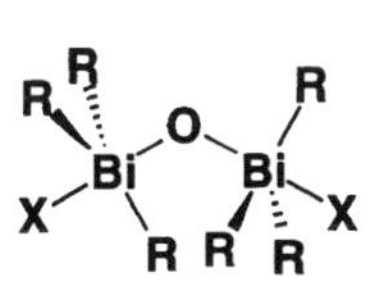

Fig. 3.2. A representative structure of an oxybis(triorganylbismuth) compound.

charged due to the less nucleophilic nature of the counter anion. This compound is readily hydrated or coordinated by donating ligand molecules to afford complexes of the type $[Ph_3Bi(L)OBi(L)Ph_3][ClO_4]_2$ (L=H_2O, Ph_3PO, Ph_3AsO, Me_2SO, pyridine *N*-oxide) [73IC944]. These bismuth complexes have been assigned unambiguously by IR spectroscopy. They show a medium intensity band at 300–405 cm^{-1}, which is assigned to the Bi–O(ligand) stretching frequency. The ν(XO) frequency (X=S, N, P or As) of the ligand is considerably reduced from the free-ligand value ($\Delta\nu$=37–115 cm^{-1}). These data indicate that the ligands are bonded to bismuth through their oxygen atoms.

Oxybis(triphenylbismuth) dichloride has been used as the reagent for oxidation and arylation in organic synthesis (Sections 5.2.4 and 5.4.3).

3.3. QUATERNARY ORGANOBISMUTH(V) COMPOUNDS

3.3.1. Methods of synthesis (Table 3.7)

3.3.1.1. From quinquenary organobismuth compounds (method A)

The cleavage of one of the Bi–C bonds in pentaarylbismuth compounds of the type Ar_5Bi or $Ar_3Ar'_2Bi$ with a protic acid such as hydrogen chloride, fluoroboric acid, carboxylic acids, sulfonic acids or triphenylsilanol, and with soft Lewis acids such as bromine, triphenylborane and phenylbismuth bis(trifluoroacetate), all result in the formation of quaternary bismuth(V) compounds of the type Ar_4BiX or $Ar_3Ar'BiX$. Many tetraarylbismuth(V) compounds have been prepared according to this methodology.

$$Ar_5Bi + HX\ (XY) \longrightarrow Ar_4BiX + ArH\ (ArY)$$

$$Ph_5Bi + Ph_3B \longrightarrow [Ph_4Bi^+][BPh_4^-]$$

Synthesis of tetraphenylbismuthonium triflate

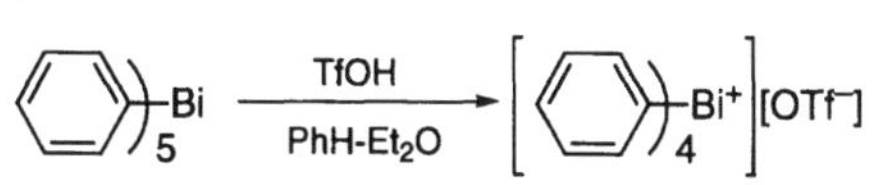

To a solution of pentaphenylbismuth (3 g, 5 mmol) in dry benzene (8 ml) under an argon atmosphere, a solution of trifluoromethanesulfonic acid (0.750 g, 5 mmol) in ether (10 ml) was added dropwise with stirring under water cooling at 15°C. Addition was continued until complete decolorization had occurred. After subsequent addition of ether, the white solid formed was filtered off, washed with cold ether, and dried in vacuo to obtain the bismuthonium trifluoromethanesulfonate (2.9 g, 90%), m.p. 205–215°C (hexane–dichloromethane) [85JCS(P1)2657].

Synthesis of tetraphenylbismuthonium tetraphenylborate

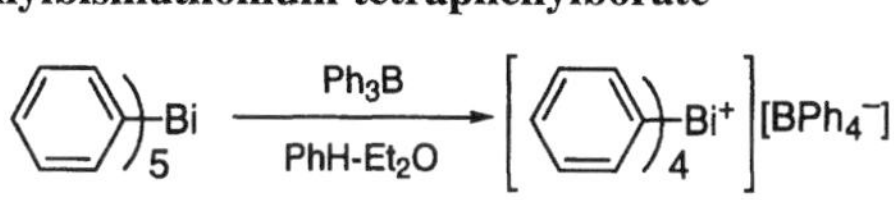

To a suspension of pentaphenylbismuth (1.6 g, 2.7 mmol) in anhydrous ether (10 ml) was added triphenylborane (0.7g, 3 mmol) in anhydrous benzene (10 ml) under a nitrogen atmosphere. The white crystals immediately precipitated were crystallized twice from nitromethane to obtain the bismuthonium tetraphenylborate as colorless needles (1.1 g, 49%), m.p. 225–228°C [52LA(578)136].

3.3.1.2. From quaternary organobismuth compounds (method B)

Treatment of tetraarylbismuthonium chloride with some metal salts such as

NaX, KX or AgX yields the corresponding tetraarylbismuth compounds of the type Ar_4BiX.

$$Ar_4BiX + MY \longrightarrow Ar_4BiY + MX$$

Tetraphenylbismuthonium nitrate, perchlorate, tetraphenylborate, cyanate, thiocyanate, perchlorate, tetrafluoroborate, hexafluorophosphate and trichloroacetate have been prepared by this metathesis method [52LA(578)136, 73JCS(D)1394]. Unsymmetrical tetraarylbismuth(V) compounds can be prepared similarly [68LA(720)198].

Synthesis of tetraphenylbismuthonium trichloroacetate

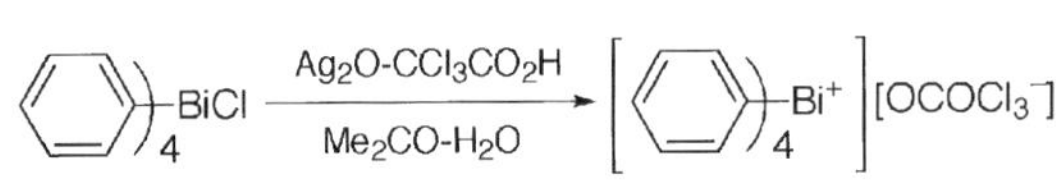

An aqueous silver trichloroacetate was prepared by dissolving a calculated amount of silver oxide in an aqueous solution of trichloroacetic acid. One equivalent of tetraphenylbismuth chloride dissolved in cold acetone (0°C) was added to this solution with constant stirring. After stirring for 15 min, the precipitated silver chloride was filtered off and the filtrate was concentrated in vacuo until it became cloudy. Upon cooling the solution in a refrigerator, the trichloroacetate was obtained as fine needles [73JCS(D)1394].

3.3.1.3. Via disproportionation of triarylbismuth dihalides (method C)

When treated with silver perchlorate or tetrafluoroborate in acetylacetone or ethyl methyl ketone, triphenylbismuth dichloride is converted to tetraphenylbismuth compounds [73JCS(D)1394].

$$Ar_3BiX_2 + AgY \longrightarrow [Ar_4Bi^+]Y^-$$

The same reaction carried out in acetone gives acetonyltriphenylbismuthonium salts as the sole products [71JCS(A)562].

Synthesis of tetraphenylbismuthonium perchlorate

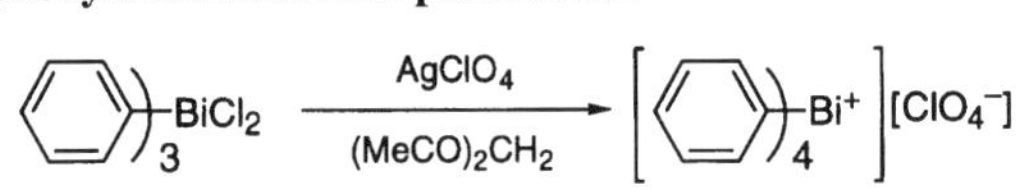

The reaction between two equiv. of silver perchlorate and triphenylbismuth dichloride in acetylacetone caused immediate precipitation of silver chloride. After stirring for 30 min, the precipitate was filtered off and the filtrate was concentrated and cooled in a refrigerator to yield the bismuthonium perchlorate as crystals, which were collected and recrystallized from a mixture of acetone and diethyl ether [73JCS(D)1394].

Synthesis of acetonyltriphenylbismuthonium perchlorate

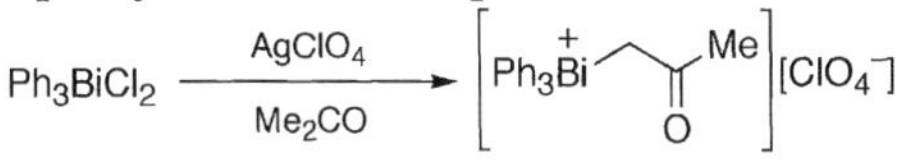

Anhydrous silver perchlorate (12 mmol) dissolved in acetone (ca. 20 ml) was added dropwise to a stirred solution of triphenylbismuth dichloride (6 mmol) in acetone (25 ml) and the resulting mixture was stirred for 1 h in the dark. Silver chloride immediately precipitated and was filtered off; the filtrate was concentrated under reduced pressure. Ether was added slowly to the concentrate until the solution became slightly turbid. It was then set aside overnight in a refrigerator. The deposited crystals were collected, washed with ether, and dried under vacuum at room temperature (ca. 60%, m.p. 122°C) [71JCS(A)562].

3.3.1.4. From triarylbismuth difluorides and organometallic reagents (method D)

Treatment of triarylbismuth difluorides with organosilicon reagents, such as silyl enol ethers [94JCS(P1)1739], siloxycyclopropanes [95JCS(P1)2543], ketene silyl acetals [99JOC6924, 95JCS(P1)2543] or allylsilanes [95TL7475], in the presence of boron trifluoride diethyl etherate or trimethylsilyl trifluoromethanesulfonate gives the corresponding 2-oxoalkyl-, 3-oxoalkyl-, alkoxycarbonylmethyl- and allyltriarylbismuthonium salts, respectively.

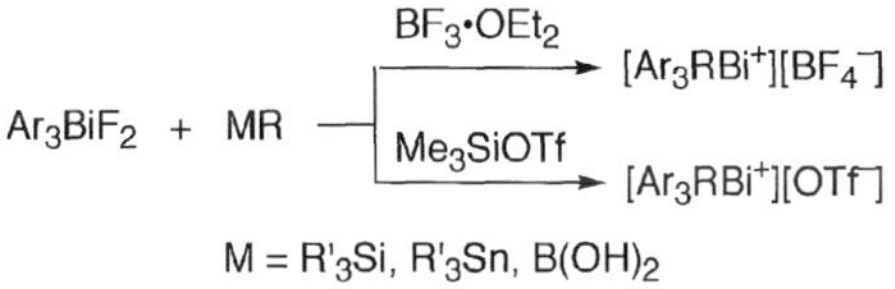

These Lewis acids play a dual important role: first, they activate the difluoride and then they are transformed into the respective counter anions, tetrafluoroborate and trifluoromethanesulfonate, which contribute to the stabilization of the resulting bismuthonium salts. Similarly organotin compounds are used for the synthesis of alkenyl- and arylbismuthonium salts [96JCS(P1)1971, 98CL127]. Tetraarylbismuthonium salts bearing the chirality at the bismuth center are successfully prepared with this methodology. When aryl- and alkenylboronic acids are used in place of the corresponding tin reagents, the Bi–C coupling proceeds more smoothly to afford aryl- and alkenylbismuthonium salts in better yields [98OM4332].

Synthesis of (3,3-dimethyl-2-oxobutyl)triphenylbismuthonium tetrafluoroborate

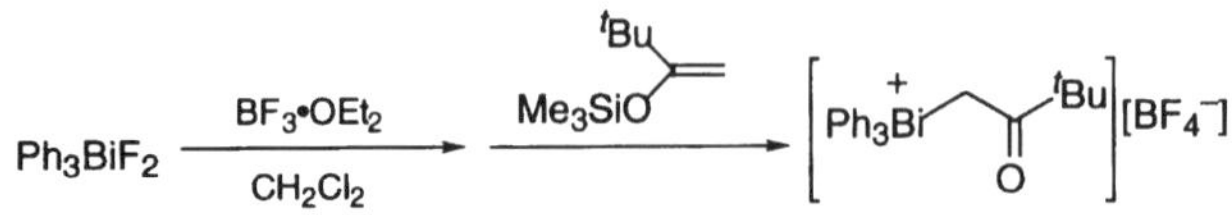

To a dichloromethane (5 ml) solution of triphenylbismuth difluoride (478 mg, 1 mmol) was added dropwise boron trifluoride diethyl etherate (0.12 ml, 1 mmol) at 0°C under argon. After stirring for 1 h, 3,3-dimethyl-2-trimethylsiloxy-1-butene (172 mg, 1 mmol) was introduced and the resulting mixture was stirred for 10 h at room temperature. Evaporation of the solvent under reduced pressure left an oily residue, which was crystallized from ether–dichloromethane (2:1) to give the bismuthonium tetrafluoroborate (601 mg, 96%) as colorless crystals (m.p. 143–145°C) [94JCS(P1)1739].

Synthesis of (4-methylphenyl)triphenylbismuthonium tetrafluoroborate

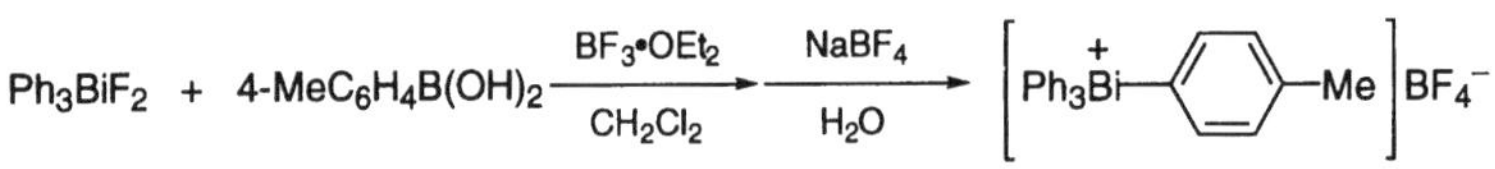

To a dichloromethane solution (5 ml) containing triphenylbismuth difluoride (239 mg, 0.50 mmol) and 4-methylphenylboronic acid (68 mg, 0.50 mmol) was added boron trifluoride diethyl etherate (65 μl, 0.50 mmol) at 0°C, and the resulting mixture was stirred at room temperature for 2 h. An aqueous solution (20 M) of sodium fluoroborate (500 mg, 4.55 mmol) was then added and the two-phase mixture was vigorously stirred for 30 min. The water phase was extracted with dichloromethane (5 ml × 2), and the combined extracts were dried over $MgSO_4$, and passed through a short silica gel column. Evaporation of the solvent under reduced pressure left an oily residue, which was crystallized from ether–dichloromethane (10:1) to yield (4-methylphenyl)triphenylbismuthonium tetrafluoroborate (300 mg, 97%) as a colorless solid [98OM4332].

3.3.1.5. From tertiary bismuthines (method E)

In marked contrast to the ease of access of lighter pnictogen congeners, quaternary bismuth compounds are not usually accessible by direct interaction of triarylbismuthines with alkylating agents due to the low nucleophilicity of the bismuthines, whose unshared electron pair is of a strong *s* character.

$$R_3Bi + R'Y \longrightarrow [R_3R'Bi^+]Y^-$$

However, there are a few known quaternary bismuth compounds that can be prepared with this methodology. Tetramethylbismuthonium triflate, which is the only known tetraalkylbismuth compound, is synthesized by the reaction between trimethylbismuthine and methyl triflate in acetonitrile [94AG(E)976]. The tritium-labeled phenyl cation, generated from the

TABLE 3.7
Tetraorganylbismuth(V) compounds

Compound		Synthetic method[a]	Yield (%)	M.p. (°C)	Physical data	Reference
Symmetrical compounds (R_4BiX)						
R	X					
Ph	Cl	A	54	–	–	[52LA(578)136]
2-$MeOC_6H_4$	Cl (**1**)[b]	E	68	195–197 (decomp.)	^1H-, ^{13}C-NMR, IR, MS	[97JCS(P1)1609]
2-$EtOC_6H_4$	Cl (**1**)[b]	E	28	212–214 (decomp.)	^1H-, ^{13}C-NMR, IR, MS	[97JCS(P1)1609]
2-$^iPrOC_6H_4$	Cl (**1**)[b]	E	39	225–227 (decomp.)	^1H-, ^{13}C-NMR, IR, MS	[97JCS(P1)1609]
4-Me-2-$MeOC_6H_3$	Cl (**2**)[b]	E	29	210–212 (decomp.)	^1H-NMR, IR, MS	[97JCS(P1)1609]
Ph	Br	A	–	–	–	[52LA(578)136]
2-$MeOC_6H_4$	Br (**1**)[b]	B	74	220–223 (decomp.)	^1H-, ^{13}C-NMR, IR, X-ray	[97JCS(P1)1609]
Ph	Br_3	A	92	–	–	[52LA(578)136]
2-$MeOC_6H_4$	I (**1**)[b]	B	74	202–204 (decomp.)	^1H-, ^{13}C-NMR, IR	[97JCS(P1)1609]
Ph	OH	A	–	–	–	[75DOK(225)581]
2-$MeOC_6H_4$	OH (**3**)[b]	B	55	155–160 (decomp.)	^1H-NMR, IR	[97JCS(P1)1609]
Ph	Ph_3SiO	A	–	145	–	[75DOK(225)581]
	NCO	B	–	–	IR	[73JCS(D)1394]
	NCS	B	–	–	IR, Raman	[73JCS(D)1394]
	NCSe	B	–	–	–	[73JCS(D)1394]
	N_3	B	–	–	–	[73JCS(D)1394]
	NO_3	B	–	–	–	[52LA(578)136]
		B	–	–	IR, Raman	[73JCS(D)1394]
	4-$O_2NC_6H_4O$	A	88	118–122	^1H-NMR, IR, MS	[85JCS(P1)2657]
	3,5-$^tBu_2C_6H_3O$	A	–	–	^1H-NMR	[85JCS(P1)2657]
	2,4,6-$(O_2N)_3C_6H_2O$	A	78	158–165	^1H-NMR, IR, MS	[87JCS(P1)251]
	C_6F_5O	A	72	122	^1H-, ^{19}F-NMR, IR, X-ray	[98OM1347]
	C_6Cl_5O	A	54	145	^1H-NMR, IR, X-ray	[98OM1347]

TABLE 3.7 *(continued)*

Compound		Synthetic method[a]	Yield (%)	M.p. (°C)	Physical data	Reference
2-$EtOC_6H_4$	HCO_2 (**1**)[b]	B	70	153–155 (decomp.)	^{1}H-NMR, IR	[97JCS(P1)1609]
Ph	$MeCO_2$	–	–	–	–	[85JCS(P1)2657]
	CF_3CO_2	A	90	120–132	^{13}C-NMR, IR, MS	[84HCA586]
	CCl_3CO_2	A	50	87–90	^{1}H-NMR, IR, MS	[87JCS(P1)251]
		B	–	–	IR, Raman	[73JCS(D)1394]
	CH_2ClCO_2	A	71	125–132	^{1}H-NMR, IR, MS	[87JCS(P1)251]
	Ph_2CHCO_2	A	67	150–158	^{1}H-NMR, IR, MS	[87JCS(P1)251]
Me	CF_3SO_3	E	73	>150	^{1}H-, ^{19}F-NMR, IR, Raman, X-ray	[94AG(E)976]
Ph	CF_3SO_3	A	90	205–215	^{1}H-, ^{13}C-NMR, IR, MS	[85JCS(P1)2657]
	4-$MeC_6H_4SO_3$	A	90	155–162	^{13}C-NMR, IR, MS, X-ray	[84HCA586]
2-$EtOC_6H_4$	4-$MeC_6H_4SO_3$	B	68	272–273	^{1}H-NMR, IR, MS	[97JCS(P1)1609]
Ph	B_3F_8	B	92	90	–	[68CB1200]
	BF_4	A	–	236	–	[71IZV2618]
		B	–	–	IR	[73JCS(D)1394]
		D	95	243–244	^{1}H-, ^{13}C-NMR, MS	[98OM4332]
2-$MeOC_6H_4$	BF_4	B	86	268–270	^{1}H-, ^{13}C-NMR, IR	[97JCS(P1)1609]
2-$EtOC_6H_4$	BF_4	B	87	>300	^{1}H-, ^{13}C-NMR, IR	[97JCS(P1)1609]
2-^{i}$PrOC_6H_4$	BF_4	B	92	>300	^{1}H-, ^{13}C-NMR, IR	[97JCS(P1)1609]
4-Me-2-$MeOC_6H_3$	BF_4	B	80	183–184	^{1}H-, ^{13}C-NMR, IR	[97JCS(P1)1609]
Ph	BPh_4	A	49	225–228	–	[52LA(578)136]
		B	–	215	–	[66JA2342]
	ClO_4	B	94	–	–	[52LA(578)136]
		C	65.7	–	IR	[66JA2342]
		B	–	–	IR, Raman	[73JCS(D)1394]
		–	–	–	X-ray	[73PS(3)33]
2-$MeOC_6H_4$	ClO_4	C	1	250–251 (decomp.)	^{1}H-, ^{13}C-NMR, IR, MS	[97JCS(P1)1609]
Ph	PF_5	B	–	–	IR	[73JCS(D)1394]
4-$Me_2NC_6H_4$	PF_5	C	90	140 (decomp.)	^{1}H-, ^{13}C-NMR, X-ray	[96OM5613]
Ph	$Ph_2Bi(OCOCF_3)_2$	A	86	142–147	^{13}C-NMR, IR, MS, X-ray	[84HCA586]

4-$Me_2NC_6H_4$	$AgCl_2$	C	–	178 (decomp.)	X-ray	[96OM5613]
Me (^{210}Bi)		–	100	–	–	[67JINC(29)869]
Unsymmetrical compounds						
R^1	R^2					
Chloride ($R^1_3R^2BiCl$)						
Ph	2-PhC_6H_4	A	76	131–133 (decomp.)	–	[68LA(720)198]
	4-ClC_6H_4	A	–	–	–	[68LA(720)198]
	R (**4**)[b]	E	13.6	110–112	–	[63JPR(22)1]
Bromide ($R^1_3R^2BiBr$)						
4-MeC_6H_4	2-OxC_6H_4 (**5**)[b]	B	70	113–115	^{1}H-NMR, IR, MS	[99OM5668]
Tetrafluoroborate ($[R^1_3R^2Bi^+][BF_4^-]$)						
Ph	$MeCOCH_2$	C	60	128 (decomp.)	^{1}H-NMR, IR	[71JCS(A)562]
		D	95	145–146	^{1}H-, ^{13}C-NMR, IR, MS	[94JCS(P1)1739]
	$^iPrCOCH_2$	D	99	156–157	^{1}H-, ^{13}C-NMR, IR, MS	[94JCS(P1)1739]
	$^tBuCOCH_2$	D	96	143–145	^{1}H-, ^{13}C-NMR, IR, MS, X-ray	[94JCS(P1)1739]
	$^sBuCOCH_2$	D	95	94–96	^{1}H-, ^{13}C-NMR, IR, MS	[94JCS(P1)1739]
	$^iBuCOCH_2$	D	95	–	^{1}H-, ^{13}C-NMR, IR, MS	[94JCS(P1)1739]
	$PhCOCH_2$	D	97	165–167	^{1}H-, ^{13}C-NMR, IR, MS	[94JCS(P1)1739]
	4-$BrC_6H_4COCH_2$	D	94	65–67	^{1}H-, ^{13}C-NMR, IR, MS	[94JCS(P1)1739]
	$^tBuCOCH_2CH_2$	D	86	144–146	^{1}H-, ^{13}C-NMR, IR, MS	[95JCS(P1)2543]
	$PhCOCH_2CH_2$	D	80	154–155	^{1}H-, ^{13}C-NMR, IR, MS	[95JCS(P1)2543]
	$^iPrO_2CCH_2$	D	45	114–115	^{1}H-NMR, IR, MS	[99JOC6924]
	PhO_2CCH_2	D	77	165–166	^{1}H-NMR, IR, MS	[99JOC6924]
	4-$MeC_6H_4O_2CCH_2$	D	65	127–129	^{1}H-NMR, IR, MS	[99JOC6924]
4-MeC_6H_4	$^tBuCOCH_2$	D	93	97–98	^{1}H-, ^{13}C-NMR, IR, MS	[94JCS(P1)1739]
	$^sBuCOCH_2$	D	90	142–144	^{1}H-, ^{13}C-NMR, IR, MS	[94JCS(P1)1739]
	$^iBuCOCH_2$	D	94	58–60	^{1}H-, ^{13}C-NMR, IR, MS	[94JCS(P1)1739]
	$PhCOCH_2$	D	87	142–144	^{1}H-, ^{13}C-NMR, IR, MS	[94JCS(P1)1739]
	$^tBuCOCH_2CH_2$	D	76	138–139	^{1}H-, ^{13}C-NMR, IR, MS	[95JCS(P1)2543]
Ph	R (**6**)[b]	E	–	–	^{1}H-NMR	[71PS(1)93]
	(*E*)-MeCH=CH	D	60	110–111	^{1}H-, ^{13}C-NMR, IR, MS	[96JCS(P1)1971]
Ph	(*E*)-PhCH=CH	D	64	–	^{1}H-, ^{13}C-NMR, IR, MS	[96JCS(P1)1971]

TABLE 3.7 *(continued)*

Compound		Synthetic method[a]	Yield (%)	M.p. (°C)	Physical data	Reference
Ph	(E)-PhCH=CH	D	95	–	–	[98OM4332]
	(*E*)-4-MeC_6H_4CH=CH	D	80	–	^1H-, ^{13}C-NMR, IR, MS	[96JCS(P1)1971]
	(*E,E*)-PhCH=CHCH=CH	D	80	–	^1H-NMR, IR, MS	[96JCS(P1)1971]
	(*Z*)-$C_6H_{13}CH$=CH	D	71	–	^1H-, ^{13}C-NMR, IR, MS	[96JCS(P1)1971]
	Me_2C=CH	D	84	147–148	^1H-, ^{13}C-NMR, IR, MS, X-ray	[96JCS(P1)1971]
	(*E*)-BuCH=CH	D	95	101–102	^1H-NMR, IR, MS	[98OM4332]
	2-MeC_6H_4	D	97	197–198	^1H-NMR, IR, MS	[98OM4332]
	4-MeC_6H_4	D	76	138–139	–	[98CL127]
		D	97	138–139	^1H-NMR, IR, MS	[98OM4332]
	2-$MeOC_6H_4$	D	85	130–131	^1H-NMR, IR, MS	[98CL127]
	4-$MeOC_6H_4$	D	77	–	^1H-NMR, IR, MS	[98CL127]
	4-FC_6H_4	D	96	172–173	^1H-NMR, IR, MS	[98OM4332]
	4-ClC_6H_4	D	97	123–124	^1H-NMR, IR, MS	[98OM4332]
	2-$MeSC_6H_4$	D	44	–	–	[98CL127]
	4-$MeSC_6H_4$	D	79	141–143	–	[98CL127]
	4-(CH_2=CH)C_6H_4	D	95	108–110	^1H-NMR, IR, MS	[98OM4332]
	4-$MeCOC_6H_4$	D	56	166–168	^1H-NMR, IR, MS	[99OM5668]
	3-$O_2NC_6H_4$	D	55	218–220	^1H-NMR, IR, MS	[98OM4332]
	2,4 6-$Me_3C_6H_2$	D	98	219–220	^1H-NMR, IR, MS	[98OM4332]
	C_6T_5	E	10.7	–	–	[89R69]
	2-C_4H_3S	D	98	181–182	^1H-NMR, IR, MS	[98OM4332]
	2-C_4H_3O	D	96	154–155	^1H-NMR, IR, MS	[98OM4332]
	1-$C_{10}H_7$	D	93	163–164	^1H-NMR, IR, MS	[98OM4332]
	4-($MeEtS^+$)C_6H_4 (**7**)[b]	–	100		–	[98CL127]
4-MeC_6H_4	Ph	D	98	149–150	^1H-NMR, IR, MS	[98OM4332]
	2-OxC_6H_4 (**5**)[b]	D	91	216–218	^1H-NMR, IR, MS, X-ray	[99OM5668]
4-$MeOC_6H_4$	Ph	D	99	–	^1H-NMR, IR, MS	[98OM4332]
4-ClC_6H_4	Ph	D	95	156–157	^1H-NMR, IR, MS	[98OM4332]
Tetraphenylborate ($[R^1_3R^2Bi^+][BPh_4^-]$)						
Ph	Me	E	–	–	–	[78IZV1643]

	2-PhC_6H_4	B	–	202–214 (decomp.)	–	[68LA(720)198]
	4-ClC_6H_4	B	20	183–188 (decomp.)	–	[68LA(720)198]
Trifluoromethanesulfonate ($[R^1_3R^2Bi^+][OSO_2CF_3^-]$)						
Ph	$^tBuCOCH_2$	D	95	152–154	^{1}H-, ^{13}C-NMR, IR, MS, X-ray	[94JCS(P1)1739]
	$^sBuCOCH_2$	D	93	117–118	^{1}H-, ^{13}C-NMR, IR, MS	[94JCS(P1)1739]
	$^iPrO_2CCH_2$	D	–	103–104	^{1}H-, ^{13}C-NMR, IR, MS	[95JCS(P1)2543]
	$^iPrO_2CCH_2CH_2$	D	47	119–120	^{1}H-, ^{13}C-NMR, IR, MS, X-ray	[95JCS(P1)2543]
	$^tBuCOCH_2CH_2$	D	12	130–131	^{1}H-, ^{13}C-NMR, IR, MS	[95JCS(P1)2543]
4-MeC_6H_4	$^tBuCOCH_2$	D	88	55–57	^{1}H-, ^{13}C-NMR, IR, MS	[94JCS(P1)1739]
	$^sBuCOCH_2$	D	86	127–128	^{1}H-, ^{13}C-NMR, IR, MS	[94JCS(P1)1739]
Ph	2-$MeOC_6H_4$	D	72	98–99	–	[98CL127]
Perchlorate ($[R^1_3R^2Bi^+][ClO_4^-]$)						
Ph	$MeCOCH_2$	C	60	122	^{1}H-NMR, IR	[71JCS(A)562]
2-$MeOC_6H_4$	$MeCOCH_2$	D	55	149–150 (decomp.)	^{1}H-, ^{13}C-NMR, IR	[97JCS(P1)1609]
Others						
$[Ph_3(2\text{-}PhC_6H_4)Bi^+][NO_3^-]$		B	–	99–101	–	[68LA(720)198]
$[Et_3MeBi^+][AlClEt_3^-]$		E	48	–	–	[67ISOC196]
$[Et_3MeBi^+][BPh_4^-]$		B	–	–	–	[67ISOC196]

TABLE 3.7 *(continued)*

Compound				Synthetic method[a]	Yield (%)	M.p. (°C)	Physical data	Reference
Chiral compounds								
$[R^1R^2R^3R^4Bi^+][BF_4^-]$								
R^1	R^2	R^3	R^4					
4-MeC_6H_4	2-$MeOC_6H_4$	4-$MeOC_6H_4$	4-ClC_6H_4	D	97	–	^{1}H-NMR, IR, MS	[99OM5668]
	2-$MeOC_6H_4$	2-^{i}$PrOC_6H_4$	4-ClC_6H_4	D	61	160–166	^{1}H-NMR, IR	[99OM5668]
	2-$MeOC_6H_4$	4-$MeSC_6H_4$	4-ClC_6H_4	D	97	–	^{1}H-NMR, IR	[99OM5668]
	4-$MeOC_6H_4$	2-^{i}$PrOC_6H_4$	4-ClC_6H_4	D	70	–	^{1}H-NMR, IR	[98CL127]
2-MeC_6H_4	4-$MeOC_6H_4$	4-ClC_6H_4	2-C_4H_3S	D	95	–	^{1}H-NMR, IR	[99OM5668]
4-MeC_6H_4	4-$MeOC_6H_4$	4-$CF_3C_6H_4$	2-OxC_6H_4 (**5**)[b]	D	97	199–200	^{1}H-NMR, IR	[98OM4332]
4-$MeOC_6H_4$	4-$CF_3C_6H_4$	2-C_4H_3S	2-OxC_6H_4 (**5**)[b]	D	72	224–226	^{1}H-NMR, IR	[99OM5668]
*$[(4\text{-}MeOC_6H_4)(4\text{-}CF_3C_6H_4)(2\text{-}C_4H_3S)(2\text{-}OxC_6H_4)Bi^+][X^-]$ (**5**)*[b]								
X								
Cl				B	58	142–144	^{1}H-NMR, IR, MS	[99OM5668]
Br				B	53	133–135	^{1}H-NMR, IR, MS	[99OM5668]
I				B	–	–	^{1}H-NMR, MS	[99OM5668]
4-$MeC_6H_4SO_3$				B	83	–	^{1}H-NMR, IR, MS	[99OM5668]
$[(4\text{-}MeC_6H_4)(2\text{-}MeOC_6H_4)(4\text{-}MeOC_6H_4)(4\text{-}ClC_6H_4)Bi^+][OSO_2R^-]$								
R								
(*D*)-Camphorsulfonate (**8**)[b]				B	63.8	–	^{1}H-NMR, IR, MS	[99OM5668] [98CL127]

[a] For notation, see Section 3.3.1.
[b] (**1**) Monohydrate; (**2**) dihydrate; (**3**) trihydrate. For other structures, see Scheme 3.5.

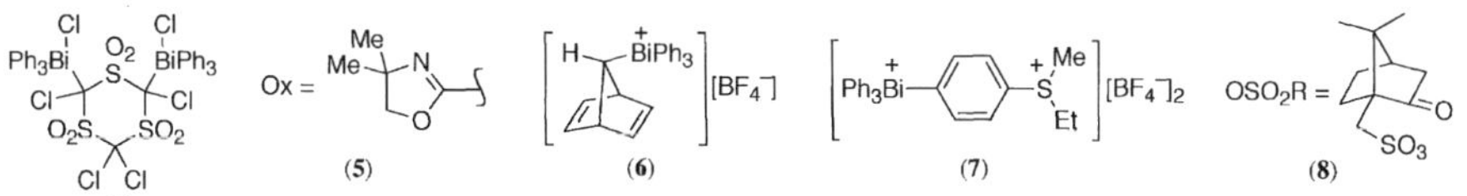

(4) (5) (6) (7) (8)

Scheme 3.5. Structures of compounds listed in Table 3.7.

β-decay of a multiple tritiated benzene, can react with triphenylbismuthine to give a mixed bismuthonium salt, $[Ph_3BiC_6T_5^+][BF_4^-]$, in 10% yield [89R69]. Tris(2-alkoxyphenyl)bismuthines react with iodosobenzene in a chlorinated solvent to give tetrakis(2-alkoxyphenyl)bismuthonium chloride [97JCS(P1)1609]. The chloride anion comes from the solvent used.

$$Ar_3Bi + PhIO \xrightarrow[CH_2Cl_2]{} [Ar_4Bi^+]Y^-$$

Synthesis of tetramethylbismuthonium triflate

$$Me_3Bi + MeOTf \xrightarrow[MeCN]{} [Me_4Bi^+][OTf^-]$$

Trimethylbismuthine (1.25 g, 4.9 mmol) in acetonitrile (2 ml) was treated with methyl trifluoromethanesulfonate (0.9 g, 5.5 mmol) under an argon atmosphere. The mixture was stirred for 6 h at room temperature and then evaporated to dryness under a high vacuum to obtain the bismuthonium triflate as a colorless solid (1.5 g), which was purified by recrystallization from acetonitrile–ether [94AG(E)976].

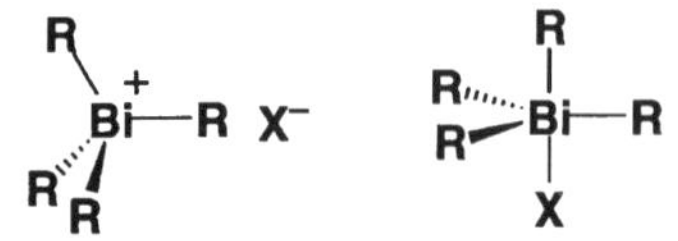

Fig. 3.3. Representative structures of tetraorganylbismuth compounds R_4BiX.

3.3.2. Properties

Thermal stability of quaternary bismuth compounds is strongly dependent on the nature of the counter anions as well as the ligand groups bound to the bismuth center. Most tetraarylbismuthonium halides are less stable comparatively to the corresponding lighter pnictogen analogs; tetraphenylbismuthonium chloride decomposes rapidly in the solid state at room temperature to yield triphenylbismuthine and chlorobenzene, although it is stable in water for several days [52LA(578)136]. By contrast, low coordinating anions enhance the stability of the bismuthonium compounds; tetraphenylbismuthonium perchlorate, tetrafluoroborate, trifluoromethanesulfonate and hexafluorophosphate are all thermally stable and do not show any sign of decomposition under ambient conditions. Tetramethylbismuthonium triflate is also stable at room temperature [94AG(E)976]. Tetraphenylbismuthonium nitrate, carboxylates, sulfonates, cyanate and thiocyanate are moderately stable, while the azide and selenocyanate are thermally unstable and decompose rapidly at room temperature [73JCS(D)1394]. On the other hand, tetrakis(2-methoxyphenyl)bismuthonium chloride, bromide and iodide are highly stabilized through the intramolecular coordination of the neighboring methoxy oxygen atoms toward the bismuth center [97JCS(P1)1609]. Triaryl(2-oxoalkyl)bismuthonium perchlorates, tetrafluoroborates and triflates are thermally stable, mainly due to the low nucleophilic nature of the counter anions [94JCS(P1)1739]. Triaryl(3-oxoalkyl)bismuthonium triflates are stable in the solid state, but in solution they slowly undergo reductive coupling to give triphenylbismuthine and the corresponding 3-oxoalkyl triflates [95JCS(P1)2543]. Alkenyltriphenylbismuthonium salts are thermally stable, while allylic bismuthonium salts decompose rapidly at room temperature, affording the allylated arenes via ligand coupling [96JCS(P1)1971, 95TL(36)7475].

Tetraorganylbismuth compounds are mostly soluble in chloroform, dichloromethane and acetonitrile, slightly soluble in benzene and THF, and almost insoluble in hexane and diethyl ether. Protic and aprotic polar solvents such as MeOH, DMF and DMSO dissolve tetraarylbismuthonium salts, but they induce decomposition of some alkylbismuthonium salts as a consequence of the nucleophilic attack of these solvent molecules on the cationic bismuth center [95JCS(P1)2543].

The crystal structures of some quaternary bismuth compounds have been elucidated by X-ray crystallographic analysis (Fig. 3.3). In general, tetraorganylbismuth(V) compounds with a soft counter anion such as perchlorate, tetrafluoroborate or triflate have a tetrahedral geometry around the bismuth center, indicating that these compounds should be described as the onium salts. For example, tetramethylbismuthonium triflate has a tetrahedral geometry with a mean C–Bi–C bond angle of 110° [94AG(E)976]. The coordination geometry of the bismuth atom in tetraphenylbismuthonium perchlorate is also tetrahedral [73PS(3)33], while that in tetraphenylbismuthonium tosylate is distorted trigonal bipyramidal with C_{eq}–Bi–C_{ax} bond angles of 98–103° and C_{eq}–Bi–C_{eq} bond angles of 113–123° [84HCA586]. The tosylate anion in Ph_4BiOTs occupies the axial position and is likely to coordinate more strongly to the bismuth center, which may therefore be better described as the pentacoordinated hypervalent compound. Thus, the weakly coordinating counter anion favors the tetrahedral geometry, whereas the strongly coordinating anion prefers the trigonal bipyramidal geometry. The central bismuth atom in 2-oxoalkyl- and 3-oxoalkyltriphenylbismuth-onium tetrafluoroborates and triflates shows a distorted tetrahedral geometry with a mean C–Bi–C bond angle of 109.1–109.2°, and the carbonyl oxygen atom is found to coordinate weakly to the bismuth center [94JCS(P1)1739, 95JCS(P1)2543]. (2-Methyl-1-propenyl)triphenylbismuthonium tetrafluoroborate also shows a distorted tetrahedral geometry with a mean C–Bi–C bond angle of 109.3° [96JCS(P1)1971]. Tetraphenylbismuth aryloxides $Ph_4Bi(OAr)$ (Ar = C_6F_5, C_6Cl_5) possess a distorted trigonal bipyramidal geometry with mean C–Bi–C_{ax} bond angles of 99.0–99.9° and mean C–Bi–C_{eq} bond angles of 117.0–117.6° [98OM1347]. The bond distance between the bismuth and axial oxygen in these aryloxides is ca. 2.54 Å. The general relationship between the coordination geometry of the bismuth atom and the nucleophilicity of the counter anion described above does not hold for tetrakis(2-methoxyphenyl)bismuthonium bromide, which consists of a distorted tetrahedral Ar_4Bi cation and a separated bromide anion in the solid state [97JCS(P1)1609]. Intramolecular coordination from four methoxy oxygen atoms toward the central bismuth atom in a tetrahedral way seems to contribute to the high thermal stabilization of this compound.

Several tetraphenylbismuth(V) compounds have been studied by IR and Raman spectroscopy as well as by molecular weight and conductance measurements [73JCS(D)1394]. The perchlorate, tetrafluoroborate and hexafluorophosphate behave as a 1:1 electrolyte in acetonitrile and nitromethane. The extent of the involvement of counter anions may be inferred from vibrational spectra. The observed wavenumbers for these bismuthonium salts clearly show the presence of the free counter anion. The nitrate and thiocyanate are also 1:1 electrolytes in these solvents, while their IR spectra indicate that the anionic groups are monodentate as ligand in the solid state. For instance, the nitrate shows strong bands at 1305 and 1450 cm^{-1}, which are assigned to the NO symmetric and asymmetric stretching frequencies, respectively. The latter band is characteristic of the monodentate nitrato-group. The trichloroacetate and cyanate are non-ionic in solution. Alkyltriphenylbismuthonium tetrafluoroborates and triflates also show strong IR absorptions due to the free counter anions at 1200–900 cm^{-1} [96BCJ2673].

Recently, unsymmetrical bismuthonium salts bearing four different aryl ligands have been prepared, and the chirality at bismuth has been investigated by ^{1}H-NMR [99OM5668]. The activation energy of isomerization at the bismuth center is strongly dependent on the nucleophilicity of the counter anions as well as the polarity of the solvents employed. This is rationalized by a pseudorotation mechanism via pentacoordinate species at the transition state.

3.3.3. Reactions

Metathesis of tetraarylbismuthonium halides with metal salts has often been used to exchange the halide anion for other anionic species (Section 3.3.1). For example, the reaction between tetrakis(2-methoxyphenyl)bismuthonium chloride and sodium or silver salts of formate, tosylate, bromide, iodide, tetrafluoroborate or perchlorate yields the corresponding bismuthonium compounds of the type $[Ar_4Bi^+]Y^-$ (Ar = 2-methoxyphenyl; Y = OCOH, OTs, Br, I, BF_4, ClO_4) [97JCS(P1)1609].

As mentioned in Section 3.3.2, both tetraphenylbismuthonium chloride and bromide decompose to yield triphenylbismuthine and the corresponding halobenzenes in the solid state at room temperature [52LA(578)136]. The tribromide undergoes another type of decomposition at room temperature to give bromodiphenylbismuthine and bromobenzene. Treatment of tetraphenylbis-

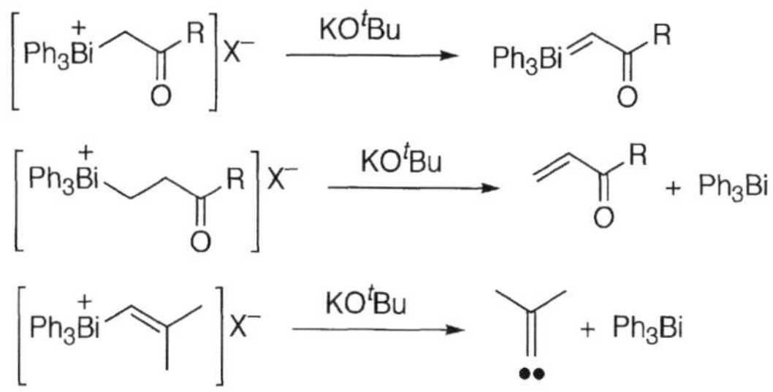

Scheme 3.6. Reactions of alkylbismuthonium salts with a base.

muthonium tetrafluoroborate with alcoholic sodium hydroxide affords triphenylbismuthine in 82% yield within a few minutes at room temperature [71IZV2618]. Photochemical reduction of tetraphenylbismuthonium tetrafluoroborate by electron transfer from an electron donor such as tetramethyl-*p*-phenylenediamine or triphenylphosphine has been studied, in which one of the phenyl groups is shifted to the donor molecule with the recovery of triphenylbismuthine [73IZV229].

The triphenylbismuth moiety in alkyl-, alkenyl- and aryltriphenylbismuthonium compounds often behaves as a good leaving group in the reactions with a variety of nucleophiles and Lewis bases [96BCJ2673]. Thus, 2-oxoalkyltriphenylbismuthonium salts readily react with triphenylphosphine, dimethyl sulfide, enolates, thiolates, phenoxides, sulfinates, piperidine and halides to give the corresponding α-substituted ketones, together with a good recovery of triphenylbismuthine [93TL8457, 94JCS(P1)1739]. 3-Oxoalkyl- and allyltriphenylbismuthonium salts also react with phosphine, sulfide, sulfinates and thiolates to transfer their alkyl moiety to the nucleophiles [95JCS(P1)2543, 95TL7475]. Alkenyltriphenylbismuthonium salts do not react with triphenylphosphine, dimethyl sulfide and piperidine, but readily transfer one of their organyl ligands to sulfinates and thiolates to give the corresponding α,β-unsaturated sulfones and sulfides, respectively [96JCS(P1)1971]. In these reactions, the alkenyl group is shifted to nucleophiles in preference to the phenyl ligand. Allyltriphenylbismuthonium compounds generated in situ from triphenylbismuth difluoride, $BF_3 \cdot OEt_2$ and allylsilanes are highly reactive and undergo the Friedel-Crafts type allylation with electron-rich arenes to give the allylated arenes. These reactions are described in Section 5.4.3.3 with reaction schemes. When treated with ethyl acrylate in the presence of a palladium(0) catalyst, alkenyltriphenylbismuthonium tetrafluoroborates effect the Heck-type alkenylation and phenylation [96JCS(P1)1971].

When treated with a base such as THF at low temperatures, 2-oxoalkyltriphenylbismuthonium salts are readily converted to bismuthonium 2-oxoalkylides, whereas 3-oxoalkyltriphenylbismuthonium salts undergo β-elimination to afford α,β-unsaturated ketones and triphenylbismuthine [94JCS(P1)2703, 95JCS(P1)2543]. Reaction of (2-methyl-1-propenyl)triphenylbismuthonium tetrafluoroborate with KO^{t}Bu in the presence of an excess of styrenes in dichloromethane at -78°C yields the corresponding alkylidene cyclopropanes and triphenylbismuthine [96JCS(P1)1971]. A Hammett study of this reaction has shown that free alkylidene carbene is generated under simultaneous elimination of the bismuthine. These base-promoted reactions are illustrated in Scheme 3.6.

Tetraphenylbismuth(V) compounds have been used as the oxidizing agents for alcohols under basic conditions (Section 5.2.4). They are also employed as the phenylating agent of alcohols, enols, amines, phenols, indoles, thiols, sulfinates, nitroalkanes, and others (Section 5.5.2). The selectivity between *O*- and *C*-arylations is dependent on the reaction conditions employed; 2-naphthol is *O*-phenylated by tetraphenylbismuthonium trifluoroacetate under acidic conditions, whereas it is *C*-phenylated under basic conditions.

Conversion of phenacylbismuthonium salt to phenacylsulfonium salt

To a stirred solution of phenacyltriphenylbismuthonium tetrafluoroborate (162 mg, 0.25 mmol) in dichloromethane (5 ml) was added dimethyl sulfide (0.18 ml, 2.5 mmol) at room temperature. Colorless crystals immediately precipitated were filtered off, washed with dichloromethane (5 ml), and dried in vacuo to obtain dimethylphenacylsulfonium tetrafluoroborate (65 mg, 97%). The filtrate was concentrated under reduced pressure to leave an oily residue, which was crystallized from methanol to give triphenylbismuthine (108 mg, 98%) [94JCS(P1)1739].

Cyclopropanation of styrene using alkenylbismuthonium salt

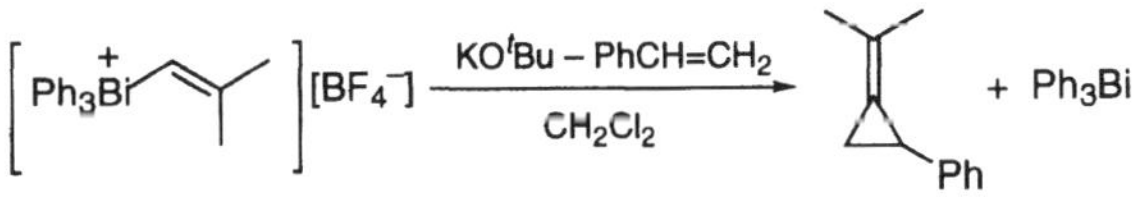

To a mixture of (2-methyl-1-propenyl)triphenylbismuthonium tetrafluoroborate (175 mg, 0.30 mmol), styrene (6 mmol) and dichloromethane (3 ml) was added potassium *tert*-butoxide (34 mg, 0.30 mmol) at -78°C. The resulting mixture was allowed to warm to room temperature and was passed through a short silica gel column to give a cyclopropane in 85% yield [96JCS(P1)1971].

3.4. QUINQUENARY ORGANOBISMUTH(V) AND RELATED COMPOUNDS

3.4.1. Methods of synthesis (Table 3.8)

3.4.1.1. From triorganylbismuth dihalides (method A)

The most convenient method for the preparation of pentaarylbismuth is the reaction between triarylbismuth dichloride and 2 equiv. of aryllithium [90B-S].

$$R_3BiX_2 + 2\ R'Li \longrightarrow R_3R'_2Bi$$

Pentamethylbismuth is prepared from trimethylbismuth dichloride and methyllithium [94AG(E)976]. Pentaorganylbismuth compounds can be transformed into lithium hexaorganylbismuthates by further action of organolithium reagent [93IC3948, 94AG(E)976].

Synthesis of bis(2-fluorophenyl)triphenylbismuth

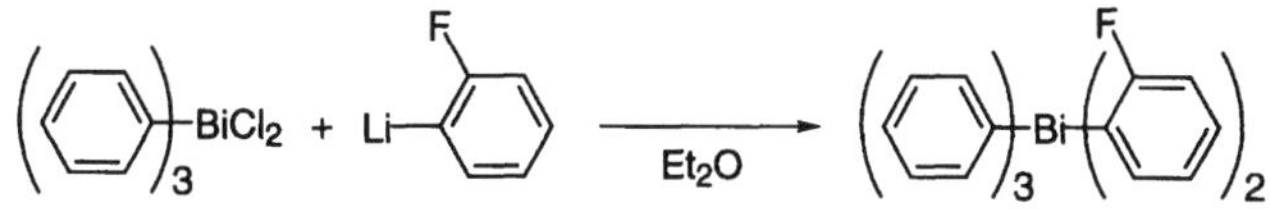

A solution of 2-fluorophenyllithium (6 mmol, freshly prepared from 1-bromo-2-fluorobenzene and butyllithium at −78°C) in anhydrous diethyl ether was fed through a small Teflon tube into a stirred suspension of triphenylbismuth dichloride (3 mmol) in the same solvent (20 ml) cooled to −78°C. The resulting suspension was gradually warmed to room temperature under stirring. After decanting from deposited lithium chloride, the clear red-violet solution was carefully evaporated. Crystallization at 0°C afforded bis(2-fluorophenyl)triphenylbismuth (45%), which is readily soluble in ether, but insoluble in pentane and sensitive to hydrolysis [90AG(E)213].

3.4.1.2. From triphenylbismuthine imide (method B)

Pentaphenylbismuth is formed in low yield by the reaction of a triphenylbismuthine imide with two equiv. of phenyllithium [64CB789].

$$Ph_3Bi{=}NSO_2Tol + 2\ PhLi \longrightarrow Ph_5Bi$$

3.4.2. Properties

Pentamethylbismuth is very unstable and explodes on rapid warming to room temperature in the crystalline state [94AG(E)976]. In solution, it decomposes to trimethylbismuthine. Although pentaphenylbismuth is thermally less stable than its phosphorus, arsenic and antimony congeners, it is still stable

TABLE 3.8
Pentaorganylbismuth and hexaorganylbismuthate compounds

Compound R^1	R^2	L(*n*)	Synthetic method[a]	Yield (%)	M.p. (°C)	Physical data	Reference
Pentaorganylbismuth ($R^1_3R^2_2Bi{\cdot}L_n$)							
Me	Me	–	A	–	–	^{1}H-NMR, X-ray	[94AG(E)976]
Ph	Ph	–	A	81	–	–	[52LA(578)136]
		–	B	26	90 (decomp.)	–	[64CB789]
		–	A	–	–	UV, X-ray	[87AG(E)1180]
	Ph (^{210}Bi)	–	A	–	–	–	[64R632]
	Ph	Py(1)	–	–	–	X-ray	[93IC3948]
	Ph	Py(3)		–	–	X-ray	[93IC3948]
	2-FC_6H_4	–	A	45	103 (decomp.)	^{19}F-NMR, IR, UV, X-ray	[90AG(E)213]
	4-ClC_6H_4	–	A	50	70–80 (decomp.)	–	[68LA(720)198]
	2,6-$F_2C_6H_3$	–	A	53	111	^{19}F-NMR, UV, X-ray	[89CB803]
	Biphenylene[b]	–	A	64	130–136	–	[68LA(720)198]
			A	–	–	X-ray	[90CB761]
4-MeC_6H_4	4-MeC_6H_4	–	A	20	111 (decomp.)	UV, X-ray	[90CB761]
	2-FC_6H_4	–	A	39	122 (decomp.)	^{19}F-NMR, IR, UV, X-ray	[90AG(E)213]
	4-$CF_3C_6H_4$	–	A	57	112 (decomp.)	^{1}H, ^{19}F-NMR, MS, UV	[89CB803]
	2,6-$F_2C_6H_3$	–	A	29	112	UV, X-ray	[90CB761]
	C_6F_5	–	A	51	114 (decomp.)	^{19}F-NMR, X-ray	[89CB803]
4-FC_6H_4	C_6F_5	–	A	29	133 (decomp.)	^{19}F-NMR, X-ray	[89CB803]
Hexaorganylbismuthate ($Li^+[R_6Bi]^-$)							
R							
Me			–	–	–	X-ray	[94AG(E)976]
Ph			–	–	–	X-ray	[93IC3948]

[a] For notation, see Section 3.4.1.

[b] 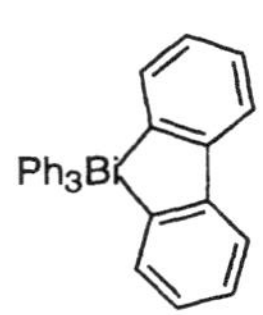

enough to be crystallized from aqueous ethereal solution at room temperature [52LA(578)136, 75DOK(225)581]. In a nitrogen atmosphere pentaphenylbismuth can be kept stable for several days, while in air it readily decomposes to a yellowish brown substance of unknown composition. When heated at around 100°C under a nitrogen atmosphere, pentaphenylbismuth decomposes exothermally to give benzene, biphenyl and triphenylbismuthine. In dry pyridine, chloroform or carbon tetrachloride, it smoothly decomposes to yield triphenylbismuthine, benzene and a tarry product. The benzyne intermediate generated in the decomposition process can be trapped with carbon monooxide, furan, ^{t}BuOH and phenol [75DOK(225)581, 78DOK(238)361, 86ZOB330, 88ZOB2305]. Recent theoretical studies on a model system predict that BiH_5 is thermodynamically less stable relative to BiH_3+H_2 by 70 kcal/mol [95JA11790, 92JA7518].

The structures of several symmetrical and unsymmetrical pentaarylbismuth compounds have been elucidated by means of X-ray crystallographic analysis (Fig. 3.4) [90B-B335, 92AOC207]. The structure of Ph_5Bi (violet-blue), $(C_6F_5)_3(4\text{-}MeC_6H_4)_2Bi$ (yellow), $(C_6F_5)_3(4\text{-}FC_6H_4)_2Bi$ (yellow), $(C_6H_5)_3(2,6\text{-}F_2C_6H_3)_2Bi$ (red), $(4\text{-}MeC_6H_4)_3(2,6\text{-}F_2C_6H_3)_2Bi$ (red) and $(C_6F_5)_3(2\text{-}FC_6H_4)_2Bi$ (violet) are all square pyramidal, while $(4\text{-}MeC_6H_4)_5Bi$ (violet) and $(4\text{-}MeC_6H_4)_3(2\text{-}FC_6H_4)_2Bi$ (orange) are trigonal bipyramidal. In the square pyramidal Ph_5Bi, four basal Bi–C bonds are significantly longer (2.32 Å) than the axial Bi–C bond (2.21 Å), while in the trigonal bipyramidal $(4\text{-}MeC_6H_4)_3(2\text{-}FC_6H_4)_2Bi$, two axial Bi–C bonds are much longer (2.39 Å) than three equatorial Bi–C bonds (2.18 Å). Pentamethylbismuth also produces violet crystals, which have been structurally characterized by X-ray crystallographic analysis at −163°C [94AG(E)976]. The central bismuth atom is trigonal bipyramidal, and the equatorial and axial Bi–C bonds are almost identical in length. Gas phase electron diffraction as well as ab initio MO calculation on the structure of pentamethylbismuth has revealed that the energy level of the trigonal bipyramidal configuration is lower than that of the square pyramidal configuration [93ACS368]. The methyl groups of pentamethylbismuth are equivalent in the ^{1}H-NMR spec-

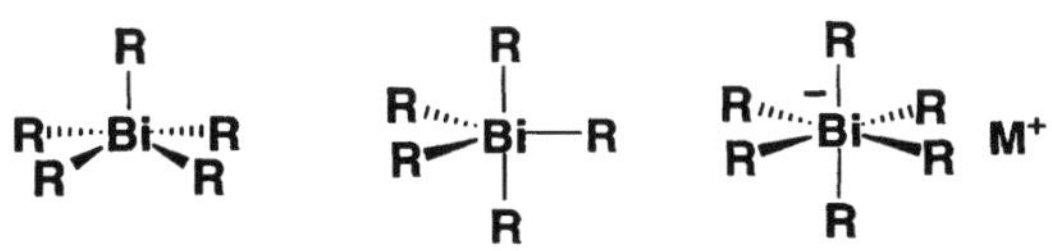

Fig. 3.4. Representative structures of pentaorganylbismuth and hexaorganylbismuthate.

trum even at -90°C, suggesting that it is a non-rigid compound [94AG(E)976].

The X-ray crystallographic analyses of yellow crystalline ate complexes, $Li^+[BiPh_6^-]$ and $Li^+[BiMe_6^-]$, have revealed that each bismuth center in these molecules has an octahedral geometry with a mean Bi–C bond length of 2.34 and 2.33 Å, respectively [94AG(E)976, 93IC3948]. The pyridine–Ph_5Bi complex has a somewhat distorted octahedral geometry with a weak interaction between the bismuth and nitrogen atoms. The relationship between the color and the coordination geometry of pentaorganylbismuth compounds has been investigated extensively by Seppelt and co-workers [90B-S].

3.4.3. Reactions

The reaction of pentaarylbismuth with 1 equiv. of bromine, hydrogen chloride, hydrogen tetrafluoroborate, triphenylsilanol and triphenylborane results in the cleavage of one of the Bi–C bonds to form tetraarylbismuth(V) compounds (Section 3.3.1).

Pentaphenylbismuth has been employed as the oxidizing and phenylating reagent in organic synthesis; isopropanol, benzyl alcohol and cholestanol are oxidized to the corresponding carbonyl compounds, whereas phenols, thiophenols and nitroalkanes are *C*- or *O*-phenylated depending on the substrates and conditions employed (Sections 5.2.4 and 5.5.2). These reactions proceed through a pentavalent organobismuth intermediate and involve no free radical species.

Pentaaryl- and pentaalkylbismuth compounds are converted by aryl- or alkyllithium reagent to the corresponding hexaaryl- and hexaalkylbismuthate complexes, respectively, at low temperatures [93IC3948, 94AG(E)976] (Scheme 3.7).

Synthesis of lithium hexaphenylbismuthate

$$Ph_5Bi + PhLi \xrightarrow[Et_2O]{} [Ph_6Bi^-]Li^+$$

To a solution of pentaphenylbismuth (0.67 mmol) in THF (12 ml) was added a solution of phenyllithium (1.4 mmol) in diethyl ether/cyclohexane (7:3) at -78°C. When the resulting yellow suspension was warmed to room temperature, the color turned blue-violet. The solution was pressed with argon through a thin Teflon tube into a glass ampule, that was then sealed off. Crystallization of the yellow bismuthinate, $Li^+[BiPh_6^-]\cdot THF$, occurred in the ampule on cooling to 0–18°C [93IC3948].

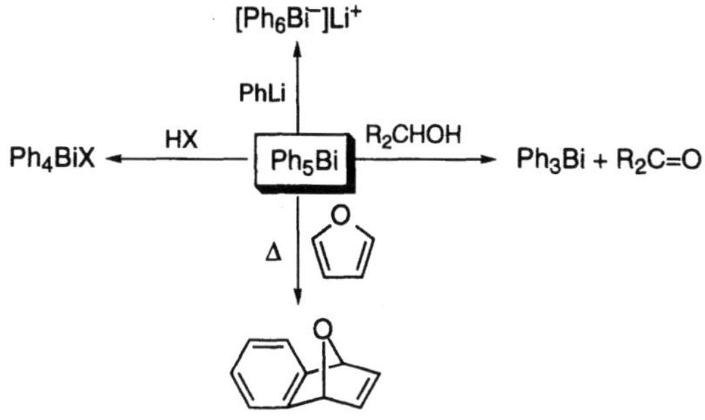

Scheme 3.7. Typical reactions of pentaphenylbismuth.

3.5. BISMUTHONIUM YLIDES

3.5.1. Methods of synthesis (Table 3.9)

3.5.1.1. From triphenylbismuthine and diazo compounds (method A)

In 1967, Lloyd et al. claimed the first synthesis of a bismuthonium ylide by heating triphenylbismuthine and diazotetraphenylcyclopentadiene at 140°C. However, experimental details confirming the product are not available [67CC1042, 88JCS(P2)1829]. Thermogravimetric analysis of the reaction shows that the carbene, cyclopentadienylidene, is formed at the initial step [72CC912].

$$Ph_3Bi + R_2C{=}N_2 \longrightarrow Ph_3Bi{=}CR_2$$

Bis(hexafluoroacytylacetonato)copper is reported to be an effective catalyst for the preparation of bismuthonium ylides from triphenylbismuthine and diazo compounds [88S319].

Synthesis of triphenylbismuthonium tetraphenylcyclopentadienylide

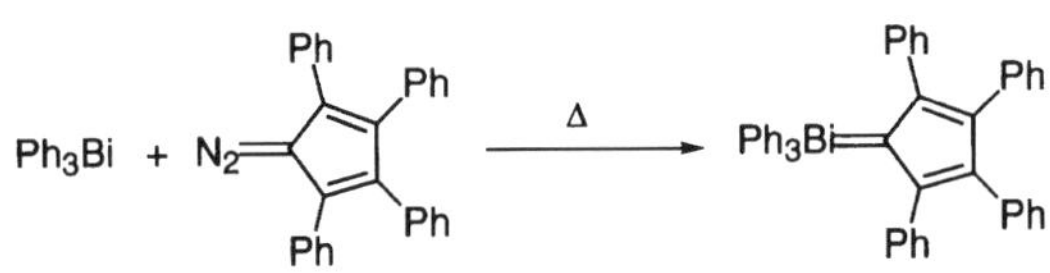

When a mixture of diazotetraphenylcyclopentadiene and triphenylbismuthine was heated to 140°C under nitrogen, triphenylbismuthonium tetraphenylcyclopentadienylide resulted as a deep blue product (m.p. 195°C, decomp.), which dissolved in ether and precipitated on addition of light petroleum [67CC1042].

3.5.1.2. From triarylbismuth(V) compounds (method B)

A stable bismuthonium ylide is formed by the reaction between triphenylbismuth carbonate and dimedone [85JCS(P1)2667]. The cyclic α,α'-dioxo structure is likely to be the main origin of high stability.

$$Ph_3BiCO_3 + (RCO)_2CH_2 \longrightarrow Ph_3Bi{=}C(COR)_2 + H_2CO_3$$

Stabilized bismuthonium ylides bearing a cyclic or acyclic α,α'-dicarbonyl or α,α'-disulfonyl skeleton can be prepared from triphenylbismuth dihalides and the corresponding 1,3-diketones or 1,3-disulfones in the presence of a suitable base [88CL847, 89CC1749, 94JFC(66)75].

$$Ph_3BiCl_2 + (RCO)_2CH_2 \xrightarrow[-\ 2\ HCl]{Base} Ph_3Bi{=}C(COR)_2$$

$$Ph_3BiCl_2 + (RSO_2)_2CH_2 \xrightarrow[-\ 2\ HCl]{Base} Ph_3Bi{=}C(SO_2R)_2$$

Synthesis of triphenylbismuthonio-4,4-dimethyl-2,6-dioxocyclohexan-1-ide

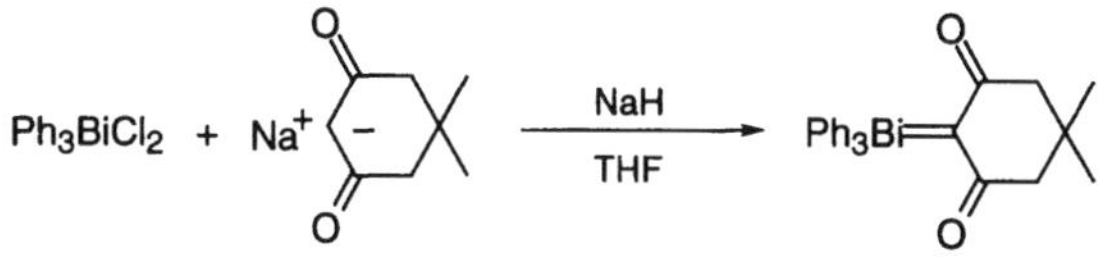

To a THF suspension (2 ml) of sodium enolate, generated from sodium hydride (ca. 80 mg) and 5,5-dimethylcyclohexane-1,3-dione (140 mg, 1.0 mmol), was added triphenylbismuth dichloride (511 mg, 1.0 mmol) at −20°C under nitrogen. The colorless suspension turned yellow and after 1 h a clear solution was obtained. Removal of the solvent followed by chromatographic purification of the product on a short silica gel column, using methanol–dichloromethane as the eluent, gave the bismuthonium ylide in 52% yield (m.p. 75°C) [89CC1749].

3.5.1.3. From bismuthonium salts and base (method C)

Abstraction of α-proton from 2-oxoalkyltriphenylbismuthonium salts by a base at low temperatures gives triphenylbismuthonium 2-oxoalkylides. The ylides cannot be isolated under ambient conditions due to thermal instability and moisture sensitivity.

$$[Ph_3BiCH_2COR]X \xrightarrow[-\ HX]{Base} Ph_3Bi{=}CHCOR$$

Synthesis of (3,3-dimethyl-2-oxobutylidene)triphenyl-λ^5-bismuthane

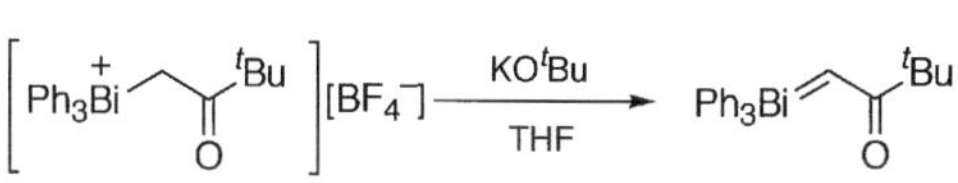

When a suspension of (3,3-dimethyl-2-oxobutyl)triphenylbismuthonium tetrafluoroborate (126 mg, 0.2 mmol) in THF (5 ml) was mixed with potassium *tert*-butoxide (24 mg, 0.2 mmol) at −78°C under argon, (3,3-dimethyl-2-oxobutylidene)triphenyl-λ^5-bismuthane was formed as a yellow solution. Attempts to isolate this ylide under ambient conditions were unsuccessful due to its thermal instability and moisture sensitivity [94JCS(P1)2703].

TABLE 3.9
Bismuthonium ylides

Compound		Synthetic method[a]	Yield (%)	M.p. (°C)	Physical data	Reference
$Ph_3Bi{=}CR_2$						
R						
MeCO		B	–	–	^{13}C-NMR	[89CC1749]
$PhSO_2$		A	22	107–109	^{1}H-NMR, IR, MS, UV	[88S319]
CF_3SO_2		B	95	108–110 (decomp.)	^{1}H-, ^{13}C-, ^{19}F-NMR	[94JFC(66)75]
$C_4F_9SO_2$		B	95	8–82	^{1}H-, ^{19}F-NMR	[94JFC(66)75]
$CHF_2CF_2CH_2OSO_2$		B	86	148–149 (decomp.)	^{1}H-, ^{19}F-NMR	[94JFC(66)75]
R_2						
PhC=CPh–CPh=CPh (**1**)[b]		A	–	195 (decomp.)	UV	[67CC1042]
		–	–	165	UV	[72T343]
$C(=O)CH_2CH_2C{=}O$ (**2**)[b]		B	–	128–132	^{1}H-NMR	[89CC1749]
		A	42	224–228	^{1}H-NMR, IR, UV	[88S319]
$C(=O)CH_2CH_2CH_2C{=}O$ (**3**)[b]		B	–	131–135	^{1}H-NMR	[89CC1749]
$C(=O)CH_2CMe_2CH_2C{=}O$ (**4**)[b]		B	75	–	^{1}H-NMR, IR, MS	[85JCS(P1)2667]
		B	83	72–75	^{1}H-, ^{13}C-NMR, IR, UV	[88CL847]
		–	–	–	X-ray	[90JCS(P1)3367]
$C(=O)OCMe_2OC{=}O$ (**5**)[b]		B	83	88–91	^{1}H-, ^{13}C-NMR, IR, UV	[88CL847]
$Ph_3Bi{=}CR^1R^2$						
R^1	R^2					
H	^{i}PrCO	C	–	–	–	[94JCS(P1)2703]
	^{t}BuCO	C	–	–	^{1}H-, ^{13}C-NMR	[94JCS(P1)2703]
	PhCO	C	–	–	–	[94JCS(P1)2703]
	$^{i}PrO_2C$	C	–	–	–	[99JOC6924]
	PhO_2C	C	–	–	–	[99JOC6924]
	$4\text{-}MeC_6H_4O_2C$	C	–	–	–	[99JOC6924]
MeCO	PhCO	B	–	–	–	[89CC1749]
	MeO_2C	B	–	–	–	[89CC1749]

[a] For notation, see Section 3.5.1.
[b] For structures, see Scheme 3.8.

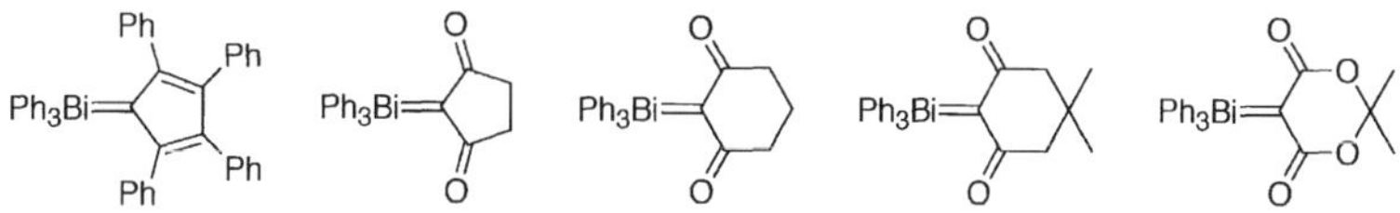

(1) (2) (3) (4) (5)

Scheme 3.8. Structures of compounds listed in Table 3.9.

3.5.2. Properties

The first bismuthonium ylide reported by Lloyd is a thermally stable colored substance, but the literature lacks details of its characterization. Several bismuthonium ylides containing a cyclic α,α′-dicarbonyl or α,α′-disulfonyl framework have been isolated as stable crystalline solids, and 4,4-dimethyl-2,6-dioxo-1-triphenylbismuthoniocyclohexane has been characterized structurally by X-ray crystallographic analysis, where the bismuth atom possesses a distorted tetrahedral geometry and interacts with one of the carbonyl oxygen atoms [90JCS(P1)3367]. The Bi–C_{ylide} bond (2.156 Å) is a bit shorter than the Bi–C_{Ph} bond (2.21–2.22 Å), suggesting little or no double bond nature of the bismuth and ylidic carbon bond (Fig. 3.5). The ylidic carbon of this class of stabilized ylide appears at δ 100–113.

Bismuthonium ylides with a non-cyclic α,α′-dicarbonyl structure are not highly stabilized and readily decompose during chromatographic treatment over silica gel [89CC1749]. Bismuthonium ylides with a non-cyclic α,α′-disulfonyl structure are colored from pale yellow to khaki to deep blue, depending on the structure of the organyl moieties attached to the sulfonyl functions [88S319, 94JFC(66)75]. They are thermally stable but decompose in open air. The moderately stabilized bismuthonium ylides with a 2-oxoalkylidene moiety can be generated only at low temperatures. The α-keto ylide of the type $Ph_3Bi{=}CHCOR$ decomposes rapidly at room temperature to give triphenylbismuthine along with small amounts of α-phenyl ketone, diacylethene, triacylcyclopropane and others. The α-ester ylide of the type

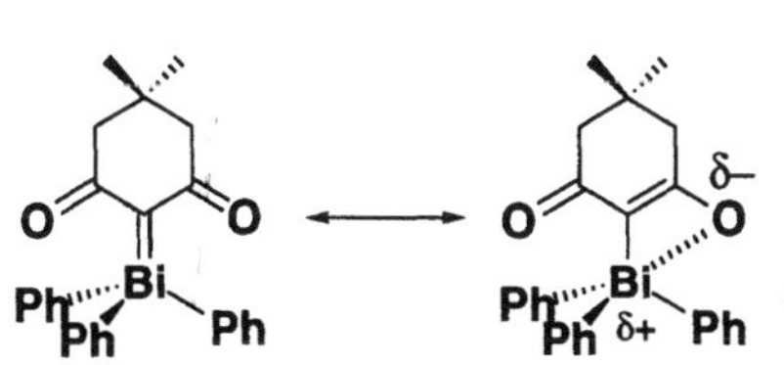

Fig. 3.5. A structural representation of a stabilized bismuthonium ylide.

$Ph_3Bi{=}CHCO_2R$ undergoes reductive dimerization at room temperature to give a good yield of triphenylbismuthine and a moderate yield of ethene $RO_2CCH{=}CHCO_2R$.

The moderately stabilized bismuthonium ylides can be characterized by low-temperature ^{13}C-NMR spectroscopy [94JCS(P1)2703]. The ylidic and carbonyl carbons of $Ph_3Bi{=}CHCO^tBu$ are observed at δ 86.4 and 187.9, respectively, at $-78°C$. The appearance of the ylidic carbon at lower field compared with the corresponding α-carbon (δ 59.6) of the parent onium salt $[Ph_3BiCH_2CO^tBu^+][BF_4^-]$ is consistent with the change in orbital hybridization mode of the carbon atom from sp^3 to sp^2. A large value for the C–H coupling constant of the ylide (J=186 Hz) as compared with the onium salt (J=147 Hz) is also a reflection of the increased s character of the ylidic C–H bond. Furthermore, the up-field appearance of the carbonyl carbon compared with that of the onium salt (δ 213.2) shows that a relatively large electron density is locarized on the relevant carbon atom. These spectral features suggest that the bismuthonium 2-oxoalkylides may well be represented as a hybrid of the canonical forms A, B and C, in which the structures B and C would be more important.

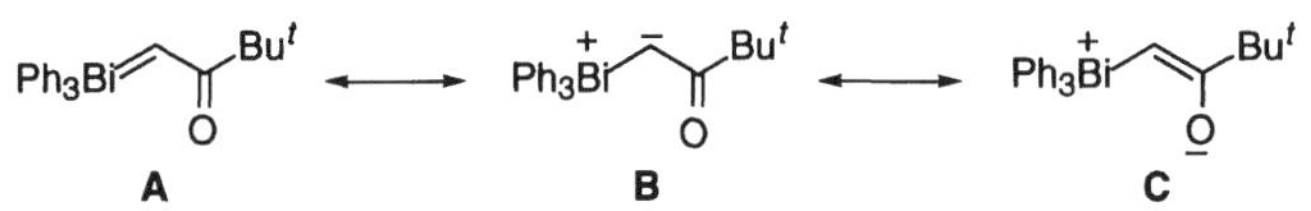

3.5.3. Reactions

Stabilized bismuthonium ylides with a cyclic α,α′-dicarbonyl structure undergo the C–C coupling reaction with aldehydes under gentle reflux in benzene or CH_2Cl_2 to give four types of product [88CL849]. The stabilized cyclic bismuthonium ylides also react with isothiocyanate [88CL847], acetylene [89CL325], phosphine and sulfide [90BCJ950] in the presence of a copper catalyst to yield the corresponding heterocyclic compounds, phosphonium and sulfonium ylides, respectively. The ylides with a non-cyclic α,α′-dicarbonyl structure fail to react with benzaldehyde but they react with a sulfene generated in situ from methanesulfonyl chloride and triethylamine to give oxathiole-*S*,*S*-dioxides [89CC1749]. These reactions are summarized in Section 5.4.3.3 with reaction schemes.

Copper(I)-catalyzed reaction of a stabilized bismuthonium ylide with phenylacetylene

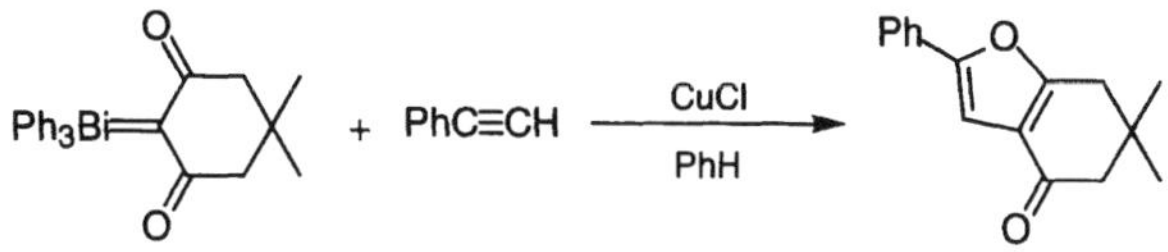

Triphenylbismuthonio-4,4-dimethyl-2,6-dioxocyclohexan-1-ide (578 mg, 1 mmol), phenylacetylene (204 mg, 2.0 mmol) and copper(I) chloride (20 mg, 0.2 mmol) were refluxed in benzene (5 ml) for 1 h. A yellow precipitate was filtered off and the filtrate was evaporated to obtain a yellow oil, which was passed through a silica gel column using benzene as an eluent to give a benzofuran derivative (31%), dimedone and triphenylbismuthine (13%) [89CL325].

The moderately stabilized triphenylbismuthonium 2-oxoalkylides readily undergo epoxidation with aldehydes to give α,β-epoxyketones and triphenylbismuthine [94JCS(P1)2703]. Although they do not react with *N*-aryl- or *N*-alkylaldimines, the bismuthonium 2-oxoalkylides react with *N*-sulfonylaldimines to afford α,β-aziridinoketones and triphenylbismuthine [95JOC4663]. The reaction mode leading to oxiranes and aziridines is characteristic of bismuth in the group 15 family; phosphonium, arsonium and stibonium 2-oxoalkylides all undergo the Wittig-type olefination with aldehydes and imines to afford α,β-unsaturated ketones. The theoretical calculations have been performed for the reaction of a series of ylides of group 15 elements, $H_3M{=}CH_2$ (M=N, P, As, Sb, Bi), with formaldehyde [94JA10080].

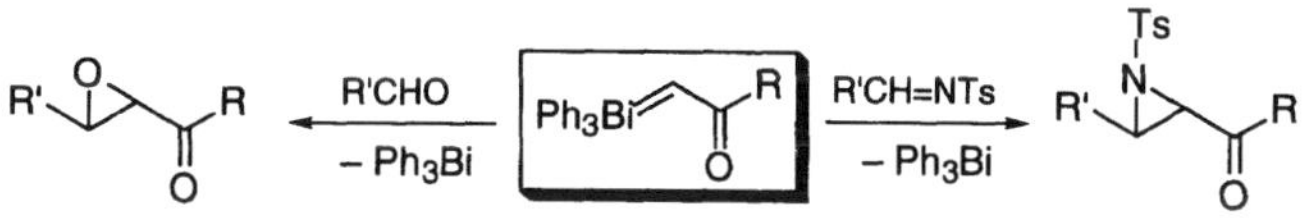

The bismuthonium 2-oxoalkylides react with acids such as dimedone, *p*-toluenesulfonic acid and benzenethiol to yield the corresponding α-substituted ketones. A catalytic amount of copper iodide or nitrosobenzene causes the reductive dimerization to form 1,2-diacylethenes and triphenylbismuthine.

Triphenylbismuthonium 2-oxoalkylides readily couple with α-dicarbonyl compounds in several different reaction modes [96CC2697, 98CC1359]. When treated with ethyl pyruvate, the epoxidation proceeds to form 2,3-difunctionalized oxiranes. The reaction with benzils gives *O*-aroylenolates derived from 1,3-diketones, while treatment with *o*-quinones affords 2-acyl-1,3-tropolone derivatives. These reaction modes are unprecedented in the ylide chemistry.

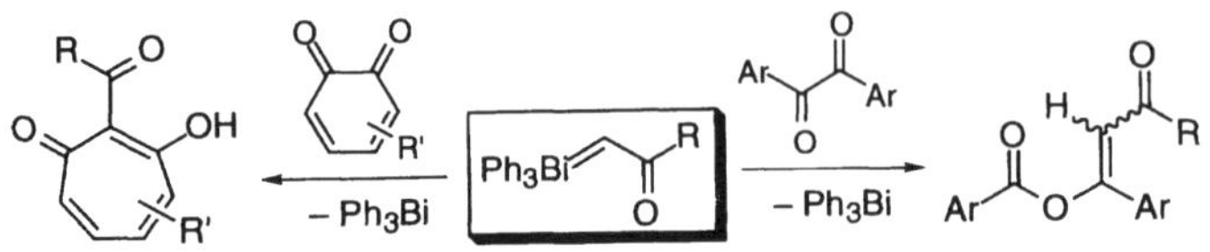

All the characteristic reactions of stabilized and moderately stabilized bismuthonium ylides are described in Section 5.4.3.3 with reaction schemes.

Conversion of aldehyde to oxirane by using bismuthonium 2-oxoalkylide

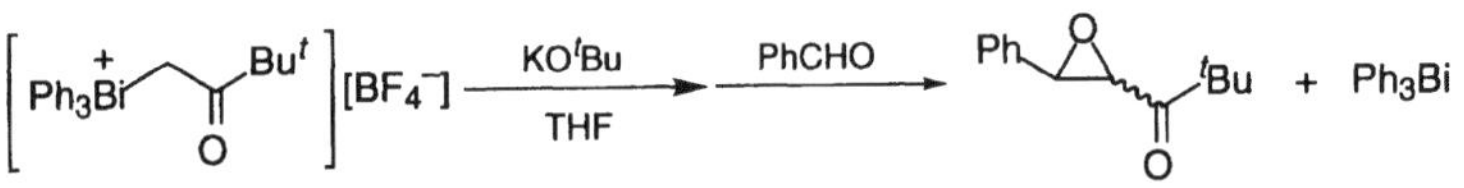

To a suspension of (3,3-dimethyl-2-oxobutyl)triphenylbismuthonium tetrafluoroborate (126 mg, 0.2 mmol) in THF (5 ml) was added potassium *tert*-butoxide (24 mg, 0.2 mmol) at −78°C. Within a few minutes, the reaction mixture turned yellow and clear. After 30 min, benzaldehyde (21 mg, 0.2 mmol) was added and the resulting mixture was allowed to warm to ambient temperature, concentrated under reduced pressure and extracted with benzene (10 ml × 2). Removal of the solvent gave an oily residue which was chromatographed on a silica gel column with hexane–ethyl acetate as the eluent to yield 1,2-epoxy-4,4-dimethyl-1-phenylpentan-3-one (90%, *cis*/*trans* 6:94) and triphenylbismuthine (89%) [94JCS(P1)2703].

3.6. BISMUTHINE IMIDES

3.6.1. Methods of synthesis (Table 3.10)

3.6.1.1. From tertiary bismuthines (method A)

A nitrene unit generated in situ from Chloramine-T® adds oxidatively to triarylbismuthines to give the corresponding triarylbismuthine *N*-tosylimides [64CB789, 91CL105]. *N*-Tosyliminoiodobenzene can also be used as a nitrene precursor [96JCR(S)24].

$$Ar_3Bi + ENTs \longrightarrow Ar_3Bi{=}NTs + E$$

Synthesis of triphenylbismuthine *N*-tosylimide

$$Ph_3Bi + TolSO_2NClNa \xrightarrow[MeCN]{} Ph_3Bi{=}NSO_2Tol + NaCl$$

To a suspension of well-dried Chloramine-T® (0.32 g, 1.4 mmol) in dry acetonitrile (20 ml) was added triphenylbismuthine (0.44 g, 1.0 mmol) and the resulting mixture was heated at 75–80°C with stirring under argon. The color of the suspension turned gradually yellow. After 3 h, the mixture was cooled to room temperature and freed from precipitated sodium chloride and unchanged Chloramine-T® by filtration. The yellowish filtrate was concentrated under reduced pressure and the residue was left in vacuo for 3 h to give *N*-tosyltriphenylbismuthimine as a pale orange syrupy oil in quantitative yield [91CL105].

3.6.1.2. From triarylbismuth dihalides and sulfonamides or amides (method B)

Crystalline bismuthine imides bearing an *N*-sulfonyl group have been prepared by the condensation of triphenylbismuth difluoride with *N,N*-bis(trimethylsilyl)sulfonamides or of triarylbismuth dichloride with sulfonamides [93JFC(63)179]. The latter combination is also applicable to the synthesis of trianylbismuthine *N*-acylimides [99OM2580].

$$Ar_3BiX_2 + H_2NSO_2R \xrightarrow[-\ 2\ HX]{Base} Ar_3Bi{=}NSO_2R$$

$$Ar_3BiX_2 + H_2NCOR \xrightarrow[-\ 2\ HX]{Base} Ar_3Bi{=}NCOR$$

Synthesis of triphenylbismuthine *N*-trifluoromethanesulfonimide

$$Ph_3BiF_2 + (Me_3Si)_2NSO_2CF_3 \xrightarrow[PhH]{} Ph_3Bi{=}NSO_2CF_3$$

A mixture of triphenylbismuth difluoride (0.478 g, 1.0 mmol), *N,N*-bis(trimethylsilyl)trifluoro-

TABLE 3.10
Triarylbismuthine imices

Compound R^1	R^2	Synthetic method[a]	Yield (%)	M.p. (°)	Physical data	Reference
Symmetrical compounds						
$R^1_3Bi{=}NSO_2R^2$						
2-$MeOC_6H_4$	Me	B	23	127–135 (decomp.)	^{1}H-NMR, IR	[UR2]
Ph	CF_3	B	100	142–144	^{1}H-, ^{19}F-NMR. MS	[93JFC(63)179]
2-MeC_6H_4	CF_3	B	97	131–138 (decomp.)	^{1}H-NMR, IR	[UR2]
2-$MeOC_6H_4$	CF_3	B	92	137–151 (decomp.)	^{1}H-NMR, IR	[UR2]
4-FC_6H_4	CF_3	B	93	136 (decomp.)	^{1}H-, ^{19}F-NMR	[93JFC(63)179]
Ph	C_4F_9	B	93	85 (decomp.)	^{1}H-, ^{19}F-NMR	[93JFC(63)179]
	4-MeC_6H_4	A	–	–	–	[64CB789]
		A	–	–	^{1}H-NMR, IR	[91CL105]
		A	–	–	–	[96JCR(S)24]
		B	97	68–70	^{1}H-NMR	[93JFC(63)179]
4-MeC_6H_4	4-MeC_6H_4	A	–	–	^{1}H-NMR, IR	[91CL105]
		A	–	–	–	[96JCR(S)24]
4-ClC_6H_4	4-MeC_6H_4	A	–	–	^{1}H-NMR, IR	[91CL105]
2,4,6-$Me_3C_6H_2$	4-MeC_6H_4	A	–	–	^{1}H-NMR, IR	[91CL105]
$R^1_3Bi{=}NCOR^2$						
2-MeC_6H_4	CF_3	B	89	108–116 (decomp.)	^{1}H-NMR, IR	[99OM2580]
2-$MeOC_6H_4$	CF_3	B	90	136–143 (decomp.)	^{1}H-NMR, IR, X-ray	[99OM2580]

TABLE 3.10 (*continued*)

Compound R^1	R^2	Synthetic method[a]	Yield (%)	M.p. (°)	Physical data	Reference
$2\text{-}MeC_6H_4$	CCl_3	B	77	113–115 (decomp.)	^{1}H-NMR, IR	[99OM2580]
$2\text{-}MeOC_6H_4$	CCl_3	B	96	112–114 (decomp.)	^{1}H-NMR, IR	[99OM2580]
$2,4,6\text{-}MeC_6H_2$	CF_3	B	86	133–134	^{1}H-NMR, IR	[99OM2580]
Unsymmetrical compounds $R^1{}_2R^2Bi{=}NSO_2CF_3$						
$4\text{-}MeC_6H_4$	$2\text{-}OxC_6H_4$[b]	B	90	188–190 (decomp.)	^{1}H-, ^{13}C-NMR, IR, X-ray	[97OM1013]
Ph	$2\text{-}OxC_6H_4$[b]	B	65	178–180 (decomp.)	^{1}H-, ^{13}C-NMR, IR	[UR4]
$4\text{-}FC_6H_4$	$2\text{-}OxC_6H_4$[b]	B	80	191–193 (decomp.)	^{1}H-, ^{13}C-NMR, IR	[UR4]

[a] For notation, see Section 3.6.1.

[b] Ox =

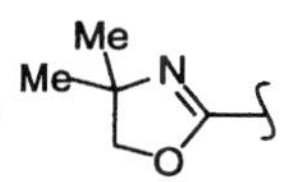

methanesulfonamide (0.293 g, 1.0 mmol) and benzene (10 ml) was stirred for about 1 h at 40°C until the evolution of fluorotrimethylsilane had ceased. The solvent was distilled off in vacuo to obtain the expected bismuthine imide in a near quantitative yield (m.p. 142–144°C) [93JFC(63)179].

Synthesis of tris(2-methylphenyl)bismuthine *N*-trifluoroacetylimide

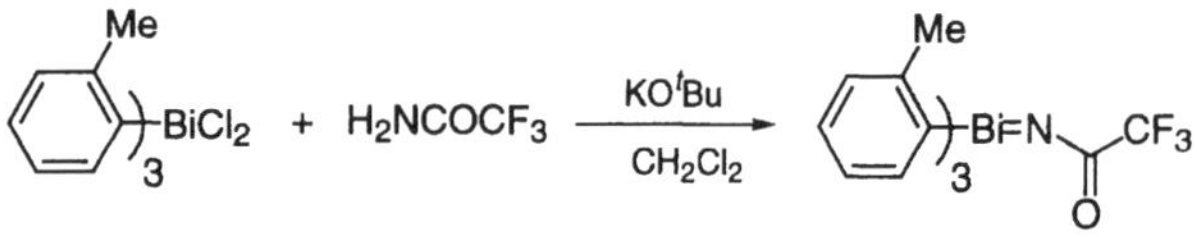

To a mixture of tris(2-methylphenyl)bismuth dichloride (2.21 g, 4.15 mmol), trifluoroacetamide (458 mg, 4.05 mmol) and dichloromethane (80 ml) was added potassium *tert*-butoxide (998 mg, 8.91 mmol) at −50°C. The resulting mixture was allowed to warm gradually to 10°C with vigorous stirring. The insoluble solid was filtered through Celite and the filtrate was concentrated under reduced pressure to leave an oily residue, which was crystallized from dichloromethane–hexane to give tris(2-methylphenyl)bismuthine *N*-trifluoroacetylimide as a pale yellow solid (2.205 g, 90%), m.p. 108–116°C (decomp.) [99OM2580].

3.6.2. Properties and reactions

Bismuthine imides containing the *N*-sulfonyl group are considerably stabilized, while the imides bearing an *N*-acyl group are rather unstable. When heated in benzene, triarylbismuthine *N*-acylimides decompose to give triarylbismuthine in moderate yield [99OM2580]. This mode of decomposition contrasts with that of triarylphosphine *N*-acylimides, where triphenylphosphine oxide and a nitrile are formed. Crystalline bismuthine imides are mostly soluble in dichloromethane, chloroform, benzene and acetonitrile, and almost insoluble in hexane. They can be kept for days under a dry atmosphere at room temperature but gradually decompose on standing in open air. However, the *ortho*-substituted triarylbismuthine imides of the type $Ar_3Bi{=}N\text{-}SO_2R$ and $Ar_3Bi{=}NCOR$, where Ar is 2-methylphenyl, 2-methoxyphenyl or 2,4,6-trimethylphenyl, are much more stable both in the solid state and in solution under atmospheric conditions when compared with the *para*-substituted ones [99OM2580]. The high stability of these imides toward moisture arises mainly from steric protection around the Bi=N bond. A coordinating imino group also stabilizes the bismuth–imido nitrogen bond; an oxazoline-substituted bismuthine imide does not show any sign of decomposition in open air [98OM1013].

Several bismuthine *N*-sulfonylimides and an bismuthine *N*-acylimide have

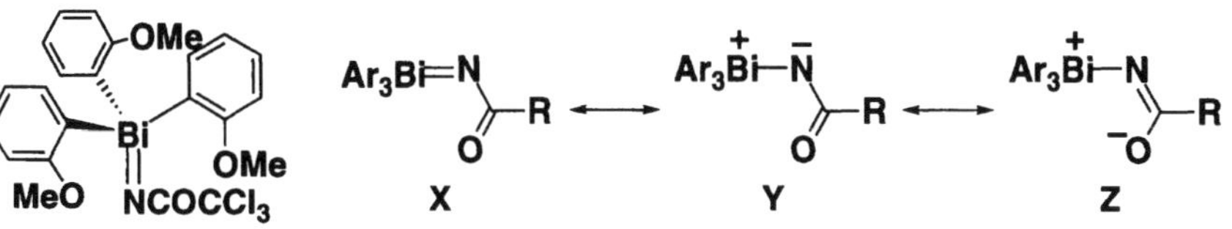

Fig. 3.6. A structural representation of a triarylbismuthine *N*-acylimide.

been characterized structurally by X-ray crystallographic analyses. The bismuth centers in $(2\text{-}MeC_6H_4)_3Bi{=}NSO_2CF_3$ and $(2\text{-}MeOC_6H_4)_3Bi{=}N\text{-}SO_2CF_3$ have a distorted tetrahedral geometry with a mean C–Bi–C bond angle of 113.4 and 111.5°, respectively [UR2]. The Bi–N bond length of 2.112(6)–2.086(6) Å lies close to the shorter end of known Bi–N single bond distances (2.10–2.24 Å), suggesting a polarized $Bi^+–N^-$ single bond nature rather than a Bi=N double bond. *N*-Acylimide, $(2\text{-}MeOC_6H_4)_3Bi{=}N\text{-}COCCl_3$, also possesses a distorted tetrahedral geometry with a mean C–Bi–C bond angle of 108.3° and a mean Bi–C bond distance of 2.20 Å [99OM2580]. The Bi–N bond length of 2.125(3) Å shows a polarized $Bi^+–N^-$ single bond character, which is consistent with a relatively short N–C bond (1.314(5) Å) and a somewhat longer C=O bond (1.255(5) Å). Thus, the canonical form Z would contribute considerably to the actual structure of this compound (Fig. 3.6). The carbonyl oxygen atom coordinates weakly to the bismuth center with a Bi–O bond distance of 2.877(3) Å. Due to this intramolecular coordination, three atoms attached to the N–C bond were located almost in the same plain with a small Bi–N–C–O torsion angle of 4.2(5)°. The theoretical calculations have been carried out for the formation and reaction of a series of imides of group 15 elements, $H_3M{=}NH$ (M=P, As, Sb, Bi) [97IC694].

The oxazoline-substituted bismuthine imide has a distorted trigonal bipyramidal structure, bearing three ipso carbon atoms at equatorial sites and two nitrogen atoms at apical sites with a Bi–N(imide) bond length of 2.13(1) Å and

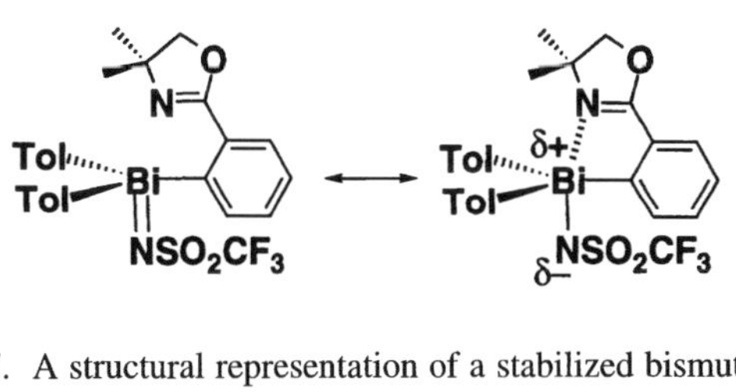

Fig. 3.7. A structural representation of a stabilized bismuthine imide.

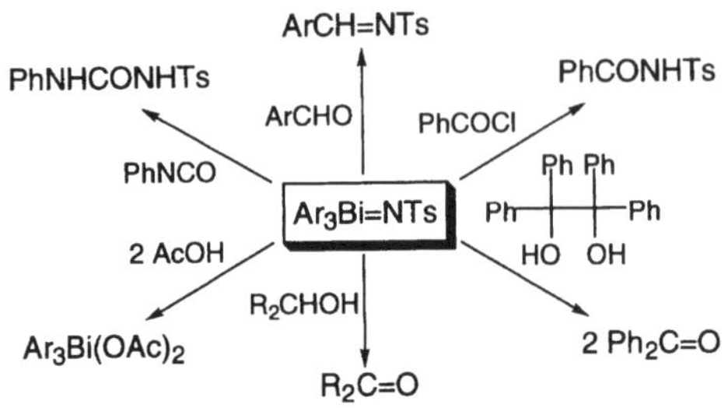

Scheme 3.9. Typical reactions of triarylbismuthine *N*-tosylimides.

a Bi–N(oxazoline) bond distance of 2.69(1) Å [98OM1013]. Intramolecular coordination of the oxazoline nitrogen atom to the bismuth center would contribute to the stabilization of this imide (Fig. 3.7).

In the IR spectra of triarylbismuthine *N*-acylimides ($Ar_3Bi{=}NCOCX_3$; X = F, Cl), a characteristic C=O stretching band is observed at considerably low frequency (1561–1593 cm^{-1}) compared with those of the parent amide (H_2NCOCX_3) or a phosphorus counterpart ($Ph_3P{=}NCOCX_3$), which suggests significant contribution of a canonical structure Z to the stabilization of this class of imide (Fig. 3.6) [99OM2580].

Triarylbismuthine *N*-tosylimides react with benzaldehyde, benzoyl chloride and phenyl isocyanate to give *N*-tosylimine, *N*-tosylamide and *N*-tosylurea derivatives, respectively [91CL105] (Section 5.5.2.3). The imides oxidize secondary and benzylic alcohols to carbonyl compounds, sometimes accompanied by the concurrent formation of a diaryl(*N*-tosylamino)bismuthine [96JCR(S)24] (Section 5.2.4). When treated with acetic acid, the imides are converted to the corresponding triarylbismuth diacetates and sulfonamides (Scheme 3.9).

Oxidation of alcohol to ketone by using triarylbismuthine imide

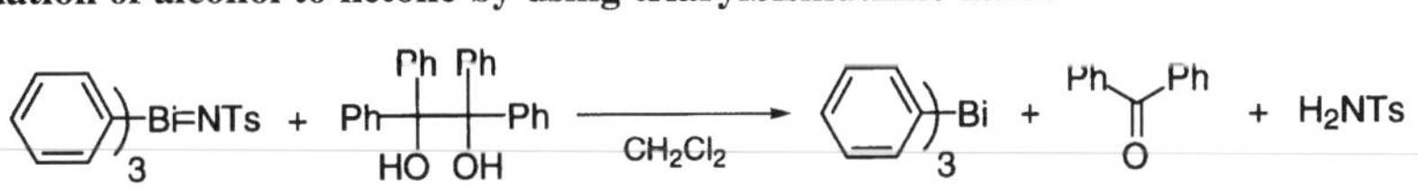

To a dichloromethane solution of triphenylbismuthine *N*-tosylimide (0.5 mmol), generated in situ from the corresponding bismuthine and iminoiodobenzene, was added 1,1,2,2-tetraphenylethane-1,2-diol (0.5 mmol) at room temperature. Within a few minutes, the reaction completed to afford triphenylbismuthine (99%), benzophenone (99%) and *p*-toluenesulfonamide (80%) [96JCR(S)24].

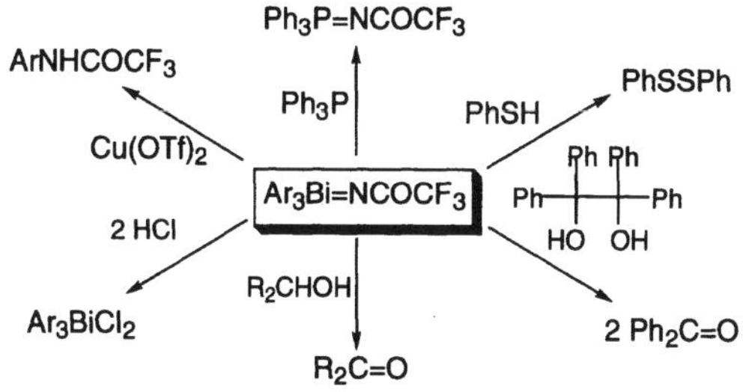

Scheme 3.10. Typical reactions of triarylbismuthine *N*-acylimide.

Triarylbismuthine *N*-acylimide transfers the nitrene moiety to triphenylphosphine to give the corresponding phosphine *N*-acylimide with a good recovery of triarylbismuthine (Scheme 3.10) [99OM2580]. The imide oxidizes secondary alcohols and benzenethiol to carbonyl compounds and diphenyl disulfide, respectively. When treated with hydrogen chloride, the bismuthine imide undergoes cleavage of the Bi=N bond to yield triarylbismuth dichloride. The *N*-arylation product is formed in the presence of a catalytic amount of copper(II) triflate.

3.7. BISMUTHINATES AND BISMUTHINIC ACIDS

3.7.1. Methods of synthesis (Table 3.11)

This class of compound has been prepared by the reaction of triarylbismuthines and Chloramine-T® in a mixture of methanol and benzene.

$$Ar_3Bi + TolSO_2NClNa + ROH \longrightarrow Ar_2Bi(=O)OR + ArH + H_2NSO_2Tol$$

Synthesis of methyl bis(4-methyl-1-naphthyl)bismuthinate

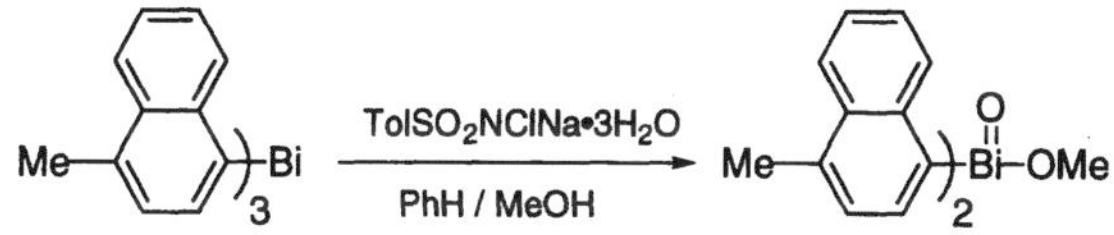

Tris(4-methyl-1-naphthyl)bismuthine (632 mg, 1 mmol) and Chloramine-T® (564 mg, 2 mmol) were heated in benzene (20 ml)/methanol (20 ml) for ca. 30 min. When the bismuthine was consumed completely, the precipitated solid was removed by filtration and the filtrate was concentrated in vacuo to ca. 20 ml and left to stand. The crystalline deposit was filtered off, washed with methanol, and recrystallized from benzene/methanol to give the dinaphthylbismuthinate as pale yellow fine needles (446 mg, 80%), m.p. 140–150°C (decomp.) [88CL2021].

TABLE 3.11
Bismuthinic acid and esters

Compound R[1]	R[2]	Yield (%)	M.p. (°C)	Physical data	Reference
$R^1_2Bi(=O)OR^2$					
Ph	H	90	–	–	[88CL2021]
2-MeC_6H_4	H	60	110–130 (decomp.)	^{1}H-NMR, IR	[88CL2021]
Ph	Me	30	110–120 (decomp.)	^{1}H-NMR, IR	[88CL2021]
4-MeC_6H_4	Me	10	–	–	[88CL2021]
3,4-$Me_2C_6H_3$	Me	20	90–110 (decomp.)	^{1}H-NMR, IR	[88CL2021]
l-$C_{10}H_7$	Me	80	135–145 (decomp.)	^{1}H-NMR, IR	[88CL2021]
4-Me-1-$C_{10}H_6$	Me	80	140–150 (decomp.)	^{1}H-NMR, IR	[88CL2021]
Ph	Et	60	140–150 (decomp.)	^{1}H-NMR, IR	[88CL2021]
	^{i}Pr	52	–	–	[88CL2021]

3.7.2. Properties and reactions

Diarylbismuthinates are mostly solids, soluble in chloroform and benzene/methanol. When a bulky group is absent at an *ortho* position of the aromatic ring, the bismuthinates readily disproportionate to bismuthines in solution. The ^{1}H-NMR spectra of methyl diarylbismuthinates show a broad signal of the methyl group around δ 3.5–3.6, suggestive of the mobility of the Bi–OMe bond. When treated with water or alcohols, the methoxy group is replaced easily by the hydroxyl or other alkoxyl groups. No crystallographic data are available for this class of compound.

Benzoin, tetraphenylethanediol, hydrazobenzene and 1,2-diphenyl-1,2-ethanedione bishydrazone are oxidized by methyl bis(1-naphthyl)bismuthinate to benzil, benzophenone, azobenzene and diphenylacetylene, respectively [88CL2021] (Section 5.2.4). Amides, thioamides, ureas and thioureas are *N*-acylated in good yield with acetic acid in the presence of the bismuthinate at room temperature [90CL1651] (Scheme 3.11).

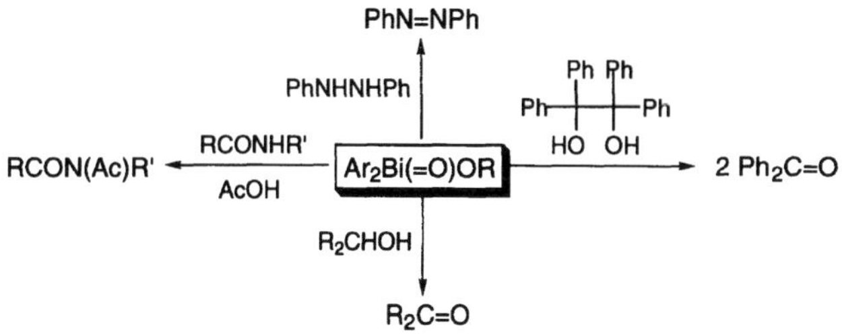

Scheme 3.11. Typical reactions of bismuthinates.

Bismuthinate-mediated *N*-acylation of amides with carboxylic acids

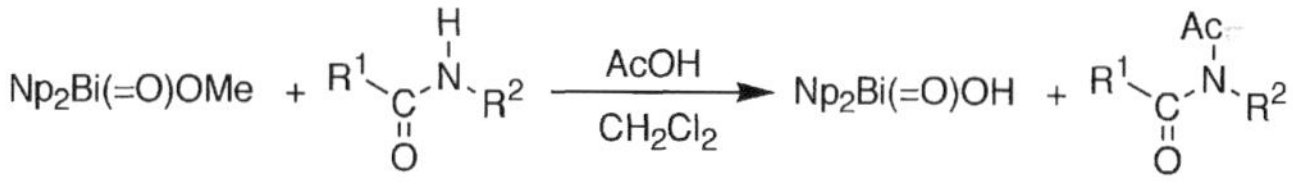

Benzamide (121 mg, 1.0 mmol) and methyl bis(1-naphthyl)bismuthinate (510 mg, 1.0 mmol) were stirred in a mixture of acetic acid (5 ml) and dichloromethane (20 ml) at room temperature. When the starting amide was consumed completely, the mixture was evaporated to dryness in vacuo and the residue was passed through a silica gel column using hexane–dichloromethane (100% hexane to 100% dichloromethane) as the eluent. *N*-Acetylbenzamide was obtained in 89% isolated yield [90CL1651].

3.8. TRIORGANYLBISMUTHINE OXIDES

In contrast with extensive works on lighter organylpnicogen oxides, little attention has hitherto been paid to the chemistry of organylbismuthine oxides. However, the oxides of triorganylbismuthines have potential as a versatile precursor for a variety of organobismuth(V) compounds.

3.8.1. Methods of synthesis

3.8.1.1. Metathesis of triorganylbismuth(V) compounds

Metathetical reaction of Ar_3BiCl_2 and silver oxide has been reported by many workers with conflicting results. A solution of triphenylbismuth dihydroxide $Ar_3Bi(OH)_2$ or oxide $Ar_3Bi{=}O$ is obtained from triphenylbismuth sulfate or dichloride by treating with ammonium hydroxide. This reacts with acetic acid to give $Ph_3Bi(O_2CMe)_2$. However, the addition of cold concentrated hydrochloric acid to a solution of this oxide apparently does not give Ph_3BiCl_2 [16JCS(109)250]. Treatment of Ph_3BiCl_2 or Ph_3BiBr_2 with aqueous potassium hydroxide gives Ph_3Bi instead of the expected dihydroxide or oxide [20JCS(117)762]. Metathesis reaction of Ar_3BiCl_2 with Ag_2O in moist acetone results in the formation of Ar_3Bi [20JCS(117)762, 22JCS(121)109], while $Ph_3Bi(OH)_2$ is reported to be formed in water [34JCS405]. The first spectral characterization of the oxide was made by Goel et al. who carried out the same metathetical reaction in a benzene–water system to obtain triphenylbismuthine oxide $Ph_3Bi{=}O$ as a white polymeric powder in 10–40% isolated yields [72JOM(36)323]. They also reported the same type of reaction between $Ph_3Bi(CN)_2$ and HgO to afford the bismuthine oxide [73JOM(50)129], although neither the chemical nature nor the yield is mentioned. Hydrolysis of triphenylbismuthine *N*-sulfonylimides yields $Ph_3Bi{=}O$ together with the corresponding sulfonamides [93JFC(63)179].

Synthesis of triphenylbismuthine oxide

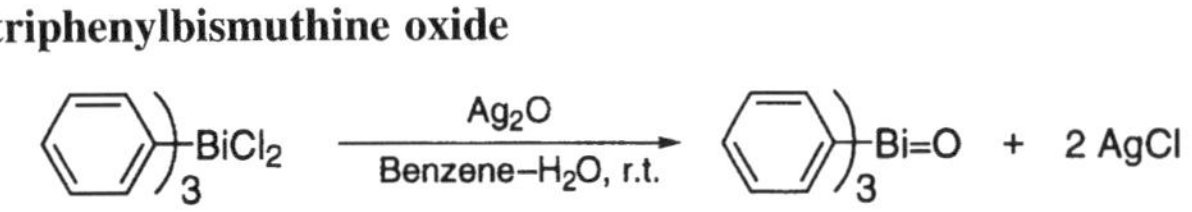

A benzene solution of triphenylbismuth dichloride (1.7 g, 3.3 mmol) was mixed with an aqueous solution of freshly prepared silver oxide (0.7725 g, 3.3 mmol) and the resulting mixture was stirred for 5 h at room temperature in the dark. The benzene layer was separated by decantation, laid on molecular sieves for 12 h, and was then filtered and concentrated by evaporation. Triphenylbismuthine oxide (40%) was precipitated from the concentrated benzene solution by adding petroleum ether slowly. Similarly, a benzene solution of triphenylbismuth dichloride (1.7 g, 3.3 mmol) was stirred with silver oxide for about 2 h and filtered. The filtrate was concentrated under vacuum and then treated with petroleum ether where triphenylbismuthine oxide was obtained as a white powder (10%) [72JOM(36)323].

Synthesis of triphenylbismuthine oxide

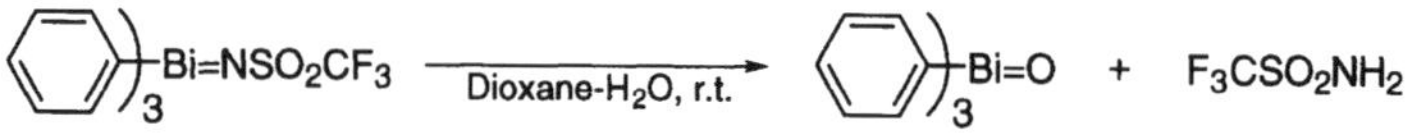

Water (1 ml) was added to a solution of triphenylbismuthine *N*-trifluoromethanesulfonylimide (0.587 g, 1.00 mmol) in dioxane (10 ml). The reaction mixture was stirred for 24 h at ambient temperature. During this period, a white solid precipitated; it was filtered off and identified as triphenylbismuthine oxide (0.492 g, 94%) from its melting point (152–154°C) [93JFC(63)179].

3.8.1.2. Direct oxidation of tertiary bismuthines

Lower trialkylbismuthines such as Me_3Bi are pyrophoric under atmospheric conditions. $(PhCH_2)_3Bi$ also decomposes with a smoke in air [57CB1176]. Controlled oxidation of trialkylbismuthines with molecular oxygen [71CJC1747, 71JCS(C)1660], *tert*-butyl hydroperoxide [71JCS(C)1660] or potassium permanganate [57CB1176] results in the cleavage of carbon–bismuth bonds. Oxidation of alkynyldiarylbismuthines with aqueous hydrogen peroxide or potassium permanganate in dry acetone does not afford the desired oxides [62ZAAC(317)54]. Tris(cyclopentadienyl)bismuthine does not form its oxide by the action of aqueous hydrogen peroxide [60CB1417].

Attempts to prepare $Ph_3Bi{=}O$ by direct oxidation of Ph_3Bi with a variety of oxygen transfer reagents have led to discordant results. Oxidation of Ph_3Bi with hydrogen peroxide [62JOC3851] or 2,4,4-trimethyl-1-pyrrolin-1-oxide [62CJC181] only leads to decomposition products, while the reaction of Ph_3Bi with selenium dioxide either gives phenylseleninic acid and bismuth selenite [77JCS(D)641] or results in the recovery of the starting materials [76JOM(116)199]. Phenyldiazonium nitrate is the main product from the reaction of Ph_3Bi with dinitrogen trioxide [39ZOB771]. Benzoyl peroxide reacts with Ph_3Bi in petroleum ether to afford the corresponding dibenzoate [27JCS209], while bis(trimethylsilyl) peroxide does not react with Ph_3Bi even after prolonged heating at 110–130°C [74JOM(73)217]. Photochemical reduction of the uranyl ion with Ph_3Bi in dioxane–water was reported to have given $Ph_3Bi{=}O$; however, evidence of the formation of the oxide is provided only by

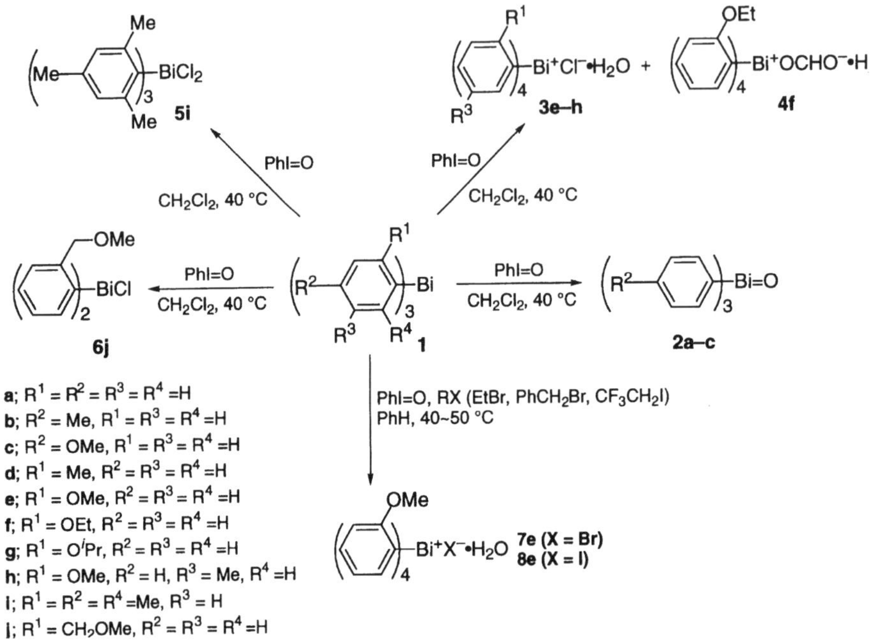

Scheme 3.12.

TLC inspection [80IJC(A)902]. Similarly, photocatalyzed air oxidation of Ph_3Bi with TiO_2 in methanol has been reported to produce the oxide in 74% yield; however, the formation of the oxide has only been claimed on the basis of GC analysis [96JOC2895]. Ph_3Bi is converted to the corresponding oxide by the action of nitric acid in acetic anhydride, though the resulting oxide is rapidly transformed into $Ph_3Bi(O_2CMe)_2$ [89G(119)545]. Conversion of Ar_3Bi into its oxide with the *N*-bromosuccinimide–hydrochloric acid–potassium bromide system is used for oxidimetric titration of Ar_3Bi [87ACA(196)351]. Ozone oxidizes triarylbismuthine to the corresponding diformate in an appropriate solvent [93JCS(P1)2411]. Suzuki et al. found that direct oxidation of triarylbismuthines with iodosobenzene under mild conditions generates triarylbismuthine oxides efficiently [94TL8197]. The formation of the oxide can be confirmed by using a trapping experiment; triarylbismuth dibenzoate and bis(4-methylbenzenesulfonate) are obtained in good yield by adding benzoic anhydride or 4-methylbenzenesulfonic acid to a solution of the oxide. Barton et al. have reported that $Ph_3Bi{=}O$ is unable to cleave trans-decaline-9,10-diol, but no mention is made on the preparation of the oxide [86T5627].

Synthesis of tris(4-methylphenyl)bismuthine oxide

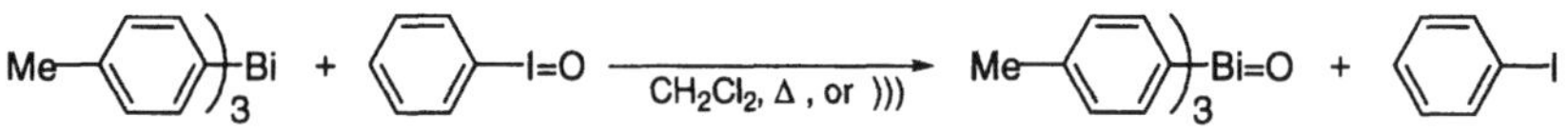

Tris(4-methylphenyl)bismuthine (0.482 g, 1 mmol) was added to a suspension of freshly prepared iodosobenzene (0.286 g, 1.3 mmol) in dry dichloromethane (30 ml) and the resulting mixture was sonicated at 35°C under argon on a commercial ultrasonic washing machine until the substrate was completely consumed (checked by TLC). After 2–3 h, the chalky suspension turned to a bright yellow solution of tris(4-methylphenyl)bismuthine oxide, which was quickly filtered through a Celite bed and used for further purposes [94TL8197].

The new oxidation reaction of triarylbismuthine by iodosobenzene is highly dependent on the structure of bismuthines and the solvent system employed. Dichloromethane or toluene are solvents of choice. Bismuthines **1a–c** are converted to a presumed oxide **2a–c**, while **1d** resists oxidation. Bismuthines **1e–h** are all similarly oxidized in dichloromethane by the present procedure to give the corresponding tetraarylbismuthonium chlorides **3e–h** in moderate yield. In the case of bismuthine **1f**, however, formate **4f** was obtained as the major product. Interestingly, bismuthine **1i** is converted to the corresponding dichloride **5i** in 80% yield, while bismuthine **1g** gives diarylchlorobismuthine **6j**. Tetraarylbismuthonium bromide **7e** and iodide **8e** are also readily available by the oxidation of **1e** with iodosobenzene in benzene in the presence of the corresponding alkyl halides [97JCS(P1)1609].

3.8.2. Properties

The physical properties of triorganylbismuthine oxides have not been examined quite so extensively. Goel et al. claim that triphenylbismuthine oxide has a five-coordinated bismuth center with a polymeric structure involving –Bi–O–Bi– units [72JOM(36)323].

Triphenylbismuthine oxide is a white solid which reportedly melts at 155°C and is moderately soluble in benzene, slightly soluble in acetonitrile but insoluble in chloroform and diethyl ether. The equivalent conductance of a 2×10^{-3} M solution in acetonitrile at 25°C is 1.5 $ohm^{-1}cm^2equiv^{-1}$. The molecular weight has not yet been determined due to its extensive decomposition in solution. The IR spectrum of solid triphenylbismuthine oxide shows no bands attributable to O–H stretching or Bi–OH bending frequencies. The bands at 630, 450 and 250 cm^{-1} can be assigned to the Bi–O–Bi asymmetric stretching, *y*- and *t*-vibrations, respectively [72JOM(36)323]. Several triarylbismuthine oxides have been obtained as a solution in dichloromethane, benzene or toluene. The ^{1}H-NMR spectrum of tris(4-methylphenyl)bismuthine oxide exhibits a methyl signal at δ 2.38, and two broad signals due to aromatic protons at δ 7.4 and 8.2; the latter two signals turn into a pair of doublets (δ 7.39 and 8.14; J = 7.6 Hz) at −75°C. By adding hexane or diethyl ether to the solution, a white to pale yellow precipitate is formed; however, it cannot be re-dissolved in dichloromethane or benzene [94TL8197].

3.8.3. Reactions

In acetone or alcohol, $Ph_3Bi{=}O$ is reduced to Ph_3Bi. It also decomposes slowly in benzene and decomposition is slow enough to allow the isolation of a relatively pure compound [72JOM(36)323]. Photolysis of a solution of $Ph_3Bi{=}O$ in the presence of TiO_2 and triethylamine produces the parent bismuthine in moderate yield. The reduction of $Ph_3Bi{=}O$ proceeds more rapidly in DMF than in methanol [96JOC2895]. Soluble bismuthine oxides generated by direct oxidation of bismuthines with iodosobenzene are converted easily to corresponding carbonates on treatment with carbon dioxide. Adding hexane to a solution of $(4\text{-}MeC_6H_4)_3Bi{=}O$, an intractable powder is formed. On stirring this powder with triphenylphosphine for 3 days, the parent bismuthine and phosphine oxide are obtained [94TL8197].

$(4\text{-}MeC_6H_4)_3Bi{=}O$ reacts with benzoic anhydride or 4-methylbenzenesulfonic acid to give $(4\text{-}MeC_6H_4)_3Bi(O_2CPh)_2$ and $(4\text{-}MeC_6H_4)_3Bi(O_3SC_6H_4\text{-}4\text{-}Me)_2$ in 66 and 70% yields, respectively. By the action of $BF_3{\cdot}OEt_2$ on this oxide at low temperatures, tetrakis(4-methylphenyl)bismuthonium tetrafluoroborate is obtained in 24% yield. Triphenylbismuthine oxide reacts with 5,5-dimethylcyclohexane-1,3-dione (dimedone) to give a stabilized bismuthonium ylide [94TL8197, 89CC1749].

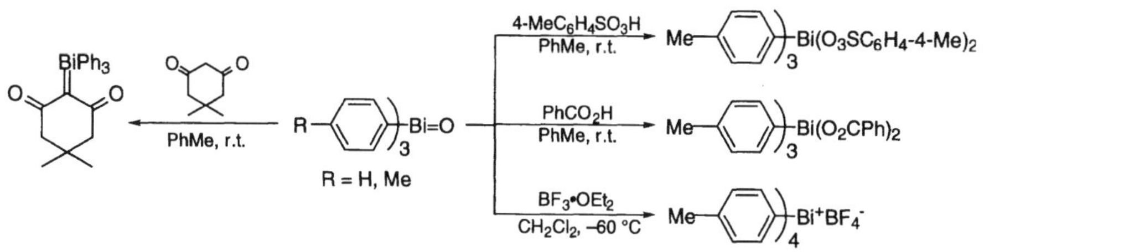

The most characteristic property of the bismuthine oxides is their oxidizing ability to convert activated alcohols, triphenylphosphine and hydrazobenzene to the corresponding ketones, oxide and azo compound, respectively. However, thioanisole and methyl octyl sulfide remain intact even after prolonged stirring with bismuthine oxide at ambient temperature [94TL8197] (Section 5.2.4).

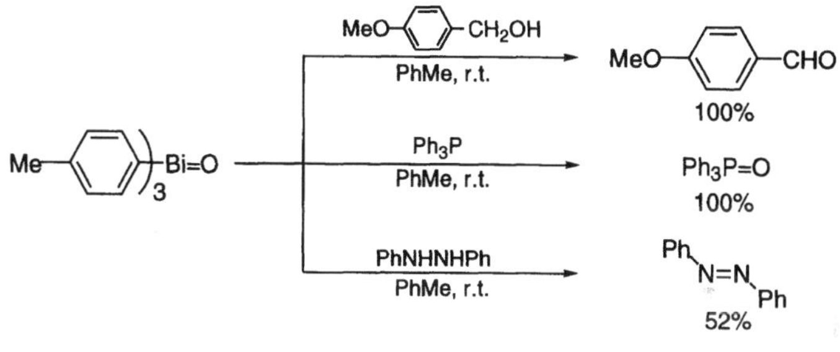

These oxidation reactions are performed under neutral conditions at ambient temperature, contrary to similar oxidations with organylantimony oxides which often need more severe conditions. The findings clearly show the difference in chemical behavior between bismuthine oxides and other organylpnicogen oxides. It is likely that the chemical nature of organylbismuthine oxides depends considerably on its preparative procedures. For some time now, a similar phenomenon has been noticed with triphenylantimony oxide that exists in a dimeric, oligomeric or polymeric form.

Chapter 4

Bismuth-Containing Heterocycles

Compared with acyclic organobismuth compounds described in the preceding chapters, the number of known ring compounds containing bismuth atom(s) is not so great. Bismuth-containing heterocycles can be classified into two groups according to ring constituting members. One is termed "bismacycles," which refers to the ring compounds containing bismuth and carbon atoms only. According to ring size, this group may be further divided into five-membered rings, six-membered rings, seven-membered rings and bicyclic compounds with bridgehead bismuth atom(s). The other group is termed "heterobismacycles," which refers to cyclic bismuth compounds containing additional hetero atom or atoms such as oxygen, sulfur or nitrogen as a ring member. Oxabismole, oxabismabenzene and azabismocin fall into this group.

4.1. BISMACYCLES

4.1.1. Five-membered rings (Table 4.1)

The bismuth analog of pyrrole is referred to as bismole. The parent 1*H*-bismole **1** is still unknown and expected to be thermally unstable due to high reactivity of the bismuth–hydrogen bond. Substituted bismoles are thermally stable and isolatable. 5-Phenyldibenzobismole **2**, stabilized by fusion with two benzene rings, as well as by the attachment of a phenyl group at the bismuth atom, is synthesized from 2,2′-dilithiobiphenyl [64CB789]. An alternative method using 2,2′-biphenylylcadmium has been reported [69JOM(17)389]. However, this method is less attractive since the organocadmium reagent needs to be prepared from 2,2′-dilithiobiphenyl.

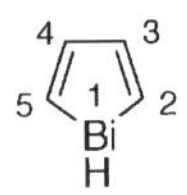

1

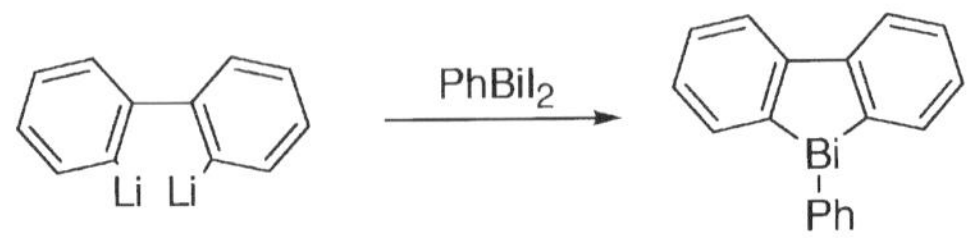

2

Norman et al. reported that the reaction between an equimolar bismuth chloride and 2,2′-dilithiobiphenyl·TMEDA affords not the expected BiCl-(biph) (biph; 2,2′-biphenylylene) **3** but $Bi_2(biph)_3$ **4** [96OM887]. The X-ray structure analysis of **4** shows the presence of some intermolecular interaction between the bismuth atoms separated by a distance of 4.7 Å. This molecule possesses an approximate C_2 symmetry and is therefore chiral. This is reflected by non-equivalence of the terminal biphenylylene groups observed in the ^{1}H- and ^{13}C-NMR spectra. The reaction of **4** with bismuth chloride gives **3**, but its isolation in a high state of purity is difficult because of the extensive contamination by bismuth chloride. Alternatively, treatment of this mixture with [PPN]Cl (PPN; $Ph_3P{=}N{=}PPh_3$) produced [PPN][$BiCl_2$(biph)] **5**. The X-ray structure analysis has revealed that this compound consists of a $[PPN]^+$ cation and a $[BiCl_2(biph)]^-$ anion. The bismuth atom adopts a trigonal bipyramidal structure, where two chlorine atoms occupy the apical positions. The equatorial plane is occupied by two carbon atoms and a lone pair of electrons. The Bi–Cl bond lengths are 2.702(2) and 2.743(2) Å.

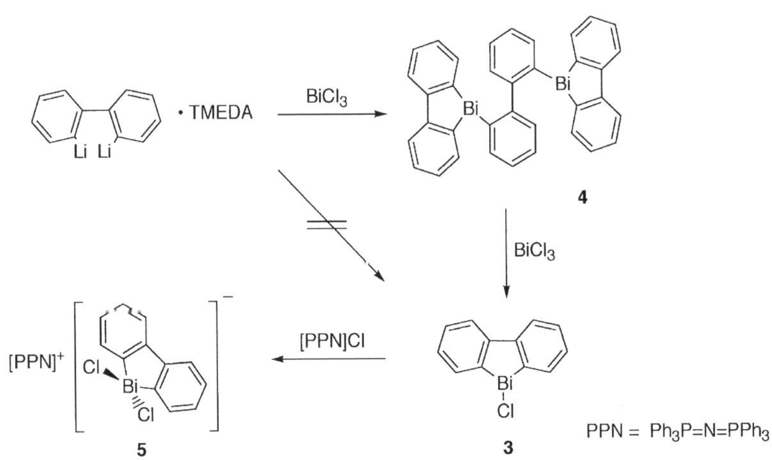

(a) Organocadmium route (Method A)

Synthesis of 5-iododibenzobismole

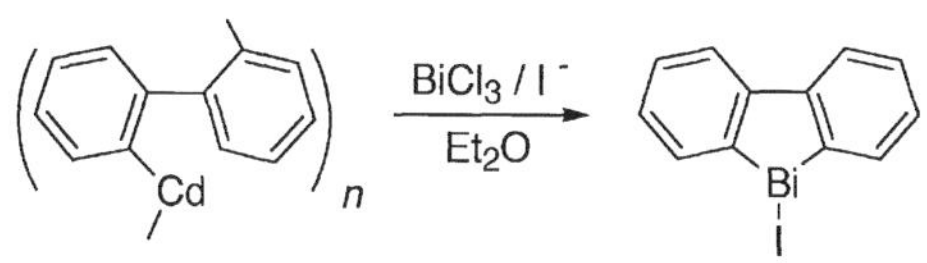

To a well-stirred solution of bismuth chloride (3.16 g, 10 mmol) in ether–THF (120 ml; 5:1) was added, at −70°C, a suspension of 2,2′-biphenylylcadmium (11 mmol) prepared from 2,2′-dilithiobiphenyl and cadmium iodide in ether–petroleum ether (2:1). An orange-red greasy substance was formed. After stirring overnight at room temperature, ethanol was added and the mixture was stirred for a further 1 h. The greasy substance changed to a yellow precipitate which was filtered off and the filtrate was concentrated to give a yellow precipitate. The combined precipitates were extracted with acetone (700 ml) and the extract was concentrated to leave crystals which were recrystallized from chloroform to give the product (3.1 g, 63%), m.p. > 250°C (decomp.). The reaction of this bismole with phenyllithium afforded compound **2** [69JOM(17)389].

Another type of bismole derivative is 5,5,5-triphenyldibenzobismole **6**, where the bismuth center is pentavalent [68LA(720)198]. Such a hypervalent bismuth compound is generally expressed by a ''10-Bi-5'' notation based on the number of total electrons in the outermost shell and bonds to the relevant heteroatom center. Compound **6** is colored orange red and dichroitic, and its structure has been clarified by Seppelt et al. [90CB761] (Section 3.4.2). The bismuth atom adopts a distorted square pyramidal structure in spite of the constraint from the 2,2′-biphenylylene ring. Compound **6** reacts with phenyllithium to form a 12-Bi-6 ate complex **7**, which dissociates at room temperature to give the original bismole and phenyllithium [68LA(720)198]. In general, pentaarylbismuth derivatives are less stable and readily undergo reductive elimination to form triarylbismuthines.

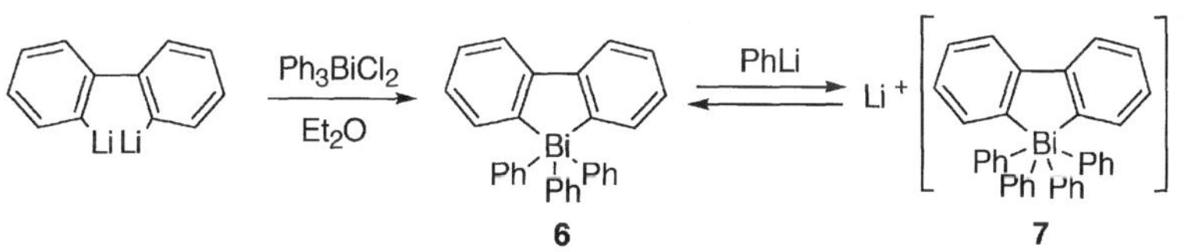

Recently, 1-phenylbenzobismole **9** was synthesized from dilithiated β-trimethylsilylstyrene via deprotection of the trimethylsilyl group of an intermediate **8** by TBAF [93CC1309]. This strategy has a broad applicability in the synthesis of *C*-unsubstituted benzometalloles of related structure. Further treatment of compound **9** with excess methyllithium leads to 1-methylbenzobismole **10** in good yield [94CPB1437].

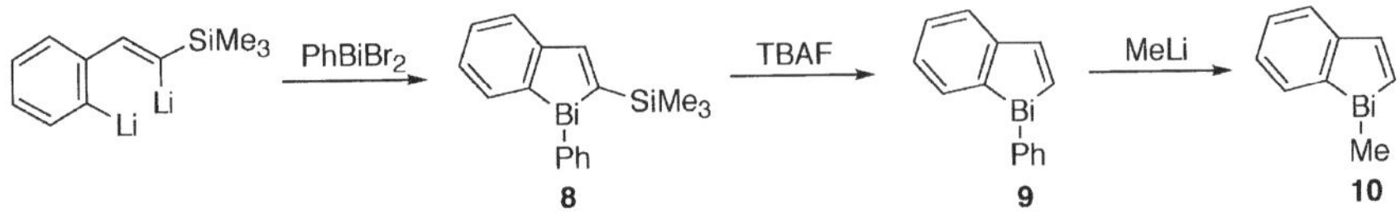

Ashe et al. succeeded in synthesizing 2,5-dimethyl-1-phenylbismole **11**, which is the least substituted bismole reported to date [84OM495, 92OM2743]. It is further converted into bibismole **12** [84OM495, 92OM2743], bismaferrocene **13** [92OM2743] and manganese carbonyl complexes **14** and **15** [93JOM(447)197].

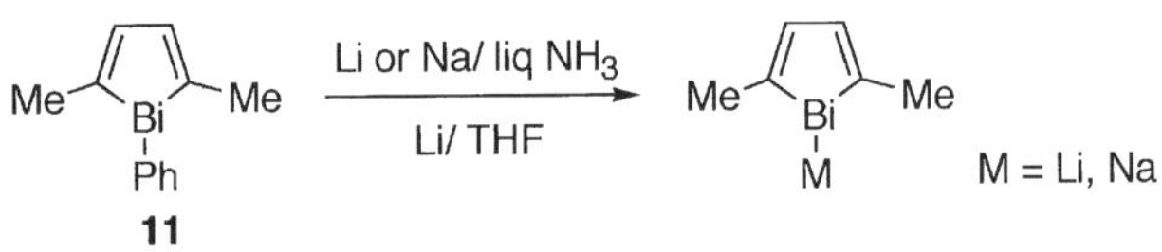

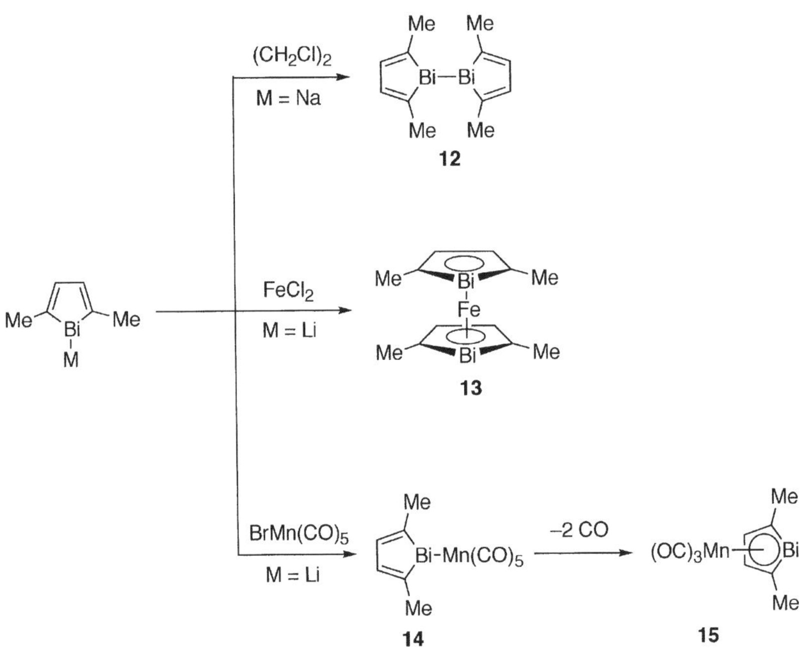

(b) Organolithium route (Method B)

Synthesis of 2,5-dimethyl-1-phenylbismole

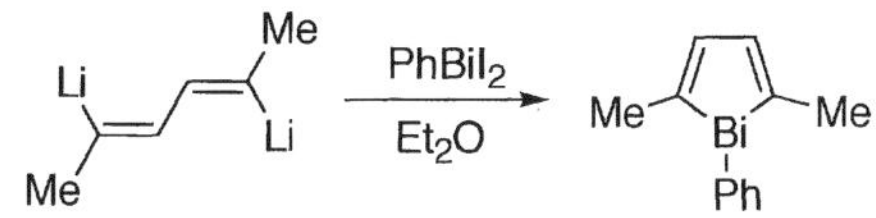

A hexane solution of 2.7 M butyllithium (2.8 ml, 7.56 mmol) was added dropwise over 10 min to a stirred solution of 2,5-diiodo-2,4-hexadiene (1.20 g, 3.60 mmol) in ether (20 ml) at −35°C. After 30 min, a precipitate was formed. The mixture was gradually warmed to 25°C and further stirred at this temperature for 2.5 h. The resulting solution of 2,5-dilithio-2,4-hexadiene was cannulated into a suspension of diiodophenylbismuthine (2.00 g, 3.71 mmol) in ether (80 ml) at −35°C. The red color was discharged over a period of 30 min, leaving a yellow-orange suspension which was stirred for an additional 3.5 h at −10°C. The solvent was evaporated in vacuo and the residue was extracted with pentane (50 ml) at 0°C. After filtration, the solvent was removed by evaporation to leave a yellow semi-solid residue. The extraction of the residue with pentane was repeated and with subsequent evaporation the combined extracts gave 2,5-dimethyl-1-phenylbismole **11** as a yellow viscous oil 1.0 g (80%) [92OM2743].

Fagan et al. have reported a more general method for the synthesis of main group heterocycles based on the transmetallation of zirconium metallacycles **16** and **18**, where 1-chloro-2,3,4,5-tetramethylbismole **17** and related bismoles are obtained in good yield [88JA2310]. This method is effective for the synthesis of substituted bismoles, especially in the case where the above dilithiobutadiene reagent is difficult to prepare. Zirconium metallacycles are readily obtained by the reduction of Cp_2ZrCl_2 with either Mg turnings or butyllithium in the presence of the corresponding alkyne [92OM3492, 92OM1491, 92JA372].

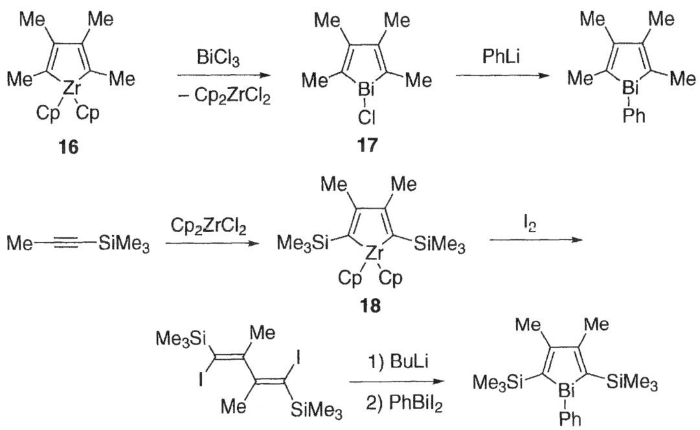

(c) Organozirconium route (Method C)

Synthesis of 1-bromo-3,4-dimethyl-2,5-bis(trimethylsilyl)bismole

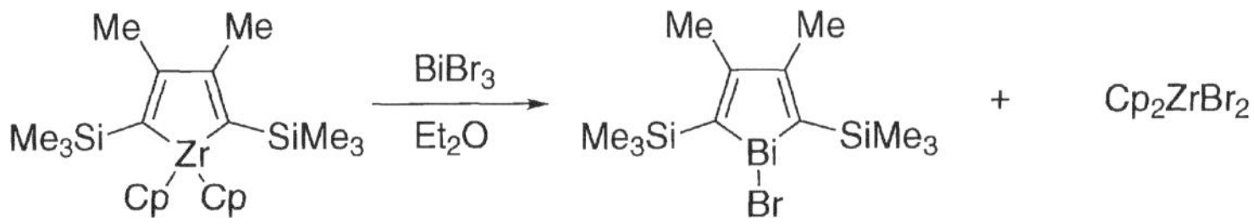

A zirconacycle was generated in situ by the sequential addition of butyllithium (24.8 ml, 1.61 M, 40 mmol) and 1-trimethylsilyl-1-propyne (7 ml, 47 mmol) to Cp_2ZrCl_2 (5.842 g, 20 mmol) in THF at −78°C and by subsequent heating of the resulting mixture to 50°C over 20 min. The volatiles were removed in vacuo and the residue was diluted with ether (60 ml) to give a red solution of zirconacycle and a white precipitate. To the reaction mixture was added a solution of bismuth bromide (8.974 g, 20 mmol) in ether (60 ml), and after 10 min the solution was cannulated from the precipitate. The ether was removed, the solid residue was washed with hexane (40 ml), and the product was again taken up in ether (160 ml). Cannula filtration to remove insoluble materials and cooling (−80°C) of the ether extract gave the product as an orange crystalline solid (4.536 g, 8.8 mmol, 44%). Recrystallization from dichloromethane gave analytically pure material, m.p. 206–208°C [92OM3492].

As illustrated above, the transmetallation of zirconium metallacycles constitutes one efficient method for the synthesis of bismoles. This is also the case with stannacycles, which undergo transmetallation with bismuth chloride to give the corresponding bismacycles [87OM1185]. Tetrahydrobismoles are synthesized from dihalobismuthine and the corresponding di-Grignard reagent [88ZN(B)739, 83OM1859]. On pyrolysis, 1,4-bis(dimethylbismuthio)butane gives 1-methylbismacyclopentane with concomitant elimination of trimethylbismuthine [88ZN(B)739]. These bismoles can be converted to bibismoles [87OM1185, 83OM1859] (Table 2.10).

(d) Organotin route (Method D)

Synthesis of 1-chloro-2,5-dimethylene-3,4-dihydrobismole

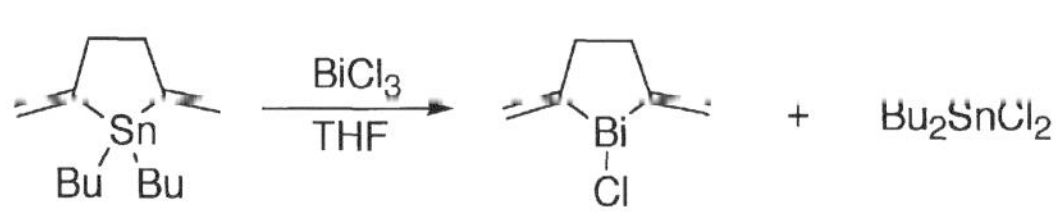

To a solution of bismuth chloride (0.2 g, 0.6 mmol) in dry THF (3 ml) was added 1,1-dibutyl-2,5-bismethylenestannolane (0.2 g, 0.6 mmol) at 0°C. The solvent was removed under vacuum at 0°C, and the residue was washed with pentane (10 ml) to remove dibutyltin dichloride. The insoluble solid was taken up in THF (10 ml) and filtered. Evaporation of the solvent at 0°C left the product (0.2 g, 99%) as yellow crystals which could be stored at 0°C, but decomposed at 25°C [87OM1185].

(e) Grignard route (Method E)

Synthesis of 1-methyltetrahydrobismole

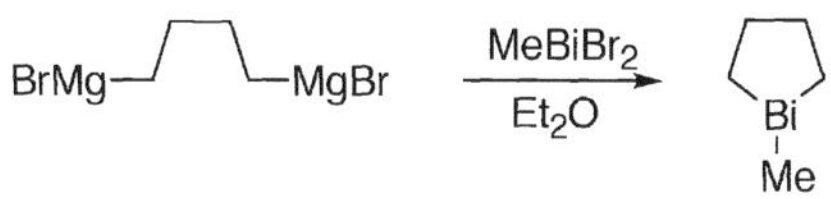

To the Grignard reagent solution prepared from magnesium (200 mmol), 1,4-dibromobutane (50 mmol) and ether was added after removal of excess magnesium, dibromomethylbismuthine (9.6 g, 25 mmol) in small portions under argon. The reaction proceeded exothermally. After stirring for 30 min at room temperature, the resulting mixture was put in a refrigerator at −20°C for 3–4 h. Excess Grignard reagent separated out as colorless crystals. The clear, colorless supernatant solution was decanted under an inert gas, and the solvent was carefully removed below 0°C to leave an air-sensitive oily liquid which was purified by vacuum distillation to give the product (1.47 g, 21%), b.p. 35°C/0.1 mmHg [88ZN(B)739].

(f) Pyrolysis route (Method F)

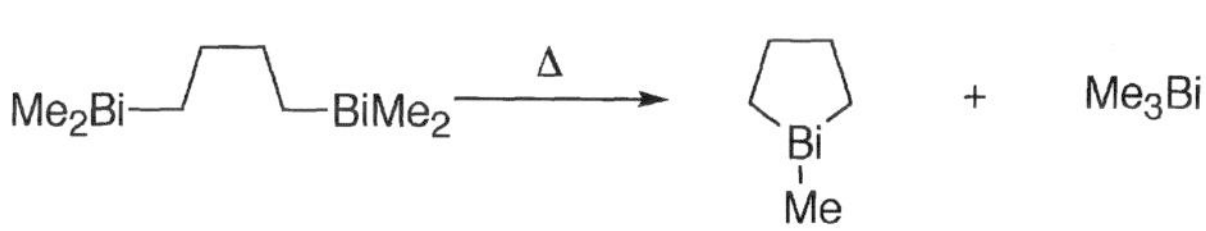

1,4-Bis(dimethylbismuthio)butane (2.08 g, 3.89 mmol) was kept at 70–90°C under vacuum conditions for 2–3 h and the volatile product (2.62 g) was collected into a cooled and dried flask. Trimethylbismuthine was removed from the distillate under vacuum conditions at 0°C to leave analytically pure 1-methyltetrahydrobismole (0.76 g, 35%) [88ZN(B)739].

4.1.2. Six-membered rings (Table 4.1)

The bismuth analog of pyridine is referred to as bismabenzene or bismin. Ashe et al. pioneered the chemistry of bismabenzene **19** [72JA7596, 76TL411, 76TL415, 82JA5693]. They were also successful in the synthesis, isolation and characterization of phosphabenzene **20** [71JA3293], arsabenzene **21** [71JA3293] and stibabenzene **22** [71JA6690]. These group 15 heterabenzenes are of considerable interest in the study of aromaticity. Compounds **20–22** are synthesized by dehydrohalogenation of **24** obtained from 1,1-dibutylstanna-

2,5-cyclohexadiene **23** and isolated by vacuum distillation. Phosphabenzene **20** is a colorless volatile liquid with a characteristic phosphine odor. It is air-sensitive but appears to be stable under an inert atmosphere. Arsabenzene **21** is a very air-sensitive liquid with a characteristic onion-like odor and a brief exposure to air makes it turn bright red. Stibabenzene **22** is a labile pale yellow-green oil. On standing at 25°C for a few minutes, it polymerizes to a resinous yellow solid that is insoluble in organic solvents. On exposure to air, it immediately forms an intractable yellow tar.

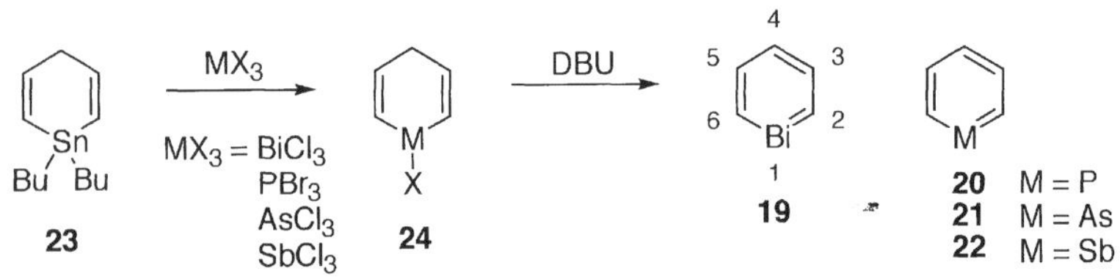

Bismabenzene **19** is generated in situ from a dihydrobismabenzene derivative, i.e. 1-chlorobisma-2,5-cyclohexadiene (**25**; R = H), by a loss of hydrogen chloride, but it is extremely labile and readily undergoes polymerization. The addition of 1 equiv. of DBU to a solution of **25** in THF at 25°C causes an immediate exothermic reaction, resulting in the formation of DBU–HCl and an intractable black precipitate [82JA5693]. On the other hand, the addition of DBU to a dilute solution of **25** at −78°C, followed by warming to 0°C, affords a golden red solution. A reaction mixture from **25** and DBU at −20°C shows a ^{1}H-NMR spectrum that is attributed to a bismabenzene dimer **26**. Warming the mixture above 0°C causes the peaks assigned to a monomer **19** to increase in intensity and become broader. The α, β and γ proton signals of **19** appear at δ 13.25 (*d*, $J = 10$ Hz), δ 9.86 (*t*, $J = 10$ Hz) and δ 7.70 (*m*), respectively. Eventually, the spectrum is lost with simultaneous formation of a black intractable tar. The introduction of a bulky alkyl group R into position 4 of **19** is effective to suppress polymerization [82JA5693]. Compound **19** (R = ^{t}Bu) is stable in solution for several hours at 0°C. Even on further cooling, no dimer of **19** can be detected. The ^{1}H-NMR spectra of group 15 heterabenzenes show the chemical shift values of α-protons in the very low field region. The α-proton (H2, H6) signals range from δ 8.6 for **20** to 13.25 for **19**, while the β- and γ-proton signals appear at lower field with reduced intensity.

The action of excess hexafluorobutyne on **19** gives a 1,4-Diels–Alder adduct **27**. The reactivity of group 15 heterabenzenes, except for pyridine as a diene, increases with the atomic number of heteroatoms. Thus, phosphabenzene **20** gives the corresponding adduct at 100°C, whereas arsabenzene **21** gives the corresponding adduct at 25°C [72JA7596]. Both stibabenzene **22** and bismabenzene **19** react rapidly at 0°C [86JA5692].

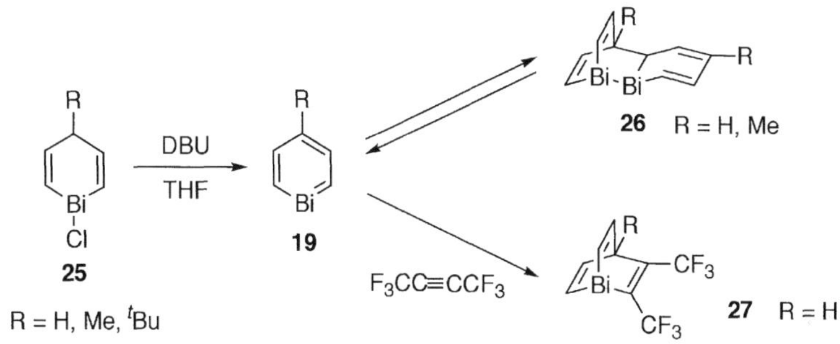

Dihydrobismabenzene (**25**; R = H) is obtained by the transmetallation of 1,1-dibutylstanna-2,5-cyclohexadiene **23** with bismuth chloride. A dibenzodihydrobismabenzene **29** stabilized by fusion with two benzene rings is synthesized from the corresponding di-Grignard reagent **28** and bismuth halide [92JCS(P1)1593].

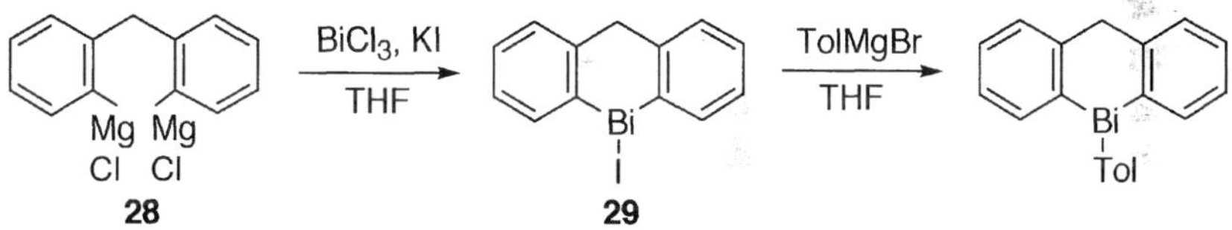

Hexahydrobismabenzene (bisminane; **30**) is obtained similarly by the reaction between the corresponding di-Grignard reagent and dihalobismuthine [15CB1473, 88ZN(B)739] or by thermolysis of 1,5-bis(dimethylbismuthio)-pentane $Me_2Bi(CH_2)_5BiMe_2$ [88ZN(B)739].

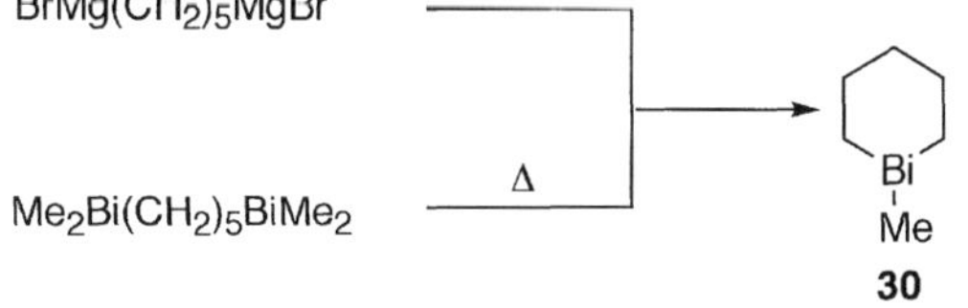

Synthesis of 4-*tert*-butylbismabenzene

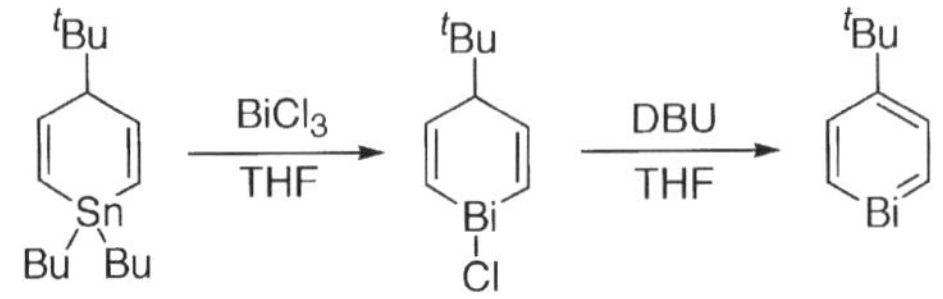

To a solution of bismuth chloride (0.5 g, 1.6 mmol) in THF (10 ml) was added a solution of 4-*tert*-butyl-1,1-dibutylstanna-2,5-cyclohexadiene (0.5 g, 1.5 mmol) in THF (10 ml) containing one drop of triethylamine. The mixture was stirred for 1 h at 25°C. The solvent was removed under reduced pressure. The residue was washed with pentane (50 ml × 3) and then dissolved in THF (20 ml) containing one drop of triethylamine. The solution was filtered and the solvent was removed under reduced pressure to leave a crude 4-*tert*-butyl-1-chlorobisma-2,5-cyclohexadiene as a golden yellow oil (0.5 g, 90%). This intermediate compound was very acid-sensitive and, therefore, used immediately without purification.

To a solution of this intermediate in THF-d_8 (3.0 ml) at −78°C was added DBU (1 ml, 6.5 mmol) via a syringe. On warming to −10°C for 10 min, the light yellow solution turned golden-red and a large amount of precipitate formed. Filtration gave 4-*tert*-butylbismabenzene as a clear golden-red solution, on which the ^{1}H- and ^{13}C-NMR measurements were made. The compound was destroyed above 45°C, and resulted in the formation of an intractable black tar. Removal of the solvent even at 0°C in vacuo led to a black tar [82JA5693].

4.1.3. Seven-membered rings (Table 4.1)

The bismuth analog of azepine is referred to as bismepine **31**. Several types of bismepine have been synthesized recently by Tsuchiya et al. The seven-membered ring system of these compounds is stabilized by fusion with benzene ring(s). Phenyl-substituted 1*H*-benzobismepine **32**, dibenzo[*b*,*d*]bismepine **33** and dibenzo[*b*,*f*]bismepine **34** are all synthesized from dibromophenylbismuthine and the corresponding dilithium reagent [93CC1817, 99CPB1108]. 3*H*-Benzobismepine (**35**; R = H) is detected at −20°C by ^{1}H-NMR but is extremely unstable thermally and cannot be isolated [96CC2183]. This compound gradually decomposes to naphthalene. The introduction of a trimethylsilyl group into the α position with respect to the bismuth atom brings kinetic stabilization. The half-lives of 1*H*- and 3*H*-benzobismepines and 2,4-bis(trimethylsilyl)-3*H*-benzobismepine, estimated from ^{1}H-NMR spectral analysis at 50°C in toluene, are 18, <1 and 34 min, respectively.

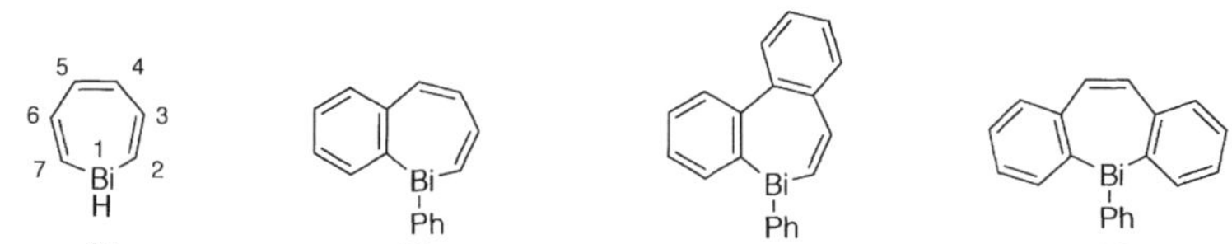

31 **32** **33** **34**

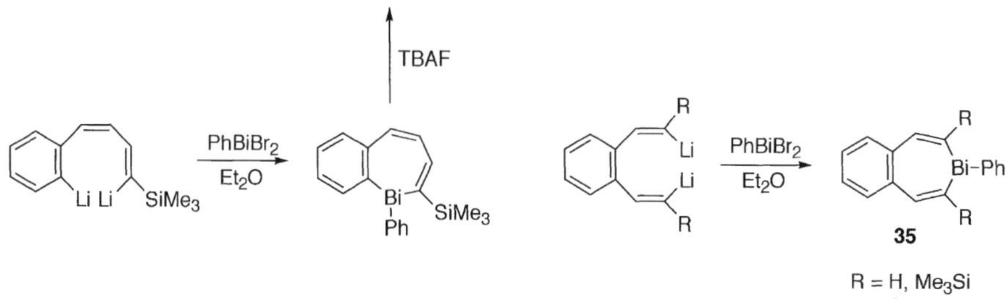

4.1.4. Bicyclic compounds with bridgehead bismuth atom or atoms (Table 4.1)

This type of cyclic bismuth compound is not common. As has been mentioned, bismabenzene is trapped by an internal alkyne to give a bridgehead compound **27**. Perfluorotriptycene **36** bearing two bismuth atoms at both bridgehead positions is obtained by heating 1,2-diiodotetrafluorobenzene and finely powdered bismuth in a sealed tube [89AOMC(3)459] or by the lithiation of $(C_6F_4Br)_3Bi$ and subsequent treatment with bismuth chloride [85JOM(287)57]. An analogous bicyclo compound **37** with a fused thiophene framework is also known [88AOMC(2)553]. However, the structures of these compounds are not well characterized yet.

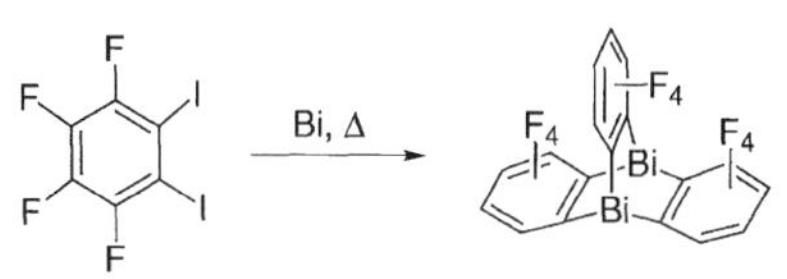

36

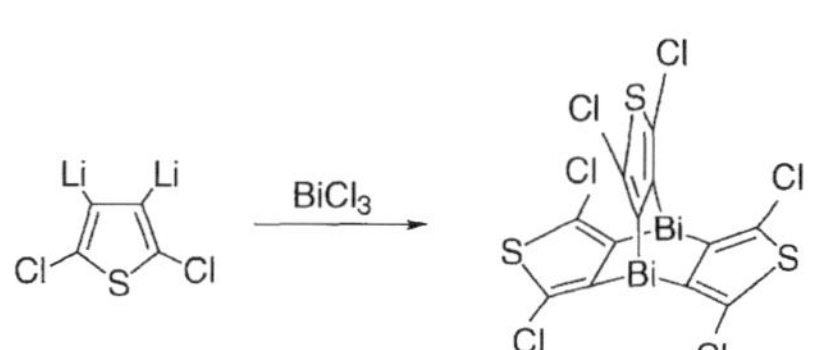

37

TABLE 4.1
Bismacycles and related compounds

Compound		Synthetic method[a]	Yield (%)	M.p. (°C)/b.p. (°C/mmHg)	Physical data	Reference
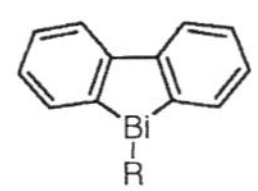	R = Ph	B	72	167–168	–	[64CB789]
		A	70	166–168	–	[69JOM(17)389]
		–	–	–	UV	[80ZFK506]
	R = Cl	B	–	–	–	[96OM887]
	R = I	A	63	250 (decomp.)	–	[69JOM(17)389]
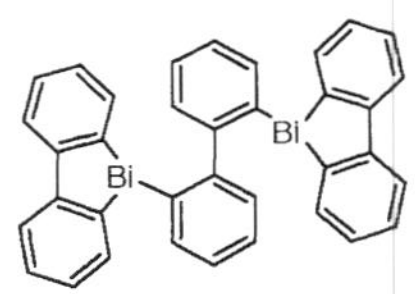		B	50	–	^{1}H-, ^{13}C-NMR, X-ray	[96OM887]
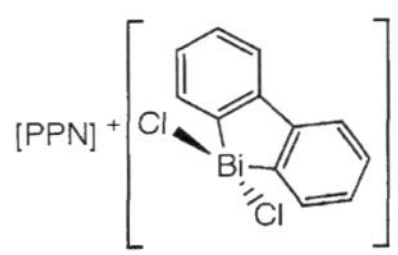	PPN = $Ph_3P{=}N{=}PPh_3$	B	70	–	^{1}H-, ^{13}C-NMR, X-ray	[96OM887]
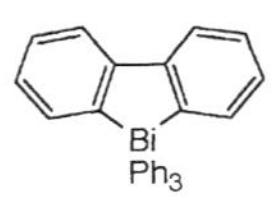		B	64	130–136	–	[68LA(720)198]
		–	–	–	X-ray	[90CB761]
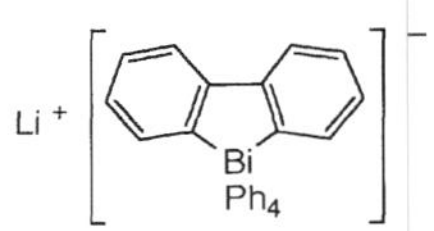		B	–	–	–	[68LA(720)198]

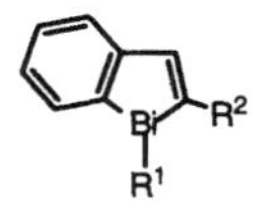	$R^1 = Ph, R^2 = Me_3Si$	B	78	73–75	–	[93CC1309]
		B	80	73–75	^{1}H-NMR, MS	[94CPB1437]
	$R^1 = Ph, R^2 = H$	B	>85	55–58	^{1}H-NMR	[93CC1309]
		B	82	55–58	^{1}H-NMR, MS	[94CPB1437]
	$R^1 = Me, R^2 = H$	B	87	–	^{1}H-NMR, MS	[94CPB1437]
		B	72	–	^{1}H-NMR, MS	[94CPB1437]
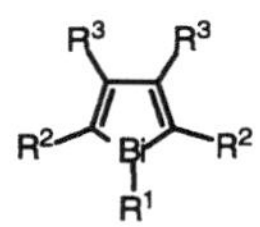	$R^1 = Ph, R^2 = Me,$	B	28	–	^{1}H-NMR, MS	[84OM495]
	$R^3 = H$	B	80	Oil	–	[92OM2743]
	$R^1 = Li, R^2 = Me,$	B	–	–	–	[92OM2743]
	$R^3 = H$	B	–	–	–	[93JOM(447)197]
	$R^1 = Na, R^2 = Me,$	B	–	–	–	[84OM495]
	$R^3 = H$	B	–	–	–	[92OM2743]
	$R^1 = Ph, R^2 = R^3 = Me$	C	95	–	–	[92OM3492]
		C	70	–	–	[88JA2310]
	$R^1 = Na, R^2 = R^3 = Me$	C	–	–	–	[92OM3492]
	$R^1 = Ph, R^2 = Me_3Si,$	B	–	–	^{1}H-, ^{13}C-NMR, MS	[92OM1491]
	$R^3 = H$	B	–	–	–	[92JA372]
	$R^1 = Br, R^2 = Me_3Si,$ $R^3 = Me$	C	44	206–208	^{1}H-, ^{13}C-NMR, IR	[92OM3492]
	$R^1 = Li, R^2 = Me_3Si,$	B	–	–	–	[92OM1491]
	$R^3 = Me$	B	–	–	–	[92JA372]
	$R^1 = Na, R^2 = Me_3Si,$ $R^3 = Me$	C	–	–	–	[92OM3492]
	$R^1 = Mn(CO)_3,$ $R^2 = Me, R^3 = H$	B	54	124–125	^{1}H-, ^{13}C-NMR, MS, X-ray	[93JOM(447)197]
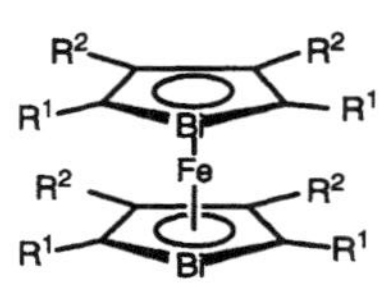	$R^1 = Me_3Si, R^2 = H$	B	37	124–125	^{1}H-, ^{13}C-NMR, MS, X-ray	[92OM2743]
	$R^1 = Me_3Si, R^2 = Me$	B	17	205 (decomp.)	^{1}H-NMR, MS	[92OM1491]

TABLE 4.1 (*continued*)

Compound		Synthetic method[a]	Yield (%)	M.p. (°C)/b.p. (°C/mmHg)	Physical data	Reference
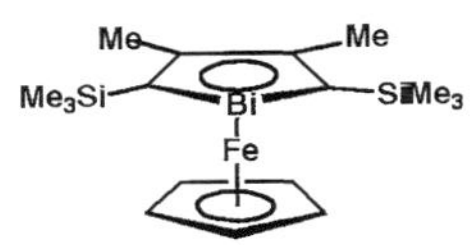		B	25	190	^{1}H-, ^{13}C-NMR, MS, UV	[92OM1491]
		B	25	–	X-ray	[92JA372]
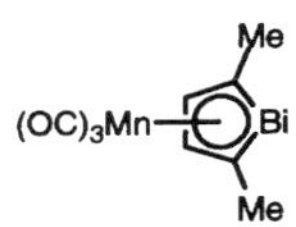		B	90	78–80	^{1}H-, ^{13}C-NMR, MS	[93JOM(447)197]
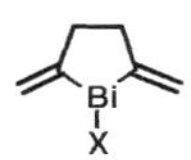	R = Cl	D	99	25 (decomp.)	^{1}H-NMR, MS	[87OM1185]
	R = Na	D	–	–	–	[87OM1185]
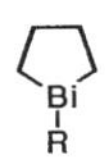	R = Me	E	21	35/0.1	^{1}H-NMR, MS	[88ZN(B)739]
		F	35	35/0.1	^{1}H-NMR, MS	[88ZN(B)739]
	R = Ph	E	49	78–80/0.1	^{1}H-NMR, IR, MS	[83OM1859]
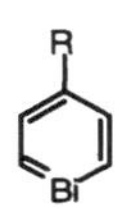	R = H	D	–	–	–	[72JA7596]
		D	–	–	–	[76TL411]
		D	–	–	^{1}H-NMR	[76TL415]
		D	–	–	^{1}H-NMR, MS	[82JA5693]
	R = Me	D	–	–	^{1}H-, ^{13}C-NMR	[82JA5693]
	R = ^{t}Bu	D	–	–	^{1}H-, ^{13}C-NMR, MS	[82JA5693]

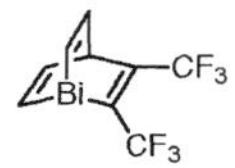		D	–	–	^{1}H-NMR	[72JA7596]
		D	–	–	^{1}H-NMR	[82JA5693]
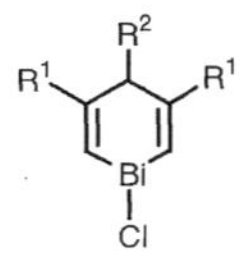	$R^1 = R^2 = H$	D	75	144–145 (decomp.)	^{1}H-NMR, MS	[72JA7596]
		D	–	–	–	[76TL415]
		D	72	144–145 (decomp.)	^{1}H-NMR, MS	[82JA5693]
	$R^1 = D, R^2 = H$	D	–	–	^{1}H-NMR	[82JA5693]
	$R^1 = H, R^2 = Me$	D	49	100 (decomp.)	^{1}H-NMR	[82JA5693]
	$R^1 = H, R^2 = {}^tBu$	D	90	Oil	^{1}H-NMR	[82JA5693]
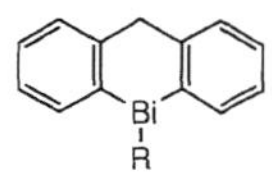	R = I	E	12	195–200 (decomp.)	^{1}H-NMR, IR, MS	[92JCS(P1)1593]
	R = Tol	E	97	98–101	^{1}H-NMR, IR, MS	[92JCS(P1)1593]
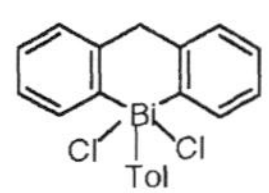		E	91	167–171 (decomp.)	^{1}H-NMR, IR, MS	[92JCS(P1)1593]
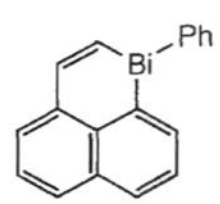						
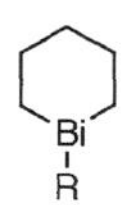		B	–	–	^{1}H-NMR	[UR3]

	R = Me	E	23	55/0.1	^{1}H-NMR, MS	[88ZN(B)739]
		F	66	55/0.1	^{1}H-NMR, MS	[88ZN(B)739]
	R = Et	E	–	108–112/18–20 (inflammable)		[15CB1473]
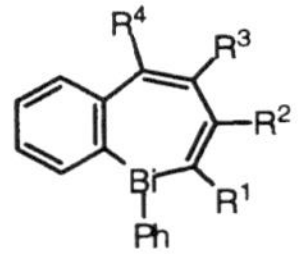	$R^1 = Me_3Si$, $R^2 = R^3 = R^4 = H$	B	40	87–88.5	–	[93CC1817] [99CPB1108]
	$R^1 = R^2 = R^3 = R^4 = H$	B	71	Oil	^{1}H-NMR	[93CC1817] [99CPB1108]
	$R^1 = R^2 = H$, $R^3, R^4 = (CH{=}CH)_2$	B	45–55	98–110 (decomp.)	^{1}H-NMR	[93CC1817]
	$R^1 = R^2 = (CH{=}CH)_2$, $R^3 = R^4 = H$	B	60–70	168–170	–	[93CC1817]
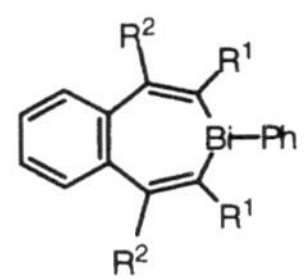	$R^1 = R^2 = H$	B	15–40	–	^{1}H-NMR	[96CC2183]
	$R^1 = Me_3Si$, $R^2 = H$	B	15	Oil	^{1}H-NMR	[96CC2183]
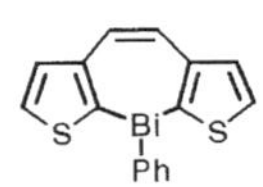		B	–	–	–	[97H1899]
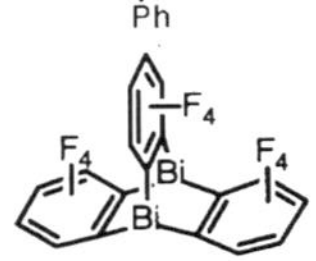		B	–	314	^{13}C-, ^{19}F-NMR, IR, MS	[85JOM(287)57]
		–	20	–	MS	[89AOMC(3)459]
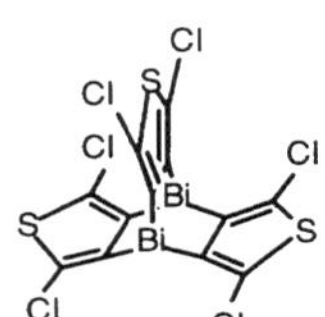		–	–	322–323 (decomp.)	MS	[88AOMC(2)553]

4.2. HETEROBISMACYCLES

4.2.1. Oxabismoles and derivatives (Table 4.2)

Oxabismoles refer to a type of bismuth alkoxide in which the bismuth–oxygen bond is incorporated into the bismole ring system. Most known oxabismoles are benzoxabismoles as reported by Akiba et al., the unique ring system being stabilized by the Martin ligand. The general synthetic pathway to benzoxabismoles and their derivatives is shown below [93OM1857].

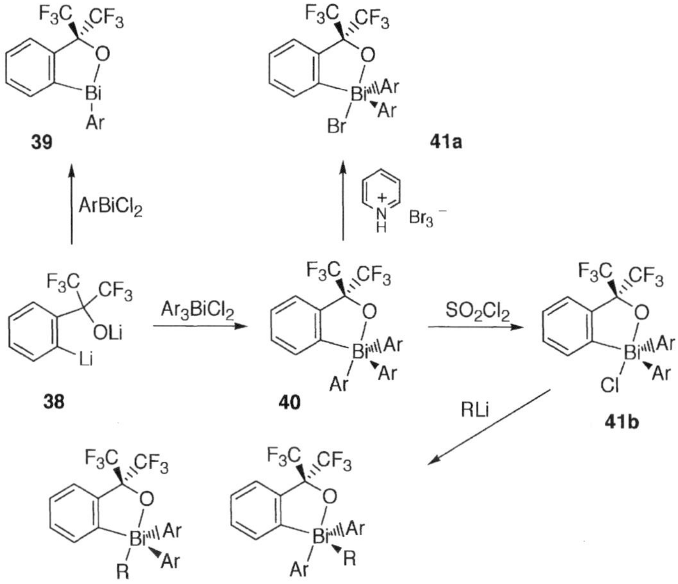

Tricoordinated benzoxabismoles **39** are obtained from the dilithiated bis(trifluoromethyl)benzyl alcohol **38** and aryldichlorobismuthine or bismuth chloride [92JA7906]. The use of triaryldichlorobismuth in place of these halides leads to pentacoordinated benzoxabismoles (**40**; 10-Bi-5) [88TL3817]. This type of compound undergoes partial halogenolysis with sulfuryl chloride and pyridinium bromide perbromide, respectively, to give the corresponding halobenzoxabismoles **41a,b** [93OM1857]. The X-ray structure studies of these pentacoordinated benzoxabismoles show that the bismuth centers possess a distorted trigonal bipyramidal geometry. Treatment of chlorobenzoxabismoles **41b** with organolithium reagent gives the substitution product, where the disposition of substituents at the bismuth atom depends on their nature [93OM1857]. Thus, when chlorobenzoxabismole **41b** (Ar = Tol) is allowed to react with 4-$CF_3C_6H_4Li$, an apical position of the bismuth center is occupied by a 4-$CF_3C_6H_4$ group in the product. On the other hand, the reaction of chlorobenzoxabismole **41b** (Ar = 4-$CF_3C_6H_4$) with TolLi directs the 4-$CF_3C_6H_4$ group from the equatorial to the apical position. These trends are consistent with the apicophilicity of electronegative substituents.

Halogenolysis of pentacoordinated benzoxabismoles substituted differently by sulfuryl chloride discloses the following reactivity order of the bismuth–carbon bonds: PhC≡C–Bi > Me–Bi > Tol–Bi > 4-$CF_3C_6H_4$–Bi [93OM1857].

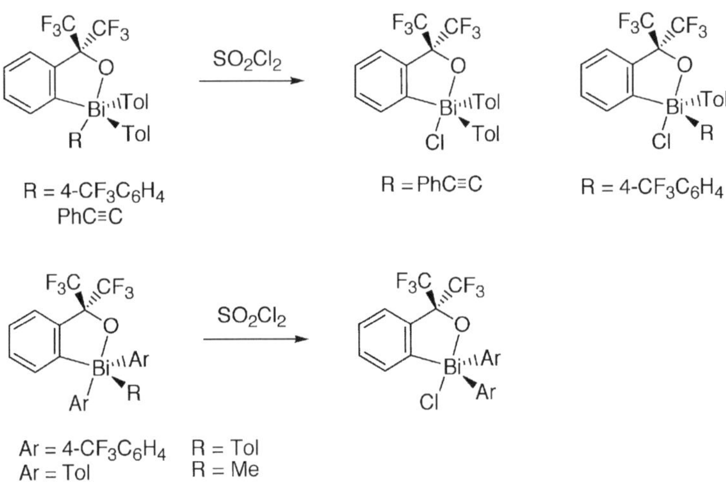

The structure of a trivalent benzoxabismole **42** bearing the intramolecular coordination between the dimethylamino group and the bismuth center has been established by X-ray crystallography [92JA7906, 95JA3922]. The bismuth atom has a distorted pseudotrigonal bipyramidal configuration, where the apical positions are occupied by the oxygen and nitrogen atoms. The equatorial plane is occupied by two carbon atoms and a lone pair of electrons. The intramolecular Bi–N distance of 2.63 Å is much shorter than the sum of the van der Waals radii (ca. 3.74 Å), indicating the formation of a hypervalent three-center-four-electron bond over the oxygen, bismuth and nitrogen atoms, i.e. of a ''10-Bi-4'' type compound. The solution studies by variable temperature dynamic ^{19}F-NMR spectra have demonstrated that the inversion at the bismuth center takes place via the edge inversion process. The intramolecular coordination of the dimethylamino group and the intermolecular coordination of nucleophilic solvents, such as DMSO or pyridine, with the bismuth atom contribute greatly to the stabilization of the transition state of the edge inversion.

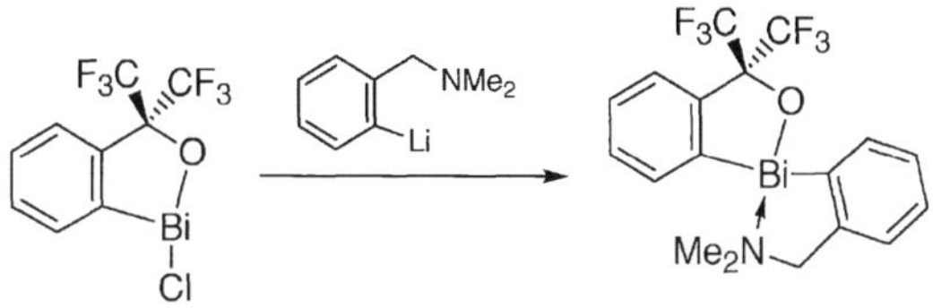

42

The structure of a hexacoordinated benzoxabismole **43** (12-Bi-6) bearing a similar coordination to the dimethylamino group has been characterized [93OM3297]. Unlike the analogous 10-Bi-4 benzoxabismole **42**, the electronegative oxygen and nitrogen atoms are in *cis* geometry. The intramolecular Bi–N distance is 2.85 Å, which is longer than that of **42** (2.63 Å).

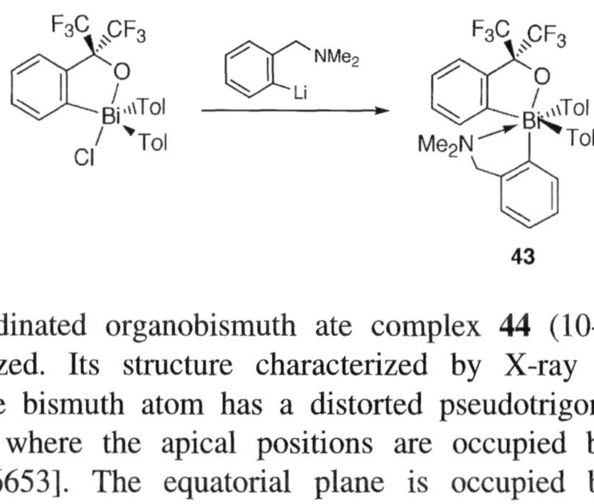

43

A tetracoordinated organobismuth ate complex **44** (10-Bi-4) has also been synthesized. Its structure characterized by X-ray crystallography shows that the bismuth atom has a distorted pseudotrigonal bipyramidal configuration, where the apical positions are occupied by two oxygen atoms [92TL6653]. The equatorial plane is occupied by two carbon atoms and a lone pair of electrons. The apical Bi–O bond lengths (2.273 and 2.306 Å) are remarkably shorter than the sum of the van der Waals radii (3.74 Å) and approximate to that of the covalent radii (2.18 Å).

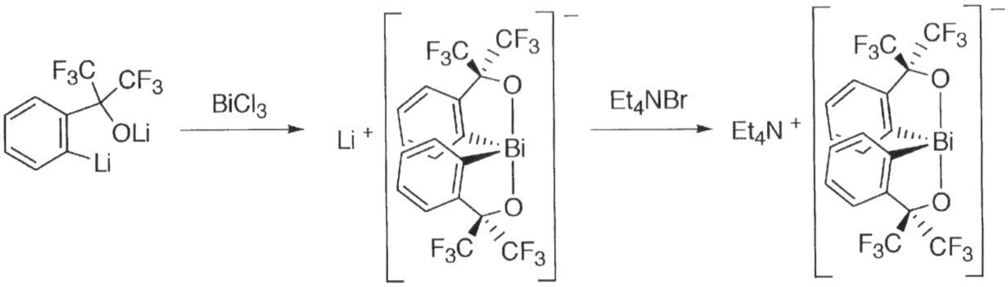

44

The reaction of potassium salt of the ate complex with sulfuryl chloride affords the pentacoordinated chlorobismuthorane **45** (10-Bi-5) almost quantitatively.

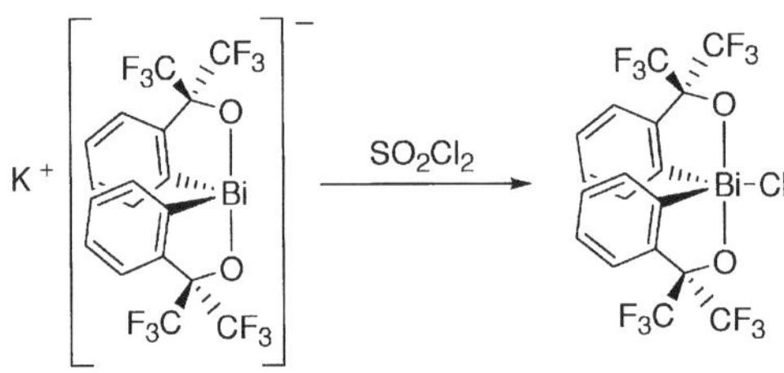

45

Hexacoordinated 12-Bi-6 ate complexes **46** have been prepared in situ by the reaction of a pentacoordinated benzoxabismole with alkyllithiums [93OM3297, 89TL953]. The formation of these complexes is confirmed by ^{19}F-NMR spectra. When methyllithium is used, protonolysis of the ate complex leads to a methyl-substituted product **47**, which is not so stable as the corresponding hexacoordinated antimonate complex and gradually decomposes at room temperature. In contrast, the reactions with ethyl- and butyllithiums do not give rise to the expected pentacoordinated benzoxabismoles even when the reaction is quenched at 78°C after 1 h. Instead, the tricoordinated benzoxabismoles **48** are obtained. This result suggests that the hexacoordinated complexes **46** initially formed are unstable and readily undergo reductive elimination to give an unstable tetracoordinated ate complex, which affords the corresponding tricoordinated benzoxabismole **48** by quenching with water.

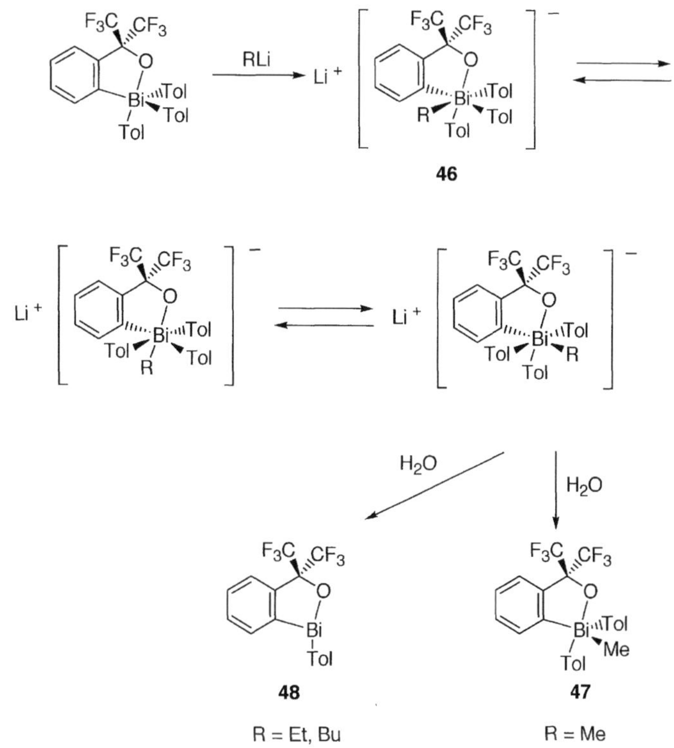

The introduction of two Martin ligands onto the bismuth center enhances considerably the stability of hexacoordinated ate complexes [89TL953, 93OM3297]. A thermally stable spirocyclic 12-Bi-6 ate complex **49** is obtained and its dynamic behavior in solution has been characterized by variable temperature ^{19}F-NMR spectra.

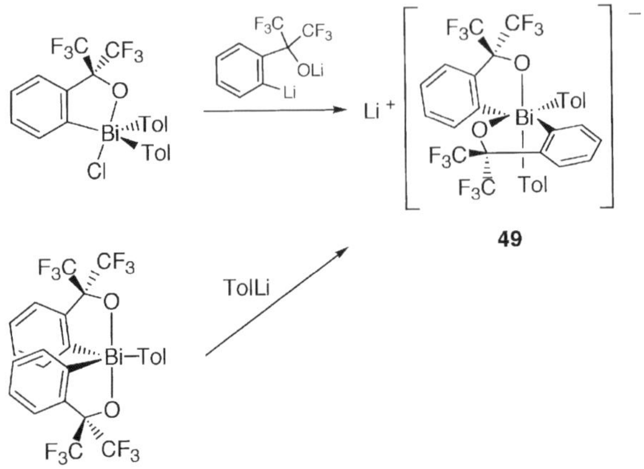

49

Oxabismole derivatives without being stabilized by the Martin ligand have been studied very little. A few examples include a cyclic bismuth carboxylate **50** which is obtained by the interaction between pentaphenylbismuth and carbon dioxide. Thermolysis of this compound leads to triphenylbismuthine, carbon dioxide and benzyne [86ZOB330].

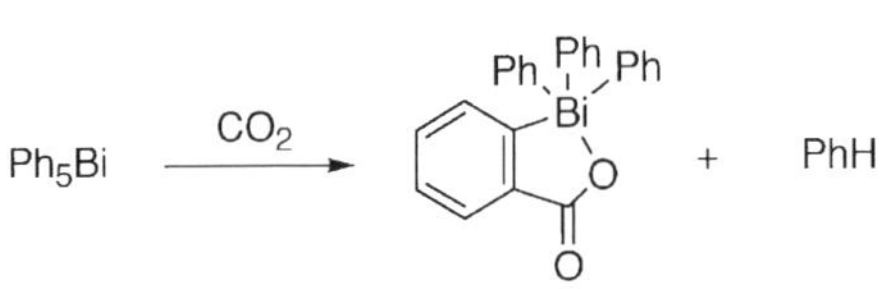

50

4.2.2. Oxabismabenzene and related compounds (Table 4.2)

Oxabismabenzene refers to dihydrobismabenzene whose methylene group is replaced with an oxygen atom. Dibenzoxabismabenzene (or phenoxabismine) **51** and related compounds **52** and **53** have been synthesized [92JCS(P1)1593]. The reaction of these compounds with benzoyl chloride in the presence of $Pd(PPh_3)_4$ affords the corresponding chlorobismuthines along with 4-methylbenzophenone. When X is SO_2, chlorobismuthine **52e** is obtained as stable crystals under ambient conditions. Although aliphatic carbon–bismuth bonds are generally known to be labile, compounds **52b–d** and **53b–d** are highly stabilized. Thus, 10-alkynyl, 10-alkenyl and 10-alkyl derivatives are all obtained in a stable crystalline form. When X is O or CH_2, the corresponding derivatives are labile under ambient conditions. The X-ray structure analysis of **52b** suggests that the transannular interaction between the bismuth and the sulfonyl oxygen atoms is responsible for the stabilization of labile chlorine–bismuth and aliphatic carbon–bismuth bonds in these systems.

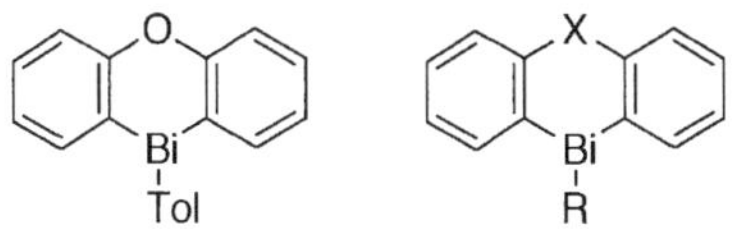

51

52: X = SO_2
53: X = S

a: R = Tol
b: R = 4-$ClC_6H_4C{\equiv}C$
c: R = $Ph_2C{=}CH$
d: R = Me
e: R = Cl

When a pentavalent bismuth dichloride **54** is refluxed in dichloromethane, reductive elimination takes place smoothly to give chlorobismuthine (**52e**) quantitatively together with 4-chlorotoluene. This process is considered to be facilitated by the transannular effect.

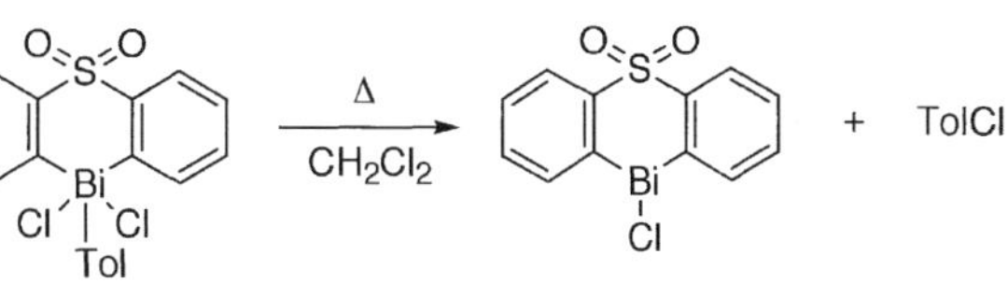

54 **52e**

4.2.3. Azabismocins (Table 4.2)

This class of compound is of interest from the viewpoint of the transannular Bi–N interaction. The reaction of bis(2-lithiobenzyl)methylamine with bismuth chloride affords 12-chlorodibenzo[*c*,*f*][1,5]azabismocin **55** (X = Cl) [89TL4841]. The presence of a 1,5-transannular Bi–N interaction is demonstrated spectroscopically; ^{13}C-NMR spectra show that the chemical shift of the N–Me carbon atom moves downfield as the electronegativity of X increases. Non-fused 5-halo-1,5-azabismocin **56** and 5-bromo-1,5-thiabismocin **57** are synthesized similarly from bismuth chloride and the corresponding di-Grignard reagent [94POL365].

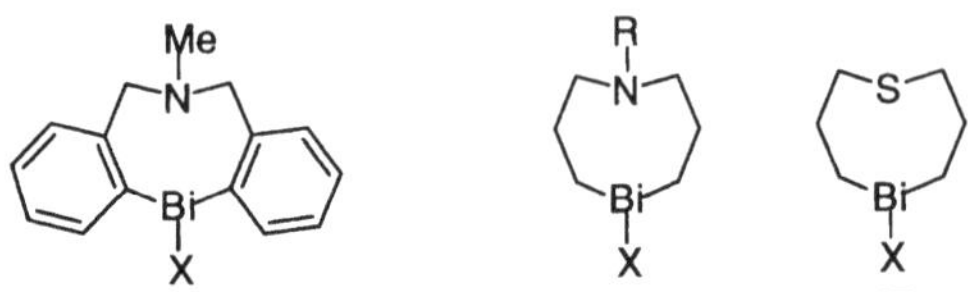

55 **56** **57**

X = F, Cl, I, Me, Ph (55); R = Me, $PhCH_2$, X = Cl, Br, I (56, 57)

4.2.4. *Miscellaneous bismuth-containing heterocycles (Table 4.2)*

This section summarizes miscellaneous heterobismacycles which contain more than two hetero atoms in one ring system. Most of such compounds are heterobismacycles whose bismuth atom is chelated by two chalcogen atoms.

4.2.4.1. Heterobismacycles containing a chalcogen–bismuth–chalcogen bond in the ring system

Triphenylbismuthine undergoes cleavage of the bismuth–carbon bond by protonolysis with dicarboxylic acid to give a cyclic bismuth dicarboxylate and benzene [80ZN(B)1000, 81ZN(B)70] (Sections 2.2.3.1 and 2.2.3.2.2).

(g) Acidolysis route (Method G)

Synthesis of phenylbismuth oxalate

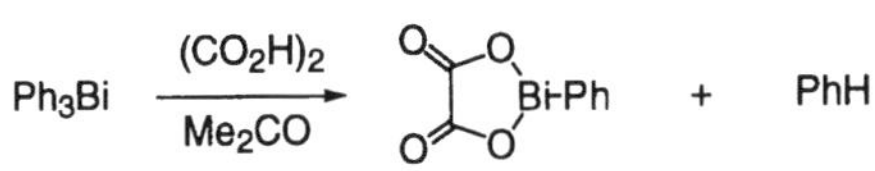

To a solution of oxalic acid (0.61 g, 6.8 mmol) in acetone (50 ml) was added a solution of triphenylbismuthine (3 g, 6.8 mmol) in the same solvent (150 ml). After refluxing at 60°C for 2 h, the product deposited as a colorless solid. It was washed repeatedly with acetone and dried (m.p. > 300°C, decomp.) [80ZN(B)1000].

Dibromomethylbismuthine reacts with lithium salt of 1,2-benzenediselenol to give the corresponding diselenabismacycle **58** [87POL1593] (Section 2.2.3.2.4).

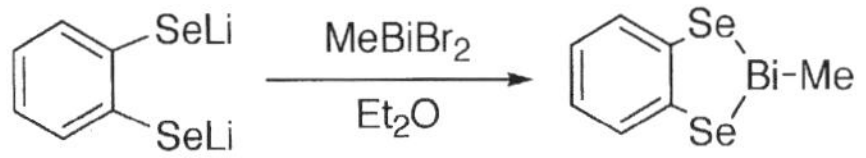

58

Dialkoxybismuthines undergo alcoholysis with 1,2-diols or 1,2-dithiols in ethanol to give the corresponding heterobismacycles [76ZAAC(423)40, 76ZAAC(423)47, 85JOM(290)133] (Sections 2.2.3.1 and 2.2.3.2.3).

(h) Alcoholysis route (Method H)

Synthesis of phenylbismuth ethanedithiolate

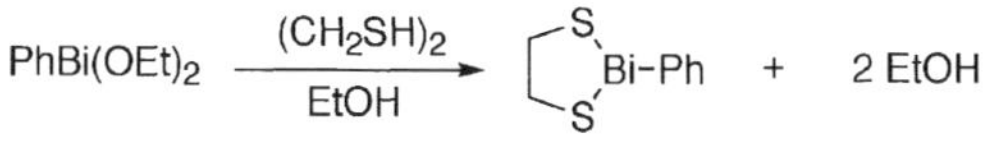

To a suspension of diethoxyphenylbismuthine, prepared from dibromophenylbismuthine (4.46 g, 10 mmol) and sodium ethoxide (20 mmol) in absolute ethanol, was added dropwise at −30 to −40°C a solution of 1,2-ethanedithiol (10 mmol) in the same solvent. The temperature was gradually raised to 0°C and the resulting yellow precipitate was filtered and dried under vacuum. The crude product was purified by recrystallization from 1,2-ethanedithiol (3.13 g, 83%) [76ZAAC(423)47].

Bismuth chloride undergoes alcoholysis with dithiols in chloroform or benzene to give the corresponding chlorodithiabismacycles **59** and **60** [74ZAAC(407)35, 74ZAAC(406)74]. Similar alcoholysis in aqueous media has also been reported. In concentrated hydrochloric or hydrobromic acids, bismuth chloride, bismuth oxycarbonate and bismuth oxychloride all undergo alcoholysis with 1,2-ethanedithiol to afford the corresponding halodithiabismacycles [68JCS(A)2587, 96JA3225] (Section 2.2.4.4).

The reaction of $Bi(OAc)_3$ with catechols in acetic acid gives bismuth catecholate **61** in good yield [90HAC233]. These compounds are insoluble in common organic solvents including DMSO and DMF; their ^{1}H-NMR spectra are measured in CF_3CO_2H. The IR spectrum of **61a** show absorption bands at 1480 and 1250 cm^{-1}, suggesting the bidentate coordination of the catechol ligand. In addition, the absorption due to the C=O bond appears at 1570 and 1448 cm^{-1}, suggesting that the bidentate acetate ligand forms a polymeric structure. The mass spectrum of this compound exhibits characteristic peaks due to M^+, $[M–OAc]^+$ and $[M–catecholate]^+$ fragments. The spectral features of **61b** and **61c** are very similar; the IR spectra show peaks at 1485 ν(CO), 1258 and 550 cm^{-1} ν(Bi–O) for the former compound, and 1410 ν(CO), 1242 and 582 cm^{-1} ν(Bi–O) for the latter [90HAC233].

Bismuth carbonate reacts with pyrocatechol in boiling aqueous sodium carbonate to give an ate complex (**62**; M = Na) [25CB891]. The use of pyridine, ammonia or potassium carbonate in place of sodium carbonate produces the corresponding ate complex.

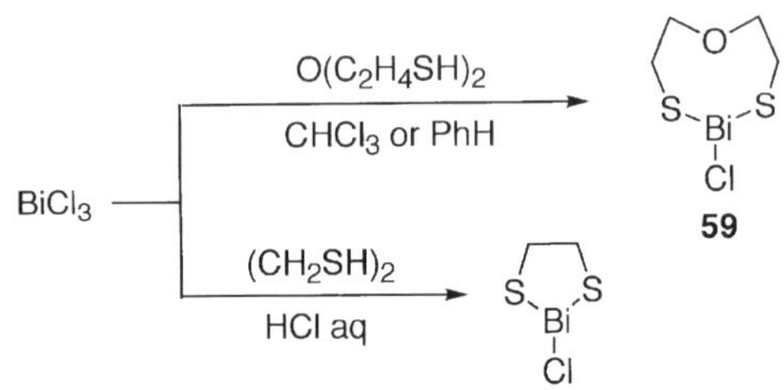

59 **60**

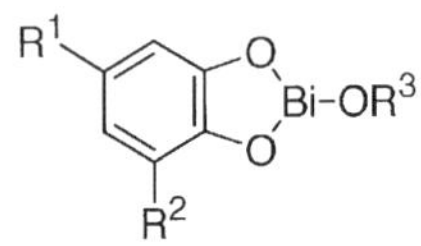

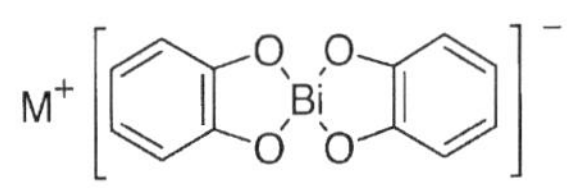

62

61 **a**; $R^1 = R^2 = H$, R^3 = COMe
b; $R^1 = {}^t$Bu, $R^2 = R^3 = H$
c; $R^1 = R^2 = {}^t$Bu, $R^3 = H$

A variety of five-membered heterobisma ring systems as shown below have been patented. However, no detailed evidence is given for the products and, therefore, their chemical identities remain to be confirmed [54BRP712828, 54BRP716647].

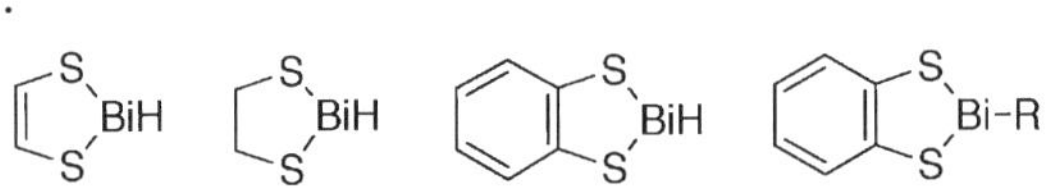

4.2.4.2. Heterobismacycles with no carbon atom ring (Table 4.2)

Tokitoh et al. have obtained a triselenatribismane **63** from lithium selenide and an aryldichlorobismuthine bearing a bulky 2,4,6-tris[bis(trimethylsilyl)-methyl]phenyl (Tbt) group. Compound **63** can be converted into a stable dibismuthene **64**, which is the first compound containing the bismuth–bismuth double bond [97SCI(277)78] (Section 2.2.2.4). Dibismuthene **64** slowly reacts with molecular oxygen in the solid state to give 1,3,2,4-dioxadibismuthene derivative **65**. A similar phenomenon is also observed in the case of antimony [98JA433].

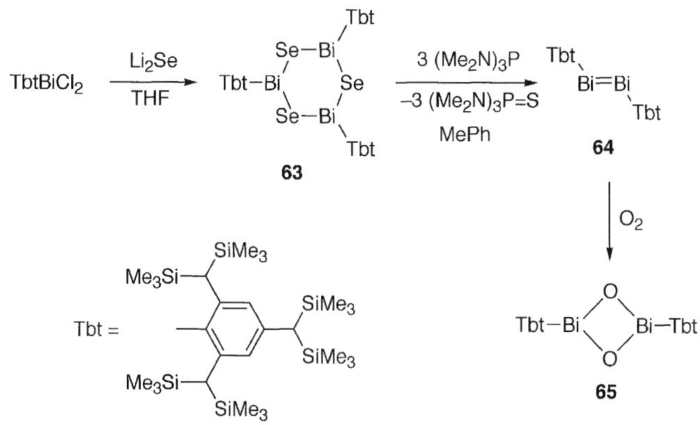

A series of bis(amino)pnictogen chlorides **66** have been synthesized [88ZAAC(557)7, 78ZN(B)1, 78ZN(B)7]. They react with Lewis acids such as $AlCl_3$, $GaCl_3$ and $InCl_3$ to form the corresponding adducts **67**, bearing a formally divalent pnictogen atom through the transfer of chloride anion to the Lewis acid [88ZAAC(559)73].

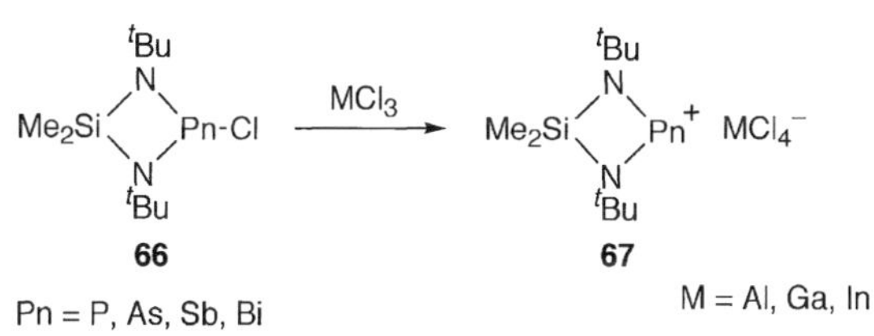

The loss of an electron at a pnictogen atom may be compensated by two different bonding mechanisms. In the case of a phosphorus derivative (Pn = P; M = Al), the electronic balance is accomplished by intramolecular backbonding from the neighboring nitrogen atoms. On the other hand, the antimony and bismuth derivatives (Pn = Sb, Bi; M = Al) involve the intermolecular coordination of a chlorine atom of the $AlCl_4$ anion with the unsaturated pnictogen atom.

Some bridgehead bismuth compounds containing the germanium or silicon atoms as three-dimensional bridgehead components are also known [80JOM(200)243, 91OM812].

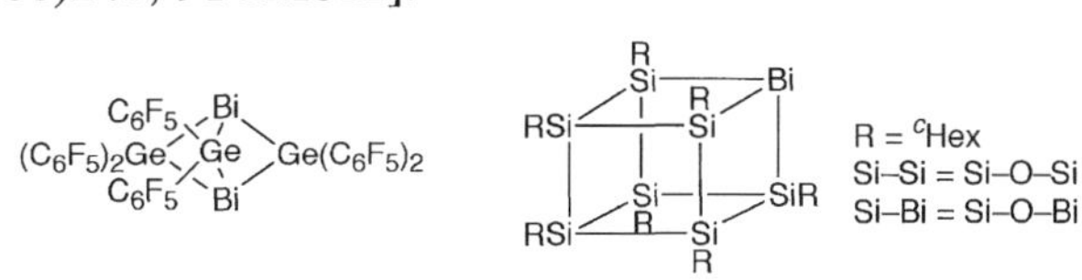

TABLE 4.2
Heterobismacycles and related compounds

Compound		Synthetic method[a]	Yield (%)	M.p. (°C)	Physical data	Reference
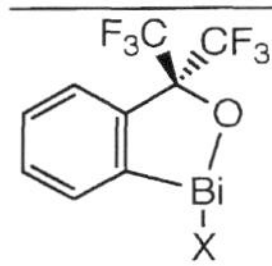	X = Cl	B	–	130 (decomp.)	–	[92JA7906]
		B	–	–	–	[93OM1857]
	X = Br	B	–	–	–	[93OM1857]
	X = OAc	B	–	267–269 (decomp.)	–	[92JA7906]
		B	95	267–269 (decomp.)	^{1}H-, ^{19}F-NMR	[95JA3922]
	X = $OCOCF_3$	B	–	185–186	–	[92JA7906]
		B	90	185–186	–	[95JA3922]
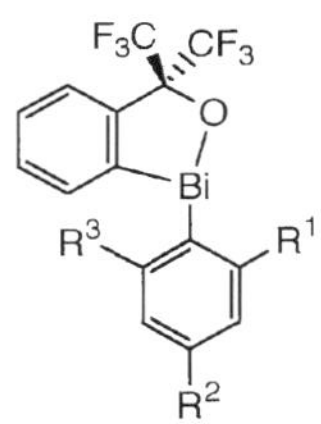	$R^1 = R^3$ = H, R^2 = Me	B	–	>300	^{1}H-, ^{19}F-NMR	[92JA7906]
		B	35	>300	^{1}H-, ^{19}F-NMR	[95JA3922]
	$R^1 = R^3$ = H, R^2 = MeO	B	34	>300	^{1}H-, ^{19}F-NMR	[95JA3922]
	$R^1 = R^3$ = H, R^2 = F	B	23	224–227	^{1}H-, ^{19}F-NMR	[93OM1857]
	$R^1 = R^3$ = H, R^2 = CF_3	B	41	223–225	^{1}H-, ^{19}F-NMR	[93OM1857]
		B	56	223–225	^{1}H-, ^{19}F-NMR	[95JA3922]
		B	–	234–235	^{1}H-, ^{19}F-NMR	[92JA7906]
	R^1 = (MeO)Me_2C, $R^2 = R^3$ = H	B	29	234–235	^{1}H-, ^{19}F-NMR	[95JA3922]
			–	–	X-ray	[95HAC293]
		B	59	234–235	^{1}H-, ^{19}F-NMR, X-ray	[95HAC293]
	R^1 = (HO)Me_2C, $R^2 = R^3$ = H	B	40	–	^{1}H-, ^{19}F-NMR	[95HAC293]
	R^1 = Me_2NCH_2, $R^2 = R^3$ = H	B	–	194–195	^{1}H-, ^{19}F-NMR, X-ray	[92JA7906]
		B	60	194–195	^{1}H-, ^{19}F-NMR, X-ray	[95JA3922]
	R^1 = $MeNHCH_2$, $R^2 = R^3$ = H	B	21	173–174	^{1}H-, ^{19}F-NMR	[95JA3922]
	R^1 = $MeOCH_2$, $R^2 = R^3$ = H	B	60	195–196	^{1}H-, ^{19}F-NMR	[95JA3922]
	$R^1 = R^3$ = Me_2NCH_2, R^2 = H	B	45	–	^{1}H-, ^{19}F-NMR, X-ray	[95JA3922]
	$R^1 = R^3$ = Me_2NCH_2, R^2 = ^{t}Bu	B	43		^{1}H-, ^{19}F-NMR	[95JA3922]

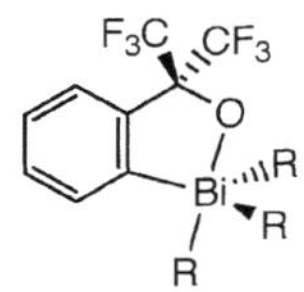	R = Tol	B	61	207–209	^{1}H-, ^{19}F-NMR	[88TL3817]
		B	55	206–209	^{1}H-, ^{19}F-NMR, X-ray	[93OM1857]
		B	–	–	–	[93OM3297]
	R = 4-$CF_3C_6H_4$	B	53	199–201	–	[88TL3817]
		B	53	199–201	^{1}H-, ^{19}F-NMR	[93OM1857]
	R = 4-FC_6H_4	B	61	161–163	^{1}H-, ^{19}F-NMR	[93OM1857]
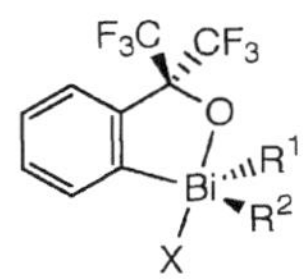	$R^1 = R^2$ = Tol,	B	>95	186–189	^{1}H-, ^{19}F-NMR	[88TL3817]
	X = Cl	B	–	186–189	^{1}H-, ^{19}F-NMR, X-ray	[93OM1857]
		B	–	–	–	[93OM3297]
	$R^1 = R^2$ = 4-$CF_3C_6H_4$,	B	>95	106 (decomp.)	–	[88TL3817]
	X = Cl	B	97	145–146	^{1}H-, ^{19}F-NMR, X-ray	[93OM1857]
	$R^1 = R^2$ = 4-FC_6H_4, X = Cl	B	88	125–127	^{1}H-, ^{19}F-NMR	[93OM1857]
	$R^1 = R^2$ = Tol, X = Br	B	82	182–185	^{1}H-, ^{19}F-NMR, X-ray	[93OM1857]
	R^1 = Tol, X = Cl,	B	>91	141–142	^{1}H-, ^{19}F-NMR	[88TL3817]
	R^2 = 4-$CF_3C_6H_4$	B	81	141–142	^{1}H-, ^{19}F-NMR, X-ray	[93OM1857]
	R^1 = Tol, R^2 = Me, X = Cl	B	–	–	–	[93OM1857]
	R^1 = Tol, X = Br, R^2 = 4-$CF_3C_6H_4$	B	88	134–136	^{1}H-, ^{19}F-NMR	[93OM1857]
	$R^1 = R^2$ = X = PhC≡C	B	10	178–181	^{1}H-, ^{19}F-NMR, X-ray	[93OM1857]
	R^1 = X = Tol,	B	70	–	–	[89TL953]
	R^2 = Me	B	–	–	–	[93OM3297]
		B	–	161–164	^{1}H-, ^{19}F-NMR, X-ray	[93OM1857]
	$R^1 = R^2$ = Tol,	B	63	182–185	^{1}H-, ^{19}F-NMR	[88TL3817]
	X = 4-$CF_3C_6H_4$	B	63	182–185	^{1}H-, ^{19}F-NMR	[93OM1857]

TABLE 4.2 *(continued)*

Compound		Synthetic method[a]	Yield (%)	M.p. (°C)	Physical data	Reference
	R^1 = X = 4-$CF_3C_6H_4$, R^2 = Tol	B	16	164–166	^{1}H-, ^{19}F-NMR	[93OM1857]
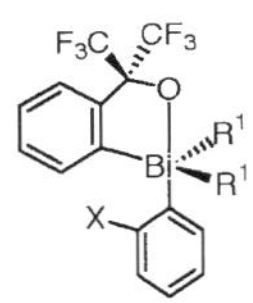	R^1 = Tol, X = Me_2NCH_2	B	84	227–229	^{1}H-, ^{19}F-NMR, X-ray	[93OM3297]
	R^1 = 4-$CF_3C_6H_4$, X = Me_2NCH_2	B	74	219–222	^{1}H-, ^{19}F-NMR, X-ray	[93OM3297]
	R^1 = Tol,	B	–	232–234	^{1}H-, ^{19}F-NMR	[89TL953]
	X = $MeO(CF_3)_2C$	B	30	232–234	^{1}H-, ^{19}F-NMR	[93OM3297]
	R^1 = 4-$CF_3C_6H_4$, X = $MeO(CF_3)_2C$	B	95	239–241	^{1}H-, ^{19}F-NMR, X-ray	[93OM3297]
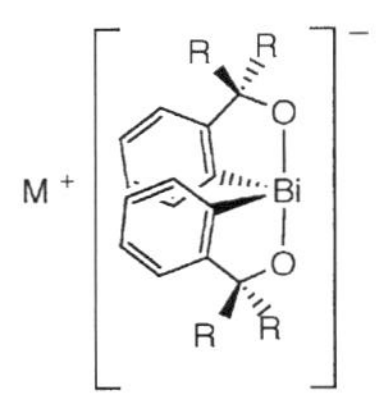	R = CF_3, M = Li	B	64	–	–	[92TL6653]
	R = CF_3, M = K	B	–	–	–	[92TL6653]
	R = CF_3, M = Et_4N	B	–	167–168	^{1}H-, ^{19}F-NMR, X-ray	[92TL6653]
		B			X-ray	[94AX(B)151]
	R = Me, M = Li	B	–	–	–	[92TL6653]
	R = Me, M = Et_4N	B	–	–	–	[92TL6653]
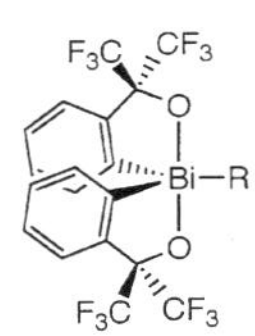	R = Cl	B	100	225–227 (decomp.)	^{1}H-, ^{19}F-NMR	[92TL6653]
	R = Tol	B	100	130–132	^{1}H-, ^{19}F-NMR	[89TL953]
		B	–	–	–	[93OM3297]

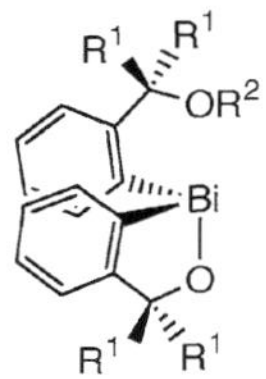	$R^1 = CF_3$, $R^2 = H$	B	–	101–102 (decomp.)	^{1}H-, ^{19}F-NMR	[92TL6653]
		B	78	100–101 (decomp.)	^{1}H-, ^{19}F-NMR	[95HAC293]
	$R^1 = CF_3$, $R^2 = OMe$	B	–	220–222 (decomp.)	^{1}H-, ^{19}F-NMR	[92TL6653]
		B	26	220–222 (decomp.)	^{1}H-, ^{19}F-NMR	[95HAC293]
	$R^1 = Me$, $R^2 = H$	B	66	227–228	^{1}H-NMR	[92TL6653]
		B	66	227–228	^{1}H-NMR	[95HAC293]
		B	–	–	X-ray	[94AX(B)151]
	$R^1 = Me$, $R^2 = OMe$	B	–	200–202	^{1}H-NMR	[92TL6653]
		B	67	200–202	^{1}H-NMR	[95HAC293]
		B	–	–	X-ray	[94AX(B)151]
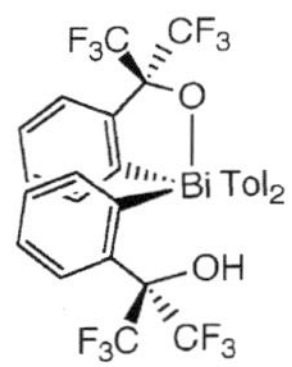		B	–	–	^{1}H-, ^{19}F-NMR	[89TL953]
		B	96	–	^{1}H-, ^{19}F-NMR	[93OM3297]
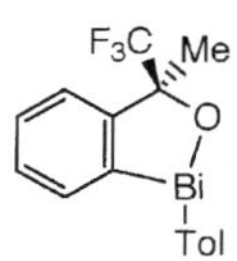		–	–	–	–	[95JA3922]
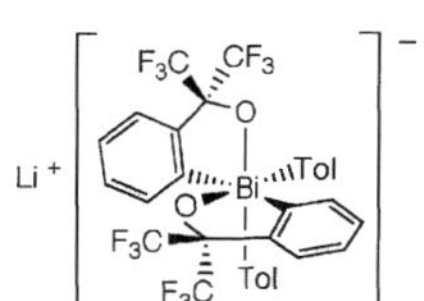	R = Me	B	–	–	–	[89TL953]
		B	–	–	–	[93OM3297]
	R = Et	B	–	–	–	[93OM3297]
	R = Bu	B	–	–	–	[93OM3297]

TABLE 4.2 (continued)

Compound		Synthetic method[a]	Yield (%)	M.p. (°C)	Physical data	Reference
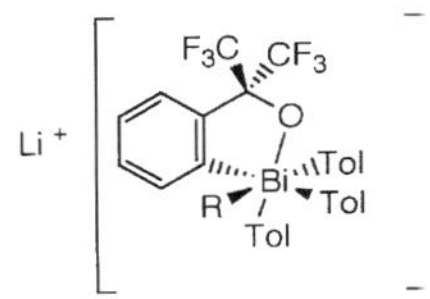	M = Li	B	–	–	^{19}F-NMR	[89TL953]
	M = (12-Crn-6)Li (Crn = crown)	B	–	–	^{19}F-NMR	[93OM3297]
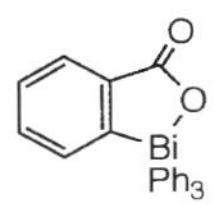		G	53	–	IR	[86ZOB330]
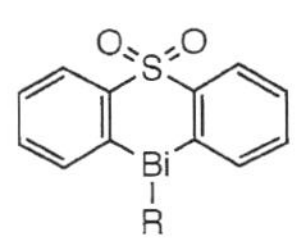	R = Tol	B	36	212–215	^{1}H-NMR, IR, MS	[92JCS(P1)1593]
	R = Ph	B	–	–	–	[92JCS(P1)1593]
	R = Cl	B	27	222–225	^{1}H-NMR, IR, MS	[92JCS(P1)1593]
	R = Me	B	86	180–183	^{1}H-NMR, IR, MS	[92JCS(P1)1593]
	R = I	B	92	277–280	^{1}H-NMR, IR, MS	[92JCS(P1)1593]
	R = $Ph_2C{=}CH$	B	92	218–220	^{1}H-NMR, IR, MS	[92JCS(P1)1593]
	R = 4-$ClC_6H_4C{\equiv}C$	B	86	229–233 (decomp.)	^{1}H-NMR, IR, MS, X-ray	[92JCS(P1)1593]
	R = OPh	B	66	203–206	^{1}H-NMR, IR	[2000OM1003]
	R = 2-$MeOC_6H_4O$	B	75	224–227	^{1}H-, ^{13}C-NMR, IR, X-ray	[2000OM1003]
	R = 4-$MeOC_6H_4O$	B	50	178–182	^{1}H-, ^{13}C-NMR, IR, UV, X-ray	[2000OM1003]
	R = 4-$MeOC_6H_4O$· 4-$MeOC_6H_4OH$	B	75	116–118	^{1}H-NMR, IR, UV, X-ray	[2000OM1003]
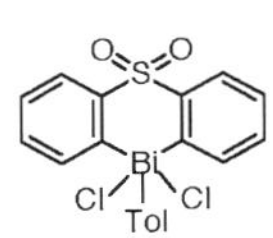		B	36	222–225 (decomp.)	^{1}H-NMR, IR, MS	[92JCS(P1)1593]

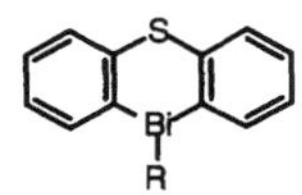	R = Tol	B	32	106–108	^{1}H-NMR, IR, MS	[92JCS(P1)1593]
	R = Me	B	35	125–127	^{1}H-NMR, IR, MS	[92JCS(P1)1593]
	R = 4-ClC$_6$H$_4$C≡C	B	35	153–156	^{1}H-NMR, IR, MS	[92JCS(P1)1593]
	R = Ph$_2$C=CH	B	58	149–151	^{1}H-NMR, IR, MS	[92JCS(P1)1593]
	R = I	B	75	187–190	^{1}H-NMR, IR, MS	[92JCS(P1)1593]
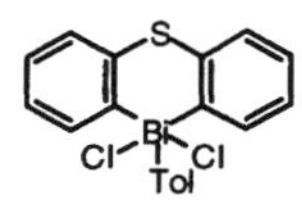		B	94	139–142	^{1}H-NMR, IR, MS	[92JCS(P1)1593]
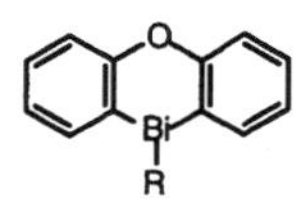	R = Tol	B	40	112–115	^{1}H-NMR, IR, MS	[92JCS(P1)1593]
	R = I	B	90	150–155	^{1}H-NMR, IR, MS	[92JCS(P1)1593]
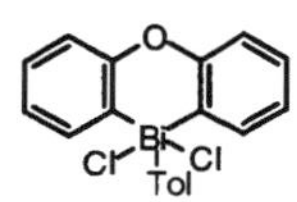		B	93	134–137 (decomp.)	^{1}H-NMR, IR, MS	[92JCS(P1)1593]
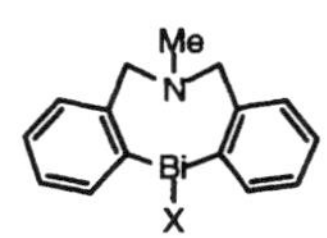	X = Cl	B	–	199–200	^{1}H-, ^{13}C-NMR	[89TL4841]
	X = F	B	–	216–219	^{1}H-, ^{13}C-NMR	[89TL4841]
	X = I	B	–	246 (decomp.)	^{1}H-, ^{13}C-NMR	[89TL4841]
	X = Me	B	–	219–222	^{1}H-, ^{13}C-NMR	[89TL4841]
	X = Ph	B	–	213–215	^{1}H-, ^{13}C-NMR	[89TL4841]
	X = Tol	B	100	152–155	^{1}H-, ^{13}C-NMR, X-ray	[99CL861]
	X = 4-CF$_3$C$_6$H$_4$	B	61	163–164	^{1}H-, ^{13}C-, ^{19}F-NMR, X-ray	[99CL861]

TABLE 4.2 *(continued)*

Compound		Synthetic method[a]	Yield (%)	M.p. (°C)	Physical data	Reference
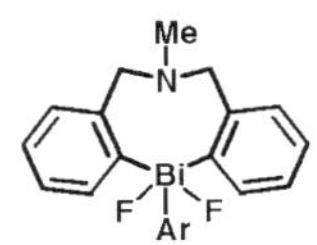	R = Tol	B	100	189–190	^{1}H-, ^{13}C-, ^{19}F-NMR, X-ray	[99CL861]
	R = 4-$CF_3C_6H_4$	B	100	187–188	^{1}H-, ^{13}C-, ^{19}F-NMR, X-ray	[99CL861]
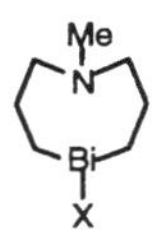	R = Me, X = Cl	E	12	202	^{13}C-NMR, IR, MS, Raman	[94POL365]
	R = Me, X = Br	E	6	223	^{13}C-NMR, IR, MS, Raman	[94POL365]
	R = Me, X = I	E	11	246	^{13}C-NMR, IR, MS, Raman	[94POL365]
	R = $PhCH_2$, X = Cl	E	16	162	^{13}C-NMR, IR, MS, Raman	[94POL365]
	R = $PhCH_2$, X = Br	E	5	179	^{13}C-NMR, IR, MS, Raman	[94POL365]
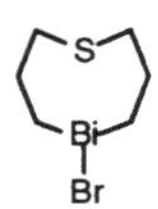		E	21	145	^{13}C-NMR, IR, MS, Raman	[94POL365]
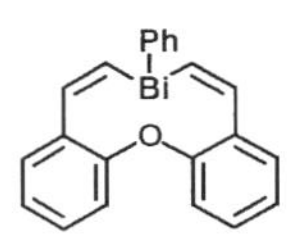		B	–	–	^{1}H-NMR, MS	[2000H in press]
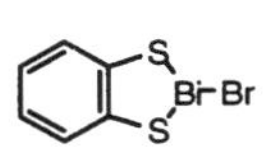		B	43	>200 (decomp.)	^{1}H-NMR, IR, MS	[87POL1593]

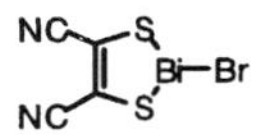		B(Na)	54	–	IR, MS	[87POL1593]
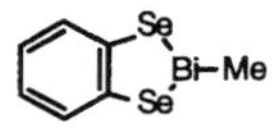		B	62	142–147 (decomp.)	^{1}H-NMR, IR, MS	[87POL1593]
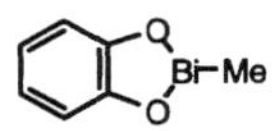		H	65	90–120	^{1}H-NMR, MS	[76ZAAC(423)40]
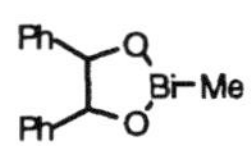		H	70	72–100	MS	[76ZAAC(423)40]
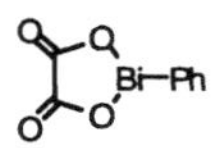		G	82	>300	MS	[80ZN(B)1000]
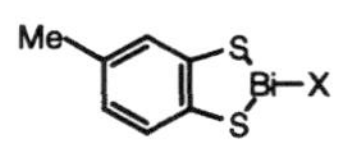	X = Me	H	82	185 (decomp.)	^{1}H-NMR, MS	[76ZAAC(423)40]
	X = Cl	H	86	233–234	–	[68JCS(A)2587]
	X = Br	H	87	222–223	–	[68JCS(A)2587]
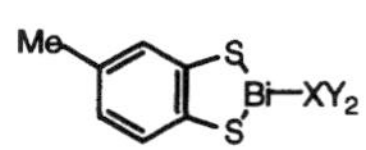	X = Cl, Y = pyridine	H	–	135–139 (decomp.)	–	[68JCS(A)2587]
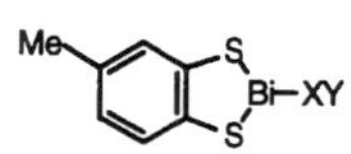	X = Cl, Y = 2,2′-bipyridine	H	98	191–192	–	[68JCS(A)2587]
	X = Br, Y = 2,2′-bipyridine	H	100	198–199 (decomp.)	–	[68JCS(A)2587]

TABLE 4.2 *(continued)*

Compound		Synthetic method[a]	Yield (%)	M.p. (°C)	Physical data	Reference
	X = Cl, Y = 1,10-phen (phen; phenanthroline)	H	100	–	–	[68JCS(A)2587]
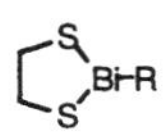	R = Me	H	78	100 (decomp.)	^{1}H-NMR, MS	[76ZAAC(423)40]
	R = Ph	H	83	137	^{1}H-NMR, MS	[76ZAAC(423)47]
	R = Cl	H	82	>120 (decomp.)	–	[68JCS(A)2587]
		–	–	–	–	[96JA3225]
		–	–	–	–	[90BSCQ165]
	R = Br	H	96	>155	–	[68JCS(A)2587]
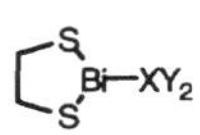	X = Cl, Y = pyridine	H	85	100–102 (decomp.)	–	[68JCS(A)2587]
		H	81	130 (decomp.)	^{1}H-NMR, IR, X-ray	[96JA3225]
	X = Br, Y = pyridine	H	93	130–132 (decomp.)	–	[68JCS(A)2587]
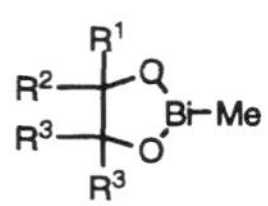	R^1 = Me, R^2 = R^3 = H	H	55	123 (decomp.)	MS	[76ZAAC(423)40]
	R^1 = R^2 = R^3 = H	H	86	168 (decomp.)	MS	[76ZAAC(423)40]
	R^1 = R^2 = R^3 = Me	H	54	88–116	MS	[76ZAAC(423)40]
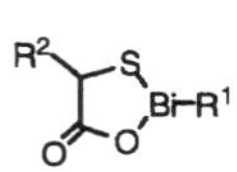	R^1 = R^2 = Me	G	70	209 (decomp.)	IR	[81ZN(B)70]
	R^1 = HO_2CCH_2S, R^2 = H	G	80	118 (decomp.)	^{1}H-NMR, IR	[81ZN(B)70]
	R^1 = Me, R^2 = H	G	72	202 (decomp.)	IR	[81ZN(B)70]
	R^1 = HS(Me)$CHCO_2$, R^2 = Me	G	80	126 (decomp.)	^{1}H-NMR, IR	[81ZN(B)70]

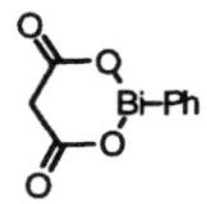		G	43	200 (decomp.)	MS	[80ZN(B)1000]
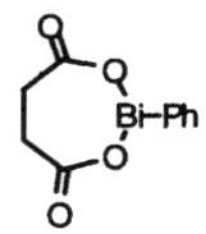		G	15	>300	MS	[80ZN(B)1000]
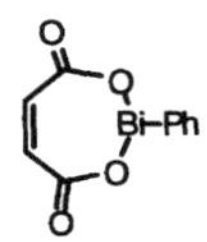		G	99	290 (decomp.)	^{1}H-NMR, MS	[80ZN(B)1000]
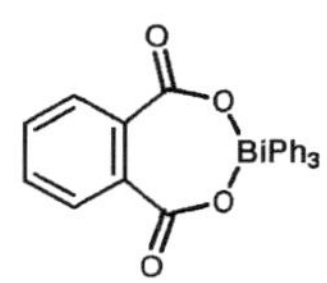		B(Na)	58	155–165	–	[51JA4470]
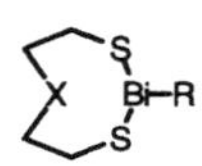	X = O, R = Ph	H	79	150 (decomp.)	^{13}C-NMR, IR, MS, Raman, X-ray	[85JOM(290)133]
	X = O, R = Me	H	69	132 (decomp.)	MS, IR, Raman	[85JOM(290)133]
	X = S, R = Ph	H	18	125 (decomp.)	^{13}C-NMR, IR, MS, Raman	[81ZN(B)70]
	X = S, R = Me	H	80	141 (decomp.)	^{13}C-NMR, IR, MS, Raman	[85JOM(290)133]
	X = S, R = Cl	H	–	120 (decomp.)	IR, MS	[74ZAAC(406)74]
		–	87	170 (decomp.)	IR, Raman, X-ray	[96JA3225]

TABLE 4.2 *(continued)*

Compound		Synthetic method[a]	Yield (%)	M.p. (°C)	Physical data	Reference
	X = O, R = Cl	H	60	75 (decomp.)	IR, MS	[74ZAAC(407)35]
		–	–	–	–	[96JA3225]
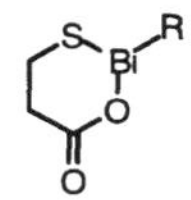	R = Ph	G	62	209 (decomp.)	IR	[81ZN(B)70]
	R = Me	G	43	172 (decomp.)	IR	[81ZN(B)70]
	R = $HO_2CC_2H_4S$	G	65	155 (decomp.)	^{1}H-NMR, IR	[81ZN(B)70]
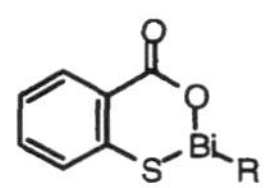	R = Ph	G	62	302 (decomp.)	IR	[81ZN(B)70]
		G	–	>250	IR	[51JA2880]
	R = Me	G	77	290 (decomp.)	IR	[81ZN(B)70]
	R = 2-$HSC_6H_4CO_2$	G	77	274 (decomp.)	^{1}H-NMR, IR	[81ZN(B)70]
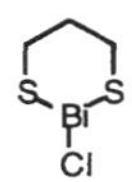		–	95	110	IR, Raman	[96JA3225]
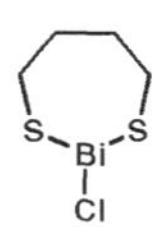		–	97	100 (decomp.)	R, Raman	[96JA3225]
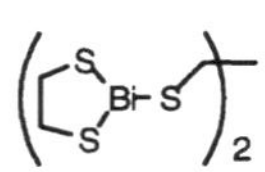		–	97	145 (decomp.)	IR, Raman, X-ray	[96JA3225]
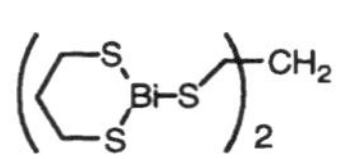		–	92	170 (decomp.)	IR, Raman	[96JA3225]

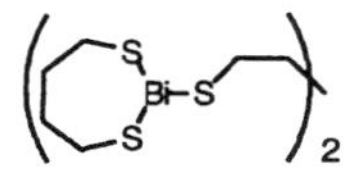		–	98	140 (decomp.)	IR, Raman	[96JA3225]
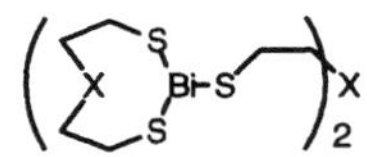	X = O	–	67	180 (decomp.)	IR, Raman, X-ray	[96JA3225]
	X = S	–	99	177–178	IR, Raman	[96JA3225]
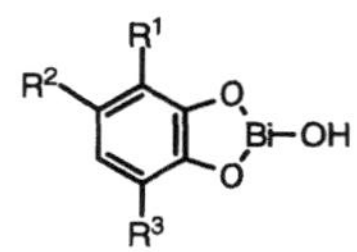	$R^1 = R^3 = H, R^2 = {}^tBu$	H	68	–	^{1}H-NMR, IR	[90HAC233]
	$R^1 = H, R^2 = R^3 = {}^tBu$	H	74	–	^{1}H-NMR, IR	[90HAC233]
	$R^1 = R^2 = R^3 = H$	H	–	–	–	[00JPHC(12)145]
	$R^1 = R^3 = H, R^2 = Me$	H	–	–	–	[00JPHC(12)145]
	$R^1 = R^3 = H,$ $R^2 = CO_2H$	H	–	–	–	[00JPHC(12)145]
		–	–	–	–	[04BSF(31)176]
	$R^1 = OH, R^2 = CO_2H,$	H	–	–	–	[00JPHC(12)145]
	$R^3 = H$	–	–	–	–	[03BSF(29)680]
	$R^1 = H, R^2 = CO_2H,$	H	–	–	–	[00JPHC(12)145]
	$R^3 = OH$	–	–	–	–	[01JPHC(14)487]
		–	–	–	–	[02ACP(25)268]
		–	–	–	–	[03BSF(29)531]
	$R^1 = H, R^2 = CO_2Me,$ $R^3 = OH$	H	–	–	–	[00JPHC(12)145]
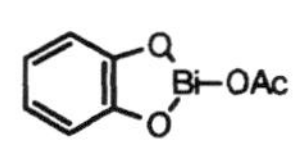		H	91	–	^{1}H-NMR, IR, MS	[90HAC233]
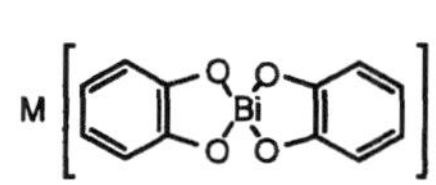	M = Na	H	–	–	–	[25CB891]
	M = K	H	–	–	–	[25CB891]
	$M = NH_4$	H	–	–	–	[25CB891]
	M = pyridinium	H	–	–	–	[25CB891]

TABLE 4.2 *(continued)*

Compound		Synthetic method[a]	Yield (%)	M.p. (°C)	Physical data	Reference
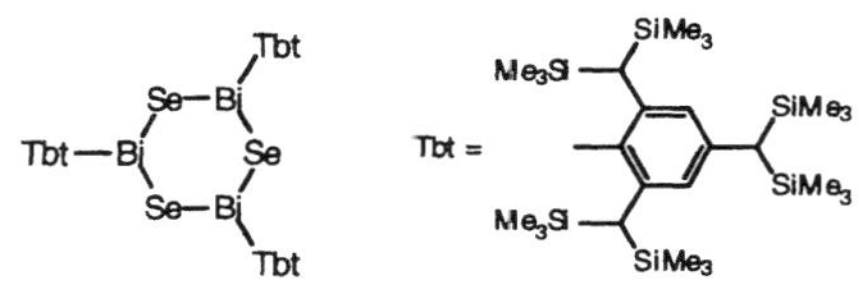		B	–	–	–	[97SCI(277)78]
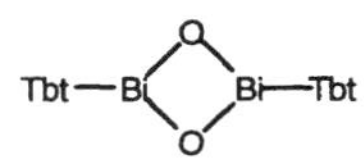		–	–	–	–	[97SCI(277)78]
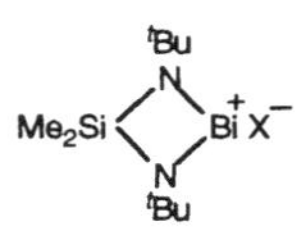	X = Cl	–	–	–	–	[88ZAAC(557)7]
		D	–	–	X-ray	[78ZN(B)1, 78ZN(B)7]
						[88ZAAC(559)73]
	X = $AlCl_4$	–	95	112	^{1}H-NMR, UV, X-ray	[88ZAAC(559)73]
	X = $GaCl_4$	–	95	94	^{1}H-NMR	[88ZAAC(559)73]
	X = $InCl_4$	–	60	120	^{1}H-NMR	
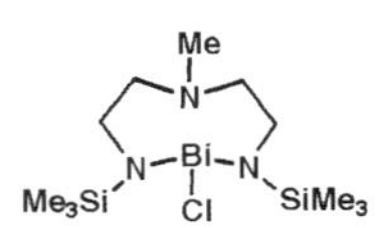		B	90	133 (decomp.)	^{1}H-, ^{13}C-NMR, X-ray	[99EJIC2295]
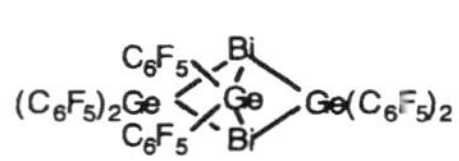		–	–	–	X-ray	[80JOM(200)243]

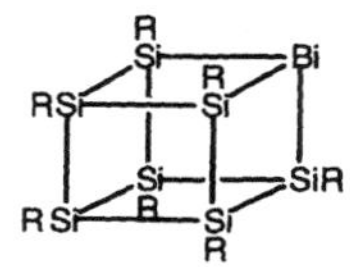	R = ^{c}Hex, Si–Si=Si–O–Si, Si–Bi=Si–O–Bi	H	–	–	^{1}H-, ^{13}C-, ^{29}Si-NMR	[91OM812]

[a] Synthetic methods A–H refer to: A, organocadmium route (p. 331); B, organolithium route (p. 333); C, organozirconium route (p. 334); D, organotin route (p. 334); E, Grignard route (p. 335); F, pyrolysis route (p. 335); G, acidolysis route (p. 352); H, alcoholysis route (p. 353).

Chapter 5

Bismuth Compounds in Organic Transformations

5.1. INTRODUCTION

The first report of organic transformation using bismuth compounds was made by Challenger in 1934 and involved the oxidation of alcohols to the corresponding carbonyl compounds with bismuth(V) compounds [34JCS405]. About two decades later, sodium bismuthate and bismuth oxide were reported by Rigby to be effective as a mild oxidant for alcohols [49NAT(164)185, 51JCS793]. During the following two decades, however, bismuth compounds were mainly studied from the promoter for organic reactions. In the 1960s, bismuth molybdate phases were used for the effective transformation of propylene to acrolein and acrylonitrile [59USP2904580, 62USP3044966], where bismuth was one of the key components of the solid phase catalysts for propane and propylene oxidations. Intensive studies have been made by Barton's and Dodonov's groups on pentavalent organobismuth compounds as oxidant and/or arylating agent during the period from the late 1970s to the 1980s. In the mid 1980s, Wada et al. reported the bismuth–based Barbier type allylation and aldol reaction [85TL4211, 88TL4719]. The chemistry of bismuthonium salts and ylides started in the late 1980s. Recently, many bismuth compounds have been gaining increasing attention as catalysts as well as reaction mediators. Typically, bismuth mandelate and triflate are effective as catalysts for molecular oxygenation and Friedel–Crafts acylation, respectively [93TL2601, 94JOM(482)119, 95T4991, 97TL8871]. Most of the bismuth compounds commercially available are not expensive, are easy to handle and non-toxic.

There are a number of review articles which deal with the topics of this chapter. Many of these reviews are devoted to organic transformations based on bismuth compounds in general [89YGK425, 96F1501, 97S249] or specific topics such as bismuthonium salts and ylides [96BCJ2673], oxidation with bismuth compounds [86B-K, 96CCR(155)127], oxidation with bismuth-containing heterogeneous phases [89CRS43, 94AIC233, 96AC(A)(145)1], arylation and oxidation with organobismuth compounds [87PAC937, 88T3039, 89CRV1487, 90B-B, 93IZV2043(93RCB1955)], and Friedel–Crafts acylation with bismuth(III) salts [96ICL15]. Progress in organobismuth chemistry has been reviewed periodically in the journal *J. Organomet. Chem. Gmelin's Handbuch der Anorganischen Chemie, volume 47* provides a comprehensive compilation of the preparation and properties of known organobismuth compounds throughout 1977. *Houben-Weyl Methoden der Organischen Chemie, Vol. XIII, Part 8*, which is mainly concerned with the preparation and reactions of organobismuth compounds, does so for 1978. *Comprehensive Organometallic Chemistry* and *Encyclopedia of Inorganic Chemistry* provide concise surveys on organic and inorganic bismuth chemistry [95COMC(11)502, 95COMC(2)321, 94EIC(1)292, 94EIC(1)280].

The present chapter deals with a variety of organic transformations based on the use of metallic bismuth, inorganic bismuth salts, bismuth-added inorganic oxides and salts, and trivalent/pentavalent organobismuth compounds. The reactions are classified according to their reaction types and are presented in the order of oxidation, reduction, carbon–carbon bond forming reactions, and carbon–heteroatom bond forming reactions.

5.2. OXIDATION

5.2.1. Metal bismuthates

5.2.1.1. Sodium bismuthate

Sodium bismuthate $NaBiO_3$ is a stable solid compound of rather ill-defined composition, which is commercially available as a yellow to light brown powder and normally used without purification. The commercial product is suggested to be a mixture of dibismuth pentaoxide, sodium carbonate and sodium peroxide. Hence, commercial samples are often variable in appearance and reactivity.

Sodium bismuthate is usually employed as a suspension in the aqueous media or organic solvents acidified with acetic or phosphoric acid. It cleaves 1,2-glycols or related alcohols to the corresponding carbonyl compounds.

Phenols and olefins are oxidized to polyphenylene oxides and *vic*-diol derivatives, respectively.

(a) Oxidative cleavage of 1,2-diols and related compounds Sodium bismuthate was first reported by Rigby in 1949 as a selective and mild agent for the oxidative cleavage of 1,2-glycols to carbonyl compounds [49NAT(164)185]. Its behavior resembles that of periodic acid and lead tetraacetate; α-hydroxylated aldehydes, ketones and carboxylic acids are converted to a pair of corresponding carbonyl–containing compounds, as shown in Scheme 5.1. The resulting aldehydes are not further oxidized with sodium bismuthate. Like lead tetraacetate, but unlike periodic acid, sodium bismuthate rapidly cleaves α-hydroxycarboxylic acids and some α-hydroxyketones.

The reaction can be carried out in the presence of acetic or phosphoric acid at room temperature. Alcohols such as methanol and ethanol can be employed as the reaction media, since they are oxidized quite slowly with sodium bismuthate. Anhydrous conditions, as required in the use of lead tetraacetate, are not necessary for sodium bismuthate [50JCS(C)1907, 66RPA185, 67AP(300)520, 70AP(303)233].

The mechanism of 1,2-cleavage of glycols with sodium bismuthate is generally considered to be similar to that with lead tetraacetate and periodic acid [40CB563], although there is still some controversy [50JCS(C)1907]. A cyclic bismuthate diester is postulated as a plausible intermediate, though the rates of oxidation observed are not so different between *cis*- and *trans*-1,2-cyclohexanediols.

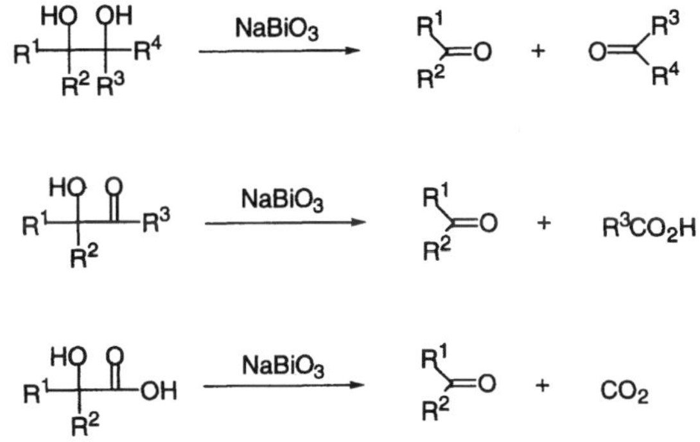

Scheme 5.1. General mode of oxidative cleavage with sodium bismuthate.

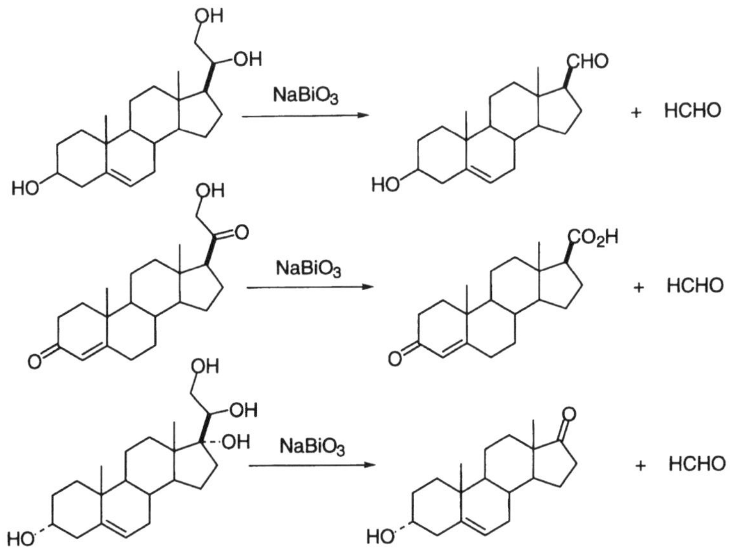

Scheme 5.2. Selective oxidative cleavage of corticosteroid side chains with sodium bismuthate [53BJ371].

The C–C bond cleavage with sodium bismuthate has been applied successfully to the conversion of cyclic *vic*-diols to dialdehydes [53HCA268] and selective oxidative cleavage of the corticosteroid side chain, as illustrated in Scheme 5.2 [53BJ371]. Although the periodate oxidation is used widely for the structure characterization of corticosteroids (cf. [43VH345]), sodium bismuthate possesses the following advantages over sodium periodate. (i) Sodium bismuthate converts corticosteroids bearing a dihydroxyacetone side chain to 17-ketosteroids, which are relatively easy to characterize and assay, whereas the periodate degrades them to α-hydroxy acids. (ii) The insolubility of sodium bismuthate in organic solvents offers technical advantages by permitting the use of a large excess of the reagent as well as simple removal of its surplus. (iii) Sodium bismuthate allows direct oxidation of urinary corticosteroids to 17-ketosteroids [52NAT(170)1074, 55A397, 57JA1130].

The reactivity of four oxidizing reagents, sodium bismuthate, sodium periodate, manganese dioxide and lead tetraacetate, has been compared in the oxidative cleavage of unsaturated alicyclic α-ketols to ω-oxo-α,β-unsaturated acids [93JOC2196]. Sodium bismuthate cleaved saturated α-ketols but not unsaturated ones (Scheme 5.3).

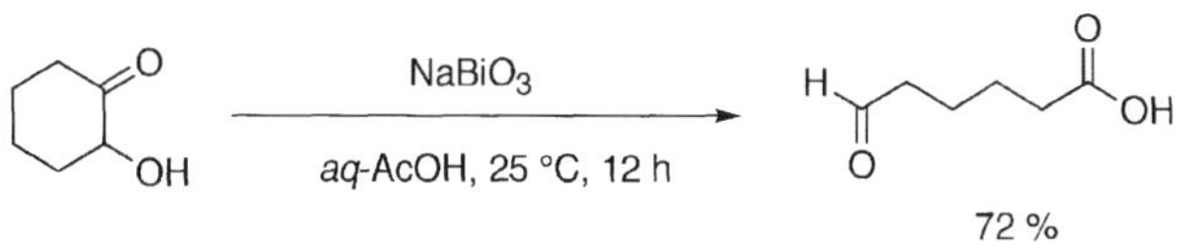

Scheme 5.3. Oxidative cleavage of 2-hydroxycyclohexanone with sodium bismuthate [93JOC2196].

Oxidative cleavage of alicyclic α-ketols with sodium bismuthate: typical procedure
To a solution of 2-hydroxycyclohexanone (114 mg, 1 mmol) in 50% aqueous acetic acid (10 ml) was added sodium bismuthate (308 mg, 1.1 mmol). The mixture was stirred at 25°C for 12 h, diluted with ethyl acetate, washed successively with 10% $NaHSO_3$, water and brine, and dried over anhydrous $MgSO_4$. The solvent was removed and the residue was chromatographed on silica gel using ethyl acetate–hexane (1:1) as the eluent to afford 6-oxohexanoic acid (93 mg, 72%) [93JOC2196].

(b) Oxidation of phenols In the oxidation of phenols with sodium bismuthate, the resulting products differ considerably depending on the solvent used. In non-polar solvents such as benzene and cyclohexane, preferential one-electron oxidation occurs to give the radical coupling product, polyphenylene oxides, as illustrated in Scheme 5.4 [76JOC1646]. Under polar conditions using acetic acid as the solvent, however, two-electron oxidation is favored to yield quinones and quinols as the major products. Representative results are summarized in Table 5.1 [65JCS2904, 71JCS(C)1750, 74ACS(B)883, 74ACS(B)888, 75JOC1515, 76JOC1646]. Monoalkyl ethers of catechol and hydroquinone often undergo dealkylation during the oxidation [68JCS(C)2770]. The oxidation of phenols with periodate [74ACS(B)883, 74ACS(B)888] and lead tetraacetate [68JCS(C)2770] also gives products similar to those from the sodium bismuthate oxidation.

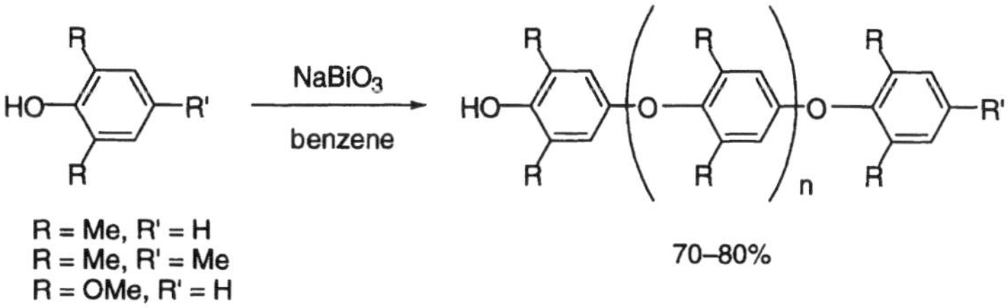

Scheme 5.4. Oxidation of substituted phenols with sodium bismuthate [75JOC1515].

TABLE 5.1
Oxidation of substituted phenols with sodium bismuthate

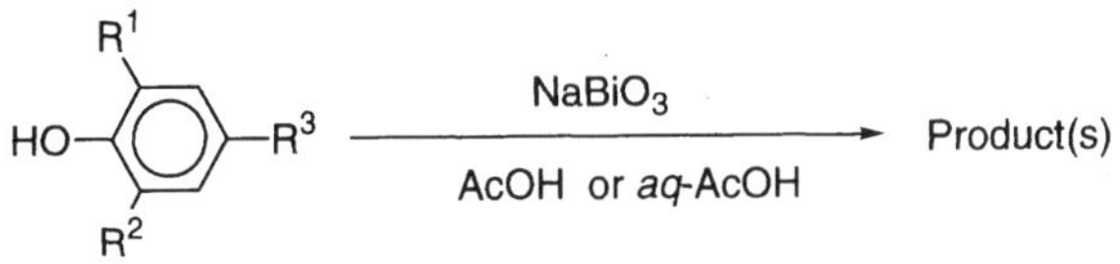

Substrate	Product(s) and yield(s) (%)
R^1 = Me, R^2 = Me, R^3 = H [76JOC1646]	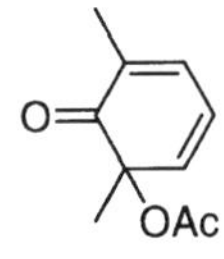 38 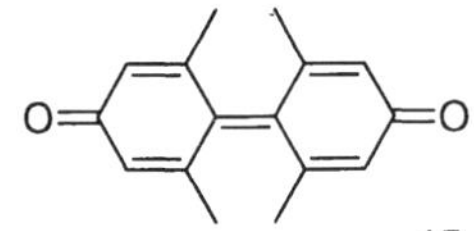 15
R^1 = Me, R^2 = Me, R^3 = CH_2OH [74ACS(B)888]	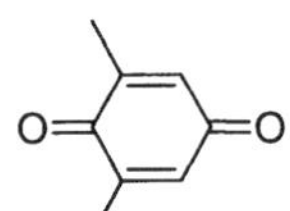 43 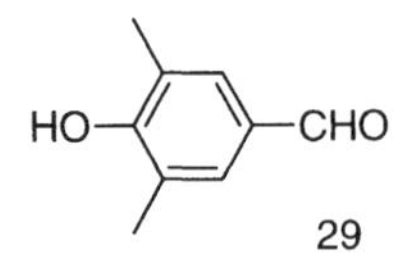 29
R^1 = Br, R^2 = ^{t}Bu, R^3 = OMe [68JCS(C)2770]	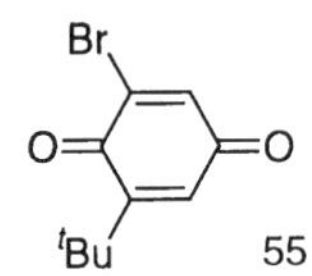 55
R^1 = ^{t}Bu, R^2 = OMe, R^3 = Br [68JCS(C)2770]	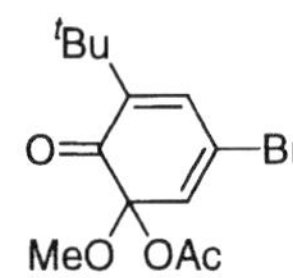 87
R^1 = Br, R^2 = OMe, R^3 = ^{t}Bu [68JCS(C)2770]	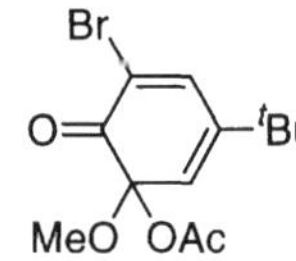 44

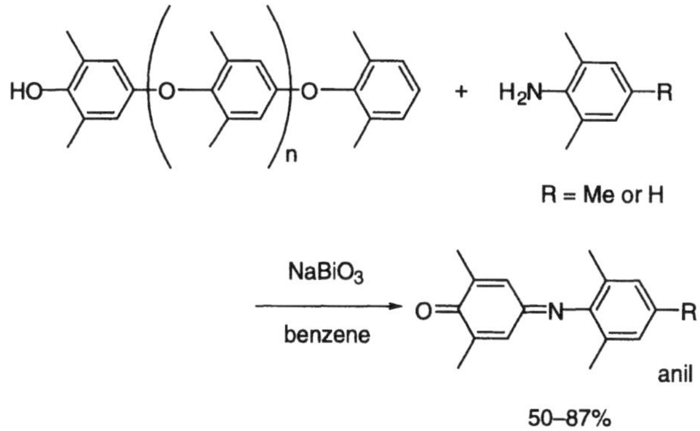

Scheme 5.5. Oxidative depolymerization of poly-1,4-phenylene oxides with sodium bismuthate [73CC562].

Oxidation of substituted phenols with sodium bismuthate: typical procedure
To a solution of 2,6-xylenol (4.1 g, 0.033 mol) in benzene (100 ml) was added sodium bismuthate (33.0 g, 0.118 mol) and the resulting mixture was refluxed for 2 h with stirring. After cooling to room temperature, the mixture was filtered and the solid residue was washed with benzene (150 ml). The combined benzene solutions were washed with 5% NaOH, dried, and concentrated to afford a crude polymer which was dissolved in chloroform and diluted with methanol (200 ml) to precipitate poly-1,4-phenylene oxide in 74% yield. From the mother liquor, a tail-to-tail coupling product 3,3′,5,5′-tetramethyldiphenoquinone was obtained in 12% yield [75JOC1515].

Sodium bismuthate can depolymerize poly-1,4-phenylene oxides in benzene in the presence of substituted aniline to give the corresponding anils in good yield (Scheme 5.5) [73CC562].

(c) Oxidation of olefins The oxidation of olefins with sodium bismuthate occurs in acetic acid to give *vic*-diol monoacetates in moderate to low yield [80JOC726]. The results are summarized in Table 5.2. The yields are highest when a tertiary carbon is present in olefinic substrates, namely, trisubstituted or geminally disubstituted olefins (entries 1–5). Monosubstituted and vicinally disubstituted olefins react sluggishly to afford hydroxy acetates in lower yields. The oxidation of olefins occurs stereospecifically; (*E*)- and (*Z*)-3-hexenes are oxidized to give *erythro*- and *threo*-4-acetoxy-3-hexanols, respectively (entries 7 and 8) and the recovered olefin retains its stereochemical integrity. Two mechanistic pathways are suggested for this

TABLE 5.2
Oxidation of olefins with sodium bismuthate [80JOC726]

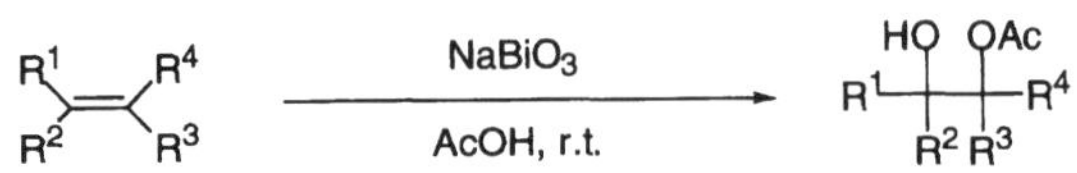

Entry	R^1	R^2	R^3	R^4	Reaction time (days)	Yield (%)[a,b]
1	Ph	$-(CH_2)_4-$		H	3	37
2	Me	$-(CH_2)_4-$		H	8	21
3	$-(CH_2)_4-$		H	H	7	40
4	Me	Ph	H	H	3	39 (62)
5	Me	Me	Me(Et)$CH(CH_2)_2$	H	7	21 (42)
6	H	$-(CH_2)_4-$		H	4	25
7	Et	H	Et	H	10	18 (41)
8	Et	H	H	Et	10	18
9	$Me(CH_2)_7$	H	H	H	21	10

[a] Isolated yield.
[b] Isolated yield (%) based on unrecovered olefin, indicated in parentheses.

Oxidation of olefins to *vic*-diol monoacetates with sodium bismuthate: typical procedure A mixture of α-methylstyrene (5.92 g, 0.05 mol), sodium bismuthate (16.62 g, 0.05 mol) and glacial acetic acid (100 ml) was stirred at room temperature in a 250-ml Morton flask for 72 h. The resulting mixture was poured into ether and washed successively with water and saturated aqueous $NaHCO_3$ until acetic acid was completely removed. The ether solution was further washed with brine, dried, filtered, and evaporated to afford crude product, which was subjected to silica gel column chromatography using hexane–dichloromethane as the solvent. 1-Acetoxy-2-phenyl-2-propanol was obtained in 39% yield [80JOC726].

type of oxidation; one is initial epoxidation followed by solvolysis in acetic acid and the other involves the process via acetoxymetallation of olefins. Since the rearranged product is predominant in the oxidation of *endo*-2-(hydroxymethyl)-5-norbornene, the latter mechanism seems to be more plausible for the bismuthate oxidation. Similarity to the behavior of sodium bismuthate in acetic acid to those of acetates of mercury(II), thallium(III) and lead(IV) may imply the in situ generation of $Bi(OAc)_5$.

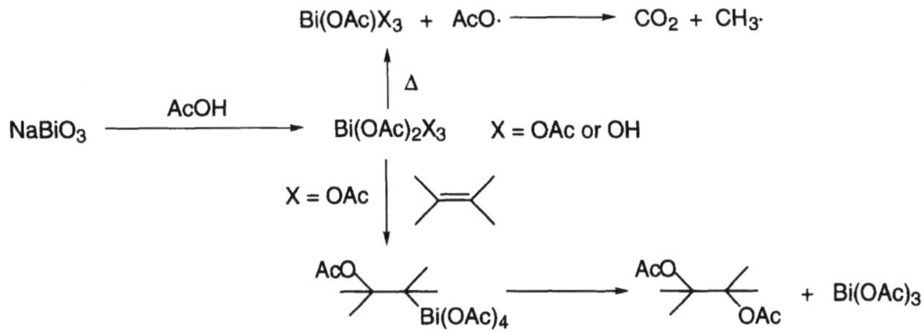

Scheme 5.6. A proposed mechanism for the sodium bismuthate oxidation of olefins [71KKZ1807].

A similar oxidation of olefins with sodium bismuthate in acetic acid at elevated temperature (100°C) leads to a mixture of *vic*-diol diacetate and carbonyl compound arising from oxidative cleavage, together with molecular oxygen and carbon dioxide [71KKZ1807]. In the absence of an olefin, sodium bismuthate works destructively on acetic acid to generate molecular oxygen, carbon dioxide, methyl acetate and a trace amount of methane, while the bismuthate is converted to bismuth acetate when the reaction is complete. The proposed mechanism is shown in Scheme 5.6.

(d) Oxidation of benzylic methylenes The oxidation of benzylic methylenes in polycyclic systems to benzylic ketones was reported quite recently with an excess amount of sodium bismuthate in acidic media [98TL7247]. Other than sodium bismuthate, it is known already that DDQ, PCC, CrO_3, ^{t}BuOOH and tetrapyridinesilver peroxydisulfate take part in the same reaction (Scheme 5.7).

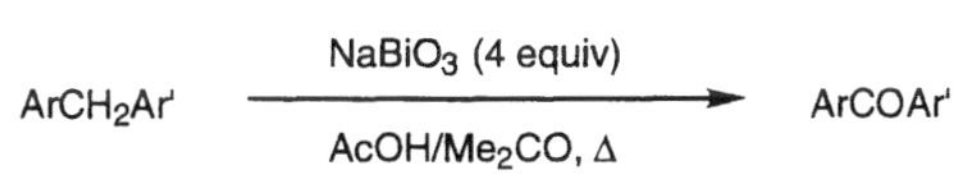

Scheme 5.7. Benzylic oxidation to aromatic ketones with sodium bismuthate [98TL7247].

Oxidation of benzylic methylenes to ketones with sodium bismuthate: general procedure To the initial hydrocarbon (20 mmol) in aqueous acetic acid (4 ml, 50% v/v) and acetone (2 ml), was added sodium bismuthate (22.4 g, 80 mmol) and the mixture was heated to reflux. At the end of the reactions as indicated by TLC, the mixture was filtered through a pad of Celite, diluted with water (10 ml) and extracted with dichloromethane (3 × 20 ml). The combined organic layer was washed with 10% sodium bicarbonate solution (3 × 10 ml) and brine (10 ml), dried over anhydrous sodium sulfate, and concentrated under reduced pressure. The crude product was purified by column chromatography over silica gel using ethyl acetate–hexanes [98TL7247].

5.2.1.2. Zinc bismuthate

Zinc bismuthate $Zn(BiO_3)_2$, readily obtained as a light brown solid from sodium bismuthate and zinc chloride, has been successfully employed for the oxidation of alcohols, thiols, thioethers and oximes (Scheme 5.8) [92BCJ1131, 94SC489]. The yields are comparable with or better than those of the barium permanganate and pyridinium chlorochromate oxidations.

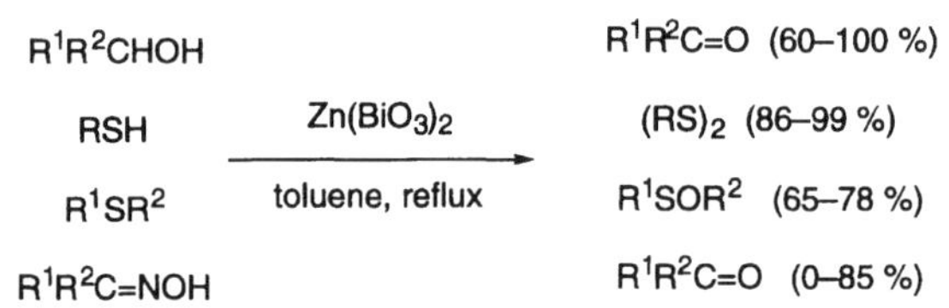

Scheme 5.8. Zinc bismuthate oxidation of alcohols, thiols, sulfides and oximes [92BCJ1131, 94SC489].

Preparation of zinc bismuthate
To a suspension of sodium bismuthate (20 g, 34.54 mmol) in water (30 ml), a 15% aqueous solution (300 ml) of zinc chloride hexahydrate was added and the resulting mixture was refluxed for 1.5 h. After cooling, the mixture was filtered off and the brown solid was washed successively with hot water (100 ml) and acetone (100 ml), and dried over calcium chloride to afford zinc bismuthate; yield 20.5 g (99%) [92BCJ1131].

Oxidation of α-hydroxyketones to α-diketones: typical procedure
In a round-bottomed flask (100 ml) equipped with a condenser was placed a solution of benzoin (1 g, 4.71 mmol) in toluene (50 ml), and zinc bismuthate (2.72 g, 471 mmol) was added in one portion. The resulting mixture was stirred magnetically under reflux for 40 min, while the progress of the reaction was monitored by TLC (eluent: $CCl_4/Et_2O = 3/1$). The reaction mixture was filtered off and the insoluble solid was washed successively with acetonitrile (50 ml) and carbon tetrachloride (50 ml). The filtrate and washings were combined and evaporated to give a residue which was chromatographed on a silica gel. Benzil was obtained as a white crystalline solid (0.96 g, 97%) [92BCJ1131].

5.2.2. Bismuth oxide and mixed bismuth/metal oxide systems as oxidant or oxidation catalyst

5.2.2.1. Oxidation with bismuth oxide

α-Ketols (acyloins) are oxidized readily and quantitatively to α-diketones

TABLE 5.3
Oxidation of acyloins to α-diketones with bismuth oxide [51JCS793]

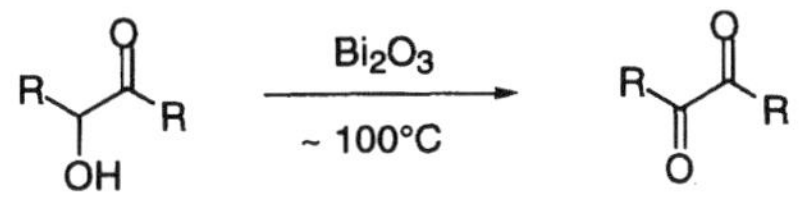

Entry	R	Reaction time (h)	Solvent	Yield (%)[a]
1	Ph	1	AcOH–$HOCH_2CH_2OEt$	95
2	Ph	0.5	AcOH	93
3	4-$MeOC_6H_4$	1	AcOH–$HOCH_2CH_2OEt$	95
4	4-$MeOC_6H_4$	0.3	AcOH	91
5	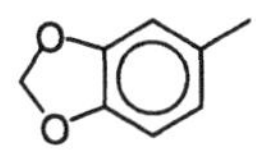	1.3	AcOH–$HOCH_2CH_2OEt$	97
6		0.8	AcOH–$HOCH_2CH_2OEt$	92
7		0.2	AcOH–H_2O	88
8	Me	0.5	AcOH	60
9	Pr	0.5	AcOH	65

[a] Isolated yield.

Oxidation of acyloins to diketones with bismuth oxide: typical procedure
Benzoin (10.6 g, 0.050 mol) and bismuth oxide (9.3 g, 0.020 mol) were stirred in a mixture of acetic acid (10 ml) and 2-ethoxyethanol (15 ml) at around 100°C for 1 h to yield benzil (10.0 g, 95%) after recrystallization [51JCS793].

with bismuth oxide in acetic acid or in a mixture of acetic acid and 2-ethoxyethanol at elevated temperatures [51JCS793]. The representative results are summarized in Table 5.3. In the oxidation of sensitive acyloins such as furoin and piperoin (entries 5–7), bismuth oxide is superior to conventional oxidizing agents. It is reduced to metallic bismuth during the reaction and easily removed by simple filtration. Hence the reaction may be used for qualitative test of acyloins, since the oxidation is rapid and specific for the acyloin group, and metallic bismuth separates as a black precipitate when the reaction is complete. Bismuth acetate is presumed as the actual oxidizing agent of acyloin and, hence, can be employed similarly in place of bismuth oxide. An enediolato complex formed from bismuth(III) and acyloin is a likely intermediate.

This oxidation methodology has often been employed for natural product transformations, which include the conversion of a five-membered α-keto-hemiketal unit in cevin or veracevine to the corresponding hydroxy-δ-lactone [55JA683], 22a,5α-spirostane-3β,12β-diol-11-one to the corresponding 11,12-dione [54JA5533], 2-hydroxypulegone to diophenolene [57JA4465], cucurbitacins B and D to cucurbitacins E and I, respectively [59CI(L)951], and 2α-hydroxyandrosta-4-ene-3,17-dione to 2-hydroxyandrosta-1,4-diene-3,17-dione [58JA1687].

Bismuth oxide is used to generate the carbon radicals from hydrocarbons in the gas phase at high temperatures (365–760°C), from methane, ethane, ethylene, [85JPC4415], propylene, [83JPC301, 83JPC3587, 85JPC4415], 1-butene, [87JPC3585] and isobutylene [83JPC3587, 87JPC3585].

5.2.2.2. Oxidation of olefins over bismuth molybdate-based catalysts

The manufacture of acrolein via the partial oxidation of propylene over cuprous oxide-based catalysts was first patented in 1948 [48USP2451485]. In the early 1960s, however, the bismuth molybdate-based catalysts developed by Standard Oil, Ohio (SOHIO) supplanted the classical cuprous oxide-based ones [59USP2904580, 62USP3044966]. The bismuth molybdate systems can also be used to effect the conversion of a propylene-ammonia mixture to acrylonitrile (Scheme 5.9). Since then, however, the bismuth molybdates have been replaced gradually by multi-component catalyst systems composed of six or more different element oxides, which include oxides of bismuth, molybdenum, iron, cobalt, nickel, manganese, phosphorous, vanadium and so on [65USP3171859, 68JCA(12)281, 69USP3454630, 72GEP2203710, 72GEP2133110, 80JCA(66)347, 84USP4442307].

Industrial catalytic formulations for the oxidation of propylene to acrolein and ammo-oxidation to acrylonitrile are based on the bismuth molybdate phases modified by the above-mentioned metal oxide components. The conversion and selectivity are excellent over the temperature range of 250–450°C, as shown in Table 5.4.

The mechanism of allylic oxidation of propylene has been investigated

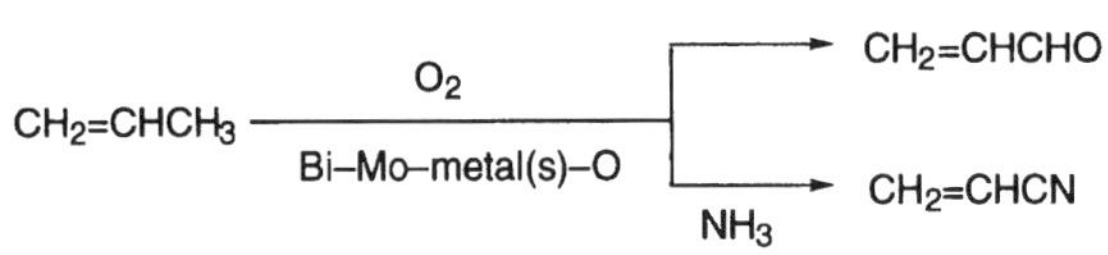

Scheme 5.9. Oxidation and ammo-oxidation of propylene over bismuth molybdate catalysts.

TABLE 5.4
Bismuth molybdate-based catalysts for the oxidation and ammo-oxidation of propylene [94AIC233, 96AC(A)(145)1]

Catalyst	Reaction temperature (°C)	Propylene conversion (%)	Selectivity to AL (%)[a]	Selectivity to AN (%)[b]	Reference
Bi–Mo–O	410–460	57	71	–	[94AIC233]
	450–500	–	–	50–65[b]	
Bi–Mo–Fe–Co–Ni–P–Mg–O	350	98	95	–	[96AC(A)(145)1]
Bi–Mo–W–Fe–Co–Ni–Si–O	250–450	96	90–93	–	[74USP3825600]
Bi–Mo–Fe–Co–Ni–Sn–K–O	300	87	97	–	[96AC(A)(145)1]
Bi–Mo–O (multicomponent)	290–350	95–99	92–96	–	[94AIC233]
	430–480	97–99	–	80–84	

[a] AL, acrolein; AN, acrylonitrile.
[b] Yield (%) of AN.

more extensively on the bismuth molybdate catalytic systems. The rate-determining abstraction of an α-methyl hydrogen to form a π-allyl intermediate has been well established in both oxidation and ammo-oxidation from the isotope effect as well as the isotopic distributions of oxygen and nitrogen insertion products. The subsequent steps can be summarized as a succession of redox and acid–base reactions. The π-allyl anion would be generated by proton abstraction on a basic active site and then oxidized to the π-allyl cation on a redox active site. Through the nucleophilic attack by a lattice O^{2-} anion, the σ-allylic species is formed, which then gives rise to acrolein by hydride abstraction in the subsequent redox step (Scheme 5.10) [89CRS43, 94AIC233, 96AC(A)(145)1].

Extensive investigations have also been undertaken on the direct use of propane for the manufacture of acrolein and acrylonitrile, since this pathway is expected to lower the cost considerably compared with those based on propylene (Scheme 5.11). Several mixed metal oxide systems such as Bi–Ba–Te–O [84USP4472314] and Ce–Te–Mo–O [84OX99] have been proposed so far, although the performance of these catalysts is not satisfactory presently for practical applications. However, successful oxidation and ammo-oxidation of propane to acrolein and acrylonitrile, respectively, have been reported recently using molecular oxygen and bismuth vanadomolybdenate catalysts of a schee-

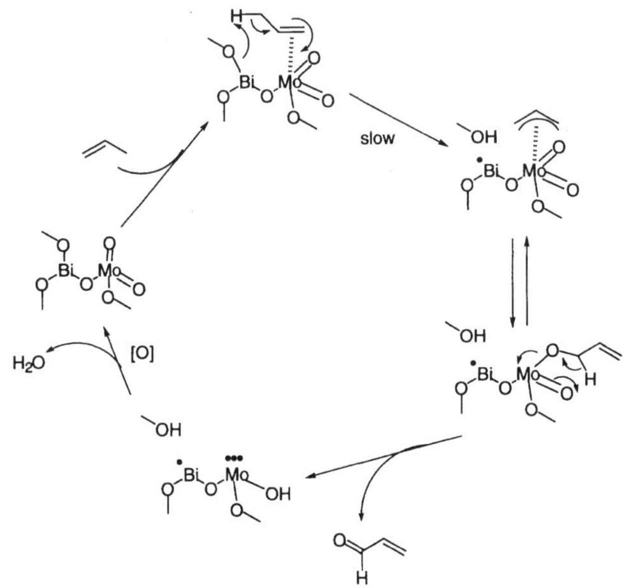

Scheme 5.10. A proposed mechanistic pathway for the allylic oxidation of propylene over bismuth molybdate catalyst.

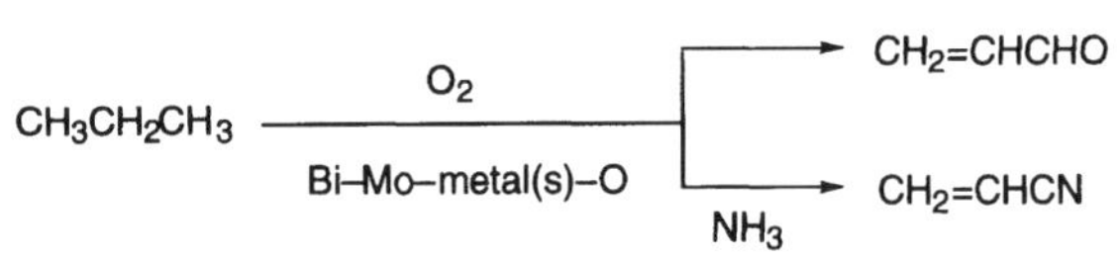

Scheme 5.11. Oxidation and ammo-oxidation of propane over bismuth molybdate catalysts.

lite structure [89CC652, 89CL531, 89CL2173, 90SSSC(55)491, 91AC175, 91AC189, 92CTO673]. The typical results are summarized in Table 5.5.

5.2.2.3. Miscellaneous oxidations using other mixed bismuth/metal oxide-based catalysts

In addition to the oxidation of propane and propylene, the following mode of oxidation over bismuth-containing mixed catalysts has been reported: (i) oxidative coupling of hydrocarbons, (ii) ammo-oxidation of toluene and xylenes, (iii) oxygenation of alkanes and alkenes, (iv) oxidation of alcohols, and (v) oxidative cleavage of carbohydrates. Representative results are summarized in Scheme 5.12.

TABLE 5.5
Bismuth molybdate-based catalysts for the oxidation and ammo-oxidation of propane [96AC(A)(145)1]

Catalyst	Reaction temperature (°C)	Propane conversion (%)	Selectivity to AL(%)[a]	Selectivity to AN(%)[a]	Reference
Bi–Mo–O	500	8.5	26	–	[92CTO673]
	515	9.7	–	51	
Bi–Mo–V–O	488	35	31	–	[91AC175]
Bi–Mo–Nb–O	500	11	58	–	[92CTO673]
	515	12	–	63	
Bi–Mo–V–Ag–O	500	13	64	–	
	515	16	–	65	
Bi–Mo–P–O/SiO_2	529	44	–	54	[96AC(A)(145)1]
Bi–Mo–Fe–Cr–W–Mg–O	470	15	–	54	[89USP4873215] [91USP5008427] [92USP5094989]
Bi–Mo–V–Sb–O/Al_2O_3	500	9	–	49	[88USP4746641] [88USP4760159] [89USP4797381]

[a] AL, acrolein; AN, acrylonitrile.

(i) Oxidative coupling of hydrocarbons

$$2\,CH_4 \xrightarrow{\text{catalyst}} CH_3CH_3$$

catalyst	reference
Bi–Mn or Bi–Mn–M^I	[92RKC(47)193]
Bi_2O_3–P_2O_5–K_2O/Sm_2O_3	[91RKC(44)103]
Bi_2O_3–P_2O_5–K_2O	[87RKC(35)361]
BiOCl–Li_2CO_3–MgO	[92CAL(13)95]

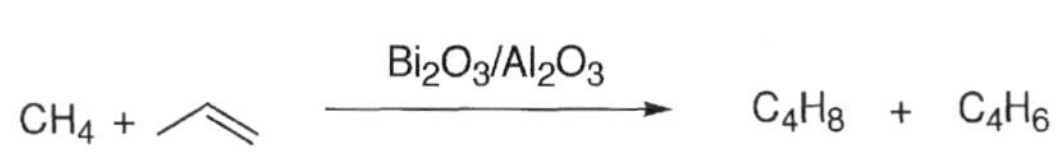

[87CL1513]

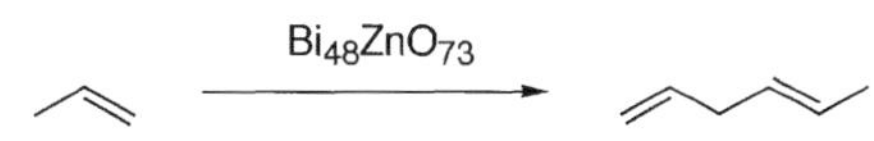

[92CAL(15)393]

(ii) Ammo-oxidation of toluene and xylenes

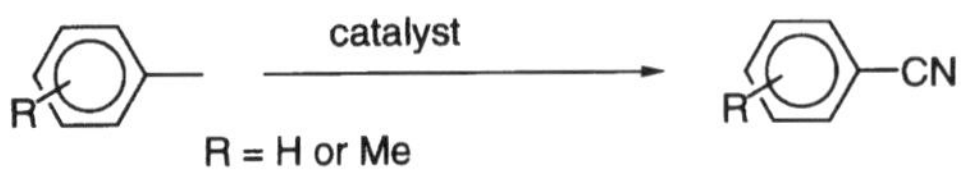

catalyst	reference
Bi–V–Sb, Bi–V–Sb–K, Bi–V–Sb–Ba	[91RKC(43)501]
Bi–V–Sb–M^{I}_2O or Bi–V–Sb–$M^{II}O$	[86RKC(31)409]
Bi–V–Sb–O	[85RKC(27)371]

(iii) Oxygenation of alkanes and alkenes

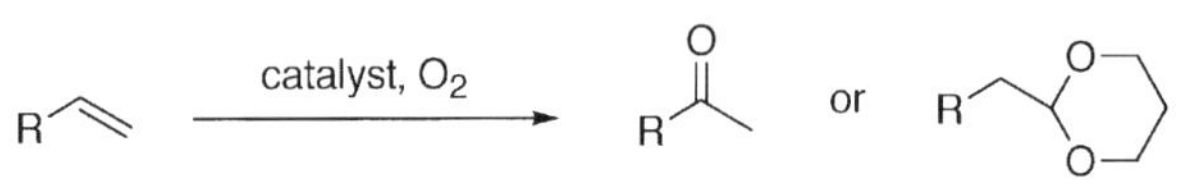

catalyst	reference
$PdCl_2(MeCN)_2$–$BiCl_3$–LiCl	[90BCJ166]
$PdCl_2$–$BiCl_3$–LiCl	[87NJC337]
$[RhCl(C_2H_4)_2]_2$–$BiCl_3$	[83OX405]
Bi_2O_3–SiO_2	[82RRC75]
Bi–Sn–Mo–O	[99AC(A)(180)299]

$$CH_4 \xrightarrow{Bi_2O_3\text{–}SnO_2,\ N_2O} HCHO$$

[85CC1455]

(iv) Oxidation of alcohols

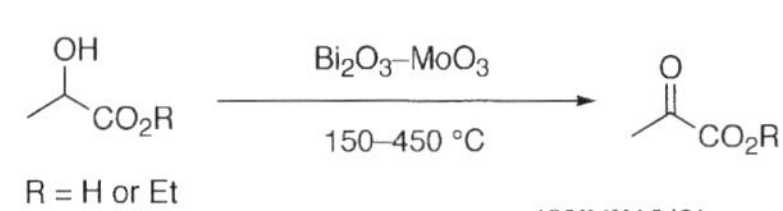

[93BCJ1542]

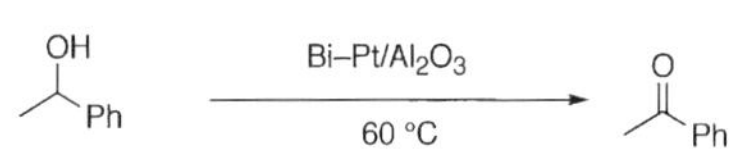

[92PAD1151]

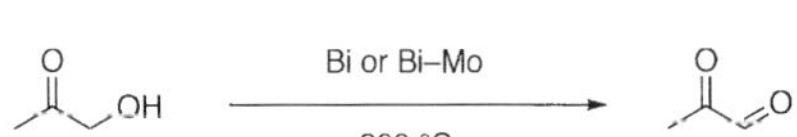

[99BCJ2143]

$$EtOH \xrightarrow[300\text{–}400\ °C]{Bi_2O_3,\ MoO_3\text{–}Bi_2O_3,\ or\ MoO_3\text{–}Bi_2O_3\text{–}V_2O_5} MeCHO + MeCO_2Et$$

[86CL1173]

$$EtOH \xrightarrow[90\text{–}230\ °C]{Bi\text{–}Pt/Al_2O_3\ or\ Bi/Al_2O_3} MeCHO + MeCO_2H$$

[85/86OX127]

(v) Oxidative cleavage of carbohydrates

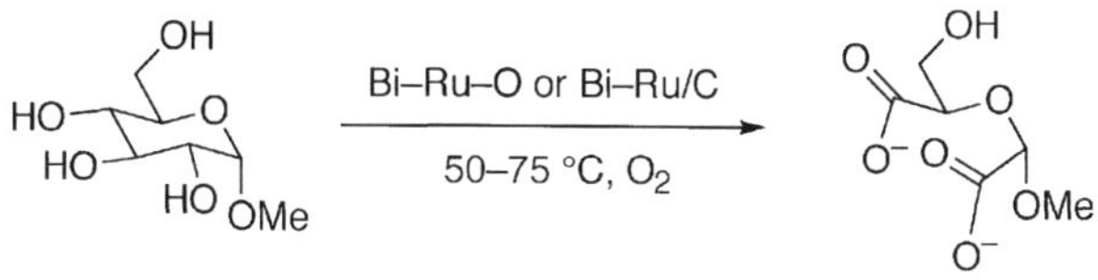

[96JCAC317, 90CAR(204)121]

Scheme 5.12. Miscellaneous oxidations of organic compounds over mixed bismuth–metal based catalysts.

5.2.3. Bismuth(III) salts

Liquid phase auto-oxidation of lower olefins proceeds in acetic acid or acetonitrile at 100°C in the presence of a catalytic amount of bismuth nitrate under pressurized oxygen, affording the corresponding epoxides and carbonyl compounds [71KKZ2072]. Under these conditions, 61.2 mol% of *cis*-2-butene is oxidized to give 33.0 mol% of the epoxide and 9.2 mol% of acetaldehyde, both based on unrecovered olefin. Catalytic performance of other bismuth salts such as bismuth chloride and sulfate is moderate as compared with bismuth nitrate, whereas bismuth acetate, carbonate, phosphate as well as bismuth oxide are almost ineffective as catalysts. A similar oxygenation of alicyclic olefins is carried out using a catalytic amount of bismuth sulfate in propionic acid at 65°C (Scheme 5.13) [76CL29]. The reaction has been found to proceed on the surface of insoluble bismuth salt, and small amounts of bismuth(III) ions present in acid solution exhibit virtually no effect on the oxidation. The mechanism of the auto-oxidation is explained in terms of the isomerization of a peroxy radical intermediate on the acidic site of solid bismuth salt. The air oxidation of lower aliphatic hydrocarbons, catalyzed by bismuth acetate and other bismuth salts at elevated temperatures under high pressure, has been patented [53USP2659746].

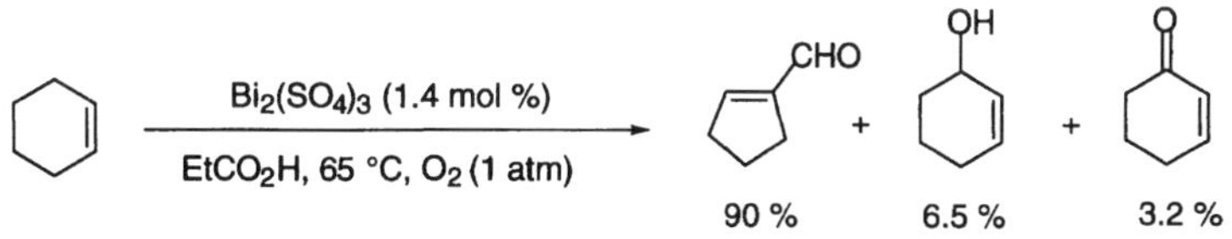

Scheme 5.13. Bismuth sulfate-catalyzed oxygenation of cyclohexene [76CL29].

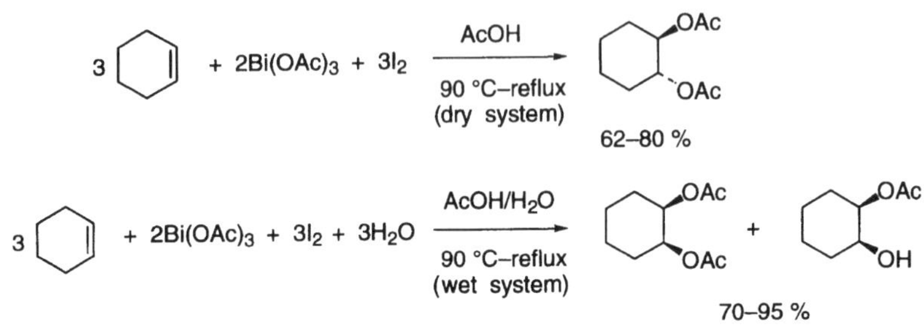

Scheme 5.14. Bismuth acetate-mediated oxidation of cyclohexene to 1,2-cyclohexanediol derivatives [89CC407].

Oxidation of olefins to *trans*-1,2-diol diacetates with bismuth acetate: typical procedure for dry system

Bismuth acetate (4.75 g, 12.3 mmol) was heated under reflux for 1 h in dry acetic acid (40 ml) containing acetic anhydride (ca. 17%). To the cooled solution, cyclohexene (3.01 g, 36.6 mmol) and iodine (9.39 g, 37.0 mmol) were added and the resulting mixture was further refluxed for 17.5 h. Usual work-up followed by distillation gave *trans*-1,2-diacetoxycyclohexane (3.56 g, 80%) with no trace of the *cis*-isomer [89CC407].

Additional examples of the bismuth-promoted oxidation include the bismuth phthalocyanine catalyzed hydroxylation of phenol to hydroquinone and catechol with hydrogen peroxide [91SSSC(66)455], and bismuth acetate-mediated oxidation (modified Prévost reaction) of cyclohexenes to 1,2-cyclohexandiol derivatives (Scheme 5.14) [89CC407].

The modified Prévost reaction uses bismuth acetate instead of silver benzoate. From a mechanistic point of view, iodine is considered to be the actual oxidant, and hence the bismuth acetate may work merely as the acetoxy source. Major advantages of this method over the previously reported analogous ones are as follows: (i) bismuth-containing wastes are non-toxic, (ii) bismuth acetate is inexpensive, and (iii) less metal salt is needed, since three acetoxyl groups can be transferred per one bismuth atom.

Oxygenation reactions using molecular oxygen have been achieved using several other bismuth salts as catalyst: the oxidative cleavage of epoxides or α-ketols to carboxylic acids with bismuth carboxylates (Scheme 5.15) [93TL2601, 94JOM(482)119, 95T4991, 96BSF913, 97NJC495], oxidation of epoxides to cyclic carbonates using bismuth bromide in DMF (Scheme

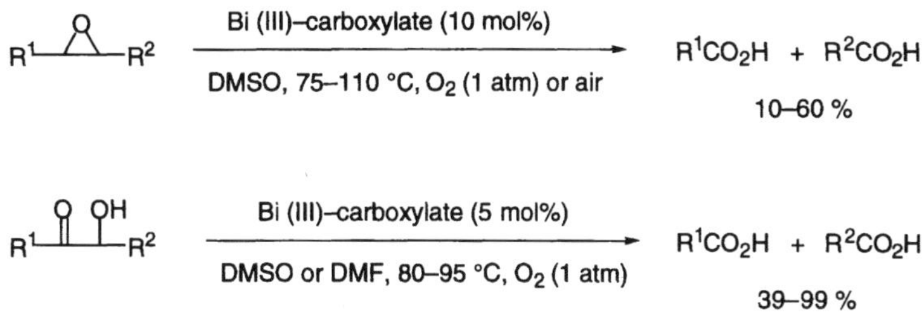

Scheme 5.15. Bismuth carboxylate-catalyzed oxidative C–C bond fission [93TL2601, 94OM(482)119, 95T4991, 96BSF913, 97NJC495].

Oxidative cleavage of epoxide to carboxylic acids with bismuth mandelate: general procedure

A solution of bismuth mandelate ($(Bi(mand)_2)_2O$; 392 mg, 0.4 mmol) in anhydrous DMSO (5 ml) was stirred at 80°C for 30 min under air. An appropriate epoxide (4 mmol) was added and the resulting mixture was stirred at 70–110°C for 22 h to 12 days. Acid hydrolysis, extraction of organic phase with ether, treatment of the organic extract with aqueous NaOH, acidification of the aqueous extract to pH 1–2, and re-extraction with ether followed by evaporation afforded the corresponding carboxylic acid [93TL2601].

5.16) [97CC95], and oxidation of organosulfur and nitrogen compounds with bismuth nitrate (Scheme 5.17) [96CC1847, 97CL1229].

For the direct conversion of epoxides to carboxylic acids with the C–C bond cleavage, bismuth mandelate prepared from bismuth oxide and mandelic acid [92SRI289] is used as an effective catalyst. Bismuth acetate and bismuth oxide are much less effective. α-Diketone, which is the sole product at an early stage of the reaction, has been suggested to be an intermediate in this oxidation. The kinetic study of some substituted styrenes affords a negative and relatively low ρ value (-1.08) in the Hammett plot, indicating the generation of the cationic center at the β-position rather than the α-position with respect to the aryl ring. It has also been suggested that the ring opening of the epoxides may occur via

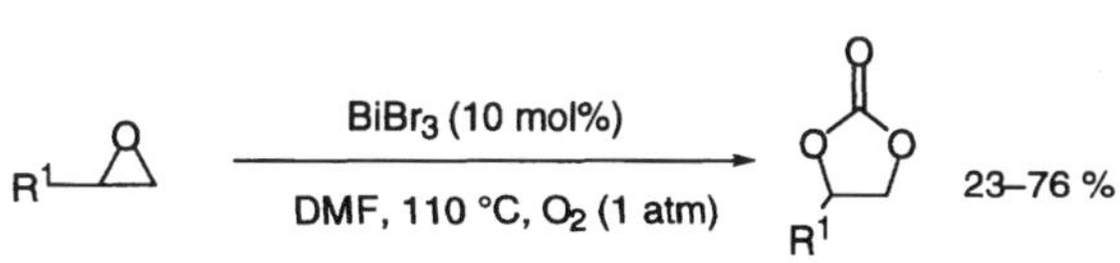

Scheme 5.16. Bismuth bromide-catalyzed formation of cyclic carbonates from epoxides and DMF [97CC95].

initial coordination of a bismuth salt, followed by attack of the sulfoxide oxygen to generate the β-cation.

The oxidative cleavage of α-ketols takes place in either DMF or DMSO, but it does not occur under nitrogen, showing that molecular oxygen is essential as the oxidant and DMSO merely acts as the solvent. Compared with zinc bismuthate and bismuth oxide which convert α-ketols to diketones, the bismuth mandelate/dioxygen system works as a stronger oxidant, leading further to the C–C bond cleavage. The bismuth mandelate catalyst exhibits high functional group selectivity in these reactions, where the bismuth salt is supposed to serve as a Lewis acid to activate both the substrate and oxidant, ruling out the possible involvement of a bismuth(V) species [69NAT(221)169].

As an extension of this oxygenative C–C bond fission, cyclic carbonates are prepared in good yield from terminal epoxides and DMF using bismuth bromide as a catalyst and molecular oxygen as an oxidant (Scheme 5.16). This is the first successful use of bismuth bromide as the catalyst for oxidative functionalization [97CC95]. Ph_3BiF_2 is examined as a catalyst for this reaction [98RCB1607].

Both bismuth nitrate-catalyzed oxidative deprotection of *S,S*-acetals with air [96CC1847] and selective air-oxidation of sulfides to sulfoxides with the bismuth bromide/bismuth nitrate binary system [97CL1229] proceed under mild aerial conditions and are therefore synthetically useful. Bismuth nitrate also mediated selective oxidation of sulfides to sulfoxides in acetic acid [98SC939] and benzyl alcohol to benzaldehyde in benzene [98NJC339]. The in situ generated NO_2 from bismuth nitrate is most likely to play a role as the oxidant for both types of oxygenation. Bismuth nitrate readily oxidizes 4-substituted Hantzsch 1,4-dihydropyridines in acetic acid at room temperature to give the parent pyridine compounds in moderate to good yield [98S713]. Bismuth nitrate can also be utilized for the nitration of aromatic compounds (Scheme 5.42 in Section 5.5.1.3) [88TL5909].

Oxidation of phosphine or arsine ligand occurs adventitiously during the preparation or recrystallization of their bismuth salt complexes. Thus, $[Bi_2Br_6(Me_2PhP)_2(Me_2PhP(=O))_2]$ is formed in the reaction of $BiBr_3$ and Me_2PhP in THF, while $[BiCl_3(Ph_2P(=O)CH_2P(=O)Ph_2)]_2$ and $[BiCl_3(Ph_2Me\text{-}As(=O))(Ph_2As(=O)CH_2CH_2As(=O)Ph_2)]_n$ are obtained during the recrystallization of $[BiCl_3(Ph_2PCH_2PPh_2)]_2$ and $[BiCl_3(Ph_2AsCH_2CH_2AsPh_2)]$, respec-

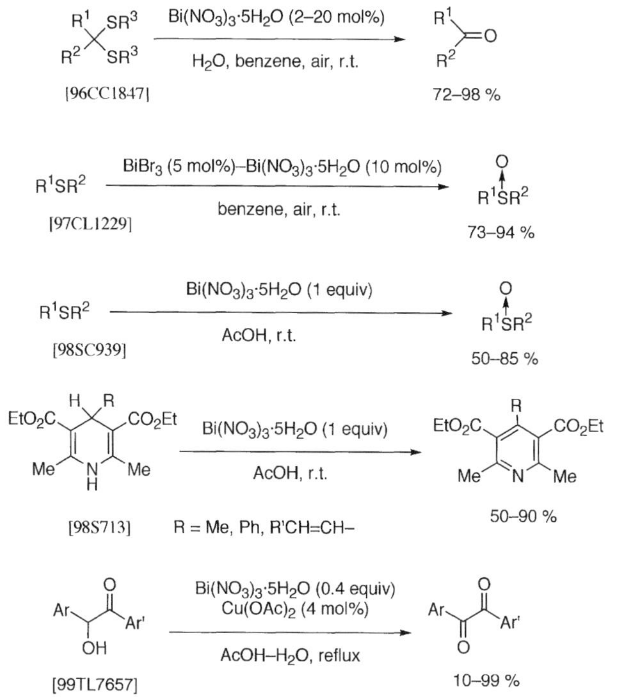

Scheme 5.17. Bismuth nitrate-catalyzed or -mediated oxidations.

Deprotection of *S,S*-acetals to carbonyl compounds: general procedure

To a solution of *S,S*-acetal (0.5 mmol) in dry benzene or dichloromethane (10 ml) was added finely powdered bismuth nitrate pentahydrate (2–20 mol%) followed by water (1 mmol). When needed, bismuth chloride (5 mol%) was also added. The mixture was magnetically stirred in a stoppered flask for an appropriate time at ambient temperature, while the progress of the reaction was intermittently monitored by GLC. The stopper was loosened at intervals in order to admit fresh air into the reaction vessel. After the reaction, the mixture was worked up as usual to give the parent carbonyl compound in 65–98% yields [96CC1847].

tively, from boiling acetonitrile/charcoal. Interestingly, the latter oxidation product results from the simultaneous C–C bond fission and oxidation of arsenic atoms [92JCS(D)1753, 95JCS(D)759]. The oxidation of ethanolamine with Bi(V) fluoride complex has been studied kinetically in $HClO_4$–HF mixture [99OX244].

5.2.4. Organobismuth(V) compounds

The mild oxidizing ability of pentavalent organic bismuth compounds was first disclosed by Challenger in 1934 in the oxidation of alcohols with $Ar_3Bi(OH)_2$ [34JCS405]. About 40 years later, Barton's and Dodonov's groups started intensive studies on a variety of pentavalent bismuth compounds for their oxidizing and arylating ability. Typical results of the bismuth(V)-based oxidations are summarized in Table 5.6. This oxidation is characterized by mild conditions of temperature and pH as well as high chemoselectivity. The reaction usually proceeds at room temperature under neutral or basic conditions. This methodology can be applied not only to the oxidation of alcohols and oxidative cleavage of *vic*-glycols but also to the oxidation of a wide variety of other functionalities, which include hydrazine, hydrazone, thiol, oxime, phenol, alkylbenzene, thiolate, alkoxide, dithiocarbamate, and phosphine. Toluene and ethylbenzene are oxidized to benzaldehyde and acetophenone, respectively, with $Ph_3Bi(OO^tBu)_2$ [88ZOB712, 92MOK1265]. Pentane and hexane are oxidized at room temperature by Ph_3Bi–^tBuOOH first to afford α-diketones, which, under the reaction conditions employed, undergo further transformation to yield carboxylic acid anhydrides [93MOK375]. Ph_4BiX and Ph_5Bi type organobismuth(V) compounds also perform the oxidation of alcohols effectively. For the oxidation of alcohols with tetraphenylbismuth trifluoroacetate, basic conditions are necessary; under neutral or slightly acidic conditions, *O*-phenylation takes place (Table 5.13 in Section 5.5.2.1) [87JCS(P1)251].

The oxidation with organobismuth(V) compounds is compatible with a variety of functionalities. The thiocarbonyl groups in xanthates, dialkylamino thiocarbamates and di-*tert*-butyl thioketone are unaffected by triphenylbismuth carbonate. Sensitive substrates such as aniline, indole, pyrrole and selenide are likewise inert under ordinary reaction conditions. Therefore, the organobismuth(V)-based oxidation has been successfully applied to the synthesis of highly functionalized natural products such as cembranolide diterpenes [79AJC1273], maytansinoids [80CC1089], 17α-methyl cardenolides [82H641], 4,6,8(14),22-ergostatetraen-3-one [85BSF849] and methyl carbapenem antibiotic precursors [93AG(E)867]. High chemoselectivity is illustrated by the examples where the allylic alcohol function is selectively oxidized to acrolein in the presence of an alcohol or a thiol function [79CC705, 81T73].

TABLE 5.6
Oxidation with organobismuth(V) compounds

Oxidant	Substrate	Condition	Product	Reference
$Np_2Bi(=O)OMe$	$[Ph_2C(OH)]_2$	Neutral	Ph_2CO	[88CL2021]
(Np = 1-naphtyl)	PhNHNHPh	Neutral	PhN=NPh	
	Benzoin	Neutral	Benzil	
$Ph_3Bi(OH)_2$	Alcohols	Neutral	Carbonyl compounds	[34JCS405]
Ph_3BiCO_3	Alcohols	Neutral	Carbonyl compounds	[79CC705] [80CC246] [81T73]
	Cholestanol	Basic	Cholestanone	[87JCS(P1)251]
	Thiols	Neutral	Disulfides	[79CC705]
	$Ph_2C{=}NNH_2$	Neutral	Ph_2CN_2	[80CC827]
	Hydrazines	Neutral	Azines	[81T73]
	Oximes	Neutral	Ketones	
	vic-Glycols	Neutral	Carbonyl compounds (C–C cleavage)	[80CC1089] [81T73] [86T5627] [93AG(E)867]
	2,6-Dimethyl-phenol	Neutral or basic	2,2′,6,6′-Tetramethyl diphenoquinone	[80CC827] [81T73]
	2,6-Di-*tert*-butylphenol	Basic	2,2′,6,6′-Tetramethyl diphenoquinone	[87T323] [88T4483]
$(Ph_3BiCl)_2O$	Alcohols	Basic	Carbonyl compounds	[78CC1099] [79AJC1273]
	vic-Glycols	Basic	Carbonyl compounds (C–C cleavage)	[81T73]
$Ph_3Bi(OO^tBu)_2$	Alcohols	Neutral	Carbonyl compounds	[88ZOB712]
	Alkyl-benzenes	Neutral	Carbonyl compounds	[98RCB659]
	Dibenzyl		Benzyl or benzoin	[97RGC927]
	vic-Glycols		Carbonyl compounds (C–C cleavage)	[97RCB1164]
	Diethyl ether		Ethoxyacet-aldehyde	[95IZV783]

TABLE 5.6 (*continued*)

Oxidant	Substrate	Condition	Product	Reference
$Ph_3Bi/^tBuOOH$	Alkanes	Neutral	Carboxylic acid anhydrides	[93MOK375]
	vic-glycols		Carbonyl compounds (C–C cleavage)	[97RCB1164]
	Diethyl ether		Ethoxyacet-aldehyde	[95IZV783]
$Ph_3Bi(OAc)_2$	Alcohols	Neutral	Mixture of carbonyl compounds, phenyl ether and acetate	[81ZOB2380] [85ZOB73]
	Alcohols	Basic	Carbonyl compounds	
	Stannylene of *vic*-glycol	Neutral	Carbonyl compounds (C–C cleavage)	[81TL2885]
	Pinacol	Neutral or basic	Acetone	[85ZOB73]
$Ph_3Bi(O_2CR)_2$ R = Me, Ph, CF_3	(−)-Carveol	Basic	(−)-Carvone	[79CC705] [81T73]
Ph_3BiCl_2, Ph_3BiBr_2	Allylic alcohols	Basic	Carbonyl compounds	[79CC705] [81T73] [87JCS(P1)251]
Ph_3BiCl_2	ROLi	Neutral	Carbonyl compounds	[84ZOB100]
	PhSNa	Basic	PhSSPh	[85JCS(P1)2667]
	Dithio-carbamates		Dithiuram disulfides	[68JOM(11)627]
	vic-Glycols	Basic	Carbonyl compounds (C–C cleavage)	[86T5627]
	2,6-Di-*tert*-butylphenol	Basic	2,2′,6,6′-tetramethyl diphenoquinone	[87T323] [88T4483]
$(Ar_3BiO)_n$	Alcohols	Neutral	Carbonyl compounds	[94TL8197]
	$[Ph_2C(OH)]_2$	Neutral	Ph_2CO	
	PhNHNHPh	Neutral	PhN=NPh	
	Ph_3P	Neutral	Ph_3PO	

TABLE 5.6 (*continued*)

Oxidant	Substrate	Condition	Product	Reference
$Ph_3Bi(NO_3)_2$	Allylic alcohols	Basic	Carbonyl compounds	[79CC705] [81T73]
$Ph_3Bi(O_2CCF_3)_2$	Cholestanol	Basic	Cholestanone	[87JCS(P1)251]
$Ph_3Bi(OTs)_2$	Cholestanol	Basic	Cholestanone	[87JCS(P1)251]
$^{cat}Ph_3Bi$, NBS	*vic*-Glycols	Basic	Carbonyl compounds	[81CC1232] [86T5627]
$Ar_3Bi{=}NR$ Ar = Ph, *o*-tolyl, *p*-tolyl	Alcohols	Neutral	Carbonyl compounds	[96JCR(S)24] [99OM2580]
	α-Ketols		Diketones	
R = Ts, CF_3CO	*vic*-Glycols		Carbonyl compounds (C–C cleavage)	
	PhSH		PhSSPh	
Ph_4BiOR R = $COCF_3$, Ts, $COCH_3$, Tf, 2,4,6-$(O_2N)_3C_6H_2$, $COCCl_3$, $COCH_2Cl$, $COCHPh_2$	Alcohols	Basic	Carbonyl compounds	[87JCS(P1)251]
Ph_5Bi	Alcohols	Neutral	Carbonyl compounds	[75DOK(225)581] [80CC827] [87JCS(P1)251]
	ArSH	Neutral	ArSPh	[80CC827]

Oxidation of alcohols with a stoichiometric amount of $(Ph_3BiCl)_2O$: general procedure
To a solution of an alcohol (0.25 mmol) and a bismuth(V) oxidant (0.20 mmol) in dichloromethane (2 ml) was added an excess of K_2CO_3 or $NaHCO_3$ (200 mg) and the resulting mixture was stirred until the oxidation was complete. The oxidation product was readily separated from the non-polar triorganylbismuth compound by chromatography on silica gel [78CC1099].

Oxidative cleavage of *vic*-glycols to carbonyl compounds with NBS in the presence of a catalytic amount of triphenylbismuthine: typical procedure
A solution of NBS (0.98 g, 5.5 mmol) in acetonitrile (55 ml) containing 1% of water was added dropwise over 10 min to a mixture of hydrobenzoin (1.07 g, 5 mmol), triphenylbismuthine (0.22 g, 0.5 mmol) and K_2CO_3 (6.9 g, 50 mmol) in the same solvent (50 ml) at room temperature in the dark. After stirring for 15 min, the mixture was filtered and the solvent is removed under vacuum. The residue was extracted with a mixture of water and ether and the organic phase was

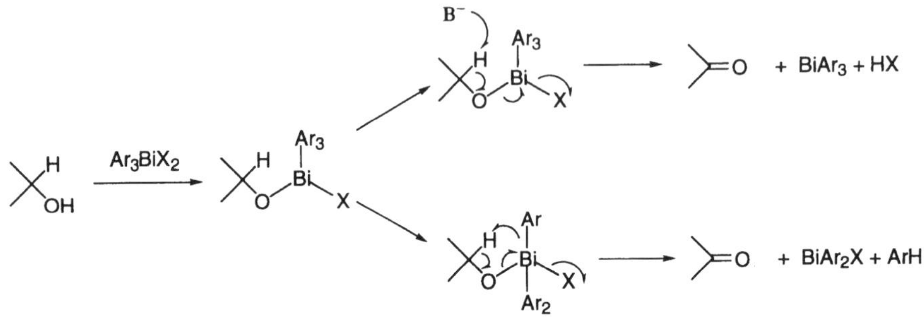

Scheme 5.18. Pathways for the oxidation of alcohols with organobismuth(V) compounds.

separated, dried over Na_2SO_4, and evaporated under vacuum. The residue was purified by preparative TLC to give benzaldehyde (0.85 g, 80%) [86T5627].

The oxidation of alcohols is proposed to proceed via a pentavalent bismuth alkoxide, which is prone to undergo reductive elimination to carbonyl compound and triorganylbismuth. From the tracer study, two mechanistic pathways are postulated as shown in Scheme 5.18 [80CC246, 81T73]. The intermediate formation of a bismuth(V) compound with the covalent Bi–O bond is confirmed by ^{1}H-NMR [87JCS(P1)251].

vic-Glycols are cleaved to carbonyl compounds by a stoichiometric use of Ph_3BiCO_3, $(Ph_3BiCl)_2O$, Ph_3BiCl_2, $(Ph_3BiO)_n$ or $Ph_3Bi(OAc)_2$, where Ph_3Bi is recovered quantitatively in the case of Ph_3BiCO_3 and $(Ph_3BiCl)_2O$. This oxidation can also be carried out using a catalytic amount of Ph_3Bi in wet acetonitrile in the presence of *N*-bromosuccinimide (NBS) as a reoxidant and potassium carbonate as a base. This is the only reported clear-cut example of the oxidation utilizing the Bi(III)–(V) redox cycle. As for the mechanism, the stoichiometric reaction is assumed to involve a metallacycle intermediate as shown in Scheme 5.19, whereas the catalytic reaction is considered to proceed via the initial formation of a glycol hypobromite which converts Ph_3Bi to a pentavalent bismuth species as shown in Scheme 5.20 [86T5627]. This mechanism is supported by the findings that the chemospe-

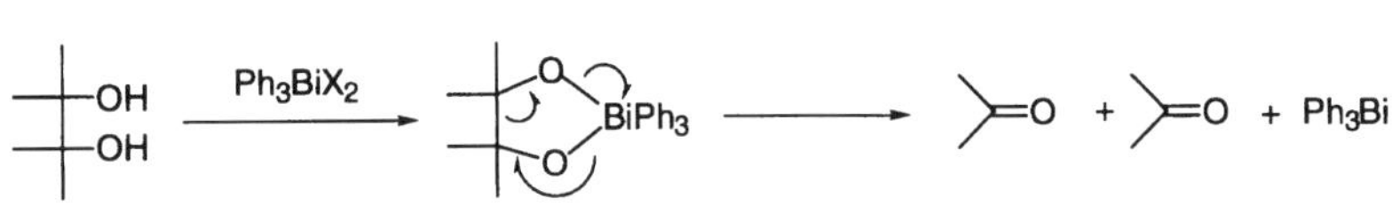

Scheme 5.19. Stoichiometric cleavage of *vic*-glycols with Ph_3BiX_2.

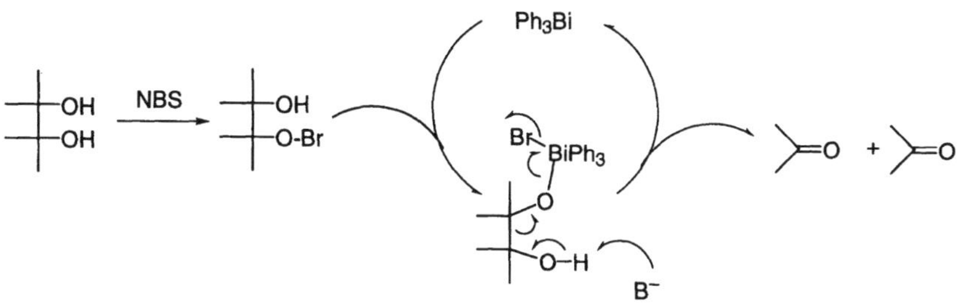

Scheme 5.20. Catalytic cleavage of *vic*-glycols with Ph_3Bi and NBS [86T5627].

cific cleavage of *cis*-diols occurs in stoichiometric reaction, while both *cis*- and *trans*-isomers are cleaved at the same rate in the catalytic process.

5.3. REDUCTION

A combination of metallic bismuth or bismuth chloride and a reducing agent has been employed for the reduction of the nitro group and C–C/C–N double bonds, as shown in Schemes 5.21 and 5.22. The reducing ability and/or selectivity of sodium borohydride is improved considerably in the presence of bismuth powder or bismuth chloride.

Aromatic and aliphatic nitro compounds as well as benzoxadiazole are reduced with a variety of combinations of bismuth compounds and reducing agents to afford the corresponding azoxy compounds, hydroxylamines or amines, as shown in Scheme 5.21. It is generally assumed that the reaction is promoted by a highly reactive metallic bismuth generated in situ. Actually, the formation of metallic bismuth as a black precipitate is accompanied by the concurrent evolution of hydrogen gas, when sodium borohydride and bismuth chloride are mixed in solution [95SC3799]. The reducing ability of the resulting bismuth(0) species may give an azoxy or amino compound as a product. The over-reduction of azoxy compounds to amines is inhibited owing to the low reactivity of metallic bismuth [94TL3167] and the in situ generated bismuth(0) from bismuth chloride and zinc dust [96SC3903]. In the presence of hydrofluoric acid and ammonium sulfate, metallic bismuth appears to be enhanced in its reactivity [95JFC(74)251, 95SC4025]. The metallic bismuth–ammonium sulfate system exhibits high chemoselectivity toward the nitro group, while the nitrile, ester, ketone, carboxylic acid and halide functionalities do not suffer any reduction [95SC4025]. A similar chemoselectivity is

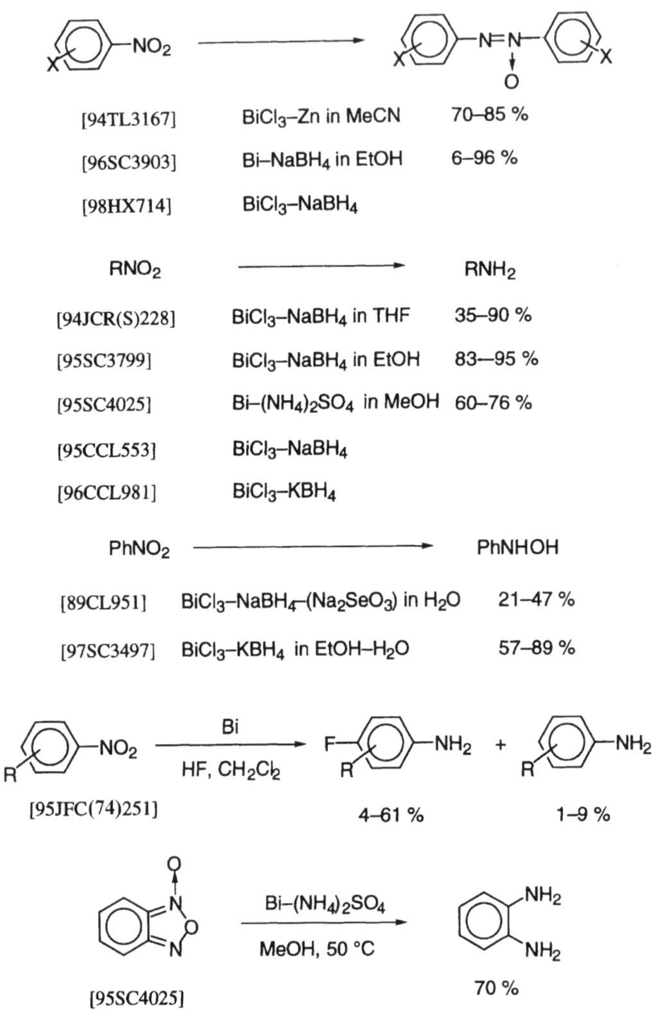

Scheme 5.21. Reduction of nitroarenes and related compounds.

Reduction of nitrobenzenes to anilines with bismuth chloride–sodium borohydride: typical procedure

Bismuth chloride (0.630 g, 2 mmol) was suspended in THF (25 ml), and sodium borohydride (0.076 g, 2 mmol) was added with stirring at ambient temperature. To the resulting black suspension was added 4-chloronitrobenzene (0.314 g, 2 mmol), followed by an additional amount of sodium borohydride (0.076 g, 2 mmol) in several portions at 60°C. The reaction mixture was stirred at 60°C for 1–2 h. After cooling, the solvent was removed under reduced pressure and water (30 ml) was added. The organic phase was extracted with ether (30 ml) three times, dried, and evaporated to leave a residue which was chromatographed on silica gel to give 4-chloroaniline (0.230 g, 90%) [94JCR(S)228].

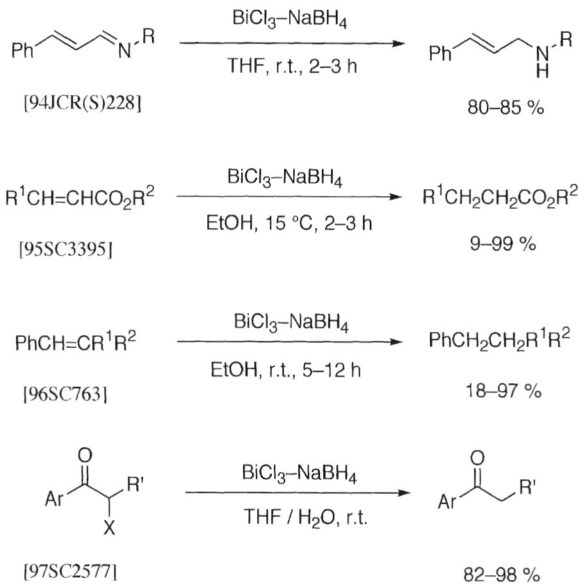

Scheme 5.22. Selective reduction of C=C, C=N and C–X bonds.

Reduction of α,β-unsaturated imines to allylic amines with bismuth chloride–sodium borohydride: typical procedure

Bismuth chloride (0.630 g, 2 mmol) was suspended in THF (20 ml) and sodium borohydride (0.076 g, 2 mmol) was added with stirring at ambient temperature. To the resulting black suspension was added *N*-cinnamylideneaniline (0.207 g, 1 mmol) in THF (5 ml), followed by sodium borohydride (0.076 g, 2 mmol) in several portions. The reaction mixture was stirred at room temperature for 2 h and then quenched with water (25 ml). The solvent was removed under reduced pressure and the residue was extracted with dichloromethane (20 ml) twice. The combined extract was dried over sodium sulfate and distilled to give the expected amine (0.178 g, 85%) [94JCR(S)228].

observed in the reduction of nitroarenes to azoxy compounds with a combination of bismuth chloride and zinc dust [94TL3167]. The reduction of nitroarenes with metallic bismuth in liquid hydrogen fluoride results in simultaneous reduction–fluorination to afford the corresponding fluoroanilines selectively, while other metals of mild reducing ability such as tin, lead, indium, cobalt, germanium and silicon are all unsatisfactory for such a purpose [95JFC(74)251]. The reductive intramolecular cyclization of *o,o'*-dinitrobiphenyl to benzo[*c*]cinnoline *N*-oxide is performed with bismuth chloride–zinc in good yield [94TL3167].

The combination of bismuth chloride–sodium borohydride is successfully applied to the selective reduction of the C=C and C=N double bonds, and the C–X bond. α,β-Unsaturated imines are selectively reduced to allylic amines with this reagent [94JCR(S)228]. It also reduces selectively the C=C double bond conjugated with an aromatic ring, ester, amide or nitrile group [95SC3395, 96SC763, 96CCL788, 97YH462]. Some nitrogenous functional groups such as nitro, cyano and diazo are reduced by polymer-supported borohydride resin under the catalysis of $BiCl_3$ [97HDX53]. A catalytic asymmetric hydrogenation of cyclic imines is performed with iridium complex in the presence of BiI_3 [98TA2415, 98TA2657]. Aromatic α-bromo and iodo ketones are reduced in water to their parent ketones. α-Chloro ketones are reduced in the presence of sodium iodide [97SC2577]. As for the reaction mechanism, it has been proposed that the in situ generated bismuth(0) adsorbs and activates the substrate toward the nucleophilic attack by borohydride [96SC763].

Mixed bismuth–metal oxychlorides with or without added alkali or alkaline earth cations show low to moderate catalytic performance in the dehydrohalogenation of *tert*-butyl halides [93BCJ347]. A Bi–Pd (Bi/Pd = 0.4–0.6) binary system supported on metal oxides is found to be an efficient catalyst for the dehydrohalogenation of 1,1,2-trichlorotrifluoroethane (FC-113) to trifluoroethene, a key intermediate to FC-134a, with 80–90% selectivity at 80–100% conversion [91CL841] (Scheme 5.23).

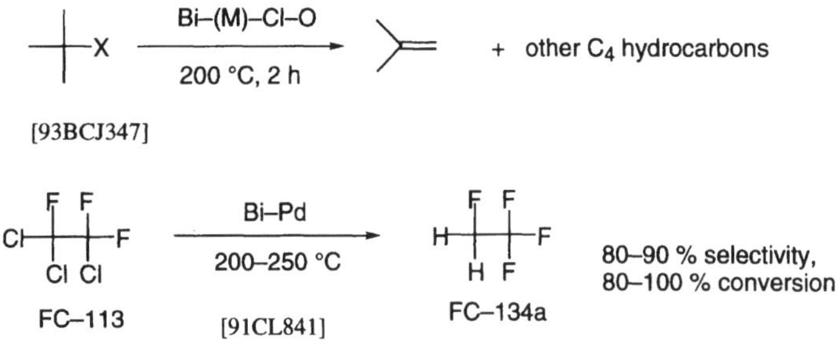

Scheme 5.23. Dehalogenation of alkyl halides over bismuth-based catalysts.

5.4. CARBON–CARBON BOND FORMING REACTIONS

5.4.1. Bi(0)-promoted carbon–carbon bond forming reactions

In 1985, Wada et al. reported that bismuth(0) can promote the Barbier-type allylation of aldehydes in DMF with high chemoselectivity over ketones [85TL4211]. Similar reactions are accomplished with bismuth(0) in situ generated from a combination of bismuth chloride and a variety of reducing agents [90YGK960, 99YGK689] such as metallic zinc, iron [86TL4771, 90BCJ1738, 97BCJ2265], aluminum [87CC708], magnesium [97TL8045], sodium borohydride [96CHJ462, 97SC2569], and also under electrolytic conditions [88CL2049] (Table 5.7). All allylations proceed with high chemo-, regio- and stereo-selectivities. The nitrile, ester, halide, alcohol and carboxylic acid functions are all intact under the conditions of this allylation. The intermediate allylic metal species discriminates aldehydes from ketones, and the carbonyl group from the carbonyl-conjugated double bond. When crotyl bromide is reacted, the addition occurs selectively at the γ-position

TABLE 5.7
Bi(0) and in situ generated Bi(0)-promoted allylation of aldehydes

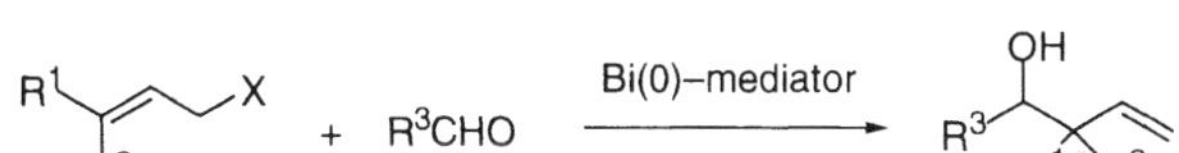

$BiCl_3$–Al	THF/H_2O, r.t., 10–20 h	30–96%	[87CC708] [90BCJ1738]
$BiCl_3$–Zn or Fe	THF, r.t., 2–7 h	45–99%	[86TL4771] [90BCJ1738] [97BCJ2265] [99YGK687]
	$THF/H_2O/HCl$ or KOH, r.t.	79–85%	[99YGK687]
$BiCl_3$–$NaBH_4$	THF/H_2O, r.t.	74–96%	[96CHJ462]
	Bu_4NBr, DMF, 50°C		[97SC2569]
$BiCl_3$–Mg	THF/H_2O, r.t.	64–90%	[97TL8045] [99YGK687]
Bi(0)	DMF, r.t., 2–12 h	53–98%	[85TL4211] [90BCS1738]
	Bu_4NBr, H_2O	70%	[96PAC919]
	MeCN, r.t.	50–91%	[99YGK687] [2000BCJ689]
$BiCl_3$–electrolysis	CH_2Cl_2/H_2O, r.t., $NaBF_4$, 5 F/mol, divided cell	55–88%	[88CL2049]

Allylation of aldehydes to homoallylic alcohols with bismuth chloride–aluminum powder: general procedure

To a mixture of commercial grade aluminum powder (0.13 g, 4.8 mmol), bismuth chloride (0.76 g, 2.4 mmol), THF (2.5 ml) and water (1 ml) was added an aldehyde (2.0 mmol) followed by an allylic halide (2.4 mmol) using a syringe under nitrogen at room temperature. After stirring for 10–20 h at room temperature, the reaction was quenched by the addition of saturated aqueous ammonium chloride, and the organic materials were extracted with ether (20 ml) three times. After drying, the ether extract was evaporated in vacuo to leave the expected product, which was isolated by preparative TLC on silica gel [87CC708, 90BCJ1738].

to afford the *erythro* product with a high diastereoselectivity. A remarkable feature of this reaction is that allylation proceeds smoothly in aqueous solvent, sometimes in the presence of a catalytic amount of bismuth chloride instead of its stoichiometric use [87CC708, 88CL2049, 90BCJ1738, 96CHJ462, 96PAC919, 99YGK689]. The dual function of water is suggested, (i) to accelerate the formation of allylbismuth reagent, and (ii) to make aldehydes more vulnerable to the attack by an allylbismuth reagent [96CHJ462].

Although the reaction mechanism is not clear at present, the intermediate formation of allylbismuth species through the oxidative addition of allylic halide to Bi(0) generated in situ has been proposed [90BCJ1738]. The generation of metallic bismuth by the reduction of bismuth chloride with metallic zinc is known [58DOK(122)614]. A plausible catalytic cycle with

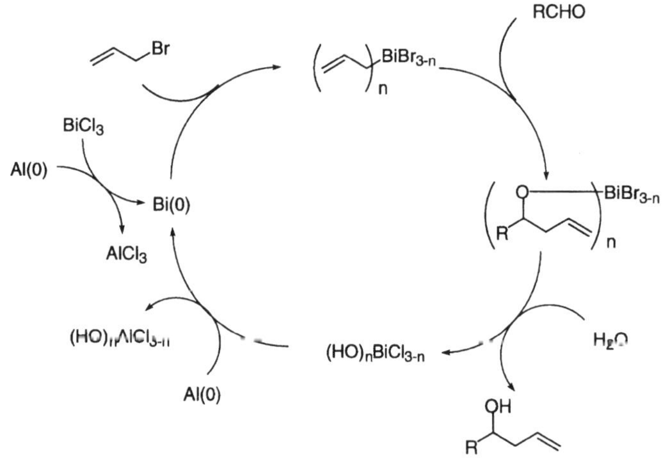

Scheme 5.24. Mechanism for the Barbier-type allylation of aldehydes catalyzed by $BiCl_3$–Al(0) in the THF–H_2O system [90BCJ1738].

the bismuth chloride–aluminum powder system is illustrated in Scheme 5.24.

The bismuth(0)-mediated allylation, especially catalytic ones in aqueous media, has stimulated further studies and several related reactions have been reported, as shown in Scheme 5.25. The bismuth chloride–aluminum powder system in aqueous media has been successfully applied to the alkylation of 1-(aminoalkyl)benzotriazoles [91TL4247, 92OM1381, 95SC2751] and the Reformatsky-type reaction of α-bromoketones [97TL2733]. The catalytic bismuth chloride–zinc dust system in aqueous media also works effectively for the reductive cross-coupling of aldehydes with α-diketones [95CL999]. The homocoupling reactions of aldimines and allylic halides proceed in methanol [89TL1261, 95TL6747]. Metallic bismuth is used for the allylation of imines [93TL7975], enamines [97CEJ1064] and sulfonyl chlorides [98SL1083] (Section 5.5.1.2) under mild conditions, propargylation–allenylation of aldehydes in H_2O/MeOH [98JOC7472], and also for the isomerization and fragmentation of allyl phenyl ether at elevated temperature [87BCJ4279].

Cross-coupling reaction between aldehydes and α-diketones with bismuth chloride and metallic zinc: typical procedure

2,3-Butanedione (0.0488 g, 0.567 mmol) and 3-phenylpropanal (0.113 g, 0.843 mmol) were added to a suspension of bismuth chloride (0.375 g, 1.19 mmol) and zinc dust (0.203 g, 3.11 mmol) in THF (10 ml) under argon at room temperature. After stirring for 40 min under reflux, the reaction was quenched by the addition of hydrochloric acid (1 M, 10 ml). Organic materials were extracted with ether three times and the combined organic extracts were washed successively with water and brine, and dried over sodium sulfate. After evaporation, the residue was subjected to thin layer chromatography on silica gel (hexane:AcOEt = 3:2) to give a diastereomeric mixture of 3,4-dihydroxy-4-methyl-1-phenyl-5-hexanone (0.107 g, 85% yield) [95CL999].

5.4.2. Bismuth(III) salt-catalyzed carbon–carbon bond forming reactions

Bismuth chloride catalyzes the Mukaiyama–aldol and Michael reactions [88TL4719, 90YGK960]. Aldehydes, ketones, acetals and α,β-unsaturated ketones all react with trimethylsilyl enol ethers under mild conditions (Table 5.8) [91BCJ990]. Catalytic activity of bismuth chloride is often enhanced remarkably by the addition of some other metal iodide [93JOC1835, 93BSF832] (Table 5.8). This mixed catalytic system was employed successfully for the Mukaiyama-aldol and Michael-type reactions of furan derivatives [92TL1053, 93BSF832], as well as for the tandem aldol-halogenation reaction [94JOC2238] (Scheme 5.43).

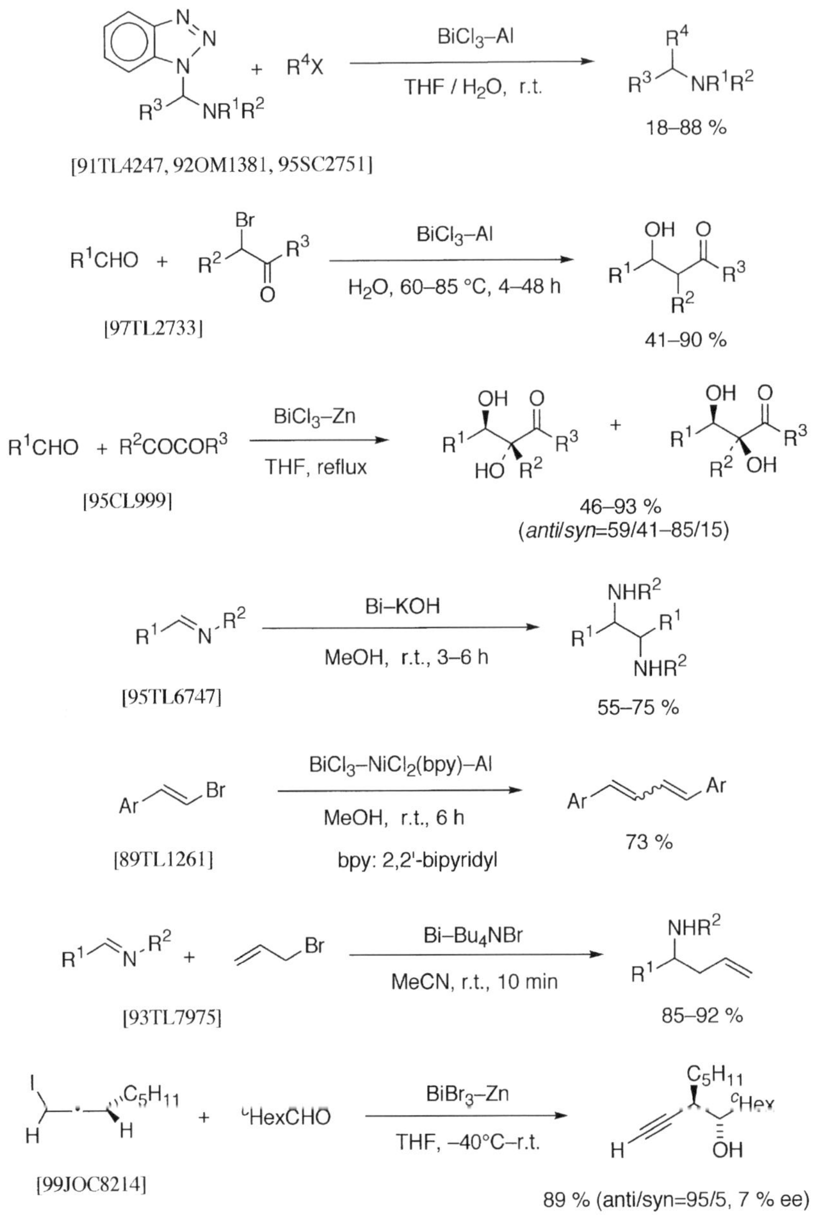

Scheme 5.25. Bi(0) and in situ generated Bi(0)-promoted coupling reactions.

TABLE 5.8
$BiCl_3$ and $BiCl_3$–metal iodide catalyzed Mukaiyama-aldol and Michael-type reactions

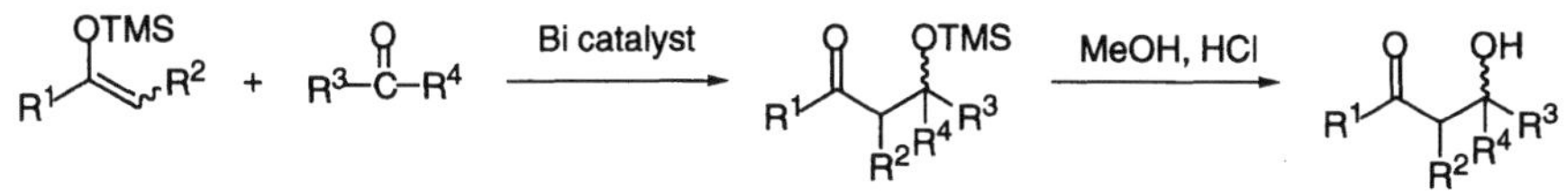

Bi catalyst	Conditions	Yield (%)	Reference
$BiCl_3$–M′I or M″I_2 (5–20 mol%)	CH_2Cl_2, −78°C–r.t., instantaneous–6 h	0–98	[93JOC1835] [93BSF832] [92TL1053]
$BiCl_3$ (5–20 mol%)	CH_2Cl_2, r.t., 25 min–40 h	0–95	[91BCJ990] [88TL4719]

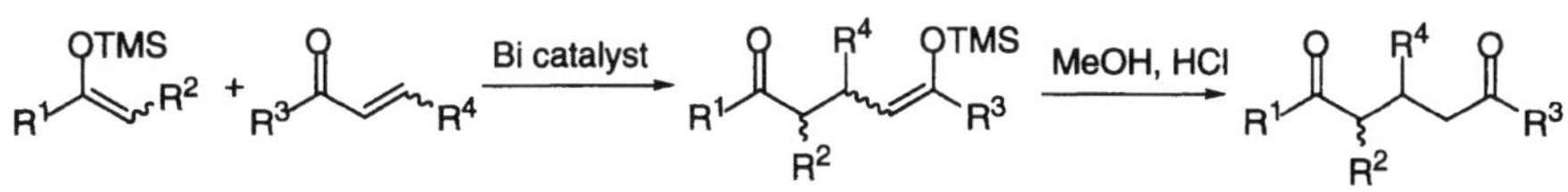

Bi catalyst	Conditions	Yield (%)	Reference
$BiCl_3$–M′I or M″I_2 (5–20 mol%)	CH_2Cl_2, r.t., 1–3 h	75–94	[93JOC1835] [93BSF832]
$BiCl_3$ (5 mol%)	CH_2Cl_2, r.t., 15 min–6 h	31–80	[91BCJ990] [88TL4719]

Bismuth chloride-catalyzed Mukaiyama-aldol reaction of aldehydes with silyl enol ethers: general procedure

Bismuth chloride (16 mg, 0.05 mmol) and dry dichloromethane (3 ml) were placed in a 50-ml two-necked flask under nitrogen. An aldehyde (1.1 mmol) and a silyl enol ether (1.0 mmol) were added sequentially and the resulting reaction mixture was stirred at room temperature. The reaction was quenched with methanol/1 M HCl (10/1, 0.1 ml), water (10 ml) is added, and the product was extracted with dichloromethane (10 ml) three times. After drying, the solvent was removed in vacuo to leave a crude product, which was purified on thin layer chromatography to afford the corresponding aldol in good yield. A similar procedure was also applicable to the Michael-type reaction [91BCJ990].

The $BiCl_3$–metal iodide catalyzed aldol reaction is suggested to proceed as depicted in Scheme 5.26 [93JOC1835]. The formation of BiI_3 from the binary

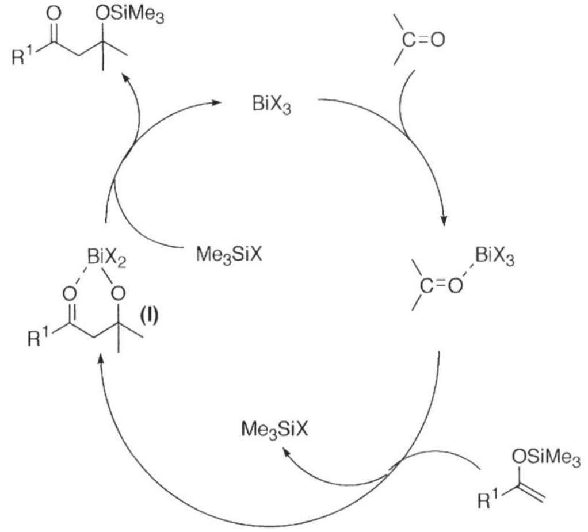

Scheme 5.26. Catalytic cycle for the aldol reaction catalyzed by the $BiCl_3$–metal iodide system [93JOC1835].

catalytic system is confirmed by X-ray powder diffraction analysis. Two proposed factors that enhance the catalytic activity are (i) carbonyl activation through the formation of a not very stable and consequently reactive Bi-chelate (I), and (ii) the catalytic key involving the decomposition of this chelate by chlorosilane formed during the activation by BiX_3 with simultaneous regeneration of Lewis acid species. $Bi(OTf)_3$ [99TL285] is also reported to be an efficient catalyst for the Mukaiyama-aldol reactions. The catalytic activity is higher than the one previously reported for the rare earth triflates $M(OTf)_3$ (M = Sc, Ln). The in situ generated trimethylsilyl trifluoromethanesulfonate via transmetallation is suggested to be a real catalyst in the catalytic system [98SL1249].

Silyl enolates and allyltrimethylsilanes can be acylated with acyl chlorides in the presence of catalytic amounts of $BiCl_3$ or $BiCl_3$–metal iodide (Scheme 5.27) [96OM4646, 96JOC3885]. With the former type of compound, *C*-acylation proceeds with high regioselectivity over *O*-acylation, while with the latter, allylation occurs exclusively at the γ-position without accompanying decarbonylation or isomerization of the double bond. The Friedel–Crafts acylation is successfully catalyzed by $Bi(OTf)_3$ [99TL285] to give the corresponding arylketones in high yields [97TL8871, 98CRA141, 99CRA455]. The catalytic activity is much higher than the one of the other metallic trifrates $M(OTf)_3$ previously reported (M = Al, Ga, Ln or Sc). Other bismuth salts such

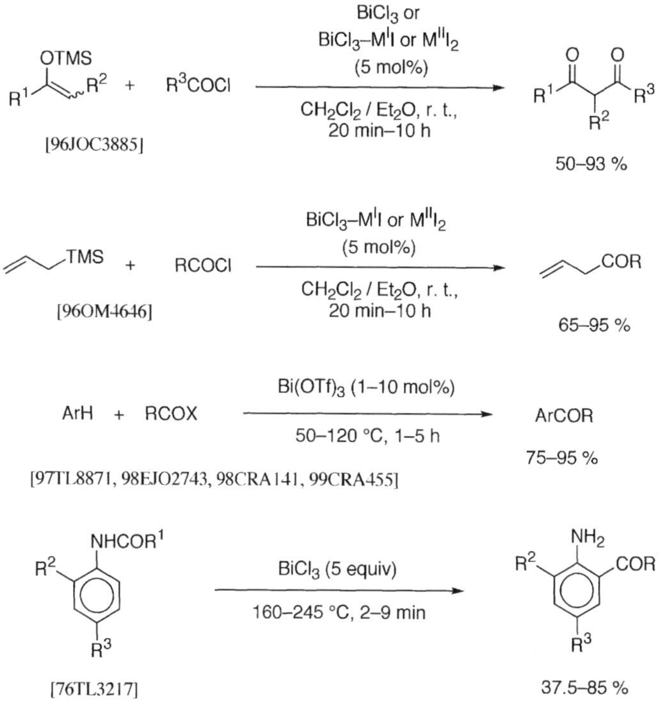

Scheme 5.27. $BiCl_3$, $BiCl_3$-metal iodide and $Bi(OTf)_3$ catalyzed acylations.

Acylation of allylsilane in the presence of $BiCl_3$–ZnI_2: typical procedure

In a flame-dried 100 ml flask with a septum inlet and a magnetic stirring bar was placed dichloromethane (50 ml) under argon. Bismuth chloride (790 mg, 2.5 mmol) and zinc iodide (1.195 g, 3.75 mmol) were transferred in a glove bag to the reaction flask. The flask was connected to an argon line. Pivaloyl chloride (6.65 g, 55 mmol) was added and the resulting suspension was stirred for 5 min at room temperature. Then, allyltrimethylsilane (5.7 g, 50 mmol) in the same solvent (10 ml) was added in one portion with vigorous stirring. The progress of the reaction was monitored by the increase of chlorotrimethylsilane (^{1}H-NMR). After stirring for 15 min at ambient temperature, the reaction mixture was quenched with saturated aqueous $NaHCO_3$ (50 ml). The organic layer was separated and the aqueous layer was washed with dichloromethane (20 ml × 3). The combined organic phase was dried over $NaSO_4$ and evaporated under reduced pressure to leave crude 2,2-dimethyl-5-hexen-3-one as a pale yellow oil (6 g, 95%). After distillation, 5.2 g (83% yield) of pure product was obtained [96OM4646].

Friedel–Crafts acylation with $Bi(OTf)_3$ as a catalyst: typical procedure

$Bi(OTf)_3 \cdot 4H_2O$ (1.16 g, 1.6 mmol), (trifluoromethoxy)benzene (5.19 g, 32 mmol) and benzoyl chloride (2.25 g, 16 mmol) were successively introduced under argon in a 50-ml flask surrounded by a condenser. The suspension was heated with an oil bath. After cooling, dichloromethane (20 ml) and water (20 ml) were added to the dark solution and the products were extracted twice with dichloromethane (20 ml). The combined organic phases were dried with $MgSO_4$ and concentrated. The crude product was purified by flash chromatography on silica gel (pentane/ether = 9/1) to give 4-(trifluoromethoxy)benzophenone (3.72 g, 87%) [97TL8871].

as $BiCl_3$, Bi_2O_3, BiOCl and bismuth carboxylates exhibit a very good catalytic activity for the Friedel–Crafts acylation of aromatic ethers [94FRP(94)10253, 96EUP698593, 98CRA141, 99CRA455]. The Friedel–Crafts reaction of phosphorous trichloride with aromatics is also catalyzed by $BiCl_3$ and $Bi(OTf)_3$ [98T10111]. Quite recently, the mechanism for the acylation was proposed in the case of benzoylation of benzene, toluene and chlorobenzene [98EJO2743]. When heated with an excess amount of $BiCl_3$ at elevated temperatures, *N*-acetylanilide undergoes migration of the acetyl group to *o*- and/or *p*-positions [76TL3217].

Additional examples of the bismuth chloride-catalyzed C–C bond forming reactions are summarized in Scheme 5.28. Active methylene compounds can react smoothly with aldehydes and α,β-unsaturated carbonyl compounds in the presence of bismuth chloride without using organic solvents [92CL1945, 97TL1449]. A catalytic amount of bismuth chloride (2 mol%) can efficiently induce the intramolecular *ene* cyclization of (+)-citronellal as well as the intermolecular *ene* reaction between (−)-β-pinene and chloral [96JOM(521)397]. The Diels–Alder reactions are catalyzed by $BiCl_3$ or $Bi(OTf)_3$ [99TL285] under very mild conditions [97JOC4880, 98TL1161, 99TL7215, 99JOM(585)253]. Both salts exhibit strong catalytic activity when compared with other catalysts previously reported. Even in the presence of water, $Bi(OTf)_3$ efficiently catalyzes hetero-Diels–Alder reaction between glyoxylic acid and dienes [98SL1138]. Allylation of benzaldehyde with crotyltributyltin is mediated by $BiCl_3$ at −78°C. Crotyldichloro-bismuthine is proposed to be an intermediate [98JOC59]. The cyanation of carbonyl compounds is catalyzed by $BiCl_3$ at room temperature or by triarylbismuthine at 80°C to give cyanohydrins [97TA3939, 99YGK689]. The reaction is applied to asymmetric cyanation with 25–58% ee by use of the catalyst prepared from $BiCl_3$ and diethyl tartrate.

Bismuth bromide is also found to be useful as a Lewis acid catalyst

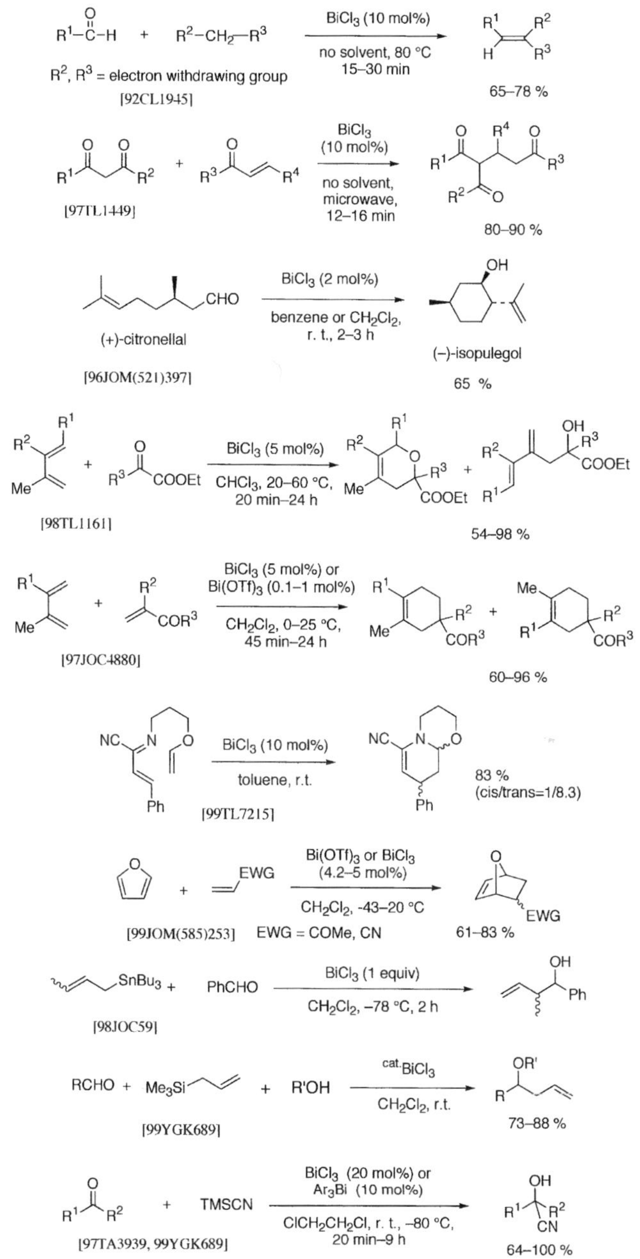

Scheme 5.28. Additional examples of $BiCl_3$-catalyzed C–C bond forming reactions.

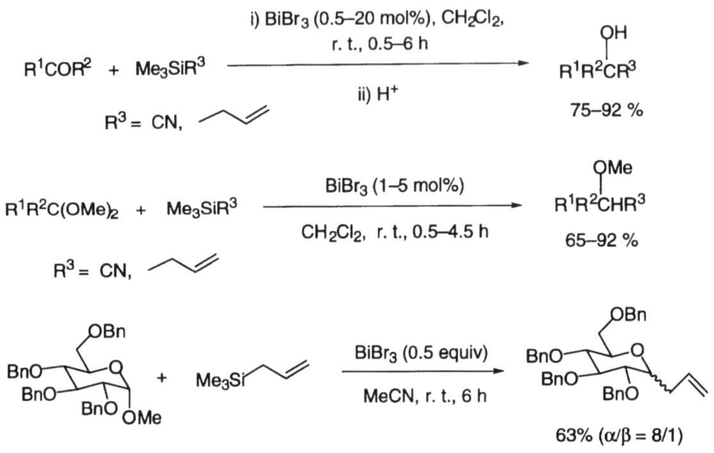

Scheme 5.29. $BiBr_3$-catalyzed C–C bond forming reactions [97TL7215].

Bismuth bromide-catalyzed allylation and cyanation of carbonyl compounds and acetals: general procedure

To a stirred suspension of bismuth bromide (0.5–20 mol%), dried in vacuo prior to use, in dry dichloromethane (2 ml) were added successively a substrate (1.0 mmol) and an organosilicon reagent (1.2 equiv) by a syringe at room temperature under an argon atmosphere. The mixture was stirred for 0.5–9 h, while the progress of the reaction was monitored intermittently by GLC. When the reaction was complete, the mixture was diluted with aqueous HCl/MeCN. Usual work-up followed by chromatographic purification of the crude product on silica gel using ethyl acetate/hexane(1/10) as the eluent gave the expected alcohol or ether in 65–92% yields [97TL7215].

(Scheme 5.29). Allylation as well as cyanation of aldehydes, ketones and acetals with organosilicon reagents proceed efficiently in the presence of a catalytic amount of bismuth bromide under mild conditions [97TL7215]. This reaction can also be employed for stereoselective *C*-glycosydation. Peracetylated α-D-ribofuranosyl bromide reacts with silyl-protected nucleoside bases in a one-pot way in the presence of bismuth bromide to give the corresponding D-nucleosides ($\alpha:\beta = 1–9:99–91$) (Scheme 5.42) [98SC603].

5.4.3. Carbon–carbon bond forming reactions using organobismuth compounds

5.4.3.1. C-Arylation

The first example of *C*-arylation with pentavalent bismuth compound is observed during the attempted oxidation of quinine with an excess of Ph_3BiCO_3 [80CC246]. Since then, many reports concerning similar *C*-arylation have been published from Barton's laboratory with tri-, tetra- and penta-arylbismuth(V) compounds. The *C*-arylation can be performed on the following types of activated organic compounds: phenols, naphthols, 1,3-dicarbonyl and related compounds, and a variety of anionic species including in situ generated ones, as summarized in Tables 5.9–5.11.

The *C*-arylation of phenols and naphthols with tri- and tetraarylbismuth(V) compounds are usually conducted under basic conditions in a solvent such as dichloromethane, benzene, toluene or THF (Table 5.9), while the similar arylation with pentaphenylbismuth can be achieved under neutral conditions. When compared with other arylbismuth(V) compounds, tetraarylbismuthonium salts allow milder conditions and shorter reaction times. They are less prone to lead to side reactions like oxidation mentioned in Section 5.2.4. Although the electronic property of substituents on aromatic rings in arylbismuth(V) compounds does not seem to influence the overall yields of arylation [86T3111], it does appear to govern the regioselectivity of arylation, especially that of phenols [87T323]. Phenols bearing an electron donating group are essentially *ortho* *C*-arylated and sometimes accompanied by small amounts of *O*-phenyl ether, whereas phenols with an electron withdrawing group are mostly and selectively *O*-arylated. Double *C*-arylation of phenol can be effected at both *ortho* positions selectively.

The *C*-arylation of 1,3-dicarbonyl compounds such as β-diketones, β-keto esters and malonic esters is readily performed with a variety of organobismuth(V) compounds under neutral or basic conditions (Table 5.10). In most cases, selective monoarylation is controlled with difficulty. However, 2,2-diarylated 1,3-dicarbonyl compounds are obtained in good yield with Ph_3BiCl_2 or tetraarylbismuthonium salts under basic conditions, and with Ph_3BiCO_3, $Ph_3Bi(OAc)_2$, or Ph_5Bi under neutral conditions. When treated with triphenylbismuth carbonate, dimedone gives an ylide, but it gives an α,α-diphenyl derivative in the reaction with tetraphenylbismuthonium salt under basic conditions [85JCS(P1)2667, 88CL847]. Similar *C*-arylation can be carried out with various triarylbismuth carbonates [86T3111]. It has also been used for the synthesis of isoflavanones and 3-aryl-4-hydroxycoumarins [88T6387].

Although non-enolizable substrates are not arylated by arylbismuth(V) compounds under neutral conditions, their derived enolate anions react easily with Ph_3BiCl_2, Ph_3BiCO_3 and tetraphenylbismuth(V) compounds to produce

TABLE 5.9
C-Arylation of phenols and naphthols with arylbismuth(V) compounds

Organobismuth	Substrate	Condition	Product	Reference
Ar_3BiCO_3 Ar = phenyl, *p*-tolyl, *p*-anisyl, *p*-$O_2NC_6H_4$	2-Naphthol	Neutral	1-Aryl-2-naphthol	[86T3111]
Ph_3BiCO_3	2-Naphthol	Basic	1-Phenyl-2-naphthol	[85JCS(P1)2657]
Ph_3BiX_2 X = O_2CCF_3, ONO_2, OTs, Cl	2-Naphthol	Basic	1-Phenyl-2-naphthol	[85JCS(P1)2657]
Ar_3BiX_2 X = O_2CCF_3, Cl	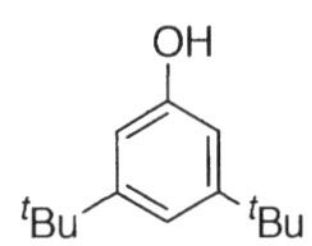	Basic	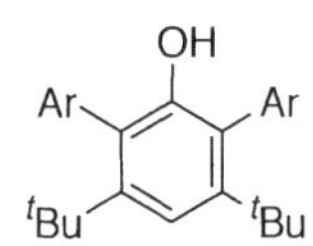	[85JCS(P1)2657]
Ar = Ph, 1-naphtyl				[99OM3016]
Ar_3BiCl_2 Ar = *o*-tolyl, 2,4-dimethyl-phenyl, mesityl	2-Naphthol	Basic	1-Aryl-2-naphthol	[99T1341]
$Ar_3Bi(OAc)_2$ Ar = 2-allyloxy-phenyl	2-Naphthol	Basic	1-Aryl-2-naphthol	[99T3377]
Ph_4BiX X = O_2CCF_3, OAc, OTf	2-Naphthol	Basic	1-Phenyl-2-naphthol	[82TL3365] [85JCS(P1)2657]
Ph_4BiOTs	2-Naphthol	Basic	1-Phenyl-2-naphthol	[82TL3365] [85JCS(P1)2657] [87JCS(P1)241]
Ph_5Bi	2-Naphthol	Neutral	1-Phenyl-2-naphthol	[80CC827]
	1-Naphthol	Neutral	2-Phenyl-1-naphthol	[85JCS(P1)2657]
	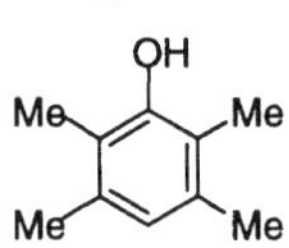	Neutral	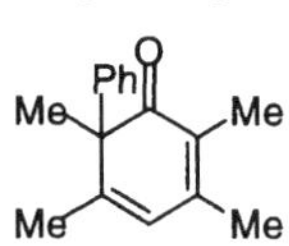	
	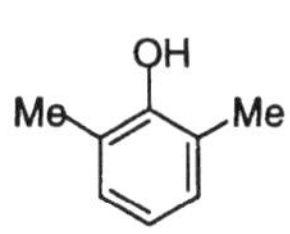	Neutral	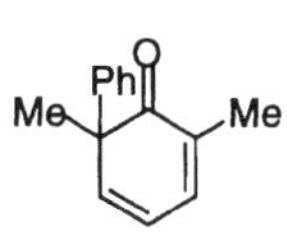	

TABLE 5.10
C-Arylation of 1,3-dicarbonyl and related compounds with arylbismuth(V) compounds

Organobismuth	Substrate	Condition	Product	Reference
Ar_3BiCO_3 Ar = phenyl, *p*-tolyl, *p*-anisyl, *p*-$O_2NC_6H_4$	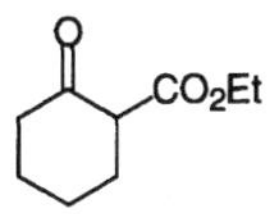	Neutral	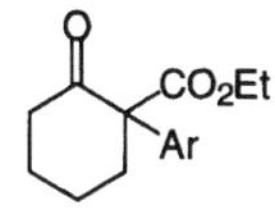	[86T3111]
Ph_3BiCO_3, Ph_3BiCl_2	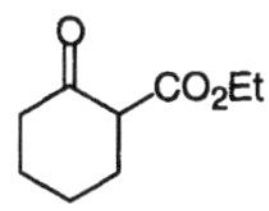	Neutral Basic	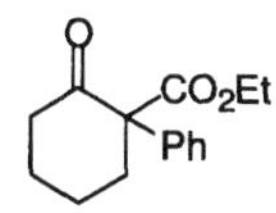	[85JCS(P1)2667]
Ph_3BiCO_3	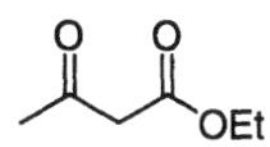	Neutral	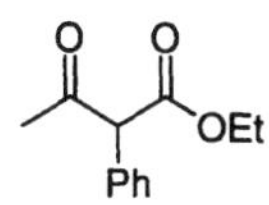	[85JCS(P1)2667]
Ph_3BiCO_3 (excess)	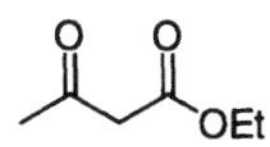	Neutral	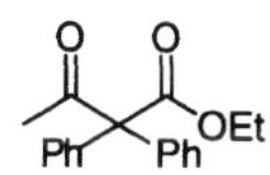 + Ph_2CHCO_2Et	
Ph_3BiCl_2	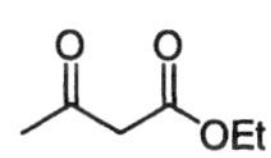	Neutral	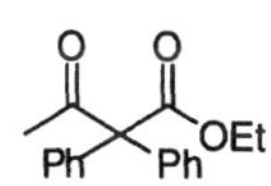	
$Ph_3Bi(OAc)_2$	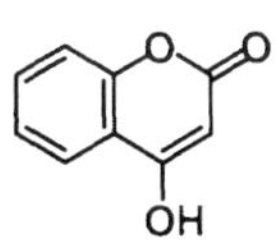	Basic Neutral	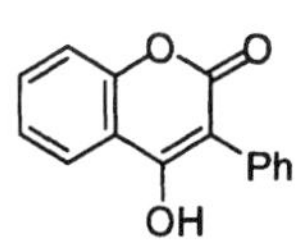	[88T6387]
Ph_3BiCO_3 (excess)	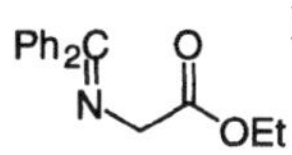	Neutral	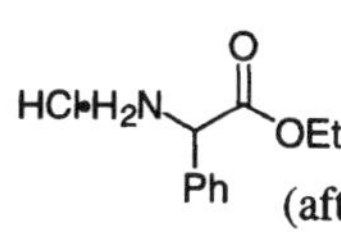 (after hydrolysis)	[89TL3909]
$Ar_3Bi(OAc)_2$ Ar = 2-allyloxy-phenyl	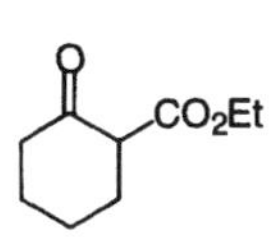	Basic	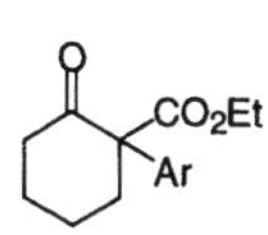	[99T3377] [99T1341]

TABLE 5.10 (*continued*)

Organobismuth	Substrate	Condition	Product	Reference
Ar_3BiCl_2 Ar = *o*-tolyl, 2,4-dimethyl-phenyl, mesityl	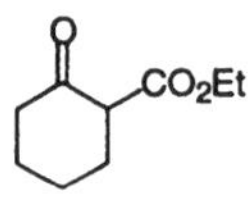	Basic	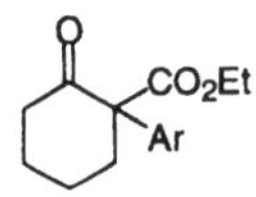	[99T3377] [99T1341]
Ph_4BiX X = O_2CCF_3, OAc, OTs, OTf	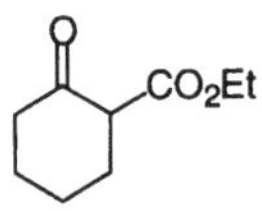	Basic	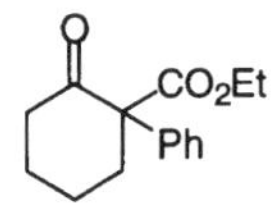	[82TL3365] [85JCS(P1)2667] [87JCS(P1)241]
Ph_5Bi	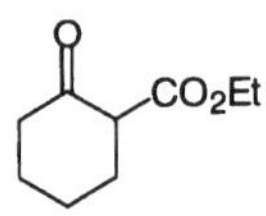	Neutral	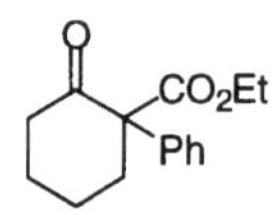	[80CC827] [85JCS(P1)2667] [87JCS(P1)241]
	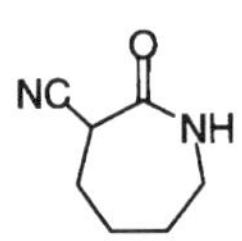	Neutral	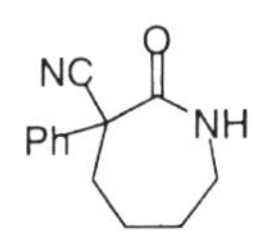	[90JOC5222]

Phenylation of 2-naphthol with tri- or tetraphenyl bismuth(V) compounds under basic conditions: typical procedure

To a stirred solution of 2-naphthol (0.14 g, 1.0 mmol) in benzene or THF (5 ml) was added tetramethyl-2-*tert*-butylguanidine (0.21 g, 1.3 mmol) at room temperature under an inert atmosphere. Triphenylbismuth bis(trifluoroacetate) (0.80 g, 1.2 mmol) was added and the resulting mixture is stirred for an appropriate time. Then the solvent was evaporated and the residue is fractionated by thick layer chromatography using ether–hexane (1:4) as the eluent to give 1-phenyl-2-naphthol in 0.15 g, 70% yield. In some cases, triphenylbismuthine was also isolated [85JCS(P1)2657].

Phenylation of 1,3-dicarbonyl compounds with various phenylbismuth(V) compounds under neutral conditions: typical procedure

A solution of acetylacetone (0.15 g, 1.5 mmol) in anhydrous benzene or dichloromethane (7.5 ml) was stirred under argon in the presence of triphenylbismuth carbonate (1.9 g, 3.8 mmol) until the reaction is complete. The reaction mixture was filtered though Celite, the filtrate was evaporated, and the residue was purified by preparative thick layer chromatography or column chromatography to give 3-phenylpentane-2,4-dione (0.10 g, 38% yield) and 3,3-diphenylpentane-2,4-dione (0.15 g, 40% yield) [85JCS(P1)2667].

TABLE 5.11
C-Arylation of organoions with arylbismuth(V) compounds

Organo-bismuth	Substrate	Base	Product	Reference
Ph_3BiCO_3	Acetophenone	KH	2,2,2-Triphenyl-acetophenone	[85JCS(P1)2667]
	Diphenylacetone	KH	Pentaphenylacetone	[85JCS(P1)2667]
	Cyclohexanone	KH	2,2,6,6-tetraphenyl-cyclohexanone	[82CC732]
	2-Nitropropane	–	2-Phenyl-2-nitropropane	[80CC827]
	Ph_3CK	–	Ph_4C	[85JCS(P1)2667]
Ph_3BiCl_2	2-Nitropropane	BTMG[a]	2-Phenyl-2-nitropropane	[87JCS(P1)241]
	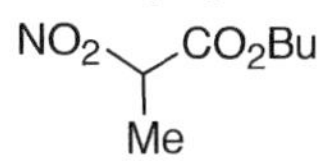	TMG[b]	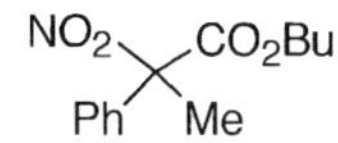	[88JOC2323]
Ph_3BiCO_3, Ph_3BiCl_2	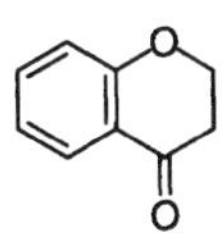	KH	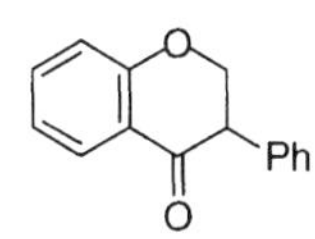	[88T6387]
Ph_4BiOTs	1,1,3,3-Tetra-phenylacetone	BTMG[a]	Pentaphenylacetone	[87JCS(P1)241]
	2-Nitropropane	BTMG[a]	2-Phenylnitropropane	
	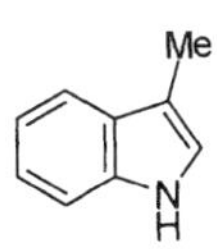	BTMG[a]	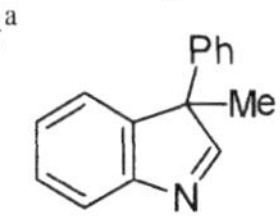	
$Ph_4BiO_2CCF_3$	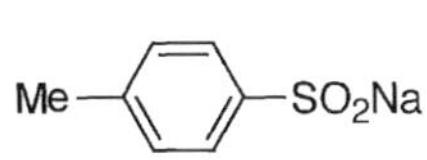	–	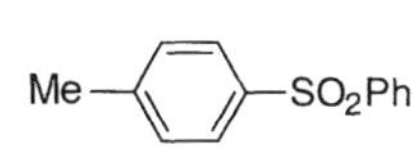	[85JCS(P1)2667]
Ph_5Bi	2-Nitropropane	–	2-Phenyl-2-nitropropane	[80CC827]

[a] BTMG, *N-tert*-butyl-*N′*, *N″*-tetramethylguanidine. [b]TMG, tetramethylguanidine.

Phenylation of enolate anions with triphenylbismuth carbonate: typical procedure
A solution of acetophenone (0.060 g, 0.50 mmol) and KH (0.090 g, 2.3 mmol) in anhydrous THF (2.5 ml) was stirred at room temperature for 15–30 min under argon. After the addition of triphenylbismuth carbonate (1.15 g, 2.3 mmol), the mixture was stirred until the reaction was complete. The reaction mixture was filtered though Celite, the filtrate evaporated, and the

residue purified by preparative thick layer chromatography or column chromatography to give 2,2,2-triphenylacetophenone (0.12 g, 69% yield) [85JCS(P1)2667].

the corresponding α-phenylation products, as shown in Table 5.11. Higher phenylation products are usually obtained as byproducts. Further examples of the anions to be phenylated include those from nitroalkanes, α-nitro carboxylic esters, triphenylmethane, indole, and alkali sulfinates.

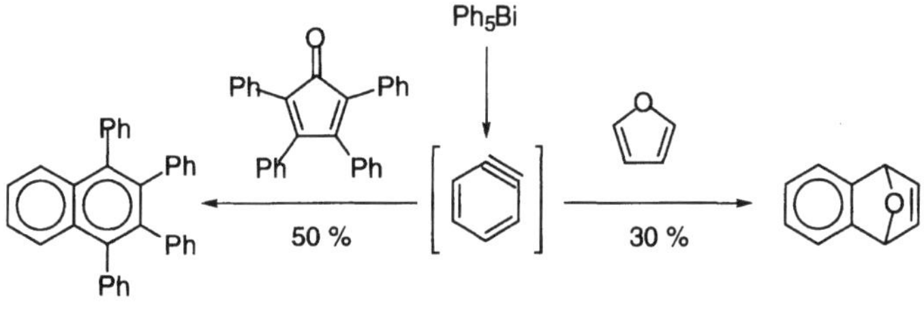

Scheme 5.30. Reaction of pentaphenylbismuth with dienes.

Pentaphenylbismuth reacts with conjugated dienes in a manner shown in Scheme 5.30. The Diels–Alder adducts are derived from the benzyne in situ generated from the thermal decomposition of pentaphenylbismuth [78DOK(238)361, 87JCS(P1)251].

The mechanistic pathway of the *C*-arylation reaction is supposed to involve a pentavalent bismuth intermediate possessing the covalent Bi–O bond, as shown in Scheme 5.31, and which has been isolated and fully characterized in some cases [85JCS(P1)2657]. Breakdown of this intermediate gives rise to the arylation product. The reductive elimination of a bismuth(III) compound is considered to proceed concertedly.

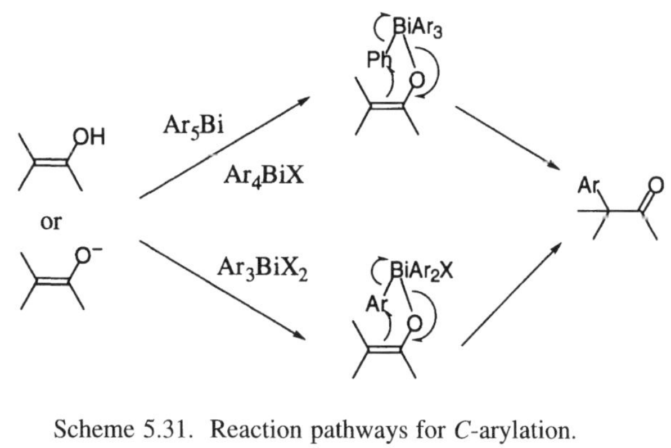

Scheme 5.31. Reaction pathways for *C*-arylation.

5.4.3.2. Transition metal-catalyzed carbon–carbon bond forming reactions

The transition metals and their salts and coordination complexes such as $RhCl_3$, $Pd(PPh_3)_4$, $Pd(OAc)_2$, Cu(0), CuCl and CuX_2 can catalyze the aryl–aryl coupling, carbonylation and phenylation reactions of organobismuth compounds with indoles, acyl chlorides, acetylenes and olefins, as illustrated in Table 5.12. Most of these reactions proceed under mild conditions.

During the course of examining arylation using various types of arylbismuth compound, Barton's group found that some transition metals effectively catalyze the aryl–aryl coupling and arylation reactions based on arylbismuth(V) compounds. The coupling reaction proceeds quantitatively with a wide variety of tri- and pentavalent arylbismuth compounds in the presence of triethylamine and a catalytic amount of palladium acetate [88T5661]. Acyl chlorides and indoles are phenylated under mild conditions in the presence of palladium and copper catalysts, respectively, to furnish the corresponding aromatic ketones and 3-arylindoles in excellent yields. Allyl, cinnamyl and propargyl bromide are also arylated with triarylbismuthine in the presence of palladium catalyst [97CCL759]. Uemura et al. reported the carbonylative aryl–aryl coupling of triarylbismuth compounds using a rhodium or palladium salt as catalyst [92CC453, 95BCJ950, 99BCJ1851]. They also found the hydrophenylation of α,β-unsaturated carbonyl compounds (conjugate addition) with sodium tetraphenylborate using palladium(0) and Lewis acid catalysts [94TL1739, 95JOC883] and phenylation of alkene (Heck-type reaction) with Ph_2BiCl and a catalytic amount of palladium(II) under air

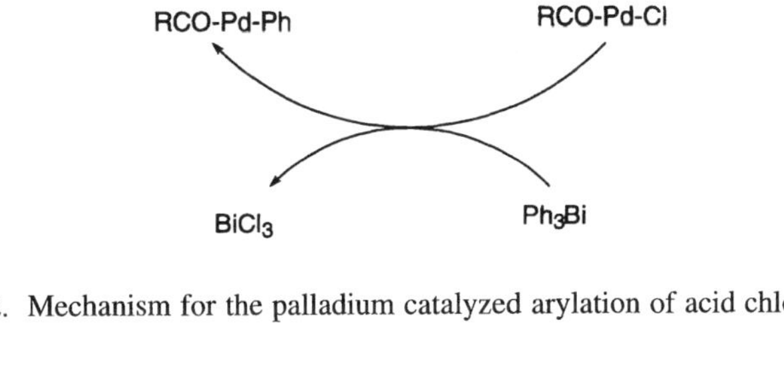

Scheme 5.32. Mechanism for the palladium catalyzed arylation of acid chlorides [88T5661].

TABLE 5.12
Transition metal-catalyzed reactions using arylbismuth compounds

Metal catalyst	Bismuth compound	Reactant	Product(s)	Reference
CuCl	Ph_3BiF_2	RC≡CH	RC≡CPh	[96TL4051]
$RhCl(H_2O)_3$	Ar_3Bi	CO (1 atm)	ArCOAr	[92CC453]
$[RhCl(CO_2)]_2$	Ar = phenyl, *p*-tolyl, *p*-anisyl,		$ArCO_2Me$	[95BCJ950]
$[RhCl(COD)]_2$	*m*-tolyl, 1-naphthyl		Ar–Ar	
$Pd(OAc)_2/K_2CO_3$				
$Pd(OAc)_2$/air	Ar_3Bi	–	Ar–Ar	[99BCJ1851]
$Pd(OAc)_2/Et_3N$	Ar_3Bi Ar = phenyl, *p*-tolyl, *p*-anisyl, $PhCH_2CH_2$	–	Ar–Ar	[88T5661]
	Ph_2BiX, X = Cl, OAc, O_2CCF_3	–	Ph–Ph	
	Ph_3BiX_2, X = Cl, OAc, O_2CCF_3, X_2 = CO_3	–	Ph–Ph	
	Ph_3Bi	RCOCl	RCOPh	
Cu, $Cu(OAc)_2$, $Cu(O_2CCF_3)_2$	$Ph_3Bi(O_2CCF_3)_2$	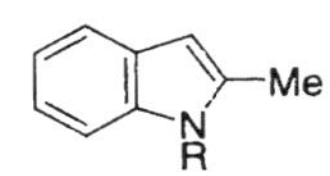 R = H or Me	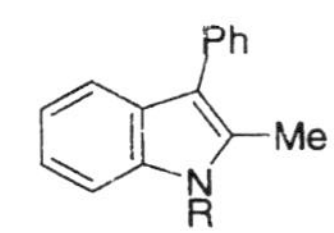	[88TL1115]
$Pd(OAc)_2$, $Pd(PPh_3)_4$	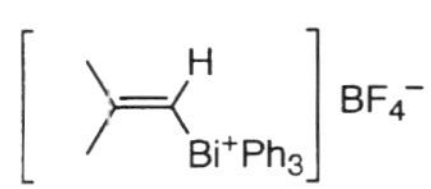	Ethyl acrylate	Ph–Ph, olefins	[96JCS(P1)1971]
$Pd(OAc)_2$	Ph_3Bi	Styrene	Ph–Ph, *trans*-stilbene	[73BCJ2910]
		1-Octene, ethyl acrylate	Ph–Ph, phenylated olefins	[77BCJ2021]

Conversion of acyl halides to aromatic ketones using triphenylbismuthine and palladium acetate: general procedure

To a stirred suspension of palladium acetate (0.05 mmol) in HMPA (1 ml) was added triethylamine (0.1 mmol) at room temperature under argon. A yellow solution resulted within 2–3 min, to which a solution of an appropriate acid chloride (5 mmol) in HMPA (2 ml) followed by triphenylbismuthine (1 mmol) in the same solvent (5 ml) was introduced slowly. The reaction mixture was heated at 65°C for 5 h, then cooled, diluted with ether (50 ml) and filtered through a short column of basic alumina. The column was washed with ether (4 × 25 ml) and the combined ethereal washings were shaken with water (3 × 250 ml). The organic phase was separated, dried over anhydrous magnesium sulfate and evaporated to give the expected ketone, which was further purified either by preparative TLC or by recrystallization [88T5661].

[99JOM(574)3]. Some unsaturated hydrocarbons undergo arylation and/or alkenylation with triarylbismuthines and alkenyltriarylbismuthonium salts in the presence of a palladium catalyst [73BCJ2910, 77BCJ2021, 96JCS(P1)1971, 96TL4051].

The mechanistic pathway for the $Pd(OAc)_2$-catalyzed formation of phenylketones from triphenylbismuthine and acyl chlorides is depicted in Scheme 5.32 [88T5661].

5.4.3.3. Carbon–carbon bond forming reactions based on bismuthonium salts and ylides

A stabilized bismuthonium ylide, readily obtainable from dimedone and triphenylbismuth carbonate [85JCS(P1)2667, 88CL847], reacts with a variety of nucleophiles as shown in Scheme 5.33. The reaction with aromatic aldehydes gives several 2:1 coupling products. The fact that cyclopropane derivatives result from the initial Wittig-type olefination of aldehydes followed by the Corey-type cyclopropanation of the resulting olefins is noteworthy. No ylide compounds which can undergo both types of reaction in such a tandem way are currently known. The reaction with a sulfene gives rise to five-membered heterocyclic sulfones [89CC1749]. In the presence of a catalytic amount of copper(I) salt, this stabilized ylide reacts with isothiocyanates [88CL847], dimethyl sulfide, triphenylarsine [90BCJ950] and terminal acetylenes [89CL325] to give the corresponding 1,3-oxazoles, sulfonium ylide, arsonium ylide and furan derivatives, respectively, in low to moderate yields.

The moderately stabilized triphenylbismuthonium 2-oxoalkylides readily undergo the Corey-type epoxidation with aldehydes to give α,β-epoxyketones and triphenylbismuthine [94JCS(P1)2703]. Competitive experiments using a series of *p*-substituted benzaldehydes have revealed that the C–C coupling proceeds through nucleophilic attack of the ylidic carbon onto the

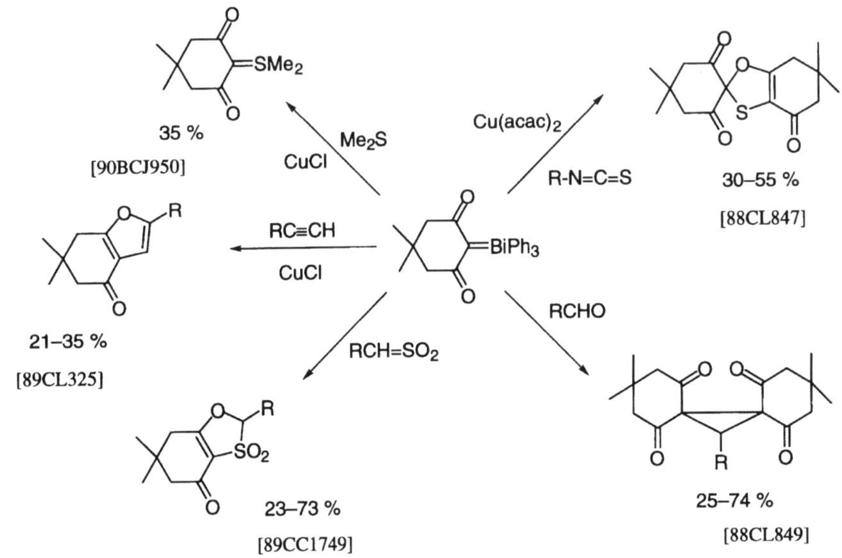

Scheme 5.33. Typical reactions of a stabilized bismuthonium ylide derived from dimedone.

carbonyl carbon atom. The ylides also react with *N*-sulfonylaldimines to afford α,β-aziridinoketones and triphenylbismuthine [95JOC4663]. The *cis/trans* geometry of the aziridine ring can be controlled by the proper choice of a base and an additive. The triphenylbismuthonio moiety always behaves as a good leaving group. The reaction mode leading to oxiranes and aziridines is in marked contrast to those observed with lighter pnictogen counterparts; phosphonium, arsonium and stibonium 2-oxoalkylides all undergo the Wittig-type olefination with aldehydes and imines to afford α,β-unsaturated ketones.

Bismuthonium 2-oxoalkylides react with dimedone, *p*-toluenesulfonic acid and benzenethiol to yield the corresponding α-substituted ketones and triphenylbismuthine. In the presence of a catalytic amount of copper iodide or nitrosobenzene, the ylides undergo reductive dimerization to form 1,2-diacylethenes. Aromatic nitrile oxide also catalyzes the dimerization but readily couples with the resulting olefin to give the corresponding isoxazoline derivative.

Although triphenylbismuthonium 2-oxoalkylides do not react with simple ketones and esters, such as benzophenone and ethyl acetate, they readily couple with α-diketones in several different reaction modes [96CC2697, 98CC1359, 99S395, 99JCS(P1)1533]. When treated with ethyl pyruvate,

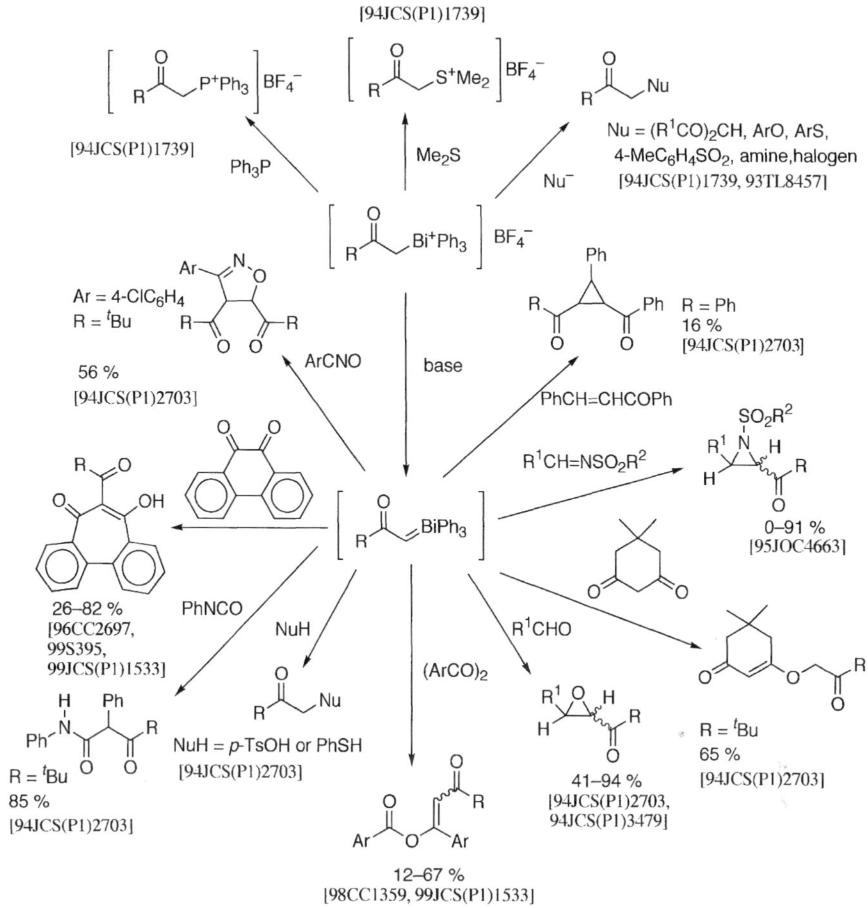

Scheme 5.34. Typical reactions of (2-oxoalkyl)triphenylbismuthonium salts and their derived ylides.

Reaction of *o*-quinones with triphenylbismuthonium 2-oxoalkylide: general procedure To a stirred suspension of (2-oxoalkyl)triphenylbismuthonium salt (0.2 mmol) in THF (5 ml) was added $NaN(SiMe_3)_2$ (0.2 mmol) at −78°C. After 10 min, *o*-quinone (0.2 mmol) was added and the resulting mixture was gradually allowed to warm to room temperature. Evaporation of the solvent under reduced pressure gave an oily residue which is diluted with benzene (15 ml) and the insoluble matter was filtered off through a Celite bed. The filtrate was concentrated under reduced pressure to leave an oily residue which was subjected to chromatographic separation on silica gel using hexane/ethyl acetate (100:0–80:20) as the solvent to elute triphenylbismuthine and 2-acyl-3-hydroxytropone, in this order [96CC2697].

2,3-difunctionalized oxiranes are formed via epoxidation. The reaction with *o*-quinones gives 2-acyl-1,3-tropolone derivatives via the ring enlargement, while the treatment with benzils affords *O*-aroylenolates derived from 1,3-diketones via the carbon-to-oxygen migration of an aryl group. Such variations of the reaction modes are unprecedented in the ylide chemistry and so characterize the bismuth from the other 15 group elements (Scheme 5.34).

Quite recently, the bismuthonium salts and ylides bearing an α-ester group have been reported to exhibit similar reactivities to their ketone analogs [99JOC6924].

(3-Oxoalkyl)triphenylbismuthonium salts behave in a similar manner as the 2-oxo analogues, when treated with such nucleophiles as sodium benzenesulfinate, potassium halides and dimethyl sulfide [95JCS(P1)2543]. In the presence of a strong base, however, β-elimination occurs to give α,β-unsaturated ketones (Scheme 5.35).

Allyltriphenylbismuthonium salts can be generated in situ, but they decompose rapidly to form allylbenzenes. In the presence of electron-rich arenes, they readily transfer the allyl moiety to arenes to produce the corresponding allylation products [95TL7475]. These bismuthonium salts may be regarded as the allyl cation equivalent for a variety of nucleophiles (Scheme 5.36).

The reactivity of (1-alkenyl)triphenylbismuthonium salts towards nucleophiles is somewhat lower as compared with alkylbismuthonium salts (Scheme 5.37) [96JCS(P1)1971]. Thus, they fail to react with triphenylphosphine and dimethyl sulfide but are capable of reacting with sulfinate and thiolate anions to give the corresponding sulfones and sulfides, respectively.

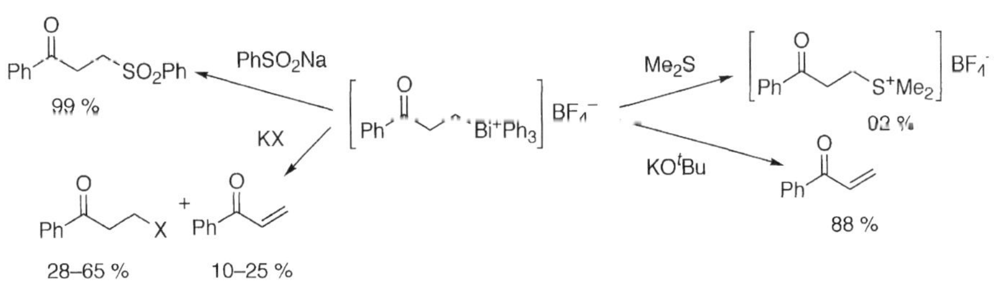

Scheme 5.35. Typical reactions of (3-oxoalkyl)triphenylbismuthonium salt [95JCS(P1)2543].

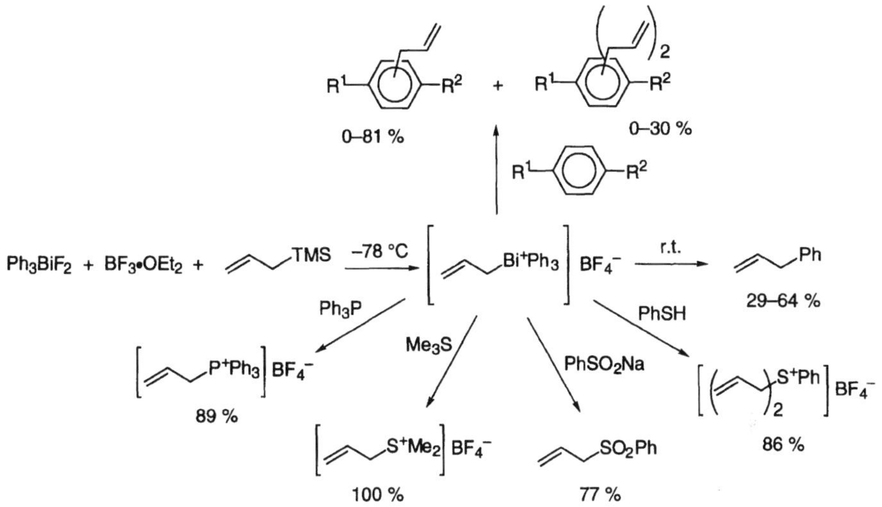

Scheme 5.36. Typical reactions of allyltriphenylbismuthonium salt [95TL7475].

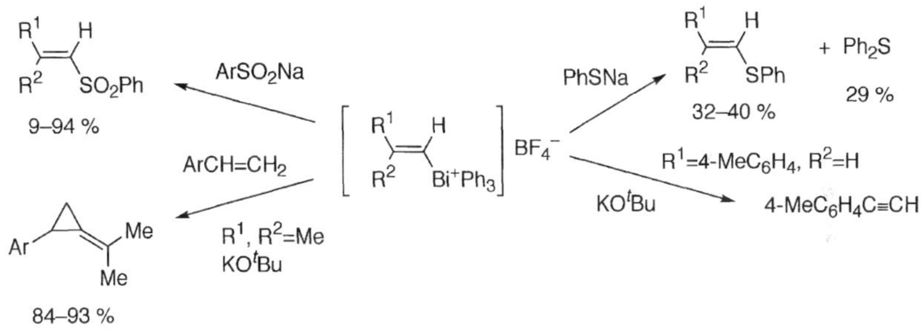

Scheme 5.37. Typical reactions of alkenyltriphenylbismuthonium salt [96JCS(P1)1971].

5.4.3.4. Miscellaneous carbon–carbon bond forming reactions

Triorganylbismuthines can mediate such C–C bond forming reactions as shown in Scheme 5.38 in the presence or absence of a transition metal salt. Tributylbismuthine promotes the allylation of carbonyl compounds with allylic bromide [91HAC297] and the olefination of diazo compounds with carbonyl compounds in the presence of catalytic amounts of a copper(I) halide [90TL5897]. A combination of triphenylbismuthine and WCl_6 promotes the metathesis of olefins, while a similar combination with $TiCl_4$ facilitates the stereoselective Diels–Alder reaction of unsaturated esters [76TMC183, 93JOC4783].

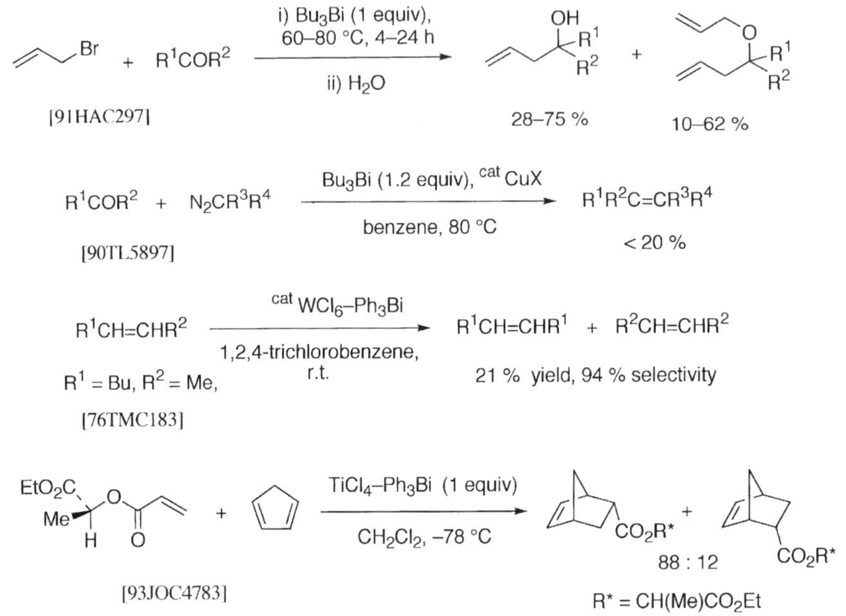

Scheme 5.38. Triorganylbismuthine-mediated C–C bond forming reactions.

The reaction of (diphenylbismuthyl)methyllithium with carbonyl compounds gives rise to α-bismuthyl alcohols which, on treatment with aqueous perchloric acid at room temperature, provide the corresponding terminal olefins [85CB1039]. Heating of alkynyldiarylbismuthines with acyl chlorides in refluxing benzene results in the ethynylation of the carbonyl function to give the corresponding acetylenic ketones [92BCJ3504] (Scheme 5.39).

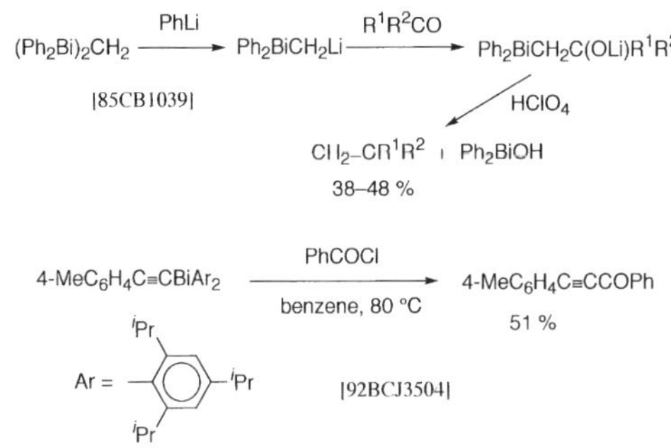

Scheme 5.39. Miscellaneous C–C bond forming reactions based on arylbismuthines.

5.5. CARBON–HETEROATOM BOND FORMING REACTIONS

5.5.1. Reactions with bismuth(III) salts

5.5.1.1. Carbon–oxygen bond forming reactions

Bismuth(III) salt-mediated C–O bond forming reactions are summarized in Scheme 5.40. The ester functionalization with simultaneous cyclopropane ring cleavage of *cis*-(−)-thujopsene proceeds under the aid of catalytic amounts of bismuth sulfate [78BCJ319]. Both alcohols and formates can be converted to acetates by heating with a slight excess of bismuth acetate at elevated temperatures [73JOC764]. Selective hydrolysis of aryl esters can be effected in the presence of a catalytic amount of bismuth mandelate in DMSO [97TL2981]. An additional presence of molecular oxygen (1 atm) slightly increases the yield of the products. Stoichiometric use of bismuth chloride as the promoter for glycosylation of 1-glycosyl dimethyl phosphite results in a good β-selectivity [93SL115]. Bismuth bromide is efficient as a catalyst for the reductive homocoupling of carbonyl compounds as well as heterocoupling between a carbonyl compound and an alkoxysilane in the presence of triethylsilane under mild conditions. This ether-forming reaction has been successfully extended to the single-step preparation of novel crownophanes of various ring sizes containing the olefinic and acetylenic linkages [97TL7219]. The rapid cleavage of oximes, hydrazones and semicarbazones to the carbonyl compounds under microwave irradiation are reported in the presence of a catalytic amount of $BiCl_3$ or on wet silica supported sodium bismuthate [97TL4267, 97SL1251, 98SC4157, 98SL1345].

Glycosylation of 1-glycosyl dimethyl phosphite with alcohols: typical procedure

To a solution of 2,3,4,6-tetra-*O*-benzyl-D-glucopyranosyl dimethyl phosphite ($\alpha/\beta = 83/17$, 37 mg, 0.058 mmol) in dichloromethane (1 ml) were added powdered molecular sieves 4A (200–300 mg), 3-phenylpropanol (7.9 mg, 0.058 mmol) and bismuth chloride (18 mg, 0.058 mmol) in this order. The resulting mixture was stirred for 20 min at room temperature, then diluted with water and filtered through a pad of Celite under suction. The filtrate was washed successively with aqueous sodium hydrogen carbonate and brine, dried and evaporated under reduced pressure. The residue was purified by preparative TLC on silica gel (AcOEt/hexane = 1/2) to give the corresponding glycoside (32 mg, 84%). The anomeric ratio, determined by a combination of ^{1}H-NMR and HPLC, was $\alpha/\beta = 2/8$ [93SL115].

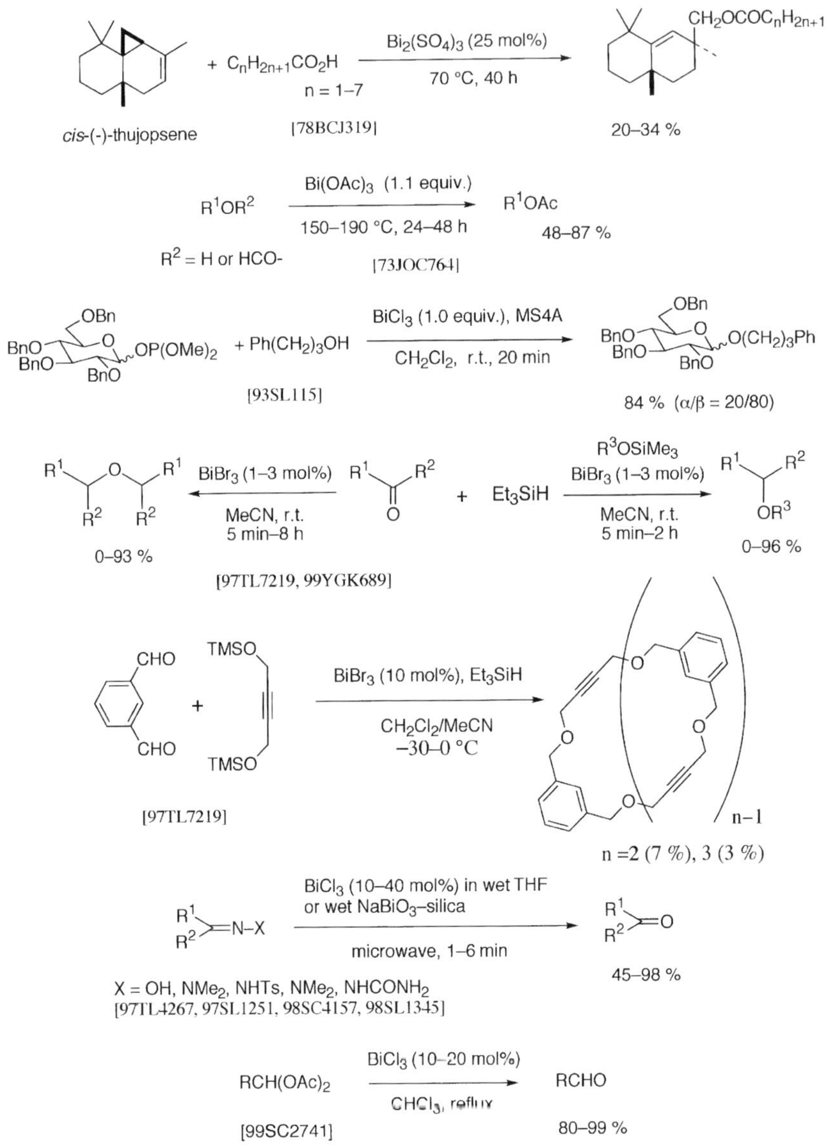

Scheme 5.40. Bismuth(III) salt-mediated C–O bond forming reactions.

Reductive homo- and hetero-coupling of carbonyl compounds: general procedure

To a stirred suspension of bismuth bromide (0.02–0.06 mmol), dried in vacuo prior to use, in dry acetonitrile (5 ml) were added successively a carbonyl compound (2.0 mmol) and, in the case of heterocoupling, an alkoxysilane (1.2 equiv), followed by triethylsilane (1.2 equiv) via a syringe at room temperature under an argon atmosphere. The resulting mixture was stirred for an appropriate time, while the progress of the reaction was monitored intermittently by GLC. After usual work-up, the crude product was chromatographed on silica gel to give the corresponding symmetrical or unsymmetrical ether in 0–96% yield [97TL7219].

5.5.1.2. Carbon–sulfur bond forming reactions

Torii et al. extended the homocoupling of allylic halides by the $BiCl_3$/Sn or $BiCl_3$/Zn redox system to the synthesis of 3-hydroxycephems and 2-*exo*-methylenepenams [90CL1867, 92SL878, 95BCJ1385, 97JOC3610, 97CL1221]. Other types of the C–S bond forming reactions are the bismuth halides and sulfate catalyzed sulfenylations of carbonyl and related compounds. Four types of sulfenylation are catalyzed efficiently: (i) dithioacetalization of aldehydes (turnover number up to 2300), (ii) dithioacetalization of ketones (up to 190), (iii) transacetalization of acetals to mono- or dithioacetals (up to 1000), and (iv) conjugate addition of thiophenol to α,β-unsaturated carbonyl compounds (up to 48) [95SL984]. Quite recently, conversion of oxiranes to thiiranes catalyzed by $BiCl_3$ [98SC3943], coupling of sulfonyl chlorides with allylic halides mediated by metallic bismuth (Section 5.4.1) [98SL1083] and transformation of benzyl bromide to benzyl sulfides and selenides promoted by $BiCl_3$/Sm [98SC4479] have been reported. These C–S bond forming reactions are illustrated in Scheme 5.41.

Dithioacetalization of carbonyl compounds with thiols: general procedure

To a suspension of bismuth salt (0.04–10 mol%) in acetonitrile (4 ml) were added successively a carbonyl compound (5.0 mmol) and a thiol (10 mmol) or a dithiol (5.0 mmol) at room temperature under aerial conditions. The mixture was magnetically stirred for an appropriate time until the reaction is complete. The mixture was evaporated on a rotary evaporator and the crude product was chromatographed on silica gel using ethyl acetate–hexane (1/10) as the eluent to give the expected thioacetal almost quantitatively [95SL984].

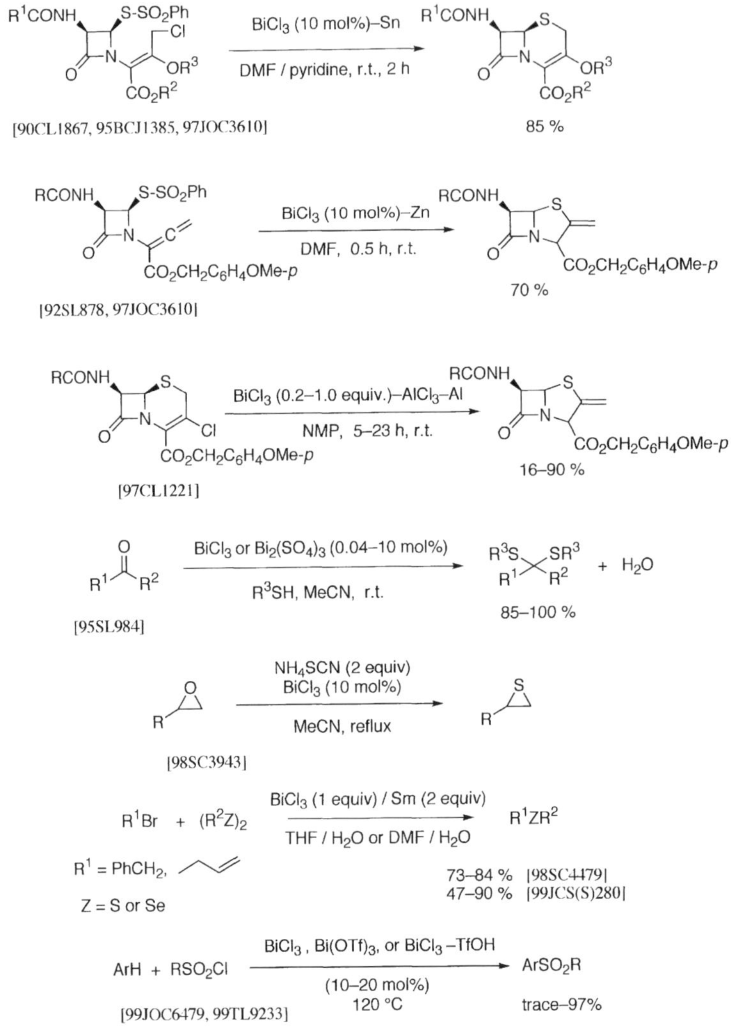

Scheme 5.41. Bismuth(III) salt-catalyzed C–S bond forming reactions.

5.5.1.3. Carbon–nitrogen bond forming reactions

Bismuth nitrate can be used for the nitration of aromatic hydrocarbons [88TL5909, 98S1724] and bismuth acetate for the acetylation of secondary amines [73JOC764], as shown in Scheme 5.42. Metallic bismuth and bismuth halides are employed as catalysts for the phosphorization and chlorination of

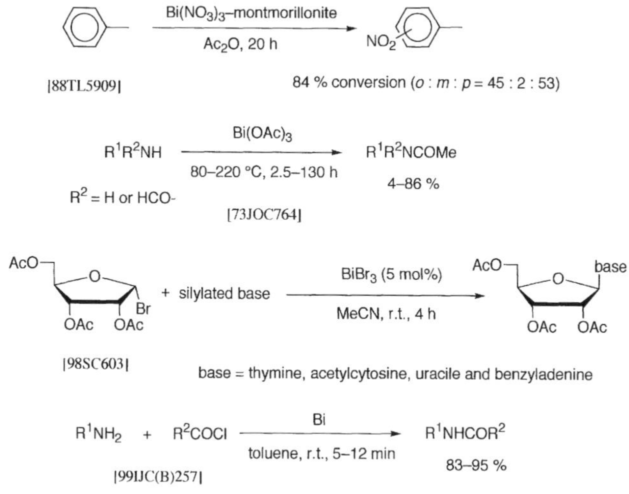

Scheme 5.42. Bismuth-mediated C–N bond forming reactions.

aromaric chlorides, respectively [90ZOB1104, 91ZOB2518]. A per-*O*-acetylated β-D-ribofuranosyl bromide reacted with silylated heterocyclic bases in the presence of $BiBr_3$ to give the corresponding β-D-nucleoside in good yield [98SC603]. Amides and related compounds are *N*-acylated with acetic acid by diarylbismuthinate (Section 3.7.2) [90CL1651].

5.5.1.4. Carbon–halogen bond forming reactions

Alcohol, silyl ether and acetate functions can be replaced by a halogen atom on treatment with trimethylsilyl halide in the presence of bismuth halide, as shown in Scheme 5.43. Most of the reactions proceed rapidly under mild conditions to afford the corresponding halides almost quantitatively [94SL723, 95BSF522]. In hydroxylic solvent, bismuth chloride is partially hydrolyzed to liberate hydrogen chloride. Thus, it can be used for selective conversion of alcohols to the corresponding chlorides [98SC1737]. Benzylic alcohols are mainly converted to dibenzyl ethers, whereas primary alcohols are inert. This type of organic transformation is applied to the tandem aldol-halogenation reaction [94JOC2238], halogenation of per-*O*-acetylated glycopyranosides [97CAR(297)175] and bromination of per-*O*-acetylated β-D-ribofuranose [98SC603]. Bismuth chloride is also used as a reagent for chlorination of alcohol as well as halogen exchange [98SC1737, 99T1971]. The exchange is not stastical, suggesting higher affinity of bromine for bismuth.

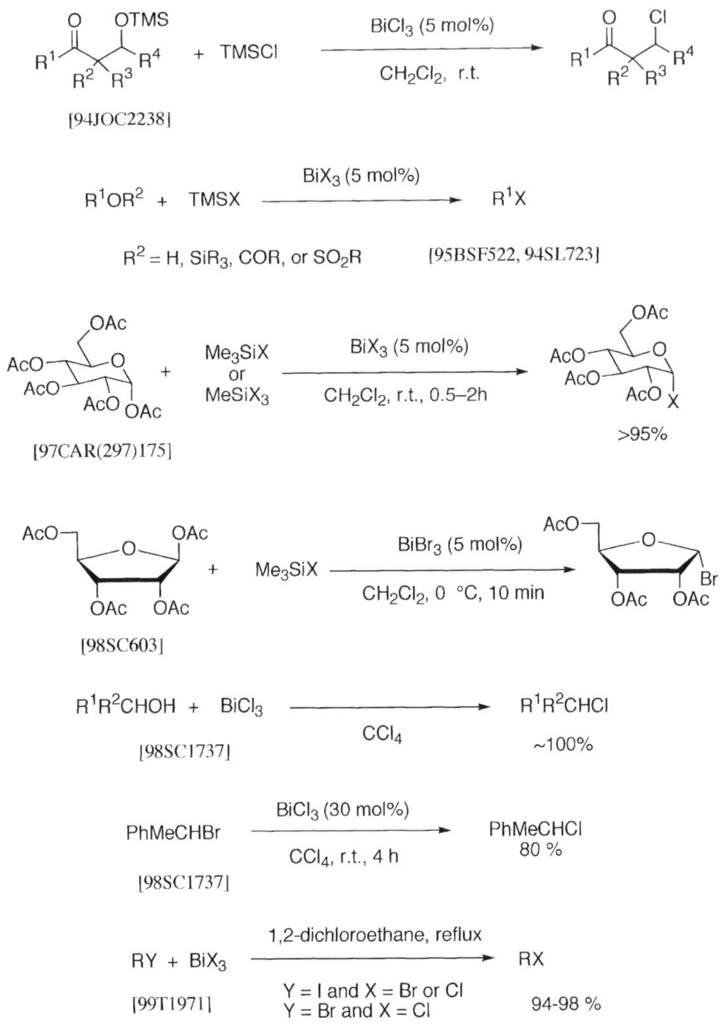

Scheme 5.43 Bismuth(III) salt-based carbon–halogen bond forming reactions.

Chlorination of alcohols with trimethylsilyl chloride in the presence of a catalytic amount of bismuth chloride: typical procedure

To a stirred mixture of cinnamyl alcohol (6.71 g, 0.05 mol) and bismuth chloride (0.79 g, 25 mmol), trimethylsilyl chloride (5.43 g, 0.05 mol) was added in small portions. After 5 min, the liquid phase was decanted and the resulting cinnamyl chloride was separated by distillation (7.25 g, 95%) [94SL723].

5.5.2. Reactions with organobismuth compounds

5.5.2.1. O-Arylation

The formation of diphenyl ether during the decomposition of pentaphenylbismuth in the presence of phenol was reported by Sharutin et al. in 1975 and constitutes the first example of the carbon–heteroatom bond forming reaction based on organobismuth compounds [75DOK(225)581, 78DOK(238)361]. Since then, a variety of alcohols, phenols, glycols, 1,3-dicarbonyl compounds and enols have been *O*-phenylated using tri-, tetra- and pentaphenylbismuth compounds in the presence or absence of a copper catalyst, as shown in Table 5.13.

As for the *O*-phenylation of phenols, tetraphenylbismuthonium trifluoroacetate gives good results, and the addition of an acid catalyst improves the yield [82TL3365, 85JCS(P1)2657]. The *O*-phenylation of phenols bearing an electron-withdrawing substituent is usually achieved with triphenylbismuth dichloride under basic conditions [87T323]. The *O*-phenylation of phenols with triphenylbismuth diacetate or tetraphenylbismuthonium trifluoroacetate is often accelerated by the presence of a catalytic amount of copper or copper salt; the yield and selectivity are also improved [86TL3619, 99T1341, 99T3377]. Triphenylbismuthine acts as the phenylating agent for phenol in the presence of an equimolar amount of copper(II) acetate and 6 equiv of triethylamine [87TL887]. Alcohols are phenylated on oxygen atom with triphenylbismuthine and 2 equiv of copper(II) acetate under neutral conditions. Atmospheric oxygen is found to improve the yields of the resulting phenyl ethers [85ZOB2630, 86ZOB2714].

Enols and 1,3-dicarbonyl compounds undergo arylation in a manner similar to phenols. Triphenylbismuth bis(trifluoroacetate) in the presence of metallic copper and tetraphenylbismuthonium trifluoroacetate under acidic conditions are both effective for the selective *O*-phenylation of enols and 1,3-dicarbonyl compounds [81CC503, 82TL3365, 85JCS(P1)2667, 86TL3619].

The *O*-phenylation of alcohols with tetraphenylbismuthonium trifluoroacetate is dependent on pH values of reaction media [87JCS(P1)251]. Thus, attempted reactions with primary or secondary alcohols under basic conditions results, mainly, in the oxidation leading to carbonyl compounds (Table 5.6), while the corresponding phenyl ethers are obtained in modest (for secondary alcohols) to good yields (for primary alcohols) under neutral or acidic conditions. *O*-Phenylation of tertiary alcohols can be performed with pentaphenylbismuth under neutral conditions [78DOK(238)361, 86ZOB330]. *tert*-Butyl

TABLE 5.13
O-Arylation of alcohols, phenols and 1,3-dicarbonyl compounds with organobismuth(V) compounds

Organobismuth	Substrate	Conditions[a]	Product	Reference
Ph_3Bi	Alcohols	$Cu(OAc)_2$	Phenyl ethers	[86ZOB2714] [85ZOB2630] [90MOK112] [90ZOB2083]
	Phenol	$Cu(OAc)_2$ basic	Diphenyl ether	[87TL887]
Ar_3Bi Ar = Ph, 4-PhC_6H_4	β-Hydroxy-esters Glycol	$Cu(O_2CC^tBu)_2$, OXONE™	*O*-Arylated compounds	[99SL1207]
R_3Bi	Alcohols	$Cu(OAc)_2$	Alkyl or aryl ethers	[93IZV2109] [95IZV156]
$Ph_3Bi(OAc)_2$	Glycols	Neutral	Phenoxyalcohols	[81TL5063] [81TL2885] [83JOC441]
	Cholestanol	Neutral	Cholestanyl phenyl ether	[86CC65]
	Glycols	$Cu(OAc)_2$	Phenoxyalcohols	
	Alcohols	$Cu(OAc)_2$, $CuCl_2$ or Cu	Phenyl ethers	[84ZOB2157] [85ZOB2514] [90MOK112] [95IZV156]
	Phenols	$Cu(OAc)_2$ or Cu	Aryl phenyl ethers	[86TL3619] [99T1341] [99T3377]

$Ph_3Bi(O_2CCF_3)_2$	${}^tBuOH$	Basic	*tert*-Butyl phenyl ether	[85ZOB73]
	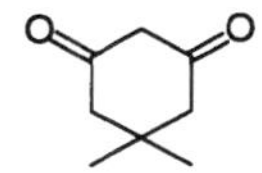	Cu	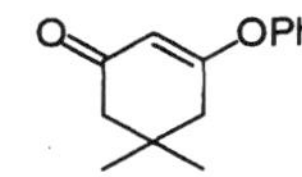	[86TL3619]
Ph_3BiCl_2	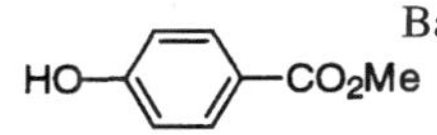	Basic	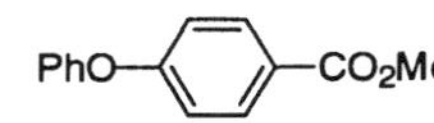	[87T323]
$Ph_4BiO_2CCF_3$	${}^tBuOLi$	Neutral	*tert*-Butyl phenyl ether	[84ZOB100]
	Phenol	Neutral	Diphenyl ether	[85JCS(P1)2657]
	2-Naphthol	Neutral	2-Phenoxynaphthol	[81CC503]
	Phenols	$Cu(OAc)_2$	Aryl phenyl ether	[86TL3619]
	Alcohols	Neutral acidic	Phenyl ethers	[87JCS(P1)251]
	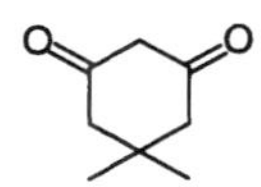	Acidic	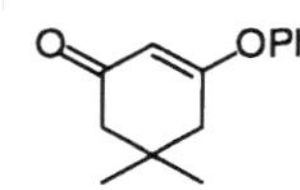	[85JCS(P1)2667] [81CC503] [82TL3365]
Ph_5Bi	Phenol	Neutral	Diphenyl ether	[75DOK(225)581]
	${}^tBuOH$	Neutral	*tert*-Butyl phenyl ether	[78DOK(238)361] [86ZOB330]

[a] Copper catalysts are generally used under neutral conditions.

phenyl ether is also obtainable in low yield from the reaction of triphenylbismuth diacetate and potassium carbonate in *tert*-butyl alcohol as well as from the reaction of triphenylbismuth dichloride and lithium *tert*-butoxide [84ZOB100, 85ZOB73].

Selective mono-*O*-phenylation of glycols is first reported by David et al. during the comparative study of the action of various oxidants toward *vic*-glycols [81TL2885, 81TL5063, 83JOC441]. Good to excellent yields are obtained for all combinations from primary-secondary to secondary-tertiary, except for ditertiary glycols; the last type of glycols afford the expected monoether only in low yield. The mono-*O*-phenylation of glycols is considerably dependent on reaction conditions and is markedly promoted by the addition of small amounts of copper(II) acetate [86CC65].

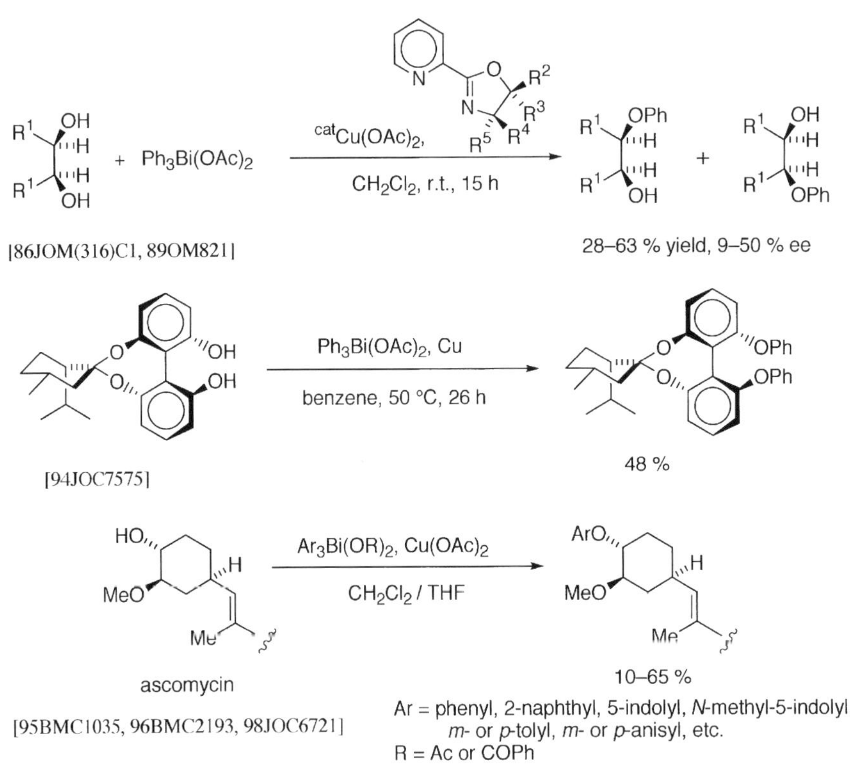

Scheme 5.44. *O*-Arylation of alcohols and phenols with $Ar_3Bi(OAc)_2$ in the presence of Cu or $Cu(OAc)_2$.

The selective monophenylation of *vic*-diols with triphenylbismuth diacetate is applied to the enantioselective preparation of α-phenoxy alcohols by using a pyridyloxazoline as a chiral source in the presence of catalytic amounts of copper(II) acetate (Scheme 5.44) [86JOM(316)C1, 89OM821]. The *O*-phenylation of phenol with triphenylbismuth diacetate under the catalysis of metallic copper is applied to the asymmetric synthesis of 6,6′-diphenoxy-2,2′-biphenyldiols [94JOC7575]. The hydroxyl group in immunosuppressive macrolides, ascomycin and L-683,742, can be *O*-arylated with triarylbismuth diacetate or dibenzoate, where the aryl group may be naphthyl, indolyl or variously substituted phenyl groups [95BMC1035, 96BMC2193, 98JOC6721].

5.5.2.2. N-Arylation and S-arylation

N-Arylation was first observed during attempted oxidation of the hydroxyl function in amino alcohols with triphenylbismuth carbonate [85T3463]. High-yield *N*-arylation has been realized in the reaction of amino alcohols with triarylbismuth diacetate in dichloromethane under reflux [86CC65]. Without the aid of copper or copper salts, amides can hardly be phenylated on the nitrogen atom by tri- and tetraarylbismuth(V) compounds, but imides and lactams can be smoothly *N*-phenylated under neutral and basic conditions, respectively [87JCS(P1)251, 90JOC5222]. Addition of a catalytic amount of copper or copper salt improves the yields of aniline derivatives in the *N*-arylation of simple aliphatic and aromatic amines. As shown in Table 5.14, a variety of aliphatic, alicyclic, aromatic and heterocyclic amines as well as hydrazines can be *N*-arylated with organobismuth(V) reagents. No reaction takes place with α-amino acids but their esters are *N*-monophenylated [89TL937]. A combination of triphenylbismuth bis(trifluoroacetate) and metallic copper effects fast and high-yielding *N*-phenylation reactions [86TL3615]. Triphenylbismuthine in the presence of a stoichiometric amount of copper(II) acetate is found to be a good *N*-phenylating agent for a variety of amines, amides, ureas, sulfonamides, carbamates and imides [87TL887, 96TL9013]. The in situ generated triarylbismuth diacetate by the reaction of triarylbismuthine with iodobenzene diacetate can be used for copper-catalyzed *N*-arylation of anilines [98T4313].

Thiols and sulfinic acids are *S*-arylated with tetraphenylbismuthonium trifluoroacetate or pentaarylbismuthines under neutral conditions to afford the corresponding sulfides and sulfones, respectively [80CC827, 81CC503].

TABLE 5.14
N- and *S*-Arylations with organobismuth compounds

Organobismuth	Substrate	Conditions[a]	Product	Reference
Ph_3Bi	Amines, amides, ureas, sulfonamides, carbamates, imides	$Cu(OAc)_2$, neutral or basic	*N*-Alkyl or arylanilines	[87TL887] [90MOK112] [96TL9013]
R_3Bi or Ph_2BiR	Amines, piperazines	$Cu(OAc)_2$	*N*-Alkyl or arylanilines	[88TL857] [97SC1669] [99T3377]
$Ph_3Bi(OAc)_2$	Amines	Cu	*N*-Alkyl or arylanilines	[86TL3615]
	Amines	$Cu(OAc)_2$	*N*-Alkyl or arylanilines	[85ZOB466] [85ZOB2514] [90MOK112]
	Piperidines	$Cu(OAc)_2$	*N*-Phenylpiperdines	[94S775]
	Amino esters	Cu, $Cu(OAc)_2$, or $Cu(O_2CCF_3)_2$	*N*-Phenylamino esters	[89TL937]
	Azoles	$Cu(OAc)_2$	*N*-Phenylazoles	[99TL2747]
$Ar_3Bi(OAc)_2$, Ar = *o*-tolyl, 2,4-dimethylphenyl, *o*-allyloxyphenyl	Anilines	$Cu(OAc)_2$	*N*-Arylanilines	[99T1341] [99T3377]
$Ar_3Bi(OAc)_2$ (Ar_3Bi + $PhI(OAc)_2$)	Anilines	$Cu(OAc)_2$	*N*-Arylanilines	[98T4313]

$Ph_3Bi(O_2CCF_3)_2$	Indoles	Cu, $Cu(OAc)_2$, or $Cu(O_2CCF_3)_2$	*N*-Phenylindoles	[88TL1115]
Ph_3BiX_2 $X = O_2CCF_3$, OTs, O_2C^tBu, etc., $X_2 = CO_3$	Amines	Cu or $Cu(O_2C^tBu)$, neutral or basic	*N*-Alkyl or arylanilines	[86TL3615] [97T4137]
Ph_3BiCl_2	Lactams	Basic	*N*-Phenyllactams	[90JOC5222]
Ph_4BiX, $X = O_2CCF_3$, OTs	Amines	Cu	*N*-Alkyl or arylanilines	[86TL3615]
$Ph_4BiO_2CCF_3$	*p*-Toluenesulphinic acid	Neutral	*p*-Tolyl phenyl sulfone	[81CC503]
	Thiols	Neutral	Sulfides	[81CC503]
	N-Phenylacetamide	Neutral	*N,N*-Diphenylacetamide	[87JCS(P1)251]
	Imides	Neutral	*N*-Phenylimides	[87JCS(P1)251]
Ar_5Bi	Amines	(Cu)	*N*-Alkyl or arylanilines	[86TL3615]
	p-Toluenesulphinic acid	Neutral	*p*-Tolyl phenyl sulfone	[81CC503]
	Thiols	Neutral	Sulfides	[80CC827]

[a] Copper catalysts are generally used under neutral conditions.

5.5.3. *Miscellaneous carbon–heteroatom bond forming reactions*

Bismuth-mediated miscellaneous reactions forming C–O and C–N bond are summarized in Schemes 5.45 and 5.46. Electron-rich triarylbismuthines are used as the selective condensation agent for ester and amide formation [93CL815, 94JCS(P1)3473]. They are effective for primary and secondary carboxylic acids but fail to work for tertiary and aromatic ones. These reagents have been successfully applied to the preparation of some macrocyclic diesters. Noteworthy is that tris(2-methoxyphenyl)bismuthine plays the dual role as a condensation agent and as a template [95CC1407]. The template effect of the organobismuth center is also utilized for the preparation of crown-type lactones [96JCS(P1)953]. Tris(4-methylphenyl)bismuth diazide reacts with isocyanates and alkynes to give the corresponding arylcarbamonyl azides and 1,2,3-triazoles, respectively [92JCR(S)34, 93TL1055]. Triarylbismuth imides transfer the imino moiety to aldehydes, benzoyl chloride, phenyl isocyanate and triphenylphosphine, giving the corresponding imines, amide, urea, and phosphine imide, respectively [91CL105, 99OM2580]. Methyl bis(1-naphthyl) bismuthinate mediates the *N*-acylation of amides, thioamides, ureas and thioureas with acetic acid [90CL1651]. All these reactions are supposed to proceed through the initial formation of some reactive organobismuth(V) intermediate.

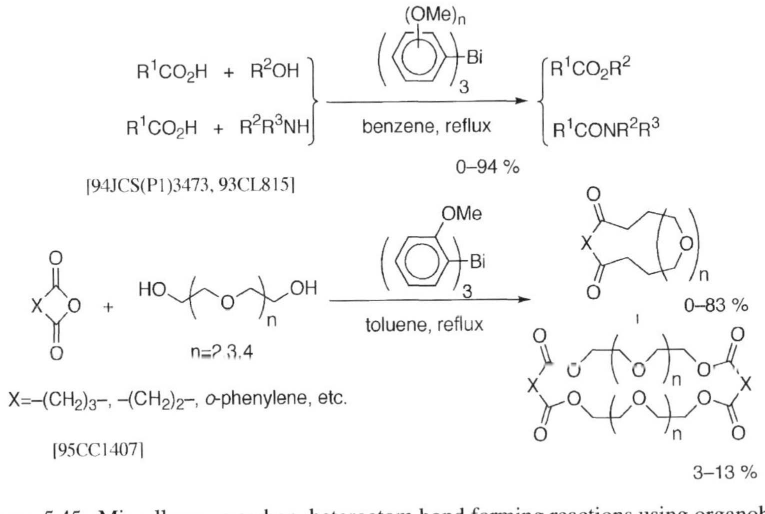

Scheme 5.45. Miscellaneous carbon–heteroatom bond forming reactions using organobismuth compounds.

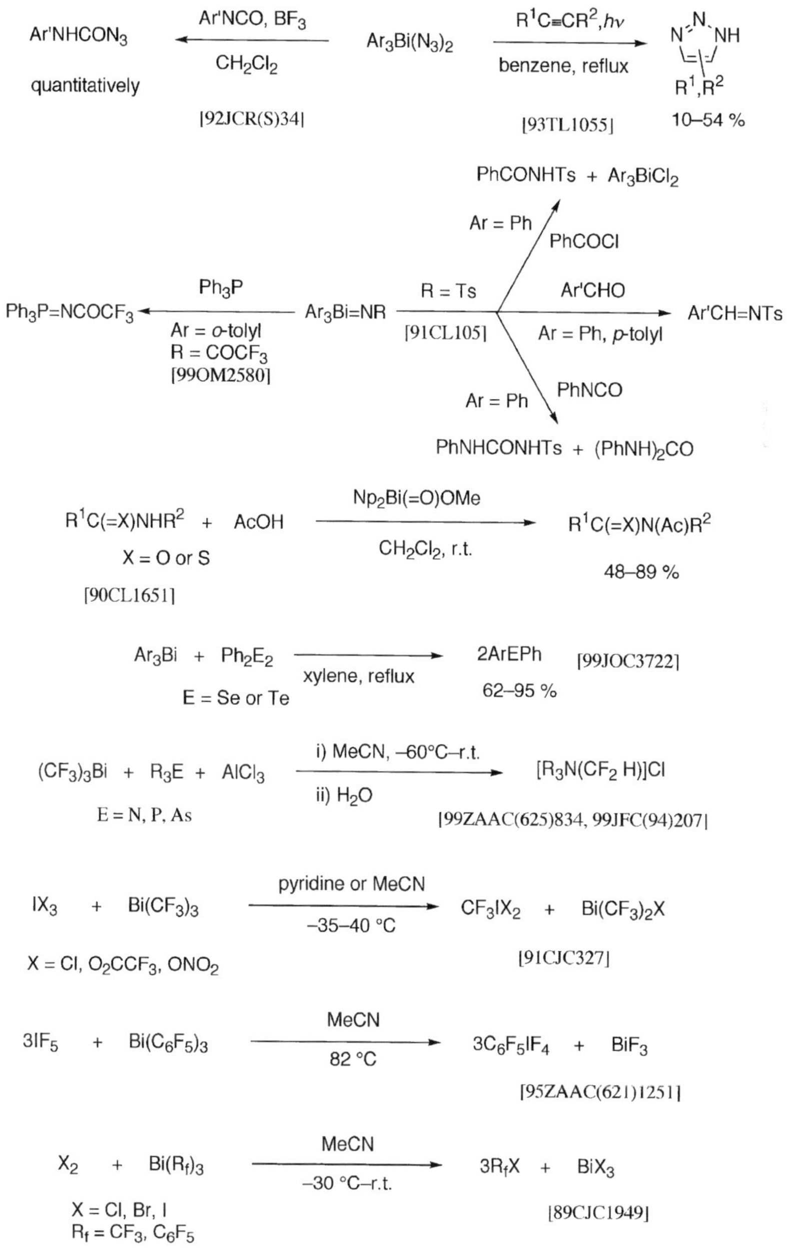

Scheme 5.45. (continued).

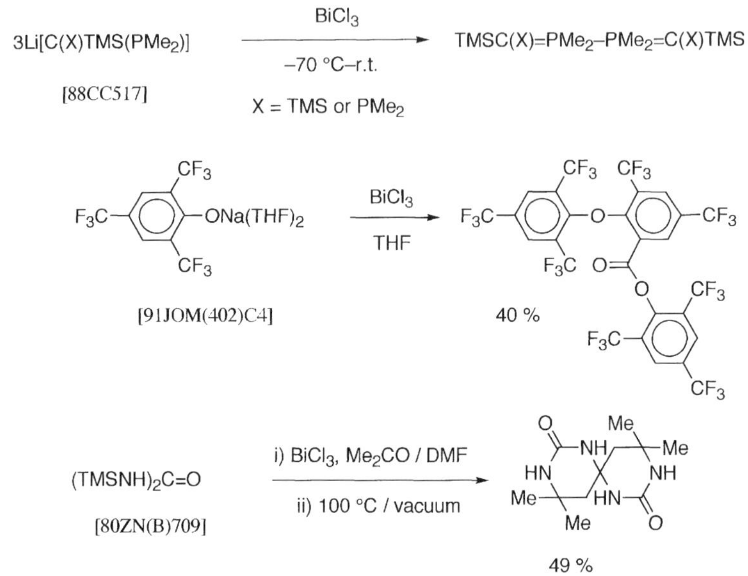

Scheme 5.46. Miscellaneous carbon–heteroatom bond forming reactions using bismuth chloride.

Chapter 6

Structural Chemistry of Organobismuth Compounds

The criterion for compounds to be included in this chapter is not strictly limited to ''organic'' bismuth compounds. Some relevant inorganic bismuth compounds are included, which may be of interest to organic chemists. Transition metal complexes are not included.

6.1. FUNDAMENTAL STRUCTURE PARAMETERS

6.1.1. Ionic radii

Reported ionic radii of pnictogen elements are tabulated in Table 6.1.

Interatomic distances (and therefore effective ionic radii) depend on the coordination number, electronic spin state, covalency as well as the presence of strong anion–anion or cation–cation repulsive forces [28ZK(67)377, 69AX(B)925]. In deriving a set of effective ionic radii listed in Table 6.1, the following assumptions are made [69AX(B)925]: (i) the principle of additivity is valid for both cation and anion radii to reproduce accurate interatomic distances, if one takes into consideration the coordination number, electronic spin, covalency and repulsive forces, (ii) the effect of covalency in shortening M–X bonds is comparable for all M–F and M–O bonds, (iii) the radii are independent of structure types, (iv) the radii of both cation and anion vary with the coordination number, (v) although individual cation–anion distances may vary in a polyhedron, the average cation–anion distances over all similar polyhedra are constant, and (vi) with a fixed anion, the volume of unit cells of isostructural series is proportional to the volume of the cations.]

TABLE 6.1
Ionic radii of pnictogen elements[a]

Ion	Coordination number	Ionic radii (Å)
N^{3+}	4	1.32
P^{3+}	6	0.58
As^{3+}	6	0.72
Sb^{3+}	6	0.90
Bi^{3+}	6	1.17
Bi^{3+}	8	1.31
N^{5+}	6	0.27
P^{5+}	4	0.31
As^{5+}	4	0.48
As^{5+}	6	0.60
Sb^{5+}	6	0.74
Bi^{5+}	6	0.90

[a] Refs.: [69AX(B)925, 76AX(A)751].

6.1.2. van der Waals radii

Mundt et al. estimated the van der Waals radii of bismuth to be 2.30 Å from an X-ray crystallographic study of tetramethyldibismuth [88ZN(B)952]. Yamamoto et al. reported the sum of van der Waals radii for Bi–N to be 3.74 Å. Abstracting the van der Waals radius of nitrogen atom (1.5 Å), a value of 2.24 Å can be attributed to the bismuth atom [92JA7906]. A value of 1.847 Å was reported for bismuth from the molar volumes of inorganic bismuth compounds [95IZV18]. The van der Waals radii of pnictogen elements are tabulated in Table 6.2.

6.2. STRUCTURES OF ORGANOBISMUTH COMPOUNDS

6.2.1. Molecular structure

In Table 6.3, measured and calculated carbon–pnictogen element (Pn) bond lengths and bond angles of acyclic trivalent and related compounds, are tabulated. In Tables 6.5 and 6.6, X-ray crystallographic structures of representative trivalent and pentavalent organic bismuth compounds, as well as some relevant inorganic bismuth compounds, are collected.

An important difference between the first and subsequent rows of main-group elements is the ability of the latter to be involved in "extended

TABLE 6.2
van der Waals radii of pnictogen elements

Element	van der Waals radii (Å)	Reference
N	1.5	[60B-P]
N	1.563[a]	[95IZV18]
P	1.9	[60B-P]
P	1.562[a]	[95IZV18]
As	2.0	[60B-P]
As	1.580[a]	[95IZV18]
Sb	2.2	[60B-P]
Sb	1.752[a]	[95IZV18]
Bi	2.30	[88ZN(B)952]
Bi	2.24	[92JA7906]
Bi	1.847[a]	[95IZV18]

[a] Data from inorganic compounds.

valence'' molecules. Such ''extended valence'' is often called ''hypervalence''. Hypervalence was first seen as the structural consequence of mixing *ns* and *np* valence atomic orbitals with *nd* valence atomic orbitals, giving hybrids such as sp^3d in PCl_5 and sp^3d^2 in PCl_6^-. However, theoretical studies have shown a very small contribution of the *d* valence atomic orbitals for these molecules [69NAT(221)1106, 69CRV(69)157, 73AG773]. Consequently, an alternative model involving the formation of three-center four-electron bonds has been presented to explain the violation of the octet rule in hypervalence [63JA112, 84JA4794, 84AG(E)272].

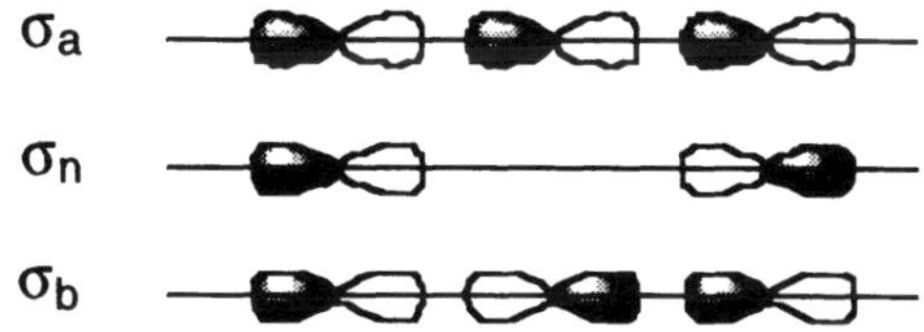

For example, in XeF_2 three MOs can be constructed as linear combinations of the $p\sigma$ valence AOs of xenon with two fluorine atoms. Of the three MOs, one is bonding (σ_b), and the others are non-bonding (σ_n) and antibonding (σ_a). Four-electrons, two from xenon and one from each of the two fluorine atoms, occupy the σ_b and σ_n MOs, so that overall bonding will occur. This three-center four-electron bonding requires that the terminal atoms be more electronegative than the central one, and that the bonds be longer than the electron paired bonds between the same elements, since there is only one bonding pair

TABLE 6.3
Measured and calculated carbon–pnictogen element (Pn) bond lengths and bond angles of acyclic trivalent and related compounds

Compound	Measured and calcd. mean bond length of C–Pn (Å)	Measured and calcd. Mean bond angle of C–Pn–C (°)	Calculated s-character of lone pair electron (%)	Reference
Me_3N	1.4535	110.40		[84AX(C)413][c]
Me_3N	1.48[a]	112.3[a]	40.7[a]	
Me_3N	1.48[b]	110.3[b]		
Me_3P	n.a.	n.a.		
Me_3P	1.87[a]	100.7[a]	78.1[a]	
Me_3P	1.88[b]	101.6[b]		
Me_3As	n.a.	n.a.		
Me_3As	1.98[a]	99.7[a]	81.1[a]	
Me_3As	1.98[b]	100.1[b]		
Me_3Sb	n.a.	n.a.		
Me_3Sb	2.16[a]	95.0[a]	94.5[a]	
Me_3Sb	2.14[b]	98.0[b]		
Me_3Bi	2.266	96.7		[73JMOS(17)429]
Me_3Bi	2.27[a]	97.0[a]	97.0[a]	
Me_3Bi	2.26[b]	96.6[b]		
Me_4Pb	2.238			[71BCJ2638]
Ph_3N	1.419	119.6		[85AX(C)967]
Ph_3N	1.47[a]	113.8[a]	34.5[a]	
Ph_3P	1.828	102.99		[64JCS3799][d]
Ph_3P	1.831	102.8		[91AX(C)345]
Ph_3P	1.86[a]	102.2[a]	78.0[a]	
Ph_3As	1.957	100.1		[83JOM(244)129, 85JCSR(15)561]
Ph_3As	1.95[a]	101.0[a]	79.4[a]	
Ph_3Sb	2.155	97.2		[90AX(C)917]
Ph_3Sb	2.11[a]	97.5[a]	94.4[a]	
Ph_3Bi	2.24	94.7		[42ZK(104)305, 66CC111, 68JCS(A)2059]
Ph_3Bi	2.23[a]	97.7[a]	96.4[a]	

[a] Calculated by MOPAC version 6.01, PM3; MOPAC version 6, J.J. Stewart, QCPE Bull. 9 (1989) 10; revised as 6.01 for Sun SPARC station by K. Nishida.

[b] Calculated by Gaussian 92, RHF/LANL2DZ; Gaussian 92, Revision E.3, M.J. Frisch et al., Gaussian, Inc., Pittsburgh, PA, 1992.

[c] At −155 °C.

[d] Photographic.

for the two bonds. The coordination number can be increased, and some further bonding may be achieved by forming the three-center four-electron bonds without expanding the valence shell [63JA112]. With this formulation,

BiX_5 has three equatorial bonds, made from the usual sp^2 hybrid orbital, and two axial bonds which use the three-center four-electron bonding, two electrons from the *p* orbital of bismuth and two electrons from the X ligands [94B-N]. In the three-center four-electron bonding, the central atom should be positively charged. This has been confirmed for MH_5 (M = P, As, Sb, Bi) with recent ab initio calculation [95JA11790].

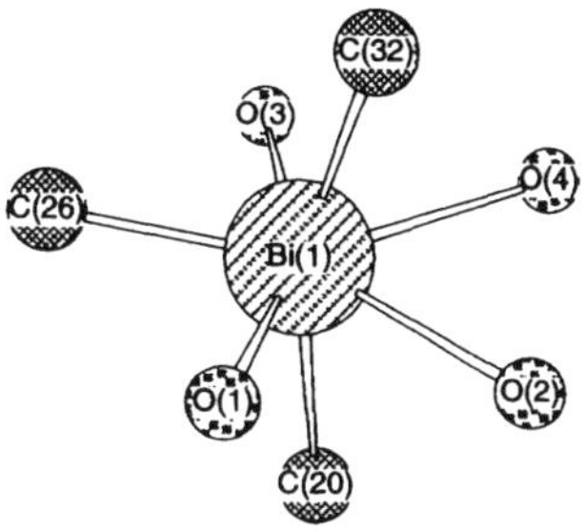

Di(carboxylato-*O*)triarylbismuth (Triarylbismuth dicarboxylate)

Although the concept of three-center four-electron bonding is useful for understanding most of the extended valence structure of organobismuth compounds, it is not necessarily applicable to all of them. As will be described below, there are many instances of extended bonding, which do not have the linear array of elements (X–Bi–Y) required to make the three-center four-electron bonding. Some of the extended bonding may be understood better with the coulombic interaction between the electronegative atoms (N, O or S) and positively charged bismuth, the charge being generated by the three-center four-electron bonding. For example, triarylbismuth dicarboxylates have, formally, seven coordinations and obviously exceed the octet rule. One set of the O–Bi bonds is significantly shorter than the other; the shorter bonds range from 2.26 to 2.30 Å, and the longer ones from 2.8 to 3.0 Å. The bonding angles O–Bi–O are as follows: O(1, short)–Bi–O(3, short), 156.7°; O(2, long)–Bi–O(4, long), 70.0°; and O(1, short)–Bi–O(4, long), 137.4°. Although the O(short)–Bi–O(short) bond may be regarded as a distorted three-center four-electron bonding, the O(long)–Bi bonds conform to this concept with difficulty, since the bond angles are too far from linearity. The O(long)–Bi bonds are probably subject to the contribution from the coulombic interaction between the electronegative carbonyl oxygen and electropositive bismuth atoms.

Perkins et al. have proposed a systematic classification scheme for the hypervalent bonding [80JA7753]. The bonding, to any atom (X) in a resonance structure in terms of the number of valence shell electrons (*N*) formally associated with that atom and the number of ligands (L) directly bonded to it, is described as an N-X-L system. With this classification scheme, Ph_3Bi is described as 8-Bi-3 (it satisfies the octet rule), and Ph_3BiCl_2 is 10-Bi-5. These 8-Bi-3 and 10-Bi-5 compounds are generally stable and a great number of compounds of these classes have been reported in the literature. Although their numbers are limited, 10-Bi-4, 12-Bi-6, 14-Bi-7 and 18-Bi-9 compounds have also been reported.

6.2.1.1. 8-Bi-3 compounds

Simple triarylbismuthines generally have a trigonal pyramidal structure with the C–Bi–C bond angles in the range of about 90–110°. The measured and calculated carbon–element bond lengths, as well as bond angles of simple trivalent pnictogenides and related compounds, are listed in Table 6.3.

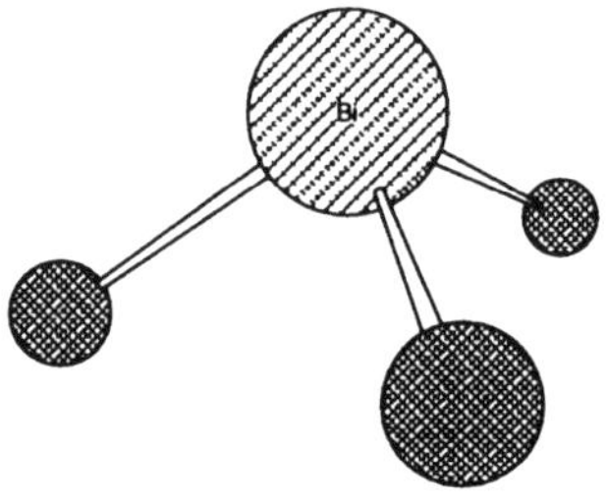

Trimethylbismuthine [73JMOS(17)429]

The molecular geometry of trimethylbismuthine has been determined by gas-phase electron diffraction. The C–Bi bond length is significantly longer than the C–Pb bond in Me_4Pb. Seeing that bismuth is located to the right of lead on the periodic table, one may expect the C–Bi bond to be shorter than the C–Pb bond [73JMOS(17)429].

The C–Bi bond length of triphenylbismuthine (2.239 ± 0.019 Å) is shorter (98.8%) than that of trimethylbismuthine (2.266 Å) and the bond angle is also a little smaller in the former compound [42ZK(104)305, 66CC111, 68JCS(A)2059]. Ab-initio calculations also show the same results. Although no data is available for triphenylbismuthine in the gas phase, these findings may be reasonably attributed to the resonance effect between the bismuth atom and phenyl rings in triphenylbismuthine. The reduction of the C–N bond length of triphenylamine (1.42 Å) relative to that of trimethylamine (1.45 Å) is about 97.9%. The larger reduction in length of the aromatic C–N bond compared with the corresponding C–Bi bond indicates that the resonance effect is less important in the C–Bi bond than in the C–N bond. It is also a general tendency of the C–Bi bond length of aromatic bismuth compounds to be a little shorter than those of aliphatic ones by the resonance effect; the average bond lengths of trivalent bismuth compounds tabulated in Table 6.5 are 2.26 ± 0.03 Å for C(aromatic)–Bi compounds, and 2.30 ± 0.05 Å for C(aliphatic)–Bi bonds, respectively.

6.2.1.2. 10-Bi-4 compounds

$[Et_4N^+][Ph_2BiX_2]^-$ (X = Cl, Br, I) and $[Ph_4As^+][Ph_nBiX_{4-n}]^-$ were first isolated in 1969 [69JOM(20)99], but the first X-ray crystallographic structure of this type of compound was reported for $[Ph_4Bi^+]$ $[Ph_2Bi(OCOCF_3)_2]^-$ two decades later, in 1984 [84HCA586]. In the anionic part of the latter compound, two trifluoroacetato groups adopt the apical positions and two phenyl rings are situated in the equatorial plane passing through the bismuth atom. The disposition leaves an empty space in the equatorial plane, which is interpreted as the remaining lone pair of a distorted trigonal bipyramid around the bismuth atom (Table 6.5, **29**).

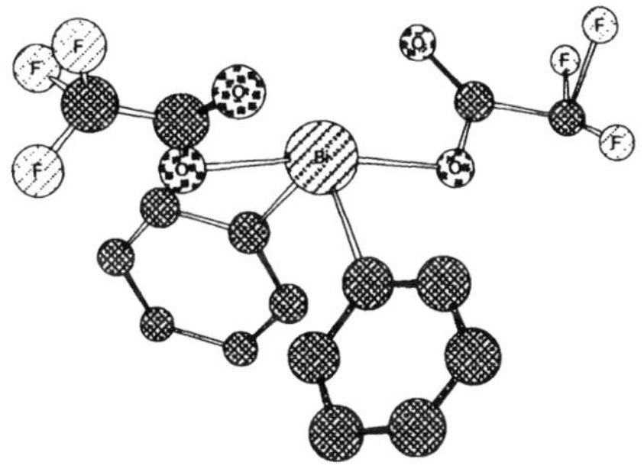

Diphenylbis(trifluoroacetato-*O*)bismuthate anion [84HCA586]

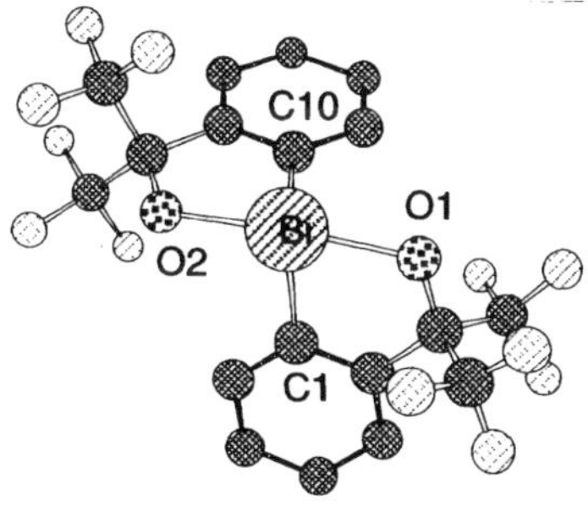

Tetraethylammonium 1,1′-spirobis[(3,3-bis(trifluoromethyl)-3*H*-2,1-benzoxabismole] [92TL6653, 95HAC293]

Organobismuth ate complexes have similar distorted trigonal bipyramidal structures with an empty space in the equatorial plane. Typical examples include tetraethylammonium 1,1′-spirobis[(3,3-bis(trifluoromethyl)-3*H*-2,1-benzoxabismole] [92TL6653, 95HAC293] (Table 6.5, **30**), [isopropyldithiocarbonato-*S*]diphenylbismuthine [80ZAAC(470)171] (Table 6.5, **32**), bis(triphenylphosphine)iminium dichloro(2,2′-biphenylene)bismuthate [96OM887] (Table 6.5, **39**), and tetraphenylphosphonium dibromodiphenylbismuthate [92JCS(D)1967] (Table 6.5, **46**).

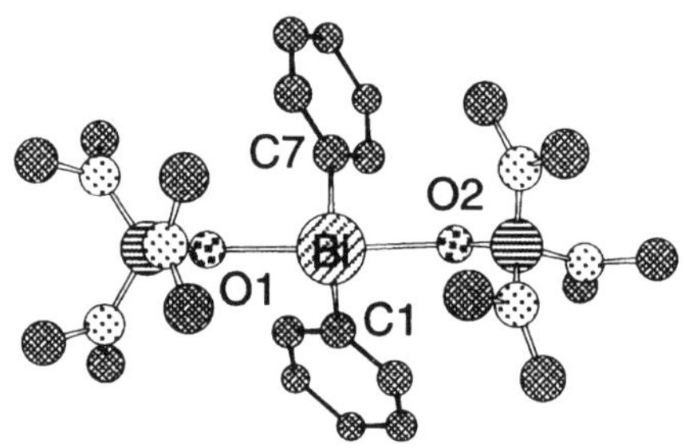

Bis(hexamethylphosphoramide-*O*) diphenylbismuthine tetrafluoroborate [93JOM(460)C22]

Diarylbismuth(III) salts of the type $[Ar_2BiX_2^+]Y^-$ have also similar distorted trigonal bipyramidal structures with an empty space in the equatorial plane. Typical examples include bis(hexamethylphosphoramide-*O*)diphenylbismuthine tetrafluoroborate [93JOM(460)C22] (Table 6.5, **20**), bis(triphenylphosphineoxide-*O*)diphenylbismuthine tetrafluoroborate (Table 6.5, **18**), bis(hexamethyl-phosphoramide-*O*)diphenylbismuthine hexafluorophosphate (Table 6.5, **19**), and bis(hexamethylphosphoramide-*O*)bis(2,4,6-trimethylphenyl)bismuthine hexafluorophosphate [96JCS(D)443] (Table 6.5, **21**).

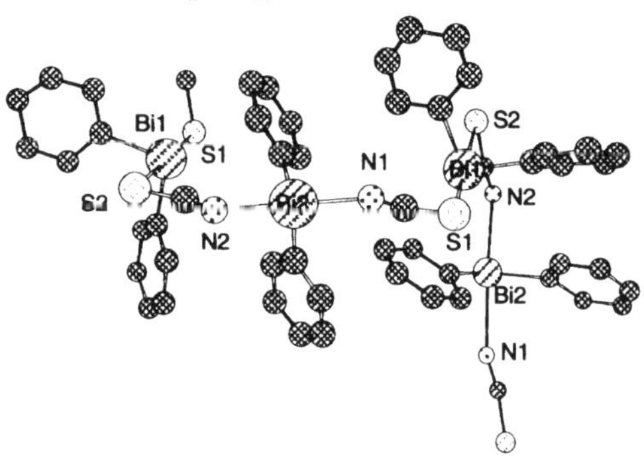

Catena[μ_2-(thiocyanato)diphenylbismuthine] chloroform solvate [95JCS(D)383]

Four coordinated organobismuth compounds with formally neutral bismuth atom have been reported to have the distorted trigonal bipyramidal structure with an empty space in the equatorial plane; catena[μ_2-(thiocyanato)diphenylbismuthine] [95JCS(D)383] (Table 6.5, **31**). Typical examples include catena[(μ_2-phenoxy)diethylbismuthine] [92CC1021] (Table 6.5, **15**), catena[(μ_2-pentafluorophenoxy)diethylbismuthine] [92CC1021] (Table 6.5, **16**), catena[(μ_2-*N*-benzoylglycinato-*O,O'*)diphenylbismuthine] [88AX(C)828] (Table 6.5, **17**), [2-(2-methoxy-2-propyl)phenyl](2-phenyl-2-propanolato]bismuthine and [2-(2-hydroxy-2-propyl)phenyl](2-phenyl-2-propanolato)bismuthine [94AX(B)151] (Table 6.5, **22** and **23**), 1-[2-(*N,N*-dimethylaminomethyl)phenyl]-3,3-bis(trifluoromethyl)-3*H*-2,1-benzoxabismole [92JA7906, 95JA3922] (Table 6.5, **24**), 1-[2-(2-methoxy-2-propyl)phenyl]-3,3-bis(trifluoromethyl)-3*H*-2,1-benzoxabismole [92JA7906, 95JA3922] (Table 6.5, **25**), chloro[bis(trimethylsilyl)(2-pyridyl)methyl-*C,N*]-[3-(2-pyridyl)-3-trimethylsilylprop-2-enyl]bismuthine [91CC1560] (Table 6.5, **35**), chlorophenyl[2-((*R*)-1-(*N,N*-dimethylamino)ethyl)phenyl-*C,N*]bismuthine [95OM1542] (Table 6.5, **36**), and chloro[2-(*N,N*-dimethylaminomethyl)phenyl-*C,N*]-(4-methylphenyl)bismuthine [93JCS(P1)2969] (Table 6.5, **37**).

6.2.1.3. 10-Bi-5 compounds

Pentaarylbismuth compounds take a square-pyramidal or a trigonal bipyramidal structure depending on the substituents on the aryl groups [90AG(E)213, 89CB803, 87AG(E)1180] (Table 6.6, **2–10**). Parameter Δ, which is defined as the difference in bond angles between two perpendicular bonding lines centered at the given element [89CB803] is often used to distinguish the structures; according to the definition, $\Delta = 0°$ means an ideal square pyramid and $\Delta = 60°$ means an ideal trigonal bipyramid.

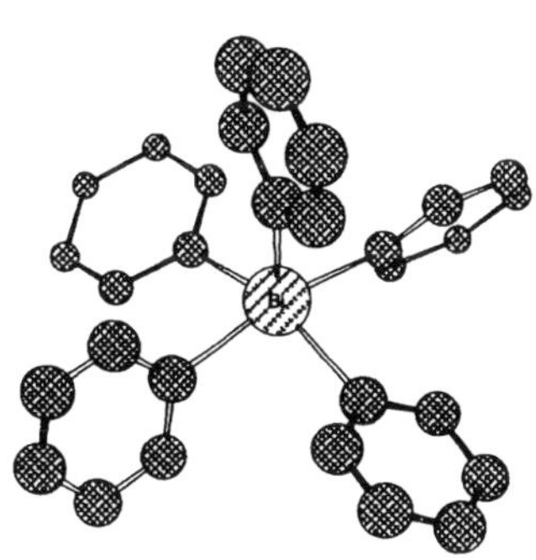

Pentaphenylbismuth [87AG(E)1180]

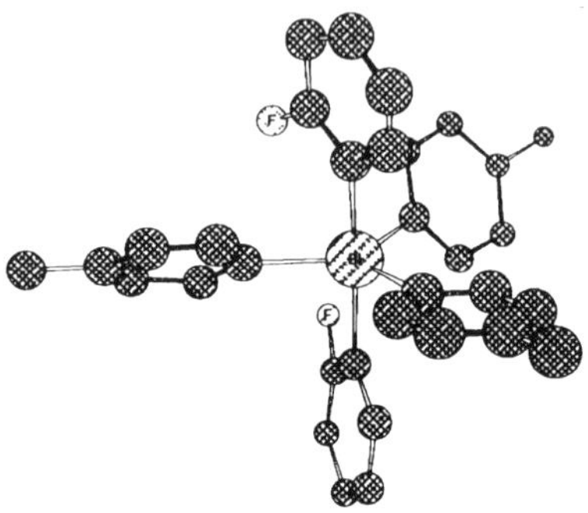

Bis(2-fluorophenyl)tris(4-methylphenyl)bismuth
[90AG(E)213]

More electronegative groups such as fluorophenyls tend to occupy the axial positions in the trigonal bipyramidal structure, as is predicted from the hypervalent theory (Table 6.4).

TABLE 6.4
Structure of pentaarylbismuths

Compound	Δ (°)	Structure
$Bi(C_6H_5)_5$	13.2	Square pyramidal
$Bi(C_6H_4CH_3\text{-}4)_3(C_6F_5)_2$	16	Square pyramidal
$Bi(C_6H_4F\text{-}4)_3(C_6F_5)_2$	5.1	Square pyramidal
$Bi(C_6H_5)_3(C_6H_3F_2\text{-}2,6)_2$	11.5	Square pyramidal
$Bi(C_6H_5)_3(C_6H_4F\text{-}2)_2$	19.2	Square pyramidal
$Bi(C_6H_4CH_3\text{-}4)_3(C_6H_4F\text{-}2)_2$	45.4	Trigonal bipyramidal

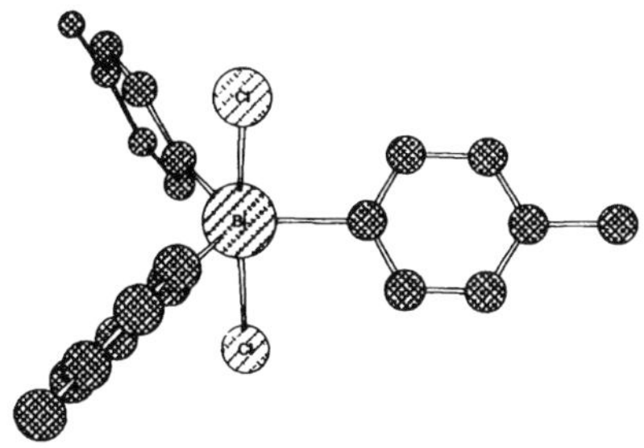

Dichlorotris(4-methylphenyl)bismuth [93OM1857]

Compounds with the formula R_3BiXY (where R denotes an aromatic or aliphatic group, and X, Y represent electronegative groups such as halogen and oxygen atoms) have a trigonal bipyramidal structure with R at equatorial positions and X, Y at axial positions. The linear X–Bi–Y bonds in these compounds are typical three-center four-electron bonds. Since the terminal atoms or groups X and Y in the three-center four-electron bonds are negatively charged, the more electronegative atoms make the more stable bonds.

6.2.1.4. 12-Bi-6 compounds

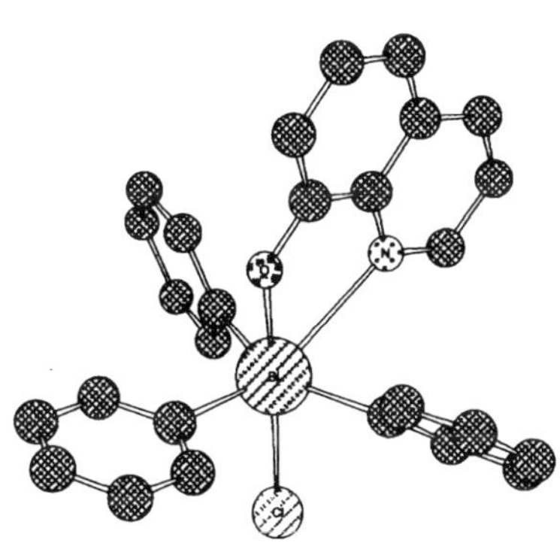

Chloro(8-hydroxylquinolinolato-*O*)triphenylbismuth
[84HCA586]

Six coordinated bismuth compounds (12-Bi-6) were detected in the reaction of Ar_5Bi with PhLi spectroscopically at low temperature [67LA(705)66,

68LA(720)198]. The first stable 12-Bi-6 compound was reported in 1984 and confirmed by X-ray crystallography [84HCA586]. In chloro(8-hydroxyquinolinato-*O*)triphenylbismuth, the nitrogen atom of quinoline moiety coordinates to the bismuth atom to form a six-coordinated structure (Table 6.6, **42**). The molecule has, essentially, a trigonal bipyramidal structure with three phenyl rings occupying the equatorial positions and the bismuth atom lies in the plane of three carbon atoms (0.03 Å). Although the coordination of the nitrogen atom to the bismuth center enlarges the bonding angle C–Bi–C of two phenyl groups to 143.1°, it is too small to be attributed to the square bipyramidal structure.

A 12-Bi-6 compound with two bidentate ligands has been reported to be stable up to 150°C [89TL953] and its X-ray crystallographic structure was elucidated [93OM3297] (Table 6.6, **24**). Now X-ray crystallographic data at low temperatures are available for a series of 12–Bi–6 compounds: Ph_5BiPy [93IC3948] (Table 6.6, **14**), Ph_6BiLi [93IC3948] (Table 6.6, **12** and **13**) and Me_6BiLi [94AG(E)976] (Table 6.6, **11**). These hexaaryl- and hexamethylbismuthate have the square bipyramidal structures where all six C–Bi bonds are almost identical in length within an error of ±0.05 Å.

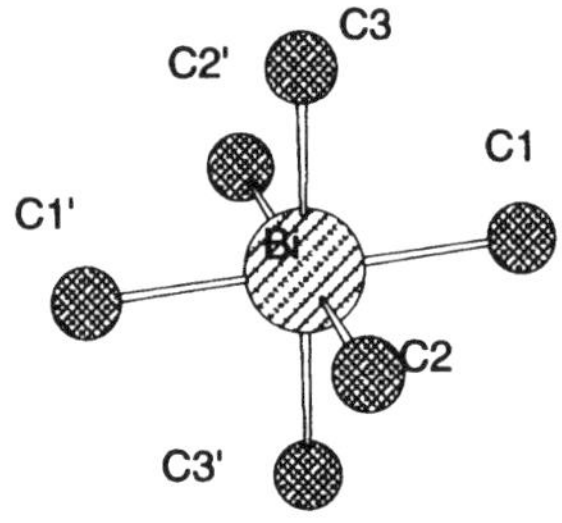

Lithium hexamethylbismuthate [94AG(E)976]

6.2.1.5. 14-Bi-7 compounds

In triarylbismuth dicarboxylates, the carboxylato groups coordinate to the central bismuth atom via two oxygen atoms for each carboxylic group, resulting in the coordination number of seven (Table 6.6, **33**–**35**). One pair of O–Bi bonds is significantly shorter than the other; the shorter bonds range from 2.26 to 2.30 Å and the longer ones range from 2.8 to 3.0 Å. The sum of the van der Waals radii of oxygen and bismuth atoms is estimated to be about 3.7 Å, so the values of 2.8–3.0 Å may be expected to fall within the bonding distance.

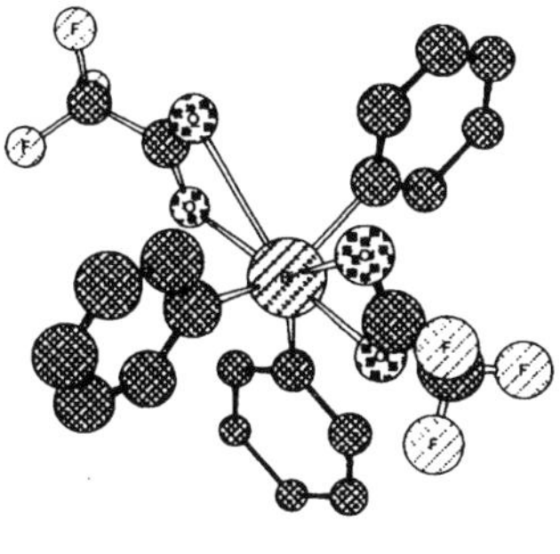

Bis(trifluoroacetato-*O*)triphenylbismuth [91JOM(419)283]

6.2.1.6. 18-Bi-9 compounds

Most phenylbismuth tetracarboxylates $PhBi(OCOR)_4$ have the square pyramidal structure with the phenyl group at the axial position and four carboxylato groups at the equatorial positions (Table 6.5, **53–58**). In these compounds, one set of the O–Bi bonds is significantly shorter than the other, and the longer ones range from 3.24 to 3.33 Å. When these distances were taken as the bonding distances, the compounds were expected to possess the 18–Bi–9 structure.

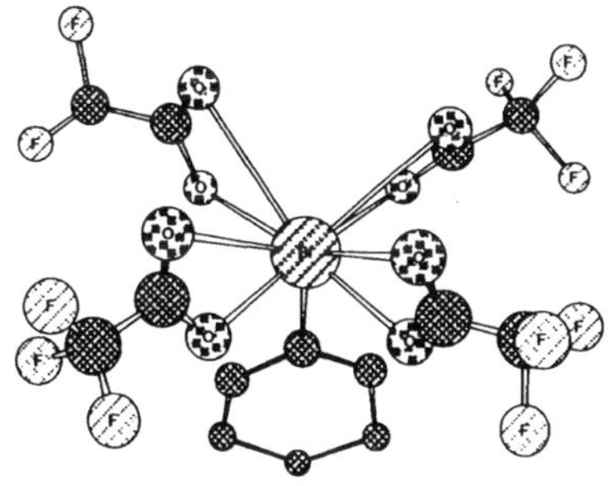

1,3-Bis(dimethylammonio)-2-propanol
tetrakis(trifluoroacetato-*O*)phenylbismuthate

N,N,N′,N′-tetramethylethylenediammonium (trifluoroacetato-*O,O′*)tris(trifluoroacetato-*O*)phenylbismuthate [93AG(E)589] (Table 6.5, **54**) has a pentagonal pyramidal structure. There are five ''short'' O–Bi bonds in this compound, and five oxygen atoms are located almost in one plane containing the bismuth atom.

N,N,N',N'-Tetramethylethylenediammonium (trifluoroacetato-*O,O'*)tris(trifluoroacetato-*O*)-phenylbismuthate [93AG(E)589]

6.2.1.7. Polymeric structure of organobismuth compounds

Significant parts of known trivalent organobismuth compounds exist in polymeric structure in the solid state, with electronegative elements working as the bridging units. Typical examples include catena[(μ_2-phenoxy)diethylbismuthine] [92CC1021] (Table 6.5, **15**), catena[(μ_2-chloro)chloro(phenyl)-(tetrahydrofuran)bismuthine] [93JCS(D)637] (Table 6.5, **65**), catena[(μ_2-bromo)bromo(phenyl)(tetrahydrofuran)(bismuthine] (Table 6.5, **70**), catena[(μ_2-iodo)iodo(phenyl)(tetrahydrofuran)bismuthine] [92JCS(D)1967] (Table 6.5, **75**). This is because the bismuth atom combines with electronegative elements such as oxygen or halogen via the hypervalent bond to form the μ-oxo or μ-halo structures. When the bond is strong, the polymeric structure does not dissociate in solution and bismuth compounds are low in solubility in organic solvents. If the intermolecular hypervalent bonding is prevented by the intramolecular hypervalent interaction or steric hindrance, the solubility in organic solvents becomes high [89CC1628, 90CC301, 90IC358].

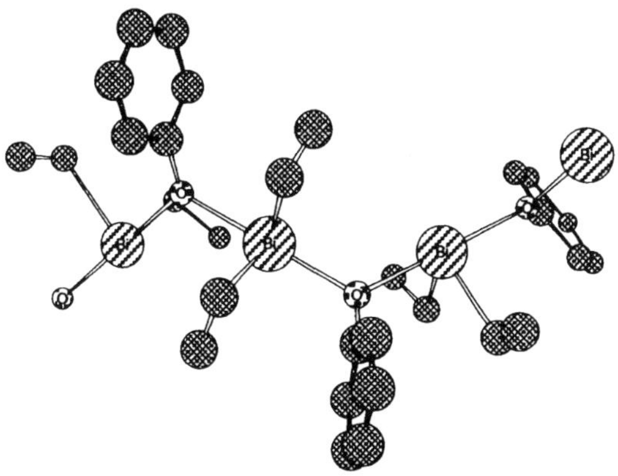

Catena[(μ_2-phenoxy)diethylbismuth] [92CC1021]

6.2.1.8. Bismuthonium ylide

Bismuthonium ylides are an example of formally double-bonded bismuth compounds whose structures were first confirmed by X-ray crystallography. The Bi–C_{ylide} bond length of 4,4-dimethyl-2,6-dioxo-1-triphenylbismuthio-cyclohexanide (Table 6.6, **26**) is 2.156(11) Å [90JCS(P1)3367] which is only 2% shorter than the Bi–C_{alkyl} bond of a bismuthonium salt (2.20(2) Å), (3,3-dimethyl-2-oxobutyl)triphenylbismuthonium tetrafluoroborate [93TL8457]. A theoretical study predicted that $H_3Bi{=}CH_2$ has a bond length of 2.117 Å and thus a single bond character [94JA10080]. A large degree of ionic character, represented as $Bi^+–C^-$, agrees with the prediction that the vacant 6*d* orbitals of bismuth do not effectively overlap the 2*p* orbitals of the carbanionic moiety.

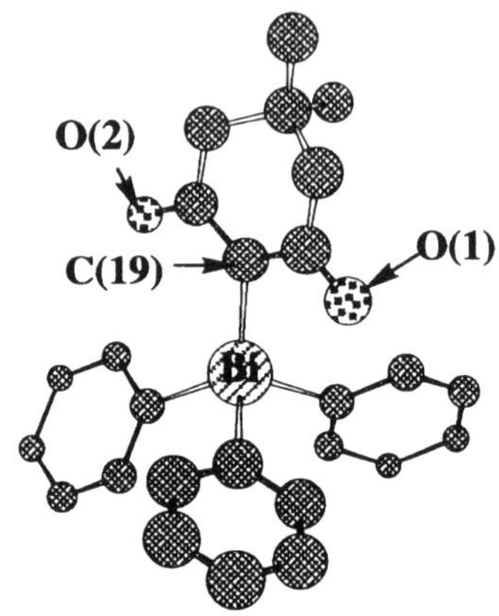

4,4-Dimethyl-2,6-dioxo-1-triphenylbismuthoniocyclohexanide [90JCS(P1)3367]

6.2.1.9. Bi–Bi double bond

Since the first isolation of a stable P=P double-bonded compound in 1981 [81JA4587], compounds featuring the double-bonding between the heavy group 15 elements have been attracting considerable interest. Although the As=As double-bonded compounds were reported soon after [83JA5506], successful synthesis of the Sb=Sb and Bi=Bi compounds was only made recently (Section 2.2.2.4).

In 1990, a theoretical study predicted the possible existence of stable distibene and dibismuthene [90CC1724] and, in 1997, the first Bi=Bi double-bonded compound was isolated and characterized by X-ray crystallography [97SCI(277)78] (Table 6.5, **77**). The Bi=Bi double bond has a length of 2.8206(8) Å, which is 6% shorter than the Bi–Bi single bond (2.990(2) Å) in $Ph_2Bi–BiPh_2$. This agrees reasonably well with the calculated bond shortening of 7% from $H_2Bi–BiH_2$ (3.009 Å) to HBi=BiH (2.795 Å). The dibismuthene is purple in hexane and shows two absorption maxima at 660 and 525 nm, which correspond to the n–π^* and π–π^* transitions of the Bi=Bi chromophore.

6.2.2. Electronic structures

Simple amines have mostly an ideal tetrahedral structure because of the almost ideal sp^3 hybrid orbitals that the central nitrogen atoms have. Going down the periodic table, the geometry of pnictogenides is gradually distorted from the ideal tetrahedral structure, and thus the bond angles of the C–Bi–C bonds in bismuthines are nearly 90°, which are almost the same as those of 6*p* orbitals. In general, these structures are determined by the combination of electronic and steric requirements. Amines use 2*s* and 2*p* orbitals to form the valence bonds. As a consequence of relatively short average bonding lengths of the orbitals, steric hindrances and electronic repulsion are comparatively significant in these compounds, and their bonds tend to part from each other to form the tetrahedral geometry. In contrast, in bismuthines which use 6*s* and 6*p* orbitals of relatively long bonding lengths, the steric and/or electronic repulsion is comparatively small. Consequently, there is no need to hybridize *s* orbitals with *p* orbitals to form a tetrahedral geometry, and the bonding electrons lie in almost pure *p* orbitals.

Of all group 15 elements, the heavy bismuth atom has the lowest tendency to form a hybrid orbital. The size difference of the valence *s* and *p* orbitals increases from nitrogen to bismuth (the significant 6*s* orbital contraction originates mainly from the relativistic effect) and thus bismuth prefers to maintain the $(6s)^2(6p)^3$ valence electron configuration. The use of three orthogonal 6*p* orbitals without significant hybridization leads to a bond angle of ~90° at the bismuth atom [97SCI(277)78].

A theoretical study on the stereochemical activity of lone pair for $MeBiI_2$ has been reported [97JA724] (Table 6.5, **74**). The compound has a one-dimensional organic/inorganic chain-like structure, in which bismuth atoms have a local square pyramidal geometry with four iodine atoms forming the base and one methyl group occupying an apical site. The molecular modeling suggests that the square pyramidal coordination at the bismuth atom arises primarily in order to reduce the Bi–C antibonding character in the HOMO. The resulting 6*s* and 6*p* hybridization in the valence band (HOMO) of bismuth leads to the stereochemically active lone pair, oriented *trans* to each methyl group.

An important outgrowth of this electronic configuration of bismuth is that trivalent bismuthines exhibit little basicity [94JCS(P1)3479]. The lack of basicity is due to the high *s*-character of the lone pair electron. All three valence electrons and two lone pair electrons occupy the sp^3 hybrid orbital in amines, whereas three valence electrons lie in almost pure *p* orbitals and lone pair electrons occupy the remaining *s* orbital in bismuthines.

TABLE 6.5
Crystallographic structure of trivalent bismuth compounds[a]

Compound			Structure	Structural parameters bond distances (Å) and bond angles (°)[b]	[Reference], **CCDB code**, and *notes*
R^1	R^2	R^3			
(**1**) Trimethylbismuthine			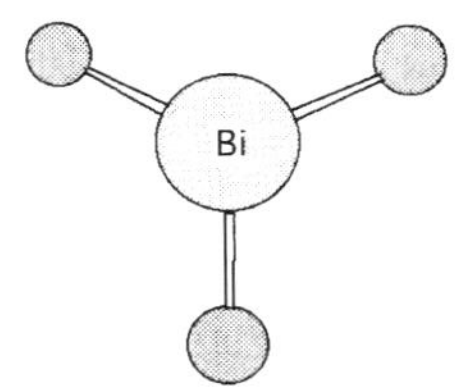	Bi–C = 2.266 C–Bi–C = 96.7	[73JMOS(17)429] *By gas-phase electron diffraction*
Me	Me	Me			
(**2**) Tris[bis(trimethylsilyl)methyl]bismuthine			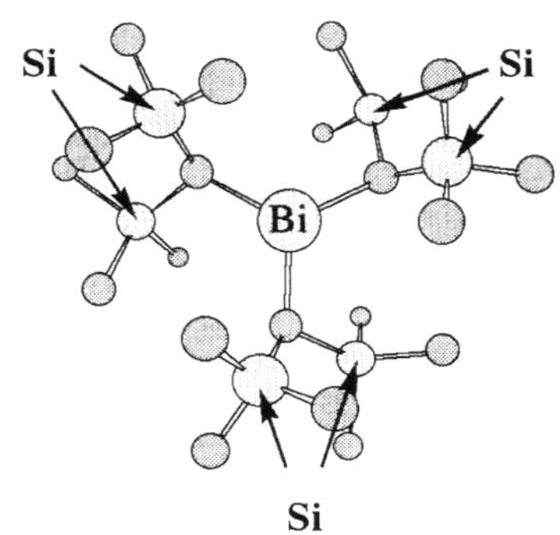	Bi–C(1) = 2.331(14), Bi–C(2) = 2.347(13), Bi–C(3) = 2.306(13) C(1)–Bi–C(2) = 102.9(5), C(1)–Bi–C(3) = 103.0(5), C(2)–Bi–C(3) = 102.7(5)	[83IC3421] **CEFCAR** *at −133°C*
$(Me_3Si)_2CH$	$(Me_3Si)_2CH$	$(Me_3Si)_2CH$			

(**3**) Tris(cyclopenta-2,4-dienyl)-bismuthine Cp Cp Cp	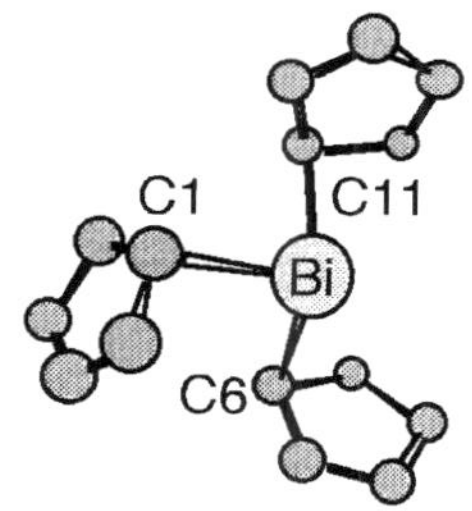	C(1)–Bi = 2.37, Bi–C(6) = 2.34, Bi–C(11) = 2.42 C(1)–Bi–C(6) = 94.3, C(1)–Bi–C(11) = 91.8, C(6)–Bi–C(11) = 98.2	[95JOM(485)149] **YOPYOR** *red modification,* *at −70°C*
(**4**) Triphenylbismuthine Ph Ph Ph	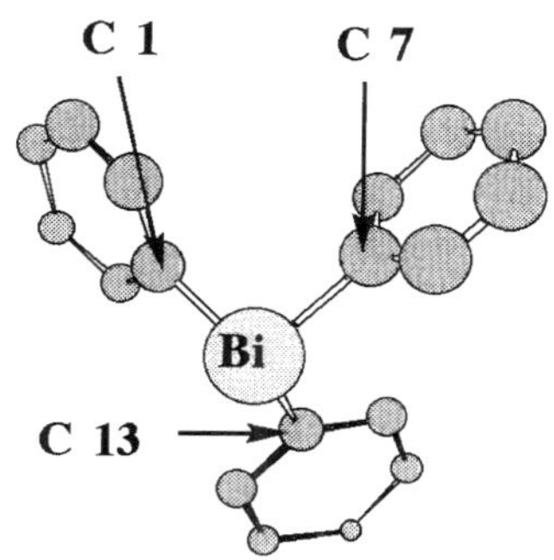	Bi–C(1) = 2.21(2), Bi–C(7) = 2.25(2), Bi–C(13) = 2.25(3) C(1)–Bi–C(7) = 92(1), C(1)–Bi–C(13) = 96(1), C(7)–Bi–C(13) = 94(1)	[42ZK(104)305; 66CC111; 68JCS(A)2059] **BITRPH10** [95ZK(210)377] **BITRPH02** *at −130°C*

TABLE 6.5 (*Continued*)

Compound R^1 R^2 R^3	Structure	Structural parameters bond distances (Å) and bond angles (°)[b]	[Reference], **CCDB code**, and *notes*
(**5**) Tris(4-methylphenyl)bismuthine Tol Tol Tol	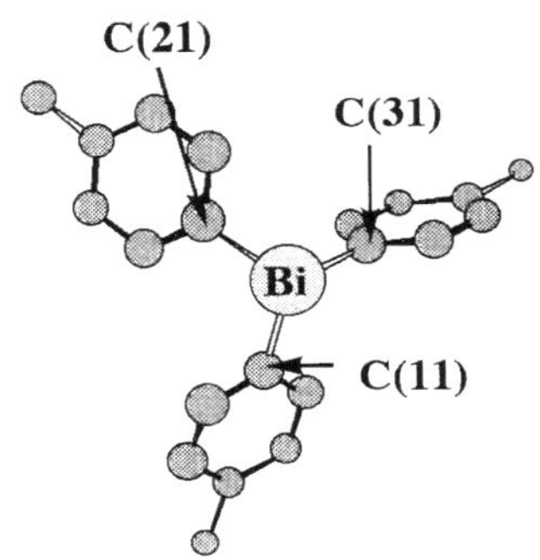	Bi–C(11) = 2.252(9), Bi–C(21) = 2.255(9), Bi–C(31) = 2.24(1) C(11)–Bi–C(21) = 94.8(3), C(21)–Bi–C(31) = 93.9(4), C(31)–Bi–C(11) = 95.3(3)	[83KK262] **BUCSOH**
(**6**) Tris(2,4,6-trimethylphenyl)-bismuthine Mes Mes Mes	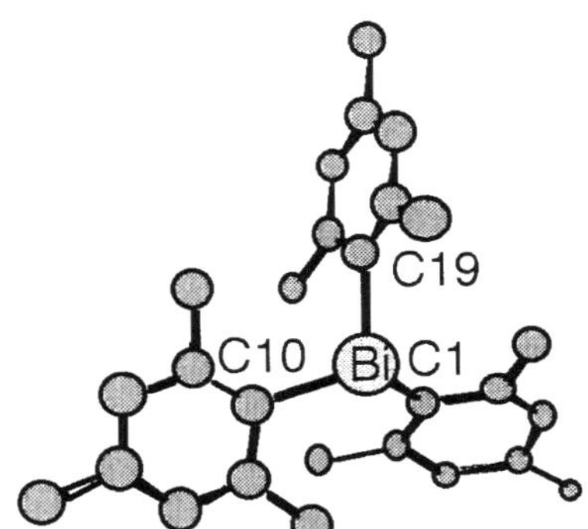	Bi–C(1) = 2.32(1), Bi–C(10) = 2.24(2), Bi–C(19) = 2.31(1) C(1)–Bi–C(10) = 107.4(5), C(1)–Bi–C(19) = 95.2(4), C(10)–Bi–C(19) = 105.5(5)	[80KK945] **MSTLBI** [94JCS(P1)3479] **MSTLBI01**

(**7**) Tris[2-(methoxy-methyl)phenyl]bismuthine 2-MeO–CH$_2$–C$_6$H$_4$ 2-MeO–CH$_2$–C$_6$H$_4$ 2–MeO–CH$_2$–C$_6$H$_4$	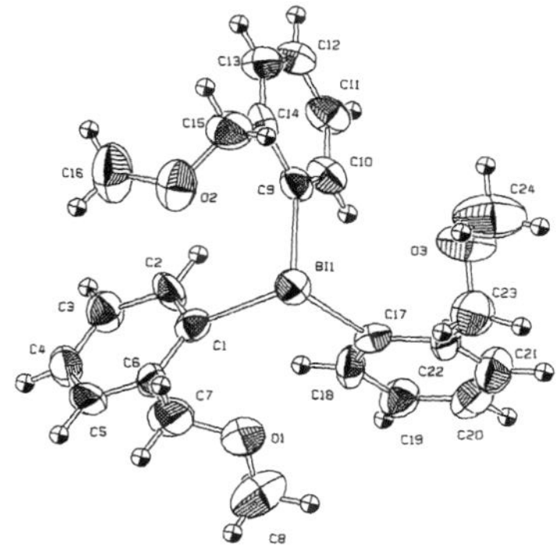	Bi–C(1) = 2.28(1), Bi–C(9) = 2.29(1), Bi–C(17) = 2.28(1), Bi...O(1) = 3.10(1), Bi...O(2) = 3.09(1), Bi...O(3) = 3.32(1) C(1)–Bi–C(9) = 93.5(5), C(1)–Bi–C(17) = 95.0(5), C(9)–Bi–C(17) = 93.1(5)	[UR4]
(**8**) Tris[2-(*N*,*N*-dimethyl-aminomethyl)phenyl]bismuthine 2-Me$_2$N–CH$_2$–C$_6$H$_4$ 2-Me$_2$N–CH$_2$–C$_6$H$_4$ 2–Me$_2$N–CH$_2$–C$_6$H$_4$	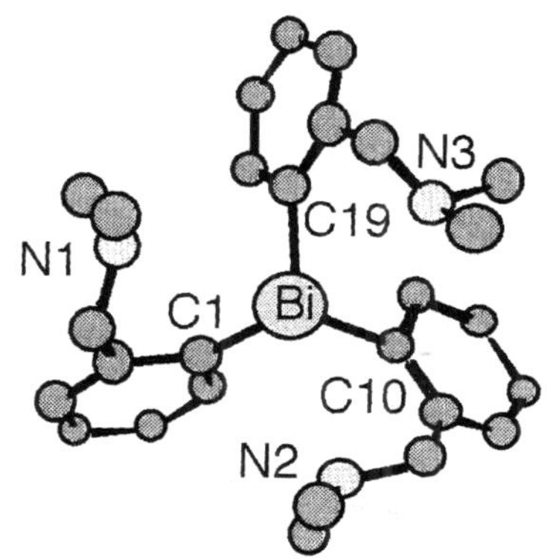	Bi–C(1) = 2.19, Bi–C(10) = 2.29, Bi–C(19) = 2.28, Bi...N(1) = 3.08, Bi...N(2) = 3.10, Bi...N(3) = 3.02 C(1)–Bi–C(10) = 91.0, C(1)–Bi–C(19) = 95.0, C(10)–Bi–C(19) = 92.2	[96IC6179] **NALNET**

TABLE 6.5 (*Continued*)

Compound			Structure	Structural parameters bond distances (Å) and bond angles (°)[b]	[Reference], **CCDB code**, and *notes*
R^1	R^2	R^3			
(**9**) Tris(2,4,6-triphenylphenyl)-bismuthine tetrahydrofuran solvate			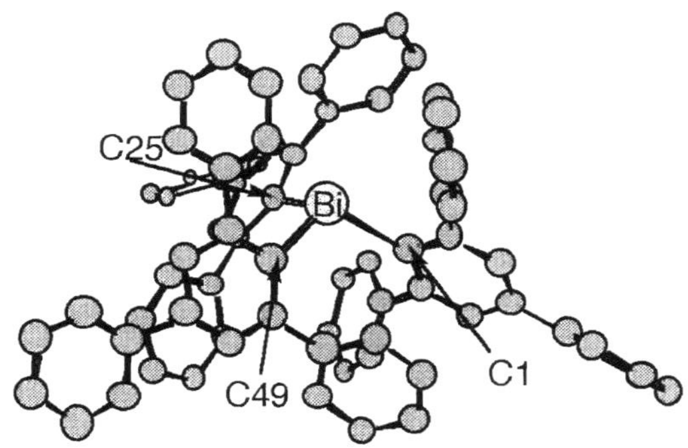	Bi–C(1) = 2.38, Bi–C(25) = 2.36, Bi–C(49) = 2.34 C(1)–Bi–C(25) = 103.7, C(1)–Bi–C(49) = 109.4, C(25)–Bi–C(49) = 104.2	[95JOM(485)141] **YITZAC**
2,4,6-Tri-phenyl-phenyl	2,4,6-Tri-phenyl-phenyl	2,4,6-Tri-phenyl-phenyl			
(**10**) (μ_2-2,2′-Biphenylene)bis(2,2′-bi-phenylene)dibismuthine tetrahydrofuran solvate			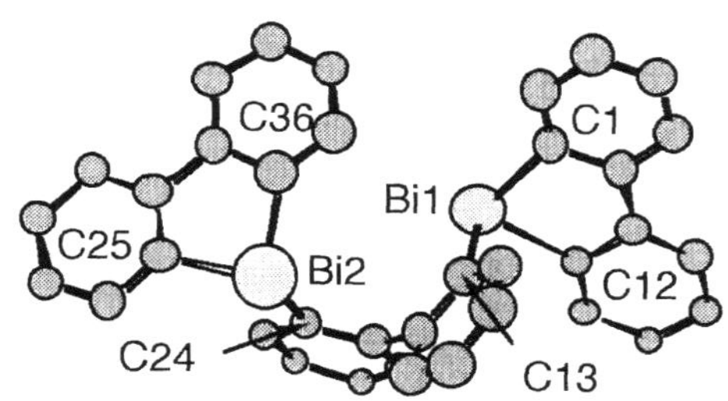	Bi–C(1) = 2.23, Bi–C(12) = 2.27, Bi–C(13) = 2.28, Bi(2)–C(24) = 2.28, Bi(2)–C(25) = 2.23, Bi(2)–C(36) = 2.23 C(1)–Bi–C(12) = 78.2, C(1)–Bi–C(13) = 92.6, C(12)–Bi–C(13) = 90.9, C(24)–Bi(2)–C(25) = 93.9, C(24)–Bi(2)–C(36) = 95.5, C(25)–Bi(2)–C(36) = 77.6	[96OM887] **ZUJQEA** *at −90°C*
Ar	Ar	Ar			

(**11**) Tris[2,4,6-tris(trifluoromethyl)phenyl]bismuthine 2,4,6-$(CF_3)_3C_6H_2$ 2,4,6-$(CF_3)_3C_6H_2$ 2,4,6-$(CF_3)_3C_6H_2$	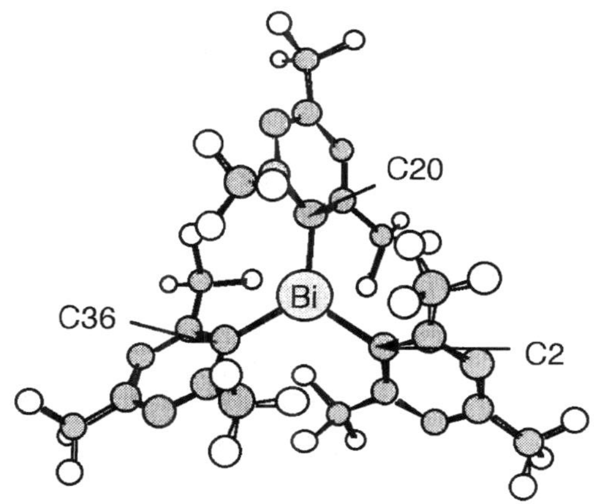 	Bi–C(2) = 2.38, Bi–C(20) = 2.37, Bi–C(36) = 2.36 C(2)–Bi–C(20) = 106.2, C(2)–Bi–C(36) = 106.5, C(20)–Bi–C(36) = 105.3	[91JOM(402)55] **KIMJIZ**
(**12**) Tris(2,6-dimethoxyphenyl)bismuthine 2,6-$(MeO)_2C_6H_3$ 2,6-$(MeO)_2C_6H_3$ 2,6-$(MeO)_2C_6H_3$	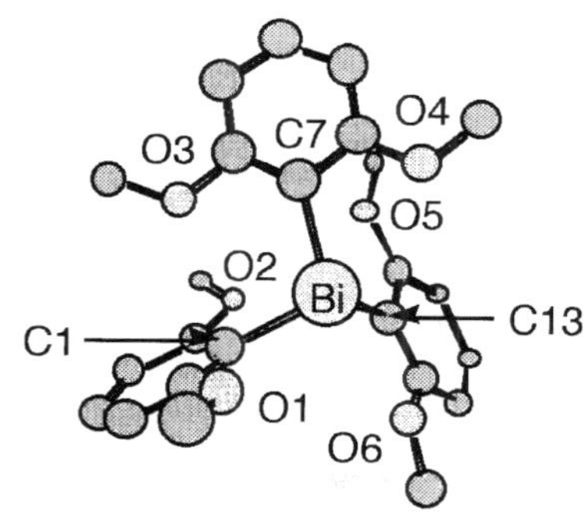 	Bi–C(1) = 2.253(7), Bi–C(13) = 2.256(7), Bi–C(7) = 2.278(8), Bi...O(1) = 3.11, Bi...O(2) = 3.33, Bi–O(3) = 3.39, Bi–O(4) = 3.05, Bi–O(5) = 3.41, Bi–O(6) = 2.98 C(1)–Bi–C(7) = 101.4(3), C(1)–Bi–C(13) = 95.7(3), C(7)–Bi–C(13) = 101.2(3)	[94JCS(P1)3479] **LIKPAW**

TABLE 6.5 (*Continued*)

Compound			Structure	Structural parameters bond distances (Å) and bond angles (°)[b]	[Reference], **CCDB code**, and *notes*
R^1	R^2	R^3			
(**13**) 10-(4-Chloro-phenylethynyl)phenothiabismin-5,5-dioxide			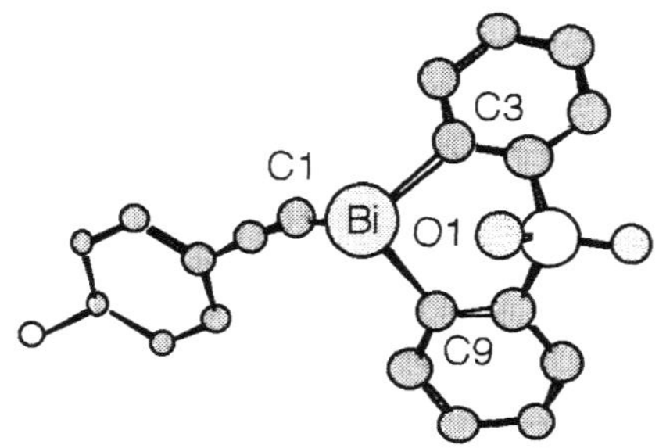	Bi–C(1) = 2.211(8), Bi–C(11) = 2.283(8), Bi–C(21) = 2.272(7), Bi...O(1) = 2.979(7) C(1)–Bi–C(11) = 95.1(3), C(1)–Bi–C(21) = 90.8(3), C(11)–Bi–C(21) = 87.0(3)	[92JCS(P1)1593] **JOXTAR**
(C)	(C)	(C) (O)			
(**14**) (2-*t*-Butyl-sulfonylphenyl)bis(4-methylphenyl)bismuthine			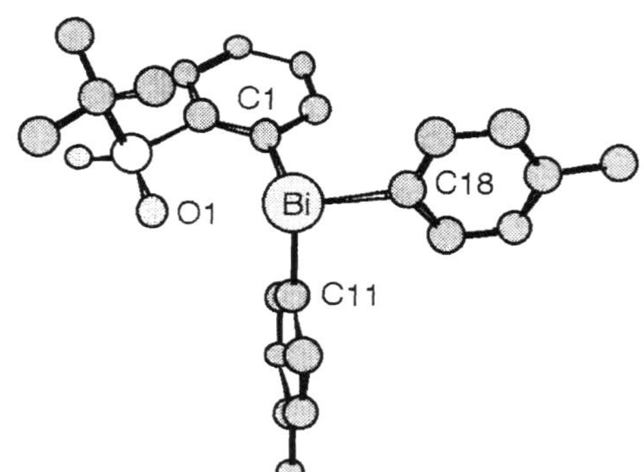	Bi–C(1) = 2.312(9), Bi–C(11) = 2.26(1), Bi–C(18) = 2.307(9), Bi...S(1) = 3.68, Bi...O(1) = 2.914(6) O(1)–Bi–C(1) = 68.1, O(1)–Bi–C(11) = 79.1, O(1)–Bi–C(18) = 158.1, C(1)–Bi–C(11) = 96.3(4), C(1)–Bi–C(18) = 92.8(3), C(11)–Bi–C(18) = 93.0(4)	[93JCS(P1)1169] **WAMPAB**
Tol	$^{t}BuSO_2$-C_6H_4	Tol (O)			

(**15**) Catena[(μ_2-phenoxy)diethyl-bismuthine]

Et Et C_6H_5O *(O)*

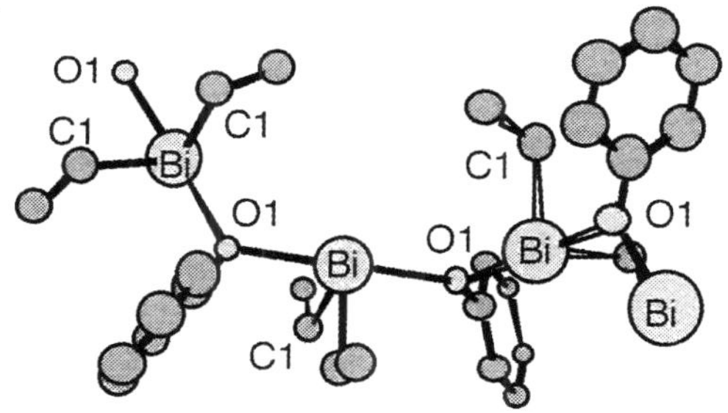

Bi–O = 2.382(7)
Bi–C = 2.35b
O–Bi–O = 179.0(3),
C–Bi–C = 81.2(9)
Exists as a chiral chain polymer in which Et2Bi groups are bridged by alkoxide

[92CC1021]
PAFWAU
at −50°C

(**16**) Catena[(μ_2-pentafluoro-phenoxy)diethylbismuthine]

Et Et C_6F_5O *(O)*

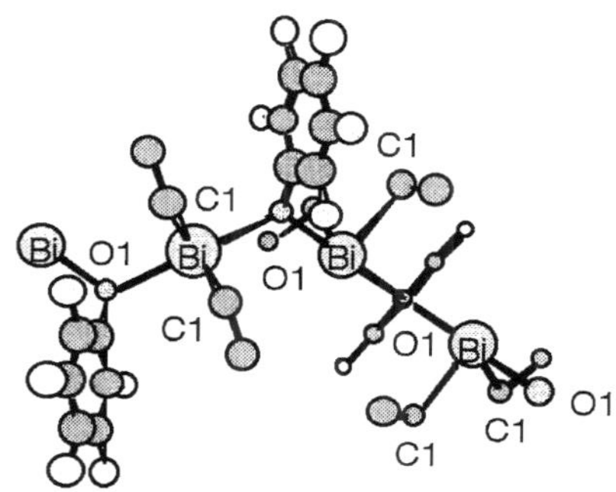

Bi–O = 2.4105(7)
O–Bi–O = 179.54(4),
C–Bi–C = 80(2)
Exists as a chiral chain polymer in which Et_2Bi groups are bridged by alkoxide

[92CC1021]
PAFVUN
at −50°C

TABLE 6.5 (*Continued*)

Compound R¹	R²	R³	Structure	Structural parameters bond distances (Å) and bond angles (°)[b]	[Reference], **CCDB code**, and *notes*
(**17**) Catena[(μ_2-*N*-benzoyl-glycinato-*O,O'*)diphenyl-bismuthine] Ph	PhC–(=O)NH CH_2C(=O)O–	Ph (*O*)	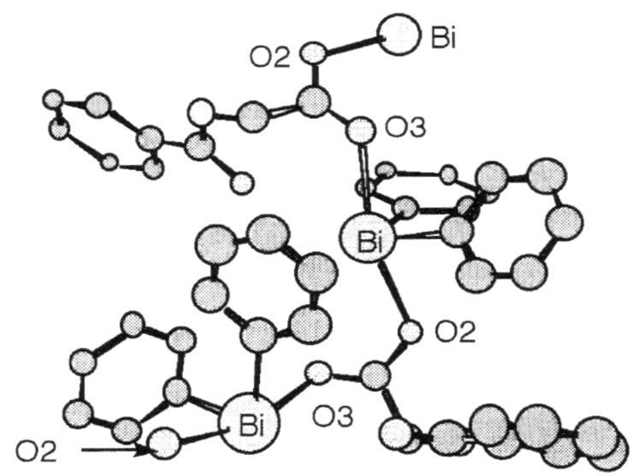	Bi–C(21) = 2.246(7), Bi–C(11) = 2.250(8), Bi–O(2) = 2.396(6), Bi–O(3′) = 2.484(6), Bi–O(3) = 3.297(6), Bi–O(1′) = 3.267(6) C(21)–Bi–C(11) = 95.10(3), C(21)–Bi–O(2) = 81.38(3), C(21)–Bi–O(3′) = 83.02(3), C(11)–Bi–O(2) = 85.85(3), C(11)–Bi–O(3′) = 84.16(3), O(2)–Bi–O(3′) = 160.63(2)	[88AX(C)828] **GEMGUA**
(**18**) Bis(triphenyl-phosphine oxide-*O*)diphenylbismuthine tetrafluoroborate dichloromethane solvate Ph	Ph	(O) (*O*)	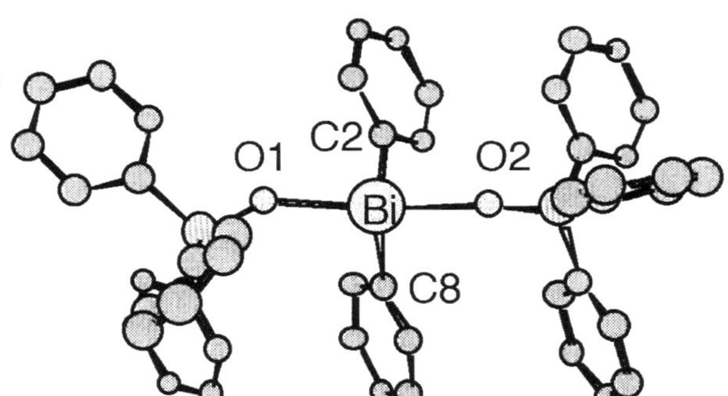	Bi–O(1) = 2.39, Bi–O(2) = 2.39, Bi–C(2) = 2.22, Bi–C(8) = 2.22 O(1)–Bi–O(2) = 173.2, O(1)–Bi–C(2) = 85.1, O(1)–Bi–C(8) = 89.4, O(2)–Bi–C(2) = 88.1, O(2)–Bi–C(8) = 90.4, C(2)–Bi–C(8) = 95.7	[96JCS(D)443] **ZUCJIQ**

(**19**) Bis(hexamethyl-phosphoramide-*O*)-diphenylbismuthine hexafluorophosphate Ph Ph (O) *(O)*	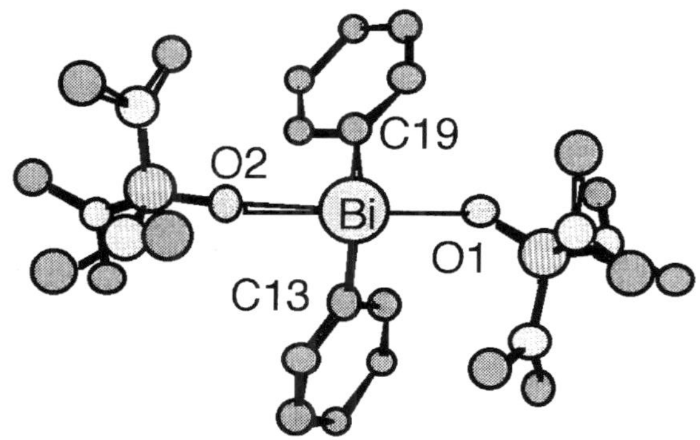	Bi–O(1) = 2.32, Bi–O(2) = 2.36, Bi–C(13) = 2.23, Bi–C(19) = 2.19 O(1)–Bi–O(2) = 172.4, O(1)–Bi–C(13) = 89.7, O(1)–Bi–C(19) = 84.4, O(2)–Bi–C(13) = 88.7, O(2)–Bi–C(19) = 88.3, C(13)–Bi–C(19) = 94.3	[96JCS(D)443] **ZUCJOW**
(**20**) Bis(hexamethyl-phosphoramide-*O*)-diphenylbismuthine tetrafluoroborate Ph Ph (O) *(O)*	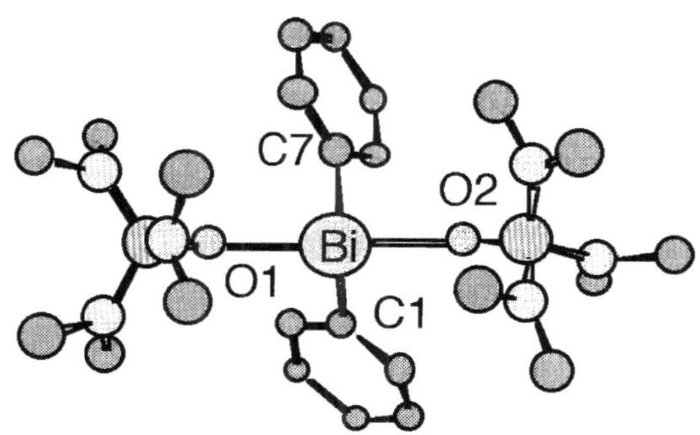	Bi–O(1) = 2.36, Bi–C(1) = 2.24, Bi–O(2) = 2.37, Bi–C(7) = 2.26 O(1)–Bi–C(1) = 87.47, O(1)–Bi–O(2) = 167.50, O(1)–Bi–C(7) = 85.32, C(1)–Bi–O(2) = 84.53, C(1)–Bi–C(7) = 93.19, O(2)–Bi–C(7) = 85.55	[93JOM(460)C22] **HAWCOX**

TABLE 6.5 (*Continued*)

Compound R[1]	R[2]	R[3]	Structure	Structural parameters bond distances (Å) and bond angles (°)[b]	[Reference], **CCDB code**, and *notes*
(**21**) Bis(hexamethyl-phosphoramide-*O*)bis(2,4,6-trimethyl-phenyl)bismuthine hexafluorophosphate 2,4,6-Me$_3$–C$_6$H$_2$	2,4,6-Me$_3$–C$_6$H$_2$	(O) *(O)*	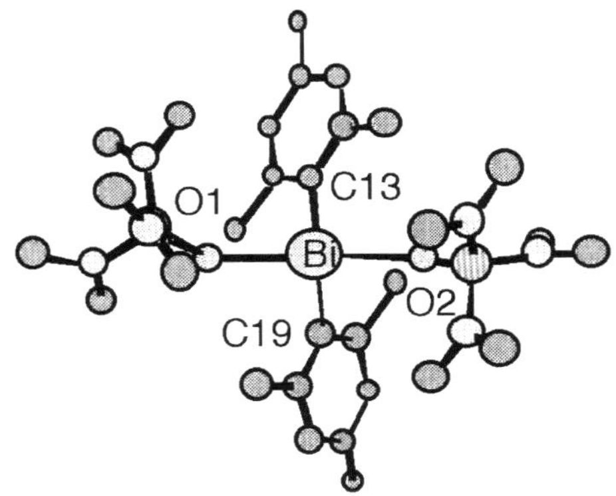	Bi–O(1) = 2.35, Bi–O(2) = 2.41, Bi–C(13) = 2.26, Bi–C(19) = 2.21 O(1)–Bi–O(2) = 177.2, O(1)–Bi–C(13) = 86.6, O(1)–Bi–C(19) = 90.7, O(2)–Bi–C(13) = 96.0, O(2)–Bi–C(19) = 87.4, C(13)–Bi–C(19) = 108.5	[96JCS(D)443] **ZUCJUC**
(**22**) [2-(2-Methoxy-2-propyl)phenyl](2-phenyl-2-propanolato)bismuthine Ar	Ar′	(O) *(O)*	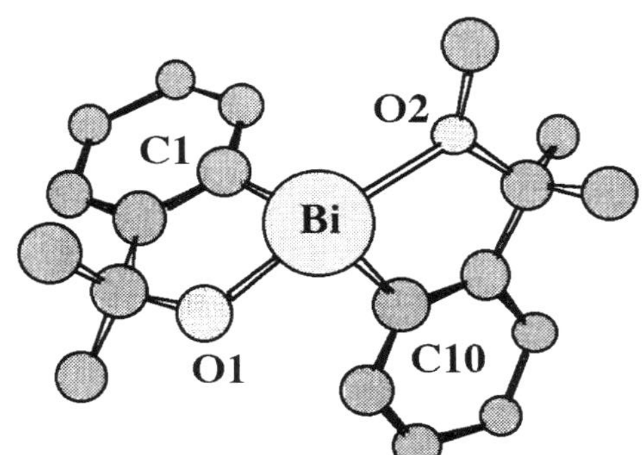	Bi–O(1) = 2.13, Bi–C(1) = 2.24, Bi–O(2) = 2.57, Bi–C(10) = 2.24 O(1)–Bi–C(1) = 77.4, O(1)–Bi–O(2) = 157.2, O(1)–Bi–C(10) = 94.5, C(1)–Bi–O(2) = 88.9, C(1)–Bi–C(10) = 92.1, O(2)–Bi–C(10) = 67.6	[94AX(B)151] **LEJHUD**

(**23**) [2-(2-Hydroxy-2-propyl)phenyl](2-phenyl-2-propanolato)bismuthine
Ar Ar′ (O) *(O)*

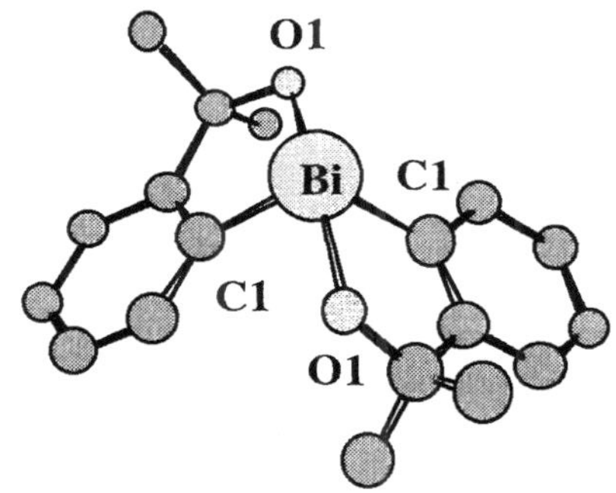

Bi–O(1) = 2.32,
Bi–C(1) = 2.17
O(1)–Bi–C(1) = 73.6,
O(1)–Bi–O(1) = 158.8,
O(1)–Bi–C(1) = 92.7,
C(1)–Bi–O(1) = 92.7,
C(1)–Bi–C(1) = 100.6,
O(1)–Bi–C(1) = 73.6

[94AX(B)151]
LEJJAL

(**24**) 1-[2-(*N,N*-dimethylaminomethyl)phenyl]-3,3-bis(trifluoromethyl)-3*H*-2,1-benzoxabismole
Ar Ar′ (O) *(N)*

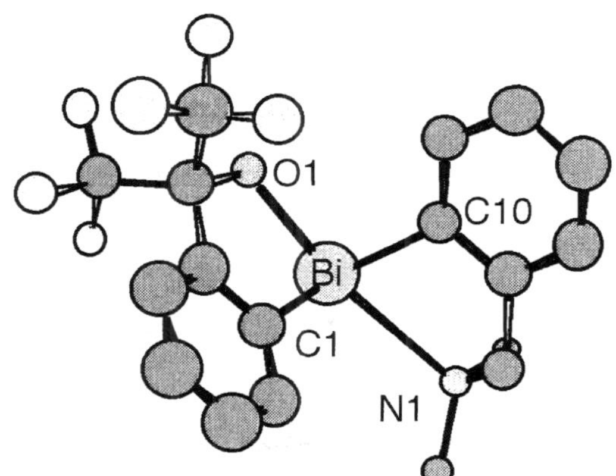

Bi–O(1) = 2.194(9),
Bi–N(1) = 2.62(1),
Bi–C(1) = 2.236(9),
Bi–C(10) = 2.22(1)
O(1)–Bi–N(1) = 160.1(3),
O(1)–Bi–C(1) = 76.6(4), O(1)–Bi–C(10) = 90.2(4),
N(1)–Bi–C(1) = 97.5(4), N(1)–Bi–C(10) = 71.0(4),
C(1)–Bi–C(10) = 93.3(4)

[92JA7906]
VUMZAE
[95JA3922]
VUMZAE10

TABLE 6.5 (*Continued*)

Compound			Structure	Structural parameters bond distances (Å) and bond angles (°)[b]	[Reference], **CCDB code**, and *notes*
R^1	R^2	R^3			
(**25**) 1-[2-(2-methoxy-2-propyl)phenyl]-3,3-bis(trifluoromethyl)-3*H*-2,1-benzoxabismole			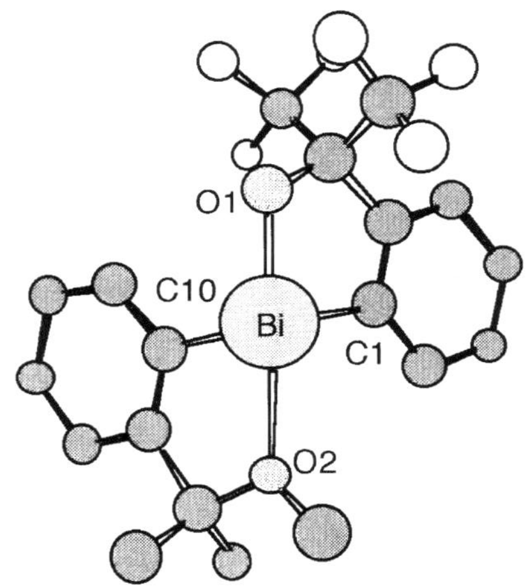	Bi–O(1) = 2.193(7), Bi–C(1) = 2.24(1), Bi–O(2) = 2.536(9), Bi–C(10) = 2.25(1) O(1)–Bi–C(1) = 76.3(3), O(1)–Bi–O(2) = 155.2(3), O(1)–Bi–C(10) = 92.3(3), C(1)–Bi–O(2) = 88.0(3), C(1)–Bi–C(10) = 94.4(4), O(2)–Bi–C(10) = 69.5(3)	[92JA7906] **VUMZEI** [95HAC293] **VUMZEI10**
Ar	Ar′	(O) *(O)*			

Compound	Structure	Data	Reference
(**26**) 1-[2,6-Bis(*N*,*N*-dimethyl-aminomethyl)phenyl]-3,3-bis(trifluoromethyl)-3*H*-2,1-benzoxabismole Ar Ar′ (O) (*N*) (*N*)	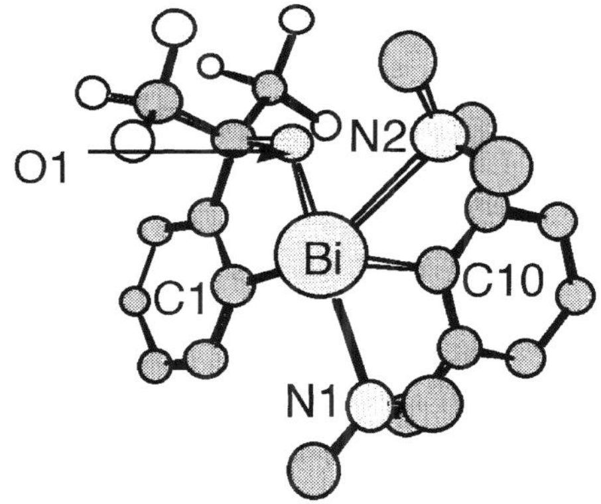	Bi–O(1) = 2.192(7), Bi–N(1) = 2.68(1), Bi–C(1) = 2.30(1), Bi–N(2) = 2.86(1), Bi–C(10) = 2.27(1) O(1)–Bi–N(1) = 158.1(4), O(1)–Bi–C(1) = 75.2(4), O(1)–Bi–N(2) = 66.5(3), O(1)–Bi–C(10) = 94.4(4), N(1)–Bi–C(1) = 90.8(4), N(1)–Bi–N(2) = 117.6(3), N(1)–Bi–C(10) = 70.0(4), C(1)–Bi–N(2) = 136.0(4), C(1)–Bi–C(10) = 96.2(4), N(2)–Bi–C(10) = 67.0(4)	[95JA3922] **YUKNOH**
(**27**) Bis(μ_2-methanedi-sulfonamide-*O*,*O*′)-tetra-phenyldibismuthine Ph Ph (O) (*O*)	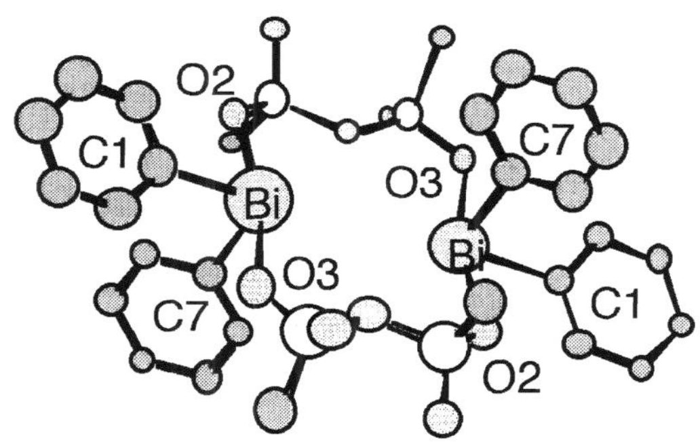	Bi–C(1) = 2.24, Bi–O(2) = 2.47, Bi–O(3) = 2.40, Bi–C(7) = 2.24 C(1)–Bi–O(2) = 83.78, C(1)–Bi–O(3) = 84.02, C(1)–Bi–C(7) = 93.04, O(2)–Bi–O(3) = 165.40, O(2)–Bi–C(7) = 86.54, O(3)–Bi–C(7) = 86.14	[95ZAAC(621)1746] **ZEMFEC** *at −130°C*

TABLE 6.5 (*Continued*)

Compound			Structure	Structural parameters bond distances (Å) and bond angles (°)[b]	[Reference], **CCDB code**, and *notes*
R^1	R^2	R^3			
(**28**) Bis[bis(2,4,6-trimethylphenyl)bismuthyl] oxide ethanol solvate			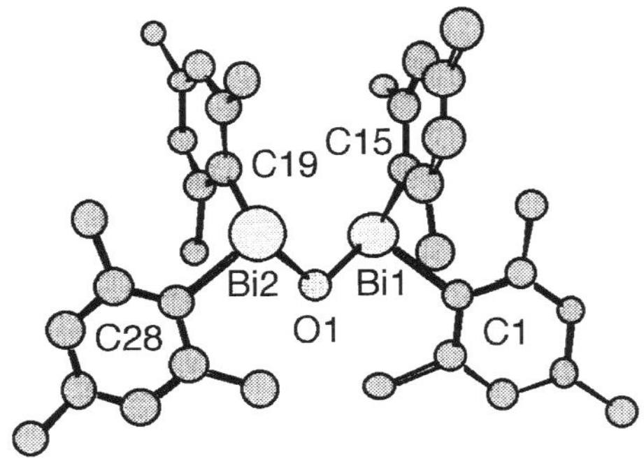	Bi(1)–O(1) = 2.10, Bi(1)–C(1) = 2.26, Bi(1)–C(10) = 2.20, O(1)–Bi(2) = 2.12, Bi(2)–C(19) = 2.30, Bi(2)–C(28) = 2.27 O(1)–Bi(1)–C(1) = 97.89, O(1)–Bi(1)–C(10) = 93.60, C(1)–Bi(1)–C(10) = 97.40, Bi(1)–O(1)–Bi(2) = 117.13, O(1)–Bi(2)–C(19) = 93.42, O(1)–Bi(2)–C(28) = 94.53, C(19)–Bi(2)–C(28) = 98.30	[95ZN(B)735] **YULYAF** *at −100°C*
Mes	Mes	(O)			

(**29**) Anionic part of tetraphenylbismuthonium diphenylbis(trifluoroacetato-*O*)-bismuthate

Ph	Ph	CF_3C–(=O)O–
		CF_3C–(=O)O–

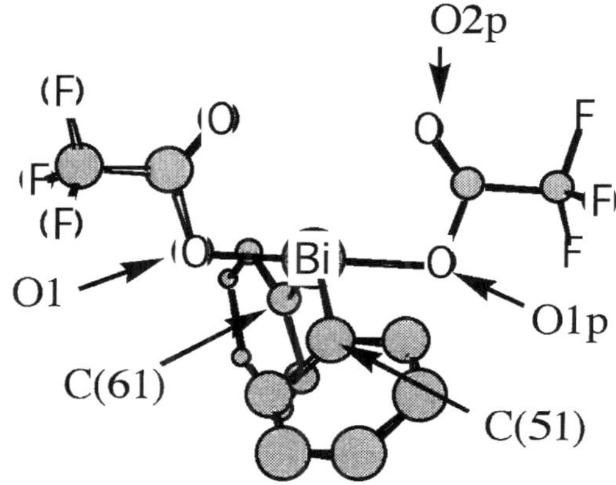

Bi–C(51) = 2.26(2),
Bi–C(61) = 2.24(2),
Bi–O1 = 2.38(2),
Bi–O1p = 2.41(2),
Bi–O2p = 3.286
C(51)–Bi–C(61) = 95,
O1–Bi–O1p = 172,
O1–Bi–C(51) = 90,
O1–Bi–C(61) = 86,
O1p–Bi–C(51) = 84,
O1p–Bi–C(61) = 88,
Bi–O1–C(1) = 116(2)

[84HCA586]
CUNFOG

(**30**) Tetraethylammonium 1,1′-spirobis[3,3-bis(trifluoromethyl)-3*H*-2,1-benzoxabismole] monohydrate

Ph	Ph	(O)
		(O)

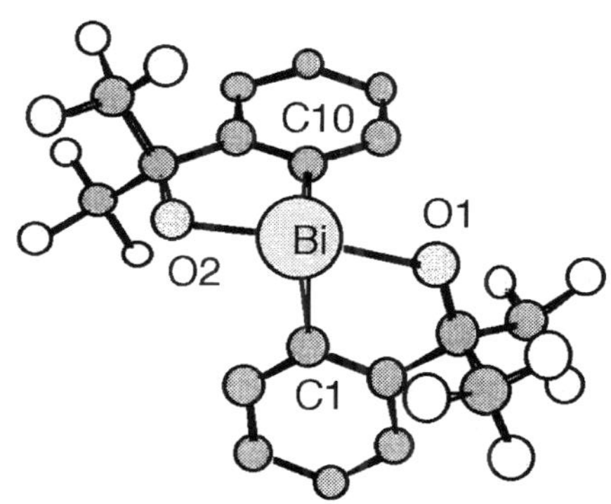

Bi–C(1) = 2.24,
Bi–O(1) = 2.31,
Bi–O(2) = 2.27,
Bi–C(10) = 2.25
C(1)–Bi–O(1) = 74.8,
C(1)–Bi–O(2) = 91.0,
C(1)–Bi–C(10) = 94.1,
O(1)–Bi–O(2) = 159.7,
O(1)–Bi–C(10) = 92.6,
O(2)–Bi–C(10) = 73.8

[92TL6653]
YAHYEL

TABLE 6.5 (*Continued*)

Compound			Structure	Structural parameters bond distances (Å) and bond angles (°)[b]	[Reference], **CCDB code**, and *notes*
R[1]	R[2]	R[3]			
(**31**) Catena[(μ_2-thiocyanato)-diphenylbismuthine] chloroform solvate			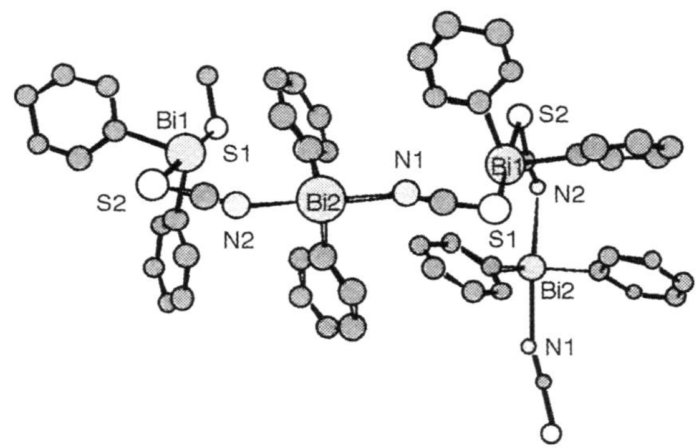	Bi(1)–C(10) = 2.23, Bi(1)–C(4) = 2.27, Bi(1)–S(1) = 2.87, Bi(1)–S(2) = 2.88, Bi(2)–C(16) = 2.20, Bi(2)–C(22) = 2.30, Bi(2)–N(2) = 2.52, N(1)–Bi(2) = 2.53 C(16)–Bi(2)–C(22) = 96.7, C(4)–Bi(1)–C(10) = 97.3, N(1)–Bi(2)–C(16) = 86.6, N(1)–Bi(2)–C(22) = 88.7, N(1)–Bi(2)–N(2) = 166.6, N(2)–Bi(2)–C(16) = 85.1, N(2)–Bi(2)–C(22) = 81.9, S(1)–Bi(1)–C(10) = 90.9, S(1)–Bi(1)–C(4) = 87.3, S(1)–Bi(1)–S(2) = 176.9, S(2)–Bi(1)–C(10) = 87.6, S(2)–Bi(1)–C(4) = 90.3	[95JCS(D)383] **YIMRAN**
Ph	Ph	(*S*) (*N*)			

(**32**) (Isopropyldithiocarbonato-*S*)diphenylbismuthine
Ph Ph ^{i}PrO–C(=S)S–

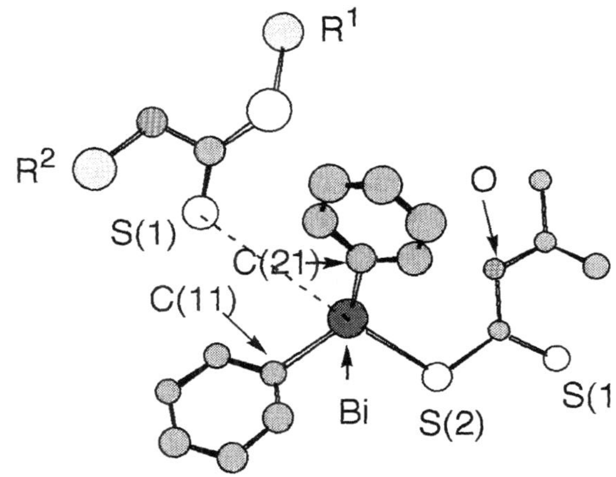

Bi–C(11) = 2.26(1), Bi–C(21) = 2.24(1), Bi–S(2) = 2.66(1), Bi...O = 3.19(1), Bi...S(1′) = 3.23(1) C(11)–Bi–C(21) = 96.5(3), C(11)–Bi–S(2) = 84.9(2), C(21)–Bi–S(2) = 90.9(1), S(2)–Bi–S(1′) = 161.9(1), S(1′)–Bi–O = 146.1(1), S(2)–Bi–O = 51.2(1)

[80ZAAC(470)171]
XNTGBI

(**33**) Bis[bis(2,4,6-trimethylphenyl)bismuthyl] sulfide
Mes Mes (S)

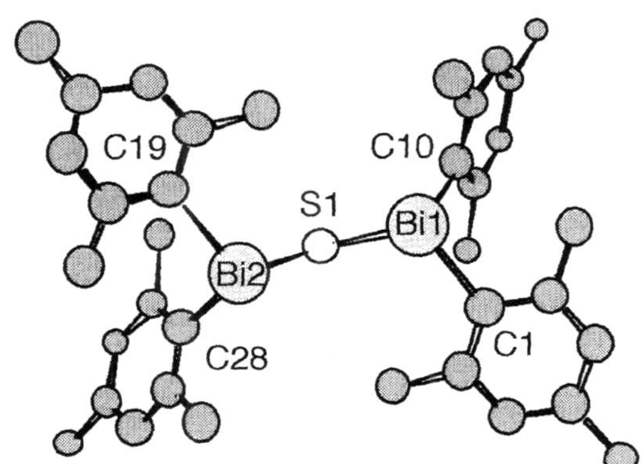

S(1)–Bi(1) = 2.52, S(1)–Bi(2) = 2.55, Bi(1)–C(1) = 2.25, Bi(1)–C(10) = 2.29, Bi(2)–C(19) = 2.33, Bi(2)–C(28) = 2.23 Bi(1)–S(1)–Bi(2) = 98.74, S(1)–Bi(1)–C(1) = 106.59, S(1)–Bi(1)–C(10) = 89.41, C(1)–Bi(1)–C(10) = 98.85, Bi(1)–C(1)–C(2) = 123.78, Bi(1)–C(1)–C(6) = 116.83, S(1)–Bi(2)–C(19) = 106.44, S(1)–Bi(2)–C(28) = 90.15, C(19)–Bi(2)–C(28) = 97.46

[95ZN(B)735]
YULYEJ
at −100°C

TABLE 6.5 (*Continued*)

Compound			Structure	Structural parameters bond distances (Å) and bond angles (°)[b]	[Reference], **CCDB code**, and *notes*
R[1]	R[2]	R[3]			
(**34**) Chlorobis[2,4,6-tris-(trifluoromethyl)-phenyl]bismuthine 2,4,6-CF_3–C_6H_2	2,4,6-CF_3–C_6H_2	Cl	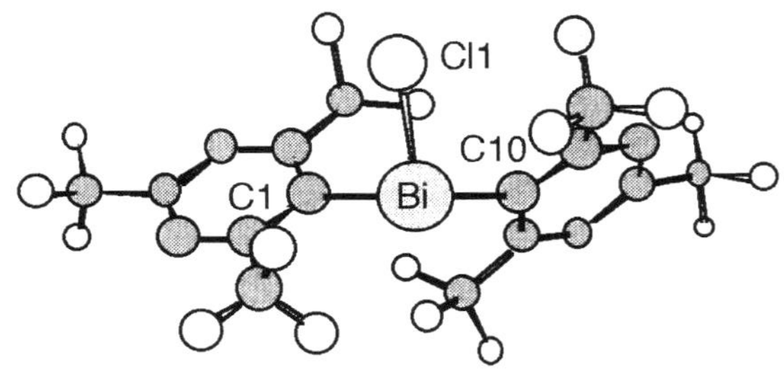	Bi–Cl(1) = 2.46, Bi–C(1) = 2.34, Bi–C(10) = 2.36 Cl(1)–Bi–C(1) = 87.77, Cl(1)–Bi–C(10) = 99.46, C(1)–Bi–C(10) = 106.83	[91JOM(402)55] **KIMJEV**
(**35**) Chloro[bis(trimethylsilyl)(2-pyridyl)methyl-*C,N*][3-(2-pyridyl)-3-trimethylsilylprop-2-enyl]bismuthine (C)	(C)	Cl (*N*)	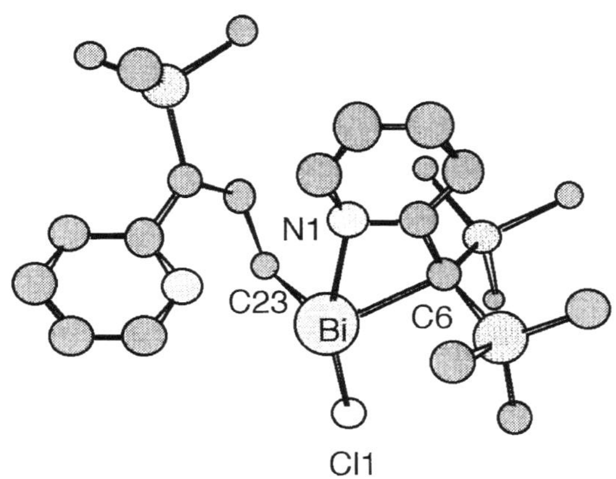	Bi–Cl(1) = 2.646(4), Bi–N(1) = 2.485(9), Bi–C(6) = 2.41(1), Bi–C(23) = 2.23(2) Cl(1)–Bi–N(1) = 156.7(3), Cl(1)–Bi–C(6) = 100.7(3), Cl(1)–Bi–C(23) = 84.9(3), N(1)–Bi–C(6) = 58.5(4), N(1)–Bi–C(23) = 91.7(4), C(6)–Bi–C(23) = 108.3(5)	[91CC1560] **SONXAU**

(**36**) Chlorophenyl[2-(*R*-1-(dimethylamino)ethyl)phenyl-*C,N*]bismuthine
Ph Ar Cl (*N*)

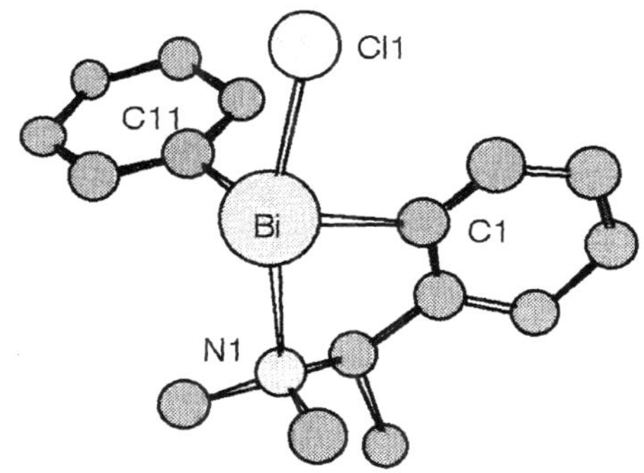

Bi–Cl(1) = 2.65, Bi–N(1) = 2.56, Bi–C(1) = 2.30, Bi–C(11) = 2.25
Cl(1)–Bi–N(1) = 162.6, Cl(1)–Bi–C(1) = 90.6, Cl(1)–Bi–C(11) = 88.2, N(1)–Bi–C(1) = 72.5, N(1)–Bi–C(11) = 89.0, C(1)–Bi–C(11) = 95.5

[95OM1542]
YOTPOM

(**37**) Chloro[2-(*N,N*-dimethyl-aminomethyl)phenyl-*C,N*](4-methylphenyl)bismuthine methanol solvate
Tol Ar Cl (*N*)

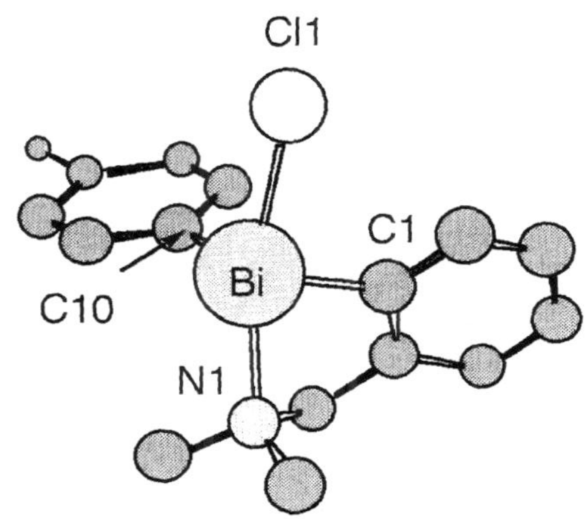

Bi–Cl(1) = 2.700(2), Bi–N(1) = 2.525(6), Bi–C(1) = 2.246(6), Bi–C(10) = 2.234(6)
Cl(1)–Bi–N(1) = 162.9(1), Cl(1)–Bi–C(1) = 90.8(2), Cl(1)–Bi–C(10) = 90.6(2), N(1)–Bi–C(1) = 72.5(2), N(1)–Bi–C(10) = 86.5(2), C(1)–Bi–C(10) = 93.7(2)

[93JCS(P1)2969]
PEZJAF

TABLE 6.5 (*Continued*)

Compound (R¹ R² R³)	Structure	Structural parameters bond distances (Å) and bond angles (°)[b]	[Reference], **CCDB code**, and *notes*
(**38**) (2-*t*-Butylsulfonylphenyl-*O*)chloro(4-methylphenyl)-bismuthine Tol Ar Cl *(O)*	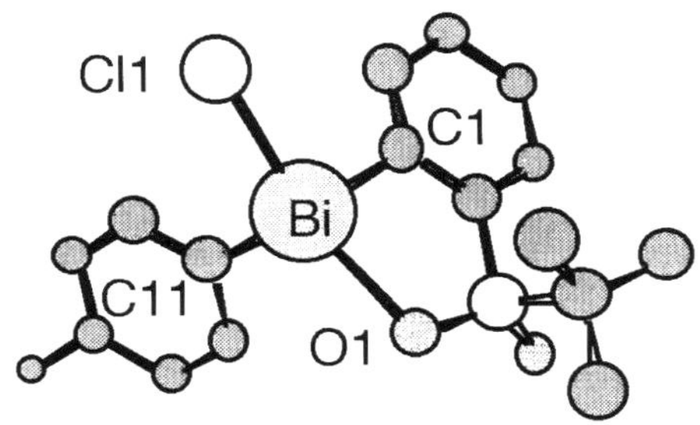 	Bi–Cl(1) = 2.556(2), Bi...O(1) = 2.592(5), Bi–C(1) = 2.288(7), Bi–C(11) = 2.247(7) Cl(1)–Bi–O(1) = 163.9(1), Cl(1)–Bi–C(1) = 91.0(2), Cl(1)–Bi–C(11) = 94.3(2), O(1)–Bi–C(1) = 73.0(2), O(1)–Bi–C(11) = 87.4(2), C(1)–Bi–C(11) = 94.3(2)	[93JCS(P1)1169] **WAMNUT**
(**39**) Bis(triphenylphosphine)-iminium dichloro(2,2′-biphenylene)bismuthate Ar Ar Cl *(Cl)*	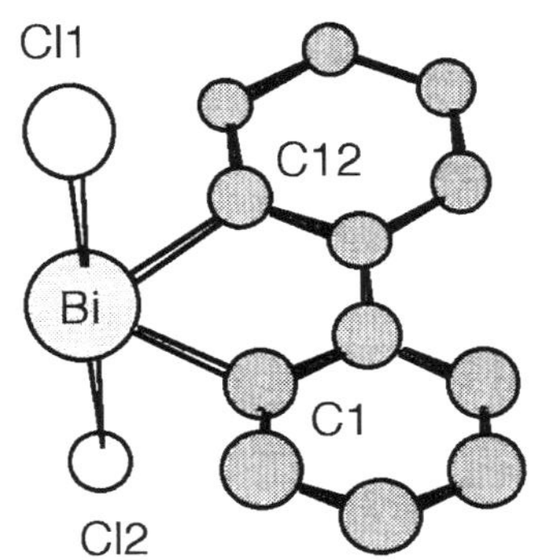 	Bi–Cl(1) = 2.70, Bi–C(1) = 2.26, Bi–Cl(2) = 2.75, Bi–C(12) = 2.23 Cl(1)–Bi–C(1) = 92.2, Cl(1)–Bi–Cl(2) = 175.1, Cl(1)–Bi–C(12) = 89.6, C(1)–Bi–Cl(2) = 90.4, C(1)–Bi–C(12) = 77.6, Cl(2)–Bi–C(12) = 87.0	[96OM887] **ZUJQIE**

TABLE 6.5 (*Continued*

Compound R^1	R^2	R^3	Structure	Structural parameters bond distances (Å) and bond angles (°)[b]	[Reference], **CCDB code**, and *notes*
(**40**) Bis[bis(2,4,6-trimethyl-phenyl)bismuthyl] selenide Mes	Mes	(Se)	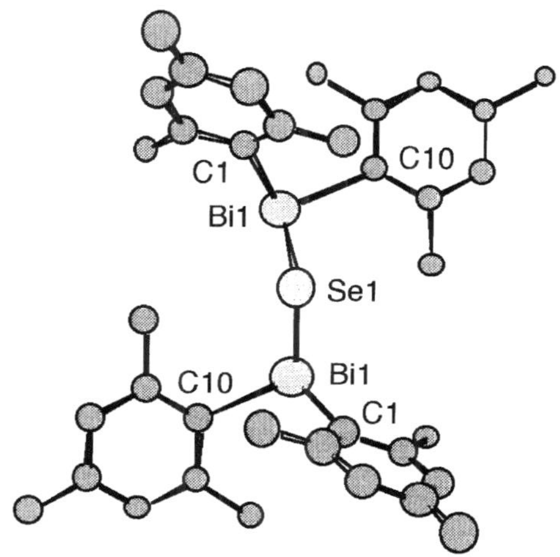	Bi(1)–Se(1) = 2.65, Bi(1)–C(1) = 2.27, Bi(1)–C(10) = 2.28, Se(1)–Bi(1) = 2.65, Bi(1)–C(1) = 2.27, Bi(1)–C(10) = 2.28 Se(1)–Bi(1)–C(1) = 88.9, Se(1)–Bi(1)–C(10) = 106.3, C(1)–Bi(1)–C(10) = 100.6, Bi(1)–Se(1)–Bi(1) = 91.3	[95ZN(B)735] **YULYIN** *at −100°C*
(**41**) Bromobis(2,4,6-trimethylphenyl)bismuthine Mes	Mes	Br	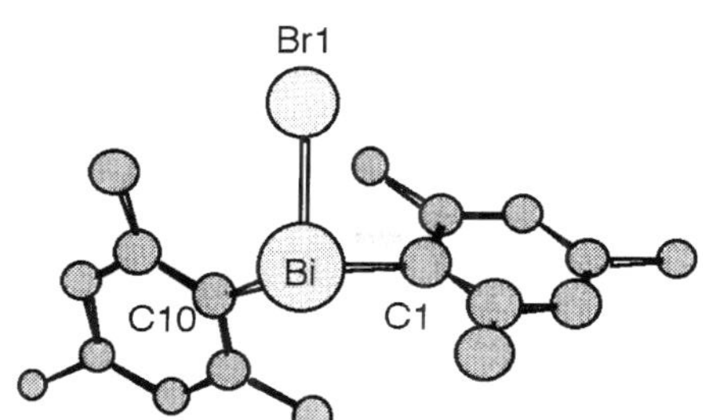	Bi–Br(1) = 2.69, Bi–C(1) = 2.27, Bi–C(10) = 2.27 Br(1)–Bi–C(1) = 89.4, Br(1)–Bi–C(10) = 106.4, C(1)–Bi–C(10) = 98.0	[94JOM(470)93] **WEXZEE** *at −120°C*

TABLE 6.5 (*Continued*)

Compound R¹	R²	R³	Structure	Structural parameters bond distances (Å) and bond angles (°)[b]	[Reference], **CCDB code**, and *notes*
(**42**) Bromobis(2,4,6-trimethyl-phenyl)(diphenylsulfoxide-*O*)-bismuthine Mes	Mes	Br (*O*)	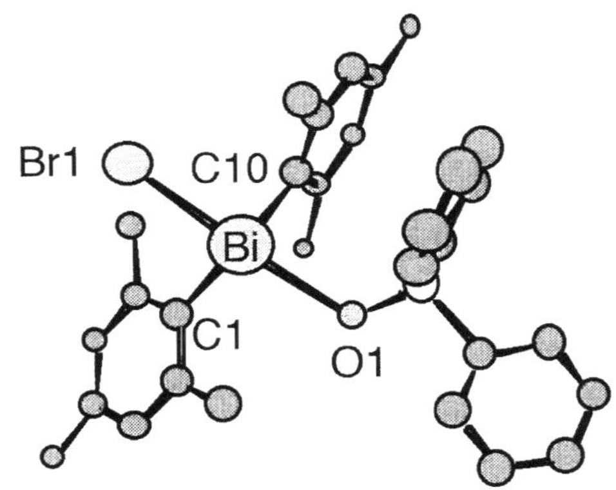	Bi–Br(1) = 2.75, Bi–O(1) = 2.64, Bi–C(1) = 2.28, Bi–C(10) = 2.25 Br(1)–Bi–O(1) = 176.1, Br(1)–Bi–C(1) = 91.9, Br(1)–Bi–C(10) = 95.1, O(1)–Bi–C(1) = 90.1, O(1)–Bi–C(10) = 81.1, C(1)–Bi–C(10) = 107.6	[95JOM(496)59] **ZASGUV**
(**43**) Bromobis(2,4,6-trimethyl-phenyl)(hexamethylphosphor-amide-*O*)bismuthine Mes	Mes	Br (*O*)	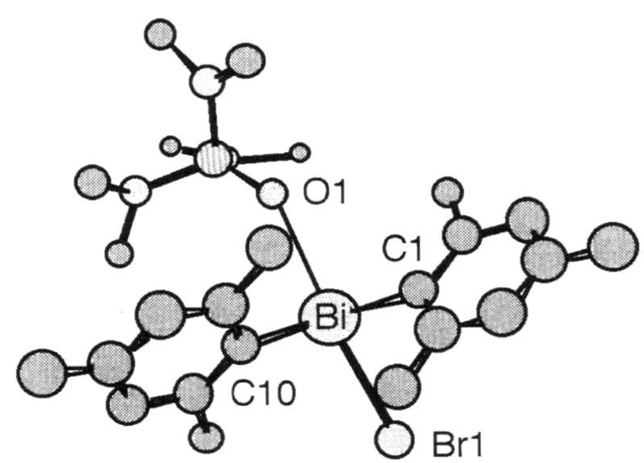	Bi–Br(1) = 2.77, Bi–O(1) = 2.63, Bi–C(1) = 2.26, Bi–C(10) = 2.24 Br(1)–Bi–O(1) = 171.7, Br(1)–Bi–C(1) = 87.8, Br(1)–Bi–C(10) = 102.0, O(1)–Bi–C(1) = 94.6, O(1)–Bi–C(10) = 85.3, C(1)–Bi–C(10) = 102.5	[95JOM(496)59] **ZASHAC**

(**44**) (2-Acetylphenyl-*C*,*O*)bromo-(4-methylphenyl)bismuthine
Tol Ar Br
(O)

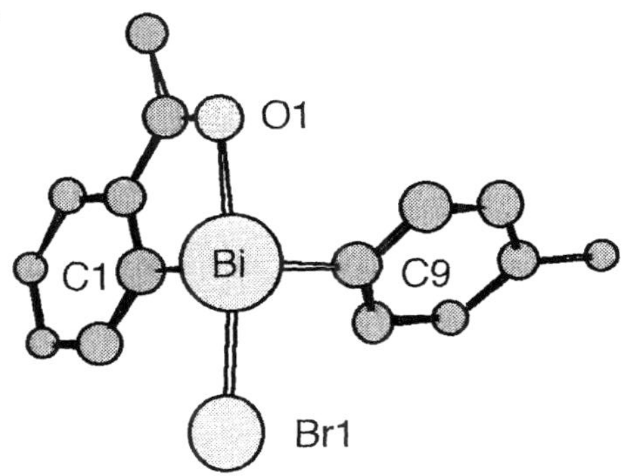

Bi–Br(1) = 2.746(1), Bi–O(1) = 2.519(7), Bi–C(1) = 2.26(1), Bi–C(9) = 2.24(1) Br(1)–Bi–O(1) = 160.8(2), Br(1)–Bi–C(1) = 90.9(3), Br(1)–Bi–C(9) = 93.0(3), O(1)–Bi–C(1) = 70.8(3), O(1)–Bi–C(9) = 83.0(3), C(1)–Bi–C(9) = 95.1(4)

[95OM3848]
ZEGMIH

(**45**) Bromodiphenyltris-(tetrahydrofuran)bismuthine
Ph Ph Br
(O)

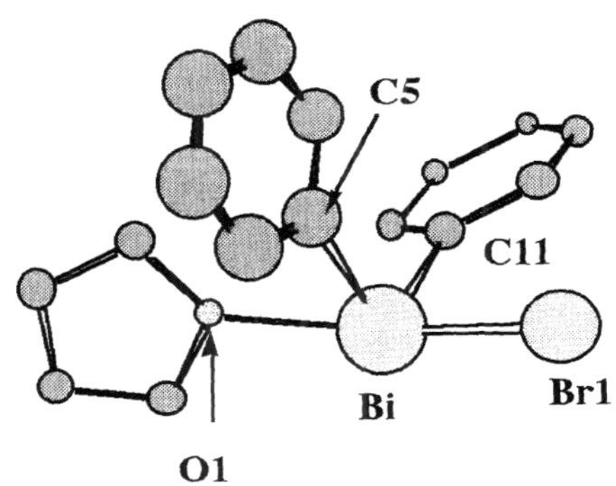

Bi–Br(1) = 2.74, Bi–O(1) = 2.59, Bi–C(5) = 2.23, Bi–C(11) = 2.27 Br(1)–Bi–O(1) = 173.5, Br(1)–Bi–C(5) = 93.2, Br(1)–Bi–C(11) = 92.7, O(1)–Bi–C(5) = 80.8, O(1)–Bi–C(11) = 85.8, C(5)–Bi–C(11) = 98.2

[92JCS(D)1967]
KUDRIK
at −33°C

TABLE 6.5 (*Continued*)

Compound			Structure	Structural parameters bond distances (Å) and bond angles (°)[b]	[Reference], **CCDB code**, and *notes*
R^1	R^2	R^3			
(**46**) Tetraphenylphosphonium dibromodiphenylbismuthate			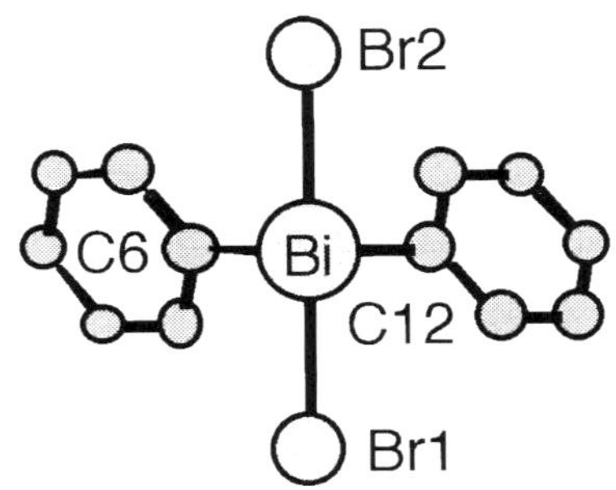	Bi–Br(1) = 2.88, Bi–Br(2) = 2.86, Bi–C(6) = 2.28, Bi–C(12) = 2.26 Br(1)–Bi–Br(2) = 177.3, Br(1)–Bi–C(6) = 91.3, Br(1)–Bi–C(12) = 91.0, Br(2)–Bi–C(6) = 90.4, Br(2)–Bi–C(12) = 91.0, C(6)–Bi–C(12) = 95.9	[92JCS(D)1967] **KUDRUW** *at −33°C*
Ph	Ph	Br *Br*			
(**47**) Tetraethylammonium diiododiphenylbismuthate			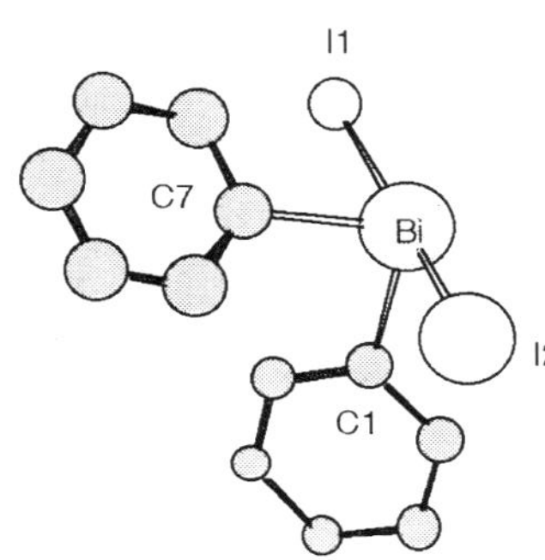	Bi–I(1) = 3.14, Bi–C(1) = 2.25, Bi–I(2) = 3.05, Bi–C(7) = 2.29 I(1)–Bi–C(1) = 89.7, I(1)–Bi–I(2) = 177.9, I(1)–Bi–C(7) = 88.1, C(1)–Bi–I(2) = 91.7, C(1)–Bi–C(7) = 93.6, I(2)–Bi–C(7) = 90.2	[93JCS(D)637] **PEBHUZ** *at −33°C*
Ph	Ph	[illegible]			

(**48**) Tetramethyldibismuthine Me Me (Bi)	$Me_2Bi\text{-}BiMe_2$	Bi–Bi′ = 3.123(7), Bi...Bi′ (inter) = 3.582(7)	[88ZN(B)952]
(**49**) 2,2′,5,5′-Tetramethyl-bibismole (C) (C) (Bi)	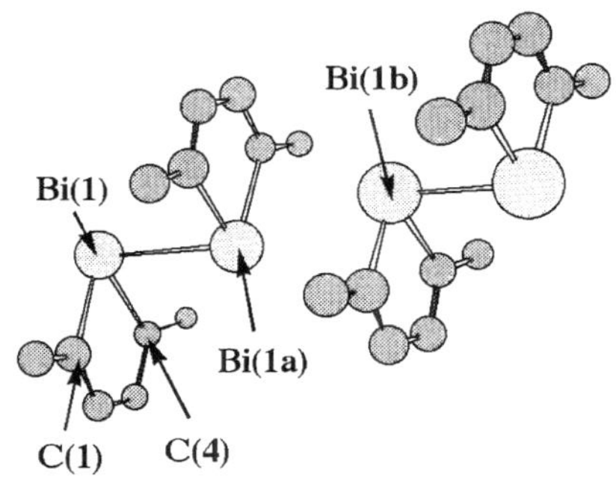	Bi–Bi(1a) = 2.9902(5), Bi(1a)–Bi(1b) = 3.6595(5), Bi–C(1) = 2.23(1), Bi–C(4) = 2.24(1) Bi–Bi(1a)–Bi(1b) = 166.53(2), C(1)–Bi–C(4) = 79.3, C(1)–Bi–Bi(1a) = 90.0(3), C(4)–Bi– Bi(1a) = 90.0(3)	[92OM2743] **JUGFAS**
(**50**) 2,2′,5,5′-Tetramethyl-1,1′-dibismaferrocene (C) (C) (Bi)	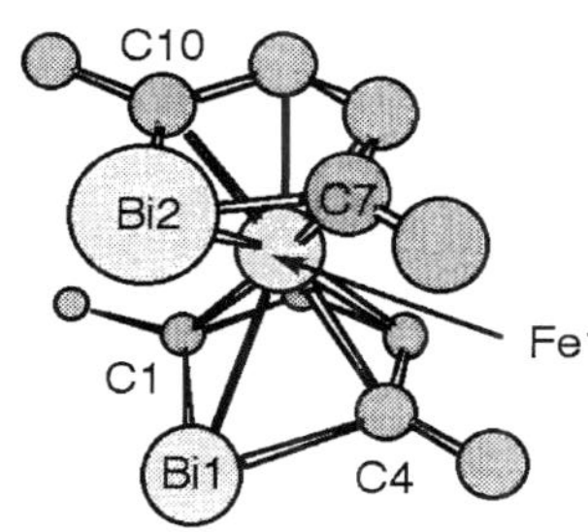	Bi–Bi(2) = 3.688(1), Fe–Bi(av) = 2.677(4), Bi–C(1), Bi–C(4) average = 2.22(2), C(1)–Bi–C(4) average = 76.4(12)	[84JCS(D)2365] **JUGFEW**

TABLE 6.5 (*Continued*)

Compound R[1]	R[2]	R[3]	Structure	Structural parameters bond distances (Å) and bond angles (°)[b]	[Reference], **CCDB code**, and *notes*
(**51a**) Tetraphenyldibismuthine Ph	Ph	(Bi)	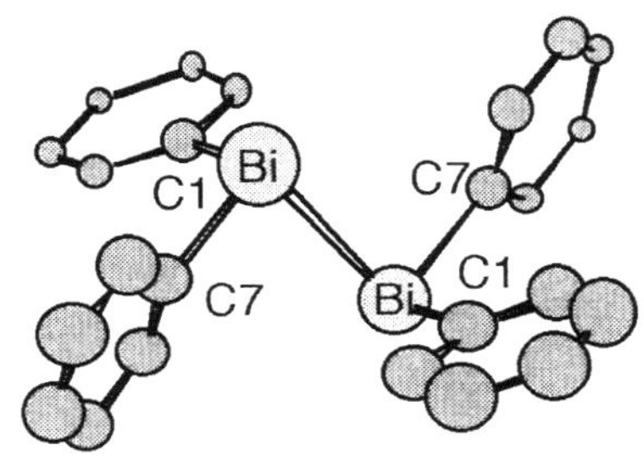	Bi–C(1) = 2.28(2), Bi–Bi = 2.990(2), Bi–C(7) = 2.26(2) Bi–Bi–C(1) = 91.6(5), Bi–Bi–C(7) = 90.9(5), C(1)–Bi–C(7) = 98.3(8)	[83CC507] **BOZVER** [84JCS(D)2365] **BOZVER10** *space group:* $P\bar{1}$
(**51b**) Tetraphenyldibismuthine Ph	Ph	(Bi)	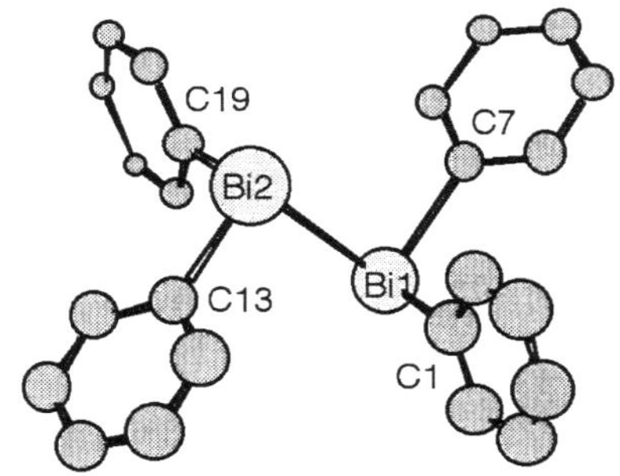	Bi(1)–C(1) = 2.24(2), Bi(1)–Bi(2) = 2.984(2), Bi(1)–C(7) = 2.27(2), Bi(2)–C(13) = 2.24(2), Bi(2)–C(19) = 2.23(2) C(1)–Bi(1)–Bi(2) = 94.8(5), C(1)–Bi(1)–C(7) = 97.7(7), Bi(2)–Bi(1)–C(7) = 91.1(5), Bi(1)–Bi(2)–C(13) = 91.7(5), Bi(1)–Bi(2)–C(19) = 93.3(5), C(13)–Bi(2)–C(19) = 95.7(8)	[92AX(C)917] **BOZVER01** *tetragonal form space group:* $I4_1/a$

(**51c**) Tetraphenyldibismuthine Ph Ph (Bi)	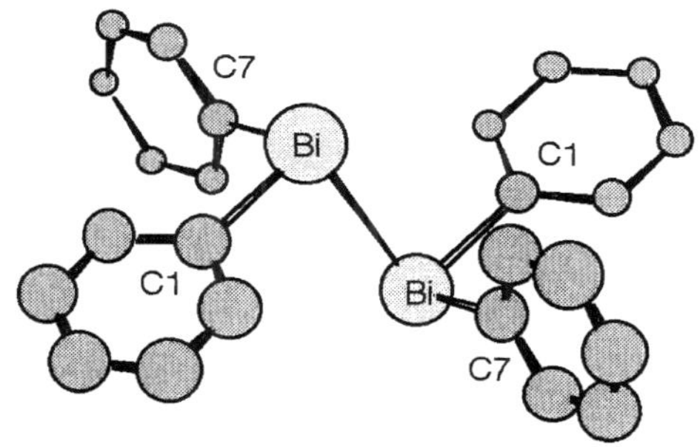 	Bi–Bi(2) = 2.988(1), Bi–C(1) = 2.25(1), Bi–C(7) = 2.26(1) Bi(2)–Bi–C(1) = 92.2(3), Bi(2)–Bi–C(7) = 95.1(3), C(1)–Bi–C(7) = 93.3(4)	[92AX(C)917] **BOZVER02** *monoclinic form space group: $P2_1/n$*
(**52**) Tetrakis(hexamethylphosphoramide-*O*)phenylbismuthine bis(hexafluorophosphate) Ph (O) (O) *(O)* *(O)*	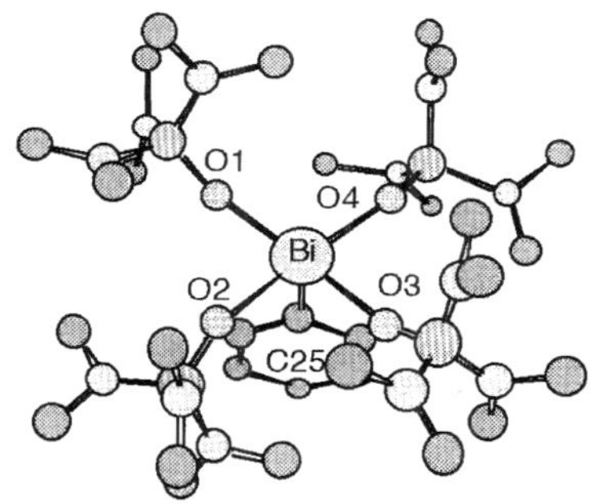 	Bi–O(1) = 2.35, Bi–O(2) = 2.32, Bi–O(3) = 2.32, Bi–O(4) = 2.36 O(1)–Bi–O(2) = 85.4, O(1)–Bi–O(3) = 174.2, O(1)–Bi–O(4) = 94.7, O(1)–Bi–C(25) = 87.5, O(2)–Bi–O(3) = 93.3, O(2)–Bi–O(4) = 172.6, O(2)–Bi–C(25) = 85.4, O(3)–Bi–O(4) = 85.8, O(3)–Bi–C(25) = 86.8, O(4)–Bi–C(25) = 87.2	[96JCS(D)443] **ZUCKAJ**

TABLE 6.5 (*Continued*)

Compound R[1]	R[2]	R[3]	Structure	Structural parameters bond distances (Å) and bond angles (°)[b]	[Reference], **CCDB code**, and *notes*
(**53**) *N,N,N′,N′*-Tetramethylpropanediammonium tetrakis(trifluoroacetato-*O*)phenylbismuthate			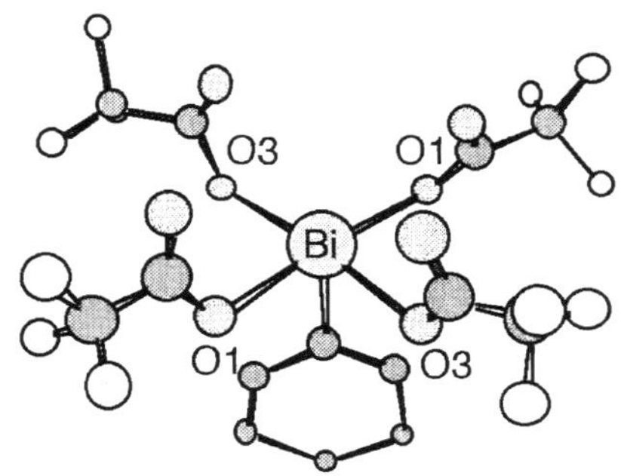	Bi–O(1) = 2.38(3), Bi–O(1′) = 3.17, Bi–O(3) = 2.42(2), Bi–O(3′) = 3.23, Bi–C(5) = 2.28(4) O(1)–Bi–O(1) = 169(1), O(1)–Bi–O(3) = 89.7, O(1)–Bi–O(3) = 88.9, O(1)–Bi–C(5) = 84.3, O(1)–Bi–C(5) = 84.3, O(3)–Bi–O(3) = 165(1)	[93AG(E)589] **SUKZAZ**
Ph	(O) *(O)*	(O) *(O)*			

(**54**) *N,N,N′,N′*-Tetramethylethylene-diammonium (trifluoroacetato-*O,O′*)tris(trifluoroacetato-*O*)-phenylbismuthate Ph (O) (O) *(O)* *(O)*	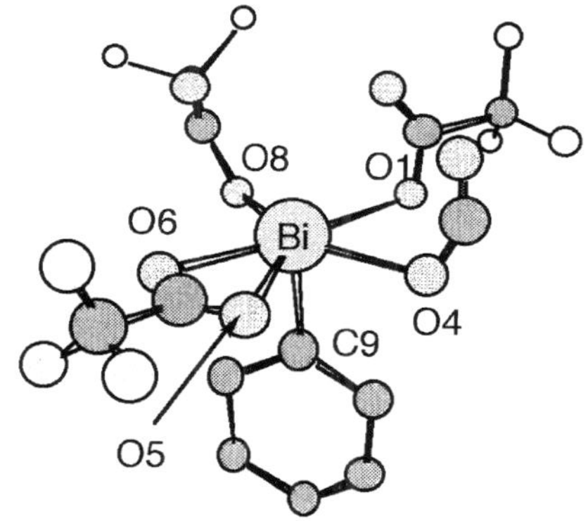	Bi–O(1) = 2.62(3), Bi–O(1′) = 3.15, Bi–O(4) = 2.50(4), Bi–O(4′) = 3.36, Bi–O(5) = 2.59(4), Bi–O(6) = 2.34(4), Bi–O(8) = 2.32(3), Bi–O(8′) = 3.22, Bi–C(9) = 2.20 O(1)–Bi–O(4) = 71(1), O(1)–Bi–O(5) = 150, O(1)–Bi–O(6) = 154, O(1)–Bi–O(8) = 71(1), O(1)–Bi–C(9) = 83, O(4)–Bi–O(5) = 78(1), O(4)–Bi–O(6) = 129, O(4)–Bi–O(8) = 142, O(4)–Bi–C(9) = 86, O(5)–Bi–O(6) = 52(1), O(5)–Bi–O(8) = 138, O(5)–Bi–C(9) = 87, O(6)–Bi–O(8) = 87(1), O(6)–Bi–C(9) = 87, O(8)–Bi–C(9) = 83	[93AG(E)589] **SUKZED** *Fluorine atoms are disordered, but the disorder could not be modelled*
(**55**) 1,3-Bis(dimethylammonio)-2-propanol tetrakis(trifluoro-acetato-*O*)phenylbismuthate Ph (O) (O) *(O)* *(O)*	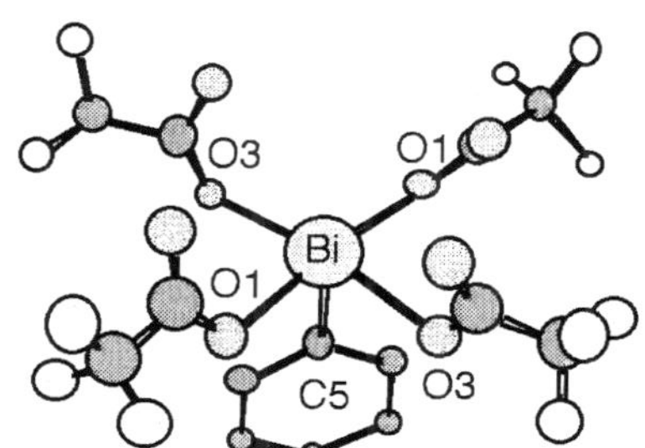	Bi–O(1) = 2.39(1), Bi–O(1′) = 3.24, Bi–O(3) = 2.40(1), Bi–O(3′) = 3.33, Bi–C(5) = 2.26(3) O(1)–Bi–O(1) = 168.2(7), O(1)–Bi–O(3) = 88.2, O(1)–Bi–O(3) = 90.4, O(1)–Bi–C(5) = 84.1, O(3)–Bi–O(3) = 165.5(7), O(3)–Bi–C(5) = 82.7	[93AG(E)589] **SUKZIH** [94JCS(D)2545] **SUKZIH10**

TABLE 6.5 (*Continued*)

Compound R[1]	R[2]	R[3]	Structure	Structural parameters bond distances (Å) and bond angles (°)[b]	[Reference], **CCDB code**, and *notes*
(**56**) *N,N,N′,N′*-Tetramethylbutane-1,4-diammonium phenyltetrakis(trifluoroacetato-*O*)bismuthate Ph	(O) (*O*)	(O) (*O*)	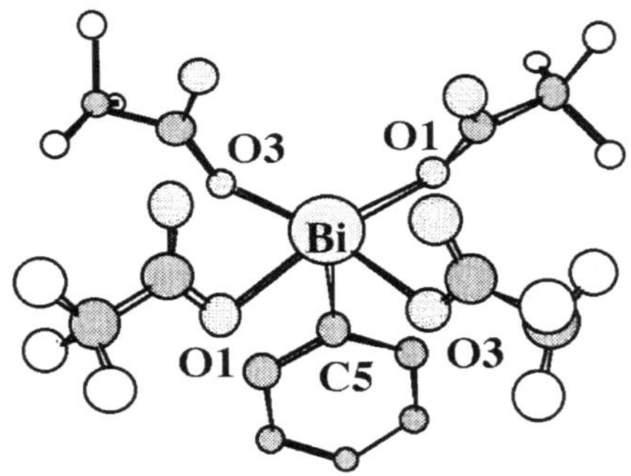	Bi–O(1) = 2.38, Bi–O(3) = 2.37, Bi–O(3) = 2.37, Bi–C(5) = 2.21 O(1)–Bi–O(1) = 165.0, O(1)–Bi–O(3) = 87.9, O(1)–Bi–O(3) = 89.8, O(1)–Bi–C(5) = 82.5, O(1)–Bi–O(3) = 87.9, O(1)–Bi–C(5) = 82.5, O(3)–Bi–O(3) = 161.9, O(3)–Bi–C(5) = 81.0	[94JCS(D)2545] **LIFGUC**
(**57**) *N,N,N′,N′*-Tetramethyl-1,4-phenylene-1,4-diammonium phenyltetrakis(trifluoroacetato-*O*)bismuthate Ph	(O) (*O*)	(O) (*O*)	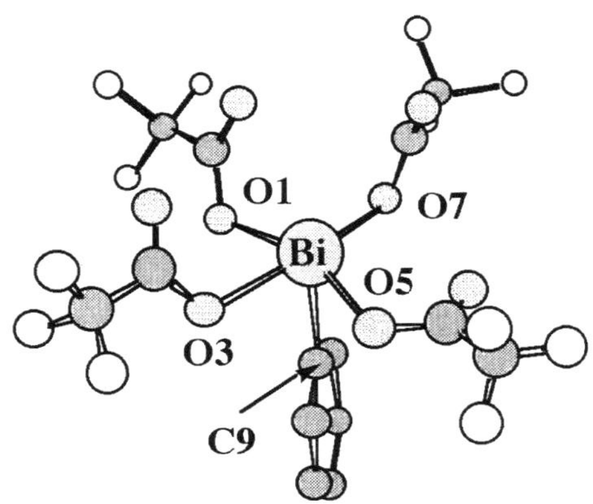	Bi–O(1) = 2.37, Bi–O(3) = 2.32, Bi–O(5) = 2.53, Bi–O(7) = 2.46, Bi–C(9) = 2.28 O(1)–Bi–O(3) = 82.9, O(1)–Bi–O(5) = 159.8, O(1)–Bi–O(7) = 74.1, O(1)–Bi–C(9) = 86.3, O(3)–Bi–O(5) = 79.8, O(3)–Bi–O(7) = 155.4, O(3)–Bi–C(9) = 80.5, O(5)–Bi–O(7) = 120.6, O(5)–Bi–C(9) = 80.7, O(7)–Bi–C(9) = 89.1	[94JCS(D)2545] **LIFHEN**

(**58**) Bis(pyridinium) phenyl-tetrakis(trifluoroacetato-*O*)-bismuthate

Ph	(O)	(O)
	(O)	*(O)*

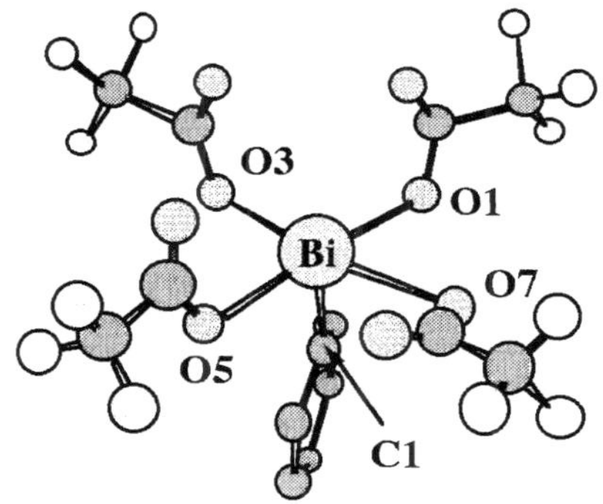

Bi–O(1) = 2.40, Bi–C(1) = 2.28, Bi–O(3) = 2.36, Bi–O(5) = 2.41, Bi–O(7) = 2.51
O(1)–Bi–C(1) = 82.2, O(1)–Bi–O(3) = 89.7, O(1)–Bi–O(5) = 168.5, O(1)–Bi–O(7) = 68.3, C(1)–Bi–O(3) = 85.7, C(1)–Bi–O(5) = 87.1, C(1)–Bi–O(7) = 80.9, O(3)–Bi–O(5) = 85.3, O(3)–Bi–O(7) = 155.6, O(5)–Bi–O(7) = 114.2

[94JCS(D)2545]
LIFHIR

(**59**) Bis(diethyldithiocarbamato-*S,S'*)methylbismuthine

Me	(S)	(S)

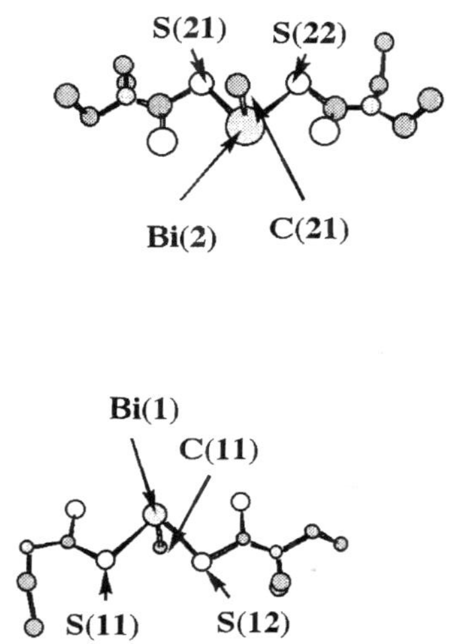

Bi–C(11) = 2.21(4), Bi–S(11) = 2.67(1), Bi–S(12) = 2.95(1), Bi–S(13) = 2.70(1), Bi–S(14) = 2.98(1), Bi–S(24) = 3.27(1), Bi(2)–C(21) = 2.27(3), Bi(2)–S(21) = 2.68(1), Bi(2)–S(22) = 2.93(1), Bi(2)–S(23) = 2.70(1), Bi(2)–S(24) = 2.96(1)
S(11)–Bi–S(12) = 63.5, S(11)–Bi–S(13) = 76.2, S(12)–Bi–S(24) = 90.7, S(13)–Bi–S(14) = 63.2, S(14)–Bi–S(22) = 89.6, S(14)–Bi–S(24) = 67.6, S(21)–Bi–S(22) = 63.3, S(21)–Bi–S(23) = 79.6, S(23)–Bi–S(24) = 62.2

[79ZN(B)1037]
MBIETC

TABLE 6.5 (*Continued*)

Compound			Structure	Structural parameters bond distances (Å) and bond angles (°)[b]	[Reference], **CCDB code**, and *notes*
R^1	R^2	R^3			
(**60**) 5-Phenyl-1,4,6,5-oxadithiabismocane			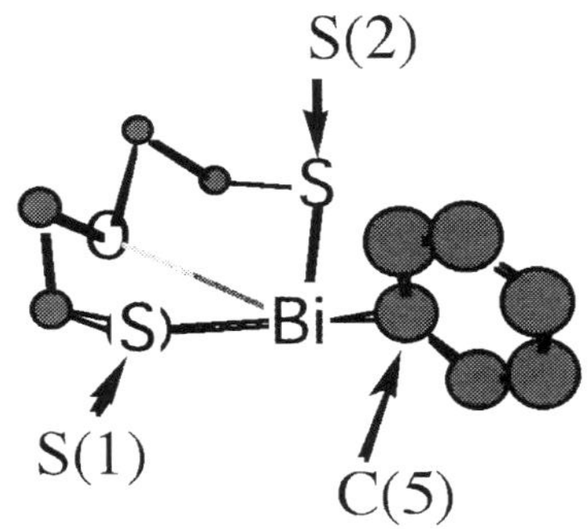	Bi–S(1) = 2.560(2), Bi–S(2) = 2.602(3), Bi–C(5) = 2.25(2), Bi–O = 2.97(1) S(2)–Bi–S(1) = 104.2(1), S(2)–Bi–C(5) = 87.2(4), S(1)–Bi–C(5) = 90.3(4)	[85JOM(290)133] **DEHWAO**
Ph	(S)	(S) *(O)*			
(**61**) Bis(methyldithiocarbonato-*S,S'*)phenylbismuthine			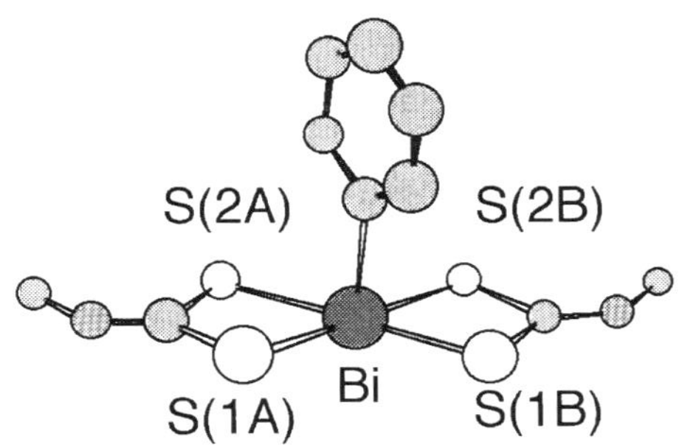	Bi–S(1A) = 2.649(3), Bi–S(1B) = 2.670(3), Bi–S(2A), 2.961(2), Bi–S(2B) = 3.079(3) S(1A)–Bi–S(2A) = 63.7(1), S(1B)–Bi–S(2B) = 61.8(1), S(1A)–Bi–S(1B) = 78.9(1), S(2A)–Bi–S(2B) = 155.1(1)	[82ZAAC(485)217] **BENXUN**
Ph	(S) *(S)*	(S) *(S)*			

(**62**) Bis(diethyldithiocarbamato-*S,S'*)phenylbismuthine Ph (S) (S) *(S)* *(S)*	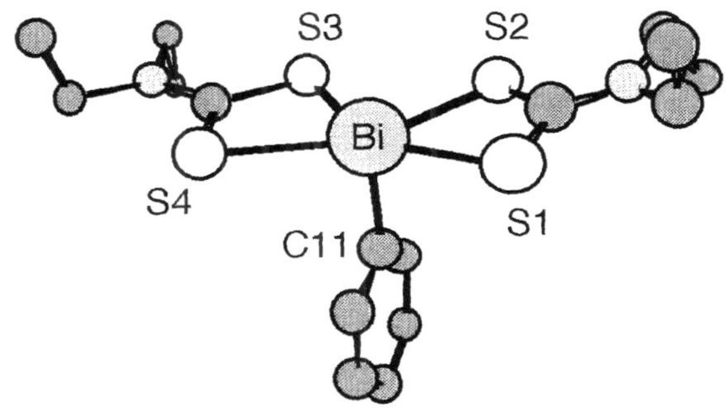	Bi–S(1) = 2.926(2), Bi–S(2) = 2.671(2), Bi–S(3) = 2.676(3), Bi–S(4) = 2.942(2), Bi–C(11) = 2.241(10) S(1)–Bi–S(2) = 63.9(1), S(1)–Bi–S(3) = 144.0(1), S(1)–Bi–S(4) = 152.5(1), S(1)–Bi–C(11) = 90.4(3), S(2)–Bi–S(3) = 80.1(1), S(2)–Bi–S(4) = 143.6(1), S(2)–Bi–C(11) = 94.6(3), S(3)–Bi–S(4) = 63.5(1), S(3)–Bi–C(11) = 92.0(3), S(4)–Bi–C(11) = 88.1(2)	[90JCS(D)899] **SEPLOO**
(**63**) Bis(diphenyldithio-phosphinato-*S,S'*)(2,4,6-trimethylphenyl)bismithine Mes (S) (S) *(S)* *(S)*	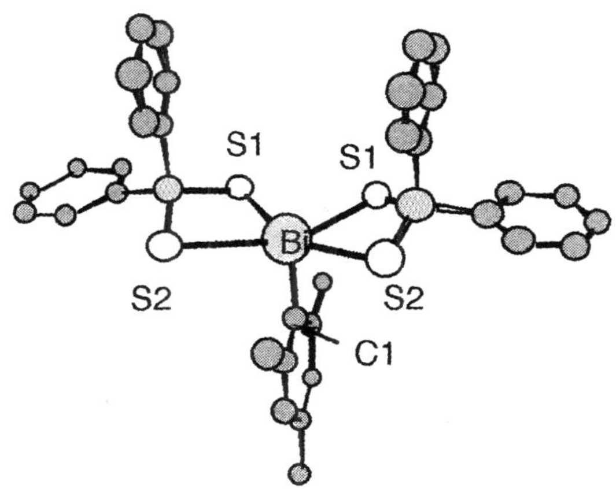	Bi–S(1) = 2.66, Bi–C(1) = 2.26, Bi–S(1) = 2.66, Bi–S(2) = 3.11, Bi–S(2) = 3.11 S(1)–Bi–C(1) = 97.9, S(1)–Bi–S(1) = 84.0, S(1)–Bi–S(2) = 70.6, S(1)–Bi–S(2) = 152.5, C(1)–Bi–S(1) = 97.9, C(1)–Bi–S(2) = 95.7, C(1)–Bi–S(2) = 95.7, S(1)–Bi–S(2) = 152.5, S(1)–Bi–S(2) = 70.6, S(2)–Bi–S(2) = 131.5	[94JOM(470)93] **WEXZII** *at −100°C*

TABLE 6.5 (*Continued*)

Compound (R¹ R² R³)	Structure	Structural parameters bond distances (Å) and bond angles (°)[b]	[Reference], **CCDB code**, and *notes*
(**64**) Bis(diethyldithiocarbamato-*S,S'*)[2-(2-pyridyl)phenyl-*C,N*]bismuthine R¹: Ar (*N*); R²: (S) (*S*); R³: (S) (*S*)	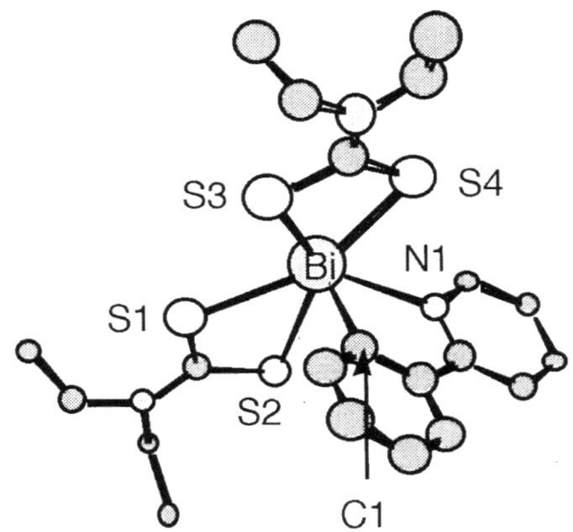	Bi–S(1) = 2.798(2), Bi–N(1) = 2.553(6), Bi–C(1) = 2.263(7), Bi–S(2) = 2.895(2), Bi–S(3) = 2.766(2), Bi–S(4) = 2.878(2) S(1)–Bi–N(1) = 137.5(1), S(1)–Bi–C(1) = 89.5(2), S(1)–Bi–S(2) = 62.9(1), S(1)–Bi–S(3) = 76.9(1), S(1)–Bi–S(4) = 139.9(1), N(1)–Bi–C(1) = 70.8(2), N(1)–Bi–S(2) = 78.4(1), N(1)–Bi–S(3) = 138.3(1), N(1)–Bi–S(4) = 78.6(1), C(1)–Bi–S(2) = 87.0(2), C(1)–Bi–S(3) = 91.6(2), C(1)–Bi–S(4) = 88.5(2), S(2)–Bi–S(3) = 139.8(1), S(2)–Bi–S(4) = 156.7(1), S(3)–Bi–S(4) = 63.1(1)	[90JCS(D)899] **SEPLII**

(**65**) Catena[(μ_2-chloro)chloro-(phenyl)(tetrahydrofuran)-bismuthine]

Ph	Cl	Cl
	Cl	*(O)*

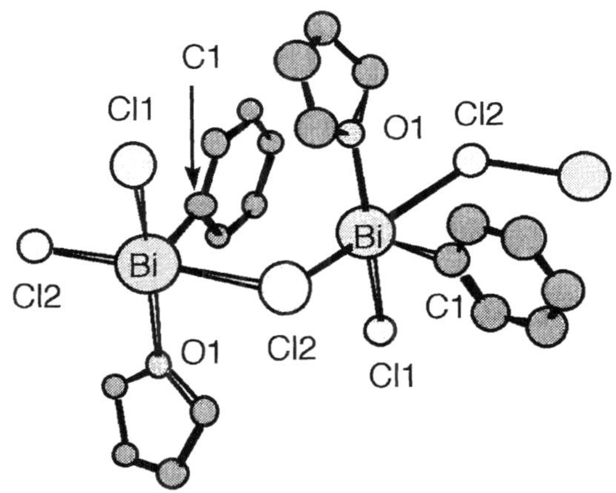

Bi–Cl(1) = 2.54, Bi–C(1) = 2.27, Bi–O(1) = 2.61, Bi–Cl(2) = 2.66, Bi–Cl(2) = 2.93
Cl(1)–Bi–C(1) = 92.6, Cl(1)–Bi–O(1) = 177.3, Cl(1)–Bi–Cl(2) = 93.3, Cl(1)–Bi–Cl(2) = 95.6, C(1)–Bi–O(1) = 88.3, C(1)–Bi–Cl(2) = 88.9, C(1)–Bi–Cl(2) = 82.8, O(1)–Bi–Cl(2) = 84.1, O(1)–Bi–Cl(2) = 87.1, Cl(2)–Bi–Cl(2) = 168.1

[93JCS(D)637]
PEBHIN
at −113°C

TABLE 6.5 (*Continued*)

Compound			Structure	Structural parameters bond distances (Å) and bond angles (°)[b]	[Reference], **CCDB code**, and *notes*
R^1	R^2	R^3			
(**66**) Catena[(ϵ3-cyclopenta-dienyl)(ϵ2-cyclopentadienyl)-bis(bis(μ_2-chloro)bismuthine)]			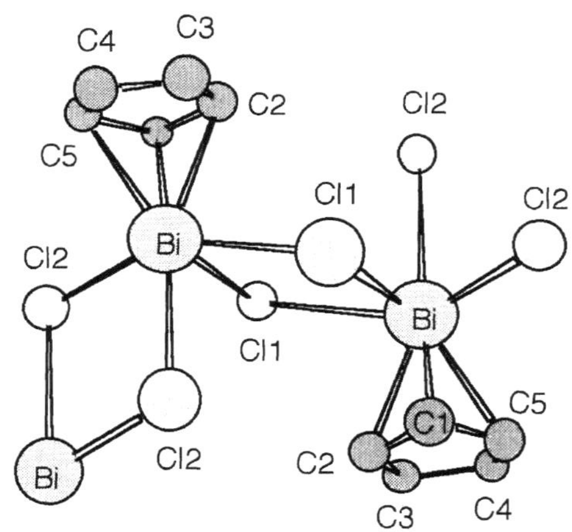	Bi–Cl(1) = 2.582(6), Bi–C(1) = 2.38(3), Bi–Cl(1) = 3.183(7), Bi–Cl(2) = 2.643(5), Bi–C(2) = 2.61(3), Bi–Cl(2) = 3.092(7), Bi–C(5) = 2.64(3), Cl(1)–Bi = 3.183(7) Cl(1)–Bi–C(1) = 87.8, Cl(1)–Bi–Cl(1) = 77.9(1), Cl(1)–Bi–Cl(2) = 86.7(1), Cl(1)–Bi–C(2) = 93.6, Cl(1)–Bi–Cl(2) = 81.4(1), Cl(1)–Bi–C(5) = 119.9, C(1)–Bi–Cl(1) = 115.5, C(1)–Bi–Cl(2) = 91.9, C(1)–Bi–C(2) = 34.5, C(1)–Bi–Cl(2) = 164.0, C(1)–Bi–C(5) = 34.2, Cl(1)–Bi–Cl(2) = 147.5(4), Cl(1)–Bi–C(2) = 83.6, Cl(1)–Bi–Cl(2) = 73.78(9), Cl(1)–Bi–C(5) = 130.0, Cl(2)–Bi–C(2) = 126.1, Cl(2)–Bi–Cl(2) = 75.8,	[90JOM(386)177] **SEYTOF** *at −20°C*
Cp	Cl	Cl			
	Cl	*Cl*			

Cl(2)–Bi–C(5) = 82.4, C(2)–Bi–Cl(2) = 157.4, C(2)–Bi–C(5) = 51.3, Cl(2)–Bi–C(5) = 148.7

(**67**) Catena[tetrakis(μ_2-bromo)diphenyldibismuthine]

Ph Br Br

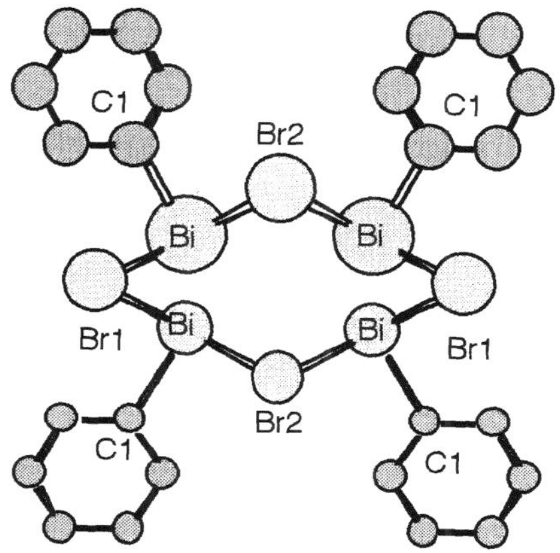

Bi–Br(1) = 2.88, Bi–Br(2) = 2.93, Bi–C(1) = 2.25 Br(1)–Bi–Br(1) = 86.0, Br(1)–Bi–Br(2) = 94.7, Br(1)–Bi–Br(2) = 178.9, Br(1)–Bi–C(1) = 89.4, Br(2)–Bi–Br(2) = 84.5, C(1)–Bi–Br(1) = 89.4, C(1)–Bi–Br(2) = 91.4

[94JMAC(4)891]
YISBIL
at −113°C

TABLE 6.5 (*Continued*)

Compound R^1	R^2	R^3	Structure	Structural parameters bond distances (Å) and bond angles (°)[b]	[Reference], **CCDB code**, and *notes*
(**68**) Catena[bis(μ_2-bromo)bromo-(2,4,6-trimethylphenyl)-bismuthine)] Mes	Br	Br	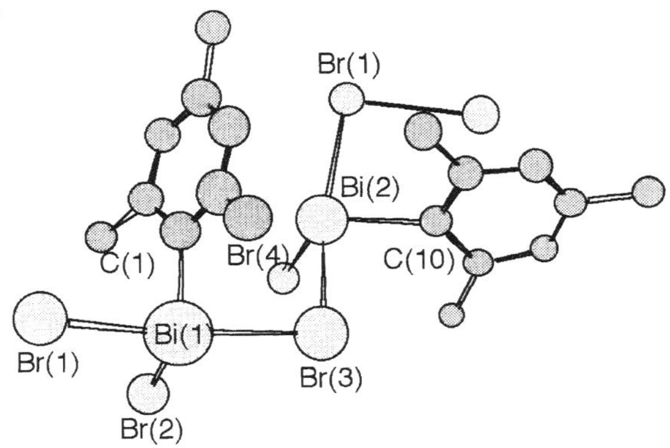	Br(1)–Bi(1) = 2.82, C(1)–Bi(1) = 2.21, C(1)–C(2) = 1.45, Bi(1)–Br(2) = 2.62, Bi(1)–Br(3) = 3.02, Br(1)–Bi(1) = 2.82, Br(1)–Bi(2) = 3.02, Bi(2)–Br(3) = 2.81, Bi(2)–Br(4) = 2.61, Bi(2)–C(10) = 2.25 Br(1)–Bi(1)–Br(2) = 92.7, Br(1)–Bi(1)–Br(3) = 172.4, Br(1)–Bi(1)–C(1) = 89.3, Br(1)–Bi(2)–Br(3) = 171.3, Br(1)–Bi(2)–Br(4) = 87.6, Br(1)–Bi(2)–C(10) = 83.3, Br(2)–Bi(1)–Br(3) = 87.4, Br(3)–Bi(2)–Br(4) = 93.0, Br(3)–Bi(2)–C(10) = 88.2, Br(4)–Bi(2)–C(10) = 106.9, C(1)–Bi(1)–Br(2) = 105.6, C(1)–Bi(1)–Br(3) = 83.4	[97ZAAC(623)941] **NASLIC** *at −100°C*

(**69**) Dibromobis(*N,N′*-dimethyl-propyleneurea-*O*)phenyl-bismuthine

Ph	Br	Br
	(O)	*(O)*

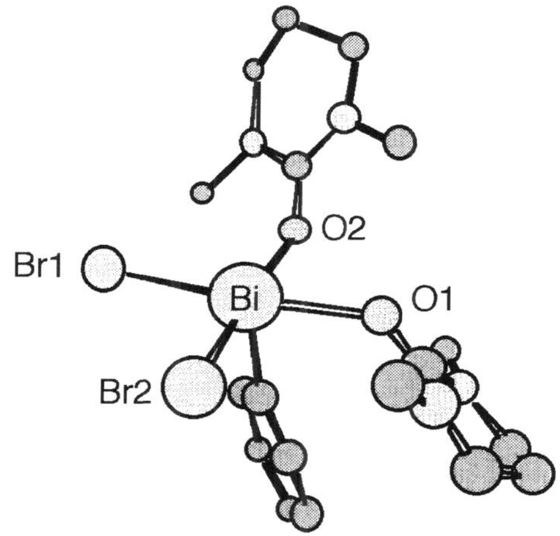

Bi–Br(1) = 2.77, Bi–C(1) = 2.25, Bi–O(1) = 2.47, Bi–Br(2) = 2.73, Bi–O(2) = 2.56
Br(1)–Bi–C(1) = 93.9, Br(1)–Bi–O(1) = 169.5, Br(1)–Bi–Br(2) = 94.4, Br(1)–Bi–O(2) = 84.5, C(1)–Bi–O(1) = 84.7, C(1)–Bi–Br(2) = 88.6, C(1)–Bi–O(2) = 83.5, O(1)–Bi–Br(2) = 96.0, O(1)–Bi–O(2) = 85.0, Br(2)–Bi–O(2) = 172.0

[95JOM(496)59]
ZASGOP

(**70**) Catena[(μ_2-bromo)bromo-(phenyl)(tetrahydrofuran)-bismuthine]

Ph	Br	Br
	(O)	*Br*

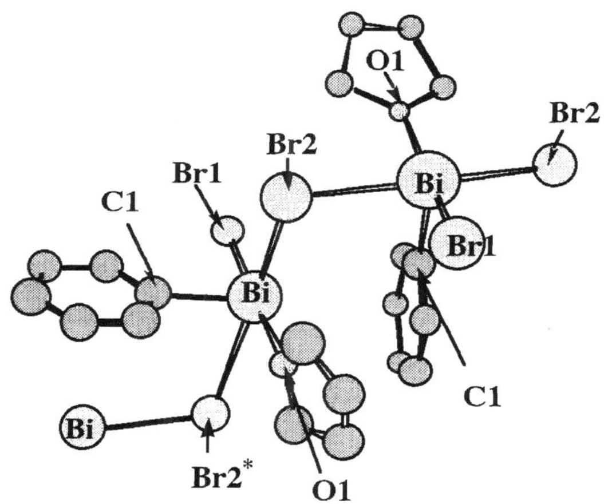

Bi–Br(1) = 2.68, Bi–C(1) = 2.28, Bi–O(1) = 2.67, Bi–Br(2) = 2.83, Bi–Br(2*) = 3.04
Br(1)–Bi–C(1) = 93.9, Br(1)–Bi–O(1) = 177.4, Br(1)–Bi–Br(2) = 93.7, Br(1)–Bi–Br(2*) = 93.5, C(1)–Bi–O(1) = 88.6, C(1)–Bi–Br(2) = 90.0, C(1)–Bi–Br(2*) = 84.4, O(1)–Bi–Br(2) = 85.4, O(1)–Bi–Br(2*) = 87.7, Br(2)–Bi–Br(2*) = 171.2

[92JCS(D)1967]
KUDPUU
at −33°C

TABLE 6.5 (*Continued*)

Compound: R^1	R^2	R^3	Structure	Structural parameters bond distances (Å) and bond angles (°)[b]	[Reference], **CCDB code**, and *notes*
(**71**) Bis[(μ_2-bromo)bromo-(phenyl)(triphenylphosphine-oxide-*O*)bismuthine] Ph	Br *(O)*	Br *Br*	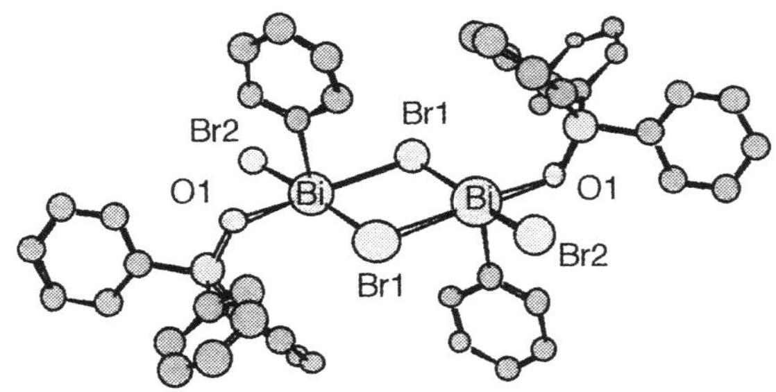	Bi–Br(1) = 3.14, Bi–O(1) = 2.39, Bi–C(1) = 2.25, Bi–Br(1) = 2.86, Bi–Br(2) = 2.69, Br(1)–Bi = 2.86, Bi–Br(1) = 3.14, Bi–O(1) = 2.39, Bi–C(1) = 2.25, Bi–Br(2) = 2.69 Br(1)–Bi–O(1) = 85.7, Br(1)–Bi–C(1) = 92.5, Br(1)–Bi–Br(1) = 84.5, Br(1)–Bi–Br(2) = 176.0, O(1)–Bi–C(1) = 83.5, O(1)–Bi–Br(1) = 167.6, O(1)–Bi–Br(2) = 93.2, C(1)–Bi–Br(1) = 89.2, C(1)–Bi–Br(2) = 91.2, Br(1)–Bi–Br(2) = 97.0, Bi–Br(1)–Bi = 95.5, Bi–O(1)–P(1) = 125.5, Bi–C(1)–C(2) = 120.4, Bi–C(1)–C(6) = 117.9	[95JOM(496)59] **ZASGIJ** *at −100°C*

(**72**) Bis[(μ_2-bromo)bromo(*N,N'*-dimethylpropyleneurea-*O*)-phenylbismuthine] Ph Br Br *(O)* *Br*	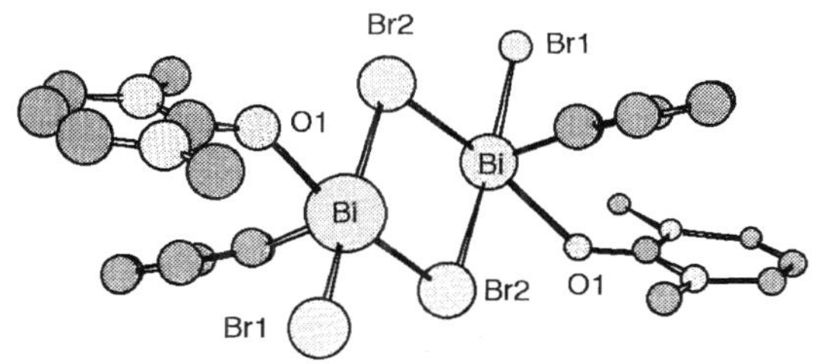 	Bi–Br(1) = 2.70, Bi–C(1) = 2.21, Bi–O(1) = 2.40, Bi–Br(2) = 3.15, Bi–Br(2) = 2.83 Br(1)–Bi–C(1) = 91.4, Br(1)–Bi–O(1) = 99.8, Br(1)–Bi–Br(2) = 176.3, Br(1)–Bi–Br(2) = 94.8, C(1)–Bi–O(1) = 85.7, C(1)–Bi–Br(2) = 91.8, C(1)–Bi–Br(2) = 89.9, O(1)–Bi–Br(2) = 82.3, O(1)–Bi–Br(2) = 164.8, Br(2)–Bi–Br(2) = 83.3	[95JOM(496)59] **ZASHEG**
(**73**) Bis(tetrabutylammonium) bis[(μ_2-bromo)dibromo-(phenyl)bismuthate] Ph Br Br *Br* *Br*	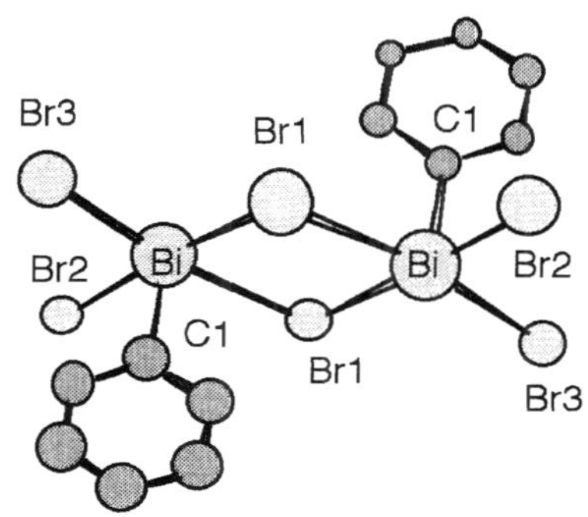 	Bi–Br(1) = 3.01, Bi–C(1) = 2.25, Bi–Br(1) = 3.06, Bi–Br(2) = 2.76, Bi–Br(3) = 2.74, Br(1)–Bi = 3.06 Br(1)–Bi–C(1) = 87.8, Br(1)–Bi–Br(1) = 84.5, Br(1)–Bi–Br(2) = 176.6, Br(1)–Bi–Br(3) = 90.3, C(1)–Bi–Br(1) = 93.2, C(1)–Bi–Br(2) = 89.6, C(1)–Bi–Br(3) = 92.8, Br(1)–Bi–Br(2) = 93.5, Br(1)–Bi–Br(3) = 171.8, Br(2)–Bi–Br(3) = 92.0	[93JCS(D)637] **PEBHOT** *at −33°C*

TABLE 6.5 (*Continued*)

Compound R[1]	R[2]	R[3]	Structure	Structural parameters bond distances (Å) and bond angles (°)[b]	[Reference], **CCDB code**, and *notes*
(**74**) Catena[bis(μ_2-iodo)(methyl)-bismuthine] Me	I	I	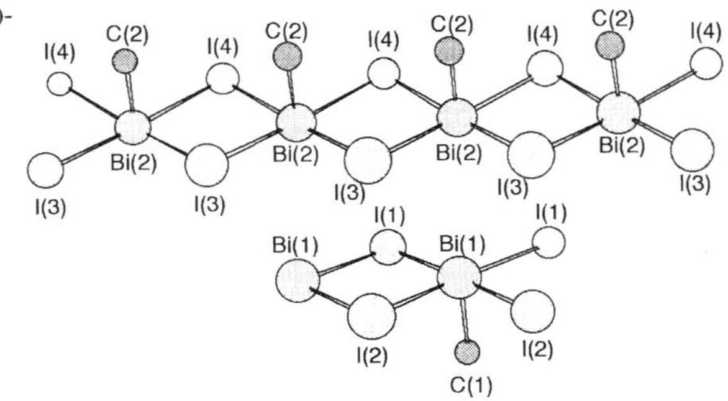	Bi(1)–I(1) = 3.128(1), Bi(1)–C(1) = 2.29(2), Bi(1)–I(2) = 3.087(1), Bi(2)–C(2) = 2.26(2), Bi(2)–I(3) = 3.122(1), Bi(2)–I(4) = 3.086(1) C(1)–Bi(1)–I(2) = 87.9(3), C(2)–Bi(2)–I(3) = 86.2(3), C(2)–Bi(2)–I(4) = 89.0(3), I(1)–Bi(1)–C(1) = 88.2(3), I(1)–Bi(1)–I(1) = 88.08(3), I(1)–Bi(1)–I(2) = 176.06(4), I(3)–Bi(2)–I(3) = 88.28(3), I(3)–Bi(2)–I(4) = 175.13(4), I(3)–Bi(2)–I(4) = 90.87(2), I(4)–Bi(2)–I(4) = 89.58(3)	[97JA724] **RELWUA**

(**75**) Catena[(μ_2-iodo)iodophenyl(tetrahydrofuran)bismuthine)]

Ph	I	I
	(O)	*I*

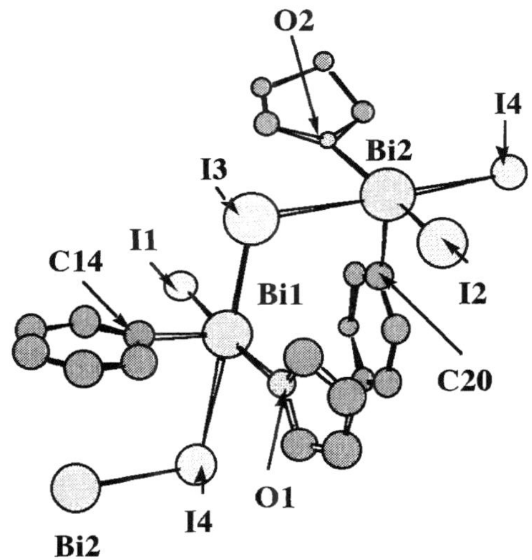

Bi–I(1) = 2.88, Bi–O(1) = 2.81, Bi–I(3) = 3.06, Bi–I(4) = 3.23, Bi–C(14) = 2.24, Bi(2)–I(2) = 2.89, Bi(2)–O(2) = 2.81, Bi(2)–I(3) = 3.21, Bi(2)–I(4) = 3.08, Bi(2)–C(20) = 2.24 I(1)–Bi–O(1) = 175.0, I(1)–Bi–I(3) = 94.8, I(1)–Bi–I(4) = 91.1, I(1)–Bi–C(14) = 94.8, O(1)–Bi–I(3) = 85.9, O(1)–Bi–I(4) = 88.5, O(1)–Bi–C(14) = 90.1, I(3)–Bi–I(4) = 172.8, I(3)–Bi–C(14) = 88.6, I(4)–Bi–C(14) = 86.7, I(2)–Bi(2)–O(2) = 174.9, I(2)–Bi(2)–I(3) = 93.4, I(2)–Bi(2)–I(4) = 93.9, I(2)–Bi(2)–C(20) = 96.7, O(2)–Bi(2)–I(3) = 83.0, O(2)–Bi(2)–I(4) = 89.8, O(2)–Bi(2)–C(20) = 86.9, I(3)–Bi(2)–I(4) = 171.9, I(3)–Bi(2)–C(20) = 87.1, I(4)–Bi(2)–C(20) = 88.6

[92JCS(D)1967]
KUDRAC
at −33°C

TABLE 6.5 (*Continued*)

Compound			Structure	Structural parameters bond distances (Å) and bond angles (°)[b]	[Reference], **CCDB code**, and *notes*
R^1	R^2	E^3			
(**76**) Bis(tetraethylammonium) bis[(μ_2-iodo)diiodo(phenyl)-bismuthate] diethyl ether solvate			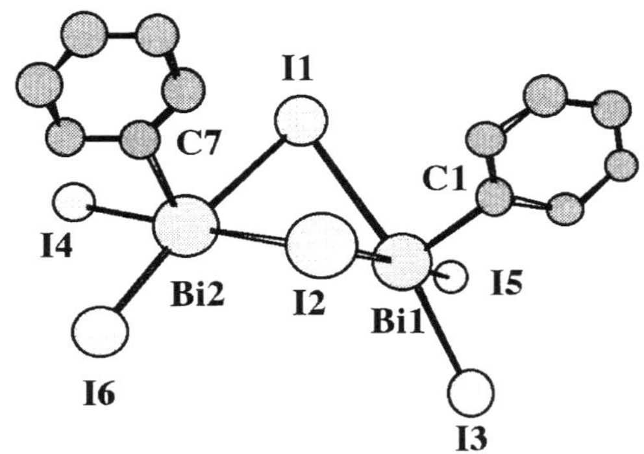	Bi–I(1) = 3.33, Bi–C(1) = 2.26, Bi–I(2) = 3.29, Bi–I(3) = 2.95, Bi–I(5) = 2.95, I(1)–Bi(2) = 3.29, Bi(2)–I(2) = 3.26, Bi(2)–I(4) = 2.97, Bi(2)–I(6) = 2.95, Bi(2)–C(7) = 2.27 I(1)–Bi–C(1) = 90.5, I(1)–Bi–I(2) = 80.3, I(1)–Bi–I(3) = 170.9, I(1)–Bi–I(5) = 96.1, C(1)–Bi–I(2) = 92.3, C(1)–Bi–I(3) = 94.3, C(1)–Bi–I(5) = 94.5, I(2)–Bi–I(3) = 91.8, I(2)–Bi–I(5) = 172.3, I(3)–Bi–I(5) = 91.1, I(1)–Bi(2)–I(2) = 81.4, I(1)–Bi(2)–I(4) = 92.8, I(1)–Bi(2)–I(6) = 171.8, I(1)–Bi(2)–C(7) = 92.3, I(2)–Bi(2)–I(4) = 173.4, I(2)–Bi(2)–I(6) = 91.4,	[92JCS(D)1967] **KUDREG**
Ph	I	I			
	I	*I*			

		I(2)–Bi(2)–C(7) = 91.5, I(4)–Bi(2)–I(6) = 94.2, I(4)–Bi(2)–C(7) = 91.8, I(6)–Bi(2)–C(7) = 91.6	
(**77**) 1,2-Bis{2,4,6-tris[bis-(trimethylsilyl)methyl]phenyl}-dibismuthene Ar Bi *	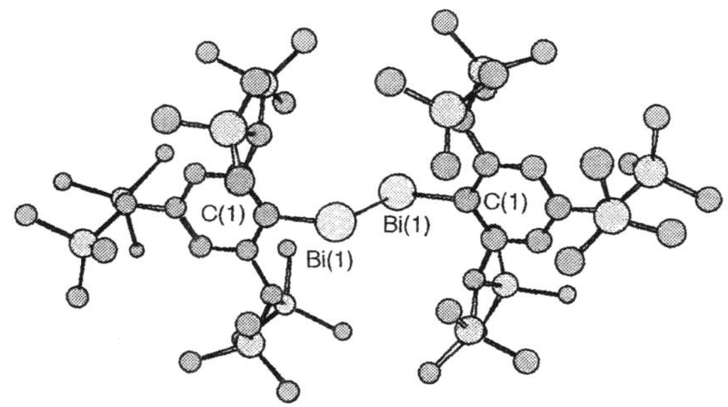 	Bi(1)–C(1) = 2.290(7), Bi(1)–Bi(1) = 2.8206(8) C(1)–Bi(1)–Bi(1) = 100.5(2)	[97SCI(277)78] **RETTIT**
(**78**) Tris(dimethylamino)-bismuthine Me_2N Me_2N Me_2N	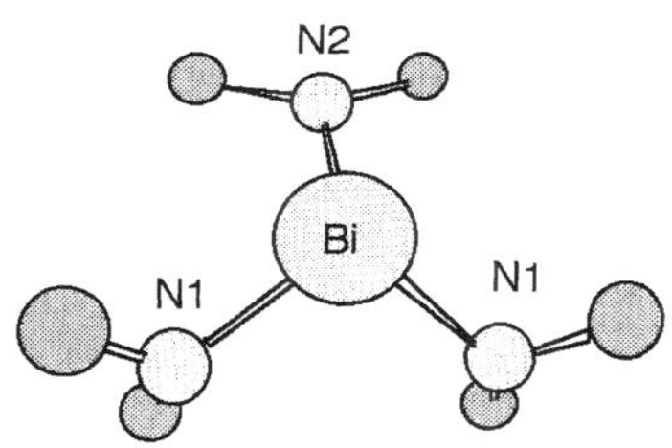 	Bi–N(1) = 2.19, Bi–N(1) = 2.19, Bi–N(2) = 2.18 N(1)–Bi–N(1) = 96.2, N(1)–Bi–N(2) = 98.3	[91IC4680] **KOMPUX** *at −33°C*

TABLE 6.5 (*Continued*)

Compound R[1] R[2] R[3]	Structure	Structural parameters bond distances (Å) and bond angles (°)[b]	[Reference], **CCDB code**, and *notes*
(**79**) *N,N'*-Di-*t*-butyl-*N,N'*-bis[(*N,N'*-di-*t*-butyldimethyl-silanediamino-*N,N'*)bismuthyl]-dimethylsilanediamine (N) (N) (N)	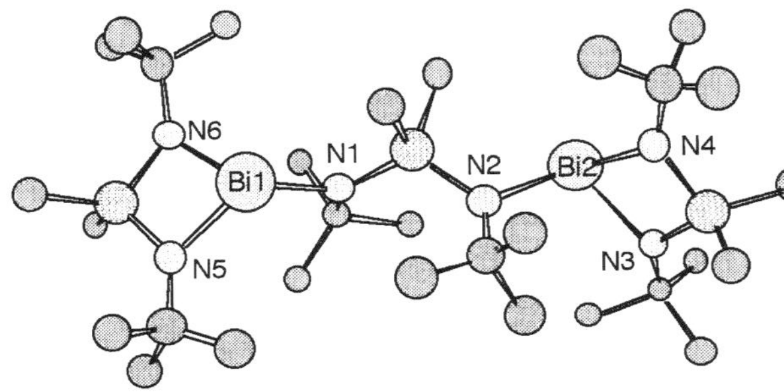	Bi(1)–N(1) = 2.202(8), Bi(1)–N(5) = 2.163(8), Bi(1)–N(6) = 2.148(9), Bi(2)–N(2) = 2.207(8), Bi(2)–N(3) = 2.179(8), Bi(2)–N(4) = 2.169(9) N(1)–Bi(1)–N(5) = 110.5(3), N(1)–Bi(1)–N(6) = 104.8(3), N(5)–Bi(1)–N(6) = 70.6(3), N(2)–Bi(2)–N(3) = 109.9(3), N(2)–Bi(2)–N(4) = 105.5(3), N(3)–Bi(2)–N(4) = 70.9(3)	[88ZAAC(557)7] **KEBSOZ**
(**80**) Catena[μ_2-chloro(*N,N'*-di-*t*-butyldimethylsilanediamino-*N,N'*)bismuthine] (N) (N) Cl *Cl*	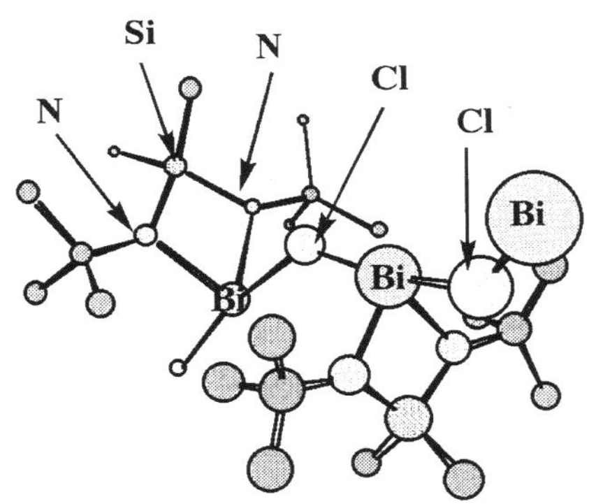	Bi–Cl = 2.748(4), Bi–N = 2.124(9) Si...Bi–Cl = 98.7(1), N–Bi–Cl = 97.8(3), N–Bi–N' = 71.8(4)	[88ZAAC(557)7] **KEBSIT**

(**81**) (*N,N*′-Di-*t*-butyldimethylsilanediamino-*N,N*′)bismuthyl tetrachloroalanate

(N) (N) Cl

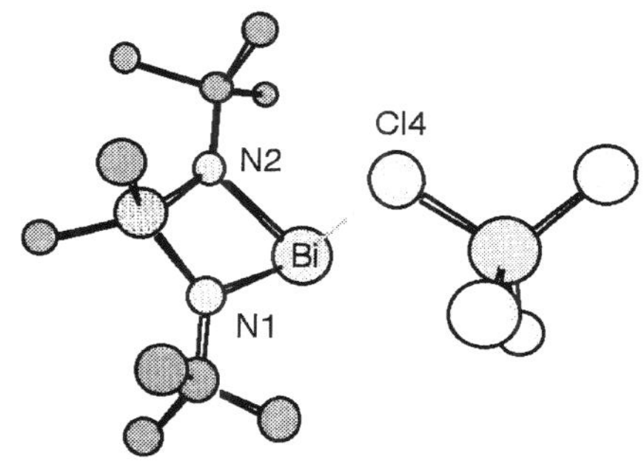

Bi–N(1) = 2.08(1), Bi–N(2) = 2.086(9), Bi–Cl(4) = 3.072(4) N(1)–Bi–N(2) = 72.9(4), N(1)–Bi–Cl(4) = 93.33, N(2)–Bi–Cl(4) = 92.33

[88ZAAC(559)73]
GAVVOO

(**82**) Catena[(bis(μ_2-methoxyethoxo)(methoxyethoxy)bismuthine]

MeO–	MeO–	MeO–
CH_2–	CH_2–	CH_2–
CH_2O	CH_2O	CH_2O

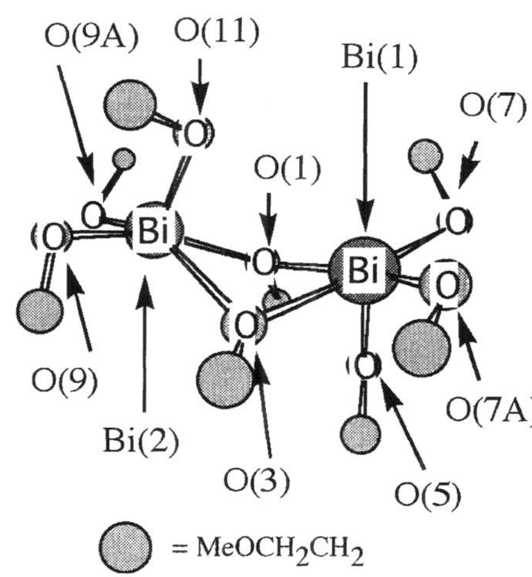

Bi–O(1) = 2.21(1), Bi–O(3) = 2.56(1), Bi–O(5) = 2.07(2), Bi–O(7) = 2.20(1), Bi–O(7A) = 2.54(1), Bi(2)–O(1) = 2.58(1), Bi(2)–O(3) = 2.21(1), Bi(2)–O(9) = 2.20(1), Bi(2)–O(9A) = 2.53(1), Bi(2)–O(11) = 2.11(1), Bi–Bi(2) = 3.641(3), Bi–Bi = 3.953(3), Bi(2)–Bi(2) = 3.975(3) O(1)–Bi–O(3) = 73.0, O(1)–Bi–O(5) = 89.4(4), O(1)–Bi–O(7) = 85.2(3), O(3)–Bi–O(7A) = 133.0(3), O(7)–Bi–O(7A) = 67.4(4),

[90CC301]
SELFEU
[90IC358]
SELFEU01

TABLE 6.5 (*Continued*)

Compound			Structure	Structural parameters bond distances (Å) and bond angles (°)[b]	[Reference], **CCDB code**, and *notes*
R^1	R^2	R^3			
				O(1)–Bi(2)–O(3) = 72.7(3), O(1)–Bi(2)–O(9A) = 129.9(3), O(1)–Bi(2)–O(11) = 72.2(4), O(3)–Bi(2)–O(9) = 88.4(4), O(9)–Bi(2)–O(9A) = 65.8(4)	
(**83**) Tris(2,6-dimethylphenoxy)-bismuthine			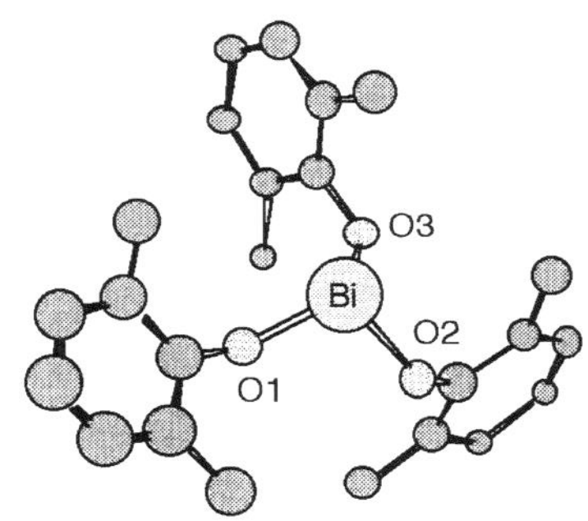	Bi–O(1) = 2.09, Bi–O(2) = 2.10, Bi–O(3) = 2.09 O(1)–Bi–O(2) = 94.2, O(1)–Bi–O(3) = 93.2, O(2)–Bi–O(3) = 88.9	[89CC1628] **KAXYUD**
2,6-Me_2–C_6H_3O	2,6-Me_2–C_6H_3O	2,6-Me_2–C_6H_3O			

(**84**) Bis[(μ_2-2,6-pyridinedicarboxylato-*N*,*O*,*O*′)(2,6-pyridinedicarboxylato-*N*,*O*)-(dimethylsulfoxide-*O*)-bismuthine] dihydrate

(*O*) (O) (O)
(*N*) (*N*)

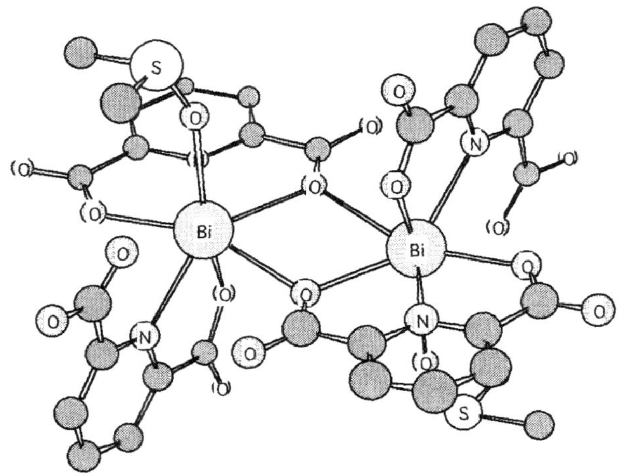

Bi–N(1) = 2.418(9), Bi–O(5) = 2.507(6), Bi–O(3) = 2.343(8), Bi–N(2) = 2.481(9), Bi–O(8) = 2.209(8), Bi–O(5′) = 2.578(4), Bi–O(2) = 2.61(1)O(2)–Bi–O(3) = 85.2(3), O(2)–Bi–O(5) = 82.6(3), O(2)–Bi–O(8) = 150.8(3), O(2)–Bi–O(9) = 77.0, O(2)–Bi–N(2) = 137.2(3), O(2)–Bi–N(1) = 77.2(3), O(3)–Bi–O(5) = 131.6(3), O(3)–Bi–O(8) = 91.2(3), O(3)–Bi–O(9) = 72.7(3), O(3)–Bi–N(2) = 72.7(3), O(3)–Bi–N(1) = 67.6(2), O(5)–Bi–O(5′) = 68.0(8), O(5)–Bi–O(8) = 78.3(3), O(5)–Bi–O(9) = 146.8(3), O(5)–Bi–N(2) = 139.1(3), O(5)–Bi–N(1) = 64.1(3), O(8)–Bi–O(9) = 129.4(3), O(8)–Bi–N(2) = 67.7(3), O(8)–Bi–N(1) = 74.7(3), O(9)–Bi–N(2) = 61.7(3), O(9)–Bi–N(1) = 133.9(3), N(2)–Bi–N(1) = 123.7(3)

[92MGMC217]
HAJBOJ

TABLE 6.5 (*Continued*)

Compound			Structure	Structural parameters bond distances (Å) and bond angles (°)[b]	[Reference], **CCDB code**, and *notes*
R^1	R^2	R^3			
(**85**) Tris-2,6-di-*t*-butyl-(4-aza-1,7-dioxahepta-2,5-diene-1,4,7-triyl)bismuthine			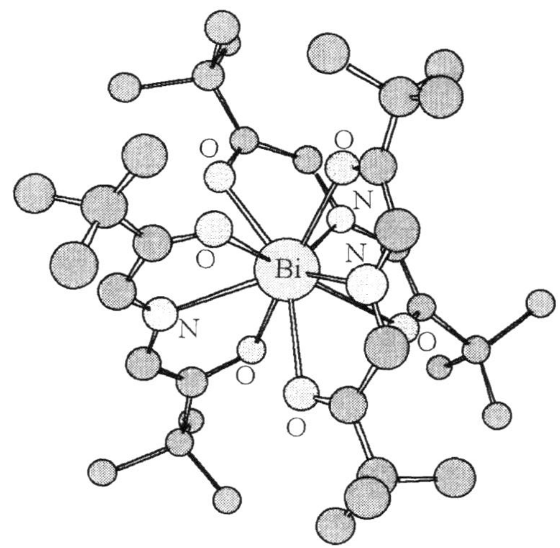	$Bi–O_{avg}$ = 2.51(3), Bi–Navg = 2.54(3) O(1)–Bi–O(2) = 128.0(0.3), O(1)–Bi–O(3) = 75.4(1.0), O(1)–Bi–O(4) = 149.8(1.5), O(2)–Bi–O(3) = 91.4(1.1), N–Bi–N = 120.0	[85JA3397] **CUYZIF** *at −100°C*
(O)	(O)	(O)			
(N)	*(N)*	*(N)*			

(**86**) Tris(triphenylsiloxy-*O*)bismuthine tetrahydrofuran solvate
(O) (O) (O)

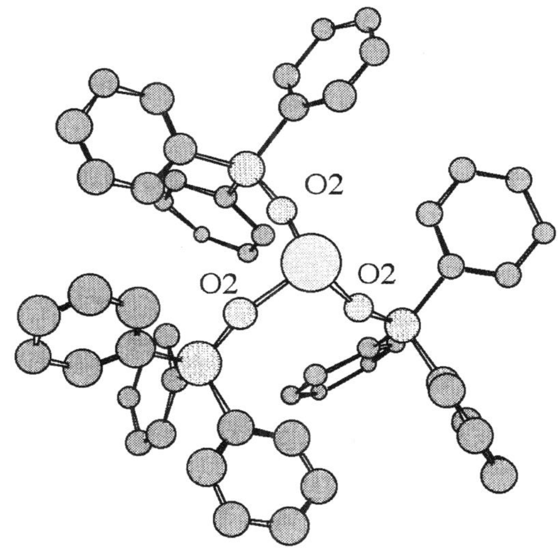

Bi–O(2) = 2.04
O(2)–Bi–O(2) = 93.0

[91POL437]
JIYLUY
at −93°C

(**87**) (1,2-Dimethoxyethane-*O,O′*)lithium bis(trimethylsilyl)-bismuthide
Me_3Si Me_3Si Li
Li

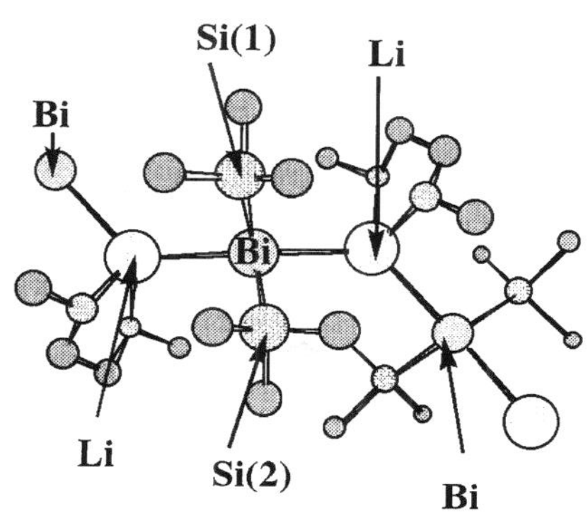

Bi–Si = 2.633(14),
Li–Bi = 2.92(3)
Si–Bi–Si′ = 91.8(4),
Li–Bi–Si = 106(1),
Li–Bi–Si′ = 96(1)

[83ZAAC(506)42]
CEBFOE
at −90°C

TABLE 6.5 (*Continued*)

Compound			Structure	Structural parameters bond distances (Å) and bond angles (°)[b]	[Reference], **CCDB code**, and *notes*
R[1]	R[2]	R[3]			
(**88**) Tetrakis(trimethylsilyl)-dibismuthine			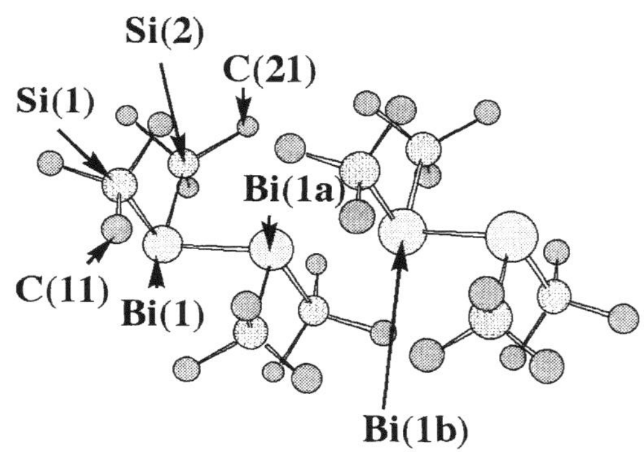	Bi–Bi(1a) (intramolecular) = 3.035(3), Bi(1a)–Bi(1b) (intermolecular) = 3.804(3), Bi–Si(1) = 2.679(13), Bi–Si(2) = 2.687(16) Bi–Bi(1a)–Si(1) = 97.4(5), Bi–Si(1)–C(11) = 107(2), Bi–Si(2)–C(21) = 103(2)	[83ZAAC(506)42] **GEPLOC**
Me_3Si	Me_3Si	(Bi)			

(**89**) Hexakis(thiourea-*S*)bismuth tetrachlorobis(thiourea-*S*) bismuthate/pentachloro(thiourea-*S*)bismuthate

(S)	(S)	(S)
(S)	*(S)*	*(S)*
Cl	Cl	Cl

$[Bi(thiourea)_6]^{3+}[Bi\{(thiourea)_{1.5}Cl_{1.5}\}Cl_3]^{1.5-}$

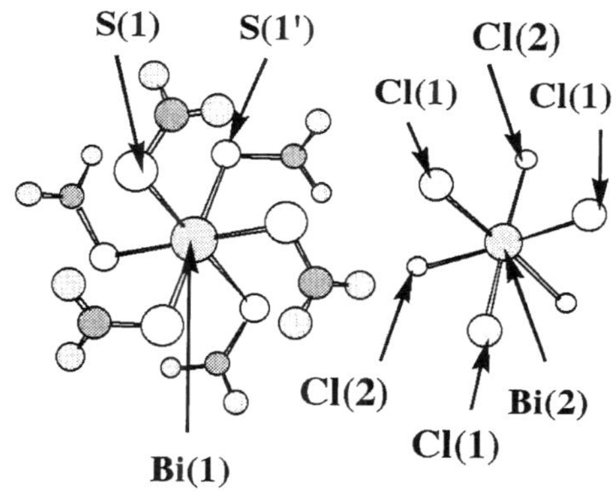

Cation part
Bi–S = 2.848(4)
S(1)–Bi–S(1′) = 91.5(1)
Anion part
Bi(2)–Cl(1) = 2.728(3), Bi–Cl(2) = 2.714(8) Cl(1)–Bi–S = 89.0, Cl(1)–Bi–Cl(2) = 89.0, Cl(1)–Bi–S(2′′)= 86.0(2), Cl(1)–Bi–Cl(2′′) = 86.0(2), Cl(1)–Bi–S(2′′′) = 177.0(3), Cl(1)–Bi–Cl(2′′′) = 177.0(3)
Both anions occupy the same position in the crystal.
Tetrachlorobis-(thiourea–*S*)bismuthate anion is not included.
2′′ means (1–y, x–y, z–1) and 2′′′ means (y–x, –x, z–1) equivalent position relative to the reference molecule at (x, y, z)

[77JCS(D)1141]
CLBITU

TABLE 6.5 (*Continued*)

Compound			Structure	Structural parameters bond distances (Å) and bond angles (°)[b]	[Reference], **CCDB code**, and *notes*
R^1	R^2	R^3			
(**90**) Tetraphenylarsonium bis(1,2-dicyanoethene-1,2-dithiolato-*S,S'*)bismuthate			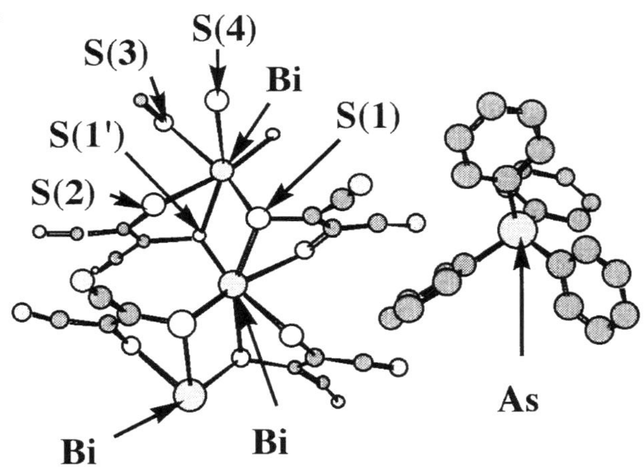	Bi–S(1) = 2.746(9), Bi–S(2) = 2.693(7), Bi–S(3) = 2.664(7), Bi–S(4) = 2.836(8), Bi–S(1′) = 3.328(10), Bi–S(4′′) = 3.182(10) S(1)–Bi–S(1′) = 103.0(2), S(1)–Bi–S(2) = 76.4(2), S(1)–Bi–S(3) = 89.4(2), S(1)–Bi–S(4′′), S(1′)–Bi–S(2) = 71.3(3), S(3)–Bi–S(1′) = 161.4(6), S(1)–Bi–S(4) = 159.8(6), S(2)–Bi–S(3) = 85.7(2), S(4)–Bi–S(4′′) = 101.5(6), S(4)–Bi–S(3) = 75.0(2), S(4)–Bi–S(2) = 89.5(2), S(4)–Bi–S(1′) = 85.6(1), S(3)–Bi–S(4′′) = 74.3(3), S(2)–Bi–S(4′′) = 161.2(6), S(1′)–Bi–S(4′′) = 132.8(3)	[83JCS(D)1067] **BUJPIF**
(S)	(S)	(S)			
(S)	*(S)*	*(S)*			

(**91**) Tris[2-(ethylamino)cyclopent-1-ene-1-dithiocarboxylato-*S*,*S*′]-bismuthine

R–	R–	R–
C(=S)	C(=S)	C(=S)
S–	S–	S–
(S)	*(S)*	*(S)*

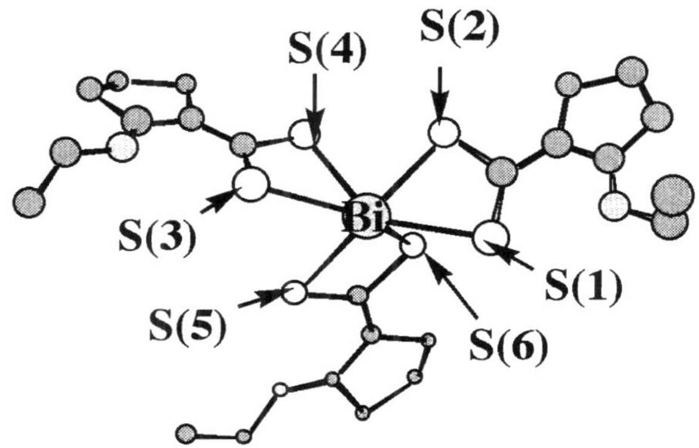

Bi–S(1) = 3.108(2), Bi–S(2) = 2.647(1), Bi–S(6) = 2.617(2), Bi–S(5) = 2.963(2), Bi–S(3) = 3.040(1), Bi–S(4) = 2.626(1) S(1)–Bi–S(2) = 61.9(4), S(5)–Bi–S(6) = 53.7(3), S(1)–Bi–S(3) = 129.8(5), S(1)–Bi–S(5) = 127.6(4), S(3)–Bi–S(4) = 63.0(5), S(3)–Bi–S(5) = 86.8(4), S(2)–Bi–S(4) = 79.4(5), S(2)–Bi–S(6) = 83.6(3), S(4)–Bi–S(6) = 93.0(4)

[87IC1453]
FEZFIZ
at −143°C

(**92**) Tris(diethyldithio-phosphinato-*S*,*S*′)bismuthine benzene solvate

Et_2P-	Et_2P-	Et_2P-
(=S)S–	(=S)S–	(=S)S–
(S)	*(S)*	*(S)*

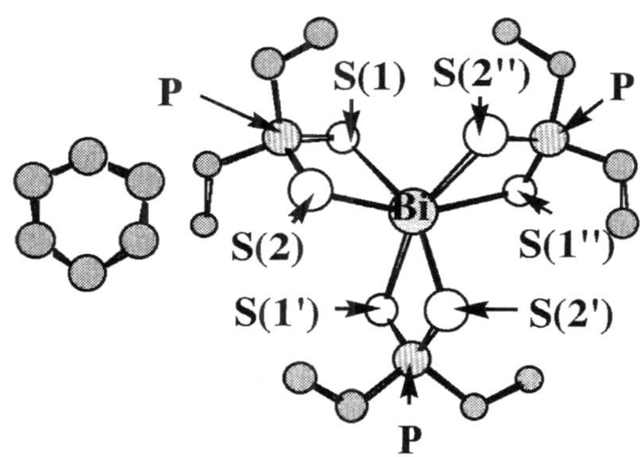

Bi–S(1) = 2.794(5), Bi–S(2) = 2.782(5) S(1)–Bi–S(2) = 73.7(1), S(1)–Bi–S(1′) = 93.0(2), S(1)–Bi–S(2′) = 158.8(2), S(1)–Bi–S(2″) = 103.8(2), S(2)–Bi–S(2′) = 93.2(2)

[87JCS(D)1257]
FESBOU

TABLE 6.5 (*Continued*)

Compound R[1]	R[2]	R[3]	Structure	Structural parameters bond distances (Å) and bond angles (°)[b]	[Reference], **CCDB code**, and *notes*
(**93**) Tris[μ_2-bis(pentafluorophenyl)germyl]dibismuthine (Ge)	(Ge)	(Ge)	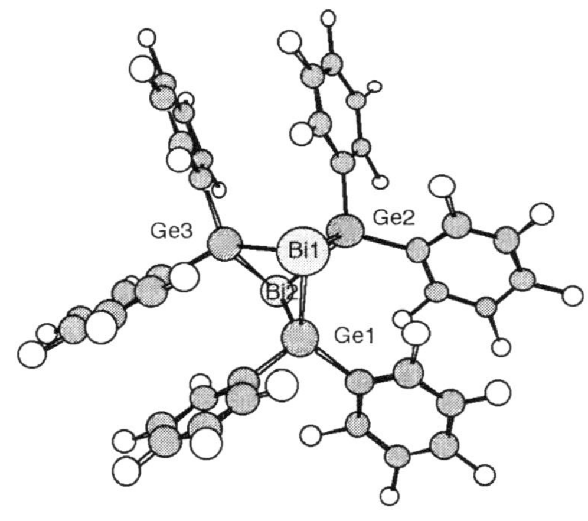	Ge(1)–Bi(1) = 2.746(1), Ge(1)–Bi(2) = 2.733(1), Ge(2)–Bi(1) = 2.749(1), Ge(2)–Bi(2) = 2.735(1), Ge(3)–Bi(1) = 2.749(1), Ge(3)–Bi(2) = 2.733(1) Ge(1)–Bi(1)–Ge(2) = 73.37(4), Ge(1)–Bi(1)–Ge(3) = 71.58(4), Ge(2)–Bi(1)–Ge(3) = 71.88(4), Ge(1)–Bi(2)–Ge(2) = 73.64(4), Ge(1)–Bi(2)–Ge(3) = 72.01(4), Ge(2)–Bi(2)–Ge(3) = 72.19(4)	[80JOM(199)205] **FGEBIA** *at −120°C*

[a] The entries of Table 6.5 are placed in the order determined by the ''*Cahn-Ingold-Prelog like*'' rule which follows: (1) substituents are listed in the order of increasing atomic number of the atom directly attached to the bismuth, (2) when two or more atoms attached to the bismuth are the same, the atomic number of the next further atom determines the order, (3) when the apparent valence is less than three the phantom atom is used to balance the valence three and is indicated by an asterisk (*). The phantom atom is given an atomic number of zero, (4) when the apparent valence is more than three the relevant extra atom is shown in *italic*, (5) all entries in Table 6.5 are listed in the order of increasing combined atomic number of all atoms directly attached to the bismuth. Thus, CCC is followed by CCO, and (6) when the atoms directly attached to the bismuth are the same, the atomic number of the next further atom determines the order. Thus, PhPhPh is followd by Ph(2-FC_6H_4)(2-FC_6H_4).

[b] The data with an error range in parentheses are taken from the original papers. The others were obtained by calculation based on the positional data without considering the error ranges.

TABLE 6.6
Crystallographic structure of pentavalent bismuth compounds[a]

Compound					Structure	Structural parameters Bond distances (Å) and bond angles (°)[b]	[Reference], **CCDB code** and *notes*
R^1	R^2	R^3	R^4	R^5			
(**1**) Pentamethylbismuth							
Me	Me	Me	Me	Me	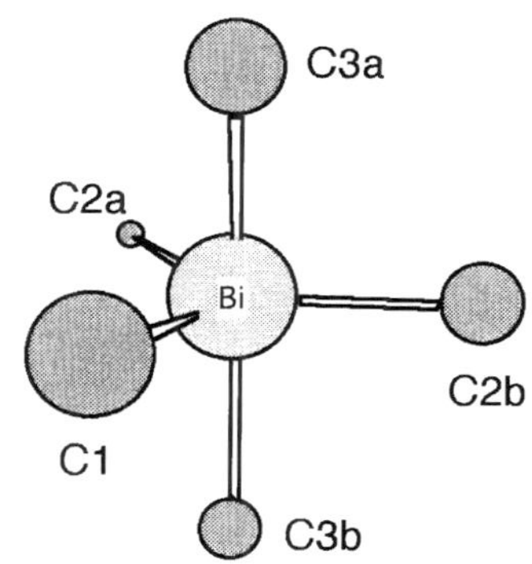 	Bi–C(1) = 2.30(2), Bi–C(2) = 2.27(1), Bi–C(3) = 2.282 C(1)–Bi–C(sa) = 119.3(5), C(1)–Bi–C(3a) = 89.9(5), C(2a)–Bi–C(2b) = 121.4(5), C(2a)–Bi–C(3a) = 90.4(6), C(2a)–Bi–C(3b) = 89.9(5), C(3a)–Bi–C(3b) = 179.8(8)	[94AG(E)976] **WEVHEK** *at −163°C*
(**2**) Pentaphenylbismuth							
Ph	Ph	Ph	Ph	Ph	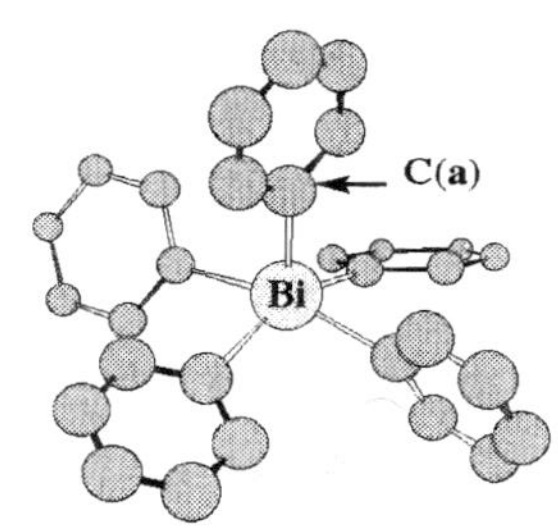 	Bi–C(a) = 2.221(9), Bi–C(e) = 2.329, 2.324, 2.327, 2.322 C(a)–Bi–C(e) = 100.1, 96.4, 103.8, 105.9(4), C(e)–Bi–C(e) = adjacent C(e): 88.1, 87.3, 88.3, 87.8; opposite C(e): 163.5, 150.3(4) C(a) and C(e) denote the axial and equatorial carbon atoms, respectively	[87AG(E)1180] **SAKZUZ** *at −96°C*

TABLE 6.6 (*continued*)

Compound					Structure	Structural parameters Bond distances (Å) and bond angles (°)	[Reference], **CCDB code** and *notes*
R^1	R^2	R^3	R^4	R^5			
(**3**) (2,2′-Biphenylene)triphenylbismuth					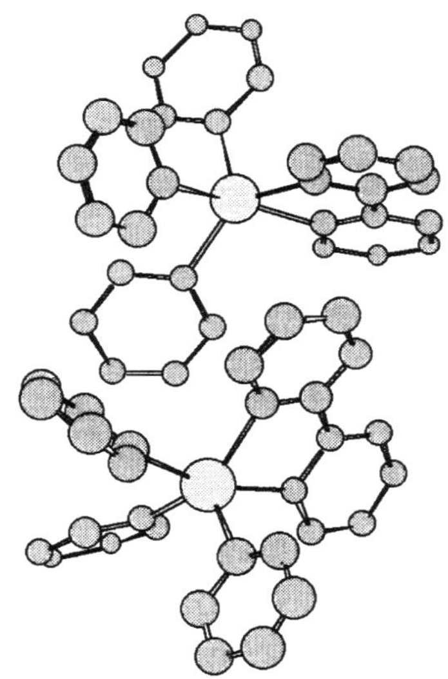	**Molecule 1** Bi(1)–C(111) = 2.14(2), Bi(1)–C(121)(B) = 2.32(2), Bi(1)–C(131)(B) = 2.32(2), Bi(1)–C(141) = 2.30(2), Bi(1)–C(151) = 2.23(2) C(121)–Bi(1)–C(141) = 166.8(6), C(131)–Bi(1)–C(151) = 138.0(6), C(121)–Bi(1)–C(131) = 76.4(6) **Molecule 2** Bi(2)–C(211) = 2.23(2), Bi(2)–C(221)(B) = 2.18(3), Bi(2)–C(231)(B) = 2.29(2), Bi(2)–C(241) = 2.28(2), Bi(2)–C(251) = 2.25(3) C(221)–Bi(2)–C(241) = 162.9(6), C(231)–Bi(2)–C(251) = 141.2(7), C(221)–Bi(2)–C(231) = 76.5(6) C(B) denotes the carbon atoms of a biphenylene ligand	[90CB761] **SETNIO** *at −138°C*
Ph	Ph	Ph	(C)	(C)			

(**4**) Bis(2-fluorophenyl)triphenylbismuth

Ph	Ph	Ph	2-F C_6H_4	2-F C_6H_4

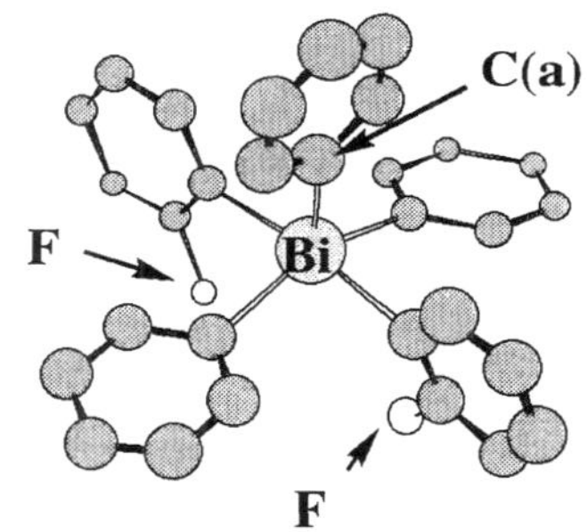

Bi–C(a) = 2.216(5),
Bi–C(basal) = 2.346(4),
2.285(7), 2.284(7), 2.363(5)
C(basal)–Bi–C(basal) =
169.2(2), 150.0(2)
C(a) denotes an axial carbon atom

[90AG(E)213]
SETYOF
at −40°C

(**5**) Bis(2,6-difluorophenyl)triphenylbismuth

Ph	Ph	Ph	2,6- $F_2C_6H_3$	2,6- $F_2C_6H_3$

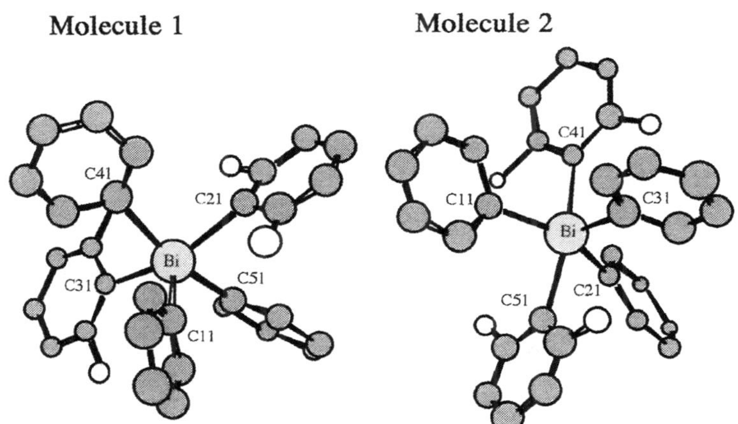

Molecule 1
Bi–C(11) = 2.26(3),
Bi–C(21) = 2.36(3),
Bi–C(31) = 2.38(2),
Bi–C(41) = 2.25(2),
Bi–C(51) = 2.27(3)
C(11)–Bi–C(21) = 106.2(9),
C(11)–Bi–C(31) = 106.3(9),
C(11)–Bi–C(41) = 101.3(8),
C(11)–Bi–C(51) = 99.6(9),
C(21)–Bi–C(31) = 147.4(8),
C(21)–Bi–C(41) = 88.0(8),
C(21)–Bi–C(51) = 83.8(8),
C(31)–Bi–C(41) = 88.5(7),
C(31)–Bi–C(51) = 88.1(8),
C(41)–Bi–C(51) = 158.9(9)
Molecule 2
Bi–C(11) = 2.25(2),
Bi–C(21) = 2.21(3),
Bi–C(31) = 2.23(2),
Bi–C(41) = 2.38(3),

[89CB803]
SARHAU
at −25°C

TABLE 6.6 (*continued*)

Compound					Structure	Structural parameters Bond distances (Å) and bond angles (°)	[Reference], **CCDB code** and *notes*
R^1	R^2	R^3	R^4	R^5			
						Bi–C(51) = 2.35(3) C(11)–Bi–C(21) = 148.3(7), C(11)–Bi–C(31) = 109.9(8), C(11)–Bi–C(41) = 85.2(8), C(11)–Bi–C(51) = 90.2(7), C(21)–Bi–C(31) = 101.8(7), C(21)–Bi–C(41) = 88.3(8), C(21)–Bi–C(51) = 88.4(8), C(31)–Bi–C(41) = 98.9(7), C(31)–Bi–C(51) = 95.8(8), C(41)–Bi–C(51) = 165.3(7)	
(**6**) Bis[(μ_2-chloro)bis(tetrahydrofuran)-lithium] bis[pentakis(4-methylphenyl)-bismuthate]					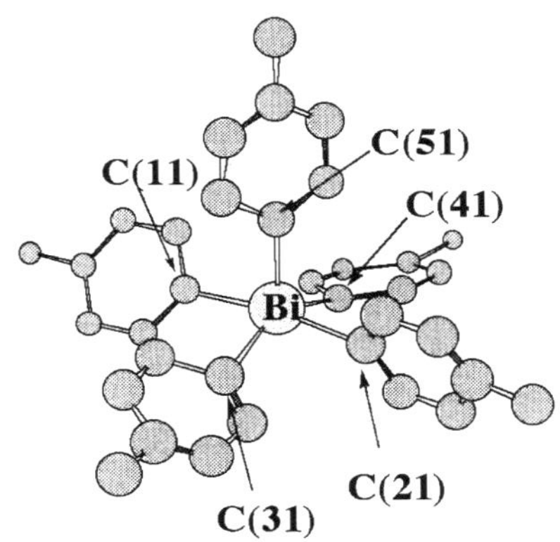	Bi–C(11) = 2.31(3), Bi–C(21) = 2.36(3), Bi–C(31) = 2.29(3), Bi–C(41) = 2.28(3), Bi–C(51) = 2.20(3) C(11)–Bi–C(21) = 171(1), C(31)–Bi–C(41) = 145(1)	[90CB761] **SETNAG** *at −145°C*
Tol	Tol	Tol	Tol	Tol			

Compound	Structure	Data	Reference
(**7**) Bis(2-fluorophenyl)tris(4-methylphenyl)bismuth Tol Tol Tol 2-F C_6H_4 2-F C_6H_4	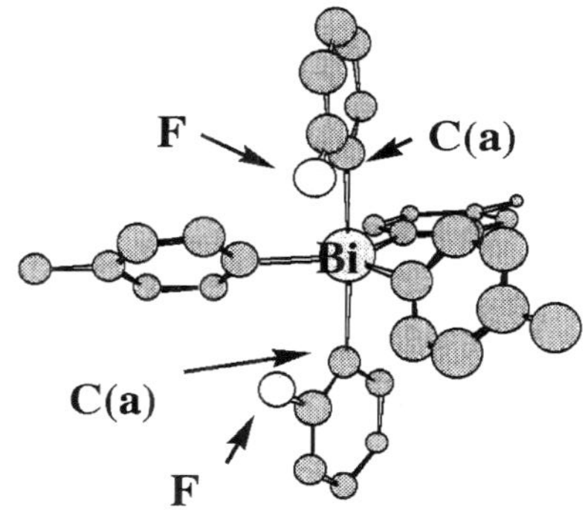	Bi–C(a) = 2.4233(16), 2.358(16), Bi–C(e) = 2.134(15), 2.201(13), 2.217(16) C(a)–Bi–C(a) = 176.3(5), C(e)–Bi–C(a) = 118.9(6), 130.6(6), 110.4(5) C(a) denotes the axial carbon atoms	[90AG(E)213] **SETYUL** *at −40°C*
(**8**) Bis(2,6-difluorophenyl)tris(4-methylphenyl)bismuth Tol Tol Tol 2,6-F_2 C_6H_3 2,6-F_2 C_6H_3	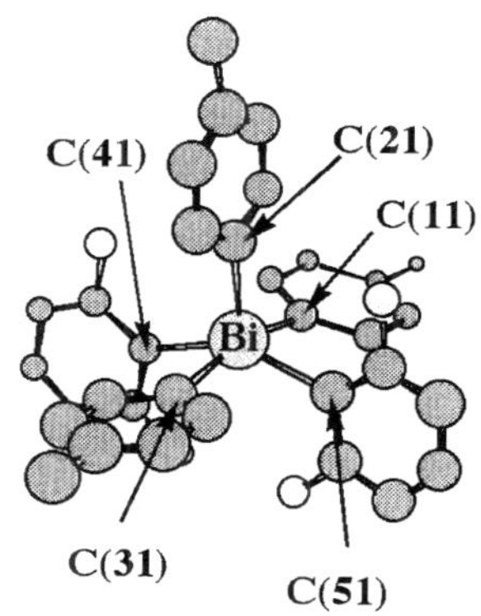	Bi–C(11) = 2.25(1), Bi–C(21) = 2.20(1), Bi–C(31) = 2.26(1), Bi–C(41)(F,F) = 2.34(1), Bi–C(51)(F,F) = 2.30(1) C(11)–Bi–C(31) = 163.1(4), C(41)–Bi–C(51) = 145.6(4)	[90CB761] **SETNEK** *at −30°C*

TABLE 6.6 (*continued*)

Compound					Structure	Structural parameters Bond distances (Å) and bond angles (°)	[Reference], **CCDB code** and *notes*
R^1	R^2	R^3	R^4	R^5			
(**9**) Tris(4-methylphenyl)bis(pentafluoro-phenyl)bismuth					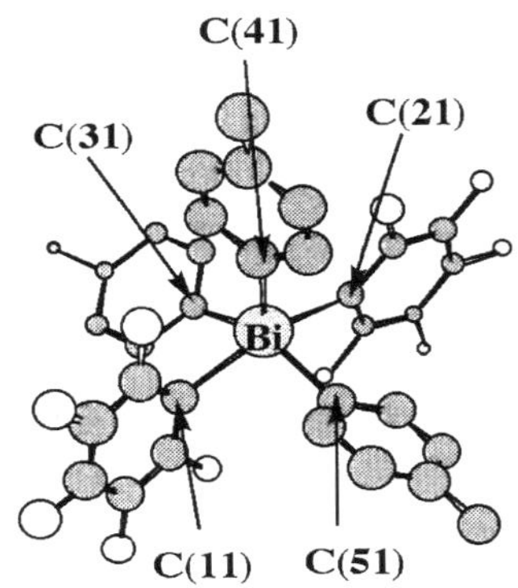	Bi–C(11) = 2.43(2), Bi–C(21) = 2.44(2), Bi–C(31) = 2.23(1), Bi–C(41) = 2.18(2), Bi–C(51) = 2.24(2) C(11)–Bi–C(21) = 162.7(4), C(11)–Bi–C(31) = 90.2(5), C(11)–Bi–C(41) = 97.9(5), C(11)–Bi–C(51) = 85.5(5), C(21)–Bi–C(31) = 85.8(4), C(21)–Bi–C(41) = 99.5(5), C(21)–Bi–C(51) = 88.6(5), C(31)–Bi–C(41) = 105.7(5), C(31)–Bi–C(51) = 146.7(4), C(41)–Bi–C(51) = 107.7(4)	[89CB803] **SARGOH**
Tol	Tol	Tol	C_6F_5	C_6F_5			
(**10**) Tris(4-fluorophenyl)bis(pentafluoro-phenyl)bismuth					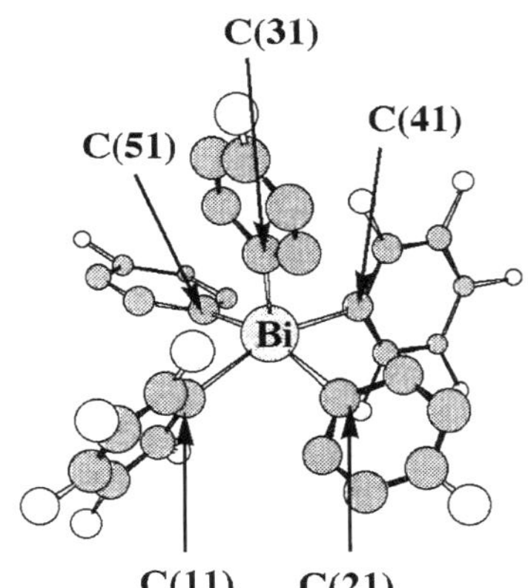	Bi–C(11) = 2.42(2), Bi–C(21) = 2.22(2), Bi–C(31) = 2.22(2), Bi–C(41) = 2.44(2), Bi–C(51) = 2.25(2) C(11)–Bi–C(21) = 99.4(5), C(11)–Bi–C(31) = 87.3(5), C(11)–Bi–C(41) = 157.9(5), C(11)–Bi–C(51) = 88.0(5), C(21)–Bi–C(31) = 103.9(6), C(21)–Bi–C(41) = 102.7(5), C(21)–Bi–C(51) = 103.3(5), C(31)–Bi–C(41) = 86.1(5), C(31)–Bi–C(51) = 152.8(6), C(41)–Bi–C(51) = 88.2(5)	[89CB803] **SARGUN** *at −25°C*
4-FC_6F_4	4-FC_6F_4	4-FC_6F_4	C_6F_5	C_6F_5			

(**11**) Tetrakis(tetrahydrofuran)lithium hexamethylbismuthate

Me Me Me Me Me *Me*

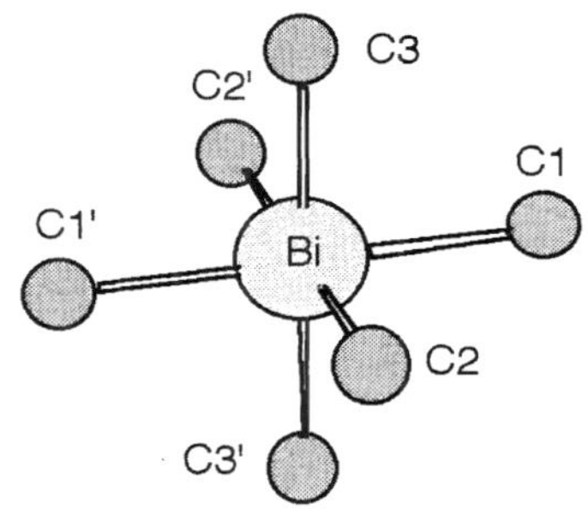

Bi–C(1) = 2.36(1),
Bi–C(2) = 2.33(1),
Bi–C(3) = 2.30(2)
C(1)–Bi–C(2) = 89.6(4),
C(1)–Bi–C(3) = 90.0,
C(2)–Bi–C(3) = 90.0,
C(3)–Bi–C(3′) = 180.0

[94AG(E)976]
WEVHAG
at −143°C

(**12**) Tetrakis(tetrahydrofuran)lithium hexaphenylbismuthate tetrahydrofuran solvate

Ph Ph Ph Ph Ph *Ph*

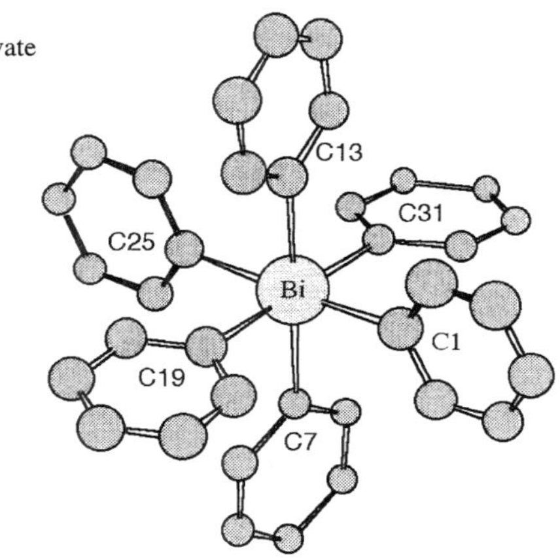

Bi–C(1) = 2.33,
Bi–C(7) = 2.36,
Bi–C(19) = 2.33,
Bi–C(25) = 2.36,
Bi–C(13) = 2.33,
Bi–C(31) = 2.35
C(1)–Bi–C(7) = 92.31,
C(1)–Bi–C(19) = 90.49,
C(1)–Bi–C(25) = 178.42,
C(1)–Bi–C(13) = 89.76,
C(1)–Bi–C(31) = 89.90,
C(7)–Bi–C(19) = 89.62,
C(7)–Bi–C(25) = 88.71,
C(7)–Bi–C(13) = 177.93,
C(7)–Bi–C(31) = 87.82,
C(19)–Bi–C(25) = 92.03,
C(19)–Bi–C(13) = 90.53,
C(19)–Bi–C(31) = 177.43,
C(25)–Bi–C(13) = 89.22,
C(25)–Bi–C(31) = 87.63,
C(13)–Bi–C(31) = 92.03

[93IC3948]
HAKVEU
at −145°C

TABLE 6.6 (*continued*)

Compound					Structure	Structural parameters Bond distances (Å) and bond angles (°)	[Reference], **CCDB code** and *notes*
R^1	R^2	R^3	R^4	R^5			
(**13**) Lithium hexapherylbismuthate diethyl etherate					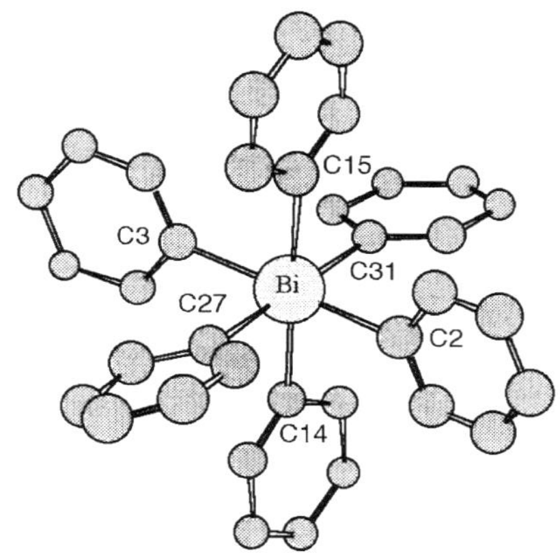	Bi–C(2) = 2.35, Bi–C(3) = 2.35, Bi–C(14) = 2.36, Bi–C(15) = 2.36, Bi–C(26) = 2.37, Bi–C(27) = 2.37 C(2)–Bi–C(3) = 180.0, C(2)–Bi–C(14) = 89.14, C(2)–Bi–C(15) = 90.86, C(2)–Bi–C(26) = 91.53, C(2)–Bi–C(27) = 88.47, C(3)–Bi–C(14) = 90.86, C(3)–Bi–C(15) = 89.14, C(3)–Bi–C(26) = 88.47, C(3)–Bi–C(27) = 91.53, C(14)–Bi–C(15) = 180.0, C(14)–Bi–C(26) = 90.25, C(14)–Bi–C(27) = 89.76, C(15)–Bi–C(26) = 89.76, C(15)–Bi–C(27) = 90.24	[93IC3948] **HAKVIY** *form 1*
Ph	Ph	Ph	Ph	Ph *Ph*			

(**14**) Pentaphenylbismuth pyridine solvate

Ph Ph Ph Ph Ph
Py

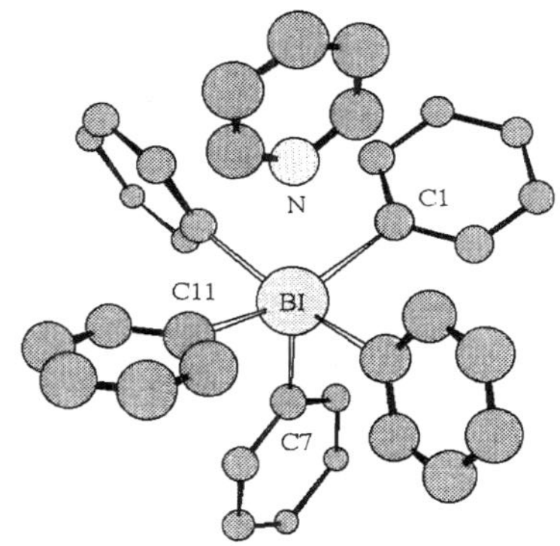

Bi–C(1) = 2.33,
Bi–C(1) = 2.33,
Bi–C(7) = 2.27,
Bi–C(11) = 2.34,
Bi–C(11) = 2.34,
Bi–N = 3.20
C(1)–Bi–C(1) = 169.9,
C(1)–Bi–C(7) = 95.0,
C(1)–Bi–C(11) = 89.0,
C(1)–Bi–C(11) = 88.8,
C(1)–Bi–C(7) = 95.0,
C(1)–Bi–C(11) = 88.8,
C(1)–Bi–C(11) = 89.0,
C(7)–Bi–C(11) = 102.6,
C(7)–Bi–C(11) = 102.6,
C(11)–Bi–C(11) = 154.8

[93IC3948]
HAKVOE
at −140°C

(**15**) 1-Methyl-1,1-bis(4-methylphenyl)-3,3-bis(trifluoromethyl)-3*H*-2,1-benzoxabismole

Me Tol Tol Ar (O)

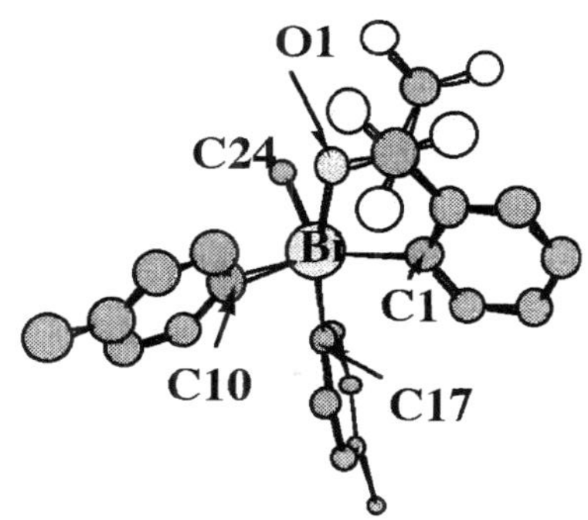

Bi–O(1) = 2.328(7),
Bi–C(1) = 2.19(1),
Bi–C(10) = 2.21(1),
Bi–C(17) = 2.25(1),
Bi–C(24) = 2.21(1)
O(1)–Bi–C(1) = 73.6(3),
O(1)–Bi–C(10) = 84.2(4)
O(1)–Bi–C(17) = 168.2(3),
O(1)–Bi–C(24) = 84.2(4),
C(1)–Bi–C(10) = 124.0(4),
C(1)–Bi–C(17) = 94.6(4),
C(1)–Bi–C(24) = 119.4(4),
C(10)–Bi–C(17) = 99.4(4),
C(10)–Bi–C(24) = 109.5(5),
C(17)–Bi–C(24) = 103.0(4)

[93OM1857]
LANGAI

TABLE 6.6 (*continued*)

Compound					Structure	Structural parameters Bond distances (Å) and bond angles (°)	[Reference], **CCDB code** and *notes*
R[1]	R[2]	R[3]	R[4]	R[5]			
(**16**) (3,3-Dimethyl-2-oxobutyl)-triphenylbismuthonium trifluoromethanesulfonate					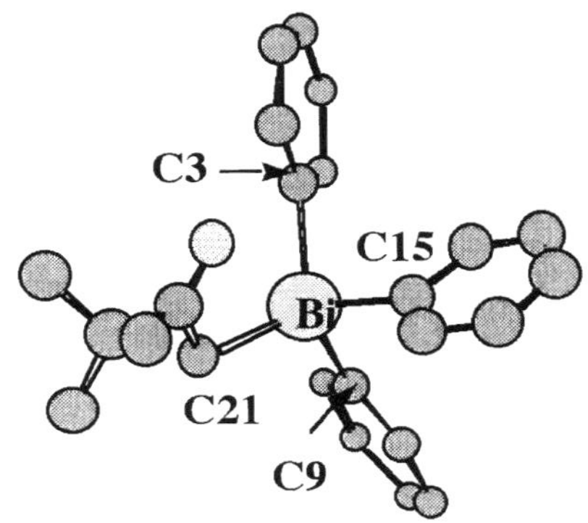	Bi–C(3) = 2.24(2), Bi–C(9) = 2.17(2), Bi–C(15) = 2.17(2), Bi–C(21) = 2.17(2), Bi–O = 2.93(1) C(3)–Bi–C(9) = 105.5(6), C(3)–Bi–C(15) = 102.6(6), C(3)–Bi–C(21) = 117.4(6), C(9)–Bi–C(15) = 105.4(6), C(9)–Bi–C(21) = 112.5(6), C(15)–Bi–C(21) = 112.2(6)	[94JCS(P1)1739] **LEVVOX**
Ph	Ph	Ph	(C)	(O)			
(**17**) (3,3-Dimethyl-2-oxobutyl)-triphenylbismuthonium tetrafluoroborate					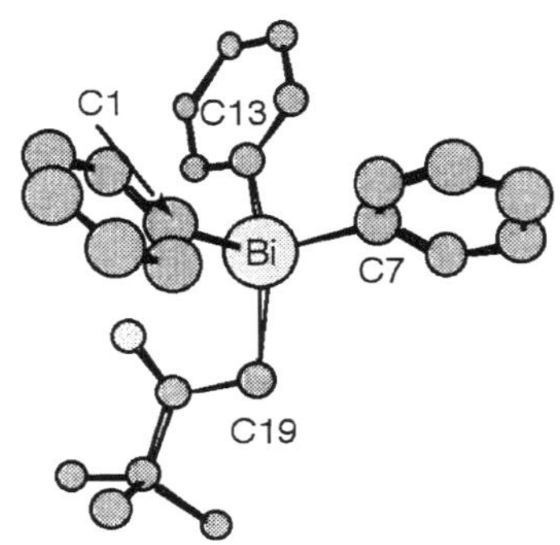	Bi–C(1) = 2.20(2), Bi–C(7) = 2.21(2), Bi–C(13) = 2.19(2), Bi–C(19) = 2.20(2), Bi–O = 2.90(1) C(1)–Bi–C(7) = 102.1(7), C(1)–Bi–C(13) = 107.0(7), C(1)–Bi–C(19) = 110.6(6), C(7)–Bi–C(13) = 105.6(8), C(7)–Bi–C(19) = 111.8(7), C(13)–Bi–C(19) = 118.4(7)	[93TL8457] [94JCS(P1)1739] **PIHFAN01**
Ph	Ph	Ph	(C)	(O)			

(**18**) 1,1,1-Tris(4-methylphenyl)-3,3-bis(trifluoromethyl)-3*H*-2,1-benzoxabismole

Tol	Tol	Tol	(C)	(O)

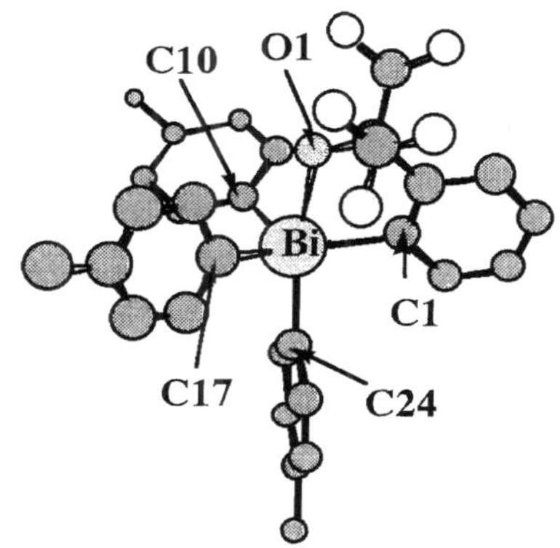

Bi–O(1) = 2.323(4),
Bi–C(1) = 2.220(6),
Bi–C(10) = 2.194(5),
Bi–C(17) = 2.209(7),
Bi–C(24) = 2.266(5)
O(1)–Bi–C(1) = 73.1(2),
O(1)–Bi–C(10) = 87.4(2),
O(1)–Bi–C(17) = 84.6(2),
O(1)–Bi–C(24) = 169.1(2),
C(1)–Bi–C(10) = 115.5(2),
C(1)–Bi–C(17) = 133.1(2),
C(1)–Bi–C(24) = 96.6(2),
C(10)–Bi–C(17) = 103.8(2),
C(10)–Bi–C(24) = 100.6(2),
C(17)–Bi–C(24) = 100.4(2)

[93OM1857]
LANFOV

(**19**) Tetraphenylbismuthonium *p*-toluenesulfonate

Ph	Ph	Ph	Ph	TsO

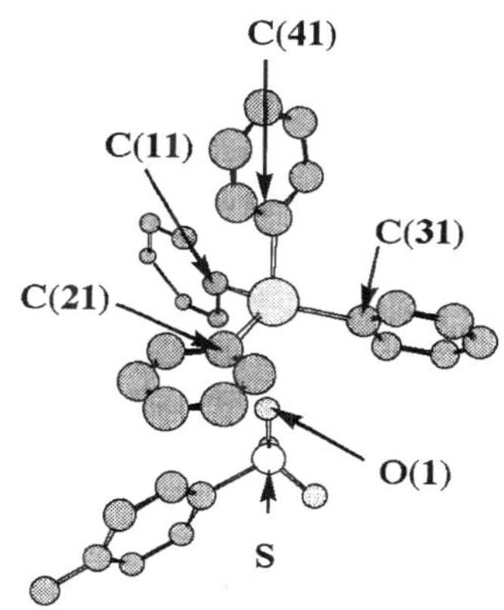

Bi–O(1) = 2.77,
Bi–C(11) = 2.24,
Bi–C(21) = 2.20,
Bi–C(31) = 2.17,
Bi–C(41) = 2.22
O(1)–Bi–C(11) = 84,
O(1)–Bi–C(21) = 75,
O(1)–Bi–C(31) = 79,
O(1)–Bi–C(41) = 177,
C(11)–Bi–C(21) = 113,
C(11)–Bi–C(31) = 123,
C(21)–Bi–C(31) = 114,
C(41)–Bi–C(11) = 98,
C(41)–Bi–C(21) = 102,
C(41)–Bi–C(31) = 103,
Bi–O–S = 166

[84HCA586]
CUNFIA

TABLE 6.6 (*continued*)

Compound R^1 R^2 R^3 R^4 R^5	Structure	Structural parameters Bond distances (Å) and bond angles (°)	[Reference], **CCDB code** and *notes*
(**20**) Cation part of tetraphenylbismuthonium diphenylbis(trifluoroacetato-*O*)bismuthate Ph Ph Ph Ph (O)	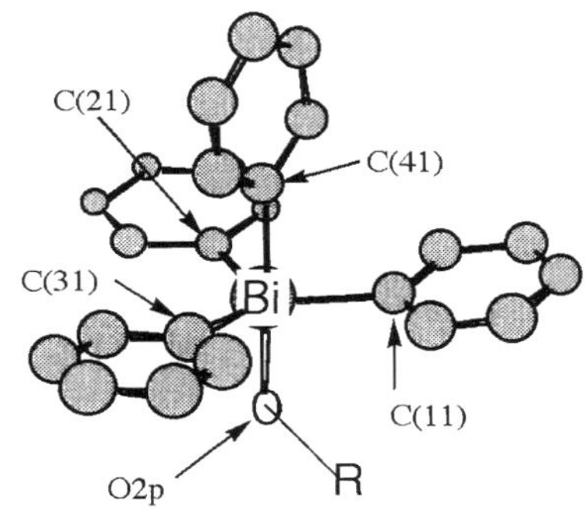	Bi–C(11) = 2.21(2), Bi–C(21) = 2.19(2), Bi–C(31) = 2.23(2), Bi–C(41) = 2.22(2), Bi–O2p = 3.11(2) C(11)–Bi–C(21) = 108(1), C(11)–Bi–C(21) = 108(1), C(11)–Bi–C(31) = 120, C(21)–Bi–C(31) = 114, C(41)–Bi–C(21) = 101, C(41)–Bi–C(31) = 103, C(41)–Bi–O2p = 177	[84HCA586] **CUNFOG**
(**21**) 1,1-Bis(4-methoxyphenyl)-1-(phenylethynyl)-3,3-bis(trifluoromethyl)-3*H*-2,1-benzoxabismole Tol Tol (C) (C) (O)	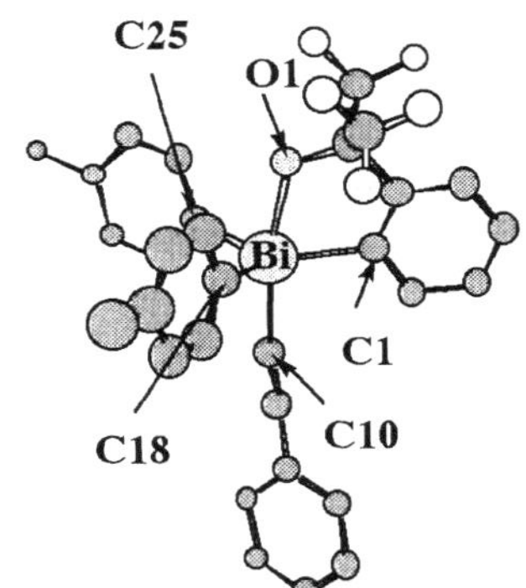	Bi–O(1) = 2.243(3), Bi–C(1) = 2.193(6), Bi–C(10) = 2.249(6), Bi–C(18) = 2.195(6), Bi–C(25) = 2.196(6) O(1)–Bi–C(1) = 75.9(2), O(1)–Bi–C(10) = 169.4(2), O(1)–Bi–C(18) = 89.8(2), O(1)–Bi–C(25) = 89.6(2), C(1)–Bi–C(10) = 93.6(2), C(1)–Bi–C(18) = 122.3(4), C(1)–Bi–C(25) = 126.8(2), C(10)–Bi–C(18) = 95.6(2), C(18)–Bi–C(25) = 108.2(2)	[93OM1857] **LANFUB**

(**22**) 1-[2-(*N,N*-dimethylaminomethyl)-phenyl]-1,1-bis(4-methylphenyl)-3,3-bis(trifluoromethyl)-3*H*-2,1-benzoxabismole
Tol Tol Ar Ar (O) (*N*)

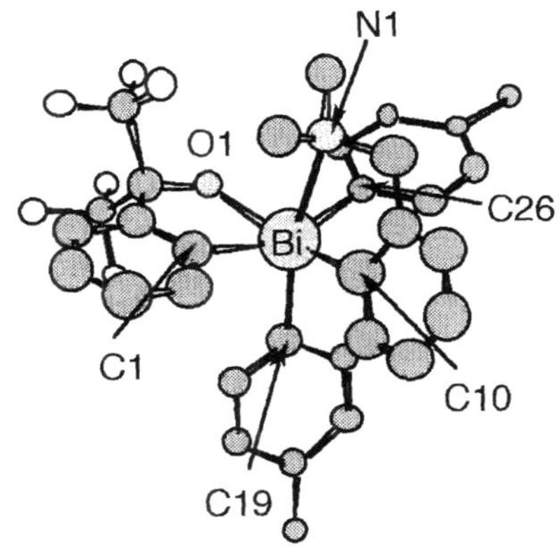

Bi–O(1) = 2.31,
Bi–N(1) = 2.85,
Bi–C(1) = 2.20,
Bi–C(10) = 2.25,
Bi–C(19) = 2.25,
Bi–C(26) = 2.22
O(1)–Bi–N(1) = 94.3,
O(1)–Bi–C(1) = 72.9,
O(1)–Bi–C(10) = 162.2,
O(1)–Bi–C(19) = 102.5,
O(1)–Bi–C(26) = 82.5,
N(1)–Bi–C(1) = 86.9,
N(1)–Bi–C(10) = 69.1,
N(1)–Bi–C(19) = 163.2,
N(1)–Bi–C(26) = 80.5,
C(1)–Bi–C(10) = 99.0,
C(1)–Bi–C(19) = 98.6,
C(1)–Bi–C(26) = 151.4,
C(10)–Bi–C(19) = 94.3,
C(10)–Bi–C(26) = 100.4,
C(19)–Bi–C(26) = 100.9

[93OM3297]
PETLAB

TABLE 6.6 (*continued*)

Compound R[1]	R[2]	R[3]	R[4]	R[5]	Structure	Structural parameters Bond distances (Å) and bond angles (°)	[Reference], **CCDB code** and *notes*
(**23**) 1-[2-(*N,N*-Dimethylaminomethyl)phenyl]-1,1-bis[4-(trifluoromethyl)phenyl]-3,3-bis(trifluoromethyl)-3*H*-2,1-benzoxabismole					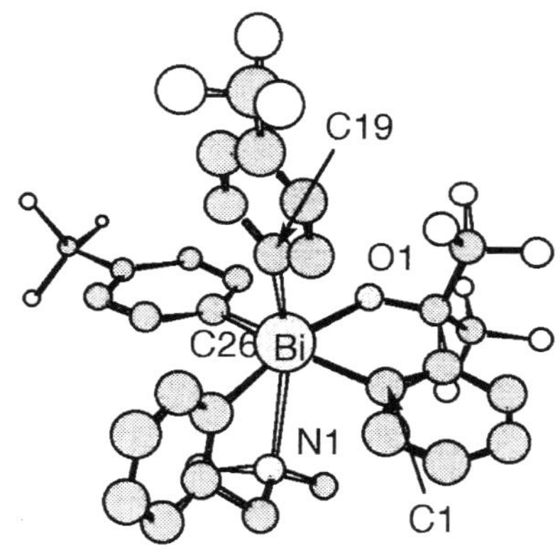	Bi–O(1) = 2.29, Bi–N(1) = 2.88, Bi–C(1) = 2.23, Bi–C(10) = 2.27, Bi–C(19) = 2.23, Bi–C(26) = 2.23 O(1)–Bi–N(1) = 97.6, O(1)–Bi–C(1) = 74.5, O(1)–Bi–C(10) = 165.5, O(1)–Bi–C(19) = 96.5, O(1)–Bi–C(26) = 82.6, N(1)–Bi–C(1) = 82.6, N(1)–Bi–C(10) = 69.0, N(1)–Bi–C(19) = 165.8, N(1)–Bi–C(26) = 80.5, C(1)–Bi–C(10) = 97.5, C(1)–Bi–C(19) = 102.3, C(1)–Bi–C(26) = 149.3, C(10)–Bi–C(19) = 97.1, C(10)–Bi–C(26) = 100.1, C(19)–Bi–C(26) = 100.2	[93OM3297] **PETLEF**
4-CF_3–C_6H_4	4-CF_3–C_6H_4	Ar	Ar	(O) (*N*)			

(**24**) 1-[2-(1,1,1,3,3,3-hexafluoro-2-oxo-2-propyl)phenyl-*C,O*]-1,1-bis[4-(trifluoromethyl)phenyl]-3,3-bis(trifluoromethyl)-3*H*-2,1-benzoxabismole

4-CF_3– C_6H_4	4-CF_3– C_6H_4	Ar	Ar	(O) *(O)*

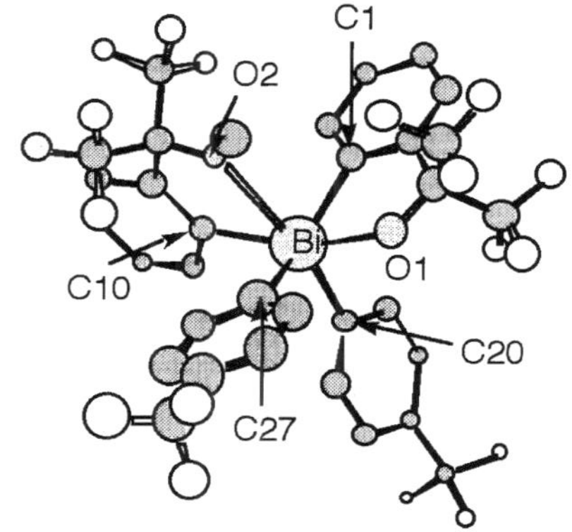

Bi–O(1) = 2.27,
Bi–C(1) = 2.22,
Bi–O(2) = 2.74,
Bi–C(10) = 2.29,
Bi –C(20) = 2.21,
Bi–C(27) = 2.22,
O(1)–Bi–C(1) = 74.2,
O(1)–Bi–O(2) = 99.9,
O(1)–Bi–C(10) = 162.6,
O(1)–Bi–C(20) = 94.9,
O(1)–Bi–C(27) = 83.0,
C(1)–Bi–O(2) = 82.7,
C(1)–Bi–C(10) = 97.1,
C(1)–Bi–C(20) = 103.7,
C(1)–Bi–C(27) = 148.2,
O(2)–Bi–C(10) = 63.6,
O(2)–Bi–C(20) = 165.1,
O(2)–Bi–C(27) = 79.5,
C(10)–Bi–C(20) = 101.9,
C(10)–Bi–C(27) = 98.3,
C(20)–Bi–C(27) = 100.2

[93OM3297]
PETKUU

(**25**)[2-(Isopropoxycarbonyl)ethyl]-triphenylbismuthonium trifluoromethanesulfonate

Ph	Ph	Ph	(C)	*

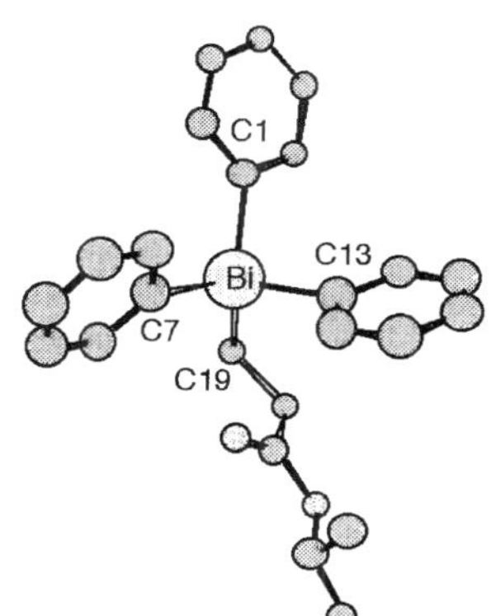

Bi–C(1) = 2.213(8),
Bi–C(7) = 2.206(9),
Bi–C(13) = 2.229(9),
Bi–C(19) = 2.235(9)
C(1)–Bi–C(7) = 108.9(3),
C(1)–Bi–C(13) = 101.0(3),
C(1)–Bi–C(19) = 115.3(3),
C(7)–Bi–C(13) = 105.5(3),
C(7)–Bi–C(19) = 115.5(3),
C(13)–Bi–C(19) = 109.1(3)

[95JCS(P1)2543]
ZEVVIF

TABLE 6.6 (*continued*)

Compound					Structure	Structural parameters Bond distances (Å) and bond angles (°)	[Reference], **CCDB code** and *notes*
R^1	R^2	R^3	R^4	R^5			
(**26**) 4,4-Dimethyl-2,6-dioxo-1-triphenylbismuthiocyclohexanide					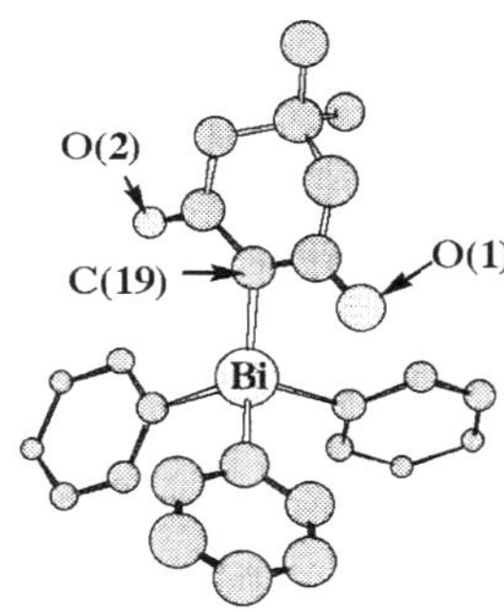	Bi–C(19) = 2.156(11), Bi–C_{ph} = 2.210, Bi...O(1) = 3.352(2), Bi...O(2) = 3.019(4), C_{ph}–Bi–C_{ph} = 101.6(4) ~ 103.1(4), C(19)–Bi–C_{ph} = 111.8(4)–119.6(4)	[90JCS(P1)3367] **JICKIP**
Ph	Ph	Ph	(C)=	*			
(**27**) Bis(phenylsulfonato-*O*)triphenylbismuth					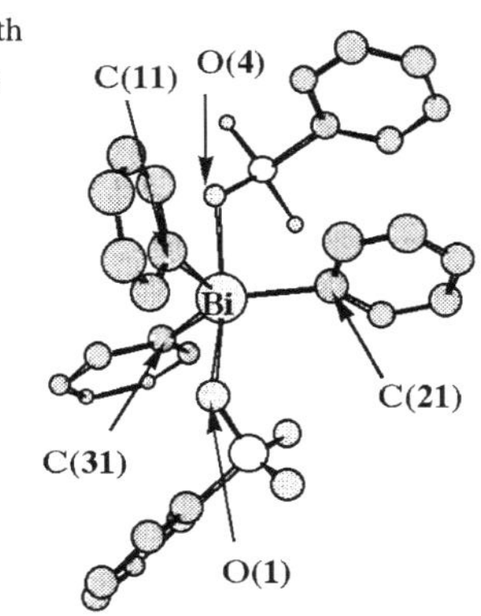	Bi–O(1) = 2.280(4), Bi–O(4) = 2.276(5), Bi–C(11) = 2.198(7), Bi–C(21) = 2.201(7), Bi–C(31) = 2.173(7) O(1)–Bi–O(4) = 171.8(2), O(1)–Bi–C(11) = 88.6(2), O(1)–Bi–C(21) = 94.9(2), O(1)–Bi–C(31) = 87.3(2), O(4)–Bi–C(11) = 86.7(2), O(4)–Bi–C(21) = 92.8(2), O(4)–Bi–C(31) = 89.0(2), C(11)–Bi–C(21) = 106.3(3), C(11)–Bi–C(31) = 117.8(3), C(21)–Bi–C(31) = 135.9(3)	[86ZAAC(539)110] **DUWNEO**
Ph	Ph	Ph	$PhSO_3$	$PhSO_3$			

(**28**) Bis(4-hydroxybutanoato-*O*)-triphenylbismuth tetrahydrofuran solvate

Ph Ph Ph (O) (O)

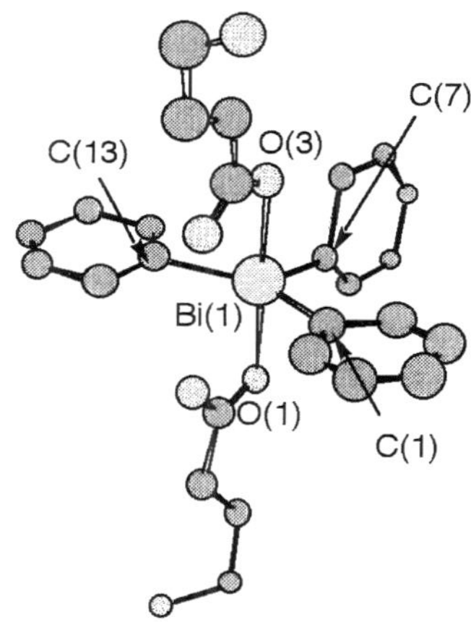

Bi(1)–O(1) = 2.28,
Bi(1)–C(1) = 2.20,
Bi(1)–O(3) = 2.32,
Bi(1)–C(7) = 2.15,
Bi(1)–C(13) = 2.24
O(1)–Bi(1)–C(1) = 91.4,
O(1)–Bi(1)–O(3) = 174.5,
O(1)–Bi(1)–C(7) = 79.7,
O(1)–Bi(1)–C(13) = 91.0,
C(1)–Bi(1)–O(3) = 91.3,
C(1)–Bi(1)–C(7) = 103.7,
C(1)–Bi(1)–C(13) = 150.0,
O(3)–Bi(1)–C(7) = 94.9,
O(3)–Bi(1)–C(13) = 89.0

[97JCS(D)2009]
NAXVEN

(**29**) Bis[3-(4-hydroxyphenyl)propanoato-*O*]triphenylbismuth tetrahydrofuran solvate

Ph Ph Ph (O) (O)

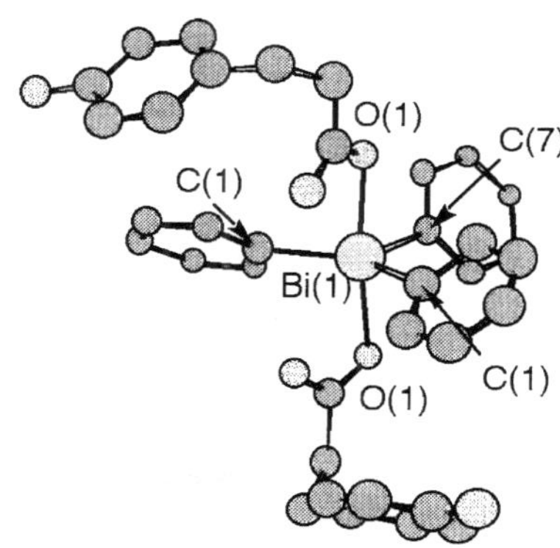

Bi(1)–C(1) = 2.23,
Bi(1)–C(7) = 2.22,
Bi(1)–O(1) = 2.25
C(1)–Bi(1)–C(1) = 145.7,
C(1)–Bi(1)–C(7) = 107.1,
C(1)–Bi(1)–O(1) = 91.9,
O(1)–Bi(1)–C(1) = 91.1,
O(1)–Bi(1)–C(7) = 85.0,
O(1)–Bi(1)–O(1) = 170.1

[97JCS(D)2009]
NAXVIR

TABLE 6.6 (*continued*)

Compound R^1 R^2 R^3 R^4 R^5	Structure	Structural parameters Bond distances (Å) and bond angles (°)	[Reference], **CCDB code** and *notes*
(**30**) Tris[4-(*N*,*N*-dimethylamino)phenyl]bis(4-hydroxybutanoato-*O*)bismuth 4-Me_2 N–C_6H_4 · 4-Me_2 N–C_6H_4 · 4-Me_2 N–C_6H_4 · (O) · (O)	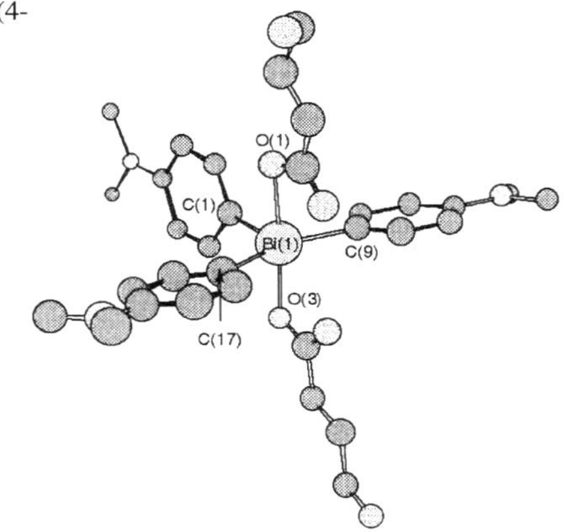	Bi(1)–O(1) = 2.32, Bi(1)–C(1) = 2.21, Bi(1)–O(3) = 2.28, Bi(1)–C(9) = 2.18, Bi(1)–C(17) = 2.18 O(1)–Bi(1)–C(1) = 87.8, O(1)–Bi(1)–O(3) = 175.7, O(1)–Bi(1)–C(9) = 91.0, O(1)–Bi(1)–C(17) = 90.2, C(1)–Bi(1)–O(3) = 88.0, C(1)–Bi(1)–C(9) = 103.8, C(1)–Bi(1)–C(17) = 103.3, O(3)–Bi(1)–C(9) = 90.8, O(3)–Bi(1)–C(17) = 90.1, C(9)–Bi(1)–C(17) = 152.9	[97JCS(D)2009] **NAXVAJ**
(**31**) Tris[4-(*N*,*N*-dimethylamino)-phenyl]bis(2-hydroxybutanoato-*O*)bismuth 4-Me_2 N–C_6H_4 · 4-Me_2 N–C_6H_4 · 4-Me_2 N–C_6H_4 · (O) · (O)	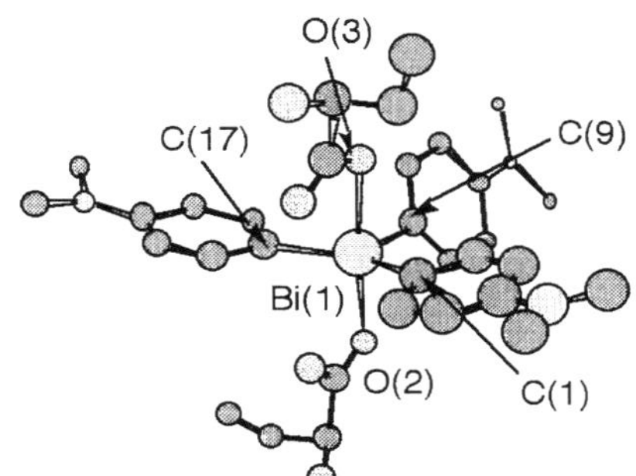	Bi(1)–C(1) = 2.15, Bi(1)–O(2) = 2.27, Bi(1)–O(3) = 2.32, Bi(1)–C(9) = 2.20, Bi(1)–C(17) = 2.16 C(1)–Bi(1)–C(17) = 144.8, C(1)–Bi(1)–C(9) = 107.5, C(1)–Bi(1)–O(2) = 96.2, C(1)–Bi(1)–O(3) = 86.0, C(9)–Bi(1)–C(17) = 107.4, O(2)–Bi(1)–C(17) = 91.4, O(2)–Bi(1)–C(9) = 84.6,	[97JCS(D)2009] **NAXVUD**

O(2)–Bi(1)–O(3) = 172.3,
O(3)–Bi(1)–C(17) = 90.9,
O(3)–Bi(1)–C(9) = 87.7

(**32**) Tris[4-(*N,N*-dimethylamino)-phenyl]bis(3-hydroxy-2-phenylpropanoato-*O*)bismuth

4-Me_2 N–C_6H_4	4-Me_2 N–C_6H_4	4-Me_2 N–C_6H_4	(O)	(O)

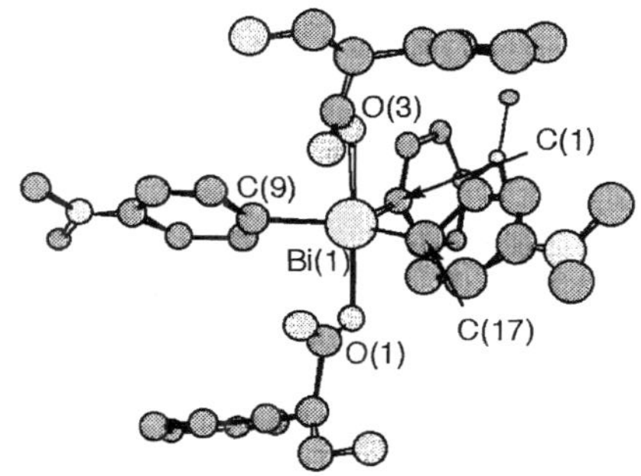

Bi(1)–O(1) = 2.31,
Bi(1)–C(1) = 2.14,
Bi(1)–O(3) = 2.20,
Bi(1)–C(9) = 2.22,
Bi(1)–C(17) = 2.30
C(1)–Bi(1)–C(17) = 105.3,
C(1)–Bi(1)–C(9) = 107.6,
C(1)–Bi(1)–O(3) = 87.3,
C(9)–Bi(1)–C(17) = 147.0,
O(1)–Bi(1)–C(1) = 89.8,
O(1)–Bi(1)–C(17) = 93.1,
O(1)–Bi(1)–C(9) = 89.1,
O(1)–Bi(1)–O(3) = 176.1,
O(3)–Bi(1)–C(17) = 90.1,
O(3)–Bi(1)–C(9) = 89.3

[97JCS(D)2009]
NAXVOX

(**33**) Bis(formato-*O*)triphenylbismuth

Ph	Ph	Ph	HC– (=O) O– *(O)*	HC– (=O) O– *(O)*

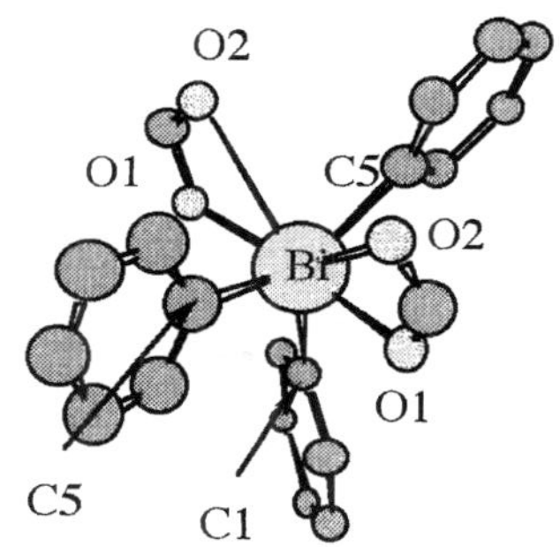

Bi–O(1) = 2.27,
Bi(1)–O(2) = 2.91,
Bi–C(1) = 2.23,
Bi–C(5) = 2.19
O(1)–Bi–C(1) = 85.8, O(1)–Bi–O(1) = 171.6, O(1)–Bi–C(5) = 90.9, O(1)–Bi–C(5) = 91.6, C(1)–Bi–O(1) = 85.8, C(1)–Bi–C(5) = 107.4, O(1)–Bi–C(5) = 91.6, O(1)–Bi–C(5) = 90.9,
C(5)–Bi–C(5) = 145.2

[93JCS(P1)2411]
LAVNUR

TABLE 6.6 (*continued*)

Compound					Structure	Structural parameters Bond distances (Å) and bond angles (°)	[Reference], **CCDB code** and *notes*
R^1	R^2	R^3	R^4	R^5			
(**34**) Bis(2-furoato-*O*)triphenylbismuth							
Ph	Ph	Ph	C_4H_3 O– C(=O) –O (*O*)	C_4H_3 O– C(=O) –O (*O*)	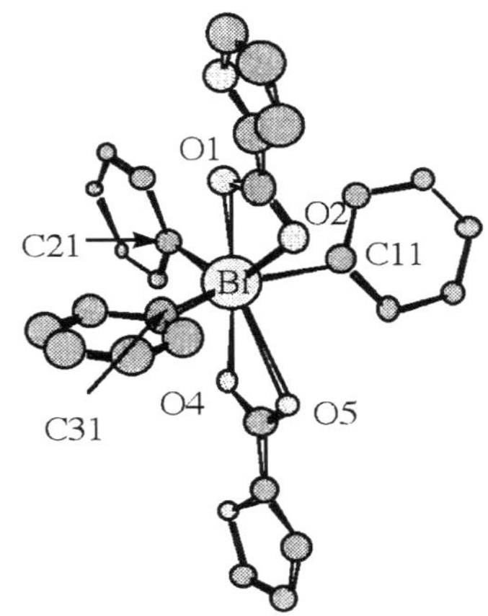	Bi–C(11) = 2.206(5), Bi–C(21) = 2.215(5), Bi–C(31) = 2.193(5), Bi–O(1) = 2.301(3), Bi–O(2) = 2.799(4), Bi–O(4) = 2.290(4), Bi–O(5) = 2.813(4) C(11)–Bi–C(21) = 104.73(2), C(11)–Bi–C(31) = 148.23(2), C(21)–Bi–C(31) = 106.79(2), O(1)–Bi–O(4) = 173.55(1), C(11)–Bi–O(1) = 90.66(2), C(11)–Bi–O(4) = 94.63(2), C(21)–Bi–O(4) = 86.91(2), C(31)–Bi–O(1) = 86.45(2), C(31)–Bi–O(4) = 90.97(2), O(1)–Bi–O(2) = 50.58(1), O(2)–Bi–O(5) = 84.40(1), O(4)–Bi–O(5) = 50.03(1)	[88AX(C)830] **GEMGOU**

(**35**) Bis(trifluoroacetato-*O*)triphenylbismuth

Ph	Ph	Ph	CF_3C– (=O)O *(O)*	CF_3C– (=O)–O *(O)*

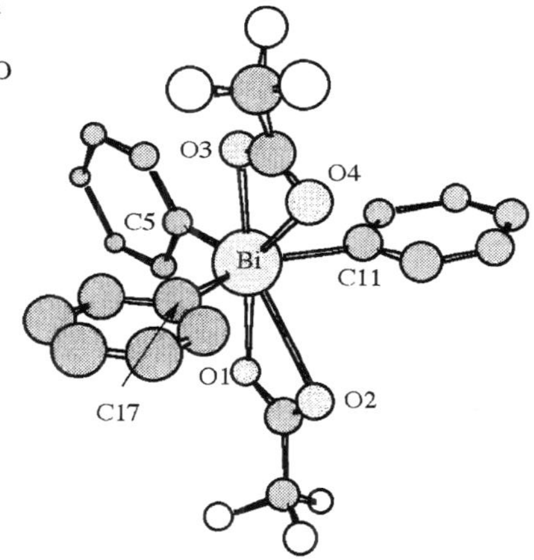

Bi–O(1) = 2.308(7),
Bi(1)–O(2) = 2.979,
Bi–O(3) = 2.309(7),
Bi(1)–O(4) = 2.981,
Bi–C(11) = 2.185(5),
Bi–C(21) = 2.138(6),
Bi–C(31) = 2.152(7)
O(1)–Bi–O(3) = 175.1(3),
O(1)–Bi–C(11) = 87.7(3),
O(1)–Bi–C(21) = 92.6(3),
O(1)–Bi–C(31) = 90.7(3),
O(3)–Bi–C(11) = 88.3(3),
O(3)–Bi–C(21) = 91.43(24),
O(3)–Bi–C(31) = 87.95(24),
C(11)–Bi–C(21) = 110.1(3),
C(11)–Bi–C(31) = 109.2(3),
C(21)–Bi–C(31) = 140.6(3)

[91JOM(419)283]
PABJEH

(**36**) Tris(4-methylphenyl)bis(tropolonato-*O*)bismuth

Tol	Tol	Tol	(O) *(O)*	(O) *(O)*

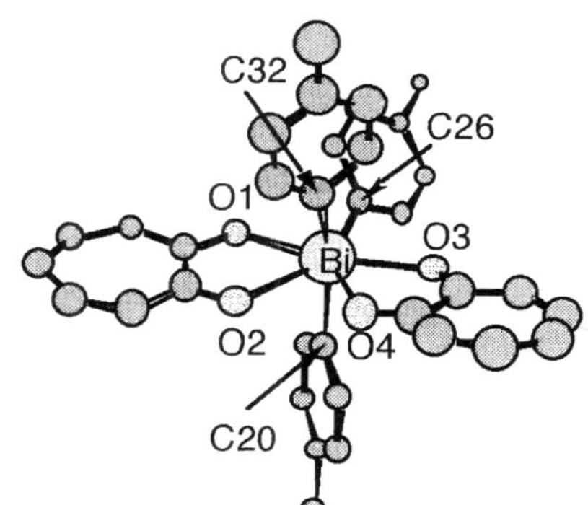

Bi–O(1) = 2.32,
Bi–O(2) = 2.50,
Bi–O(3) = 2.31,
Bi–O(4) = 2.46,
Bi–C(20) = 2.21,
Bi–C(26) = 2.16,
Bi–C(32) = 2.12
O(1)–Bi–O(2) = 67.6,
O(1)–Bi–O(3) = 156.7,
O(1)–Bi–O(4) = 137.4,
O(1)–Bi–C(20) = 90.3,
O(1)–Bi–C(26) = 79.4,
O(1)–Bi–C(32) = 93.6,
O(2)–Bi–O(3) = 134.7,
O(2)–Bi–O(4) = 70.0,

[96AG(E)67]
ZOGBIG

TABLE 6.6 (*continued*)

Compound					Structure	Structural parameters Bond distances (Å) and bond angles (°)	[Reference], **CCDB code** and *notes*
R^1	R^2	R^3	R^4	R^5			
						O(2)–Bi–C(20) = 79.9, O(2)–Bi–C(26) = 147.0, O(2)–Bi–C(32) = 82.4, O(3)–Bi–O(4) = 65.7, O(3)–Bi–C(20) = 88.4, O(3)–Bi–C(26) = 78.1, O(3)–Bi–C(32) = 96.0, O(4)–Bi–C(20) = 86.2, O(4)–Bi–C(26) = 142.7, O(4)–Bi–C(32) = 76.9, C(20)–Bi–C(26) = 101.7, C(20)–Bi–C(32) = 158.8, C(26)–Bi–C(32) = 99.5	

(**37**) Bis(5-nitrotropolonato-*O*)triphenyl-bismuth

Ph	Ph	Ph	(O)	(O)
			(O)	*(O)*

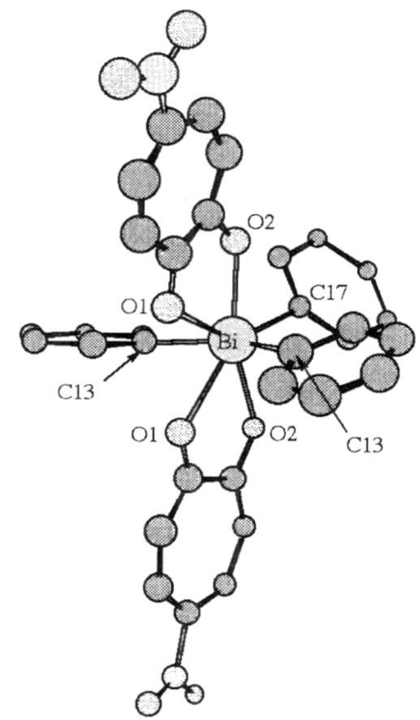

Bi–O(1) = 2.59,
Bi–O(1) = 2.59,
Bi–O(2) = 2.32,
Bi–C(13) = 2.14,
Bi–C(17) = 2.39
O(1)–Bi–O(1) = 72.3,
O(1)–Bi–O(2) = 64.7,
O(1)–Bi–O(2) = 137.0,
O(1)–Bi–C(13) = 76.5,
O(1)–Bi–C(13) = 83.2,
O(1)–Bi–C(17) = 143.8,
O(2)–Bi–O(2) = 158.2,
O(2)–Bi–C(13) = 87.9,
O(2)–Bi–C(13) = 96.8,
O(2)–Bi–C(17) = 79.1,
O(2)–Bi–C(13) = 87.9,
C(13)–Bi–C(13) = 154.8,
C(13)–Bi–C(17) = 102.6

[96AG(E)67]
ZOGBOM

TABLE 6.6 (*continued*)

Compound R^1 R^2 R^3 R^4 R^5	Structure	Structural parameters Bond distances (Å) and bond angles (°)	[Reference], **CCDB code** and *notes*
(**38**) 1-Chloro-1,1-bis(4-methylphenyl)-3,3-bis(trifluoromethyl)-3*H*-2,1-benzoxabismole Tol Tol Ar (O) Cl	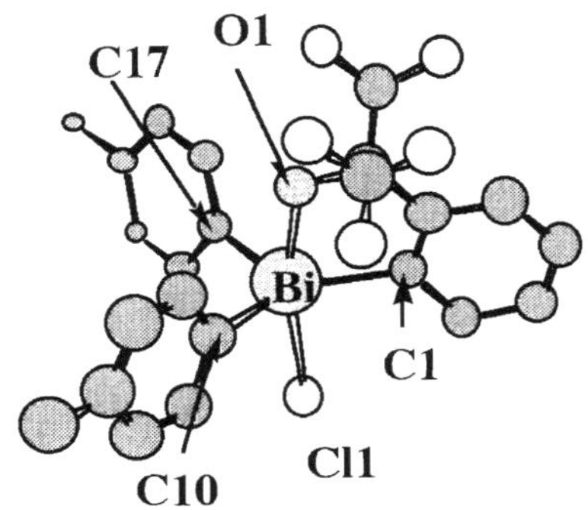	Bi–Cl(1) = 2.562(3), Bi–O(1) = 2.179(6), Bi–C(1) = 2.196(8), Bi–C(10) = 2.201(8), Bi–C(17) = 2.191(8) Cl(1)–Bi–O(1) = 170.0(2), Cl(1)–Bi–C(1) = 92.3(2), Cl(1)–Bi–C(10) = 94.4(2), Cl(1)–Bi–C(17) = 93.9(2), O(1)–Bi–C(1) = 77.7(2), O(1)–Bi–C(10) = 91.6(3), O(1)–Bi–C(17) = 91.5(2), C(1)–Bi–C(10) = 124.2(3), C(1)–Bi–C(17) = 124.8(3), C(10)–Bi–C(17) = 109.8(3)	[93OM1857] **LANGEM**
(**39**) 1-Chloro-1-(4-methylphenyl)-3,3-bis(trifluoromethyl)-1-(4-trifluoro-methylphenyl)-3*H*-2,1-benzoxabismole Tol 4-CF_3-C_6H_4 Ar (O) Cl	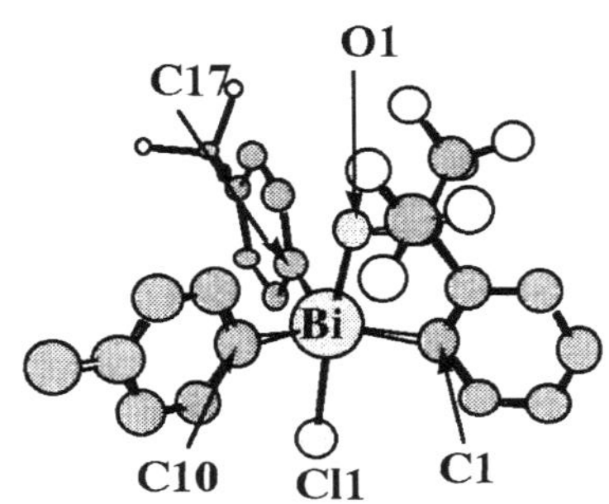	Bi–Cl(1) = 2.547(4), Bi–O(1) = 2.186(9), Bi–C(1) = 2.19(1), Bi–C(10) = 2.21(1), Bi–C(17) = 2.20(1) Cl(1)–Bi–O(1) = 171.2(3), Cl(1)–Bi–C(1) = 93.7(4), Cl(1)–Bi–C(10) = 94.8(3), Cl(1)–Bi–C(17) = 94.6(4), O(1)–Bi–C(1) = 77.5(4), O(1)–Bi–C(10) = 90.5(4), O(1)–Bi–C(17) = 90.0(4), C(1)–Bi–C(10) = 123.1(5), C(1)–Bi–C(17) = 124.4(5), C(10)–Bi–C(17) = 110.7(5)	[93OM1857] **LANHAJ**

(**40**) 1-Chloro-3,3-bis(trifluoromethyl)-1,1-bis(4-trifluoromethylphenyl)-3*H*-2,1-benzoxabismole

4-CF_3-C_6H_4	4-CF_3-C_6H_4	Ar	(O)	Cl

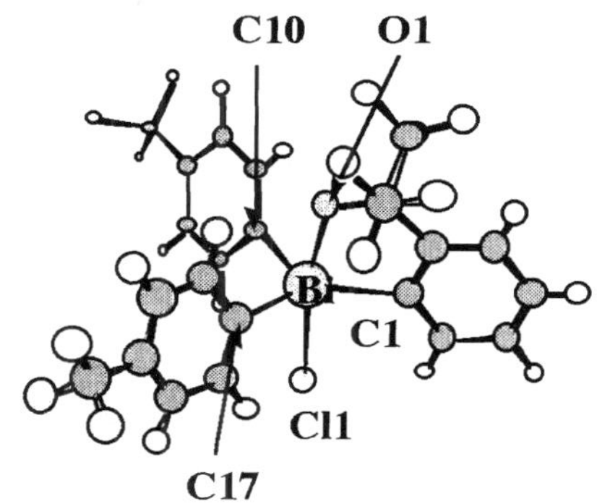

Bi–Cl(1) = 2.558(4),
Bi–O(1) = 2.17(1),
Bi–C(1) = 2.20(1),
Bi–C(10) = 2.22(1),
Bi–C(17) = 2.19(1)
Cl(1)–Bi–O(1) = 170.7(2),
Cl(1)–Bi–C(1) = 92.9(4),
Cl(1)–Bi–C(10) = 92.7(3),
Cl(1)–Bi–C(17) = 93.8(4),
O(1)–Bi–C(1) = 77.9(4),
O(1)–Bi–C(10) = 91.6(4),
O(1)–Bi–C(17) = 92.9(4),
C(1)–Bi–C(10) = 129.1(5),
C(1)–Bi–C(17) = 123.0(5),
C(10)–Bi–C(17) = 107.0(4)

[93OM1857]
LANGOW

(**41**) Chloro(2-methyl-8-quinolinolato-*O*)-triphenylbismuth

Ph	Ph	Ph	(O)	Cl *(N)*

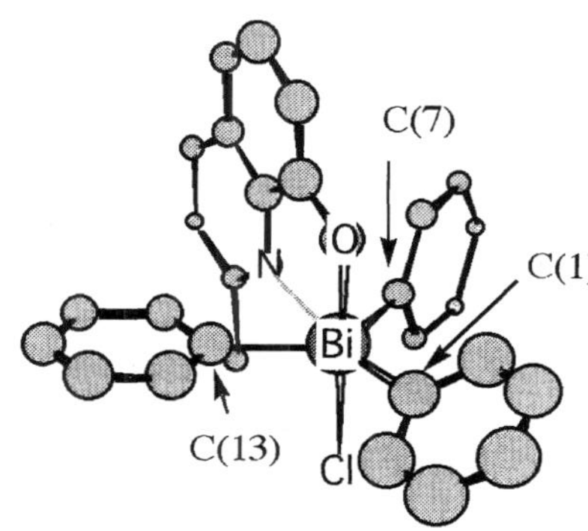

Bi–Cl = 2.67(1),
Bi–O = 2.19(2),
Bi–N = 2.71(2),
Bi–C(1) = 2.16(2),
Bi–C(7) = 2.29(2),
Bi–C(13) = 2.25(2)
N–Bi–O = 66.3(6),
N–Bi–Cl = 119.0(5),
N–Bi–C(7) = 81.5(4),
N–Bi–C(13) = 73.1(4),
N–Bi–C(1) = 150.5(5),
O–Bi–Cl = 174.5(4),
O–Bi–C(1) = 84.7(4),
O–Bi–C(7) = 106.0(1),
O–Bi–C(13) = 91.9(4),
C(1)–Bi–C(7) = 106.0(1),
C(1)–Bi–C(13) = 150.0(1),
C(7)–Bi–Cl = 87.3(2),
C(13)–Bi–Cl = 91.0(2)

[83JOM(253)317]
CASVUN

TABLE 6.6 (*continued*)

Compound					Structure	Structural parameters Bond distances (Å) and bond angles (°)	[Reference], **CCDB code** and *notes*
R[1]	R[2]	R[3]	R[4]	R[5]			
(**42**) Chloro(8-hydroxyquinolinolato-*O*)-triphenylbismuth					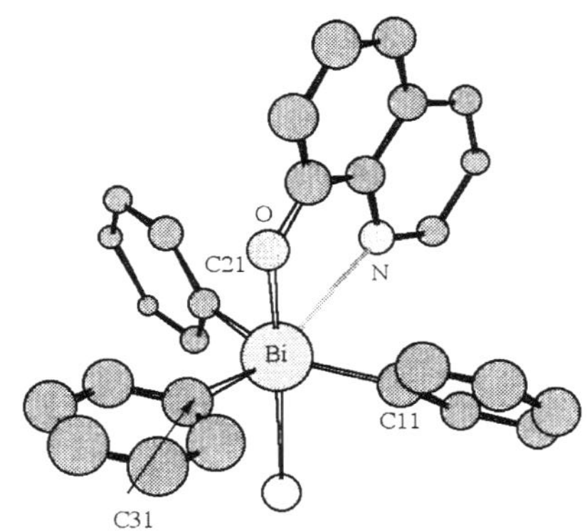	Bi–Cl = 2.657(3), Bi–O = 2.175(7), Bi–N = 2.807(10), Bi–C(11) = 2.168(10), Bi–C(21) = 2.209(11), Bi–C(31) = 2.214(11) Cl–Bi–O = 175.1(2), Cl–Bi–N(9) = 116.8(2), Cl–Bi–C(11) = 85.9(3), Cl–Bi–C(21) = 90.6(3), Cl–Bi–C(31) = 92.3(3), O–Bi–N = 67.8(3), O–Bi–C(11) = 94.4(3), O–Bi–C(21) = 92.1(3), O–Bi–C(31) = 83.0(3), N–Bi–C(11) = 72.8(3), N–Bi–C(21) = 76.2(3), N–Bi–C(31) = 150.8(3), C(11)–Bi–C(21) = 143.1(4), C(11)–Bi–C(31) = 110.6(4), C(21)–Bi–C(31) = 106.2(4), Bi–O–C(1) = 124.(6)	[84HCA586] **CUNFEW**
Ph	Ph	Ph	(O)	Cl (*N*)			

(**43**) 1-Bromo-1,1-bis(4-methylphenyl)-3,3-bis(trifluoromethyl)-3*H*-2,1-benzoxabismole
Tol Tol Ar (O) Br

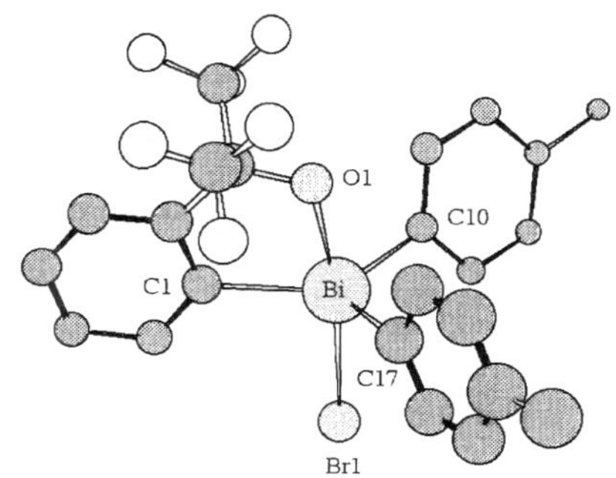

Bi–Br(1) = 2.708(2),
Bi–O(1) = 2.184(8),
Bi–C(1) = 2.21(1),
Bi–C(10) = 2.20(1),
Bi–C(17) = 2.21(1)
Br(1)–Bi–O(1) = 170.1(2),
Br(1)–Bi–C(1) = 92.8(3),
Br(1)–Bi–C(10) = 94.5(2),
Br(1)–Bi–C(17) = 94.8(3),
O(1)–Bi–C(1) = 77.5(3),
O(1)–Bi–C(10) = 90.0(3),
O(1)–Bi–C(17) = 92.1(3),
C(1)–Bi–C(10) = 126.9(3),
C(1)–Bi–C(17) = 122.6(4),
C(10)–Bi–C(17) = 109.0(4)

(**44**) μ_2-Oxobis(triphenylbismuth) diperchlorate
Ph Ph Ph O–BiPh$_3$ (ClO$_4$) *

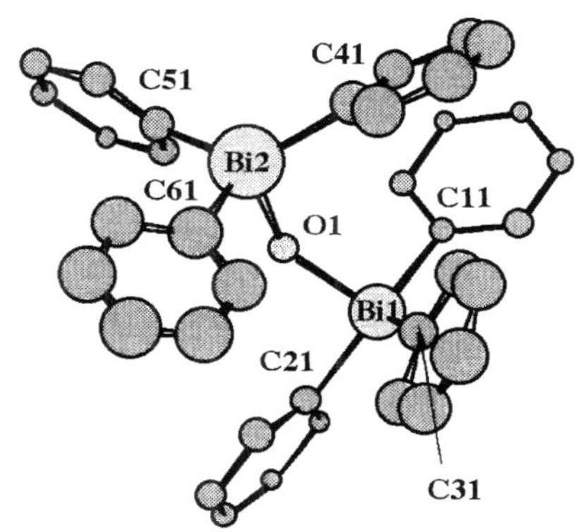

Bi(1)–O(1) = 2.065(15),
Bi(2)–O(1) = 2.062(14),
Bi(1)–C(11) = 2.13(2),
Bi(1)–C(21) = 2.19(2),
Bi(1)–C(31) = 2.19(2),
Bi(2)–C(41) = 2.19(2),
Bi(2)–C(51) = 2.26(2),
Bi(2)–C(61) = 2.15(2),
Bi(1)–O(11a) = 2.53(3),
Bi(1)–O(11b) = 2.76(3),
Bi(2)–O(21) = 2.655(15)
Bi(1)–O(1)–Bi(2) = 142.2(7),
O(1)–Bi(1)–C(11) = 98.2(7),
O(1)–Bi(1)–C(21) = 92.1,
O(1)–Bi(1)–C(31) = 99.7(7),
O(11a)–Bi(1)–C(11) = 87.9(10), O(11b)–Bi(1)–C(11) = 71.2(9), O(11a)–Bi(1)–C(21) = 86.0(10),

[75JCS(D)1291]
TPBIOP10

TABLE 6.6 (*continued*)

Compound R^1	R^2	R^3	R^4	R^5	Structure	Structural parameters Bond distances (Å) and bond angles (°)	[Reference], **CCDB code** and *notes*
						O(11b)–Bi(1)–C(21) = 84.9(9), O(11a)–Bi(1)–C(31) = 75.5(9), O(11b)–Bi(1)–C(31) = 96.1(9), C(11)–Bi(1)–C(21) = 126.6(8), C(11)–Bi(1)–C(31) = 114.7(8), C(21)–Bi(1)–C(31) = 114.8(8), O(1)–Bi(1)–O(11a) = 173.5(8), O(1)–Bi(1)–O(11b) = 163.7(7), O(1)–Bi(2)–C(41) = 103.2(7), O(1)–Bi(2)–C(51) = 93.4(7), O(1)–Bi(2)–C(61) = 96.1(7), O(21)–Bi(2)–C(41) = 81.4(7), O(21)–Bi(2)–C(51) = 87.5(6), O(21)–Bi(2)–C(61) = 77.6(7), C(41)–Bi(2)–C(51) = 125.7(8), C(41)–Bi(2)–C(61) = 116.7(8), C(51)–Bi(2)–C(61) = 112.2(9), O(1)–Bi(2)–O(21) = 173.5(5)	

(**45**) μ_2-Oxobis(triphenylbismuth) bis(trifluoromethanesulfonate)

Ph Ph Ph (O) *

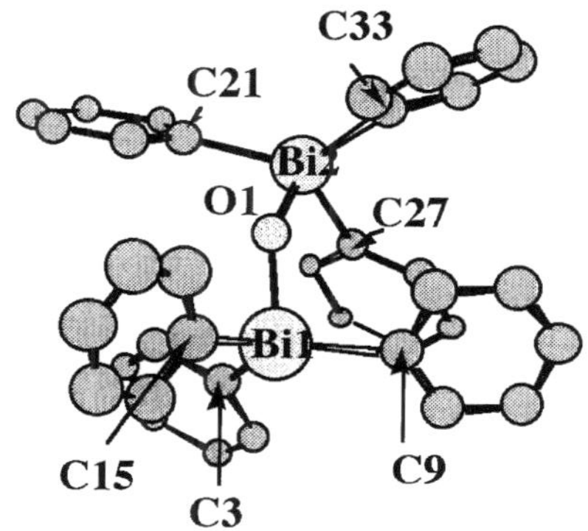

Bi(1)–O(1) = 2.039(8), Bi(1)–C(3) = 2.20(1), Bi(1)–C(9) = 2.20(2), Bi(1)–C(15) = 2.22(1), O(1)–Bi(2) = 2.091(8), Bi(2)–C(21) = 2.19(1), Bi(2)–C(27) = 2.20(1), Bi(2)–C(33) = 2.21(1) O(1)–Bi(1)–C(3) = 98.9(5), O(1)–Bi(1)–C(9) = 97.7(5), O(1)–Bi(1)–C(15) = 92.4(5), C(3)–Bi(1)–C(9) = 119.3(7), C(3)–Bi(1)–C(15) = 112.7(6), C(9)–Bi(1)–C(15) = 124.4(5), Bi(1)–O(1)–Bi(2) = 134.0(4), O(1)–Bi(2)–C(21) = 93.2(5), O(1)–Bi(2)–C(27) = 100.4(4), O(1)–Bi(2)–C(33) = 92.6(4), C(21)–Bi(2)–C(27) = 113.9(6), C(21)–Bi(2)–C(33) = 130.5(5), C(27)–Bi(2)–C(33) = 113.3(5)

[94JCS(P1)1739]
LEVVUD

TABLE 6.6 (*continued*)

Compound					Structure	Structural parameters Bond distances (Å) and bond angles (°)	[Reference], **CCDB code** and *notes*
R^1	R^2	R^3	R^4	R^5			
(**46**) Difluorotris(pentafluorophenyl)bismuth bis[tris(pentafluorophenyl)bismuthine]					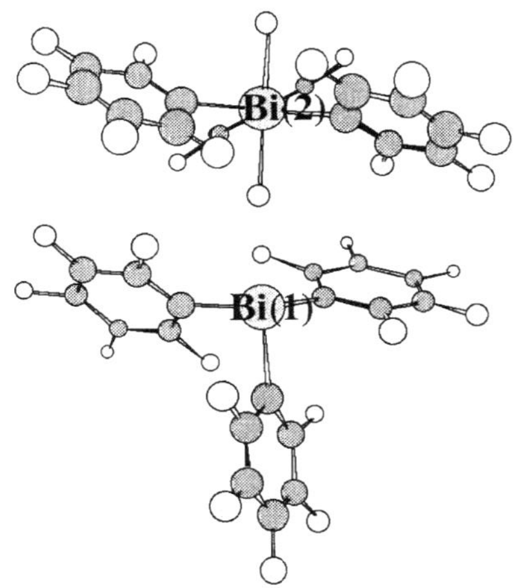	**$(C_6F_5)_3BiF_2$:** Bi(2)–C(211) = 2.18(2), Bi(2)–C(221) = 2.19(2), Bi(2)–F(1) = 2.088(8) F(1)–Bi(2)–C(211) = 91.6(2), F(1)–Bi(2)–C(221) = 87.2(5), C(211)–Bi(2)–C(221) = 118.3(4), F(1)–Bi(2)–F(1′) = 180 **$Bi(C_6F_5)_3$:** Bi(1)–C(111) = 2.27(2), Bi(1)–C(121) = 2.24(2), Bi(1)–C(131) = 2.26(2), Bi(1)…F(1) = 2.759(8) C(111)–Bi(1)–C(121) = 93.4(6), C(111)–Bi(1)–C(131) = 93.7(6), C(121)–Bi(1)–C(131) = 98.3(6)	[90CB761] **SETNOU**
C_6F_5	C_6F_5	C_6F_5	F	F			

(**47**) Dichlorotrimethylbismuth acetone solvate

Me	Me	Me	Cl	Cl

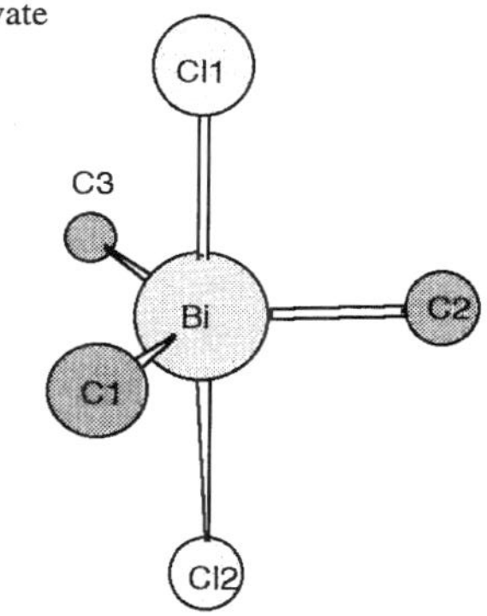

Bi–Cl(1) = 2.602(2),
Bi–C(1) = 2.189(7),
Bi–Cl(2) = 2.617(2),
Bi–C(2) = 2.196(6),
Bi–C(3) = 2.207(6)
Cl(1)–Bi–C(1) = 88.684,
Cl(1)–Bi–Cl(2) = 177.8(1),
Cl(1)–Bi–C(2) = 88.8,
Cl(1)–Bi–C(3) = 90.8,
C(1)–Bi–Cl(2) = 90.3,
C(1)–Bi–C(2) = 117.8(1),
C(1)–Bi–C(3) = 127.3(2),
Cl(2)–Bi–C(2) = 90.0,
Cl(2)–Bi–C(3) = 91.4,
C(2)–Bi–C(3) = 115.3(2)

[94AG(E)976]
WEVGUZ
at −143°C

(**48**) Dichlorotriphenylbismuth

Ph	Ph	Ph	Cl	Cl

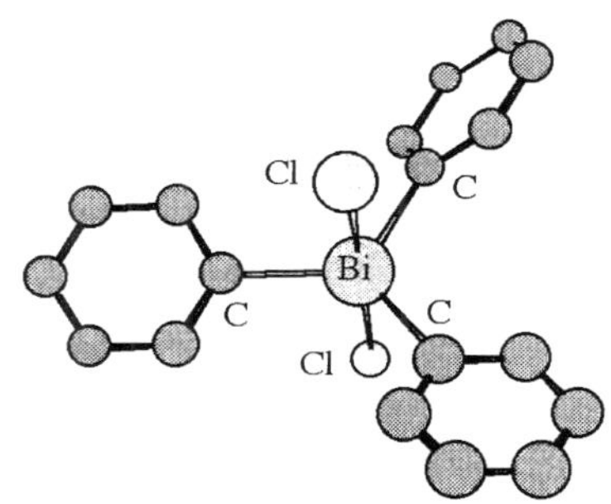

Bi–C = 2.24,
Bi–Cl = 2.605

[66CC111]

TABLE 6.6 (*continued*)

Compound					Structure	Structural parameters Bond distances (Å) and bond angles (°)[b]	[Reference], **CCDB code** and *notes*
R[1]	R[2]	R[3]	R[4]	R[5]			
(**49**) Dichlorotris(4-methylphenyl)bismuth					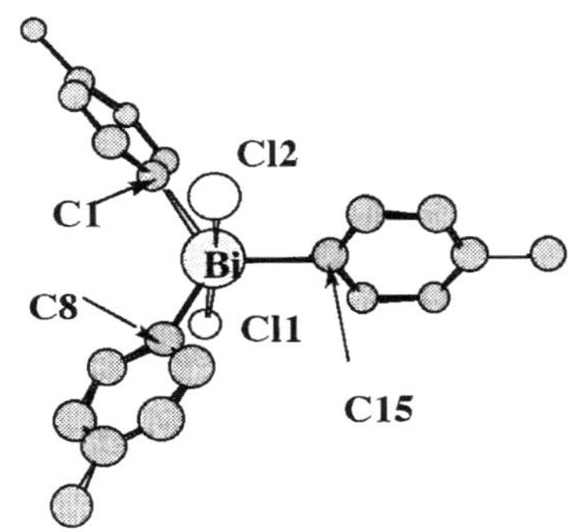	Bi–Cl(1) = 2.590(4), Bi–C(1) = 2.23(2), Bi–Cl(2) = 2.595(4), Bi–C(8) = 2.19(1), Bi–C(15) = 2.22(1) Cl(1)–Bi–C(1) = 90.0(4), Cl(1)–Bi–Cl(2) = 176.0(1), Cl(1)–Bi–C(8) = 88.7(4), Cl(1)–Bi–C(15) = 91.4(4), C(1)–Bi–Cl(2) = 90.0(4), C(1)–Bi–C(8) = 121.6(5), C(1)–Bi–C(15) = 119.1(4), Cl(2)–Bi–C(8) = 87.9(4), Cl(2)–Bi–C(15) = 92.1(4), C(8)–Bi–C(15) = 119.2(4)	[93OM1857] **LANFIP**
Tol	Tol	Tol	Cl	Cl			

[a] The entries of Table 6.6 are placed in the order determined by the ''*Cahn-Ingold-Prelog like*'' rule which follows: (1) substituents are listed in the order of increasing atomic number of the atom directly attached to the bismuth, (2) when two or more atoms attached to the bismuth are the same, the atomic number of the next further atom determines the order, (3) when the apparent valence is less than five, the phantom atom is used to balance the valence five and is indicated by an asterisk (*). An atomic number of zero is given to the phantom atom, (4) when the apparent valence is more than five, the relevant extra atom are is shown in *italic*, (5) all entries in Table 6.6 are listed in the order of increasing combined atomic number of all atoms directly attached to the bismuth. Thus, they appear in the order CCCCC > CCCCCC > CCCCO > CCCC* > CCCOO > CCCO*, and (6) when the atoms directly attached to the bismuth are the same, the atomic number of the next further atom determines the order. Thus, PhPhPhPhPh is followed by PhPhPh(2-FC_6H_4)(2-FC_6H_4)

[b] The data with an error range in parentheses are taken from the original papers. The others were obtained by calculation based on the positional data without considering the error ranges.

Formula Index for X-Ray Data

Formula	Name	Number
CH_3BiI_2	Diiodo(methyl)bismuthine	6.5–74
$(CH_3BiI_2)_n$	Catena[bis(μ$_2$-iodo)(methyl)bismuthine]	6.5–74
C_3H_9Bi	Trimethylbismuthine	6.5–1
$C_3H_9BiCl_2 \cdot C_3H_6O$	Dichlorotrimethylbismuth (*trimethylbismuth dichloride*) acetone solvate	6.6–47
$C_4H_{12}Bi_2$	Tetramethyldibismuthine	6.5–48
$C_5H_5BiCl_2$	Dichloro(cyclopentadienyl)bismuthine	6.5–66
$C_5H_{15}Bi$	Pentamethylbismuth	6.6–1
$C_6H_5BiCl_2$	Dichloro(phenyl)bismuthine	6.5–65
$C_6H_5BiBr_2$	Dibromo(phenyl)bismuthine	6.5–69, 70, 71, 72
$C_6H_5BiI_2$	Diiodo(phenyl)bismuthine	6.5–75
$C_6H_{18}BiN_3$	Tris(dimethylamino)bismuthine	6.5–78
$C_6H_{18}BiLi$	Lithium hexamethylbismuthate	6.6–11
$C_6H_{18}Bi \cdot C_{16}H_{32}LiO_4$	Tetrakis(tetrahydrofuran)lithium hexamethylbismuthate	6.6–11
$C_6H_{24}BiN_{12}S_6 \cdot C_2H_8BiCl_4N_4S_2 \cdot CH_4BiCl_5N_2S$	Hexakis(thiourea-*S*)bismuth tetrachlorobis(thiourea-*S*)bismuthate/ pentachloro(thiourea-*S*)bismuthate	6.5–89
$(C_9H_{11}BiBr_2)_n$	Catena[bis(μ$_2$-bromo)bromo(2,4,6-trimethylphenyl)bismuthine]	6.5–68
$C_9H_{21}BiO_6$	Tris(methoxyethoxy)bismuthine	6.5–82
$(C_9H_{21}BiO_6)_n$	Catena[(bis(μ$_2$-methoxyethoxo)methoxyethoxy)bismuthine]	6.5–82
$C_{10}H_{15}BiO$	Diethyl(phenoxy)bismuthine	6.5–15
$C_{10}H_{10}BiF_5O$	Diethyl(pentafluorophenoxy)bismuthine	6.5–16
$(C_{10}H_{10}Bi_2Cl_4)_n$	Catena[(ε_3-cyclopentadienyl)(ε_2-cyclopentadienyl)bis(bis(μ$_2$-chloro)-bismuthine)]	6.5–66
$(C_{10}H_{24}BiClN_2Si)_n$	Catena[(μ$_2$-chloro)(*N,N′*-di-*t*-butyldimethylsilanadiamino-*N,N′*)-bismuthine]	6.5–80
$C_{10}H_{11}BiO_2S_4$	Bis(methyldithiocarbonato-*S,S′*)phenylbismuthine	6.5–61

Formula	Name	Number
$C_{10}H_{13}BiOS_2$	5-Phenyl-1,4,6,5-oxadithiabismocane	6.5–60
$C_{10}H_{24}BiN_2Si{\cdot}AlCl_4$	(*N,N*′-Di-*t*-butyldimethylsilanadiamino-*N,N*′)bismuthyl tetrachloroalanate	6.5–81
$C_{11}H_{23}BiN_2S_4$	Bis(diethyldithiocarbamato-*S,S*′)methylbismuthine	6.5–59
$C_{12}H_8BiCl{\cdot}Cl{\cdot}C_{36}H_{30}NP_2$	Bis(triphenylphosphine)iminium dichloro(2,2′-biphenylene)-bismuthate	6.5–39
$C_{12}H_{10}BiBr$	Bromo(diphenyl)bismuthine	6.5–45
$C_{12}H_{10}BiBr_2{\cdot}C_{24}H_{20}P$	Tetraphenylphosphonium dibromodiphenylbismuthate	6.5–46
$C_{12}H_{10}BiI_2{\cdot}C_8H_{20}N$	Tetraethylammonium diiododiphenylbismuthate	6.5–47
$(C_{12}H_{10}Bi_2Br_4)_n$	Catena[tetrakis(μ_2-bromo)diphenyldibismuthine]	6.5–67
$C_{12}H_{10}Bi_2Br_6{\cdot}2C_{16}H_{36}N$	Bis(tetrabutylammonium) bis[(μ_2-bromo)dibromo(phenyl)bismuthate]	6.5–73
$C_{12}H_{10}Bi_2I_6{\cdot}2C_8H_{20}N{\cdot}C_4H_{10}O$	Bis(tetraethylammonium) bis[(μ_2-iodo)diiodo(phenyl)bismuthate] diethyl ether solvate	6.5–76
$C_{12}H_{16}Bi_2$	2,2′,5,5′-Tetramethylbibismole	6.5–49
$C_{12}H_{16}Bi_2Fe$	2,2′,5,5′-Tetramethyl-1,1′-dibismaferrocene	6.5–50
$C_{12}H_{30}BiP_3S_6{\cdot}C_6H_6$	Tris(diethyldithiophosphinato-*S,S*′)bismuthine benzene solvate	6.5–92
$C_{12}H_{36}Bi_2Si_4$	Tetrakis(trimethylsilyl)dibismuthine	6.5–88
$(C_{13}H_{10}BiNS)_n{\cdot}0.5(CHCl_3)_n$	Catena[(μ_2-thiocyanato)diphenylbismuthine] chloroform solvate	6.5–31
$C_{14}H_5BiF_{12}O_8{\cdot}C_5H_6N$	Bis(pyridinium) phenyltetrakis(trifluoroacetato-*O*) bismuthate	6.5–58
$C_{14}H_5BiF_{12}O_8{\cdot}C_6H_{18}N_2$	*N,N,N*′,*N*′-Tetramethylethylenediammonium (trifluoroacetato-*O,O*′)tris(trifluoroacetato-*O*)phenylbismuthate	6.5–54
$C_{14}H_5BiF_{12}O_8{\cdot}C_7H_{20}N_2$	*N,N,N*′,*N*′-Tetramethylpropanediammonium tetrakis(trifluoro-acetato-*O*)phenylbismuthate	6.5–53
$C_{14}H_5BiF_{12}O_8{\cdot}C_7H_{20}N_2O_2$	1,3-Bis(dimethylammonio)-2-propanol tetrakis(trifluoroacetato-*O*)-phenylbismuthate	6.5–55
$C_{14}H_5BiF_{12}O_8{\cdot}C_8H_{22}N_2$	*N,N,N*′,*N*′-Tetramethylbutane-1,4-diammonium phenyltetrakis-(trifluoroacetato-*O*)bismuthate	6.5–56

$C_{14}H_5BiF_{12}O_8 \cdot C_{10}H_{18}N_{22} \cdot C_4H_8O$	*N,N,N′,N′*-Tetramethyl-1,4-phenylene-1,4-diammonium phenyltetrakis-(trifluoroacetato-*O*)bismuthate	6.5–57
$C_{15}H_{14}BiBrO$	(2-Acetylphenyl-*C,O*)bromo(4-methylphenyl)bismuthine	6.5–44
$C_{15}H_{15}Bi$	Tris(cyclopenta-2,4-dienyl)bismuthine	6.5–3
$(C_{16}Bi_2N_8S_8)_n \cdot 2(C_{24}H_{20}As)_n$	Tetraphenylarsonium bis(1,2-dicyanoethene-1,2-dithiolato-*S,S′*)bismuthate	6.5–90
$C_{16}H_{10}BiF_6O_4$	Tetraphenylbismuthonium diphenylbis(trifluoroacetato-*O*)bismuthate (anion part)	6.5–29
$C_{16}H_{17}BiOS_2$	(Isopropyldithiocarbonato-*S*)diphenylbismuthine	6.5–32
$C_{16}H_{18}BiBrO$	Bromodiphenyltris(tetrahydrofuran)bismuthine	6.5–45
$C_{16}H_{19}BiClN$	Chlorophenyl[2-((*R*)-1-(*N,N*-dimethylamino)ethyl-phenyl-*C,N*)]bismuthine	6.5–36
$C_{16}H_{19}BiClN \cdot CH_4O$	Chloro[2-(*N,N*-dimethylaminomethyl)phenyl-*C,N*]-(4-methylphenyl)-bismuthine methanol solvate	6.5–37
$C_{16}H_{25}BiN_2S_4$	Bis(diethyldithiocarbamato-*S,S′*)phenylbismuthine	6.5–62
$C_{17}H_{20}BiClO_2S$	(2-*t*-Butylsulfonylphenyl-*O*)chloro(4-methylphenyl)bismuthine	6.5–38
$C_{18}BiF_{17} \cdot 2C_{18}BiF_{15}$	Difluorotris(pentafluorophenyl)bismuth bis[tris(pentafluorophenyl)-bismuthine]	6.6–46
$C_{18}H_4BiClF_{18}$	Chlorobis[2,4,6-tris(trifluoromethyl)phenyl]bismuthine	6.5–34
$C_{18}H_8BiF_{12}O_2 \cdot C_8H_{20}N \cdot H_2O$	Tetraethylammonium 1,1′-spirobis[3,3-bis(trifluoromethyl)-3*H*-2,1-benzoxabismole] monohydrate	6.5–30
$C_{18}H_{15}Bi$	Triphenylbismuthine	6.5–4
$C_{18}H_{15}BiCl_2$	Dichlorotriphenylbismuth (*triphenylbismuth dichloride*)	6.6–48
$C_{18}H_{16}BiF_6NO$	1-[2-(*N,N*-Dimethylaminomethyl)phenyl]-3,3-bis(trifluoromethyl)-3*H*-2,1-benzoxabismole	6.5–24
$C_{18}H_{21}BiO_2$	[2-(2-Hydroxy-2-propyl)phenyl](2-phenyl-2-propanolato)bismuthine	6.5–23
$C_{18}H_{22}BiBr$	Bromobis(2,4,6-trimethylphenyl)bismuthine	6.5–41, 42, 43
$(C_{18}H_{22}Bi_2Br_4)_n$	Catena[bis(μ_2-bromo)bromo(2,4,6-trimethylphenyl)bismuthine)]	6.5–68
$C_{18}H_{29}BiBr_2N_4O_2$	Dibromobis(*N,N′*-dimethylpropyleneurea-*O*)phenylbismuthine	6.5–69
$C_{19}H_{17}BiF_6O_2$	1-[2-(2-Methoxy-2-propyl)phenyl]-3,3-bis(trifluoromethyl)-3*H*-2,1-benzoxabismole	6.5–25

Formula	Name	Number
$C_{19}H_{23}BiO_2$	[2-(2-Methoxy-2-propyl)phenyl](2-phenyl-2-propanolato)bismuthine	6.5–22
$C_{20}H_{12}BiClO_2S$	10-(4-Chlorophenylethynyl)phenothiabismin-5,5-dioxide	6.5–13
$C_{20}H_{17}BiO_4$	Bis(formato-*O*)triphenylbismuth (*triphenylbismuth diformate*)	6.6–33
$(C_{20}H_{26}Bi_2Cl_4O_2)_n$	Catena[(μ_2-chloro)chloro(phenyl)(tetrahydrofuran)bismuthine]	6.5–65
$(C_{20}H_{26}Bi_2Br_4O_2)_n$	Catena[(μ_2-bromo)bromo(phenyl)(tetrahydrofuran)bismuthine]	6.5–70
$(C_{20}H_{26}Bi_2I_4O_2)_n$	Catena[(μ_2-iodo)iodo(phenyl)(tetrahydrofuran)bismuthine]	6.5–75
$(C_{20}H_{56}Bi_2Li_2O_4Si_4)_r$	(1,2-Dimethoxyethane-*O*,*O*′)lithium bis(trimethylsilyl)bismuthide	6.5–87
$C_{21}H_{21}Bi$	Tris(4-methylphenyl)bismuthine	6.5–5
$C_{21}H_{21}BiCl_2$	Dichlorotris(4-methylphenyl)bismuth (*tris(4-methylphenyl)-bismuth dichloride*)	6.6–49
$C_{21}H_{23}BiF_6N_2O$	1-[2,6-Bis(*N*,*N*-dimethylaminomethyl)phenyl]-3,3-bis(trifluoromethyl)-3*H*-2,1-benzoxabismole	6.5–26
$C_{21}H_{28}BiN_3S_4$	Bis(diethyldithiocarbamato-*S*,*S*′)[2-(2-pyridyl)phenyl-*C*,*N*]bismuthine	6.5–64
$C_{21}H_{57}BiSi_6$	Tris[bis(trimethylsilyl)methyl]bismuthine	6.5–2
$C_{22}H_{15}BiF_6O_4$	Bis(trifluoroacetato-*O*)triphenylbismuth (*triphenylbismuth bis(trifluoro-acetate)*)	6.6–35
$C_{23}H_{12}BiClF_{12}O$	1-Chloro-3,3-bis(trifluoromethyl)-1,1-bis[4-(trifluoromethyl)phenyl]-3*H*-2,1-benzoxabismole	6.6–40
$C_{23}H_{15}BiCl_1F_9O$	1-Chloro-1-(4-methylphenyl)-3,3-bis(trifluoromethyl)-1-[4-(trifluoro-methyl)phenyl]-3*H*-2,1-benzoxabismole	6.6–39
$C_{23}H_{18}BiBrF_6O$	1-Bromo-1,1-bis(4-methylphenyl)-3,3-bis(trifluoromethyl)-3*H*-2,1-benzoxabismole	6.6–43
$C_{23}H_{18}BiClF_6O$	1-Chloro-1,1-bis(4-methylphenyl)-3,3-bis(trifluoromethyl)-3*H*-2,1-benzoxabismole	6.6–38
$C_{23}H_{38}BiClN_2Si_3$	Chloro[bis(trimethylsilyl)(2-pyridyl)methyl-*C*,*N*]-[3-(2-pyridyl)-3-trimethylsilylprop-2-enyl]bismuthine	6.5–35
$C_{24}H_{20}Bi$	Tetraphenylbismuthonium diphenylbis(trifluoroacetato-*O*)bismuthate (cation part)	6.6–20
$C_{24}H_{20}Bi{\cdot}C_7H_7O_3S$	Tetraphenylbismuthonium *p*-toluenesulfonate	6.6–19

$C_{24}H_{20}Bi_2$	Tetraphenyldibismuthine	6.5–51a, 51b, 51c
$C_{24}H_{21}BiF_6O$	1-Methyl-1,1-bis(4-methylphenyl)-3,3-bis(trifluoromethyl)-3*H*-2,1-benzoxabismole	6.6–15
$C_{24}H_{26}BiO{\cdot}BF_4$	(3,3-Dimethyl-2-oxobutyl)triphenylbismuthonium tetrafluoroborate	6.6–17
$C_{24}H_{26}BiO{\cdot}CF_3O_3S$	(3,3-Dimethyl-2-oxobutyl)triphenylbismuthonium trifluoromethanesulfonate	6.6–16
$C_{24}H_{26}BiO_2{\cdot}CF_3O_3S$	[2-(Isopropoxycarbonyl)ethyl]triphenylbismuthonium trifluoromethanesulfonate	6.6–25
$C_{24}H_{27}BiO_2S$	(2-*t*-Butylsulfonylphenyl)bis(4-methylphenyl)bismuthine	6.5–14
$C_{24}H_{27}BiO_3$	Tris(2,6-dimethylphenoxy)bismuthine	6.5–83
$C_{24}H_{27}BiO_3$	Tris[2-(methoxymethyl)phenyl]bismuthine	6.5–7
$C_{24}H_{27}BiO_6$	Tris(2,6-dimethoxyphenyl)bismuthine	6.5–12
$C_{24}H_{34}Bi_2Br_4N_4O_2$	Bis[(μ_2-bromo)bromo(*N,N'*-dimethylpropyleneurea-*O*)phenylbismuthine]	6.5–72
$C_{24}H_{36}BiN_3S_6$	Tris[2-(ethylamino)cyclopent-1-ene-1-dithiocarboxylato-*S,S'*]-bismuthine	6.5–91
$C_{24}H_{40}BiBrN_3OP$	Bromobis(2,4,6-trimethylphenyl)(hexamethylphosphoramide-*O*)-bismuthine	6.5–43
$C_{24}H_{46}BiN_6O_2P_2{\cdot}BF_4$	Bis(hexamethylphosphoramide-*O*)diphenylbismuthine tetrafluoroborate	6.5–20
$C_{24}H_{46}BiN_6O_2P_2{\cdot}PF_6$	Bis(hexamethylphosphoramide-*O*)diphenylbismuthine hexafluorophosphate	6.5–19
$C_{26}H_{25}BiO_2$	4,4-Dimethyl-2,6-dioxo-1-triphenylbismuthiocyclohexanide	6.6–26
$C_{26}H_{29}BiO_6{\cdot}C_4H_8O$	Bis(4-hydroxybutanoato-*O*)triphenylbismuth tetrahydrofuran solvate	6.6–28
$C_{27}H_6BiF_{27}$	Tris[2,4,6-tris(trifluoromethyl)phenyl]bismuthine	6.5–11
$C_{27}H_{21}BiClNO$	Chloro(8-hydroxyquinolinolato-*O*)triphenylbismuth	6.6–42
$C_{27}H_{33}Bi$	Tris(2,4,6-trimethylphenyl)bismuthine	6.5–6
$C_{27}H_{36}BiN_3$	Tris[2-(*N,N*-dimethylaminomethyl)phenyl]bismuthine	6.5–8
$C_{28}H_{21}BiO_6$	Bis(2-furoato-*O*)triphenylbismuth (*triphenylbismuth di(2-furoate)*)	6.6–34
$C_{28}H_{23}BiClNO$	Chloro(2-methyl-8-quinolinolato-*O*)triphenylbismuth	6.6–41
$C_{28}H_{32}Bi_2N_2O_8S_4$	Bis(μ_2-methanedisulfonamide-*O,O'*)tetraphenyldibismuthine	6.5–27
$C_{30}H_{12}BiF_{13}$	Tris(4-fluorophenyl)bis(pentafluorophenyl)bismuth	6.6–10

Formula	Name	Number
$C_{30}H_{21}BiF_4$	Bis(2,6-difluorophenyl)triphenylbismuth	6.6–5
$C_{30}H_{23}Bi$	(2,2′-Biphenylene)triphenylbismuth	6.6–3
$C_{30}H_{23}BiF_2$	Bis(2-fluorophenyl)triphenylbismuth	6.6–4
$C_{30}H_{25}Bi$	Pentaphenylbismuth	6.6–2, 14
$C_{30}H_{25}Bi{\cdot}C_5H_5N$	Pentaphenylbismuth pyridine solvate	6.6–14
$C_{30}H_{25}BiF_6O$	1,1,1-Tris(4-methylphenyl)-3,3-bis(trifluoromethyl)-3*H*-2,1-benzoxabismole	6.6–18
$C_{30}H_{25}BiO_6S_2$	Bis(phenylsulfonato-*O*)triphenylbismuth (*triphenylbismuth bis(phenylsulfonate)*)	6.6–27
$(C_{30}H_{30}Bi_3F_{15}O_3)_n$	Catena[(μ$_2$-pentafluorophenoxy)diethylbismuthine]	6.5–16
$C_{30}H_{32}BiBrOS$	Bromobis(2,4,6-trimethylphenyl)(diphenylsulfoxide-*O*)bismuthine	6.5–42
$(C_{30}H_{45}Bi_3O_3)_n$	Catena[(μ$_2$-phenoxy)diethylbismuthine]	6.5–15
$C_{30}H_{58}BiN_6O_2P_2{\cdot}PF_6$	Bis(hexamethylphosphoramide-*O*)bis(2,4,6-trimethylphenyl)bismuthine hexafluorophosphate	6.5–21
$C_{30}H_{72}Bi_2N_6Si_3$	*N,N′*-Di-*t*-butyl- *N,N′*-bis[(*N,N′*-di-*t*-butyldimethylsilanadiamino-*N,N′*)bismuthyl]dimethylsilanadiamine	6.5–79
$C_{30}H_{77}BiN_2O_4P_4{\cdot}2FF_6$	Tetrakis(hexamethylphosphoramide-*O*)phenylbismuthine bis(hexafluorophosphate)	6.5–52
$C_{31}H_{23}BiF_6O$	1,1-Bis(4-methylphenyl)-1-(phenylethynyl)-3,3-bis(trifluoromethyl)-3*H*-2,1-benzoxabismole	6.6–21
$C_{32}H_{23}BiN_2O_8$	Bis(5-nitrotropolonato)triphenylbismuth	6.6–37
$C_{32}H_{24}BiF_{12}NO$	1-[2-(*N,N*-Dimethylaminomethyl)phenyl]-1,1-bis[4-(trifluoromethyl)-phenyl]-3,3-bis(trifluoromethyl)-3*H*-2,1-benzoxabismole	6.6–23
$C_{32}H_{26}Bi_2N_4O_8S_2{\cdot}2H_2O$	Bis[(μ$_2$-2,6-pyridinedicarboxylato-*N,O,O′*)(2,6-pyridinedicarboxylato-*N,O*)(dimethylsulfoxide-*O*)bismuthine] dihydrate	6.5–84
$C_{32}H_{30}BiF_6NO$	1-[2-(*N,N*-Dimethylaminomethyl)phenyl] -1,1-bis-(4-methylphenyl)-3,3-bis(trifluoromethyl)-3*H*-2,1-benzoxabismole	6.6–22
$C_{32}H_{44}BiN_3O_6$	Tris[4-(dimethylamino)phenyl]bis(4-hydroxybutanoato-*O*)bismuth	6.6–30

$C_{32}H_{44}BiN_3O_6$	Tris[4-(dimethylamino)phenyl]bis(2-hydroxybutanoato-*O*)bismuth	6.6–31
$C_{33}H_{19}BiF_{18}O_2$	1-[2-(1,1,1,3,3,3-Hexafluoro-2-oxy-2-propyl)phenyl-*C,O*]-1,1-bis[4-(trifluoromethyl)phenyl]-3,3-bis(trifluoromethyl)-3*H*-2,1-benzoxa-bismole	6.6–24
$C_{33}H_{21}BiF_{10}$	Tris(4-methylphenyl)bis(pentafluorophenyl)bismuth	6.6–9
$C_{33}H_{27}BiF_4$	Bis(2,6-difluorophenyl)tris(4-methylphenyl)bismuth	6.6–8
$C_{33}H_{29}BiF_2$	Bis(2-fluorophenyl)tris(4-methylphenyl)bismuth	6.6–7
$C_{33}H_{31}BiP_2S_4$	Bis(diphenyldithiophosphinato-*S,S'*)(2,4,6-trimethylphenyl)bismuthine	6.5–63
$C_{35}H_{31}BiO_4$	Tris(4-methylphenyl)bis(tropolonato-*O*)bismuth	6.6–36
$C_{35}H_{35}Bi{\cdot}0.5(C_{16}H_{32}Cl_2Li_2O_4)$	Bis[(μ_2-chloro)bis(tetrahydrofuran)lithium] bis[pentakis(4-methyl-phenyl)bismuthate]	6.6–6
$C_{36}Bi_2F_{30}Ge_3$	Tris[μ_2-bis(pentafluorophenyl)germyl]dibismuthine	6.5–93
$C_{36}H_{24}Bi_2{\cdot}C_4H_8O$	(μ_2-2,2′-Biphenylene)bis(2,2′-biphenylene)dibismuthine tetrahydro-furan solvate	6.5–10
$C_{36}H_{30}BiLi$	Lithium hexaphenylbismuthate	6.6–12, 13
$C_{36}H_{30}Bi_2O{\cdot}2CF_3O_3S$	μ_2-Oxobis(triphenylbismuth) bis(trifluoromethanesulfonate)	6.6–45
$C_{36}H_{30}Bi_2O{\cdot}2ClO_4$	μ_2-Oxobis(triphenylbismuth) diperchlorate	6.6–44
$C_{36}H_{30}Bi{\cdot}C_{16}H_{32}LiO_4{\cdot}C_4H_8O$	Tetrakis(tetrahydrofuran)lithium hexaphenylbismuthate tetrahydrofuran solvate	6.6–12
$C_{36}H_{30}Bi{\cdot}Li{\cdot}C_4H_{10}O$	Lithium hexaphenylbismuthate diethyl etherate	6.6–13
$C_{36}H_{33}BiO_6{\cdot}C_4H_8O$	Bis(3-(4-hydroxyphenyl)propanoato-*O*)triphenylbismuth tetrahydrofuran solvate	6.6–29
$C_{36}H_{44}Bi_2O{\cdot}0.5C_2H_6O$	Bis[bis(2,4,6-trimethylphenyl)bismuthyl] oxide ethanol solvate	6.5–28
$C_{36}H_{44}Bi_2S$	Bis[bis(2,4,6-trimethylphenyl)bismuthyl] sulfide	6.5–33
$C_{36}H_{44}Bi_2Se$	Bis[bis(2,4,6-trimethylphenyl)bismuthyl] selenide	6.5–40
$C_{36}H_{60}BiN_3O_6$	Tris(2,6-di-*t*-butyl-4-aza-1,7-dioxahepta-2,5-diene-1,4,7-triyl)-bismuthine	6.5–85
$(C_{42}H_{36}Bi_2N_2O_6)_n$	Catena[(μ_2-*N*-benzoylglycinato-*O,O'*)diphenylbismuthine]	6.5–17
$C_{42}H_{48}BiN_3O_6$	Tris[4-(*N,N*-dimethylamino)phenyl]bis(3-hydroxy-2-phenylpropanoato-*O*)-bismuth	6.6–32

Formula	Name	Number
$C_{48}H_{40}BiO_2P_2 \cdot BF_4 \cdot CH_2Cl_2$	Bis(triphenylphosphineoxide-*O*)diphenylbismuthine tetrafluoroborate dichloromethane solvate	6.5–18
$C_{48}H_{40}Bi_2Br_4O_2P_2$	Bis[(μ_2-bromo)bromo(phenyl)(triphenylphosphineoxide-*O*)-bismuthine]	6.5–71
$C_{54}H_{45}BiO_3Si_3 \cdot 3C_4H_8O$	Tris(triphenylsiloxy-*O*)bismuthine tetrahydrofuran solvate	6.5–86
$C_{54}H_{118}Bi_2Si_{12}$	1,2-Bis{2,4,6-tris[bis(trimethylsilyl)methyl]phenyl}dibismuthene	6.5–77
$C_{72}H_{51}Bi \cdot 4.8C_4H_8O$	Tris(2,4,6-triphenylphenyl)bismuthine tetrahydrofuran solvate	6.5–9

Abbreviations and Symbols

The abbreviations in some instances may differ from those commonly used in other branches of chemistry.

Ac	acetyl
acac	acetylacetonato (2,4-pentanedionato)
AIBN	2,2′-azobis(isobutyronitrile)
AL	acrolein
AN	acrylonitrile
aq	aqueous
Ar	aryl
atm	atmospheric pressure
b.p.	boiling point
bpy	2,2′-bipyridyl
BTMG	*N-tert*-butyl-*N*′,*N*′,*N*″,*N*″-tetramethylguanidine
Bu	butyl
^{i}Bu	*iso*-butyl
^{t}Bu	*tert*-butyl
c-	cyclo-
cat.	catalytic
^{c}Hex	cyclohexyl
CIMS	chemical ionization mass spectra
cit	citrato
conc.	concentrated
Cp	η^5-cyclopentadienyl or 2,4-cyclopentadienyl
d	density
dam	$Me_2N(CH_2)_nNMe_2$ ($n = 2,3$)
DBU	1,8-diazabicyclo[5.4.0]undec-7-ene
decomp.	decomposes
dil.	diluted
dipic	pyridine-2,6-dicarboxylato (2,6-dipicolinato)

DMF	*N,N*-dimethylformamide
dmpu	*N,N'*-dimethylpropylene urea
DMSO	dimethyl sulfoxide
dpm	dipivaloylmethanato (1,3-di-*tert*-butylpropane-1,3-dionato)
EC	electric conductance
Et	ethyl
equiv.	equivalent
(g)	gaseous
GLC	gas liquid chromatography
glc	glycolato
GPC	gas permeation chromatography
GS	glutathionato
HMPA (hmpa)	hexamethylphosphoramide
i-	iso-
IR	infrared spectra
LAH	lithium aluminum hydride
LD_{50}	lethal dosage 50
LDA	lithium diisopropylamide
liq.	liquid or liquified
mand	mandelato
Mc	mercaptocarboxylato
MCPBA	4-chloroperbenzoic acid
Me	methyl
Mes	2,4,6-trimethylphenyl (mesityl)
m/e	mass number of an ion
mmHg	millimeters of mercury
m.p.	melting point
MS	mass spectra
NBS	*N*-bromosuccinnimide
NHE	normal hydrogen electrode
NMR	nuclear magnetic resonance
Nu	nucleophile
Ox	4,4-dimethyloxazolin 2 yl
Ph	phenyl
Pn	pnictogen element
Piv	2,2-dimethylpropanoyl (pivaloyl)
PPN	$Ph_3P{=}N{=}PPh_3$
Pr	propyl

^{i}Pr	*iso*-propyl
Py (py)	2-pyridyl or pyridyl
Pz (pz)	2-pyrazyl
quant.	quantitative
QO	8-hydroxyquinolinolato
R_f	perfluoroalkyl
r.t.	room temperature
sec-	secondary
sublim.	sublimes
sym-	symmetrical
t-	tertiary
$t_{1/2}$	half life
TBAF	tetrabutylammonium fluoride
Tbt	2,4,6-tris[bis(trimethylsilyl)methyl]phenyl
tert-	tertiary
Tf	trifluoromethanesulfonyl (triflyl)
TfOH	trifluoromethanesulfonic acid (triflic acid)
TGA	thermogravimetric analysis
THF	tetrahydrofuran
TLC	thin layer chromatography
TMEDA	tetramethylethylenediamine
TMG	tetramethylguanidine
TMS	trimethylsilyl
Tol	4-methylphenyl (*p*-tolyl)
Ts	4-methylphenylsulfonyl (*p*-toluenesulfonyl, tosyl)
Tsal	thiosalicylato
UV	ultraviolet spectra
UV-Vis	ultraviolet and visible spectra

Procedure Index

PREPARATION OF SELECTED ORGANOBISMUTH COMPOUNDS

Representative examples for the preparation of selected organobismuth compounds are illustrated in the text by experimental procedures. All compounds are listed in alphabetical order, except a few which have complicated names and therefore are shown by chemical formulae at the end of this list.

Organobismuth compounds

Organic transformations based on bismuth compounds

Bismuth-based organic transformations, illustrated in the text by experimental procedures, are listed below in alphabetical order.

Reference Code Index

Journals are in most cases abbreviated according to the CASSI (Chemical Abstracts Service Source Index) system and listed below in alphabetical order. Dictionaries composed of more than one volume are assigned the specific codes.

A	*Analyst*
AC	*Appl. Catal.*
AC(A)	*Appl. Catal. A*
ACA	*Anal. Chim. Acta*
ACP	*Ann. Chim. Phys.*
ACR	*Acc. Chem. Res.*
ACS	*Acta Chem. Scand.*
ACS(A)	*Acta Chem. Scand. Ser. A*
ACS(B)	*Acta Chem. Scand. Ser. B*
AG	*Angew. Chem.*
AG(E)	*Angew. Chem., Int. Ed. Engl.*
AIC	*Adv. Catal.*
AJC	*Aust. J. Chem.*
ANSP	*Akad. Nauk SSSR Proc. Sect. Chem.*
AOC	*Adv. Organomet. Chem.*
AOMC	*Appl. Organomet. Chem.*
AP	*Arch. Pharm.*
AS	*Appl. Spectr.*
AST	*Anal. Sci. Technol.*
AX(A)	*Acta Crystallogr. Part A*
AX(B)	*Acta Crystallogr. Part B*
AX(C)	*Acta Crystallogr. Part C*
BAU	*Bull. Acad. Soc. USSR, Div. Chim. Sci.*
BBI	*The Bulletin of the Bismuth Institute*
BCJ	*Bull. Chem. Soc. Jpn.*

BJ	*Biochem. J.*
BM	*Biol. Met.*
BMC	*Bioorg. Med. Chem. Lett.*
BPIP	*Bull. Pharm. Inst. Pol.*
BRP	*Brit. Pat.*
BSCQ	*Bol. Soc. Chil. Quim.*
BSF	*Bull. Chim. Soc. Fr.*
BTI	*B. Trav. Inst. Pharm. Etat*
CAL	*Cat. Lett.*
CAR	*Carbohydr. Res.*
CB	*Chem. Ber. (Ber. Deut. Chem. Ges.)*
CC	*J. Chem. Soc., Chem. Commun.*
CCC	*Comprehensive Coordination Chemistry*
CCL	*Chin. Chem. Lett.*
CCR	*Coord. Chem. Rev.*
CEJ	*Chem. Eur. J.*
CHJ	*Chin. J. Chem.*
CI(L)	*Chem. Ind. (London)*
CJC	*Can. J. Chem.*
CL	*Chem. Lett.*
CM	*Chem. Mater.*
COMC	*Comprehensive Organometallic Chemistry*
CPB	*Chem. Pharm. Bull.*
CRA	*C. R. Acad. Sci., Ser. IIc: Chim.*
CRS	*Catal. Rev. Sci. Eng.*
CRV	*Chem. Rev.*
CSR	*Chem. Soc. Rev.*
CT	*Chemtracts*
CTO	*Cat. Today*
DOK	*Dokl. Akad. Nauk SSSR*
EA	*Electrochim. Acta*
EAS	*Encyclopedia of Analytical Science*
EDM	*Elsevier's Encyclopaedic Dictionary of Medicines*
EIC	*Encyclopedia of Inorganic Chemistry*
EJIC	*Eur. J. Inorg. Chem.*
EJO	*Eur. J. Org. Chem.*
EL	*Elektrokhimiya*
EOHS	*Encyclopedia of Occupational Health and Safety*

EUP	*Eur. Pat. Appl.*
F	*Farumashia*
FRP	*Fr. Pat.*
G	*Gazz. Chim. Ital.*
GEP	*Ger. Pat.*
GH	*Gmelins Handbuch der Anorganischen Chemie*
H	*Heterocycles*
HAC	*Heteroatom Chem.*
HCA	*Helv. Chim. Acta*
HDX	*Hubei Daxue Xuebao, Ziran Kexueban*
HX	*Huaxue Xuebao*
IC	*Inorg. Chem.*
ICA	*Inorg. Chim. Acta*
ICL	*Ind. Chem. Libr.*
IJC	*Ind. J. Chem.*
IJC(A)	*Ind. J. Chem. Sect. A*
IJC(B)	*Ind. J. Chem. Sect. B*
INCL	*Inorg. Nucl. Chem. Lett.*
ISOC	*3rd Int. Symp. Organomet. Chem., München*
IZV	*Izv. Akad. Nauk SSSR, Ser. Khim.*
JA	*J. Am. Chem. Soc.*
JAP(K)	*Jpn. Kokai*
JAPS	*J. Appl. Polym. Sci.*
JC	*J. Chromatogr.*
JCA	*J. Catal.*
JCAC	*J. Carbohydr. Chem.*
JCC	*J. Coord. Chem.*
JCE	*J. Chem. Ed.*
JCP	*J. Chem. Phys.*
JCR(S)	*J. Chem. Res. (S)*
JCS	*J. Chem. Soc.*
JCS(A)	*J. Chem. Soc. (A)*
JCS(B)	*J. Chem. Soc. (B)*
JCS(C)	*J. Chem. Soc.(C)*
JCS(D)	*J. Chem. Soc., Dalton Trans.*
JCS(P1)	*J. Chem. Soc., Perkin Trans. 1*
JCS(P2)	*J. Chem. Soc., Perkin Trans. 2*
JCSR	*J. Cryst. Spectr. Res.*

OM	*Organometallics*
OMC	*Organomet. Chem.*
OMR	*Organomet. React.*
OMS	*Org. Mass Spectrom.*
OX	*Oxidation Commun.*
PAC	*Pure Appl. Chem.*
PAD	*Prepr. Am. Chem. Soc., Div. Pet. Chem.*
PCS	*Proc. Chem. Soc.*
PM	*Polymer*
POL	*Polyhedron*
PPS	*Proc. Phys. Soc.*
PR	*Phys. Rev.*
PS	*Phosphorus and Sulfur*
R	*Radiokhimiya*
RA	*Radiochim. Acta*
RACE	*Rev. Acad. Cienc. Exact. Fis. Quim. Nat. Zaragoza*
RCB	*Russ. Chem. Bull.*
RGC	*Russ. J. Gen. Chem.*
RKC	*React. Kinet. Catal. Lett.*
RPA	*Rev. Pure Appl. Chem.*
RRC	*Rev. Roum. Chim.*
RZC	*Rocz. Chem.*
S	*Synthesis*
SA(A)	*Spectrochimica Acta A*
SAG	*Soobshch. Akad. Nauk Gruz. SSR*
SC	*Synth. Commun.*
SCI	*Science*
SK(B)	*Suomen Kemistilehti B*
SL	*Synlett*
SRI	*Synth. React. Inorg. Met.-Org. Chem.*
SSOK	*Sbornik Stat. Obshch. Khim.*
SSSC	*Stud. Surf. Sci. Catal.*
T	*Tetrahedron*
TA	*Tetrahedron: Asymmetry*
TCC	*Top. Curr. Chem.*
TFS	*Trans. Faraday Soc.*
TL	*Tetrahedron Lett.*
TMC	*Transition Met. Chem.*

TTFI	*Tr. Tashkent. Farm. Inst.*
UE	*Ullmann's Encyclopedia of Industrial Chemistry*
UKZ	*Uzbek. Khim. Zh.*
UR	*Unpublished result*
USP	*US Pat.*
YGK	*Yuki Gosei Kagaku Kyokaishi*
VH	*Vitam. Horm.*
VLU	*Vestn. Leningrad. Univ. Fiz. Khim.*
VMU	*Vestn. Moscov. Univ. Fiz.-Mat. Estestven. Nauk*
YH	*Youji Huaxue*
ZAAC	*Z. Anorg. Allg. Chem.*
ZAK	*Z. Anal. Khim.*
ZFK	*Zh. Fiz. Khim.*
ZK	*Z. Kristallogr.*
ZN(B)	*Z. Naturforsch. Teil B*
ZNK	*Zh. Neorg. Khim.*
ZOB	*Zh. Obshch. Khim.*
ZPC(B)	*Z. Phys. Chem. Teil B*
ZSK	*Zh. Strukt. Khim.*
ZVKO	*Zh. Vses. Khim. Obshch.*

References

All references for the six chapters are collected together and listed here in chronological order. They are shown according to the number-lettering code system that has been used in *Comprehensive Heterocyclic Chemistry*, eds A. R. Katritzky, C. W. Rees and E. F. V. Scriven, Elsevier Science, Oxford, UK (1996). The references are indicated in square brackets each time they appear in the text, where the first two numbers denote the year of publication (tens and units), the next two to four letters denote the journal, and the final numbers denote the page. For the last century references, the year is given in full in the code. For journals that have more than one volume per year, the volume number is included in parentheses immediately after the journal code letters. If there are more than one references from the same volume of the journal in question, they are listed in increasing page order. Journals are mostly abbreviated according to the Chemical Abstracts Service Source Index (CASSI) system.

Monographs, handbooks, and data collections are indicated by a letter code, B, preceded by the year and followed by the capital letter(s) of author(s)' or editor(s)' family name(s), and the page number if necessary. Dictionaries composed of more than one volume are given the specific abbreviation code, followed by a volume number in parentheses and the page number, if necessary. Patents are shown by a letter code P, which is preceded by the year and followed by a registration number. For some publications that are not readily accessible, the corresponding *Chemical Abstracts* references are given in addition to the original journal. Organobismuth chemistry has been reviewed on an annual basis in the 1980s and irregularly in the 1970s and 1990s in *Journal of Organometallic Chemistry* and *Organometallic Chemistry*. For these as well as general references and specific topics on bismuth chemistry, refer to the explanatory guide on p. 2. Unpublished results are given a specific code UR and collected at the end of the list.

1850LA(75)315	Löwig, C., Schweizer, E. *Liebigs Ann. Chem.* **1850**, *75*, 315.
1854LA(92)371	Dünhaupt, F. *Liebigs Ann. Chem.* **1854**, *92*, 371.
1876BSF(25)183	Claesson, P. *Bull. Soc. Chim. Fr.* **1876**, *25*, 183.
1877JPR(15)209	Claesson, P. *J. Prakt. Chem.* **1877**, *15*, 209.
1887CB54	Michaelis, A., Polis, A. *Ber. Deut. Chem. Ges.* **1887**, *20*, 54.
1887CB1516	Marquardt, A. *Ber. Deut. Chem. Ges.* **1887**, *20*, 1516.
1888CB2035	Marquardt, A. *Ber. Deut. Chem. Ges.* **1888**, *21*, 2035.
1888CB2038	Marquardt, A. *Ber. Deut. Chem. Ges.* **1888**, *21*, 2038.
1889LA(251)323	Michaëlis, A., Marquardt, A. *Liebigs Ann. Chem.* **1889**, *251*, 323.
1897CB2843	Gillmeister, A. *Ber. Deut. Chem. Ges.* **1897**, *30*, 2843.
00JPHC(12)145	Richard, E. *J. Pharm. Chim.* **1900**, *12*, 145.
01JPHC(14)487	Thibault, P. *J. Pharm. Chim.* **1901**, *14*, 487.
02ACP(25)268	Thibault, P. *Ann. Chim. Phys.* **1902**, *25*, 268.
03BSF(29)531	Thibault, P. *Bull. Soc. Chim. Fr.* **1903**, *29*, 531.
03BSF(29)680	Thibault, P. *Bull. Soc. Chim. Fr.* **1903**, *29*, 680.
04BSF(31)176	Thibault, P. *Bull. Soc. Chim. Fr.* **1904**, *31*, 176.
04CB4620	Pfeiffer, P., Heller, I., Pietsch, H. *Ber. Deut. Chem. Ges.* **1904**, *37*, 4620.
13CB1675	Hilpert, S., Gruttner, G. *Ber. Deut. Chem. Ges.* **1913**, *46*, 1675.
14CB3257	Grüttner, G. *Ber. Deut. Chem. Ges.* **1914**, *47*, 3257.
14JCS(105)2210	Challenger, F. *J. Chem. Soc.* **1914**, *105*, 2210.
15CB1473	Grüttner, G., Wiernik, M. *Ber. Deut. Chem. Ges.* **1915**, *48*, 1473.
15JCS(107)16	Challenger, F., Allpress, C.F. *J. Chem. Soc.* **1915**, *107*, 16.
16JCS(109)250	Challenger, F. *J. Chem. Soc.* **1916**, *109*, 250.
17CB21	Vanino, L., Mussgnug, F. *Chem. Ber.* **1917**, *50*, 21.
17ZAAC(100)249	Schaefer, K., Hein, F. *Z. Anorg. Allg. Chem.* **1917**, *100*, 249.
20CB2267	Classes, A., Ney, O. *Ber. Deut. Chem. Ges.* **1920**, *53*, 2267.
20JCS(117)762	Challenger, F., Goddard, A.E. *J. Chem. Soc.* **1920**, *117*, 762.
21JA2367	Denham, H.G. *J. Am. Chem. Soc.* **1921**, *43,* 2367.
21JCS(119)913	Challenger, F., Allpress, C.F. *J. Chem. Soc.* **1921**, *119*, 913.
21ZAAC(115)253	Classen, A., Ney, O. *Z. Anorg. Allg. Chem.* **1921**, *115*, 253.
22BSF(31)545	Volmer, M., Chardeyron, M. *Bull. Chim. Soc. Fr.* **1922**, *31*, 545.
22JCS(121)36	Goddard, A.E. *J. Chem. Soc.* **1922**, *121*, 36.
22JCS(121)91	Challenger, F., Wilkinson, J.F. *J. Chem. Soc.* **1922**, *121*, 91.
22JCS(121)104	Challenger, F., Ridgway, L.R. *J. Chem. Soc.* **1922**, *121*, 104.
22JCS(121)109	Challenger, F., Ridgway, L.R. *J. Chem. Soc.* **1922**, *121*, 109.
22JCS(121)256	Goddard, D., Goddard, A.E. *J. Chem. Soc.* **1922**, *121*, 256.
22JCS(121)978	Goddard, A.E., Ashley, J. N., Evans, R.B. *J. Chem. Soc.* **1922**, *121*, 978.
23JCS(123)1046	Challenger, F., Smith, A.L., Paton, F.J. *J. Chem. Soc.* **1923**, *123*, 1046.
24JCS(125)854	Wilkinson, J. F., Challenger, F. *J. Chem. Soc.* **1924**, *125*, 854.
24JCS(125)864	Challenger, F., Pritchard, F. *J. Chem. Soc.* **1924**, *125*, 864.
25CB891	Rosenheim, A, Baruttschisky, I. *Ber. Deut. Chem. Ges.* **1925**, *58*, 891.
25CB1893	Vorlander, D. *Ber. Deut. Chem. Ges.* **1925**, *58*, 1893.
25RZC(5)298	Supniewski, J.V. *Rocz. Chem.* **1925**, *5*, 298.
26JA507	Supniewski, J.V., Adams, R. *J. Am. Chem. Soc.* **1926**, *48*, 507.

26JCS1648	Challenger, F., Peters, A.T., Halévy, J. *J. Chem. Soc.* **1926**, 1648.
26RZC(6)97	Supniewski, J.V. *Rocz. Chem.* **1926**, *6*, 97.
27BPIP1	Fabrykant, F. *Bull. Pharm. Inst. Pol.* **1927**, 1.
27BTI3	Rozenblumowna, S., Weil, S. *B. Trav. Inst. Pharm. Etal* **1927**, 3.
27JCS209	Challenger, F., Wilson, V.K. *J. Chem. Soc.* **1927**, 209.
28ZK(67)377	Paulling, L. *Z. Kristallogr.* **1928**, *67*, 377.
29CB1710	Krause, E., Renwanz, G. *Ber. Deut. Chem. Ges.* **1929**, *62*, 1710.
30CB1110	Ipatiew, W., Rasuwajew, G. *Ber. Deut. Chem. Ges.* **1930**, *63*, 1110.
31BSF(49)187	Davies, W.C., Norrick, I., Jones, W.J. *Bull. Soc. Chim. Fr.* **1931**, *49*, 187.
31JA1025	Blicke, F.F., Oakdale, U.O., Smith, F.D. *J. Am. Chem. Soc.* **1931**, *53*, 1025.
32CB777	Krause, E., Renwantz, G. *Ber. Deut. Chem. Ges.* **1932**, *65*, 777.
32ZPC(B)401	Bergmann, E., Schütz, W. *Z. Phys. Chem. B* **1932**, *19*, 401.
34JCS405	Challenger, F., Richards, O.V. *J. Chem. Soc.* **1934**, 405.
34JCS1258	Challenger, F., Rothstein, E. *J. Chem. Soc.* **1934**, 1258.
34TFS(30)179	Paneth, F.A. *Trans. Faraday Soc.* **1934**, *30*, 179.
35JCS366	Paneth, F.A., Loleit, H. *J. Chem. Soc.* **1935**, 366.
35NAT(135)1038	Norvick, I. *Nature* **1935**, *135*, 1038.
36JA1820	Worrall, D.E. *J. Am. Chem. Soc.* **1936**, *58*, 1820.
37JA935	Gilman, H., Nelson, J.F. *J. Am. Chem. Soc.* **1937**, *59*, 935.
37ZOB143	Makarova, L.G. *Zh. Obshch. Khim.* **1937**, *7*, 143.
38ZOB1831	Talalaeva, T.V., Kocheshkov, K.A. *Zh. Obshch. Khim.* **1938**, *8*, 1831.
38ZOB1839	Zhitkova, L.A., Sheverdina, N.I., Kocheshkov, K.A. *Zh. Obshch. Khim.* **1938**, *8*, 1839.
39JA1170	Gilman, H., Yablunky, H.L., Svigoon, A.C. *J. Am. Chem. Soc.* **1939**, *61*, 1170.
39JA3586	Gilman, H., Svigoon, A.C. *J. Am. Chem. Soc.* **1939**, *61*, 3586.
39JOC162	Gilman, H., Apperson, L.D. *J. Org. Chem.* **1939**, *4*, 162.
39JPH(67)17	Sollmann, T., Seifter, J. *J. Pharmacol.* **1939**, *67*, 17.
39ZOB771	Makarove, L.G., Nesmeyanov, K.A. *Zh. Obshch. Khim.* **1939**, *9*, 771.
39ZOB2283	Koton, M.M. *Zh. Obshch. Khim.* **1939**, *9*, 2283.
40CB563	Criegee and Büchner, *Chem. Ber.* **1940**, *73*, 563.
40JA665	Gilman, H., Yablunky, H.L. *J. Am. Chem. Soc.* **1940**, *62*, 665.
40JCP366	Rosenbaum, E.J., Rubin, D.J., Sandberg, C.R. *J. Chem. Phys.* **1940**, *8*, 366.
41JA207	Gilman, H., Yablunky, H.L. *J. Am. Chem. Soc.* **1941**, *63*, 207.
41JA212	Gilman, H., Yablunky, H.L. *J. Am. Chem. Soc.* **1941**, *63*, 212.
41JA949	Gilman, H., Yablunky, H.L. *J. Am. Chem. Soc.* **1941**, *63*, 949.
41JOC421	Smyth, C.P. *J. Org. Chem.* **1941**, *6*, 421.
41ZAAC(248)84	Hein, F., Pobloth, H. *Z. Anorg. Allg. Chem.* **1941**, *248*, 84.
41ZOB379	Koton, M.M. *Zh. Obshch. Khim.* **1941**, *11*, 379.
42CRV(30)281	Gilman, H., Yale, H.L. *Chem. Rev.* **1942**, *30*, 281.
42JA173	Oesper, P.F., Smyth, C.P. *J. Am. Chem. Soc.* **1942**, *64*, 173.
42JA392	Calingaert, G., Soroos, H., Hnizda, V. *J. Am. Chem. Soc.* **1942**, *64*, 392.

42ZK(104)305	Wetzel, J. *Z. Kristallogr.* **1942**, *104*, 305.
43VH345	Reichstein, T., Shoppee, C.W. *Vitam. Horm.* **1943**, *1*, 345.
43ZAAC(250)257	Jensen, K.A. *Z. Anorg. Allg. Chem.* **1943**, *250*, 257.
45BAU522	Nesmeyanov, A.N., Kocheskov, K.A. *Bull. Acad. Sci. USSR* **1945**, 522.
46ZOB891	Kozminskaya, T.K., Nad, M.M., Kocheshkov, K.A. *Zh. Obshch. Khim.* **1946**, *16*, 891.
46ZOB897	Nad, M.M., Kozminskaya, T.K., Kocheshkov, K.A. *Zh. Obshch. Khim.* **1946**, *16*, 897.
47ZOB1307	Koton, M.M. *Zh. Obshch. Khim.* **1947**, *17*, 1307.
48USP2451485	Hearne, G.W., Adams, M.L. *US Patent* **1948**, 2,451,485; *Chem. Abstr.* **1949**, *43,* 2222f.
48ZOB936	Koton, M.M. *Zh. Obshch. Khim.* **1948**, *18*, 936.
49NAT(164)185	Rigby, W. *Nature* **1949**, *164*, 185.
49ZOB1675	Koton, M.M., Moskvina, E.P., Florinskii, F.S. *Zh. Obshch. Khim.* **1949**, *19*, 1675.
50JA8	Gilman, H., Yale, H.L. *J. Am. Chem. Soc.* **1950**, *72*, 8.
50JCS617	Lile, W.J., Menzies, R.C. *J. Chem. Soc.* **1950**, 617.
50JCS1907	Rigby, W. *J. Chem. Soc.* **1950**, 1907.
50PR(78)471	Proctor, G., Yu, F.C. *Phys. Rev.* **1950**, *78*, 471.
50ZOB2093	Koton, M.M., Moskvina, E.P., Florinskii, F.S. *Zh. Obshch. Khim.* **1950**, *20*, 2093.
51JA2880	Gilman, H., Yale, H.L. *J. Am. Chem. Soc.* **1951**, *73*, 2880.
51JA4470	Gilman, H., Yale, H.L. *J. Am. Chem. Soc.* **1951**, *73*, 4470.
51JCS793	Rigby, W. *J. Chem. Soc.* **1951**, 793.
51ZOB93	Manulkin, Z.M., Tatarenko, A.N. *Zh. Obshch. Khim.* **1951**, *21*, 93.
52LA(578)136	Wittig, G., Clauss, K. *Liebigs Ann. Chem.* **1952**, *578*, 136.
52NAT(170)1074	Norymberski, J.K. *Nature* **1952**, *170*, 1074.
52ZOB643	Koton, M.M. *Zh. Obshch. Khim.* **1952**, *22*, 643.
52ZOB1528	Yakubovich, A.Y., Makarov, S.P. *Zh. Obshch. Khim.* **1952**, *22*, 1528.
53BJ371	Brooks, C.J.W., Norymberski, J.K. *Biochem. J.* **1953**, *55*, 371.
53HCA268	Stoll, A., Lindenmann, A., Jucker, E. *Helv. Chim. Acta***1953**, *36*, 268.
53PR(89)595	Ng, Y., Williams, D. *Phys. Rev.* **1953**, *89*, 595.
53PR(89)1305	Robinson, H.G., Dehmelt, H.G., Gordy, W. *Phys. Rev.* **1953**, *89*, 1305.
53SSOK1308	Manulkin, Z.M., Tatarenko, A.N., Yusupov, F. *Sbornik Stat. Obshch. Khim.* **1953**, *2*, 1308.
53USP2659746	Morgan, C.S., Robertson, N.C. *US Patent* **1953**, 2,659,746; *Chem. Abstr.* **1954**, *49,* 1099e.
53VMU119	Reutov, O.A. *Vestn. Moscov. Univ. Fiz.-Mat. Estestven. Nauk* **1953**, 119.
54BRP712828	Friedheim, E.A.H. *Br. Patent* 712,828, *Chem. Abstr.* **1955**, *49*, 15946e.
54BRP716647	Friedheim, E.A.H. *Br. Patent* 716,647, *Chem. Abstr.* **1955**, *49*, 12530d.
54JA5533	Djerassi, C., Ringold, H.J., Rosenkranz, G. *J. Am. Chem. Soc.* **1954**, *76*, 5533.
55A397	Smith, D.C., Tomsett, S.L. *Analyst* **1955**, *80*, 397.
55AG233	Hartmann, H., Habenicht, G., Reiss, W. *Angew. Chem.* **1955**, *67*, 233.

55JA683	Kupchan, S.M., Lavie, D. *J. Am. Chem. Soc.* **1955**, *77*, 683.
55JA4262	Freedman, L.D., Doak, G.O., Petit, E.L. *J. Am. Chem. Soc.* **1955**, *77*, 4262.
56ZN(B)755	Wiberg, E., Mödritzer, K. *Z. Naturforsch.* **1956**, *11B*, 755.
57AG205	Fischer, E.O., Schreiner, S. *Angew. Chem.* **1957**, *69*, 205.
57CB1176	Bahr, G., Zoche, G. *Chem. Ber.* **1957**, *90*, 1176.
57JA1130	Fried, J., Sabo, E.F. *J. Am. Chem. Soc.* **1957**, *79*, 1130.
57JA4465	Reitsema, R.H. *J. Am. Chem. Soc.* **1957**, *79*, 4465.
57JA5884	Maier, L., Seyferth, D., Stone, F.G.A., Rochow, E.G. *J. Am. Chem. Soc.* **1957**, *79*, 5884.
57TTFI(1)321	Solomakhina, F.K. *Tr. Tashkent. Farm. Inst.* **1957**, *1*, 321.
57ZN(B)132	Wiberg, V.E., Mödritzer, K. *Z. Naturforsch.* **1957**, *12B*, 132.
57ZN(B)263	Maier, L., Seyferth, D., Stone, F.G.A., Rochow, E. *Z. Naturforsch.* **1957**, *12B*, 263.
58B-P(11)	Domange, L., in *Nouveau Traite de Chimie Minerale*, ed. Pascal, P., Masson, Paris, **1958**, vol. 11.
58DOK(122)614	Nesmeyanov, A.N., Tolstaya, T.P., Isaeva, L.S. *Dokl. Akad. Nauk SSSR* **1958**, *122*, 614.
58DOK(122)1032	Reutov, O.A., Ptitsyna, O.A., Styazhkina, N.B. *Dokl. Akad. Nauk SSSR* **1958**, *122*, 1032.
58JA1336	Seyferth, D. *J. Am. Chem. Soc.* **1958**, *80*, 1336.
58JA1687	Baran, J.S. *J. Am. Chem. Soc.* **1958**, *80*, 1687.
58JIC(35)573	Parab, N.K., Desai, D.M. *J. Indian Chem. Soc.* **1958**, *35*, 573.
58UKZ41	Manulkin, Z.M. *Uzbek. Khim. Zh.* **1958**, 41; *Chem. Abstr.* **1959**, *53*, 9112.
59BRP824944	Ramsden, H.E. *Br. Patent* **1959**, 824,944; *Chem. Abstr.* **1960**, *54*, 17238.
59CI(L)951	Lavie, D., Shvo, Y., Willner, D. *Chem. Ind. (London)* **1959**, 951.
59DOK(125)330	Nesmeyanov, A.N., Tolstaya, T.P., Isaeva, L.S. *Dokl. Akad. Nauk SSSR* **1959**, *125*, 330.
59DOK(125)1265	Nesmeyanov, A.N., Reutov, O.A., Tolstaya, T.P., Ptitsyna, O.A., Isaeva, L.S., Turchinskii, M.F., Bochkareva, G.P. *Dokl. Akad. Nauk SSSR* **1959**, *125*, 1265.
59IZV1942	Zakharkin, L.I., Okhlobystin, O.Y. *Izv. Akad. Nauk SSSR, Ser. Khim.* **1959**, 1942.
59PPS945	Flynn, C.P., Seymens, E.F.W. *Proc. Phys. Soc.* **1959**, *73*, 945.
59TL1	Maier, L. *Tetrahedron Lett.* **1959**, *6*, 1.
59USP2904580	Idol, J.D. *US Patent* **1959**, 2,904,580; *Chem. Abstr.* **1960**, *54*, 5470f.
60B-P	Pauling, L., *The Nature of the Chemical Bond*, Cornell Univ. Press, Ithaca, **1960**.
60CB1417	Fischer, E.O., Schreiner, S. *Chem. Ber.* **1960**, *93*, 1417.
60DOK(133)123	Mirin, A.N., Nefedov, V.D., Zaizev, V.M., Grachev, S.A. *Dokl. Akad. Nauk SSSR* **1960**, *133*, 123.
60TTFI(2)317	Solomakhina, F.K. *Tr. Tashkent. Farm. Inst.* **1960**, *2*, 317.
61CB1447	Amberger, E. *Chem. Ber.* **1961**, *94*, 1447.

61CJC171 Rao, C.N.R., Ramachandran, J., Balasubramanian, A. *Can. J. Chem.* **1961**, *39*, 171.
61JCS1238 Peach, M.E., Waddington, T.C. *J. Chem. Soc.* **1961**, 1238.
61JINC(16)213 Maier, L., Rochow, E.G., Fernelius, W.C. *J. Inorg. Nucl. Chem.* **1961**, *16*, 213.
61JINC(17)43 Allred, A.L., Hensley, A.L., Jr. *J. Inorg. Nucl. Chem.* **1961**, *17*, 43.
61ZAAC(308)33 Becke-Goehring, M., Thielemann, H. *Z. Anorg. Allg. Chem.* **1961**, *308*, 33.
61ZVKO110 Sterlin, R.N., Dubov, S.S., Li, W.-K., Vakhomchik, L.P., Knunyants, I.L. *Zh. Vses. Khim. Obshch.* **1961**, *6*, 110.
62AJC710 Dickson, R.S., West, B.O. *Aust. J. Chem.* **1962**, *15*, 710.
62CJC181 Agolini, F., Bonnett, R. *Can. J. Chem.* **1962**, *40*, 181.
62IZV638 Ptitsyna, O.A., Reutov, O.A., Ovodor, Y.S. *Izv. Akad. Nauk SSSR, Ser. Khim.* **1962**, 638.
62JCS2854 Nuttall, R.H., Roberts, E.R., Sharp, D.W.A. *J. Chem. Soc.* **1962**, 2854.
62JCS3746 Jackson, G.D.F., Sasse, W.H.F. *J. Chem. Soc.* **1962**, 3746.
62JOC3851 Monagle, J.J. *J. Org. Chem.* **1962**, *27*, 3851.
62LA(652)1 Ziegler, K., Steudel, O.W. *Liebigs Ann. Chem.* **1962**, *652*, 1.
62PCS224 Bell, T.N., Pullman, B.J., West, B.O. *Proc. Chem. Soc.* **1962**, 224.
62USP3044966 Callahan, J.L., Foreman, R.W., Veatch, F. *US Patent* **1962**, 3,044,966; *Chem. Abstr.* **1962**, *57*, 9270b.
62ZAAC(317)54 Hartmann, H., Habenicht, G., Reiss, W. *Z. Anorg. Allg. Chem.* **1962**, *317*, 54.
62ZVKO117 Sterlin, R.N., Dubov, S.S., Li, W.-K., Vakhomchik, L.P., Knunyants, I.L. *Zh. Vses. Khim. Obshch.* **1962**, *7*, 117; *Chem. Abstr.* **1962**, *57*, 294.
63AJC636 Bell, T.N., Pullman, B.J., West, B.O. *Aust. J. Chem.* **1963**, *16*, 636.
63BSF1583 Benlian, D., Bigorgne, M. *Bull. Soc. Chim. Fr.* **1963**, 1583.
63IZV1507 Borisov, A.E., Osipova, M.A., Nesmeyanov, A.N. *Izv. Akad. Nauk SSSR, Ser. Khim.* **1963**, 1507.
63JA112 Rundle, R.E. *J. Am. Chem. Soc.* **1963**, *85*, 112.
63JCS1739 Aroney, M.J., Le Févre, R.J.W., Saxby, J. D. *J. Chem. Soc.* **1963**, 1739.
63JCS5612 Hey, D.H., Shingleton, D.A., Williams, G.H. *J. Chem. Soc.* **1963**, 5612.
63JINC(25)318 Popplewell, D.S. *J. Inorg. Nucl. Chem.* **1963**, *25*, 318.
63JPR(22)1 El-Hewehi, V.Z., Hempel, D. *J. Prakt. Chem.* **1963**, *22*, 1.
63MAC(62)183 Braun, B., Diamon, H., Becker, G. *Macromol. Chem.* **1963**, *62*, 183.
63USP3109851 Ramsden, H.E. *US Patent* **1963**, 3,109,851; *Chem. Abstr.* **1964**, *60*, 3,015.
64AG99 Stamm, W., Breindel, A. *Angew. Chem.* **1964**, *76*, 99.
64AG(E)316 Schumann, H., Schmidt, M, *Angew. Chem., Int. Ed. Engl.* **1964**, *3*, 316.
64ANSP(156)383 Nesmeyanov, A.N., Anisimov, K.N., Kolobova, N.E., Khandozhko, V.N. *Akad. Nauk SSSR Proced. Sec. Chem.* **1964**, *156*, 383.
64AOC(2)49 Skinner, H.A. *Adv. Organomet. Chem.* **1964**, *2*, 49.
64CB789 Wittig, G., Hellwinkel, D. *Chem. Ber.* **1964**, *97*, 789.
64CB2306 Kuchen, W., Metten, J., Judat, A. *Chem. Ber.* **1964**, *97*, 2306.

64DOK(156)383	Nesmeyanov, A.N., Anisimov, K.N., Kolobova, N.E., Khandozhko, V.N. *Dokl. Akad. Nauk SSSR* **1964**, *156*, 383.
64GH(19)	Meyer, R.J., Pietsch, E.H.E., in *Gmelins Handbuch der Anorganischen Chemie*, ed. Kotowski, A., Verlag Chemie, Weinheim, **1964**, vol. 19.
64IC500	Roper, W.R., Wilkins, C.J. *Inorg. Chem.* **1964**, *3*, 500.
64JC(15)70	Adloff-Bacher, M., Adloff, J.P. *J. Chromatogr.* **1964**, *15*, 70.
64JCS3799	Daly, J.J. *J. Chem. Soc.* **1964**, 3799.
64JMSP(14)320	Jackson, J.A., Nielsen, J.R. *J. Mol. Spectrosc.* **1964**, *14*, 320.
64JOM(2)59	Heitz, C., Adloff, J.P. *J. Organomet. Chem.* **1964**, *2*, 59.
64NEP6405308	M&T Chemicals Inc. *Neth. Appl.* **1964**, 64-05308; *Chem. Abstr.* **1965**, *63*, 1,897.
64NEP6405309	M&T Chemicals Inc. *Neth. Appl.* **1964**, 64-05309; *Chem. Abstr.* **1965**, *62*, 16,300.
64R632	Nefedov, V.D., Zhuravlev, V.E., Toropova, M.A., Levchenko, A.V. *Radiokhimiya* **1964**, *6*, 632.
64ZOB2787	Vyazankin, N.S., Razuvaev, G.A., Korneva, S.P. *Zh. Obshch. Khim.* **1964**, *34*, 2787.
65AG130	Schrauzer, G.N., Kratel, G. *Angew. Chem.* **1965**, *77*, 130.
65CJC1879	Desnoyers, J., Rivest, R. *Can. J. Chem.* **1965**, *43*, 1879.
65JCS2904	Hewgill, F.R., Kennedy, B.R., Kilpin, D. *J. Chem. Soc.* **1965**, 2904.
65JCS3411	Deacon, G.B., Green, J.H.S., Nyholm, R.S. *J. Chem. Soc.* **1965**, 3411.
65JGU392	Kruglaya, O.A., Vyazankin, N.S., Razuvaev, G.A. *J. Gen. Chem. USSR (Engl. Trans.)*, **1965**, *35*, 392.
65JINC(27)1431	Olszewski, E.J., Albinak, M.J. *J. Inorg. Nucl. Chem.* **1965**, *27*, 1431.
65JOM(3)420	Considine, W.J., Ventura, J.J. *J. Organomet. Chem.* **1965**, *3*, 420.
65JPC2880	Peterson, D.B., Arakawa, T., Walmsley, D.A.G., Burton, M. *J. Phys. Chem.* **1965**, *69*, 2880.
65RA(4)104	Nefedov, V.D., Vobecky, M., Borak, J. *Radiochim. Acta* **1965**, *4*, 104.
65USP3171859	Sennewald, K., Gehrmann, K., Vogt, W., Schaefer, S. *US Patent* **1965**, 3,171,859; *German Patent* **1962**, 1,125,901; *Chem. Abstr.* **1962**, *57*, 12325b.
65ZOB394	Kruglaya, O.A., Vyazankin, N.S., Razuvaev, G.A. *Zh. Obshch. Khim.* **1965**, *35*, 394.
65ZOB481	Razuvaev, G.G., Petukhov, G.G., Titov, V.A., Druzhkov, O.N. *Zh. Obshch. Khim.* **1965**, *35*, 481.
66BCJ1823	Okawara, R., Yasuda, K., Inoue, M. *Bull. Chem. Soc. Jpn.* **1966**, *39*, 1823.
66CB1609	Müller, R., Dathe, C. *Chem. Ber.* **1966**, *99*, 1609.
66CC111	Hawley, D.M., Ferguson, G., Harris, G.S. *J. Chem. Soc., Chem. Commun.* **1966**, 111.
66CC228	Shaw, R.A., Smith, B.C., Thakur, C.P. *J. Chem. Soc., Chem. Commun.* **1966**, 228.
66DOK(166)99	Vyazankin, N.S., Kruglaya, O.A., Razuvaev, G.A., Semchikova, G.S. *Dokl. Akad. Nauk SSSR* **1966**, *166*, 99.

66DOK(169)598	Koton, M.M., Florinskii, F.S. *Dokl. Akad. Nauk SSSR* **1966**, *169*, 598.
66DOK(170)333	Egorochkin, A.N., Byazankin, N.S., Razuvaev, G.A., Kruglaya, O.A., Bochkarev, M.N. *Dok. Akad. Nauk SSSR,* **1966**, *170*, 333.
66IC2119	Houk, L.W., Dobson, G.R. *Inorg. Chem.* **1966**, *5*, 2119.
66IJC537	Mehrotra, R.C., Rai, A.K. *Indian. J. Chem.* **1966**, *4*, 537.
66JA467	Dessy, R.E., Chivers, T., Kitching, W. *J. Am. Chem. Soc.* **1966**, *88*, 467.
66JA2342	Doak, G.O., Long, G.G., Kakar, S.K., Freedman, L.D. *J. Am. Chem. Soc.* **1966**, *88*, 2342.
66JA4856	Muetterties, E.L., Roesky, H., Wright, C.M. *J. Am. Chem. Soc.* **1966**, *88,* 4856.
66JA5117	Dessy, R.E., Weissman, P.M., Pohl, R.L. *J. Am. Chem. Soc.* **1966**, *88,* 5117.
66JA5121	Dessy, R.E., Pohl, R.L., King, R.B. *J. Am. Chem. Soc.* **1966**, *88,* 5121.
66JCS(A)520	Beveridge, A.D., Harris, G.S., Inglis, F. *J. Chem. Soc. (A)* **1966**, 520.
66JOM(6)259	Schere, O.J., Hornig, P., Schmidt, M. *J. Organomet. Chem.* **1966**, *6*, 259.
66JOM(6)474	Vyazankin, N.S., Razuvaev, G.A., Kruglaya, O.A., Semchikova, G.S. *J. Organomet. Chem.* **1966**, *6*, 474.
66R98	Nefedov, V.D., Kirin, I.S., Gracheva, L.M., Grachev, S.A. *Radiokhimiya* **1966**, *8*, 98.
66R376	Nefedov, V.D., Grachev, S.A., Gracheva, L.M., Petrov, L.N. *Radiokhimiya* **1966**, *8*, 376.
66RPA185	Pickering, W.F. *Rev. Pure Appl. Chem.* **1966**, *16*, 185.
66USP3285890	Aftandilian, V.D. *US Patent* **1966**, 3,285,890; *Chem. Abstr.* **1967**, *66*, 11,289.
66ZOB2308	Egorochkin, A.N., Khorshev, S.Ya., Vyazankin, N.S., Bochkarev, M.N., Kruglaya, O.A., Semchikova, G.S. *Zh. Obshch. Khim.* **1966**, *37*, 2308.
67AP(300)520	Möhrle, H., Hang, W. *Arch. Pharm.* **1967**, *300*, 520.
67CC1042	Lloyd, D., Singer, M.I.C. *J. Chem. Soc., Chem. Commun.* **1967**, 1042.
67CJC161	Barbeau, C. *Can. J. Chem.* **1967**, *45*, 161.
67DOK(177)1352	Nesmeyanov, A.N., Sazonova, V.A., Sazonova, N.S., Plyukhina, V.N. *Dokl. Akad. Nauk SSSR* **1967**, *177*, 1352.
67ISOC196	Dötzer, R. *3rd Intern. Symp. Organomet. Chem.*, München, **1967**, Abstr., p. 196.
67IZV1629	Kostyanovskii, R.G., Chervin, I.I., Yakshin, V.V., Stepanyants, A.U. *Izv. Akad. Nauk SSSR, Ser. Khim.* **1967**, 1629.
67JINC(29)869	Duncan, J.F., Thomas, F.G. *J. Inorg. Nucl. Chem.* **1967**, *29*, 869.
67JOM(8)491	Wittig, G., Maercker, A. *J. Organomet. Chem.* **1967**, *8*, 491.
67LA(705)66	Hellwinkel, D., Kilthau, G. *Liebigs Ann. Chem.* **1967**, *705*, 66.
67M(98)731	Zorn, H., Schindlbauer, H., Hammer, D. *Monatsh. Chem.* **1967**, *98*, 731.
67R738	Gracheva, L.M., Nefedov, V.D., Grachev, S.A. *Radiokhimiya* **1967**, *9*, 738.
67ZAAC(349)328	Schmitz-DuMont, O., Ross, B. *Z. Anorg. Allg. Chem.* **1967**, *349*, 328.
68CB1200	Amberger, E., Gut, E. *Chem. Ber.* **1968**, *101*, 1200.

68CC1474	El Sheikh, S.I.A., Smith, B.C. *J. Chem. Soc., Chem. Commun.* **1968**, 1474.
68CJC2699	Peach, M.E. *Can. J. Chem.* **1968**, *46*, 2699.
68JCA(12)281	Aykan, K. *J. Catal.* **1968**, *12*, 281.
68JCS(A)1195	Dalton, J., Paul, I., Smith, J.G., Stone, F.G.A. *J. Chem. Soc. (A)* **1968**, 1195.
68JCS(A)2059	Hawley, D.M., Ferguson, G. *J. Chem. Soc. (A)* **1968**, 2059.
68JCS(A)2539	Hawley, D.M., Ferguson, G. *J. Chem. Soc. (A)* **1968**, 2539.
68JCS(A)2587	Powell, P. *J. Chem. Soc. (A)* **1968**, 2587.
68JCS(C)2770	Adderley, C.J.R., Hewgill, F.R. *J. Chem. Soc. (C)* **1968**, 2770.
68JOM(11)627	Kupchik, E.J., Theisen, C.T. *J. Organomet. Chem.* **1968**, *11*, 627.
68JOM(13)529	Reichle, W.T. *J. Organomet. Chem.* **1968**, *13*, 529.
68LA(713)30	Shaw, R.A., Smith, B.C., Thakur, C.P. *Liebigs Ann. Chem.* **1968**, *713*, 30.
68LA(720)198	Hellwinkel, D., Bach, M. *Liebigs Ann. Chem.* **1968**, *720*, 198.
68OCR(B)426	Doak, G.O., Freedman, L.D. *J. Organometal. Chem. Rev. B* **1968**, *4,* 426.
68OMS127	Wszolek, P.C., McLafferty, F.W., Brewster, J.H. *Org. Mass. Spectrom.* **1968**, *1*, 127.
68ZAAC(363)84	Schmidbaur, H., Belgfeld, M. *Z. Anorg. Allg. Chem.* **1968**, *363,* 84.
68ZOB205	Vyazankin, N.S., Kalinina, G.S., Kruglaya, O.A., Razuvaev, G.A. *Zh. Obshch. Khim.* **1968**, *38*, 205.
68ZOB276	Egorochkin, A.N., Khorshev, S.Y., Vyazankin, N.S., Gladyshev, E.N., Bychkov, V.T., Kruglaya, O.A. *Zh. Obshch. Khim.* **1968**, *38*, 276.
69ACS1525	Stålhandske, C.-I. *Acta Chem. Scand.* **1969**, *23*, 1525.
69AS(23)12	Shobatake, K., Postmus, C., Ferraro, J.R., Nakamoto, K. *Appl. Spectr.* **1969**, *23*, 12.
69AX(B)925	Shannon, R.D., Prewitt, C.T. *Acta Crystallogr.* **1969**, *B25*, 925.
69CRV(69)157	Mitchell, K.A.R. *Chem. Rev.* **1969**, *69*, 157.
69FRP1559505	Mueller, R., Frey, H., Dathe, C. *Fr. Patent* **1969**, 1,559,505; *Chem. Abstr.* **1970**, *72*, 43,877.
69IC83	Angelici, R.J., Ingemanson, C. M. *Inorg. Chem.* **1969**, *8*, 83.
69IZV1827	Nesmeyanov, A.N., Sazonova, N.S., Sazonova, V.A., Meskhi, L.M. *Izv. Akad. Nauk SSSR, Ser. Khim.* **1969**, 1827.
69JCE413	Gavin, R.M. *J. Chem. Ed.* **1969**, *46*, 413.
69JCS(A)2138	Deganello, G., Dolcetti, G., Giustiniani, M., Belluco, U. *J. Chem. Soc. (A)* **1969**, 2138.
69JOM(17)389	Hellwinkel, D., Bach, M. *J. Organomet. Chem.* **1969**, *17*, 389.
69JOM(20)99	Faraglia, G. *J. Organomet. Chem.* **1969**, *20*, 99.
69JOM(20)211	Wizemann, T., Müller, H., Seybold, D., Dehnicke, K. *J. Organomet. Chem.* **1969**, *20,* 211.
69NAT(221)169	Burstein, G.T., Wright, G. A. *Nature* **1969**, *221*, 169.
69NAT(221)1106	Coulson, A. *Nature* **1969**, *221*, 1106.
69OCR(B)186	Freedman, L.D., Doak, G.O. *Organometal. Chem. Rev. B* **1969**, *5*, 186.
69RACE119	Royo, P., Uson, R. *Rev. Acad. Cienc. Exact. Fis. Quim. Nat. Zaragoza* **1969**, *24*, 119.

69TL111 Zeeh, B., Thomson, J.B. *Tetrahedron Lett.* **1969**, *10*, 111.
69USP3454630 Nippon Kayaku Co. Ltd. *US Patent* **1969**, 3,454,630; *Neth. Appl.* **1966**, 6,512,930; *Chem. Abstr.* **1966**, *65*, 5369c.
69VLU167 Petrov, L.N., Kyuntsel, I.A., Grechiskhin, V.S. *Vestn. Leningrad. Univ. Fiz. Khim.* **1969**, 167.
69ZAAC(368)113 Müller, R., Frey, H.J. *Z. Anorg. Allg. Chem.* **1969**, *368*, 113.
70AG(E)400 Rauk, A., Allen, L.C., Mislow, K. *Angew. Chem., Int. Ed. Engl.* **1970**, *9*, 400.
70AP(303)233 Berka, A. *Arch. Pharm.* **1970**, *303*, 233.
70B-D Dub, M. *Organometallic Compounds*, Springer, New York, **1970**.
70B-DF Doak, G.O., Freedman, L.D. *Organometallic Compounds of Arsenic, Antimony, and Bismuth*, Wiley, New York, **1970**.
70BRP1211595 Goldman, M., Johnson, A.L. *Br. Patent* **1970**, 1,211,595; *Chem. Abstr.* **1971**, *74*, 81,794.
70CB799 Deubzer, B., Elian, M., Fischer, E.O., Fritz, H.P. *Chem. Ber.* **1970**, *103*, 799.
70CC1621 Gansow, O.A., Kimura, B.Y. *J. Chem. Soc., Chem. Commun.* **1970**, 1621.
70CJC2488 Goel, R.G., Prasad, H.S. *Can. J. Chem.* **1970**, *48*, 2488.
70IC408 Kolling, O.W., Mawdsley, E.A. *Inorg. Chem.* **1970**, *9*, 408.
70INCL71 Manoussakis, G.E., Karayannidis, P. *Inorg. Nucl. Chem. Lett.* **1970**, *6*, 71.
70JA266 Gimarc, B.M. *J. Am. Chem. Soc.* **1970**, *92*, 266.
70JCS(A)2258 Casey, M., Manning, A.R. *J. Chem. Soc. (A)* **1970**, 2258.
70JCS(B)735 Davies, A.G., Hook, S.C.W. *J. Chem. Soc. (B)* **1970**, 735.
70JINC(32)3249 Selbin, J., Ahmad, N., Pribble, M.J. *J. Inorg. Nucl. Chem.* **1970**, *32*, 3249.
70OCR(A)283 Davis, D.D., Gray, C.E. *Organometal. Chem. Rev. A* **1970**, *6*, 283.
70OCR(B)656 Freedman, L.D., Doak, G.O. *Organometal. Chem. Rev. B* **1970**, *6*, 656.
70USP3504005 Moedritzer, K., Van Wazer, J.R., Weingarten, H.I., Johnson, A.L. *US Patent* **1970**, 3504005; *Chem. Abstr.* **1970**, *72*, 121,707.
71BCJ2638 Oyamada, T., Iijima, T., Kimura, M. *Bull. Chem. Soc. Jpn.* **1971**, *44*, 2638.
71CB1892 Müller, A., Rao, V.V.K., Klinksiek, G. *Chem. Ber.* **1971**, *104*, 1892.
71CJC1747 Winchester, R.V. *Can. J. Chem.* **1971**, *49*, 1747.
71CJC2529 Goel, R.G., Prasad, H.S. *Can. J. Chem.* **1971**, *49*, 2529.
71DOK(198)590 Nesmeyanov, A.N., Sazonova, V.A., Sedova, N.N. *Dokl. Akad. Nauk SSSR* **1971**, *198*, 590.
71IC1080 White, J.F., Farona, M.F. *Inorg. Chem.* **1970**, *10*, 1080.
71IZV2618 Ptitsyna, O.A., Gurskii, M.E., Mariorova, T.D., Reutov, O.A. *Izv. Akad. Nauk SSSR, Ser. Khim.* **1971**, 2618.
71JA593 Gimarc, B.M. *J. Am. Chem. Soc.* **1971**, *93*, 593.
71JA815 Gimarc, B.M. *J. Am. Chem. Soc.* **1971**, *93*, 815.
71JA822 Connor, J., Young, P.J., Strausz, O.P. *J. Am. Chem. Soc.* **1971**, *93*, 822.
71JA3293 Ashe, A.J., III *J. Am. Chem. Soc.* **1971**, *93*, 3293.
71JA6690 Ashe, A.J., III *J. Am. Chem. Soc.* **1971**, *93*, 6690.

71JCS(A)562	Goel, R.G., Prasad, H.S. *J. Chem. Soc. (A)* **1971**, 562.
71JCS(C)1660	Davies, A.G., Hook, S.C.W. *J. Chem. Soc. (C)* **1971**, 1660.
71JCS(C)1750	Hewitt, D.G. *J. Chem. Soc. (C)* **1971**, 1750.
71JINC(33)892	Brown, R.A., Dobson, G.R. *J. Inorg. Nucl. Chem.* **1971**, *33*, 892.
71JOM(26)C10	Deacon, G.B., Fallon, G.D., Felder, P. W. *J. Organomet. Chem.* **1971**, *26*, C10.
71JOM(27)149	Marshall, R.A.G., Pollard, D.R. *J. Organomet. Chem.* **1971**, *27*, 149.
71JOM(30)365	Deketelaere, R.F., Delbeke, F.T., VanderKelen, G.P. *J. Organomet. Chem.* **1971**, *30*, 365.
71JOM(32)C11	Smith, B.C., Waller, C.B. *J. Organomet. Chem.* **1971**, *32*, C11.
71JOM(33)C81	Crociani, B., Nicolini, M., Boschi, T. *J. Organomet. Chem.* **1971**, *33*, C81.
71JOM(48)292	Freedman, L.D., Doak, G.O. *J. Organomet. Chem.* **1971**, *48*, 292.
71KKZ1807	Tsuchiya, F., Suzuki, S., Ikawa, T. *Kogyo Kagaku Zassi* **1971**, *74*, 1807.
71KKZ2072	Tsuchiya, F., Sumi, T., Ikawa, T. *Kogyo Kagaku Zassi* **1971**, *74*, 2072.
71OCR(B)247	Freedman, L.D., Doak, G.O. *Organometal. Chem. Rev. B* **1971**, *8*, 247.
71PS(1)93	Schipper, P., Buck, H.M. *Phosphorus* **1971**, *1*, 93.
71SAG77	Bregadze, V.I., Sadzhaya, D.N., Okhlobystin, O.Y. *Soobshch. Akad. Nauk Gruz. SSR* **1971**, *63*, 77.
71SK(B)317	Vaska, L., Peone, J. *Suomen Kemistilehti B* **1971**, *44*, 317; *Chem. Abstr.* **1971**, *75*, 157,736.
71VLU113	Petrov, L.N., Sivkov, V.P., Avrorin, V.V., Zhuraviev, V.E. *Vestn. Leningrad. Univ. Fiz. Khim.* **1971**, 113.
71ZAAC(381)103	Müller, J. *Z. Anorg. Allg. Chem.* **1971**, *381*, 103.
72AJC2107	Deacon, G.B., Fallon, G.D. *Aust. J. Chem.* **1972**, *25*, 2107.
72CC912	Freeman, B.H., Harris, G.S., Kennedy, B.W., Lloyd, D. *J. Chem. Soc., Chem. Commun.* **1972**, 912.
72GEP2133110	Shiraishi, T., Kishiwada, S., Shimizu, Y. *German Patent* **1972**, 2,133,110; *Chem. Abstr.* **1972**, *76*, 58981f.
72GEP2203710	Grasselli, R.K., Miller, A.F., Hardman, H.F. *Ger. Patent* **1972**, 2,203,710; *Chem. Abstr.* **1972**, *77*, 139423y.
72INCL(8)271	Deacon, G.B., Johnson, I.K. *Inorg. Nucl. Chem. Lett.* **1972**, *8*, 271.
72JA7596	Ashe, A.J., III, Gordon, M.D. *J. Am. Chem. Soc.* **1972**, *94*, 7596.
72JCS(D)1120	Curry, J.D., Jandacek, R.J. *J. Chem. Soc., Dalton Trans.* **1972**, 1120.
72JINC(34)2523	Paul, R.C., Paul, K.K., Malhortra, K.C. *J. Inorg. Nucl. Chem.* **1972**, *34*, 2523.
72JINC(34)3913	Radheshwar, P.V., Dev, R., Cady, G.H. *J. Inorg. Nucl. Chem.* **1972**, *34*, 3913.
72JOM(36)323	Goel, R.G., Prasad, H.S. *J. Organomet. Chem.* **1972**, *36*, 323.
72JOM(38)91	Faraglia, G., Rivarola, E., Di Bianca, F. *J. Organomet. Chem.* **1972**, *38*, 91.
72JOM(44)317	Faleschini, S., Zanella, P., Doretti, L., Faraglia, G. *J. Organomet. Chem.* **1972**, *44*, 317.

72OMR73	Reutov, O.A., Ptitsyna, O.A., in *Organometallic Reactions*, eds. Becker, E.I., Tsutsui, M., **1972**, vol. 4, p. 73.
72OMS1183	Kostyanovsky, R.G., Plekhanov, V.G. *Org. Mass. Spectrom.* **1972**, *6*, 1183.
72R436	Nefedov, V.D., Petrov, L.N., Avrorin, V.V. *Radiokhimiya* **1972**, *12*, 436.
72T343	Freeman, B.H., Lloyd, D., Singer, M.I.C. *Tetrahedron* **1972**, *28*, 343.
72ZN(B)591	Keck, J.M., Klar, G. *Z. Naturforsch.* **1972**, *27B*, 591.
72ZSK(13)701	Borisova, N.P., Petrov, L.N. *Zh. Strukt. Khim.* **1972**, *13*, 701.
73AG773	Bock, H., Ramsey, B.G. *Angew. Chem.* **1973**, *85*, 773.
73B-BENT547	Smith, J.D., in *Comprehensive Inorganic Chemistry*, eds. Bailar, J.C., Jr., Emelius, H.J., Nyholm, R., Trotman-Dickenson, A.F., Perganom, Oxford, **1973**, p. 547.
73BCJ2910	Asano, R., Moritani, I., Fujiwara, Y., Teranishi, S. *Bull. Chem. Soc. Jpn.* **1973**, *46*, 2910.
73CC562	Kon, E., McNelis, E. *J. Chem. Soc., Chem. Commun.* **1973**, 562.
73IC944	Beaumont, R.E., Goel, R.G., Prasad, H.S. *Inorg. Chem.* **1973**, *12*, 944.
73IC3015	Manoussakis, G.E., Tsipis, C.A., Christophides, A.G. *Inorg. Chem.* **1973**, *12*, 3015.
73IZV229	Ptitsyna, O.A., Gurskii, M.E., Reutov, O.A. *Izv. Akad. Nauk SSSR, Ser. Khim.* **1973**, 229.
73JA5870	Shaw, R.W. Jr., Carroll, T.X., Thomas, T.D. *J. Am. Chem. Soc.* **1973**, *95*, 5870.
73JCS(D)1394	Beaumont, R.E., Goel, R.G. *J. Chem. Soc., Dalton Trans.* **1973**, 1394.
73JMOS(17)429	Beagley, B., McAloon, K.T. *J. Mol. Struct.* **1973**, *17*, 429.
73JOC764	Reese, A.L., McMartin, K., Miller, D., Wickham, P.P. *J. Org. Chem.* **1973**, *38*, 764.
73JOM(47)367	Nesmeyanov, A.N., Kravtosov, D.N., Krasov, B.A., Pombrik, S.I., Fedin, E.I. *J. Organomet. Chem.* **1973**, *47*, 367.
73JOM(50)129	Goel, R.G., Prasad, H.S. *J. Organomet. Chem.* **1973**, *50,* 129.
73JP(2)471	Scaiano, J. *J. Photochem.* **1973/74**, *2*, 471.
73PS(3)33	Bordner, J., Freedman, L.D. *Phosphorus* **1973**, *3*, 33.
73S328	Doak, G.O., Freedman, L.D. *Synthesis*, **1973**, 328.
74ACS(B)883	Adler, E., Holmberg, K., Ryrfors, L.-O. *Acta Chem. Scand.* **1974**, *28B*, 883.
74ACS(B)888	Adler, E., Holmberg, K., Ryrfors, L.-O. *Acta Chem. Scand.* **1974**, *28B*, 888.
74BBI(suppl.)	Arellano, M.E., Guzman, L.A. *The Bullen of the Bismuth Institute*, **1974**, Suppl.
74CL733	Nakayama, H., Nishijima, C., Tachiyashiki, S. *Chem. Lett.* **1974**, 733.
74EL(10)1424	Chernykh, I.N., Tomilov, A.P. *Elektrokhimiya* **1974**, *10*, 1424.
74IC135	Lawton, S.L., Fuhrmeister, C.J., Haas, R.G., Jarman, C.S. Jr., Lohmeyer, F.G. *Inorg. Chem.* **1974**, *13*, 135.
74IZV234	Kravtsov, D.N., Krasov, B.A., Pombrik, S.I., Fedin, E.I. *Izv. Akad. Nauk SSSR, Ser. Khim.* **1974**, 234.

74IZV927 Kravtsov, D.N., Krasov, B.A., Pombrik, S.I., Fedin, E.I. *Izv. Akad. Nauk SSSR, Ser. Khim.* **1974**, 927.
74JA112 Rossi, R.A., Bunnett, J.F. *J. Am. Chem. Soc.* **1974**, *96*, 112.
74JINC(36)472 Müller, A., Rao, V.V.K., Christophliemk, P. *J. Inorg. Nucl. Chem.* **1974**, *36*, 472.
74JMOS(23)329 van der Kelen, G.P., de Ketelaere, R.F. *J. Mol. Struc.* **1974**, *23*, 329.
74JOM(68)411 Freedman, L.D., Doak, G.O. *J. Organomet. Chem.* **1974**, *68*, 411.
74JOM(73)217 Brandes, D., Blaschette, A. *J. Organomet. Chem.* **1974**, *73*, 217.
74JOM(73)251 Deketelaere, R.F., Van der Kelen, G.P. *J. Organomet. Chem.* **1974**, *73*, 251.
74PS(4)25 Padma, D.K., Shaw, R.A., Murthy, A.R., Woods, M. *Phosphorus* **1974**, *4*, 25.
74USP3825600 Ohara, T., Ueshima, M., Nishinomiya, H., Yanagisawa, I. *US Patent* **1974**, 3,825,600; *Ger. Patent* 2,125,032; *Chem. Abstr.* *76*, 58980e.
74ZAAC(406)74 Engler, R. *Z. Anorg. Allg. Chem.* **1974**, *406*, 74.
74ZAAC(407)35 Engler, R. *Z. Anorg. Allg. Chem.* **1974**, *407*, 35.
75CC164 Levason, W., McAuliffe, C.A., Murray, S.G. *J. Chem. Soc., Chem. Commun.* **1975**, 164.
75CJC1530 Manoussakis, G.E., Tsipis, C.A., Hadjikostas, C.C. *Can. J. Chem.* **1975**, *53*, 1530.
75DOK(225)581 Razuvaev, G.A., Osanova, N.A., Sharutin, V.V. *Dokl. Akad. Nauk. SSSR* **1975**, *225*, 581.
75IC1058 Elian, M., Hoffmann, R. *Inorg. Chem.* **1975**, *14*, 1058.
75IC1722 Garner, C.D., Hughes, B. *Inorg. Chem.* **1975**, *14*, 1722.
75JCS(D)1291 March, F.C., Ferfuson, G. *J. Chem. Soc., Dalton Trans.* **1975**, 1291.
75JINC(37)2011 Ando, F., Hayashi, T., Ohashi, K., Koketsu, J. *J. Inorg. Nucl. Chem.* **1975**, *37*, 2011.
75JINC(37)2347 Ouchi, A., Uehiro, T., Yoshino, Y. *J. Inorg. Nucl. Chem.* **1975**, *37*, 2347.
75JINC(37)2559 Ouchi, A., Honda, H., Kitazima, S. *J. Inorg. Nucl. Chem.* **1975**, *37*, 2559.
75JOC1515 Kon, E., McNelis, E. *J. Org. Chem.* **1975**, *40*, 1515.
75JOM(86)383 Kravtsov, D.N., Kvasov, B.A., Pombrik, S.I., Fedin, E.I. *J. Organomet. Chem.* **1975**, *86*, 383.
75JOM(87)83 Schumann, H., Breunig, H.J. *J. Organomet. Chem.* **1975**, *87*, 83.
75JOM(88)329 Krommes, P., Lorberth, J. *J. Organomet. Chem.* **1975**, *88*, 329.
75JOM(89)183 Freedman, L.D., Doak, G.O. *J. Organomet. Chem.* **1975**, *89*, 183.
75JOM(93)339 Krommes, P., Lorberth, J. *J. Organomet. Chem.* **1975**, *93*, 339.
75TL4361 Lema, R.H., Scaiano, J.C. *Tetrahedron Lett.* **1975**, *16*, 4361.
75ZAAC(417)235 Manoussakis, G.E., Tsipis, C.A., Christophides, A.G. *Z. Anorg. Allg. Chem.* **1975**, *417*, 235.
76AX(A)751 Shannon, R.D. *Acta Crystallogr.* **1976**, *A32*, 751.
76AJC559 Winter, G. *Aust. J. Chem.* **1976**, *29*, 559.
76CB1140 Fischer, E.O., Richter, K. *Chem. Ber.* **1976**, *109*, 1140.
76CL29 Suzuki, S., Moro-oka, Y., Ikawa, T. *Chem. Lett.* **1976**, 29.
76IZV335 Smirnov, S.G., Rodionov, A.N., Rogozhin, K.L., Syutkina, O.P., Panov,

	E.M., Shigorin, D.N., Kocheshkov, K. A. *Izv. Akad. Nauk SSSR, Ser. Khim.* **1976**, 335.
76IZV338	Rogozhin, K.L., Rodionov, A.N., Smirnov, S.G., Shigorin, D.N., Syutkina, O.P., Lodochnikova, V.I., Kocheshkov, K.A. *Izv. Akad. Nauk SSSR, Ser. Khim.* **1976**, 338.
76JCS(D)791	Raston, C.L., White, A.H. *J. Chem. Soc., Dalton Trans.* **1976**, 791.
76JOC1646	Kon, E., McNelis, E. *J. Org. Chem.* **1976**, *41*, 1646.
76JOM(106)255	Freedman, L.D., Doak, G.O. *J. Organomet. Chem.* **1976**, *106*, 255.
76JOM(111)C1	Burns, R.P., Levason, W., McAuliffe, C.A. *J. Organomet. Chem.* **1976**, *111*, C1.
76JOM(116)199	Glidewell, C.C. *J. Organomet. Chem.* **1976**, *116*, 199.
76JOM(122)351	Levason, W., McAuliffe, C.A., Sedgwick, R.D. *J. Organomet. Chem.* **1976**, *122*, 351.
76OMS1019	Spalding, T.R. *Org. Mass. Spectrom.* **1976**, *11*, 1019.
76PRB896	Poole, R.T., Liesegang, J., Leckey, R.C.G., Jenkin, J.G., Peel, J.B. *Phys. Rev. B* **1976**, *13*, 896.
76SA(A)569	Goel, R.G., Prasad, H.S. *Spectrochim. Acta* **1976**, *32A*, 569.
76TL411	Bastide, J., Heilbronner, E., Maier, J.P., Ashe, A.J., III *Tetrahedron Lett.* **1976**, *6*, 411.
76TL415	Ashe, A.J., III *Tetrahedron Lett.* **1976**, *6*, 415.
76TL3217	Basha, A., Ahmed, S.S., Farooqui, T.A. *Tetrahedron Lett.* **1976**, 3217.
76TMC183	Ichikawa, K., Watanabe, O., Fukuzumi, K. *Transition Met. Chem.* **1976**, *1*, 183.
76ZAAC(423)40	Wieber, M., Baudis, U. *Z. Anorg. Allg. Chem.* **1976**, *423*, 40.
76ZAAC(423)47	Wieber, M., Baudis, U. *Z. Anorg. Allg. Chem.* **1976**, *423*, 47.
77ACS(A)775	Ahrland, S., Berg, T., Trinderup, P. *Acta Chem. Scand.* **1977**, *31A*, 775.
77BCJ2021	Kawamura, T., Kikukawa, K., Takagi, M., Matsuda, T. *Bull. Chem. Soc. Jpn.* **1977**, *50*, 2021.
77CB3430	Stelzer, O., Unger, E. *Chem. Ber.* **1977**, *110*, 3430.
77DOK(237)852	Razuvaev, G.A., Domrachev, G.A., Sharutin, V.V., Suvorova, O.N. *Dokl. Akad. Nauk. SSSR* **1977**, *237*, 852.
77GH(47)	Wieber, M. *Gmelins Handbuch der Anorganischen Chemie, Bismut-Organische Verbindungen*, Springer, Berlin, **1977**, vol. 47.
77IC1823	Morrison, J.A., Lagow, R.J. *Inorg. Chem.* **1977**, *16*, 1823.
77IZV1035	Rogozhin, K.L., Rodionov, A.N., Smirnov, S.G., Shigorin, D.N., Panov, E.M., Kocheshkov, K.A. *Izv. Akad. Nauk SSSR, Ser. Khim.* **1977**, 1035.
77JCS(D)641	El Cheikh, S.I.A., Patel, M.S., Smith, B.C., Waller, C.B. *J. Chem. Soc., Dalton Trans.* **1977**, 641.
77JCS(D)711	Levason, W., McAuliffe, C.A., Murray, S.G. *J. Chem. Soc., Dalton Trans.* **1977**, 711.
77JCS(D)1141	Battaglia, L.P., Corradi, A.B., Pelizzi, G., Vidoni, M.E. *J. Chem. Soc., Daton Trans.* **1977**, 1141.
77JINC(39)565	Peach, M.E. *J. Inorg. Nucl. Chem.* **1977**, *39*, 565.

77JMOS(40)89	Claeys, E.G., Van der Kelen, G.P., De Ketelaere, R.F. *J. Mol. Struct.* **1977**, *40*, 89.
77JOM(129)55	Gruning, R., Lorberth, J. *J. Organomet. Chem.* **1977**, *129*, 55.
77JOM(129)163	Wuyts, L.F., Van de Vondel, D.F., Van der Kelen, G.P. *J. Organomet. Chem.* **1977**, *129*, 163.
77JOM(134)C32	Panster, P., Malisch, W. *J. Organomet. Chem.* **1977**, *134*, C32.
77JOM(136)185	Pombrik, S.I., Kravtosov, D.N., Krasov, B.A., Fedin, E.I. *J. Organomet. Chem.* **1977**, *136*, 185.
77MJ(22)442	Bishara, S.W., El-Samman, F.M. *Microchem. J.* **1977**, *22*, 442.
77ZAAC(436)101	Wieber, M., Baudis, U. *Z. Anorg. Allg. Chem.* **1977**, *436*, 101.
78ACS(A)933	Ahrland, S., Berg, T., Bläuenstein, P. *Acta Chem. Scand.* **1978**, *32A*, 933.
78AG(E)214	Ruppert, I., Bastian, V. *Angew. Chem., Int. Ed. Engl.* **1978**, *17*, 214.
78B-LS	Lederer, C.M., Shirley, V. S. *Table of Isotopes*, Wiley, New York, **1978**, 7th ed.
78BCJ319	Abe, K., Ito, M. *Bull. Chem. Soc. Jpn.* **1978**, *51*, 319.
78CB1978	Talay, R., Rehder, D. *Chem. Ber.* **1978**, *111*, 1978.
78CC1099	Barton, D.H.R., Kitchin, J.P., Motherwell, W.B. *J. Chem. Soc., Chem. Commun.* **1978**, 1099.
78DOK(238)361	Razuvaev, G.A., Osanova, N.A., Sharutin, V.V., Sorokin, A.I., Okhlopkova, I.E. *Dokl. Akad. Nauk. SSSR* **1978**, *238*, 361.
78IC2261	Hall, M.B. *Inorg. Chem.* **1978**, *17*, 2261.
78IJC(16A)778	Bhattacharya, S.N., Singh, M. *Indian J. Chem.* **1978**, *16A*, 778.
78IZV1643	Arbuzov, B.A., Belkin, Y.V., Polezhaeva, N.A., Buslaeva, G.E. *Izv. Akad. Nauk SSSR, Ser. Khim.* **1978**, 1643.
78JA5717	Palke, W.E., Kirtman, B. *J. Am. Chem. Soc.* **1978**, *100*, 5717.
78JMOC(4)231	Uson, R., Oro, L.A., Claver, C., Garralda, M.A., Moreto, J.M. *J. Mol. Catal.* **1978**, *4*, 231.
78JOM(162)289	Bochkarev, M.N., Gur'ev, N.I., Razuvaev, G.A. *J. Organomet. Chem.* **1978**, *162*, 289.
78JOM(163)141	Doak, G.O., Freedman, L.D. *J. Organomet. Chem.* **1978**, *163*, 141.
78KOE(3)912	Carapella, S.C. Jr., Howe, H.E., in *Kirk-Othmer Encyclopedia of Chemical Technology*, Wiley, New York, 3rd ed., **1978**, vol. 3, p. 912.
78MOC(13-8)	Samaan, S., in *Methoden der Organischen Chemie, Metalloorganischen Verbindungen, As, Sb, Bi*, Georg Thieme, Stuttgart, **1978**, Vol. XIII, Part 8.
78ZAAC(439)134	Wieber, M., Baudis, U. *Z. Anorg. Allg. Chem.* **1978**, *439,* 134.
78ZAAC(439)139	Wieber, M., Baudis, U. *Z. Anorg. Allg. Chem.* **1978**, *439,* 139.
78ZN(B)1	Veith, M. *Z. Naturforsch.* **1978**, *33b*, 1.
78ZN(B)7	Veith, M. *Z. Naturforsch.* **1978**, *33b*, 7.
79AJC1273	Ahond, A., Coll, B.F.B.C., Fourneron, J.D., Mitchell, S.J. *Aust. J. Chem.* **1979**, *32*, 1273.
79B-WSA1111	Poller, R., in *Comprehensive Organometallic Chemistry*, eds. Wilkinson, G., Stone, F.G., Abel, E.W., Pergamon, **1979**, p. 1111.

79CC639 Carty, A.J., Taylor, N.J., Coleman, A.W., Lappert, M.F. *J. Chem. Soc., Chem. Commun.* **1979**, 639.

79CC705 Barton, D.H.R., Lester, D.J., Motherwell, W.B., Papoula, M.T.B. *J. Chem. Soc., Chem. Commun.* **1979**, 705.

79JAP(K)83435 Sakai, K., Hashimoto, M., Sasaki, M., Ohta, M., Tsutsui, K., Kazami, T. *Jpn. Kokai* **1979**, 83,435, *Chem. Abstr.* **1980**, *92*, 102,295.

79JCT(11)187 Steele, W.V. *J. Chem. Thermodyn.* **1979**, *11*, 187.

79JES191 Hoste, S., van de Vondel, D.F., van der Kelen, G.P. *J. Electron. Spectrosc. Relat. Phenom.* **1979**, *17*, 191.

79JINC(41)1390 Peach, M.E. *J. Inorg. Nucl. Chem.* **1979**, *41*, 1390.

79JOM(166)365 Allman, T., Goel, R.G., Prasad, H.S. *J. Organomet. Chem.* **1979**, *166*, 365.

79JOM(171)333 Otero, A., Royo, P. *J. Organomet. Chem.* **1979**, *171*, 333.

79JOM(180)143 Doak, G.O., Freedman, L.D. *J. Organomet. Chem.* **1979**, *180*, 143.

79JOM(182)C49 Alberti, A., Hudson, A. *J. Organomet. Chem.* **1979**, *182*, C49.

79JPR(321)370 Dahlmann, J., Winsel, K. *J. Prakt. Chem.* **1979**, *321*, 370.

79SA(A)339 Goel, R.G., Prasad, H.S. *Spectrochim. Acta, Part A*, **1979**, *35*, 339.

79ZAAC(448)89 Wieber, M., Basel, A. *Z. Anorg. Allg. Chem.* **1979**, *448,* 89.

79ZN(B)1037 Burschka, C., Wieber, M. *Z. Naturforsch.* **1979**, *34B*, 1037.

79ZOB1185 Anishchenko, N.I., Panov, E.M., Syutkina, O.P., Kocheshkov, K.A. *Z. Obshch. Khim.* **1979**, *49*, 1185.

79ZOB2257 Lopatin, M.A., Kuznetsov, V.A., Egorochkin, A.N., Bochkarev, M.N. *Zh. Obshch. Khim.* **1979**, *49*, 2257.

80AG(E)723 Steinseifer, F., Kauffmann, T. *Angew. Chem., Int. Ed. Engl.* **1980**, *19*, 723.

80B-O *Remington's Pharmaceutical Sciences*, ed. Osol, A., Mack Publishing, Easton, PA, **1980**, 16th ed.

80CC246 Barton, D.H.R., Lester, D.J., Motherwell, W.B., Papoula, M.T.B. *J. Chem. Soc., Chem. Commun.* **1980**, 246.

80CC827 Barton, D.H.R., Blazejewski, J.-C., Charpiot, B., Lester, D.J., Motherwell, W.B., Papoula, M.T.B. *J. Chem. Soc., Chem. Commun.* **1980**, 827.

80CC1089 Barton, D.H.R., Gero, S.D., Maycock, C.D. *J. Chem. Soc., Chem. Commun.* **1980**, 1089.

80IC3863 Klemperer, W.G., Liu, R.S. *Inorg. Chem.* **1980**, *19*, 3863.

80IJC(A)902 Brar, A.S., Sarpal, A.S., Sandhu, S.S. *Indian J. Chem.* **1980**, *19A*, 902.

80INCL189 Williams, D.J. *Inorg. Nucl. Chem. Lett.* **1980**, *16*, 189.

80JA7753 Perkins, C.W., Martin, J.C., Arduengo, A.J., III, Lau, W., Algria, A., Kochi, J.K. *J. Am. Chem. Soc.* **1980**, *102*, 7753.

80JCA(66)347 Brazdil, J.F., Suresh, D.D., Grasselli, R.K. *J. Catal.* **1980**, *66*, 347.

80JCS(D)2428 Gynane, M.J.S., Hudson, A., Lappert, M.F., Power, P.P., Goldwhite, H. *J. Chem. Soc., Dalton Trans.* **1980**, 2428.

80JOC726 Truesdale, L.K., Reuman, M.E. *J. Org. Chem.* **1980**, *45*, 726.

80JOM(199)205 Bochkarev, M.N., Razuvaev, G.A., Zakharov, L.N., Struchkov, Yu. T. *J. Organomet. Chem.* **1980**, *199*, 205.

80JOM(200)243	Razuvaev, G.A. *J. Organomet. Chem.* **1980**, *200*, 243.
80JOM(203)359	Doak, G.O., Freedman, L.D. *J. Organomet. Chem.* **1980**, *203*, 359.
80KK805	Zakharov, L.N., Struchkov, Y.T., Sharutin, V.V., Suvorova, O.N. *Koord. Khim.* **1980**, *6*, 805.
80KK945	Sobolev, A.N., Romm, I.P., Bel'skii, V.K., Gur'yanova, E.N. *Koord. Khim.* **1980**, *6*, 945.
80ZAAC(470)171	Wieber, M., Rüdling, H.G., Burschka, C. *Z. Anorg. Allg. Chem.* **1980**, *470,* 171.
80ZFK506	Davydov, S.N., Rodionov, A.N., Shigorin, D.N., Syutkina, O.P., Krasnova, T. L. *Zh. Fiz. Khim.* **1980**, *54*, 506.
80ZN(B)709	Georgiades, A., Schilling, G., Latscha, H.P. *Z. Naturforsch.* **1980**, *35B*, 709.
80ZN(B)1000	Georgiades, A., Latscha, H.P. *Z. Naturforsch.* **1980**, *35B,* 1000.
81B-CC1558	*Patty's Industrial Hygiene and Toxicology*, eds. Clayton, G.D., Clayton, F.E, Wiley, New York, **1981**, 3rd. ed., p. 1558.
81CC503	Barton, D.H.R., Blazejewski, J.-C., Charpiot, B., Motherwell, W.B. *J. Chem. Soc., Chem. Commun.* **1981**, 503.
81CC1232	Barton, D.H.R., Motherwell, W.B., Stobie, A. *J. Chem. Soc., Chem. Commun.* **1981**, 1232.
81IZV1874	Aladin, A.A., Artemov, A.N., Razuvaev, G.A. *Izv. Akad. Nauk SSSR, Ser. Khim.* **1981**, 1874.
81JA4587	Yoshifuji, Y., Shima, I., Inamoto, N., Hirotsu, K., Higuchi, T. *J. Am. Chem. Soc.* **1981**, *103*, 4587.
81JINC(43)1469	Janzen, A.F., Vaidya, O,C., Willis, C.J. *J. Inorg. Nucl. Chem.* **1981**, *43*, 1469.
81JOM(212)C7	Bras, P., Herwijer, H., Wolters, J. *J. Organomet. Chem.* **1981**, *212*, C7.
81T73	Barton, D.H.R., Kitchin, J.P., Lester, D.J., Motherwell, W.B., Papoula, M.T.B. *Tetrahedron* **1981**, *37(Suppl. 1)*, 73.
81TL2885	David, S., Thieffry, A. *Tetrahedron Lett.* **1981**, *22*, 2885.
81TL5063	David, S., Thieffry, A. *Tetrahedron Lett.* **1981**, *22*, 5063.
81ZN(B)70	Praeckel, U., Huber, F. *Z. Naturforsch.* **1981**, *36B,* 70.
81ZN(B)451	Talay, R., Rehder, D. *Z. Naturforsch.* **1981**, *36B*, 451.
81ZOB2380	Dodonov, V.A., Brilkina, T.G., Gushchin, A.V. *Zh. Obshch. Khim.* **1981**, *51*, 2380.
82AG(E)439	Breunig, H.J., Müller, D. *Angew. Chem., Int. Ed. Engl.* **1982**, *21*, 439.
82AG448	Breunig, H.J., Müller, D. *Angew. Chem.* **1982**, *94*, 448.
82AX(B)2140	Niven, M.L., Irving, H.M.N.H., Nassimbeni, L.R.N., Hutton, A.T. *Acta Crystallogr.* **1982**, *B38*, 2140.
82CC732	Barton, D.H.R., Papoula, M.T.B., Guilhem, J., Motherwell, W.B., Pascard, C., Dau, E.T.H. *J. Chem. Soc., Chem. Commun.* **1982**, 732.
82COMC(2)681	Wardell, J.L., in *Comprehensive Organometallic Chemistry*, eds. Abel, E.W., Stone, F.G.A., Wilkinson, G., Pergamon, Oxford, **1982**, vol. 2, p. 681.
82CRV15	Freedman, L.D., Doak, G.O. *Chem. Rev.* **1982**, *82*, 15.

82H641 Atwal, K.S., Sahoo, S.P., Tsai, T.Y.R., Wiesner, K. *Heterocycles* **1982**, *19*, 641.

82IC2097 Williams, D.J., Quicksall, C.O., Barkigia, K.M. *Inorg. Chem.* **1982**, *21*, 2097.

82JA5693 Ashe, A.J., III, Diephouse, T.R., El-Sheikh, M.Y. *J. Am. Chem. Soc.* **1982**, *104*, 5693.

82JCC(12)53 Levason, W., Sheikh, B., McCullough, F.P. *J. Coord. Chem.* **1982**, *12*, 53.

82JOM(228)C29 Kaul, H.-A., Greissinger, D., Luksza, M., Malisch, W. *J. Organomet. Chem.* **1982**, *228*, C29.

82JOM(237)51 Doak, G.O., Freedman, L.D. *J. Organomet. Chem.* **1982**, *237*, 51.

82JOM(240)C45 Praeckel, U., Huber, F. *J. Organomet. Chem.* **1982**, *240*, C45.

82OM1408 Ashe, A.J., III, Ludwig, E.G., Jr. *Organometallics* **1982**, *1*, 1408.

82OMC129 Wardell, J.L. *Organomet. Chem.* **1982**, *10*, 129.

82RRC75 Ionescu, N.I., Caldararu, M. *Rev. Roum. Chim.* **1982**, *27*, 75.

82TL3365 Barton, D.H.R., Charpiot, B., Motherwell, W.B. *Tetrahedron Lett.* **1982**, *23*, 3365.

82ZAAC(485)217 Burschka, C. *Z. Anorg. Allg. Chem.* **1982**, *485*, 217.

82ZAK(37)2079 Shushunova, A.F., Prokhorova, L.P., Aladin, A.A., Dodonov, V.A. *Z. Anal. Khim.* **1982**, *37*, 2079.

82ZN(B)91 Becker, G., Rössler, M. *Z. Naturforsch.* **1982**, *37B*, 91.

82ZN(B)815 Huber, F., Bock, S. *Z. Naturforsch.* **1982**, *37B*, 815.

83CC507 Calderazzo, F., Morvillo, A., Pelizzi, G., Poli, R. *J. Chem. Soc., Chem. Commun.* **1983**, 507.

83EOHS(1)291 *Encyclopedia of Occupational Health and Safety*, ed. Parmeggiani, L., 3rd ed., **1983**, vol. 1, p. 291.

83IC3421 Murray, B., Hvoslef, J., Hope, H., Power, P.P. *Inorg. Chem.* **1983**, *22*, 3421.

83ICA(73)141 Alonzo, G. *Inorg. Chim. Acta* **1983**, *73*, 141.

83IJC(A)814 Singh, S., Verma, A.D., Verma, R.D. *Indian. J. Chem.* **1983**, *22A*, 814.

83JA5506 Cowley, A.H., Lasch, J.G., Norman, N.C., Pakulski, M. *J. Am. Chem. Soc.* **1983**, *105*, 5506.

83JCS(D)1067 Hunter, G., Weakley, T.J. R. *J. Chem. Soc., Dalton Trans.* **1983**, 1067.

83JMR139 Morgan, K., Sayer, B.G., Schrobilgen, G.J. *J. Mag. Res.* **1983**, *52*, 139.

83JOC441 David, S., Thieffry, A. *J. Org. Chem.* **1983**, *48*, 441.

83JOM(243)305 Bodner, G.M., Gagnon, C., Whittern, D.N. *J. Organomet. Chem.* **1983**, *243*, 305.

83JOM(244)129 Sobolev, A.N., Belsky, V.K., Chernikova, N. Yu., Akhmadulina, F. Yu. *J. Organomet. Chem.* **1983**, *244*, 129.

83JOM(252)C23 Greissinger, D., Malisch, W., Kaul, H.A. *J. Organomet. Chem.* **1983**, *252*, C23.

83JOM(253)317 Falaglia, G., Graziani, R., Volponi, L., Casellato, U. *J. Organomet. Chem.* **1983**, *253*, 317.

83JOM(253)C21 Breunig, H.J., Muller, D. *J. Organomet. Chem.* **1983**, *253*, C21.

83JOM(256)C1 Bras, P., VanderGen, A., Wolters, J. *J. Organomet. Chem.* **1983**, *256*, C1.

83JPC301 Driscoll, D.J., Lunsford, J.H. *J. Phys. Chem.* **1983**, *87*, 301.

83KK262 Sobolev, A.A., Bel'skii, V.K., Romm, I.P. *Koord. Khim.* **1983**, *9*, 262.

83OM1859 Ashe, A.J., III., Ludwig, E.G., Jr., Oleksyszyn, J. *Organometallics* **1983**, *2,* 1859.

83OMC157 Wardell, J.L. *Organomet. Chem.* **1983**, *11*, 157.

83OX405 Dahlmann, J., Höft, E. *Oxidation Commun.* **1983**, *5*, 405.

83PS(14)253 Horner, L., Simons, G. *Phosphorus and Sulfur* **1983**, *14*, 253.

83PS(18)121 Hellwinkel, D., Stahl, H., Gaa, H.G., Gottfried, R. *Phosphorus and Sulfur* **1983**, *18*, 121.

83TCC1 Hellwinkel, D. *Top. Curr. Chem.* **1983**, *109*, 1.

83ZAAC(505)138 Wieber, M., Wirth, D., Hess, K. *Z. Anorg. Allg. Chem.* **1983**, *505*, 138.

83ZAAC(505)147 Wieber, M., Rüdling, H.G. *Z. Anorg. Allg. Chem.* **1983**, *505*, 147.

83ZAAC(505)150 Wieber, M., Rüdling, H.G. *Z. Anorg. Allg. Chem.* **1983**, *505*, 150.

83ZAAC(506)42 Mundt, O., Becker, G., Rössler, M., Witthauer, C. *Z. Anorg. Allg. Chem.* **1983**, *506*, 42.

83ZN(B)125 Breunig, H.J., Müller, D. *Z. Naturforsch.* **1983**, *38B,* 125.

84AG(E)272 Kutzelnigg, W. *Angew. Chem., Int. Ed. Engl.* **1984**, *23*, 272.

84AJC527 Deacon, G.B., Jackson, W.R., Pfeiffer, J.M. *Aust. J. Chem.* **1984**, *37*, 527.

84AX(C)413 Blake, A.J., Ebsworth, E.A.V., Welch, A.J. *Acta Crystallogr.* **1984**, *C40*, 413.

84B-GE Greenwood, N.N., Earnshaw, A. *Chemistry of the Elements*, Pergamon, Oxford, **1984**.

84CB1999 Herberhold, M., Ehrenreich, W., Guldner, K. *Chem. Ber.* **1984**, *117*, 1999.

84HCA586 Barton, D.H.R., Charpiot, B., Dau, E.T.H., Motherwell, W.B., Pascard, C., Pichon, C. *Helv. Chim. Acta* **1984**, *67*, 586.

84ICA(81)L33 Hoskins, B.F., Tiekink, E.R.T., Winter, G. *Inorg. Chim. Acta* **1984**, *81,* L33.

84JA4794 Trinquier, G., Daudey, J.-P., Caruana, G., Madaule, Y. *J. Am. Chem. Soc.* **1984**, *106*, 4794.

84JAP(K)124347 *Jpn. Kokai* **1984**, 124347; *Chem. Abstr.* **1985**, *102*, 36678r.

84JC(290)377 Ermoshkin, A.E., Makarenko, N.P., Sakodynskii, K.I. *J. Chromatogr.* **1984**, *290*, 377.

84JCS(D)2365 Calderazzo, F., Poli, R., Pelizzi, G. *J. Chem. Soc., Dalton Trans.* **1984**, 2365.

84JCS(D)2535 Calderazzo, F., Poli, R., Pelizzi, G. *J. Chem. Soc., Dalton Trans.* **1984**, 2535.

84JOM(261)59 Doak, G.O., Freedman, L.D. *J. Organomet. Chem.* **1984**, *261*, 59.

84OM495 Ashe, A.J., III, Drone, F.J. *Organometallics* **1984**, *3*, 495.

84OMC127 Wardell, J.L. *Organomet. Chem.* **1982**, *12*, 127.

84OMS647 Gielen, M., Bras, P., Wolters, J. *Org. Mass Spectrom.* **1984**, *19*, 647.

84OX99 Giordano, N., Bart, J.C.J., Vitarelli, P., Cavallaro, S. *Oxidation Commun.* **1984**, *7*, 99.

84POL1337 Chauhan, H.P.S., Srivastava, G., Mehrotra, R.C. *Polyhedron* **1984**, *3*, 1337.

84USP4442307 Lewis R.M., Ryan, R.C., Slaugh, L.H. *US Patent* **1984**, 4,442,307; *Chem. Abstr.* **1984**, *100*, 209170h.

84USP4472314 Conner, W.C., Soled, S.L., Signorelli, A.J., DeRites, B.A. *US Patent* **1984**, 4,472,314; *Chem. Abstr.* **1985**, *102*, 5681g.

84ZAAC(510)104 Breunig, H.J., Gräfe, U. *Z. Anorg. Allg. Chem.* **1984**, *510*, 104.

84ZN(B)887 Wieber, M., Sauer, I. *Z. Naturforsch.* **1984**, *39B,* 887.

84ZN(B)1668 Wieber, M., Sauer, I. *Z. Naturforsch.* **1984**, *39B,* 1668.

84ZOB100 Dodonov, V.A., Gushchin, A.V., Grishin, D.F., Brilkina, T.G. *Zh. Obshch. Khim.* **1984**, *54*, 100.

84ZOB2157 Dodonov, V.A., Gushchin, A.V., Brilkina, T.G. *Zh. Obshch. Khim.* **1984**, *54*, 2157.

85AX(C)967 Sobolev, A.N., Belsky, V.K., Romm, I.P., Chernikova, N.Yu., Guryanova, E.N. *Acta Crystallogr.* **1985**, *C41*, 967.

85BSF849 Barton, D.H.R., Lusinchi, X., Sandoval, R. *Bull. Soc. Chim. Fr.* **1985**, *122*, 849.

85CB1031 Kauffmann, T., Steinseifer, F. *Chem. Ber.* **1985**, *118*, 1031.

85CB1039 Kauffmann, T., Steinseifer, F., Klas, N. *Chem. Ber.* **1985**, *118*, 1039.

85CC1455 Solymosi, F., Tombacz, I., Kutsan, G. *J. Chem. Soc., Chem. Commun.* **1985**, 1455.

85ICA(99)177 Hoskins, B.F., Tiekink, E.R.T., Winter, G. *Inorg. Chim. Acta* **1985**, *99,* 177.

85JA1062 Arif, A.M., Cowley, A.H., Norman, N.C., Pakulski, M. *J. Am. Chem. Soc.* **1985**, *107*, 1062.

85JA3397 Stewart, C.A., Calabrese, J.C., Arduengo, A.J., III *J. Am. Chem. Soc.* **1985**, *107*, 3397.

85JCS(D)629 Boyar, E.B., Robinson, S.D. *J. Chem. Soc., Dalton Trans.* **1985**, 629.

85JCS(P1)2657 Barton, D.H.R., Bhatnagar, N.Y., Blazejewski, J.-C., Charpiot, B., Finet, J.-P., Lester, D.J., Motherwell, W.B., Papoula, M.T.B., Stanforth, S.P. *J. Chem. Soc., Perkin Trans. I* **1985**, 2657.

85JCS(P1)2667 Barton, D.H.R., Blazejewski, J.-C., Charpiot, B., Finet, J.-P., Motherwell, W.B., Papoula, M.T.B., Stanforth, S.P. *J. Chem. Soc., Perkin Trans. I* **1985**, 2667.

85JCSR(15)561 Mazhar-Ul-Haque, Tayim, H.A., Ahmed, J., Horne, W. *J. Cryst. Spectr. Res.* **1985**, *15*, 561.

85JOM(287)57 Al-Jabar, N.A.A., Massey, A.G. *J. Organomet. Chem.* **1985**, *287*, 57.

85JOM(288)145 Al-Jabar, N.A.A., Massey, A.G. *J. Organomet. Chem.* **1985**, *288*, 145.

85JOM(290)133 Dräger, M., Schmidt, B.M. *J. Organomet. Chem.* **1985**, *290*, 133.

85JOML(17)353 Doak, G.O., Freedman, L.D. *J. Organomet. Chem. Libr.* **1985**, *17*, 353.

85JPC4415 Driscoll, D.J., Lunsford, J.H. *J. Phys. Chem.* **1985**, *89*, 4415.

85KK1171 Imyanitov, N.S. *Koord. Khim.* **1985**, *11*, 1171.

85OM641 Nakamura, E., Shimada, J., Kuwajima, I. *Organometallics* **1985**, *4*, 641.

85OMC133 Wardell, J.L. *Organomet. Chem.* **1985**, *13*, 133.

85/86OX127	Ismagilov, Z.R., Dobrynkin, N.M., Shepelin, K.P., Zhdan, P.A., Popovskii, V.V., Szabó, S., Nagy, F. *Oxidation Commun.* **1985/86**, *8*, 127.
85RKC(27)371	Guseinov, A.B., Mamedov, E.A., Rizaev, R.G. *React. Kinet. Catal. Lett.* **1985**, *27*, 371.
85T3463	Walts, A.E., Roush, W.R. *Tetrahedron* **1985**, *41*, 3463.
85TL4211	Wada, M., Akiba, K.-y. *Tetrahedron Lett.* **1985**, *26*, 4211.
85UE(A4)171	Krüger, J., Winkler, P., Lüderitz, E., Lück, M., in *Ullmann's Encyclopedia of Industrial Chemistry*, ed. Gerhartz, W., VCH, Weinheim, **1985**, 5th ed., vol. A4, p. 171.
85ZAAC(520)59	Wieber, M., Fetzer-Kremling, I., Wirth, D., Rüdling, H.G. *Z. Anorg. Allg. Chem.* **1985**, *520*, 59.
85ZN(B)848	du Mont, W.-W., Severengiz, T., Breunig, H.J., Müller, D. *Z. Naturforsch.* **1985**, *40B,* 848.
85ZN(B)1320	Müller, J., Müller, U., Loss, A., Lorberth, J., Donath, H., Massa, W. *Z. Naturforsch.* **1985**, *40B,* 1320.
85ZN(B)1476	Breunig, H.J., Müller, D. *Z. Naturforsch.* **1985**, *40B*, 1476.
85ZOB73	Dodonov, V.A., Grishin, A.V., Brilkina, T.G. *Zh. Obshch. Khim.* **1985**, *55*, 73.
85ZOB466	Dodonov, V.A., Grishin, A.V., Brilkina, T.G. *Zh. Obshch. Khim.* **1985**, *55*, 466.
85ZOB586	Dodonov, V.A., Starostina, T.I., Grishin, A.V., Zelenova, N.A. *Zh. Obshch. Khim.* **1985**, *55*, 586.
85ZOB1767	Pankratov, L.V., Lineva, A.N., Latyaeva, V.N., Cherkasov, V.K., Bochkarev, M.N., Razuvaev, G.A. *Zh. Obshch. Khim.* **1985**, *55*, 1767.
85ZOB2514	Dodonov, V.A., Grishin, A.V., Brilkina, T.G. *Zh. Obshch. Khim.* **1985**, *55*, 2514.
85ZOB2630	Grishin, A.V., Brilkina, T.G., Dodonov, V.A. *Zh. Obshch. Khim.* **1985**, *55*, 2630.
86ACR406	Huttner, G., Evertz, K. *Acc. Chem. Res.* **1986**, *19*, 406.
86AG(E)757	Schier, A., Wallis, J.M., Müller, G., Schmidbaur, H. *Angew. Chem., Int. Ed. Engl.* **1986**, *25*, 757.
86B-K	Kitchin, J.P. in *Organic Synthesis by Oxidation with Metal Compounds*, eds. Mijs, W.J., Jong, C.R.H.I., Plenum, New York, **1986**, p. 817.
86CC65	Barton, D.H.R., Finet, J.-P., Pichon, C. *J. Chem. Soc., Chem. Commun.* **1986**, 65.
86CL1173	Wang, L., Eguchi, K., Arai, H., Seiyama, T. *Chem. Lett.* **1986**, 1173.
86EUP173563	Philip, Jr. J.B. *Eur. Patent* **1986**, 173,563; *Chem. Abstr.* **1986**, *105*, 105,702.
86IC4836	Arif, A.M., Cowley, A.H., Norman, N.C., Pakulski, M. *Inorg. Chem.* **1986**, *25*, 4836.
86ICA(113)43	Deacon, G.B., Felder, P.W., Domagala, M., Huber, F., Rüther, R. *Inorg. Chim. Acta* **1986**, *113*, 43.
86JOM(298)67	Doak, G.O., Freedman, L.D. *J. Organomet. Chem.* **1986**, *298*, 67.

86JOM(303)197	Ashe, A.J., III, Ludwig, E.G., Jr. *J. Organomet. Chem.* **1986**, *303,* 197.
86JOM(308)289	Ashe, A.J., III, Ludwig, E.G., Jr. *J. Organomet. Chem.* **1986**, *308,* 289.
86JOM(316)C1	Brunner, H., Obermann, U., Wimmer, P. *J. Organomet. Chem.* **1986**, *316*, C1.
86OCS(3)636	Breunig, H.J., Müller, D. *Organomet. Synth.* **1986**, *3*, 636.
86OCS(3)638	Breunig, H.J., Müller, D. *Organomet. Synth.* **1986**, *3*, 638.
86OMC141	Wardell, J.L. *Organomet. Chem.* **1986**, *14*, 141.
86POL1513	Casassas, E., Visa, T. *Polyhedron* **1986**, *5*, 1513.
86RKC(31)409	Suleimanov, T.E., Vislovskii, V.P., Mamedov, E.A., Rizaev, R.G. *React. Kinet. Catal. Lett.* **1986**, *31*, 409.
86T3111	Barton, D.H.R., Bhatnagar, N.Y., Finet, J.-P., Motherwell, W.B. *Tetrahedron* **1986**, *42*, 3111.
86T5627	Barton, D.H.R., Finet, J.-P., Motherwell, W.B., Pichon, C. *Tetrahedron* **1986**, *42*, 5627.
86TL3615	Barton, D.H.R., Finet, J.-P., Khamsi, J. *Tetrahedron Lett.* **1986**, *27*, 3615.
86TL3619	Barton, D.H.R., Finet, J.-P., Khamsi, J., Pichon, C. *Tetrahedron Lett.* **1986**, *27*, 3619.
86TL4771	Wada, M., Ohki, H., Akiba, K.-y. *Tetrahedron Lett.* **1986**, *27*, 4771.
86ZAAC(539)110	Rüther, R., Huber, F., Preut, H. *Z. Anorg. Allg. Chem.* **1986**, *539*, 110.
86ZN(B)1129	Breunig, H.J., Müller, D. *Z. Naturforsch.* **1986**, *41B,* 1129.
86ZOB330	Sharutin, V.V., Bychkov, V.T., Bolotova, O.P., Kuzina, V.I. *Zh. Obshch. Khim.* **1986**, *56*, 330.
86ZOB2714	Dodonov, V.A., Gushchin, A.V., Brilkina, T.G., Muratova, L.V. *Zh. Obshch. Khim.* **1986**, *56*, 2714.
87ACA(196)351	Gawargious, Y.A., Hassouna, M.E., Hassan, H.N.A. *Anal. Chim. Acta* **1987**, *196*, 351.
87AG(E)74	Frank, W., Weber, J., Fuchs, E. *Angew. Chem., Int. Ed. Engl.* **1987**, *26*, 74.
87AG(E)1180	Schmuck, A., Buschmann, J., Fuchs, J., Seppelt, K. *Angew. Chem., Int. Ed. Engl.* **1987**, *26*, 1180.
87AJC743	Snow, M.R., Tiekink, E.R.T. *Aust. J. Chem.* **1987**, *40*, 743.
87BCJ1564	Ando, F., Kohmura, Y., Koketsu, J. *Bull. Chem. Soc. Jpn.* **1987**, *60*, 1564.
87BCJ4279	Ozawa, S., Sasaki, Y., Ogino, Y. *Bull. Chem. Soc. Jpn.* **1987**, *60*, 4279.
87CB1829	Schmidbaur, H., Nowak, R., Schier, A., Wallis, J.M., Huber, B., Müller, G. *Chem. Ber.* **1987**, *120*, 1829.
87CB1837	Schmidbaur, H., Wallis, J.M., Nowak, R., Huber, B., Müller, G. *Chem. Ber.* **1987**, *120*, 1837.
87CC708	Wada, M., Ohki, H., Akiba, K.-y. *J. Chem. Soc., Chem. Commun.* **1987**, 708.
87CCC(3)237	McAuliffe, C.F., in *Comprehensive Coordination Chemistry*, eds. Wilkinson, G., Gillard, R.D., McCleverty, J.A., Abel, E.G., Pergamon, Oxford, **1987**, vol. 3, p. 237.
87CL1513	Sodesawa, T., Matsubara, M., Satoh, S., Nozaki, F. *Chem. Lett.* **1987**, 1513.

87IC1453	Bharadwaj, P.K., Musker, W.K. *Inorg. Chem.* **1987**, *26,* 1453.
87IZV187	Sharutin, V.V., Ermoshkin, A.E. *Izv. Akad. Nauk SSSR, Ser. Khim.* **1987**, 187.
87JA627	Arduengo, A.J., III, Stewart, C.A., Davidson, F., Dixon, D.A., Becker, J.Y., Culley, S.A., Mizen, M.B. *J. Am. Chem. Soc.* **1987**, *109*, 627.
87JCC(16)315	Williams, D.J., Travis, J.B., Bergbauer, K.L. *J. Coord. Chem.* **1987**, *16*, 315.
87JCS(D)145	Begley, M.J., Sowerby, D.B., Haiduc, I. *J. Chem. Soc., Dalton Trans.* **1987**, 145.
87JCS(D)1257	Sowerby, D.B., Haiduc, I. *J. Chem. Soc., Dalton Trans.* **1987**, 1257.
87JCS(P1)241	Barton, D.H.R., Finet, J.-P., Giannotti, C., Halley, F. *J. Chem. Soc., Perkin Trans. 1* **1987**, 241.
87JCS(P1)251	Barton, D.H.R., Finet, J.-P., Motherwell, W.B., Pichon, C. *J. Chem. Soc., Perkin Trans. 1* **1987**, 251.
87JOM(319)59	Humphries, R.E., Al-Jabar, N.A.A., Bowen, D., Massey, A.G., Deacon, G.B. *J. Organomet. Chem.* **1987**, *319*, 59.
87JOM(324)39	Doak, G.O., Freedman, L.D. *J. Organomet. Chem.* **1987**, *324*, 39.
87JOM(331)299	Klapötke, T. *J. Organomet. Chem.* **1987**, *331*, 299.
87JOM(334)323	Naumann, D., Tyrra, W. *J. Organomet. Chem.* **1987**, *334*, 323.
87JPC3585	Driscoll, D.J., Martir, W., Lunsford, J.H. *J. Phys. Chem.* **1987**, *91*, 3585.
87NJC337	Brégeault, J.-M., Faraj, M., Martin, J., Martin, C. *New J. Chem.* **1987**, *11*, 337.
87OM1185	Ashe, A.J., III, Kausch, C.M., Eisenstein, O. *Organometallics* **1987**, *6*, 1185.
87OMC138	Wardel, J.L. *Organometallic Chemistry*, **1987**, *15*, 138.
87PAC937	Barton, D.H.R., Finet, J.-P. *Pure Appl. Chem.* **1987**, *59*, 937.
87POL1593	Klapötke, T. *Polyhedron* **1987**, *6*, 1593.
87RKC(35)361	Keulks, G.W., Yu, M. *React. Kinet. Catal. Lett.* **1987**, *35*, 361.
87SRI(17)559	Singhal, K., Raj, P., Jee, F. *Synth. React. Inorg. Met.-Org. Chem.* **1987**, *17*, 559.
87T323	Barton, D.H.R., Y-Bhatnager, N., Finet, J.-P., Khamsi, J., Motherwell, W.B., Stanforth, S.P. *Tetrahedron* **1987**, *43*, 323.
87TL887	Barton, D.H.R., Finet, J.-P., Khamsi, J. *Tetrahedron Lett.* **1987**, *28*, 887.
87ZN(B)695	Wieber, M., Sauer, I. *Z. Naturforsch.* **1987**, *42B,* 695.
87ZN(B)940	Klapötke, T., Gowik, P. *Z. Naturforsch.* **1987**, *42B,* 940.
87ZOB2034	Yablokov, V.A., Zelyaev, I.A., Makarov, E.I., Lokhov, N.S. *Zh. Obshch. Khim.* **1987**, *57*, 2034.
88AX(C)828	Huber, F., Domagala, M., Preut, H. *Acta Crystallogr.* **1988**, *C44*, 828.
88AX(C)830	Domagala, M., Preut, H., Huber, F. *Acta Crystallogr.* **1988**, *C44*, 830.
88AOMC(2)553	Al-Soudani, A.R., Massey, A.G. *Appl. Organomet. Chem.* **1988**, *2*, 553.
88B-CW	Cotton, F.A., Wilkinson, G. *Advanced Inorganic Chemistry*, Wiley, New York, **1988**, 5th ed.
88BM(1)69	Klapötke, T. *Biol. Met.* **1988**, *1*, 69.

88CC517 Karsch, H.H., Deugbelly, B., Müller, G. *J. Chem. Soc., Chem. Commun.* **1988**, 517.

88CL847 Suzuki, H., Murafuji, T., Ogawa, T. *Chem. Lett.* **1988**, 847.

88CL849 Ogawa, T., Murafuji, T., Suzuki, H. *Chem. Lett.* **1988**, 849.

88CL2021 Ogawa, T., Murafuji, T., Iwata, K., Suzuki, H. *Chem. Lett.* **1988**, 2021.

88CL2049 Minato, M., Tsuji, J. *Chem. Lett.* **1988**, 2049.

88CSR269 Norman, N.C. *Chem. Soc. Rev.* **1988**, *17*, 269.

88IC3730 Calderazzo, F., Morvillo, A., Pelizzi, G., Poli, R., Ungari, F. *Inorg. Chem.* **1988**, *27*, 3730.

88ICA(150)217 Baker, P.K., Fraser, S.G., Matthews, T.M. *Inorg. Chim. Acta* **1988**, *150*, 217.

88ICA(152)49 Köpf-Maier, P., Klapötke, T. *Inorg. Chim. Acta* **1988**, *152*, 49.

88JA2310 Fagan, P.J., Nugent, W.A. *J. Am. Chem. Soc.* **1988**, *110*, 2310.

88JAP(K)174926 Imazaki, H., Fujikawa, M., Konishi, T., Kawaguchi, Y. *Jpn. Kokai* **1988**, 174,926; *Chem. Abstr.* **1989**, *111*, 154,111j.

88JCS(D)191 Vezzosi, I.M., Zanoli, A.F., Battaglia, L.P., Corradi, A.B. *J. Chem. Soc., Dalton Trans.* **1988**, 191.

88JCS(P2)1829 Ferguson, G., Glidewell, C., Gosney, I., Lloyd, D., Metcalfe, S., Lumbroso, H. *J. Chem. Soc., Perkin Trans. 2* **1988**, 1829.

88JFC(38)367 Hergett, S.C., Peach, M.E. *J. Fluorine Chem.* **1988**, *38*, 367.

88JOC2323 Lalonde, J.J., Bergbreiter, D.E., Wong, C.-H. *J. Org. Chem.* **1988**, *53*, 2323.

88JOM(342)185 Rüther, R., Huber, F., Preut, H. *J. Organomet. Chem.* **1988**, *342*, 185.

88JOM(347)27 Hasseler, K., Seidl, S. *J. Organomet. Chem.* **1988**, *347*, 27.

88JOM(351)63 Doak, G.O., Freedman, L.D. *J. Organomet. Chem.* **1988**, *351*, 63.

88M1317 Klapötke, T. *Monatsh. Chem.* **1988**, *119*, 1317.

88MJ(37)51 Hassan, H.N.A., Hassouna, M.E.M., Gawargious, Y.A. *Microchem. J.* **1988**, *37*, 51.

88OMC142 Wardel, J.L. *Organomet. Chem.* **1988**, *16*, 142.

88PR(B)3745 Mattheiss, L.F., Gyorgy, D.W., Johnson, D.W. *Phys. Rev. B: Condens. Matter* **1988**, *37*, 3745.

88R537 Severin, A.V., Batalov, A.P. *Radiokhimiya* **1988**, *30*, 537.

88S319 Glidewell, C., Lloyd, D., Metcalfe, S. *Synthesis* **1988**, 319.

88T3039 Abramovitch, R.A., Barton, D.H.R., Finet, J.-P. *Tetrahedron* **1988**, *44*, 3039.

88T4483 Barton, D.H.R., Finet, J.-P., Giannotti, C., Halley, F. *Tetrahedron* **1988**, *44*, 4483.

88T5661 Barton, D.H.R., Ozbalik, N., Sarma, J.C. *Tetrahedron* **1988**, *44*, 5661.

88T6387 Barton, D.H.R., Donnelly, D.M.X., Finet, J.-P., Stenson, P.H. *Tetrahedron* **1988**, *44*, 6387.

88TL857 Barton, D.H.R., Ozbalik, N., Ramesh, M. *Tetrahedron Lett.* **1988**, *29*, 857.

88TL1115 Barton, D.H.R., Finet, J.-P., Khamsi, J. *Tetrahedron Lett.* **1988**, *29*, 1115.

88TL3817	Akiba, K.-y., Ohdoi, K., Yamamoto, Y. *Tetrahedron Lett.* **1988**, *29*, 3817.
88TL4719	Ohki, H., Wada, M., Akiba, K.-y. *Tetrahedron Lett.* **1988**, *29*, 4719.
88TL5909	Cornélis, A., Delaude, L., Gerstmans, A., Laszlo, P. *Tetrahedron Lett.* **1988**, *29*, 5909.
88USP4746641	Guttmann, A.T., Grasselli, R.K., Brazdil, J.F. *US Patent* **1988**, 4,746,641; *Chem. Abstr.* **1988**, *109*, 171076c.
88USP4760159	Suresh, D.D., Orndoff, D.A., Brazdil, J.F., Glaeser, L.C. *US Patent* **1988**, 4,760,159; *Chem. Abstr.* **1988**, *109*, 171074a.
88ZAAC(557)7	Veith, M., Bertsch, B. *Z. Anorg. Allg. Chem.* **1988**, *557*, 7.
88ZAAC(559)73	Veith, M., Bertsch, B., Huch, V.A. *Z. Anorg. Allg. Chem.* **1988**, *559*, 73.
88ZN(B)739	Wieber, M., Rudolph, K. *Z. Naturforsch.* **1988**, *43B*, 739.
88ZN(B)952	Mundt, O., Riffel, H., Becker, G., Simon, A. *Z. Naturforsch.* **1988**, *43B*, 952.
88ZOB712	Dodonov, V.A., Zinov'eva, T.I., Osadchaya, N.N. *Zh. Obshch. Khim.* **1988**, *58*, 712.
88ZOB1270	Dodonov, V.A., Gushchin, A.V., Ezhova, M.B. *Zh. Obshch. Khim.* **1988**, *58*, 1270.
88ZOB2305	Sharutin, V.V. *Zh. Obshch. Khim.* **1988**, *58*, 2305.
89ACS511	Svensson, G., Albertsson, J. *Acta Chem. Scand.* **1989**, *43*, 511.
89AOMC(3)459	Al-Jabar, N.A.A., Jones, J.B., Brown, D.S., Colligan, A.H., Massey, A.G., Miller, J.M., Nye, J.W. *Appl. Organomet. Chem.* **1989**, *3*, 459.
89B-FD(5)397	Freedman, L.D., Doak, G.O., in *The Chemistry of the Metal-Carbon Bond*, ed. Hartley, F.R., Wiley, Chichester, **1989**, vol. 5, p. 397.
89CB287	Jutzi, P., Schwartzen, K.H. *Chem. Ber.* **1989**, *122*, 287.
89CB803	Schmuck, A., Seppelt, K. *Chem. Ber.* **1989**, *122*, 803.
89CC407	Campi, E.M., Deacon, G.B., Edward, G.L., Fitzroy, M.D., Giunta, N., Jackson, W.R., Trainor, R. *J. Chem. Soc., Chem. Commun.* **1989**, 407.
89CC652	Kim, Y.-C., Ueda, W., Moro-oka, Y. *J. Chem. Soc., Chem. Commun.* **1989**, 652.
89CC1628	Evans, W.J., Hain, J.H., Jr., Ziller, J.W. *J. Chem. Soc., Chem. Commun.* **1989**, 1628.
89CC1749	Ogawa, T., Murafuji, T., Suzuki, H. *J. Chem. Soc., Chem. Commun.* **1989**, 1749.
89CJC1949	Tyrra, W., Naumann, D. *Can. J. Chem.* **1989**, *67*, 1949.
89CL325	Ogawa, T., Murafuji, T., Iwata, K., Suzuki, H. *Chem. Lett.* **1989**, 325.
89CL531	Kim, Y.-C., Ueda, W., Moro-oka, Y. *Chem. Lett.* **1989**, 531.
89CL951	Yanada, K., Yamaguchi, H., Yanada, R., Meguri, H., Uchida, S. *Chem. Lett.* **1989**, 951.
89CL2173	Kim, Y.-C., Ueda, W., Moro-oka, Y. *Chem. Lett.* **1989**, 2173.
89CRS43	Snyder, T.P., Hill, C.G., Jr. *Catal. Rev.-Sci. Eng.* **1989**, *31*, 43.
89CRV1487	Finet, J.-P. *Chem. Rev.* **1989**, *89*, 1487.
89G(119)545	Maccarone, E., Passerini, A., Passerini, R., Tassone, G. *Gazz. Chim. Ital.* **1989**, *119*, 545.

89IC3164 Whitmire, K.H., Shieh, M., Cassidy, J. *Inorg. Chem.* **1989**, *28*, 3164.

89JOM(360)297 Doak, G.O., Freedman, L.D. *J. Organomet. Chem.* **1989**, *360*, 297.

89JOM(379)251 Oetting, I.T., Klapotke, T. *J. Organomet. Chem.* **1989**, *379*, 251.

89OM821 Brunner, H., Obermann, U., Wimmer, P. *Organometallics* **1989**, *8*, 821.

89OMC130 Wardell, J.L. *Organomet. Chem.* **1982**, *17*, 130.

89POL483 Iglesias, M., Del Pino, C., Martínez-Carrera, S. *Polyhedron* **1989**, *8*, 483.

89POL1579 Clegg, W., Compton, N.A., Errington, R.J., Norman, N.C., Wishart, N. *Polyhedron* **1989**, *8*, 1579.

89R69 Nefedov, V.D., Toropova, M.A., Shchepina, N.E., Avrorin, V.V., Zhuravlev, V.E., Trofimova, N.I. *Radiokhimiya* **1989**, *31*, 69.

89T2615 Barton, D.H.R., Bridon, D., Zard, S.Z. *Tetrahedron* **1989**, *45*, 2615.

89TL937 Barton, D.H.R., Finet, J.-P., Khamsi, J. *Tetrahedron Lett.* **1989**, *30*, 937.

89TL953 Akiba, K.-y., Ohdoi, K., Yamamoto, Y. *Tetrahedron Lett.* **1989**, *30*, 953.

89TL1261 Tanaka, H., Kosaka, A., Yamashita, S., Morisaki, K., Torii, S. *Tetrahedron Lett.* **1989**, *30*, 1261.

89TL3909 O'Donnell, M.J., Bennett, W.D., Jacobsen, W.N., Ma, Y., Huffman, J.C. *Tetrahedron Lett.* **1989**, *30*, 3909.

89TL4841 Ohkata, K., Takemoto, S., Ohnishi, M., Akiba, K.-y. *Tetrahedron Lett.* **1989**, *30*, 4841.

89USP4797381 Bartek, J.P., Guttmann, A.T. *US Patent* **1989**, 4,797,381; *Chem. Abstr.* **1989**, *110*, 142305s.

89USP4873215 Guttmann, A.T., Grasselli, R.K., Brazdil, J.F. *US Patent* **1989**, 4,873,215; *Chem. Abstr.* **1988**, *109*, 171076c.

89YGK425 Wada, M., Ohki, H. *Yuki Gosei Kagaku Kyokaishi* **1989**, *47*, 425.

90AX(C)917 Adams, E.A., Kolis, J.W., Pennington, W.T. *Acta Crystallogr.* **1990**, *C46*, 917.

90AG(E)213 Schmuck, A., Pyykko, P., Seppelt, K. *Angew. Chem., Int. Ed. Engl.* **1990**, *29*, 213.

90AOC(30)77 Ashe, A.J., III *Adv. Organomet. Chem.* **1990**, *30*, 77.

90AOC(31)91 Compton, N.A., Errington, R.J., Norman, N.C. *Adv. Organomet. Chem.* **1990**, *31*, 91.

90B-B Barton D.H.R. in *Heteroatom Chemistry*, ed. Block, E., VCH, New York, **1990**, Chap. 5.

90B-S Seppelt, K. in *Heteroatom Chemistry*, ed. Block, E., VCH, New York, **1990**, Chap. 19.

90BCJ166 Hosokawa, T., Ataka, Y., Murahashi, S.-I. *Bull. Chem. Soc. Jpn.* **1990**, *63*, 166.

90BCJ950 Suzuki, H., Murafuji, T. *Bull. Chem. Soc. Jpn.* **1990**, *63*, 950.

90BCJ1738 Wada, M., Ohki, H., Akiba, K.-y. *Bull. Chem. Soc. Jpn.* **1990**, *63*, 1738.

90BSCQ165 Cea-Olivares, R., Hernandez, A., Rios, E., Wingartz, J., Valdes-Martinez, J. *Bol. Soc. Chil. Quim.* **1990**, *35*, 165.

90CAR(204)121 Hendriks, H.E. J., Kuster, B.F.M., Martin, G.B. *Carbohydr. Res.* **1990**, *204*, 121.

90CB761	Schmuck, A., Leopold, D., Wallenhauer, S., Seppelt, K. *Chem. Ber.* **1990**, *123*, 761.
90CC301	Massiani, M.-C., Papiernik, R., Hubert-Pfalzgraf, L.G., Daran, J.-C. *J. Chem. Soc., Chem. Commun.* **1990**, 301.
90CC1724	Nagase, S., Suzuki, S., Kurakake, T. *J. Chem. Soc., Chem. Commun.* **1990**, 1724.
90CL1651	Ogawa, T., Miyazaki, K., Suzuki, H. *Chem. Lett.* **1990**, 1651.
90CL1867	Tanaka, H., Taniguchi, M., Kameyama, Y., Monnin, M., Sasaoka, M., Shiroi, T., Nagao, S., Torii, S. *Chem. Lett.* **1990**, 1867.
90EDM(D)	*Elsevier's Encyclopaedic Dictionary of Medicine. Part D*, ed. Dorian, A.F., Elsevier, Amsterdam, **1990**.
90EUP353461	Neuman, G.A., Bloss, K.H. *Eur. Patent* **1990**, 353461; *Chem. Abstr.* **1990**, *112*, 184,639.
90HAC233	Guilhaume, N., Postel, M. *Heteroatom Chem.* **1990**, *1*, 233.
90IC358	Matchett, M.A., Chiang, M.Y., Buhro, W.E. *Inorg. Chem.* **1990**, *29*, 358.
90JAPS(40)835	Delaviz, Y., Zhang, Z.X., Cabasso, I., Smid, J. *J. Appl. Polym. Sci.* **1990**, *40*, 835.
90JCS(D)899	Ali, M., McWhinnie, W.R., West, A.A., Hamor, T.A. *J. Chem. Soc., Dalton Trans.* **1990**, 899.
90JCS(D)1675	Battaglia, L.P., Corradi, A.B., Vezzosi, I.M., Zanoli, A.F. *J. Chem. Soc., Dalton Trans.* **1990**, 1675.
90JCS(P1)3367	Yasui, M., Kikuchi, T., Iwasaki, F., Suzuki, H., Murafuji, T., Ogawa, T. *J. Chem. Soc., Perkin Trans. 1* **1990**, 3367.
90JFC(48)421	Niyogi, D.G., Singh, S., Gill, S., Verma, R.D. *J. Fluorine Chem.* **1990**, *48*, 421.
90JOC5222	Akhtar, M.S., Brouillette, W.J., Waterhous, D.V. *J. Org. Chem.* **1990**, *55*, 5222.
90JOM(380)35	Doak, G.O., Freedman, L.D. *J. Organomet. Chem.* **1990**, *380*, 35.
90JOM(384)263	Hassler, H., Seidl, S. *J. Organomet. Chem.* **1990**, *384*, 263.
90JOM(386)177	Frank, W. *J. Organomet. Chem.* **1990**, *386*, 177.
90JOM(399)C21	Clegg, W., Compton, N.A., Errington, R.J., Fisher, G.A., Norman, N.C., Wishart, N. *J. Organomet. Chem.* **1990**, *399*, C21.
90OMC124	Wardell, J.L. *Organomet. Chem.* **1990**, *18*, 124.
90MOK112	Dodonov, V.A., Gushchin, A.V. *Metalloorg. Khim.* **1990**, *3*, 112.
90SSSC(55)491	Kim, Y.-C., Ueda, W., Moro-oka, Y. *Stud. Surf. Sci. Catal.* **1990**, *55*, 491.
90TL5897	Liao, Y., Huang, Y.-Z. *Tetrahedron Lett.* **1990**, *31*, 5897.
90YGK960	Wada, M. *Yuki Gosei Kagaku Kyokaishi* **1990**, *48*, 960.
90ZOB1104	Zhun, V.I., Tsvetkov, A.L., Sukhikh, A.S., Sheludyakov, V.D. *Zh. Obshch. Khim.* **1990**, *60*, 1104.
90ZOB2083	Dodonov, V.A., Gushchin, A.V., Bolotova, O.P. *Zh. Obshch. Khim.* **1990**, *60*, 2083.
91AC175	Kim, Y.-C., Ueda, W., Moro-oka, Y. *Appl. Catal.* **1991**, *70*, 175.
91AC189	Kim, Y.-C., Ueda, W., Moro-oka, Y. *Appl. Catal.* **1991**, *70*, 189.
91AX(C)345	Dunne, B.J., Orpen, A.G. *Acta Crystallogr.* **1991**, *C47*, 345.

91B-E	Emsley, J. *The Elements*, Clarendon Press, Oxford, 2nd ed., **1991**.
91BCJ990	Wada, M., Takeichi, E., Matsumoto, T. *Bull. Chem. Soc. Jpn.* **1991**, *64*, 990.
91CC1560	Jones, C., Engelhardt, L.M., Junk, P.C., Hutchings, D.S., Patalinghug, W.C., Raston, C.L., White, A.H. *J. Chem. Soc., Chem. Commun.* **1991**, 1560.
91CJC327	Tyrra, W., Naumann, F. *Can. J. Chem.* **1991**, *69*, 327.
91CL105	Suzuki, H., Nakaya, C., Matano, Y., Ogawa, T. *Chem. Lett.* **1991**, 105.
91CL841	Ohnishi, R., Suzuki, I., Ichikawa, M. *Chem. Lett.* **1991**, 841.
91G(121)93	Calderazzo, F., Marchetti, F., Ungari, F., Wieber, M. *Gazz. Chim. Ital.* **1991**, *121*, 93.
91HAC297	Huang, Y.-Z., Liao, Y. *Heteroatom Chem.* **1991**, *2*, 297.
91IC1274	Calderazzo, F., Juris, A., Poli, R., Ungari, F. *Inorg. Chem.* **1991**, *30*, 1274.
91IC2579	Herrmann, W.A., Herdtweck, E., Pajdla, L. *Inorg. Chem.* **1991**, *30*, 2579.
91IC2788	Cassidy, J.M., Whitmire, K.H. *Inorg. Chem.* **1991**, *30*, 2788.
91IC4210	Asato, E., Driessen, W.L., de Graaff, R.A.G., Hulsbergen, F.B., Reedijk, J. *Inorg. Chem.* **1991**, *30*, 4210.
91IC4680	Clegg, W., Compton, N.A., Errington, R.J., Fisher, G.A., Green, M.E., Hockless, D.C.R., Norman, N.C. *Inorg. Chem.* **1991**, *30*, 4680.
91IC4784	Block, E., Ofori-Okai, G., Kang, H., Wu, J., Zubieta, J. *Inorg. Chem.* **1991**, *30*, 4784.
91JMAR1595	Wasa, K., Adachi, H., Hirochi, K., Ichikawa, Y., Matsushima, T., Setsune, K. *J. Mater. Res.* **1991**, *6*, 1595.
91JOM(402)55	Whitmire, K.H., Labahn, D., Roesky, H.W., Noltemeyer, M., Sheldrick, G.M. *J. Organomet. Chem.* **1991**, *402*, 55.
91JOM(402)C4	Whitmire, K.H., Roesky, H.W., Brooker, S., Sheldrick, G.M. *J. Organomet. Chem.* **1991**, *402*, C4.
91JOM(403)183	Schumann, H., Eguren, L. *J. Organomet. Chem.* **1991**, *403*, 183.
91JOM(404)87	Doak, G.O., Freedman, L.D. *J. Organomet. Chem.* **1991**, *404*, 87.
91JOM(417)C47	Pasenok, S., Naumann, D., Tyrra, W. *J. Organomet. Chem.* **1991**, *417*, C47.
91JOM(419)283	Ferguson, G., Kaitner, B., Glidewell, C., Smith, S. *J. Organomet. Chem.* **1991**, *419*, 283.
91KK909	Troyanov, S.I., Pisarevskii, A.P. *Koord. Khim.* **1991**, *17*, 909.
91NJC927	Zevaco, T., Guilhaume, N., Postel, M. *New J. Chem.* **1991**, *15*, 927.
91OM812	Feher, F.J., Budzichowski, T.A. *Organometallics* **1991**, *10*, 812.
91POL437	Massiani, M.-C., Papiernik, R., Hubert-Pfalzgraf, L.G., Daran, J.-C. *Polyhedron* **1991**, *10*, 437.
91POL665	Schumann, H., Enemark, J.H., Labarre, M.J., Bruck, M., Wexler, P. *Polyhedron* **1991**, *10*, 665.
91RKC(43)501	Guseinova, E.G., Suleimanov, T.E., Kalinkin, A.V. *React. Kinet. Catal. Lett.* **1991**, *43*, 501.
91RKC(44)103	Parida, K.M., Rao, S.B. *React. Kinet. Catal. Lett.* **1991**, *44*, 103.

91SSSC(66)455	Masri, Y., Hronec, M. *Stud. Surf. Sci. Catal.* **1991**, *66*, 455.
91TL4247	Katritzky, A.R., Shobana, N., Harris, P.A. *Tetrahedron Lett.* **1991**, *32*, 4247.
91USP5008427	Brazdil, J.F., Jr., Toft, M.A., Glaeser, L.C. *US Patent* **1991**, 5,008,427; *Chem. Abstr.* **1991**, *115*, 50487c.
91ZN(B)1319	Rudolph, K., Wieber, M. *Z. Naturforsch.* **1991**, *46B*, 1319.
91ZOB2518	Petrov, K.A., Kryukova, L.Y., Treshchalina, L.V. *Zh. Obshch. Khim.* **1991**, *61*, 2518.
92AG(E) 451	Jones, C.M., Burkart, M.D., Whitmire, K.H. *Angew. Chem., Int. Ed. Engl.* **1992**, *31*, 451.
92AOC207	Seppelt, K. *Adv. Organomet. Chem.* **1992**, *34*, 207.
92AX(C)917	Whitmire, K.H, Cassidy, J.M. *Acta Crystallogr.* **1992**, *C48,* 917.
92B-L	*CRC Handbook of Chemistry and Physics*, ed. Lide, D.R., CRC Press, Boca Raton, 73rd ed., **1992**.
92BCJ1131	Firouzabadi, H., M-Baltork, I. *Bull. Chem. Soc. Jpn.* **1992**, *65*, 1131.
92BCJ3504	Matano, Y., Kinoshita, M., Suzuki, H. *Bull. Chem. Soc. Jpn.* **1992**, *65*, 3504.
92CAL(13)95	Khan, A.Z., Ruckenstein, E. *Cat. Lett.* **1992**, *13*, 95.
92CAL(15)393	Ramaroson, E., Blanchard, G., Che, M., Tatibouët, J.-M. *Cat. Lett.* **1992**, *15*, 393.
92CB43	Kiprof, P., Scherer, W., Pajdla, L., Herdtweck, E., Herrmann, W.A. *Chem. Ber.* **1992**, *125*, 43.
92CC453	Cho, C.S., Ohe, T., Itoh, O., Uemura, S. *J. Chem. Soc., Chem. Commun.* **1992**, 453.
92CC1021	Whitmire, K.H., Hutchison, J.C., McKnight, A.L., Jones, C.M. *J. Chem. Soc., Chem. Commun.* **1992**, 1021.
92CC1143	Suzuki, H., Murafuji, T. *J. Chem. Soc., Chem. Commun.* **1992**, 1143.
92CC1638	Jones, C.M., Burkart, M.D., Whitmire, K.H. *J. Chem. Soc., Chem. Commun.* **1992**, 1638.
92CL1945	Prajapati, D., Sandhu, J.S. *Chem. Lett.* **1992**, 1945.
92CL1967	Asato, E., Katsura, K., Mikuriya, M., Fujii, T., Reedijk, J. *Chem. Lett.* **1992**, 1967.
92CTO673	Kim, Y.-C., Ueda, W., Moro-oka, Y. *Cat. Today* **1992**, *13*, 673.
92ICA(198-200)271	Atwood, D.A., Cowley, A.H., Ruiz, J. *Inorg. Chim. Acta* **1992**, *198-200*, 271.
92JA372	Ashe, A.J., III, Kamph, J.W., Al-Taweel, S.M. *J. Am. Chem. Soc.* **1992**, *114*, 372.
92JA2960	Rogers, R.D., Bond, A.H., Aguinaga, S. *J. Am. Chem. Soc.* **1992**, *114*, 2960.
92JA7518	Schwerdtfeger, P., Heath, G.A., Dolg, M., Bennett, M. *J. Am. Chem. Soc.* **1992**, *114*, 7518.
92JA7906	Yamamoto, Y., Chen, X., Akiba, K.-y. *J. Am. Chem. Soc.* **1992**, *114*, 7906.
92JCR(S)34	Suzuki, H., Nakaya, C., Matano, Y. *J. Chem. Res. (S)* **1992**, 34.

92JCS(D)1753 Clegg, W., Errington, R.J., Flynn, R.J., Green, M.E., Hockless, D.C.R., Norman, N.C., Gibson, V.C., Tavakkooli, K. *J. Chem. Soc., Dalton Trans.* **1992**, 1753.

92JCS(D)1967 Clegg, W., Errington, R.J., Fisher, G.A., Hockless, D.C.R., Norman, N.C., Orpen, A.G., Stratford, S.E. *J. Chem. Soc., Dalton Trans.* **1992**, 1967.

92JCS(P1)1593 Suzuki, H., Murafuji, T., Azuma, N. *J. Chem. Soc., Perkin Trans. 1* **1992**, 1593.

92JCSR(22)231 Tiekink, E.R.T. *J. Cryst. Spectr. Res.* **1992**, *22*, 231.

92JOM(442)61 Doak, G.O., Freedman, L.D. *J. Organomet. Chem.* **1992**, *442*, 61.

92L90 Steiner, U.B., Neuenschwander, P., Caseri, W.R., Suter, U.W., Stucki, F. *Langmuir* **1992**, *8*, 90.

92M391 Cea-Olivares, R., Toscano, R.A., Carreón, G., Martínez, J.V. *Monatsh. Chem.* **1992**, *123*, 391.

92MGMC217 Zevaco, T., Postel, M., Benali-Cherif, N. *Main Group Met. Chem.* **1992**, *15*, 217.

92MOK1265 Dodonov, V.A., Zinovyeva, T.I. *Metalloorg. Khim.* **1992**, *5*, 1265.

92OM1381 Katritzky, A.R., Shobana, N., Harris, P.A. *Organometallics* **1992**, *11*, 1381.

92OM1491 Ashe, A.J., III, Kamph, J.W., Al-Taweel, S.M. *Organometallics* **1992**, *11*, 1491.

92OM2743 Ashe, A.J., III, Kamph, J.W., Puranik, D.B., Al-Taweel, S.M. *Organometallics* **1992**, *11*, 2743.

92OM3492 Spence, R.E.v.H., Hsu, D.P., Buchwald, S.L. *Organometallics* **1992**, *11*, 3492.

92PAD1151 Mallat, T., Bodnar, Z., Baiker, A. *Prepr. Am. Chem. Soc., Div. Pet. Chem.* **1992**, *37*, 1151.

92PM(33)1724 Ignatious, F., Sein, A., Delaviz, Y., Cabasso, I. *Polymer* **1992**, *33*, 1724.

92POL1219 Mishra, A.K., Gupta, V.D., Linti, G., Nöth, H. *Polyhedron* **1992**, *11*, 1219.

92RKC(47)193 Vislovskii, V.P., Baidikova, I.V., Mamedov, E.A., Rizayev, R.G. *React. Kinet. Catal. Lett.* **1992**, *47*, 193.

92SL878 Tanaka, H., Kameyama, Y., Torii, S. *Synlett* **1992**, 878.

92SRI289 Zevaco, T., Postel, M. *Synth. React. Inorg. Met.-Org. Chem.* **1992**, *22*, 289.

92TL1053 Roux, C.L., Maraval, M., Borredon, M.E., G-Iloughmane, H., Dubac, J. *Tetrahedron Lett.* **1992**, *33*, 1053.

92TL6653 Chen, X., Yamamoto, Y., Akiba, K.-y., Yoshida, S., Yasui, M., Iwasaki, F. *Tetrahedron Lett.* **1992**, *33*, 6653.

92USP5094989 Lynch, C.S., Glaeser, L.C., Brazdil, J.F., Toft, M.A. *US Patent* **1992**, 5,094,989; *Chem. Abstr.* **1992**, *116*, 201954b.

93ACS368 Haaland, A., Hammel, A., Rypdal, K., Swang, O., Brunvoll, J., Gropen, O., Greune, M., Weidlein, J. *Acta Chem. Scand.* **1993**, *47*, 368.

93AG(E)589 Breeze, S.R., Wang, S. *Angew. Chem., Int. Ed. Engl.* **1993**, *32*, 589.

93AG(E)867	Anaya, J., Barton, D.H.R., Gero, S.D., Grande, M., Martin, N., Tachdijian, C. *Angew. Chem., Int. Ed. Engl.* **1993**, *32*, 867.
93AOMC137	Ali, M., McWhinnie, W.R. *Appl. Organomet. Chem.* **1993**, *7*, 137.
93BBI(suppl.)	Palmieri, Y. *The Bulletin of the Bismuth Institute,* **1993**, Suppl.
93BCJ347	Ueda, W., Yamazaki, M., Horikawa, Y. *Bull. Chem. Soc. Jpn.* **1993**, *66*, 347.
93BCJ1542	Sugiyama, S., Shigemoto, N., Masaoka, N., Suetoh, S., Kawami, H. *Bull. Chem. Soc. Jpn.* **1993**, *66*, 1542.
93BSF832	Roux, C.L., G-Iloghmane, H., Dubac, J. *Bull. Soc. Chim. Fr.* **1993**, *130*, 832.
93CB51	Herrmann, W.A., Herdtweck, E., Scherer, W., Kiprof, P., Pajdla, L. *Chem. Ber.* **1993**, *126*, 51.
93CC335	Troyanov, S.I, Pisarevsky, A.P. *J. Chem. Soc., Chem. Commun.* **1993**, 335.
93CC799	Frank, W., Schneider, J., Becker, S.M. *J. Chem. Soc., Chem. Commun.* **1993**, 799.
93CC1309	Kurita, J., Ishii, M., Yasuike, S., Tsuchiya, T. *J. Chem. Soc., Chem. Commun.* **1993**, 1309.
93CC1817	Yasuike, S., Ohta, H., Shiratori, S., Kurita, J., Tsuchiya, T. *J. Chem. Soc., Chem. Commun.* **1993**, 1817.
93CL815	Ogawa, T., Hikasa, T., Ikegami, T., Ono, N., Suzuki, H. *Chem. Lett.* **1993**, 815.
93IC2212	Shieh, M., Liou, Y., Peng, S.-M., Lee, G.-H. *Inorg. Chem.* **1993**, *32*, 2212.
93IC2972	Atwood, D.A., Cowley, A.H., Hernandez, R.D., Jones, R.A., Rand, L.L., Bott, S.G., Atwood, J.L. *Inorg. Chem.* **1993**, *32*, 2972.
93IC3948	Wallenhauer, S., Leopold, D., Seppelt, K. *Inorg. Chem.* **1993**, *32*, 3948.
93IC5136	Jones, C.M., Burkart, M.D., Bachman, R.E., Serra, D.L., Hwu, S.-J., Whitmire, K.H.J. *Inorg. Chem.* **1993**, *32*, 5136.
93IC5322	Asato, E., Katsura, K., Mikuriya, M., Fujii, T., Reedijk, J. *Inorg. Chem.* **1993**, *32*, 5322.
93IZV2043	Dodonov, V.A., Gushchin, A.V. *Izv. Akad. Nauk, Ser. Khim.* **1993**, 2043.
93IZV2109	Dodonov, V.A., Starostina, T.I., Belukhina, E.V., Vorob'eva, N.V. *Izv. Akad. Nauk, Ser. Khim.* **1993**, 2109.
93JCS(D)637	Clegg, W., Errington, R.J., Fisher, G.A., Flynn, R.J., Norman, N.C. *J. Chem. Soc., Dalton Trans.* **1993**, 637.
93JCS(P1)1169	Suzuki, H., Murafuji, T., Azuma, N. *J. Chem. Soc., Perkin Trans. 1* **1993**, 1169.
93JCS(P1)2411	Suzuki, H., Ikegami, T., Matano, Y., Azuma, N. *J. Chem. Soc., Perkin Trans. 1* **1993**, 2411.
93JCS(P1)2969	Suzuki, H., Murafuji, T., Matano, Y., Azuma, N. *J. Chem. Soc., Perkin Trans. 1* **1993**, 2969.
93JFC(63)179	Pasenok, S.V., Kirij, N.V., Yagupolskii, Y.L., Naumann, D., Tyrra, W. *J. Fluorine Chem.* **1993**, *63*, 179.

93JOC1835 Roux, C.L., G-Iloghmane, H., Dubac, J. *J. Org. Chem.* **1993**, *58*, 1835.

93JOC2196 Floresca, R., Kurihara, M., Watt, D.S. *J. Org. Chem.* **1993**, *58*, 2196.

93JOC4783 Suzuki, I., Yamamoto, Y. *J. Org. Chem.* **1993**, *58*, 4783.

93JOM(447)197 Ashe, A.J., III, Kampf, J.W., Puranik, D.B. *J. Organomet. Chem.* **1993**, *447*, 197.

93JOM(460)C22 Carmalt, C.J., Norman, N.C., Orpen, A.G., Stratford, S.E. *J. Organomet. Chem.* **1993**, *460*, C22.

93MOK375 Dodonov, V.A., Zinovyeva, T.I. *Metalloorg. Khim.* **1993**, *6*, 375.

93OM343 Lohr, L.L., Ashe, A.J., III *Organometallics* **1993**, *12,* 343.

93OM1857 Chen X., Ohdoi, K., Yamamoto, Y., Akiba, K.-y. *Organometallics* **1993**, *12*, 1857.

93OM3297 Yamamoto, Y., Ohdoi, K., Chen, X., Kitano, M., Akiba, K.-y. *Organometallics* **1993**, *12*, 3297.

93OM4926 Shieh, M., Liou, Y., Jeng, B.-W. *Organometallics* **1993**, *12*, 4926.

93RCB1955 Dodonov, V.A., Gushchin, A.V. *Russ. Chem. Bull.* **1993**, *42*, 1955.

93SL115 Watanabe, Y., Nakamoto, C., Ozaki, S. *Synlett* **1993**, 115.

93TL1055 Suzuki, H., Nakaya, C., Matano, Y. *Tetrahedron Lett.* **1993**, *34*, 1055.

93TL2601 Zevaco, T., Duñach, E., Postel, M. *Tetrahedron Lett.* **1993**, *34*, 2601.

93TL7975 Bhuyan, P.J., Prajapati, D., Sandhu, J.S. *Tetrahedron Lett.* **1993**, *34*, 7975.

93TL8457 Matano, Y., Azuma, N., Suzuki, H. *Tetrahedron Lett.* **1993**, *34*, 8457.

93ZAAC(619)1073 Müller-Becker, S., Frank, W., Schneider, J. *Z. Anorg. Allg. Chem.* **1993**, *619*, 1073.

93ZNK(38)1205 Fukin, G.K., Pisarevskii, A.P., Yanovskii, A.I., Struchkov, Y.T. *Zh. Neorg. Khim.* **1993**, *38*, 1205.

94AG(E)976 Wallenhauer, S., Seppelt, K. *Angew. Chem., Int. Ed. Engl.* **1994**, *33*, 976.

94AIC233 Moro-oka, Y., Ueda, W. *Adv. Catal.* **1994**, *40*, 233.

94AX(B)151 Yoshida, S., Yasui, M., Iwasaki, F., Yamamoto, Y., Chen, X., Akiba, K. *Acta Crystallogr.* **1994**, *B50*, 151.

94B-AY Akiba, K.-y., Yamamoto, Y., in *The Chemistry of Organic Arsenic, Antimony, and Bismuth Compounds*, ed. Patai, S., Wiley, New York, **1994**, Chapter 20.

94B-N Nagase, S. in *The Chemistry of Organic Arsenic, Antimony, and Bismuth Compounds*, ed. Patai, S., Wiley, New York, **1994**, Chapter 1.

94CB1335 Sitzmann, H., Wolmershäuser, G. *Chem. Ber.* **1994**, *127*, 1335.

94CCR(133)115 Champness, N.R., Levason, W. *Coord. Chem. Rev.* **1994**, *133*, 115.

94CL2123 Asato, E., Katsura, K., Arakaki, T., Mikuriya, M., Kotera, T. *Chem. Lett.* **1994**, 2123.

94CPB1437 Kurita, J., Ishii, M., Yasuike, S., Tsuchiya, T. *Chem. Pharm. Bull.* **1994**, *42,* 1437.

94EIC(1)280 Whitmire, K.H., in *Encyclopedia of Inorganic Chemistry*, ed. King, R.B., Wiley, Chichester, **1994**, vol. 1, p. 280.

94EIC(1)292 Whitmire, K.H., in *Encyclopedia of Inorganic Chemistry*, ed. King, R.B., Wiley, Chichester, **1994**, vol. 1, p. 292.

94FRP(94)10253 Dubac, J., Labrouillére, M., Laporterie, A., Desmurs, J.R. *French Patent* **1994**, *94*, 10253; *Chem. Abstr.* **1996**, *124*, 316758y.

94IC4607 Wirringa, U., Roesky, H.W., Noltemeyer, M., Schmidt, H.-G. *Inorg. Chem.* **1994**, *33*, 4607.

94IJC(A)687 Asthana, A. *Indian J. Chem.* **1994**, *33A*, 687.

94IZV171 Miller, A.O., Firin, G.G. *Izv. Akad. Nauk SSSR Ser. Khim.* **1994**, 171.

94JA10080 Naito, T., Nagase, S., Yamataka, H. *J. Am. Chem. Soc.* **1994**, *116*, 10080.

94JCR(S)228 Borah, H.N., Prajapati, D., Sandhu, J.S. *J. Chem. Res. (S)* **1994**, 228.

94JCS(D)2545 Breeze, S.R., Chen, L., Wang, S. *J. Chem. Soc., Dalton Trans.* **1994**, 2545.

94JCS(P1)1739 Matano, Y., Azuma, N., Suzuki, H. *J. Chem. Soc., Perkin Trans. 1* **1994**, 1739.

94JCS(P1)2703 Matano, Y. *J. Chem. Soc., Perkin Trans. 1* **1994**, 2703.

94JCS(P1)3473 Ogawa, T., Hikasa, T., Ikegami, T., Ono, N., Suzuki, H. *J. Chem. Soc., Perkin Trans. 1* **1994**, 3473.

94JCS(P1)3479 Ogawa, T., Ikegami, T., Hikasa, T., Ono, N., Suzuki, H. *J. Chem. Soc., Perkin Trans. 1* **1994**, 3479.

94JFC(66)75 Kirij, N. V., Pasenok, S.V., Yagupolskii, Y.L., Naumann, D., Tyrra, W. *J. Fluorine Chem.* **1994**, *66*, 75.

94JFC(66)79 Naumann, D., Schlengermann, R., Tyrra, W. *J. Fluorine Chem.* **1994**, *66*, 79.

94JFC(69)219 Kirij, N.V., Pasenok, S.V., Yagupolskii, Y.L., Naumann, D., Tyrra, W. *J. Fluorine Chem.* **1994**, *69*, 219.

94JMAC(4)891 Clegg, W., Elsegood, M.R.J., Errington, R.J., Fisher, G.A., Norman, N.C. *J. Mater. Chem.* **1994**, *4*, 891.

94JOC2238 Roux, C.L., G-Iloghmane, H., Dubac, J. *J. Org. Chem.* **1994**, *59*, 2238.

94JOC7575 Harada, T., Ueda, S., Yoshida, T., Inoue, A., Takeuchi, M., Ogawa, N., Oku, A. *J. Org. Chem.* **1994**, *59*, 7575.

94JOM(470)93 Ebert, K.H., Schluz, R.E., Breunig, H.J., Silvestru, C., Haiduc, I. *J. Organomet. Chem.* **1994**, *470*, 93.

94JOM(477)31 Doak, G.O., Freedman, L.D. *J. Organomet. Chem.* **1994**, *477*, 31.

94JOM(482)119 Boisselier, V.L., Duñach, E., Postel, M. *J. Organomet. Chem.* **1994**, *482*, 119.

94MGMC727 Tiekink, E.R.T. *Main Group Met. Chem.* **1994**, *17*, 727.

94POL365 Bräu, E., Falke, R., Ellener, A., Beuter, M., Kolb, U., Dräger, M. *Polyhedron* **1994**, *13*, 365.

94POL547 Edelmann, F.T., Noltemeyer, M., Haiduc, I., Silvestru, C., Cea-Olivares, R. *Polyhedron* **1994**, *13*, 547.

94S775 Banfi, A., Bartoletti, M., Bellora, E., Bignotti, M., Turconi, M. *Synthesis* **1994**, 775.

94SC489 Firouzabadi, H., M-Baltork, I. *Synth. Commun.* **1994**, *24*, 489.

94SC989 Liming, J., Xian, H. *Synth. Commun.* **1994**, *24*, 989.

94SL723 Labrouillére, M., Roux, C.L., G-Iloughmane, H., Dubac, J. *Synlett* **1994**, 723.

94TL1739 Cho, C.S., Motofusa, S., Uemura, S. *Tetrahedron Lett.* **1994**, *35*, 1739.

94TL3167 Borah, H.N., Prajapati, D., Sandhu, J.S., Ghosh, A. C. *Tetrahedron Lett.* **1994**, *35*, 3167.

94TL8197 Suzuki, H., Ikegami, T., Matano, Y. *Tetrahedron Lett.* **1994**, *35*, 8197.

94ZOB1051 Sharutin, V.V., Sharutina, O.K., Pavlov, K.V., Shcherbimin, V.V. *Zh. Obshch. Khim.* **1994**, *64*, 1051.

95AG(E) 2187 Herrmann, W.A., Huber, N.W., Runte, O. *Angew. Chem., Int. Ed. Engl.* **1995**, *34*, 2187.

95AG(E) 2416 Frank, W., Reiss, G.J., Schneider, J. *Angew. Chem., Int. Ed. Engl.* **1995**, *34*, 2416.

95BBI(68) Endriss, H. *The Bulletin of the Bismuth Institute*, **1995**, No. 68.

95BCJ950 Cho, C.S., Yoshimori, Y., Uemura, S. *Bull. Chem. Soc. Jpn.* **1995**, *68*, 950.

95BCJ1385 Tanaka, H., Taniguchi, M., Kameyama, Y., Monnin, M., Torii, S., Sasaoka, M., Shiroi, T., Nagao, S., Yamada, T., Tokumaru, Y. *Bull. Chem. Soc. Jpn.* **1995**, *68*, 1385.

95BMC1035 Sinclair, P.J., Wong, F., Wyvratt, M., Staruch, M.J., Dumont, F. *Bioorg. Med. Chem. Lett.* **1995**, *5*, 1035.

95BSF522 Labrouillére, M., Roux, C.L., Oussaid, A., G-Iloughmane, H., Dubac, J. *Bull. Soc. Chim. Fr.* **1995**, *132*, 522.

95CB335 Diemer, R., Keppler, B.K., Dittes, U., Nuber, B., Seifried, V., Opferkuch, W. *Chem. Ber.* **1995**, *128*, 335.

95CC1407 Ogawa, T., Yoshikawa, A., Wada, H., Ogawa, C., Ono, N., Suzuki, H. *J. Chem. Soc., Chem. Commun.* **1995**, 1407.

95CCL553 Ren, P.-D., Pan, S.F., Dong, T.W., Wu, S.H. *Chin. Chem. Lett.* **1995**, *6*, 553.

95CL999 Miyoshi, N., Fukuma, T., Wada, M. *Chem. Lett.* **1995**, 999.

95COMC(2)321 Wardell, J.L., in *Comprehensive Organometallic Chemistry*, eds. Abel, E.W., Stone, F.G.A., Wilkinson, G., Pegamon, New York, **1995**, vol. 2, p. 321.

95COMC(11)502 Huang, Y.Z., Zhau, Z.L., in *Comprehensive Organometallic Chemistry*, eds. Abel, E.W., Stone, F.G.A., Wilkinson, G., Pegamon, New York, **1995**, vol. 11, p. 502.

95CT169 Whitmire, K.H. *Chem. Tracts: Inorg. Chem.* **1995**, *7*, 169.

95EAS(1)365 Chattopadhyay, P., in *Encyclopedia of Analytical Science*, ed. Townshend, A., Academic, London, **1995**, vol. 1, p. 365.

95HAC293 Chen, X., Yamamoto, Y., Akiba, K. *Heteroatom Chem.* **1995**, *6*, 293.

95IC2447 Asato, E., Katsura, K., Mikuriya, M., Turpeinen, U., Mutikainen, I., Reedijk, J. *Inorg. Chem.* **1995**, *34*, 2447.

95ICA(237)169 Battaglia, L.P., Cramarossa, M.R., Vezzosi, I.M. *Inorg. Chim. Acta* **1995**, *237*, 169.

95IZV18 Batsanov, S.S. *Izv. Akad. Nauk, Ser. Khim.* **1995**, 18.

95IZV156 Dodonov, V.A., Starostina, T.I., Kuznetsova, Y.L., Gushchin, A.V. *Izv. Akad. Nauk, Ser. Khim.* **1995**, 156.

95IZV783	Dodonov, V.A., Zinov'eva, T.I., Dolganova, N.V. *Izv. Akad. Nauk, Ser. Khim.* **1995**, 783.
95IZV964	Grishin, A.V., Dyomina, E.E., Dodonov, V.A. *Izv. Akad. Nauk, Ser. Khim.* **1995**, 964.
95JA3922	Yamamoto, Y., Chen, X., Kojima, S., Ohdoi, K., Kitano, M., Doi, Y., Akiba, K.-y. *J. Am. Chem. Soc.* **1995**, *117*, 3922.
95JA11790	Moc, J., Morokuma, K. *J. Am. Chem. Soc.* **1995**, *117*, 11790.
95JCS(D)383	Forster, G.E., Begley, M.J., Sowerby, D.B. *J. Chem. Soc., Dalton Trans.* **1995**, 383.
95JCS(D)759	Willey, G.R., Rudd, M.D., Samuel, C.J., Drew, M.G.B. *J. Chem. Soc., Dalton Trans.* **1995**, 759.
95JCS(D)1163	Farrugia, L.J., Lawlor, F.J., Norman, N.C. *J. Chem. Soc., Dalton Trans.* **1995**, 1163.
95JCS(D)1649	Bochmann, M., Song, X., Hursthouse, M.B., Karaulov, A. *J. Chem. Soc., Dalton Trans.* **1995**, 1649.
95JCS(D)2129	Clegg, W., Elsegood, M.R.J., Farrugia, L.J., Lawlor, F.J., Norman, N.C., Scott, A.J. *J. Chem. Soc., Dalton Trans.* **1995**, 2129.
95JCS(P1)2543	Matano, Y., Azuma, N., Suzuki, H. *J. Chem. Soc., Perkin Trans. 1* **1995**, 2543.
95JFC(70)237	Niyogi, D.G., Singh, S., Verma, R.D. *J. Fluorine Chem.* **1995**, *70*, 237.
95JFC(74)251	Tordeux, M., Wakselman, C. *J. Fluorine Chem.* **1995**, *74*, 251.
95JOC883	Cho, C.S., Motofusa, S., Ohe, K., Uemura, S. *J. Org. Chem.* **1995**, *60*, 883.
95JOC4663	Matano, Y., Yoshimune, M., Suzuki, H. *J. Org. Chem.* **1995**, *60*, 4663.
95JOM(485)1	Doak, G.O., Freedman, L.D. *J. Organomet. Chem.* **1995**, *485*, 1.
95JOM(485)141	Li, X.-W., Lorberth, J., Massa, W., Wocadlo, S. *J. Organomet. Chem.* **1995**, *485*, 141.
95JOM(485)149	Lorberth, J., Massa, W., Wocadlo, S., Sarraje, I., Shin, S.-H., Li, X.-W. *J. Organomet. Chem.* **1995**, *485*, 149.
95JOM(496)1	Doak, G.O., Freedman, L.D. *J. Organomet. Chem.* **1995**, *496*, 1.
95JOM(496)59	Carmalt, C.J., Cowley, A.H., Decken, A., Norman, N.C. *J. Organomet. Chem.* **1995**, *496*, 59.
95OM1542	Murafuji, T., Azuma, N., Suzuki, H. *Organometallics* **1995**, *14*, 1542.
95OM3848	Murafuji, T., Mutoh, T., Satoh, K., Tsunenari, K., Azuma, N., Suzuki, H. *Organometallics* **1995**, *14*, 3848.
95POL311	Farrugia, L.J., Lawlor, F.J., Norman, N.C. *Polyhedron* **1995**, *14*, 311.
95SA(A)2019	Ludwig, C., Götze, H.-J. *Spectrochim. Acta, Part A* **1995**, *51*, 2019.
95SC2751	Katritzky, A.R., Allin, S.M. *Synth. Commun.* **1995**, *25*, 2751.
95SC3395	Ren, P.-D., Pan, S.-F., Dong, T.-W., Wu, S.-H. *Synth. Commun.* **1995**, *25*, 3395.
95SC3799	Ren, P.-D., Pan, S.-F., Dong, T.-W., Wu, S.-H. *Synth. Commun.* **1995**, *25*, 3799.
95SC4025	Prajapati, D., Borah, H.N., Sandhu, J.S. *Synth. Commun.* **1995**, *25*, 4025.
95SL984	Komatsu, N., Uda, M., Suzuki H. *Synlett* **1995**, 984.

95T4991 Boisselier, V.L., Coin, C., Postel, M., Duñach, E. *Tetrahedron* **1995**, *51*, 4991.

95TL6747 Baruah, B., Prajapati, D., Sandhu, J. S. *Tetrahedron Lett.* **1995**, *36*, 6747.

95TL7475 Matano, Y., Yoshimune, M., Suzuki, H. *Tetrahedron Lett.* **1995**, *36*, 7475.

95ZAAC(621)1251 Frohn, H.J., Georg, S., Henkel, G., Laege, M. *Z. Anorg. Allg. Chem.* **1995**, *621*, 1251.

95ZAAC(621)1288 Wieber, M., Clarius, T. *Z. Anorg. Allg. Chem.* **1995**, *621*, 1288.

95ZAAC(621)1403 Wieber, M., Schroepf, M. *Z. Anorg. Allg. Chem.* **1995**, *621*, 1403.

95ZAAC(621)1746 Weitze, A., Henshel, D., Blaschette, A., Jones, P.G. *Z. Anorg. Allg. Chem.* **1995**, *621*, 1746.

95ZK(210)377 Jones P.G., Blaschette, A., Henschel, D., Weitze, A. *Z. Kristallogr.* **1995**, *210*, 377.

95ZN(B)735 Breunig, H.J., Ebert, K.H., Schulz, R.E., Wieber, M., Sauer, I. *Z. Naturforsch.* **1995**, *50B,* 735.

95ZOB345 Sharutin, V.V., Sharutina, O.K., Viktorov, N.A. *Zh. Obshch. Khim.* **1995**, *65*, 345.

96AC(A)(145)1 Bettahar, M.M., Costentin, G., Savary, L., Lavalley, J.C. *Appl. Catal. A* **1996**, *145*, 1.

96AG(E)67 Dittes, U., Keppler, B.K., Nuber, B. *Angew. Chem., Int. Ed. Engl.* **1996**, *35*, 67.

96BCJ2673 Matano, Y., Suzuki, H. *Bull. Chem. Soc. Jpn.* **1996**, *69*, 2673.

96BMC2193 Sinclair, P.J., Wong, F., Staruch, M.J., Wiederrecht, G., Parsons, W.H., Dumont, F., Wyvratt, M. *Bioorg. Med. Chem. Lett.* **1996**, *6*, 2193.

96BSF(133)913 Coin, C., Zevaco, T., Duñach, E., Postel, M. *Bull. Soc. Chim. Fr.* **1996**, *133*, 913.

96CC1693 Murafuji, T., Mutoh, T., Sugihara, Y. *Chem. Commun.* **1996**, 1693.

96CC1847 Komatsu, N., Taniguchi, A., Uda, M., Suzuki, H. *Chem. Commun.* **1996**, 1847.

96CC2183 Yasuike, S., Kiharada, T., Kurita, J., Tsuchiya, T. *Chem. Commun.* **1996**, 2183.

96CC2697 Matano, Y., Suzuki, H. *Chem. Commun.* **1996**, 2697.

96CCL788 Ren, P.-D., Pan, S.F., Dong, T.S. *Chin. Chem. Lett.* **1996**, *7*, 788.

96CCL981 Pan, S.F., Ren, P.D., Dong, T.W. *Chin. Chem. Lett.* **1996**, *7*, 981.

96CCR(155)127 Postel, M., Duñach, E. *Coord. Chem. Rev.* **1996**, *155*, 127.

96CEJ701 Sadler, P.J., Sun, H., Li, H. *Chem. Eur. J.* **1996**, *2*, 701.

96CHJ462 Ren, P.-D., Pan, S.-F., Dong, T.-W., Wu, S.-H. *Chin. J. Chem.* **1996**, *14*, 462.

96EUP698593 Dubac, J., Labrouillére, M., Laporterie, A., Desmurs, J.R. *Eur. Patent Appl.* **1996**, 698,593; *Chem. Abstr.* **1996**, *124*, 316758y.

96F1501 Suzuki, H. *Farumashia* **1996**, *32*, 1501.

96IC3948 García-Montalvo, V., Cea-Olivares, R., Williams, D.J., Espinosa-Pérez, G. *Inorg. Chem.* **1996**, *35*, 3948.

96IC4013	Burford, N., Macdonald, C.L.B., Robertson, K.N., Cameron, T.S. *Inorg. Chem.* **1996**, *35*, 4013.
96IC6179	Kamepalli, S., Carmalt, C.J., Culp, R.D., Cowley, A.H., Jones, R.A., Norman, N.C. *Inorg. Chem.* **1996**, *35*, 6179.
96ICA(248)9	Edwards, A.J., Beswick, M.A., Galsworthy, J.R., Paver, M.A., Raithby, P.R., Rennie, M.-A., Russell, C.A., Verhorevoort, K.L., Wright, D.S. *Inorg. Chim. Acta* **1996**, *248*, 9.
96ICA(248)263	Farrugia, L.J., Carmalt, C.J., Norman, N.C. *Inorg. Chim. Acta* **1996**, *248*, 263.
96ICL15	Desmurs, J.-R., Labrouillere, M., Dubac, J., Laporterie, A., Gaspard, H., Metz, F. *Ind. Chem. Libr.* **1996**, *8*, 15.
96JA3225	Agocs, L., Burford, N., Cameron, T.S., Curtis, J.M., Richardson, J.F., Robertson, K.N., Yhard, G.B. *J. Am. Chem. Soc.* **1996**, *118*, 3225.
96JCAC317	Arts, S.J.H.F., Rantwijk, F.V., Sheldon, R.A. *J. Carbohydr. Chem.* **1996**, *15*, 317.
96JCR(S)24	Suzuki, H., Ikegami, T. *J. Chem. Res. (S)* **1996**, 24.
96JCS(D)443	Carmalt, C.J., Farrugia, L.J., Norman, N.C. *J. Chem. Soc., Dalton Trans.* **1996**, 443.
96JCS(D)2023	Wullens, H., Devillers, M., Tinant, B., Declercq, J.-P. *J. Chem. Soc., Dalton Trans.* **1996**, 2023.
96JCS(D)2417	Barrie, P.J., Djuran, M.I., Mazid, M.A., McPartlin, M., Sadler, P.J., Scowen, I.J., Sun, H. *J. Chem. Soc., Dalton Trans.* **1996**, 2417.
96JCS(P1)953	Habata, Y., Fujishiro, F., Akabori, S. *J. Chem. Soc., Perkin Trans. 1* **1996**, 953.
96JCS(P1)1971	Matano, Y., Yoshimune, M., Azuma, N., Suzuki, H. *J. Chem. Soc., Perkin Trans. 1* **1996**, 1971.
96JOC2895	Somasundaram, N., Srinivasan, C. *J. Org. Chem.* **1996**, *61*, 2895.
96JOC3885	Roux, C.L., Mandrou, S., Dubac, J. *J. Org. Chem.* **1996**, *61*, 3885.
96JOM(515)245	Oldenburg, K., Vogler, A. *J. Organomet. Chem.* **1996**, *515*, 245.
96JOM(521)397	Peidro, L., Roux, C.L., Laporterie, A., Dubac, J. *J. Organomet. Chem.* **1996**, *521*, 397.
96MI	*Merck Index*, 12nd ed., Merck & Co., **1996**.
96MJR	Fickling, M. *Mining Journal Annual Review*, **1996**.
96OM887	Carmalt, C.J., Cowley, A.H., Decken, A., Lawson, Y.G., Norman, N.C. *Organometallics* **1996**, *15*, 887.
96OM1951	Matano, Y., Miyamatsu, T., Suzuki, H. *Organometallics* **1996**, *15*, 1951.
96OM4646	Roux, C.L., Dubac, J. *Organometallics* **1996**, *15*, 4646.
96OM5613	Hassan, A., Breeze, S.R., Courtenay, S., Deslippe, C., Wang, S. *Organometallics* **1996**, *15*, 5613.
96PAC919	Chan, T., Isaac, M. *Pure Appl. Chem.* **1996**, *68*, 919.
96SC763	Ren, P.-D., Pan, S.-F., Dong, T.-W., Wu, S.-H. *Synth. Commun.* **1996**, *26*, 763.
96SC3903	Ren, P.-D., Pan, S.-F., Dong, T.-W., Wu, S.-H. *Synth. Commun.* **1996**, *26*, 3903.

96SC4569 Combes, S., Finet, J.-P. *Synth. Commun.* **1996**, *26*, 4569.
96TL4051 Lermontov, S.A., Rakov, I.M., Zefirov, N.S., Stang, P.J. *Tetrahedron Lett.* **1996**, *37*, 4051.
96TL9013 Chan, D.M.T. *Tetrahedron Lett.* **1996**, *37*, 9013.
96ZAAC(622)923 Ehleiter, Y., Wolmershäuser, G., Sizmann, H., Boese, R. *Z. Anorg. Allg. Chem.* **1996**, *622*, 923.
96ZAAC(622)2009 Lewe, T., Naumann, D., Nowicki, G., Schneider, H., Tyrra, W., Gilles, T., Tebbe, K.-F. *Z. Anorg. Allg. Chem.* **1996**, *622*, 2009.
97BBI(70) Erkens, L.J.H., Vos, L.J. *The Bullentin of the Bismuth Institute*, **1997**, No. 70.
97BCJ2265 Wada, M., Honma, M., Kuramoto, Y., Miyoshi, N. *Bull. Chem. Soc. Jpn.* **1997**, *70*, 2265.
97CAR(297)175 Montero, J.-L., Winum, J.-Y., Leydet, A., Kamel, M., Pavia, A.A., Roque, J.-P. *Carbohydr. Res.* **1997**, *297*, 175.
97CB669 Sun, H., Li, H., Sadler, P.J. *Chem. Ber./Recueil* **1997**, *130*, 669.
97CC95 Boisselier, V.L., Postel, M., Duñach, E. *Chem. Commun.* **1997**, 95.
97CC2295 Suzuki, H., Kurata, H., Matano, Y. *Chem. Commun.* **1997**, 2295.
97CCL759 Huang, X., Wu, J.L. *Chin. Chem. Lett.* **1997**, *8*, 759.
97CCR(163)345 Dittes, U., Vogel, E., Keppler, B.K., *Coord. Chem. Rev.* **1997**, *163*, 345.
97CEJ1064 Tussa, L., Lebreton, C., Mosset, P. *Chem. Eur. J.* **1997**, *3*, 1064.
97CL1221 Tanaka, H., Tokumaru, Y., Kameyama, Y., Torii, S. *Chem. Lett.* **1997**, 1221.
97CL1229 Komatsu, N., Uda, M., Suzuki, H. *Chem. Lett.* **1997**, 1229.
97CM1335 Tamber, H., Smid, J., Cabasso, I. *Chem. Mater.* **1997**, *9*, 1335.
97EA(42)1979 Fuchigami, T., Miyazaki, M. *Electrochim. Acta* **1997**, *42*, 1979.
97H1899 Yasuike, S., Nakashima, F., Kurita, J., Tsuchiya, T. *Heterocycles* **1997**, *45*, 1899.
97HAC317 Cea-Olivares, R., Ebert, K.H., Silaghi-Dumitrescu, L., Haiduc, I. *Heteroatom Chem.* **1997**, *8*, 317.
97HDX53 Yang, G., Chen, J. *Hubei Daxue Xuebao, Ziran Kexueban* **1997**, *19*, 53; *Chem. Abstr.* **1998**, 128, 166993n.
97IC694 Koketsu, J., Ninomiya, Y., Suzuki, Y., Koga, N. *Inorg. Chem.* **1997**, *36*, 694.
97IC2770 Carmalt, C.J., Cowley, A.H., Culp, R.D., Jones, R.A., Kamepalli, S., Norman, N.C. *Inorg. Chem.* **1997**, *36*, 2770.
97IZV1212 Dolganova, N.V., Dodonov, V.A., Zinoveva, T.I., Prezhbog, I.G. *Izv. Akad. Nauk, Ser. Khim.* **1997**, 1212.
97JA724 Wang, S.M., Mitzi, D.B., Landrum, G.A., Genin, H., Hoffmann, R. *J. Am. Chem. Soc.* **1997**, *119*, 724.
97JCS(D)2009 Hassan, A., Wang, S. *J. Chem. Soc., Dalton Trans.* **1997**, 2009.
97JCS(P1)1609 Suzuki, H., Ikegami, T., Azuma, N. *J. Chem. Soc., Perkin Trans. 1* **1997**, 1609.
97JFC(84)69 Naumann, D., Tyrra, W., Lewe, T. *J. Fluorine Chem.* **1997**, *84*, 69.

97JOC3610 Tanaka, H., Sumida, S., Nishioka, Y., Kobayashi, N., Tokumaru, Y., Kameyama, Y., Torii, S. *J. Org. Chem.* **1997**, *62*, 3610.

97JOC4880 Garrigues, B., Gonzaga, F., Robert, H., Dubac, J. *J. Org. Chem.* **1997**, *62*, 4880.

97JOM(529)151 Karsch, H.H., Witt, E. *J. Organomet. Chem.* **1997**, *529*, 151.

97JOM(530)71 Avtomonov, E.V., Li, X.-W., Lorberth, J. *J. Organomet. Chem.* **1997**, *530*, 71.

97JOM(545)111 Holmes, N.J., Levason, W., Webster, M. *J. Organomet. Chem.* **1997**, *545-546*, 111.

97JOM(548)223 Wada, M., Natsume, S., Suzuki, A., Uo, A., Nakamura, M., Hayase, S., Erabi, T. *J. Organomet. Chem.* **1997**, *548*, 223.

97NJC495 Kallal, K., Coin, C., Duñach, E., Postel, M. *New J. Chem.* **1997**, *21*, 495.

97OM1655 Mlynek, P.D., Dahl, L.F. *Organometallics* **1997**, *16*, 1655.

97OM3565 Rahman, M.M., Matano, Y., Suzuki, H. *Organometallics* **1997**, *16*, 3565.

97OM3597 Carmalt, C.J., Walsh, D., Cowley, A.H., Norman, N.C. *Organometallics* **1997**, *16*, 3597.

97POL1211 Venkatachalam, V., Ramalingam, K., Casellato, U., Graziani, R. *Polyhedron* **1997**, *16*, 1211.

97RCB1164 Dolganova, N.V., Dodonov, V.A., Zinov'eva, T.I., Prezhbog, I.G. *Russ. Chem. Bull.* **1997**, *46*, 1164.

97RGC927 Dodonov, V.A., Zabrudaeva, E.A., Dolganonva, N.V., Stepovik, L.P., Zinov'eva, T.I. *Russ. J. Gen. Chem.* **1997**, *67*, 927.

97S249 Suzuki, H., Ikegami, T., Matano, Y. *Synthesis* **1997**, 249.

97SA(A)2363 Ludwig, C., Solney, M., Götze, H.-J. *Spectrochim. Acta, Part A* **1997**, *53*, 2363 .

97SC1669 Vassileva, E., Shopova, M., Fugier, C., H.-Basch, E. *Synth. Commun.* **1997**, *27*, 1669.

97SC2569 Ren, P.-D., Shao, D., Dong, T.-W. *Synth. Commun.* **1997**, *27*, 2569.

97SC2577 Ren, P.-D., Jin, Q.-H., Yao, Z.-P. *Synth. Commun.* **1997**, *27*, 2577.

97SC3497 Ren, P.-D., Pan, X.-W., Jin, Q.-H., Yao, Z.-P. *Synth. Commun.* **1997**, *27*, 3497.

97SCI(277)78 Tokitoh, N., Arai, Y., Okazaki, R., Nagase, S. *Science* **1997**, *277*, 78.

97SL1251 Boruah, A., Baruah, B., Prajapati, D., Sandhu, J.S. *Synlett* **1997**, 1251.

97T4137 Arnauld, T., Barton, D.H.R., Doris, E. *Tetrahedron* **1997**, *53*, 4137.

97TA3939 Wada, M., Takahashi, T., Domae, T., Fukuma, T., Miyoshi, N., Smith, K. *Tetrahedron: Asymmetry* **1997**, *8*, 3939.

97TL365 Arnauld, T., Barton, D.H.R., Doris, E. *Tetrahedron Lett.* **1997**, *38*, 365.

97TL1449 Baruah, B., Baruah, A., Prajapati, D., Sandhu, J.S. *Tetrahedron Lett.* **1997**, *38*, 1449.

97TL2733 Shen, Z., Zhang, J., Zou, H., Yang, M. *Tetrahedron Lett.* **1997**, *38*, 2733.

97TL2981 Boisselier, V.L., Postel, M., Duñach, E. *Tetrahedron Lett.* **1997**, *38*, 2981.

97TL4267 Boruah, A., Baruah, B., Prajapati, D., Sandhu, J.S. *Tetrahedron Lett.* **1997**, *38*, 4267.

97TL7215 Komatsu, N., Uda, M., Suzuki, H., Takahashi, T., Domae, T., Wada, M. *Tetrahedron Lett.* **1997**, *38*, 7215.

97TL7219 Komatsu, N., Ishida, J., Suzuki, H. *Tetrahedron Lett.* **1997**, *38*, 7219.

97TL8045 Wada, M., Fukuma, T., Morioka, M., Takahashi, T., Miyoshi, N. *Tetrahedron Lett.* **1997**, *38*, 8045.

97TL8871 Desmurs, J.R., Labrouillére, M., Roux, C.L., Gaspard, H., Laporterie, A., Dubac, J. *Tetrahedron Lett.* **1997**, *38*, 8871.

97YH462 Ren, P.-D., Shao, D., Dong, T.-W. *Youji Huaxue* **1997**, *17*, 462; *Chem. Abstr.* **1997**, *127*, 330953b.

97ZAAC(623)122 Lewe, T., Naumann, D., Nowicki, G., Schneider, H., Tyrra, W. *Z. Anorg. Allg. Chem.* **1997**, *623*, 122.

97ZAAC(623)941 Becker, G., Egner, J., Meiser, M., Mundt, O., Weidlein, J. *Z. Anorg. Allg. Chem.* **1997**, *623*, 941.

97ZN(B)149 Hillwig, R., Kunkel, F., Harms, K., Neumüller, B., Dehnicke, K. *Z. Naturforsch.* **1997**, *52B*, 149.

97ZN(B)398 Sitzmann, H., Wolmershäuser, G. *Z. Naturforsch.* **1997**, *52B*, 398.

98AG(E)3175 Breunig, H.J., Rösler, R., Lork, E. *Angew. Chem., Int. Ed. Engl.* **1998**, *37*, 3175.

98AOC(42)1 Whitmire, K.H. *Adv. Organomet. Chem.* **1998**, *42*, 1.

98AST(11)42 Lorberth, J., Shin, S.-H., Kong, Y.K. *Anal. Sci. Technol.* **1998**, *11*, 42.

98B-A *Aldrich Catalog/Handbook of Fine Chemicals* **1998–1999** (ALDRICH Chemical Company, Inc., 1001, W. Saint Paul Avenue, Milwaukee, WI 53233, USA).

98BBI(71) Karol, T.J. *The Bulletin of the Bismuth Institute*, **1998**, No. 71.

98B-CN Carmalt, C.J., Norman, N.C., in *Chemistry of Arsenic, Antimony and Bismuth*, ed. Norman, N.C., Blackie Academic & Professional, London, **1998**, Chap. 1.

98B-R Reglinski, J., in *Chemistry of Arsenic, Antimony and Bismuth*, ed. Norman, N.C., Blackie Academic & Professional, London, **1998**, Chap. 8.

98B-SM Suzuki, H., Matano, Y., in *Chemistry of Arsenic, Antimony and Bismuth*, ed. Norman, N.C., Blackie Academic & Professional, London, **1998**, Chap. 6.

98B-W *Wako Catalogue* **1998** (Wako Pure Chemical Industries, Ltd., 3-1-2, Doshomachi, Chuo-ku, Osaka 540-8605, Japan).

98CC1359 Rahman, M.M., Matano, Y., Suzuki, H. *Chem. Commun.* **1998**, 1359.

98CL127 Matano, Y., Miyamatsu, T., Suzuki, H. *Chem. Lett.* **1998**, 127.

98CRA141 Laporte, C., Baules, P., Laporterie, A., Desmurs, J.R., Dubac, J. *C. R. Acad. Sci. Paris, Ser. IIc: Chim.* **1998**, *1*, 141.

98EJO2743 Répichet, S., Roux, C.L., Dubac, J., Desmurs, J.R. *Eur. J. Org. Chem.* **1998**, 2743.

98ET(1)174 *Encyclopedia of Toxicology*, ed. Wexler, P., Academic, San Diego, CA, **1998**, vol. 1, p. 174.

98HX714 Ren, P.-D., Pan, S.-F., Dong, T.-W., Wu, S.-H. *Huaxue Xuebao* **1998**, *56*, 714; *Chem. Abstr.* **1998**, *129*, 161370.

98ICA(275/276)340 Armelao, L., Bandoli, G., Casarin, M., Depaoli, G., Tondello, E., Vittadini, A. *Inorg. Chim. Acta* **1998**, *275/276*, 340.

98JA433 Tokitoh, N., Arai, Y., Sasamori, T., Okazaki, R., Nagase, S., Uekusa, H., Ohashi, Y. *J. Am. Chem. Soc.* **1998**, *120*, 433.

98JCS(P1)2511 Matano, Y., Aratani, Y., Miyamatsu, T., Kurata, H., Miyaji, K., Sasako, S., Suzuki, H. *J. Chem. Soc., Perkin Trans. 1* **1998**, 2511.

98JOC59 Serre, S.L., Guillemin, J.-C., Karapti, T., Soos, L., Nyulkgzi, L., Veszprémi, T. *J. Org. Chem.* **1998**, *63*, 59.

98JOC6721 Brands, K.M. J., Dolling, U.-H., Jobson, R.B., Marchesini, G., Reamer, R.A., Williams, J.M. *J. Org. Chem.* **1998**, *63*, 6721.

98JOC7472 Yi, X.-H., Meng, Y., Hua, X.-G., Li, C.-J. *J. Org. Chem.* **1998**, *63*, 7472.

98JOM(560)211 Li, X.-W., Lorberth, J., Ebert, K.H., Massa, W., Wocadlo, S. *J. Organomet. Chem.* **1998**, *560*, 211.

98NJC339 Bekassy, S., Cseri, T., Horvath, M., Farkas, J., Figueras, F. *New J. Chem.* **1998**, *22*, 339.

98NJC973 Wenkin, M., Touillaux, R., Devillers, M. *New J. Chem.* **1998**, *22*, 973.

98OM1013 Ikegami, T., Suzuki, H. *Organometallics* **1998**, *17*, 1013.

98OM1347 Hoppe, S., Whitmire, K.H. *Organometallics* **1998**, *17*, 1347.

98OM1711 Murafuji, T., Satoh, K., Sugihara, Y., Azuma, N. *Organometallics* **1998**, *17*, 1711.

98OM4049 Matano, Y., Kurata, H., Murafuji, T., Azuma, N., Suzuki, H. *Organometallics* **1998**, *17*, 4049.

98OM4332 Matano, Y., Begum, S.R., Miyamatsu, T., Suzuki, H. *Organometallics* **1998**, *17*, 4332.

98RCB659 Zinov'eva, T.I., Dolganova, N.V., Dodonov, V.A., Prezhbog, I.G. *Russ. Chem. Bull.* **1998**, *47*, 659.

98RCB1607 Lermontov, S.A., Shkavrov, S.V., Lermontov, A.S., Zaorin, S.I. *Russ. Chem. Bull.* **1998**, *47*, 1607.

98S713 Mashraqui, S.H., Karnik, M.A. *Synthesis* **1998**, 713.

98S1724 Cao, Z., Armstrong, K., Shaw, M., Petry, E., Harris, N. *Synthesis* **1998**, 1724.

98SC603 Winum, J.-Y., Kamal, M., Barragan, V., Leydet, A., Montero, J.-L. *Synth. Commun.* **1998**, *28*, 603.

98SC939 Mashraqui, S.H., Mudalier, C.D., Karnik, M.A. *Synth. Commun.* **1998**, *28*, 939.

98SC1737 Boyer, B., Keramane, E.M., Montero, J.-L., Roque, J.-P. *Synth. Commun.* **1998**, *28*, 1737.

98SC3943 Mohammadpoor-Baltork, I., Aliyan, H. *Synth. Commun.* **1998**, *28*, 3943.

98SC4157 Baruah, M., Prajapati, D., Sandhu, J.S. *Synth. Commun.* **1998**, *28*, 4157.

98SC4479 Lu, G., Zhang, Y. *Synth. Commun.* **1998**, *28*, 4479.

98SL1083 Baruah, M., Boruah, A., Prajapati, D., Sandhu, J.S. *Synlett* **1998**, 1083.

98SL1138 L.-Robert, H., Roux, C.L., Dubac, J. *Synlett* **1998**, 1138.

98SL1249 Roux, C.L., Ciliberti, L., L.-Robert, H., Laporterie, A., Dubac, J. *Synlett* **1998**, 1249.
98SL1345 Mitra, A.K., De, A., Karchaudhuri, N. *Synlett* **1998**, 1345.
98T4313 Combes, S., Finet, J.-P. *Tetrahedron* **1998**, *54*, 4313.
98T10111 Simeon, F., Jaffres, P.-A., Villemin, D. *Tetrahedron* **1998**, *54*, 10111.
98TA2415 Zhu, G., Zhang, X. *Tetrahedron: Asymmetry* **1998**, *9*, 2415.
98TA2657 Satoh, K., Inenaga, M., Kanai, K. *Tetrahedron: Asymmetry* **1998**, *9*, 2657.
98TL1161 Robert, H., Garrigues, B., Dubac, J. *Tetrahedron Lett.* **1998**, *39*, 1161.
98TL7247 Banik, B.K., Venkatraman, M.S., Mukhopadhyay, C., Becker, F.F. *Tetrahedron Lett.* **1998**, *39*, 7247.
99AC(A)299 Moens, L., Ruiz, P., Delmon, B., Devillers, M. *Appl. Catal. A* **1999**, *180*, 299.
99AG(E)967 Schulz, S., Nieger, M. *Angew. Chem. Int. Ed. Engl.* **1999**, *38*, 967.
99BBI(73) Brown, R.D. *The Bulletin of the Bismuth Institute*, **1999**, No. 73.
99BBI(74) Fickling, M. *The Bulletin of the Bismuth Institute*, **1999**, No. 74.
99BCJ1851 Ohe, T., Tanaka, T., Kuroda, M., Cho, C.S., Ohe, K., Uemura, S. *Bull. Chem. Soc. Jpn.* in press.
99BCJ2143 Ai, M., Ohdan, K. *Bull. Chem. Soc. Jpn.* **1999**, *72*, 2143.
99CL861 Minoura, M., Kanamori, Y., Miyake, A., Akiba, K.-y. *Chem. Lett.* **1999**, 861.
99CPB1108 Yasuike, S., Shiratori, S.-i., Kurita, J., Tsuchiya, T. *Chem. Pharm. Bull.* **1999**, *47*, 1108.
99CRA455 Laporte, C., Marquie, J., Laporterie, A., Desmurs, J.R., Dubac, J. *C. R. Acad. Sci. Paris, Ser. IIc: Chim.* **1999**, *2*, 455.
99CRV2601 Briand, G.G., Burford, N. *Chem. Rev.* **1999**, *99*, 2601.
99CRV3277 Silvestru, C., Breunig, H. J., Althaus, H. *Chem. Rev.* **1999**, *99*, 3277.
99EJIC2295 Faure, J.-L., Gornitzka, H., Reau, R., Stalke, D., Bertrand, G. *Eur. J. Inorg. Chem.* **1999**, 2295.
99IJC(B)257 Gadhwal, S., Sandhu, J.S. *Indian J. Chem., Sect. B* **1999**, *38B*, 257.
99JA3357 Twamley, B., Sofield, C.D., Olmstead, M.M., Power, P.P. *J. Am. Chem. Soc.* **1999**, *121*, 3357.
99JA4409 Stark, J.L., Harms, B., Guzman-Jimenez, I., Whitmire, K.H., Gautier, R., Halet, J.-F., Saillard, J.-Y. *J. Am. Chem. Soc.* **1999**, *121*, 4409.
99JCR(S)280 Zhan, Z., Lu, G., Zhang, Y. *J. Chem. Res. (S)* **1999**, 280.
99JCS(D)657 Tyrra, W., Naumann, D., Kirij, N.V., Kolomeitsev, A.A., Yagupolskii, Y.L. *J. Chem. Soc., Dalton Trans.* **1999**, 657.
99JCS(D)2837 James, S.C., Norman, N.C., Orpen, A.G. *J. Chem. Soc., Dalton Trans.* **1999**, 2837.
99JCS(P1)1533 Rahman, M.M., Matano, Y., Suzuki, H. *J. Chem. Soc., Perkin Trans. 1* **1999**, 1533.
99JFC(93)103 Lermontov, S.A., Rakov, I.M., Zefirov, N.S., Stang, P.J. *J. Fluorine Chem.* **1999**, *93*, 103.

99JFC(94)207 Kirij, N.V., Pasenok, S.V., Yagupolskii, Y.L., Fitzner, A., Tyrra, W., Naumann, D. *J. Fluorine Chem.* **1999**, *94*, 207.

99JOC3722 Arnauld, T., Barton, D.H.R., Normant, J.-F. *J. Org. Chem.* **1999**, *64*, 3722.

99JOC6479 Répichet, S., Roux, C.L., Hernandez, P., Dubac, J. *J. Org. Chem.* **1999**, *64*, 6479.

99JOC6924 Matano, Y., Rahman, M.M., Yoshimune, M., Suzuki, H. *J. Org. Chem.* **1999**, *64*, 6924.

99JOC8214 Marshall, J.A., Grant, C.M. *J. Org. Chem.* **1999**, *64*, 8214.

99JOM(574)3 Matoba, K., Motofusa, S.-I., Cho, C.S., Ohe, K., Uemura, S. *J. Organomet. Chem.* **1999**, *574*, 3.

99JOM(584)179 Holmes, N.J., Levason, W., Webster, M. *J. Organomet. Chem.* **1999**, *584*, 179.

99JOM(585)253 Garrigues, B., Oussaid, A. *J. Organomet. Chem.* **1999**, *585*, 253.

99OM328 Althaus, H., Breunig, H.J., Rösler, R., Lork, E. *Organometallics* **1999**, *18*, 328.

99OM2580 Matano, Y., Nomura, H., Shiro, M., Suzuki, H. *Organometallics* **1999**, *18*, 2580.

99OM3016 Riley, P.N., Thorn, M.G., Vilardo, J.S., Lockwood, M.A., Fanwick, P.E., Rothwell, I.P. *Organometallics* **1999**, *18*, 3016.

99OM5668 Matano, Y., Begum, S.A., Miyamatsu, T., Suzuki, H. *Organometallics*, **1999**, *18*, 5668.

99OX244 Mishra, P.M., Mishra, P., Jha, C.S. *Oxidation Commun.* **1999**, *22*, 244.

99S395 Rahman, M.M., Matano, Y., Suzuki, H. *Synthesis* **1999**, 395.

99SC2741 Mohammadpoor-Baltork, I., Aliyan, H. *Synth. Commun.* **1999**, *29*, 2741.

99SL1207 Sheppard, G.S. *Synlett* **1999**, 1207.

99T1341 Fedorov, A., Combes, S., Finet, J.-P. *Tetrahedron* **1999**, *55*, 1341.

99T1971 Boyer, B., Keramane, E.M., Arpin, S., Montéro, J.-L., Roque, J.-P. *Tetrahedron* **1999**, *55*, 1971.

99T3377 Combes, S., Finet, J.-P. *Tetrahedron* **1999**, *55*, 3377.

99TL2747 Fedorov, A., Finet, J.-P. *Tetrahedron Lett.* **1999**, *40*, 2747.

99TL285 Labrouillére, M., Roux, C.L., Gaspard, H., Laporterie, A., Dubac, J., Desmurs, J. R. *Tetrahedron Lett.* **1999**, *40*, 285.

99TL7215 Motorina, I.A., Grierson, D.S. *Terahedron Lett.* **1999**, *40*, 7215.

99TL7657 Tymonko, S.A., Nattier, B.A., Mohan, R.S. *Terahedron Lett.* **1999**, *40*, 7657.

99TL9233 Répichet, S., Roux, C.L., Dubac, J. *Terahedron Lett.* **1999**, *40*, 9233.

99YGK689 Wada, M., Miyoshi, N. *Yuki Gosei Kagaku Kyokaishi* **1999**, *57*, 689.

99ZAAC(625)629 Schumann, H., Mühle, S.H. *Z. Anorg. Allg. Chem.* **1999**, *625*, 629.

99ZAAC(625)834 Pasenok, S.V., Kirij, N.V., Yagupolskii, Y.L., Naumann, D., Tyrra, W., Fitzner, A. *Z. Anorg. Allg. Chem.* **1999**, *625*, 834.

99ZAAC(625)1354 Üffing, C., Ecker, A., Baum, E., Schnöckel, H. *Z. Anorg. Allg. Chem.* **1999**, *625*, 1354.

2000B-L(2)521	Lewis, R.J., Sr. ed. *Sax's Dangerous Properties of Industrial Materials*, vol. 2, 10th ed., Wiley, New York, **2000**, p. 521.
2000B-L(3)3585	Lewis, R.J., Sr. ed. *Sax's Dangerous Properties of Industrial Materials*, vol. 3, 10th ed., Wiley, New York, **2000**, pp. 3585, 3610.
2000BCJ689	Miyoshi, N., Nishio, M., Murakami, S., Fukuma, T., Wada, M. *Bull. Chem. Soc. Jpn.* **2000**, *73*, 689.
2000H in press	Yasuike, S., Tsukada, S., Kurita, J., Tsuchiya, T., Tsuda, Y., Kiuchi, H., Hosoi, S. *Heterocycles* **2000**, in press.
2000OM in press	Murafuji, T., Nagasue, M., Tashiro, Y., Suguhara, Y., Azuma, N. *Organometallics* **2000**, in press.
UR1	Sasamori, T., Takeda, N., Tokitoh, N. unpublished results.
UR2	Matano, Y., Nomura, H., Suzuki, H. unpublished results.
UR3	Yasuike, S., Niwa, M., Kurita, J., Tsuchiya, T., Yamaguchi, K. unpublished results.
UR4	Ikegami, T. unpublished results.

Subject Index

FSC
www.fsc.org
MIX
Papier | Fördert gute Waldnutzung
FSC® C083411

Kolibri 360 GmbH
Ferdinand-Jühlke-Straße 7
99095 Erfurt, Deutschland
produktsicherheit@kolibri360.de